Reference Elements

CLOSE-UP

USING HYPHENS

Compound adjectives that contain words ending in *-ly* are not hyphenated, even when they precede the noun.

Many upwardly mobile families consider items like home computers, microwave ovens, and videocassette recorders to be necessities.

(2)   With certain prefixes or suffixes

Use a hyphen between a prefix and a proper noun or an adjective formed from a proper noun.

mid-July   pre-Columbian

See ◄
Pt.9
Use a hyphen to connect the prefixes *all-*, *ex-*, *half-*, *quarter-*, *quasi-*, and *self-* and the suffixes *-elect* and *-odd* to a noun.

all-pro   quasi-serious
ex-senator   self-centered
half-pint   president-elect
quarter-moon   thirty-odd

NOTE: The words *selfhood*, *selfish*, and *selfless* do not include hyphens. In these cases *self* is the root, not a prefix.

(3)   For clarity

Hyphenate to prevent misreading one word for another.

In order to reform criminals, we must re-form our ideas about prisons.

Hyphenate to avoid combinations that are hard to read such as two *i*'s (*semi-illiterate*) or more than two of the same consonant (*shell-less*).

Hyphenate in most cases between a capital initial and a word when the two combine to form a compound.

A-frame   T-shirt

But check your dictionary; some letter- or numeral-plus-word compounds do not require hyphens.

B flat   F major

550

Left running heads indicate chapter title. Right running heads indicate section titles.

Tabs show abbreviated but descriptive chapter title and section number.

Blue cross-references direct users to related discussions in other sections.

Blue arrows and underlining highlight grammatical concepts under discussion

Yellow screens highlight punctuation marks under discussion.

Works Cited Format: MLA

Author's last name   First name   Underlined title (all major words capitalized)
      ↓ [1]   ↓ [2]   ↓ [2]   ↓ [1]
   Dyson, Freeman.   Disturbing the Universe.   New York:

Double →   [1]
space   Harper, 1979.
              ↑   ↑   ↑   ↑
           Publisher Date   Period   City

Annotated models illustrate MLA, APA, CMS, and CBE documentation styles.

The
Holt Handbook
FOURTH EDITION

Publisher   Ted Buchholz
Acquisitions Editor   Michael Rosenberg
Developmental Editor   Camille Adkins
Project Editor   Barbara Moreland
Production Manager   Tad Gaither
Art Director   Nick Welch
Permissions Editor   Sheila Shutter

ISBN: 0-15-501633-4

Library of Congress Catalog Card Number: 93-81250

Requests for permission to make copies of any part of the work should be mailed to: Permissions Department, Harcourt Brace & Company, 8th Floor, Orlando, Florida 32887.

Address for Editorial Correspondence: Harcourt Brace College Publishers, 301 Commerce Street, Suite 3700, Fort Worth, TX 76102.

Address for Orders: Harcourt Brace & Company, 6277 Sea Harbor Drive, Orlando, FL 32887. 1-800-782-4479, or 1-800-433-0001 (in Florida).

(Copyright acknowledgments begin on page L1, which constitutes a continuation of this copyright page.)

Printed in the United States of America

4 5 6 7 8 9 0 1 2 3   039   9 8 7 6 5 4 3 2 1

The
Holt Handbook

FOURTH EDITION

Laurie G. Kirszner

Philadelphia College of Pharmacy and Science

Stephen R. Mandell

Drexel University

Harcourt Brace College Publishers

Fort Worth   Philadelphia   San Diego   New York   Orlando   Austin   San Antonio
Toronto   Montreal   London   Sydney   Tokyo

Preface to the Annotated Instructor's Edition

We are pleased to introduce the Annotated Instructor's Edition of *The Holt Handbook*, Fourth Edition. It is our hope that this special edition, which supplements the material contained in *The Holt Handbook*, will provide interesting and thought-provoking material for in-class discussion.

The features contained in the Annotated Instructor's Edition, which appear as annotations in the margins of the text, are designed to make your teaching easier and more enjoyable.

1. *Answers to Exercises* The Annotated Instructor's Edition contains answers to the many exercises contained in the *Handbook*. This feature eliminates the need to bring a separate manual to class or to write answers in the classroom text.
2. *Further Reading* Abstracts of articles presenting scholarly information about writing across the curriculum, collaborative learning, critical thinking, and composition theory serve as a starting point for those instructors who want to explore these subjects in more detail. Together, the abstracts form an annotated bibliography of practical and useful articles on composition and learning.
3. *Computer Tips* Designed to supplement the "Guide to Writing with Computers," these tips offer practical advice to instructors who have access to a computer writing center, enabling them to give helpful hints to students who compose on computers. Included are tips on the various stages of the writing process, sug-gested lab activities, and discussions on using computers to enhance the writing process.
4. *In The Classroom* This feature suggests in-class activities for students to help reinforce the topics presented in the *Handbook*. Both individual and group activities are designed to lead to classroom discussion.
5. *Quotations* Selected quotations drawn from the work of writers and scholars can serve as jumping-off points for classroom discussion. In addition, they can show students that the concerns they have as beginning writers have also been shared by experienced writers.
6. *Students Working Together* This feature offers collaborative learning activities to bring students together in all stages of the writing process.
7. *Teaching Strategy* These strategies offer suggestions for teaching *Handbook* material and highlight areas that are particularly troublesome to students. Included are suggestions for classroom activities, additional exercises, and hints to help instructors better explain the material being discussed.

In addition to the above features, we have included seven complete scholarly articles chosen to represent recent research in the areas of: the multicultural classroom, teaching composition, peer editing, writing across the curriculum, critical thinking and writing, composing with computers, and responding to students' writing. Following the articles is a bibliography of recommended readings in the area of composition. We hope you find the articles and bibliography both informative and thought-provoking.

We hope you share our enthusiasm for the Annotated Instructor's Edition of *The Holt Handbook*, Fourth Edition, and invite you to comment on the features listed above and submit any of your own teaching strategies for inclusion in the fifth edition. Send these to Laurie Kirszner and Stephen Mandell ℅ English Editor, Harcourt Brace, 301 Commerce Street, Suite 3700, Fort Worth, Texas, 76102.

Contents

Following the recommended reading list is the complete student edition of *The Holt Handbook*, Fourth Edition. The Annotated Instructor's Edition of *The Holt Handbook* contains annotations not included in the student edition. A description of these annotations can be found in the Preface to the Instructor's Edition.

Brief Contents of Student Edition

Annotated Instructor's Edition

Multicultural Classrooms, Monocultural Teachers

Terry Dean
The University of California, Davis

> Remember gentlemen, John Chrysostom's exquisite story about the day he entered the rhetorician Libanius' school in Antioch. Whenever a new pupil arrived at his school, Libanius would question him about his past, his parents, and his country.
>
> Renan, *La Réforme intellectuelle et morale*

Sometimes more than others, I sense the cultural thin ice I walk on in my classrooms, and I reach out for more knowledge than I could ever hope to acquire, just to hang on. With increasing cultural diversity in classrooms, teachers need to structure learning experiences that both help students write their way into the university and help teachers learn their way into student cultures. Now this is admittedly a large task, especially if your students (like mine) are Thai, Cambodian, Vietnamese, Korean, Chinese, Hmong, Laotian (midland Lao, lowland Lao), Salvadoran, Afro-American, Mexican, French, Chicano, Nicaraguan, Guatemalan, Native American (Patwin, Yurok, Hoopa, Wintu), Indian (Gujarati, Bengali, Punjabi), Mexican-American, Jamaican, Filipino (Tagalog, Visayan, Ilocano), Guamanian, Samoan, and so on. It may take a while for the underpaid, overworked freshman composition instructor to acquire dense cultural knowledge of these groups. But I have a hunch that how students handle the cultural transitions that occur in the acquisition of academic discourse affects how successfully they acquire that discourse. The very least we can do, it seems to me, is to educate ourselves so that when dealing with our students,

Dean, Terry. "Multicultura Classrooms, Monocultural Teachers." College Composition and Communication 40, no. 1 (February 1989): 23–37. Copyright 1989 by the National Council of Teachers of English. Reprinted with permission.

Terry Dean has taught ESL and Basic Writing courses for regular and special admission EOP/SAA students for twelve years at The University of California, Davis. The author can be reached at The Learning Skills Center, Basement, South Hall.

in the words of Michael Holzman, "We should stop doing harm if we can help it" (31).

Some would question how much harm is being done. If enough students pass exit exams and the class evaluations are good, then everything is OK. Since we want our students to enter the mainstream, all we need worry about is providing them the tools. Like opponents of bilingual education, some would argue that we need to concern ourselves more with providing student access to academic culture, not spending time on student culture. But retention rates indicate that not all students are making the transition into academic culture equally well. While the causes of dropout are admittedly complex, cultural dissonance seems at the very least to play an important role. If indeed we are going to encounter "loss, violence, and compromise" (142) as David Bartholomae describes the experience of Richard Rodriguez, should we not be directing students to the counseling center? And if the attainment of biculturalism in many cases is painful and difficult, can we be assured, as Patricia Bizzell suggests, that those who do achieve power in the world of academic discourse will use it to argue persuasively for preservation of the language and the culture of the home world view? (299) This was not exactly Richard Rodriguez's response to academic success, but what if, after acquiring the power, our students feel more has been lost than gained? I think as teachers we have an obligation to raise these issues. Entering freshmen are often unaware of the erosion of their culture until they become seniors or even later. Like Richard Rodriguez, many students do not fully realize what they have lost until it is too late to regain it. Let me briefly outline the problem as I see it and offer some possible solutions.

The Problem

A lot is being asked of students. David Bartholomae describes the process: "What our beginning students need to learn is to extend themselves, by successive approximations, into the commonplace, set phrases, rituals and gestures, habits of mind, tricks of persuasion, obligatory conclusions, and necessary connections that determine 'what might be said' and constitute knowledge within the various branches of our academic community" (146). "Rituals and gestures, tricks of persuasion" mean taking on much more than the surface features of a culture. Carried to an extreme, students would have to learn when it is appropriate to laugh at someone slipping on a banana peel. When we teach composition, we are teaching culture. Depending on students' backgrounds, we are teaching at least academic culture, what is acceptable evidence, what persuasive strategies work best, what is taken to be a demonstration of "truth" in different disciplines. For students whose home culture is distant from mainstream culture, we are also teaching how, as a people, "mainstream" Americans view the world. Consciously or unconsciously, we do this, and the responsibility is frightening.

In many situations, the transitions are not effective. Several anthropological and social science studies show how cultural dissonance can affect learning. Shirley Brice Heath examines the ways in which the natural language environments of working-class Black and white children can interfere with their success in schools designed primarily for children from middle-class mainstream culture. The further a child's culture is from the culture of the school, the less chance for success. Classroom environments that do not value the home culture of the students lead to decreased motivation and poor academic performance (270–72). In a study of Chicano and Black children in Stockton, California schools, John Ogbu arrives at a similar conclusion. Susan Urmston Philips analyzes the experiences of Warm Springs Native American children in a school system in Oregon where the administrators, teachers, and even some parents thought that little was left of traditional culture. But Philips shows that "children who speak English and who live in a material environment that is overwhelmingly Western in form can still grow up in a world where by far the majority of their enculturation experience comes from their interaction with other Indians. Thus school is still the main source of their contact with mainstream Anglo culture" (11). Philips describes the shock that Warm Springs Native American children experience upon entering a school system designed for the Anglo middle-class child. Because of differences in the early socialization process of Native American children (especially in face-to-face interaction), they feel alienated in the classroom and withdraw from class activities (128).

Pierre Bourdieu and Jean-Claude Passeron examine how the cultural differences of social origin relate directly to school performance (8–21). Educational rewards are given to those who feel most at home in the system, who, assured of their vocations and abilities, can pursue fashionable and exotic themes that pique the interest of their teachers, with little concern for the vocational imperatives of working-class and farm children. Working-class and farm children must struggle to acquire the academic culture that has been passed on by osmosis to the middle and upper classes. The very fact that working-class and farm children must laboriously acquire what others come by naturally is taken as another sign of inferiority. They work hard because they have no talent. They are remedial. The further the distance from the mainstream culture, the more the antipathy of mainstream culture, the more difficulty students from outside that culture will have in acquiring it through the educational system (which for many is the only way):

> Those who believe that everyone would be given equal access to the highest level of education and the highest culture, once the same economic means were provided for all those who have the requisite "gifts," have stopped halfway in their analysis of the obstacles; they ignore the fact that the abilities measured by the scholastic criteria stem not so much from natural "gifts" (which must remain hypothetical so long as the educational inequalities can be traced to other causes), but from the greater or lesser affinity between class cultural habits and the demands of the educational system or the criteria which define success within it. [Working-class children] must assimilate a whole set of knowledge and techniques which are never completely separable from social values often contrary to those of their class origin. For the children of peasants, manual workers, clerks, or small shopkeepers, the acquisition of culture is acculturation. (Bourdieu and Passeron 22)

Social and cultural conditions in the United States are not the same as in France, but the analyses of Bourdieu, Passeron, Heath, Ogbu, and others suggest interesting lines of inquiry when we look at the performance of students from different cultures and classes in U.S. schools. Performance seems not so much determined by cultural values (proudly cited by successful groups), but by class origins, socio-economic mobility, age at time of immigration, the degree of trauma experienced by immigrants or refugees, and the acceptance of student culture by the mainstream schools. Stephen Steinberg argues in *The Ethnic Myth* that class mobility precedes educational achievement in almost all immigrant groups (131–32). I really do not believe that Black, Native American, and Chicano cultures place less emphasis on the importance of education than Chinese, Jewish, Vietnamese, or Greek cultures do. We do not have over one hundred Black colleges in the United States because Blacks don't care about education. I have never been to a Native American Studies Conference or visited a rancheria or reservation that did not have newsletters, workshops, and fund-raisers in support of education. Bourdieu and Passeron's analysis suggests that educational success depends to a large extent on cultural match, and if an exact match is not possible, there must at least be respect and value of the culture children bring with them. Acculturation (assimilation) is possible for some, but it is not viable for all.

Acculturation itself poses problems. Jacquelyn Mitchell shows how cultural conflict affects the preschool child, the university undergraduate, the graduate student, and the faculty member as well. Success brings with it, for some people, alienation from the values and relationships of the home culture. "In fulfilling our academic roles, we interact increasingly more with the white power structure and significantly less with members of our ethnic community. This is not without risk or consequence; some minority scholars feel in jeopardy of losing their distinctive qualities" (38). The question Mitchell poses is, "How can blacks prepare themselves to move efficiently in mainstream society and still maintain their own culture?" (33) Jacqueline Fleming, in a cross-sectional study of Black students in Black colleges and predominantly white colleges, found black colleges more effective despite the lack of funding because Black colleges are more "supportive" of students (194). Long before the recent media coverage of racial incidents on college campuses, Fleming noted that "all is not well with Black students in predominantly white colleges" (162). And in California, the dropout rate of "Hispanics" (a term that obscures cultural diversity much as the term "Asian" does) is greater than that of any other group except possibly Native Americans. But despite gloomy statistics, there is hope.

Theoretical Models for Multicultural Classrooms

Several theoretical models exist to help students mediate between cultures. In "Empowering Minority Students: A Framework for Intervention," James Cummins provides one:

> The central tenet of the framework is that students from "dominated" societal groups are "empowered" or "disabled" as a direct result of their interactions with educators in the schools. These interactions are mediated by the implicit or explicit role definitions that educators assume in relation to four institutional characteristics of schools. These characteristics reflect the extent to which (1) minority students' language and culture are incorporated into the school program; (2) minority community participation is encouraged as an integral component of children's education; (3) the pedagogy promotes intrinsic motivation on the part of students to use language actively in order to generate their own knowledge; and (4) professionals involved in assessment become advocates for minority students rather than legitimizing the location of the "problem" in the students. For each of these dimensions of school organization the role definitions of educators can be described in terms of a continuum, with one end promoting the empowerment of students and the other contributing to the disabling of students. (21)

Like Bourdieu, Cummins sees bicultural ambivalence as a negative factor in student performance. Students who have ambivalence about their cultural identity tend to do poorly whereas "widespread school failure does not occur in minority groups that are positively oriented towards both their own and the dominant culture, [that] do not perceive themselves as inferior to the dominant group, and [that] are not alienated from their own cultural values" (22). Cummins argues that vehement resistance to bilingual education comes in part because "the incorporation of minority languages and cultures into the school program confers status and power (jobs for example) on the minority group" (25). But for Cummins, it is precisely this valuing of culture within the school that leads to academic success because it reverses the role domination of students by the school.

Shirley Brice Heath's model is similar to Cummins'. The main difference is that she focuses on ethnography as a way for both teachers and students to mediate between home and school cultures. A consideration of home culture is the only way students can succeed in mainstream schools, increase scores on standardized tests, and be motivated to continue school: "Unless the boundaries between classrooms and communities can be broken, and the flow of cultural patterns between them encouraged, the schools will continue to legitimate and reproduce communities of townspeople who control and limit the potential progress of other communities who themselves remain untouched by other values and ways of life" (369). Like Cummins, Heath aims for cultural mediation. As one student stated: "Why should my 'at home' way of talking be 'wrong' and your standard version be 'right?' . . . Show me that by adding a fluency in standard dialect, you are adding something to my language and not taking something away from me. Help retain my identity and self-respect while learning to talk 'your' way" (271). Paulo Freire, on the other hand, wants those from the outside to totally transform mainstream culture, not become part of it: "This, then, is the great humanistic and historical task of the oppressed: to liberate themselves and their oppressors as well" (*Pedagogy* 28). Yet in almost all other respects, Freire's model, like Cummins' and Heath's, is grounded in a thorough knowledge of the home culture by teachers, and actively learned, genuine knowledge by the student. Each of these models has different agendas. Teachers should take from them whatever suits their teaching style, values, and classroom situations. I would simply encourage an inclusion of the study of the wide diversity within student cultures.

Teaching Strategies for Multicultural Classes

Cultural Topics

Culturally oriented topics are particularly useful in raising issues of cultural diversity, of different value systems, different ways of problem solving. Several successful bridge programs have used comparison of different cultural rituals (weddings, funerals, New Year's) as a basis for introducing students to analytic academic discourse. Loretta Petrie from Chaminade University in Hawaii has a six-week summerschool curriculum based on this. Students can use their own experience, interview relatives, and read scholarly articles. Reading these papers to peer response groups gives students additional insights into rituals in their own culture as well as making them aware of similarities and differences with other groups. I have used variations of Ken Macrorie's I-Search paper (Olson 111–22) to allow students to explore part of their cultural heritage that they are not fully aware of. One Vietnamese student, who was three years old when she came to America, did a paper on Vietnam in which she not only interviewed relatives to find out about life there but sought out books on geography and politics; she literally did not know where Vietnam was on the map and was embarrassed when other students would ask her about life there. Several students whose parents were from the Philippines did research that was stimulated by the desire to further understand family customs and to explain to themselves how the way they thought of themselves as "American" had a unique quality to it. One student wrote:

> For over eighteen years I have been living in the United States. Since birth I have been and still am a citizen of this country. I consider myself a somewhat typical American who grew up with just about every American thing you can think of; yet, at home I am constantly reminded of my Filipino background. Even at school I was reminded of my Filipino culture. At my previous school, the two other Filipinos in my class and I tried to get our friends to learn a little about our culture.

But classmates were not always open to cultural diversity, and their rejection raised the central question of just how much you have to give up of your culture to succeed in the mainstream society:

> During the Philippine presidential election, there were comments at my school that we Filipinos were against fair democracy and were as corrupt as our ex-President Marcos. Also it was said that Filipinos are excessively violent barbaric savages. This is partly due to our history of fighting among ourselves, mostly one group

that speaks one dialect against another group of a different language within the islands. Also maybe we are thought savages because of the food we eat such as "chocolate meat" and "balut," which is sort of a salted egg, some of which may contain a partially developed chicken. So, to be accepted into society you must give up your old culture.

"You must give up your old culture" is misleading. The student had to be careful about sharing home culture with peers, but he isn't giving up his culture; he is gaining a greater understanding of it. His essay ended with:

> I now have a better understanding of why I was doing a lot of those things I didn't understand. For example, whenever we visited some family friends, I had to bow and touch the older person's hand to my forehead. My mother didn't really explain why I had to do this, except it was a sign of respect. Also my mother says my brother should, as a sign of respect, call me "manong" even though he is only two years younger than I. At first I thought all this was strange, but after doing research I found out that this practice goes back a long way, and it is a very important part of my Filipino culture.

Richard Rodriguez's widely anthologized "Aria" (from *Hunger of Memory*) allows students to analyze his assertion that loss of language and culture is essential to attain a "public voice." Although the student above seems to agree in part with Rodriguez, most students find Rodriguez's assertions to be a betrayal of family and culture:

> I understand Rodriguez's assertion that if he learned English, he would lose his family closeness, but I think that he let paranoia overcome his senses. I feel that the lack of conversation could have been avoided if Rodriguez had attempted to speak to his parents instead of not saying anything just because they didn't. I am sure that one has to practice something in order to be good at it and it was helpful when his parents spoke in English to them while carrying on small conversation. To me, this would have brought the family closer because they would be helping each other trying to learn and grow to function in society.

Instead of feeling left out at home and in his society, Rodriguez could've been included in both.

I have used this topic or variations on it for a half dozen years or so, and most of my freshmen (roughly 85% of them) believe they do not have to give up cultural and home values to succeed in the university. I find quite different attitudes among these students when they become juniors and seniors. More and more students graduate who feel that they have lost more than they gained. Raising this issue early provides students more choices. In some cases it may mean deciding to play down what seem to be unacceptable parts of one's culture (no balut at the potluck); in other cases it may lead to the assertion of positive values of the home culture such as family cohesiveness and respect for parents and older siblings. Courageous students will bring the balut to the potluck anyway and let mainstream students figure it out. I often suspect that some of the students who drop out of the university do so because they feel too much is being given up and not enough is being received. Dropping out may be a form of protecting cultural identity.

Cultural topics are equally important, if not more so, for students from the mainstream culture. Many mainstream students on predominantly white campuses feel inundated by Third World students. Their sense of cultural shock can be as profound as that of the ESL basic writer. One student began a quarter-long comparison/contrast essay on the immigration experiences of his Italian grandparents with the experiences of Mexicans and Vietnamese in California. As the quarter went on, the paper shifted focus as the student became aware that California was quickly becoming nonwhite. It scared him. The essay was eventually titled "Shutting the Doors?" and ended with:

> I have had some bad experiences with foreigners. On a lonely night in Davis, three other friends and I decided to go to a Vietnamese dance. When we got there, I couldn't believe how we were treated. Their snobbishness and arrogance filled the air. I was upset. But that was only one incident and possibly I am over reacting. I often reflect on my high school teacher's farewell address. He called for our acceptance of the cultural and religious background of

each other. But after long days thinking, I, like many others, am unable to answer. My only hope is that someone has a solution.

The student had grown up in Richmond, California, a culturally diverse East Bay community, and was friendly with students from many cultures (it was his idea to go to the dance). The very recognition on the part of the student of what it feels like to be surrounded by difference at the dance is a beginning step for him to understand what it means to try to be who he is in the midst of another culture. I see this student quite often. He is not racist. Or if he is, he at least has the courage to begin asking questions. The solution for which the student yearned was not immediately forthcoming: his yearning for one is worth writing about.

Language Topics

It is not unusual for ESL errors to persist in the writing or the pronunciation of highly educated people (doctors, lawyers, engineers, professors) because, consciously or unconsciously, those speech patterns are part of the person's identity and culture. The same can be true of basic writers. Language-oriented topics are one way to allow students to explore this kind of writing block. Assignments that require students to analyze their attitudes toward writing, their writing processes, and the role that writing plays in their lives can make these conflicts explicit. For example, one student wrote:

> Moreover, being a Chinese, I find myself in a cultural conflict. I don't want to be a cultural betrayer. In fact, I want to conserve my culture and tradition. I would be enjoyable if I wrote my mother language. For example I like to write letter in Chinese to my friends because I can find warmth in the letter. The lack of interest in writing and the cultural conflict has somehow blocked my road of learning English.

Overcoming this block may, however, cause problems in the home community. In response to Rodriguez, one student wrote:

> I can relate when Rodriguez say that his family closeness was broken. Even though I speak the language that is understood and is comfortable at home, when I speak proper English around my friends at home, they accuse me of trying to be something I'm not. But what they don't realize is that I have to talk proper in order to make it in the real world.

Jacquelyn Mitchell writes of returning to her community after college:

> My professional attire identified me in this community as a "middle-class," "siddity," "uppity," "insensitive" school teacher who had made her way out of the ghetto, who had returned to "help and save," and who would leave before dark to return to the suburbs. My speech set me up as a prime candidate for suspicion and distrust. Speaking standard English added to the badges that my role had already pinned on me. (31)

In some way problems like this affect us all. In the small town that I grew up in, simply going to college was enough to alienate you from your peers. Although my parents encouraged me to get an education, they and their peers saw college as producing primarily big egos—people who thought they knew everything. No easy choices here. It took my uncle, who made the mistake of becoming a Franciscan priest, forty years to be accepted back into the family by my mother.

Another way to help students with cultural transitions is to make the home language the subject of study along with the different kinds of academic discourse they will be required to learn. Suzy Groden, Eleanor Kutz, and Vivian Zamel from the University of Massachusetts have developed an extensive curriculum in which students become ethnographers and analyze their language patterns at home, at school, and in different social situations, using techniques developed by Shirley Brice Heath. This approach takes time (several quarters of intensive reading and writing), but such a curriculum has great potential for helping students acquire academic discourse while retaining pride and a sense of power in the discourse they bring with them.

Peer Response Groups

Peer response groups encourage active learning and help students link home and university cultures. The

Puente Project, in affiliation with the Bay Area Writing Project, combines aggressive counseling, community mentors, and English courses that emphasize active peer response groups. The Puente Project has turned what used to be a 50–60% first-year dropout rate into a 70–80% retention rate in fifteen California community colleges. All of the students are academically high risk (meaning they graduated from high school with a D average), they are Chicanos or Latinos, and all have a past history of avoidance of English classes and very low self-confidence when it comes to writing. Writing response groups give the students a sense of belonging on campus. As students make the transition from home to school, the groups become, in the students' own words, "una familia":

> Now after two quarters of Puente, it's totally different. My writing ability has changed to about 110%. I might not be the best speller in the world, but I can think of different subjects faster and crank out papers like never before . . . having Latinos in a class by themselves is like sun to a rose. This is the only class where I know the names of every student and with their help I decide what to write. (*Puente* 20)

Joan Wauters illustrates how structured non-confrontational editing can make peer response more than a support group. Students work in pairs on student essays with specific training and instructions on what to look for, but the author of the essay is not present. The author can later clarify any point she wishes with the response group, but Wauters finds that the non-confrontational approach allows students to be more frank about a paper's strengths and weaknesses and that it is "especially valuable for instructors who work with students from cultures where direct verbal criticism implies 'loss of face'" (159). Wauters developed these techniques with Native American students, but they apply equally well to other cultures.

Response groups do not have to be homogeneous. Any small group encourages participation by students who may not feel comfortable speaking up in class for whatever reason. They provide a supportive environment for exploring culturally sensitive issues that students might hesitate to bring up in class or discuss with the teacher. The following paragraph was read by a Black student to a group consisting of a Filipino, a Chinese, and two Chicano students:

> I am black, tall, big, yet shy and handsome. "I won't hurt you!" Get to know who I am first before you judge me. Don't be scared to speak; I won't bite you. My size intimidates most people I meet. I walk down my dormitory hallway and I can feel the tension between me and the person who's headed in my direction. A quick "Hi!" and my response is "Hello, how are you doing?" in a nice friendly way. It seems that most of the guys and girls are unsure if they should speak to me. I walk through the campus and eyes are fixed on me like an eagle watching its prey. A quick nod sometimes or a half grin. Do I look like the devil? No, I don't. Maybe if I shrink in size and lightened in color they wouldn't be intimidated. Hey, I'm a Wild and Crazy Guy too!

This small group discussion of what it felt like to be an outsider spilled over into the class as a whole, and students that normally would not have participated in class discussion found themselves involved in a debate about dorm life at Davis. I know that peer response groups have limitations, need structure, and can be abused by students and teachers alike. But I have never heard complaints from teachers using peer groups about how difficult it is to get ESL students to participate. In some cases the problem is to shut them up.

Class Newsletters

Class newsletters encourage students to write for an audience different from the teacher, and they generate knowledge about multicultural experiences. I use brief 20-minute in-class writing assignments on differences between the university and home, or how high school is different from the university, or ways in which the university is or is not sensitive to cultural differences on campus. Sometimes I simply have students finish the statement, "The university is like. . . ." These short paragraphs serve as introductions to issues of cultural transition, and when published, generate class discussions and give ideas for students who are ready to pursue the topic in more detail. Newsletters can be done in a variety of formats from ditto masters to desktop publishing. Students who feel comfortable discussing ethnic or cultural tensions establish a forum for those students whose initial response would be one of denial. For example, the story of a Guamanian and a Black student who thought they were not invited to a white fraternity party led to an extended class discussion of whether this kind of experience was typical and whether they had not been invited for ethnic reasons. The next time I had students write, a Chicana student articulated her awakening sense of cultural conflict between the university and her family:

> I was so upset about leaving home and coming to Davis. I was leaving all my friends, my boyfriend, and my family just to come to this dumb school. I was angry because I wanted to be like all my other friends and just have small goals. I was resentful that I had to go away just to accomplish something good for me. I felt left out and angry because it seemed that my family really wanted me to go away and I thought it was because they did not love me. I was not studying like I should because I wanted to punish them. My anger grew when I realized I was a minority at Davis. My whole town is Mexican and I never thought of prejudice until I came to Davis.

Family is central to Chicano/Latino/Mexican-American students. The pull toward home can create ambivalence for students about their school commitments. In this case, the family was aware of this pull, and encouraged the student to give college the priority. The daughter interpreted this as a loss of love. The student's ambivalence about home and school put her on probation for the first two quarters at Davis. She is now a junior, doing well as a pre-med student, and her chances for a career in medicine look good. I don't think just writing about these issues made the difference. The class discussion generated by her article helped her realize that her situation was shared by others who were experiencing the same thing but had not quite articulated it. She was not alone.

Bringing Campus Events into the Classroom

I recently assigned a paper topic for a quarter-long essay that made reading of the campus newspapers mandatory. I was surprised to find that many students did not read the campus newspapers on a daily basis and in many cases were quite unaware of campus issues that directly concerned them, for example, the withdrawal of funding from the Third World Forum (a campus newspaper that deals with Third World issues), compulsory English examinations for international graduate teaching assistants in science and math classes, and increasing incidents of hostility toward Asians (as reflected in bathroom graffiti, "Lower the curve: kill a chink"). Admission policies at UC Berkeley, particularly as reported in the press, have pitted Blacks and Chicanos against Asians, and quite often students find themselves in dorm discussions without having enough specific knowledge to respond. The more articulate students can be about these issues, the greater the chance the students will feel integrated in the university.

Assignments can make mandatory or strongly encourage students to attend campus events where cultural issues will be discussed within the context of campus life—issues such as the self-images of women of color or how Vietnamese students feel they are perceived on campus. A panel entitled "Model Minority Tells the Truth" called for Vietnamese students to become more involved in campus life partly in order to overcome misperceptions of some students:

> Vietnamese students feel inferior to Americans because Americans do not understand why we are here. We are refugees, not illegal immigrants. There are a lot of unspoken differences between Vietnamese and Americans because the memory of the Vietnam war is so fresh, and it is difficult for Americans to be comfortable with us because we are the conflict. Vietnamese students also suffer an identity crisis because the Vietnamese community has not established itself yet in America as other Asian groups have.

Anecdotes

Anecdotes about oneself and former class experiences are another way to generate discussion and raise issues of cultural transition and identity. The teacher's own curiosity and experience of cultural diversity will often give students ideas for other topics. Cultural identity does not depend on a Spanish surname alone nor does it reside in skin color. Richard Rodriguez, for example, does not consider himself a Chicano and was insulted when he was so identified at Berkeley during the sixties. I mentioned this in class and described the experiences of several former students whose parents were from different cultures (so-called "rainbow children"). Several weeks later, a student whose mother was Mexican and whose father was Anglo wrote:

> Someone once told me that I'd have to fight everyday to prove to everyone that I wasn't "another stupid Mexican." He was convinced that the whole American population was watching his every move, just waiting for him to slip and make a mistake. Having it emphasized that he is a minority certainly won't help his attitude any. It will just remind him that he is different. All of my life I never considered myself a minority. I didn't speak Spanish, I didn't follow Mexican customs, and I hung around with "American" kids. It is real hard for me to understand why minorities get so much special treatment.

This student found the existence of different student cultural groups on campus to be disturbing:

> When I came to Davis, instead of seeing a melting pot like the one I expected, I saw distinct cultural groups. When I first heard about CHE (Chicanos in Health Education) I was furious. I could not see any need for a special club just for Hispanic students interested in health careers. It seemed that the students in CHE were segregating themselves from the real world. They should actually be interacting with everyone else proving that they weren't different.

After interviewing members of CHE and VSA (Vietnamese Student Association), this student was able to see how some people benefited from these clubs:

> I didn't think I could find some positive aspects about these clubs, but I found some. Some clubs help immigrants assimilate into the Western culture. They provide the member with a sense of pride about who they are and they strengthen cultural bonds. If students attend classes and become discouraged by lower grades than they expected, they can go to a CHE meeting and see "one of their own" explain how they made it through the bad times and how they came back to beat the odds. I asked Trinh why she joined the VSA. She said she joined to learn more

about her culture and to improve her language. But others join to help themselves assimilate into Western culture. I was wondering what was so important about her culture that Trinh couldn't retain unless she went to these meetings. She said, "I can't explain it." But there is an atmosphere there that she can't get anywhere else. And if this gives her a good feeling, then more power to her.

This student still has reservations about cultural differences on campus (primarily because she does not want to see herself as "different"), but the movement from "I was furious about CHE" to "more power to her" is a step toward recognizing her own cultural diversity. Cultural identity is not always simple. I have seen second-generation Vietnamese, Indian, Korean, and Chinese students who saw themselves primarily as "American" (no hyphens), and in some cases as white. One Chinese student, from a Black neighborhood in Oakland, grew up wanting to be Black. It was the cool thing to be.

For some students, examining home culture and the culture of the university can cause anxiety. Teenagers often do not relish the idea of being "different." They have enough difficulty keeping their grades up, forming peer relationships, adjusting to being in a new environment. My sixteen-year-old stepson argues constantly that he is just like his peers. He certainly tries to be. But all you have to do is walk on the school grounds, look around a few minutes, and it is impossible to miss the six-foot-tall, dark-skinned, handsome boy bobbing up and down amongst a sea of white faces: definitely Indian (other Indians can identify him on sight as Telegu). It is becoming increasingly difficult on predominantly white campuses for students to deny differences in culture. What is important to learn is that while differences between home and school can lead to conflict, differences in themselves do not inherently cause conflict. The home culture can be a source of strength which can enable the student to negotiate with the mainstream culture. One of the major factors of success of students coming from cultures least valued by society is the ability of the family to help the student maintain a positive self-image that allows her to withstand rejection and insensitivity of mainstream peers. Occasionally, I have discovered some parents who did not want their children in school at all and did everything in their power to deter the education of their children. But most often, it is not the home culture that causes problems, but a fear on the part of students that elements of that culture will not be accepted in the university environment.

Implications for Teacher Training and Classroom Research

These topics and assignments not only help students mediate between school and home cultures, they provide windows for the teacher into the diversity within each of the cultures that students bring with them. They can serve as a base for ongoing teacher research into the ways in which home and university cultures interact. There simply is no training program for teachers, and can be no definitive research study that will ever account for the realities our students bring with them. Change is constant. Each generation is different. Given the lack of homogeneity in our classes, given the incredible diversity of cultures we are being exposed to, who better to learn from than our students? The culture and language topics I have described here comprise roughly 30–50% of the assignments I give in a ten-week course that meets for two hours twice a week. Some quarters I find myself using more cultural topics than others; it depends on the students. The course is English A, a four-unit course (two units counting toward graduation) with English Department administered, holistically-graded diagnostic, midterm, and final exams. The point is that if one can begin to integrate cultures under these constraints, one should be able to do it anywhere.

The cultural transitions we ask of our students are by no means easy. Cultural transition is ultimately defined by the student, whether she decides to assimilate and leave her culture behind, or attempts to integrate her world view with the academic world view. As composition teachers we are offered a unique opportunity to make these transitions easier for students, and at the same time increase our skills in moving between cultures. Clifford Geertz puts it this way: "The primary question, for any cultural institution anywhere, now that nobody is leaving anybody else alone and isn't ever again going to, is not whether everything is going to come seamlessly together or whether, contrariwise, we are all going to exist sequestered in our separate prejudices. It is whether hu-

man beings are going to be able, in Java or Connecticut, through law, anthropology, or anything else, to imagine principled lives they can practically lead" (234). But we may find, and this has been my experience, that in helping students make cultural transitions, we learn from them how to make transitions ourselves.

Works Cited

Bartholomae, David. "Inventing the University." *When a Writer Can't Write*. Ed. Mike Rose. New York: Guilford, 1985. 134–65.

Bizzell, Patricia. "What Happens When Basic Writers Come to College?" *College Composition and Communication* 37 (October 1986): 294–301.

Bourdieu, Pierre, and Jean-Claude Passeron. *The Inheritors: French Students and Their Relation to Culture*. Chicago: U of Chicago P, 1979.

Cummins, James. "Empowering Minority Students: A Framework for Intervention." *Harvard Educational Review* 56 (February 1986): 18–36.

Fleming, Jacqueline. *Blacks in College: A Comparative Study* of *Students' Success in Black and in White Institutions*. San Francisco: Jossey-Bass, 1985.

Freire, Paulo. *Pedagogy* of *the Oppressed*. New York: Continuum, 1982.

———. *The Politics of Education: Culture, Power, and Liberation*. South Hadley, MA: Bergin & Garvey, 1985.

Geertz, Clifford. *Local Knowledge*. New York: Basic Books, 1983.

Groden, Suzy, Eleanor Kutz, and Vivian Zamel. "Students as Ethnographers: Investigating Language Use as a Way to Learn Language." *The Writing Instructor* 6 (Spring–Summer 1987): 132–40.

Heath, Shirley Brice. *Ways with Words: Language, Life, and Work in Communities and Classrooms*. New York: Cambridge UP, 1983.

Holzman, Michael. "The Social Context of Literacy Education." *College English* 48 (January 1986): 27–33.

Mitchell, Jacquelyn. "Reflections of a Black Social Scientist: Some Struggles, Some Doubts, Some Hopes." *Harvard Educational Review* 52 (February 1982): 27–44.

Ogbu, John. *The Next Generation: An Ethnography of Education in an Urban Neighborhood*. New York: Academic Press, 1974.

Olson, Carol Booth, ed. *Practical Ideas for Teaching Writing as a Process*. Sacramento: California State Department of Education, 1986.

Petrie, Loretta. "Pulling Together the Multicultural Composition Class." CCCC Convention. New Orleans, March 1986.

Philips, Susan Urmston. *The Invisible Culture: Communication in the Classroom and Community on the Warm Springs Indian Reservation*. New York: Longman, 1983.

The Puente Project: Building Bridges. Berkeley: Bay Area Writing Project, 1985.

Rodriguez, Richard. *Hunger of Memory*. Boston: Bantam Books, 1982.

Steinberg, Stephen. *The Ethnic Myth: Race, Ethnicity, and Class in America*. New York: Atheneum, 1981.

Wauters, Joan. "Non-Confrontational Critiquing Pairs: An Alternative to Verbal Peer Response Groups." *The Writing Instructor* 7 (Spring/Summer 1988): 156–66.

Flux: The Shifting Role of the Composition Teacher

Dabney Gray
University of Alabama

I was a farmer by birth, as was my father. We came from generations of farmers who toiled in the sun to glean the often meager living the earth afforded. Farming, the actual work of preparing land, sowing seed, caring for a crop through unpredictable spring weather and summer heat, and then trying to harvest it in the fall before the rain and wind ruined the year's work, can dissipate the strength and break the will of the heartiest individual. The physical labor takes a toll on the body, and the anxiety of such a huge investment exposed to the forces of capricious nature preys on the mind. This is the most visible side of farming, the one concerned with survival, of lasting through the current year and hoping that the next one will be more forgiving. But farming has another side, a capacity for revitalizing the dispirited, a potential for rejuvenation seemingly unconnected to the mundane, day-to-day tedium of working the fields. What gives farming this hidden value is the sense of universal proprietorship, a kind of mutual holding of the land and its potential for life in trust that one acquires when he works closely with nature.

And here is the most important lesson my father taught me about his profession. For him the greatest reward from tilling the soil came not at the end of the year when he harvested his cotton and soybeans, selling them for what was seldom enough to cover the expenses incurred over seasons of unpredictable cold snaps, rain storms, and dry, dusty Augusts. For my Dad the true value of farming had little to do with tangible gain but came from a sense of personal worth he received from participating in a process larger than himself, because he was part of an immense, perpetually enduring cycle in a very personal way. A farmer functions as protector

Gray, Dabney, "Flux: The Shifting Role of the Composition Teacher." *Freshman English News* 17, no. 1 (1988): 29–33. Copyright 1988 by *Freshman English News*. Reprinted with permission.

of and in communion with nature in its broadest context. His province is the nurturing of life.

Teaching composition parallels closely the process the farmer follows to prepare for his crop, for the magic of life that springs from the seeds he plants. By plowing his soil, forming a seedbed, and waiting until the time and conditions in the field are just right to receive the seed, the farmer does all he can to insure that the process on which he depends will begin anew each year. As composition teachers, we, too, attempt to prepare as carefully as possible by creating an environment in which the process of writing can flourish. Like the farmer, we have no guarantees, no real control over the changing conditions that might ruin our crop at any time. We can only hope that once our seeds are planted the magic will somehow happen again.

When l left farming and entered graduate school, l found that what my father had taught me about farming applied to my new job as a teacher of freshman composition. The sense of worth derived from farming parallels that found by those teachers truly dedicated to helping freshmen learn. In many ways the farmer's physical work of nurturing life in the field is inextricably tied to his reward because his labor contributes to the greater cycle of life. Likewise, the composition teacher's goals and rewards are bound together if he or she is truly concerned with helping students become literate, independent thinkers, and not just with imposing on them a set of preconceived standards. Our province is the nurturing of potential. Like the farmer whose careful planting, watering, weeding, and eventual harvesting of his crop helps insure not only his financial survival, but his personal fulfillment as well, the concerned teacher's informed and sensitive application of pedagogy allows the individual student to flourish and in turn contributes to the teacher's own sense of accomplishment.

Insights into teaching, like those in farming, are best revealed through experience. What I have discovered about the field of freshman composition is that, for anyone who actively seeks to teach freshmen to write, helpful pedagogy is in a constant state of flux, and that this flow of contemporary ideas often takes tangential and in many cases opposite directions to established, more traditional methods of teaching. And even where pedagogical theories may seem a helpful

foundation on which to build a personal teaching strategy, they can only suggest guidelines, broad, rough frameworks, over which each teacher must stretch the fabric of his or her own style, molding and shaping the pedagogy until it takes on that teacher's unique features. Seen in this context, it is quite often the pedagogue and not the pedagogy that matters most in the classroom.

Too often beginning teachers become jaded or burned out because the methods they were taught and those they are expected to use instructing freshmen are out of sync with the way they, as writers, create their own compositions. Textbooks as well as suggested departmental syllabi often advocate a linear movement from word, sentence, paragraph, essay and prescribe writing tasks with the same lack of awareness of the writing process. I faced this problem my first semester as a graduate teaching assistant. The syllabus supplied from my English department required a research paper in the course and included a suggested timetable for students to dutifully complete each step in the "process" of "writing" the paper. Bewildered by such supposed organization in the process of writing, I made an honest effort to use this syllabus with its exhortations to collect bibliographic cards, outlines, and note cards. Then, I gave up. Never having written a paper with such things, I had no idea how to teach someone else to do it. The only time I had ever written a formal outline or note cards had been for my freshman English instructor, and those I wrote after finishing the paper— because I was required to turn them in as evidence of work completed in preparation to writing.

For my inability to teach "the term paper" the way it "should" be taught, the "correct" way, I felt like a failure, a traitor to the ranks of "real" English teachers. Had anyone suggested to me that there is no fixed method, no "right" way, to teach freshman composition the anxiety I experienced could have been greatly alleviated. Surely, my case is not an isolated instance of ignorance or inability. What I needed was not a prescribed, unbending syllabus, but enough exposure to current trends in composition pedagogy to have some awareness that no one teaching method is the only acceptable way to teach writing.

Now I know better how to deal with the incongruities between suggested methodologies and the way writers write. I have learned that vigilance and flexibility in teaching help my students most. I must allow learning to take place by having the judgment not to interfere while the process is working smoothly, but be aware of what my students are experiencing and the difficulties they face to more successfully act as a support when demands threaten to overwhelm them. Like the farmer who stays in touch with a crop by walking the fields through the mud or the dust to experience the choking weeds and entangling vines that threaten what he has so tenderly nurtured, the teacher must have the wisdom to know when circumstances require him to sometimes advise, sometimes correct, or sometimes only watch the burgeoning crop. This is a difficult thing to learn. Nurturing students is so often a matter of timing.

Metaphors abound to explain this concept. For Peter Elbow the teacher becomes a coach, an ally, who encourages students to take risks, to deform themselves and their material alike in order to experience the "violence" and exhilaration of learning (331). Carolyn Matalene encourages the teacher to become an editor in the most professional sense of the word, giving support or censure to student writers, depending on which one she deems necessary in a particular situation (4). Nancy Sommers, taking her lead from Knoblauch and Brannon, warns teachers not to become appropriators of a student text by forcing it to conform to any preconceptions the teacher might have, but to listen to the student's voice in each text and to focus on what the student, not the teacher, is trying to say (149). Donald Murray emphasizes that before students can be introduced to the finer points of critical thinking they must first confront questions of content, meaning, or focus in their compositions. Only then can they, as writers, begin to make significant decisions about language and clarity of expression ("Teaching" 145). He sees himself as a wilderness guide "who doesn't lead so much as stand behind the young explorer, pointing out alternatives only at the moment of panic" ("Teaching" 142).

All of these writers, I believe, are pursuing the same thing. In attempting to demonstrate a concern for the student writer as an individual whose thoughts and perceptions have validity, they are also reserving a significant role for the teacher, one that is less certain than in the traditional teacher-student relationship and, at the same time, more exacting because of its lack of struc-

ture. But this new role for the teacher need not seem threatening. Once a teacher accepts this extension of the traditional view of his or her job, the current emphasis on post-structuralism meshes with pedagogy and translates into informed writing instruction, obligating each composition teacher to meet students on a common ground, to aid them in what Ann Berthoff calls the "making of meaning," to show them that the act of composing is "the work of an active mind and is thus within their natural capacity" (*Mean*ing 67). Doing this implies a commitment by the teacher to plastic methods, to tailoring teaching practices for the individual student in so far as that is possible.

Jay Robinson delineates this new direction for teaching composition by describing the attitude he brings to his own writing class:

> When I teach a composition class, I must remember that my cultural frames are not my students' or theirs mine. I must remember that their experiences are neither mine, nor something I want to appropriate by investing them with my meanings. And yet, I want them to learn; and yes, I want to work with them toward common meanings, meanings that we can share, meanings that will make possible the possibility of a common language, a public discourse made of and constitutive both of self and community. (495)

Clearly, Robinson exhibits the influence of a directional shift in composition pedagogy. The thrust here is to empower the student, to locate and develop what Donald Murray calls "the other self," the ability in each student to discover his own hidden potential for creating and evaluating a personal text. No longer does the traditional model of all-knowing teacher and ignorant student dominate the classroom. The role of teacher and that of student are both redefined.

What Robinson proposes is a compromise between theorists concerned with defining composition as solely a mode of thinking, of making sense of the world, and practitioners who regard composition mainly as a mode of communication. In the freshman classroom, space for both these concepts exists. For any instructor concerned with maintaining the student at the center of the curriculum, Robinson's concern for allowing students to make their own meaning while at the same time working for a common language suggests a difficult but challenging role for the student and the teacher. In a Robinson styled classroom the emphasis would be twofold: first, to help each student find within himself or herself the potential for making sense of the world; second, to enable that student to communicate individual perceptions and thoughts in a way that will be readily understood and appreciated in the broader discourse of the university. In short, Robinson is suggesting that we as teachers of composition need to allow our students the liberty of individual expression so long as that expression remains grounded in literacy and critical thinking.

But, by the use of the term literacy I do not mean to imply that upon first entering the composition classroom all or even most students can be considered literate, with the ability to think critically. The desired goal is the development of the kind of literacy defined by Robinson as "an outcome, not a skill, and not (even) a competency. It is something that is achieved when competencies are enabled through exercise of the human capacity to make meaning" (485). The composition classroom is a place to nurture literacy, not to find it full grown.

Literacy comes to students at different speeds; some may not acquire it at all. And for those to whom it does come, it is not necessary for them to be aware of their gain. They can learn the application of an intellectual process without insight into the process itself Often such insight arrives with sudden and unpredictable impact, the burst of light aptly named "allatonceness" by Ann Berthoff. This, I believe, is what Robinson means when he calls literacy an "outcome," the culmination of energizing the various, subtle capacities involved in moving a student from writer to critical thinker, from emulator to creator.

This new model for the composition teacher does not imply a rejection of the traditional evaluative function of teaching, but it suggests that this role be fulfilled with compassion and insight, attune to the individual student as the focus of the composition process. Peter Elbow has some useful insights into the dual and seemingly conflicting duties of teacher as both facilitator and evaluator. He contends that a good teacher, one concerned with student learning, must function unflinchingly in both capacities, and he offers as examples worthy of emulation the teaching methods of Socrates and Christ, the "archetypal good teachers—archetypal in being paradoxical." Elbow credits them with doing violence to what they taught in order to get their message across while at the same time insisting that "learners bend and transform themselves in order to become fit receptacles" for what they had to impart (333).

The formula Elbow recommends for this implies a dichotomy in the role of teacher. He suggests that as a facilitator a teacher should:

1. assume students are intelligent and capable.
2. let them know you are on their side by "showing them that the perplexity or ignorance they reveal . . . will not be used against them in tests, grading, or certifying."
3. take the stance of advocate in support of the student regardless of how outrageously that student might have violated your trust in the past.
4. demonstrate that you too are still learning, having doubts, seeking help, and needing support, that you are open to new ideas, because neither you nor they have all the answers (332).

For the role of evaluator Elbow also proposes four techniques:

1. Insist on standards that are high in the sense of demonstrating that there exists "a 'real world' of truth, of good reasoning, of good writing. . . ."
2. Approach all student performance with a critical eye in an attempt to reveal those students who either cannot or will not learn.
3. Refuse to become attached to students, to assume their sensibilities, their viewpoints.
4. Identify yourself with the culture and the academy and not with the students you teach (333).

For Elbow the key to putting these two roles into practice is to clearly delineate his evaluative persona at the beginning of the semester and each time he makes an assignment. Otherwise, the persona of friend, advisor, and facilitator has precedence. By allowing students to realize that the teacher does have two jobs, but that they can and will be separated at different phases of the course, both the students and the teachers are spared some of the trauma of dealing with seemingly split and opposing teacher obligations.

My own search through the flow of pedagogical ideas has taken my classroom through several metamorphoses. When I began teaching, I lectured because I thought that was the way a teacher taught. That changed because it seemed I could not talk my students into a good essay. After trying various other methods—peer evaluation, short in-class assignments, rough draft workshops in which I tried to read all their drafts at one succession—I started having conferences with individual students, one-on-one. This method seems to produce the best writing from my students and enables me to apply Elbow's ideas of a dual teaching role directly and personally to each writer. In these face-to-face meetings, I get to know the individuals I am dealing with on a level not possible in a classroom situation. And although these sessions do not last as long as a class period, usually 10–15 minutes, the results I have seen have convinced me that I may be on the right track.

In these conferences I try to adopt the techniques proposed by Donald Murray who suggests teaching less and giving up some controls over the student and his text in an effort to allow the student to learn more. Murray believes that by supplying the appropriate questions about a draft at the right time a teacher can motivate, not intimidate, a student into better writing ("Listening Eye" 14). He offers six questions for the teacher to pose and for the student to consider as a prompt to future drafts:

1. What did you learn from this piece of writing?
2. What do you intend to do in the next draft?
3. What surprised you in the draft?
4. Where is this piece of writing taking you?
5. What do you like best in this piece of writing?
6. What questions do you have of me? ("Listening Eye" 15)

These questions, or variations of them, work primarily because they put emphasis on the student's writing and do not allow the conference to become too teacher oriented.

Of course, a student too eager for approval, or unsure of what role he is expected to play in a conference setting, can often return with a second draft and say that he now has a much better paper because all the changes I suggested have been made. When this happens I know I have failed for the present in reaching this writer, in helping him discover the "other self" and become a thinker and not just an emulator. But even this failure is a sign of progress. At least the second or third draft exists, and we have something to work with. Then, at other times a student will leave a conference with a parting comment suggesting that on the next draft he will "figure me out," as if a certain combination exists, and as soon as he finds it, he will crack this thing called composition. When this happens I always feel that I have had a small success, that at least for the present I have dodged the bullet of figurability.

Conference teaching has two distinct advantages over the methods I formerly employed and for these reasons alone it is a viable alternative to the traditional classroom situation. First, my students are producing multiple drafts, drafts that I have the chance to see take shape. Conferences make it possible for me to see writing in progress more than I ever did in the classroom where because of time limitations I could only get a general idea of how each paper was progressing. Second, conferences allow each student to move at his or her own speed, providing a way for more advanced writers to stretch the limits and not be bound by the constraints of a class that may not be up to their level. And conferences enable weaker students to receive the special attention they need in a situation more personal and direct than the classroom. Often knowing a teacher is interested in them not just as writers but as individuals sparks these students to put forth the effort necessary to achieve far greater gains than they ever thought themselves capable. For students with less enthusiasm, conferences bring them face to face with a teacher and require of them more than minimal effort. In the typical classroom setting these students are often ignored by teachers whose time restrictions may prompt them to take a greater interest in their classmates who show more enthusiasm.

This system is not perfect, however, at least not the way I have been able to implement it. Some students still slip through, as they always will, and do not perform up to the hoped for standards. I view this as a fault of the conference method of teaching that puts the burden of learning on the student. It requires a certain amount of maturity that some have not attained. Another limitation of this system is the lack of perspective the students are allowed in appraising their writing. I have found that at times having another student read over a paper and confront the writer with a vague sentence or an awkward phrase is far more productive than my doing it from what is usually perceived as an adversary point of view. In a conference this, of course, is a difficult situation to arrange. The only way I have been able to do it effectively is to schedule a group of students to come for conferences at the same time and use those waiting their turns as my readers. This small group situation furnishes a pool of readers that can be beneficial in reaffirming or rejecting any particularly critical comments I might make about a paper. The limitations of this are the same as with other peer evaluations, however, in that the whole process depends on abilities and attitudes of the readers.

Overall, though, I have been more satisfied with the conference system than any other method I have tried. The most satisfying aspect is watching as my students grapple with the writing process and feeling that I have a real input into their attempts to work through the problems they face as writers. There is a more practical consideration, however. This method allows me to read through and evaluate papers holistically and eliminates the need for so many written comments. By encouraging successive drafts to be turned in for a grade at any time, the pressure of grading a complete set of papers at any one time is alleviated because the papers do not all come in at once. In this respect the work load is not lessened, but it is spread over the entire semester and the work required seems not so monumental as grading an entire set of papers often does.

In the future I would like to try a couple of techniques that might enhance my conference teaching, or at least alter it somewhat. I believe that the use of portfolio grading might lend itself naturally to the conference system; indeed, even though I now give a set number of assignments, by requiring successive drafts and by not accepting any paper until the student and I both feel it is ready for a grade, I have created a portfolio situation without calling it such. Something else I would like to try if the facilities were available is to have students save their work on computer disks and grade for grammar and mechanics by computer with some sort of style checker. This could differentiate for the student the job of editing from that of revision. It would also let conference time be spent more productively than if we allowed ourselves to become bogged down in error correction.

But these are problems for next semester or next year. Or perhaps I will change my methods, completely abandon conferences, and again enter the flux of ideas swirling around me, hoping as every time before, this time to grab and hold the "right" one. And I know before I start that I am fortunately doomed to failure, understanding that I can never reach closure with pedagogy because each situation offers different circumstances and suggests a new approach. Another year may require a different set of methods, a new design for success.

As I write this I know it is not exactly what I want to say about teaching. My ideas remain romantic and fluid. I feel like Faulkner must have when asked to justify his style. He claimed that he wanted to say everything worth saying in one sentence. Failing to do so in one sentence, he attempted the same thing in one book—book after book. Teaching composition is like that, an addictive endeavor. There is no end. Like Faulkner, I will keep on trying to get it right and knowing that I never will. Faulkner, himself a farmer, knew that the end never comes, the answer remains elusive but that the value of his endeavor was in the attempt. The beauty, the worth, in teaching and in farming comes from personal fulfillment.

Works Cited

Berthoff, Ann E. *The Making of Meaning.* Upper Montclair, N.J.: Boynton/Cook, 1981.

Elbow, Peter. "Embracing Contraries in the Teaching Process." *College English* 45 (1983): 327–39.

Matalene, Carolyn. "The Teacher as Editor." *Journal of Teaching Writing* 5 (1986): 3–15.

Murray, Donald M. "The Listening Eye: Reflections on the Writing Conference." *College English* 41 (1979): 13–18.

———. "Teaching the Other Self." *College Composition and Communication* 33 (1982): 140–47.

Robinson, Jay L. "Literacy in the Department of English." *College English* 47 (1985): 482–98.

Sommers, Nancy. "Responding to Student Writing." *College Composition and Communication* 33 (1982): 148–156.

Improving Students' Responses to their Peers' Essays

Nancy Grimm
Michigan Technological University

As any composition teacher knows, dividing a class into groups of four or five people does not automatically insure that everyone will receive useful response. Even students complain that too often peer-group work feels like the blind leading the blind. When I first started using groups, my intention to give students a sense of direction led me to compose critique sheets on ditto masters with questions such as "What do you think the main point of the paper is?" "What examples does the writer provide to support the main point?" In spite of the good intentions behind the response sheets, the purple dittoed pages called up the fill-in-the-blank syndrome. Students mechanically filled in the empty spaces without the analytical thought or oral exchange that a good response session demands. If a question asked them to identify the paper's thesis, the students obediently copied the last sentence in the first paragraph. The real thesis might have lain buried in the last paragraph, but discussion of that possibility didn't occur. The critique sheet turned response into busywork. Students responded to the teacher's questions instead of to their own.

Grimm, Nancy. "Improving Students' Responses to their Peers' Essays." *College Composition and Communication* 38 (1986): 91–94. Copyright 1986 by the National Council of Teachers of English. Reprinted with permission.

After several years of experimentation and observation, I have replaced the critique sheets with the guidelines printed below. I discovered that students do need guidance through the response process, but they need a framework rather than a blueprint. The guidelines, in addition to providing students with a familiar procedure, also suggest a form of group etiquette that I derived from watching successful groups operate.

I give my students a copy of the guidelines the class period before an early draft is due. We discuss the process outlined and review it periodically throughout the term. The periodic review is important because it deepens their awareness of the process in the same way that mirrors in a dance studio develop a self-consciousness of movement. For this article, I follow each point in the guidelines with comments on how or why the step evolved.

Guidelines for Writing Groups

1. You will always have *2 days* for group work on a paper. Divide the time equally among group members. Distribute copies of the paper and begin quickly. Keep a pile of extra copies handy for me.

Lack of adequate time contributes to superficial response. With four to five students per group and fifty-minute class periods, I set aside two days for group work on early drafts. Knowing that it takes me at least fifteen minutes to read and respond to a paper, I allow students about thirty minutes per paper to allow for their lack of experience.

2. Classwork should be oral. For homework, do *written responses*. Give these responses to one another on the second day of responding. Remember to turn them in with the final draft.

The expressive talk that occurs during group discussion often helps the writer find a stronger focus and develop his or her points more specifically. However, because talk is ephemeral and writing is more responsible and committed, I ask students to write responses to the drafts outside of class. These written responses can often suggest solutions to problems that the oral exchange has uncovered and named.

3. Always begin by having the writer read his or her piece *aloud* while you follow along on your copy, marking places you want to discuss.

This step has value for both the writer and the readers. The sound of her own voice gives the writer clues about what needs to be revised, and the slower pace of an oral reading allows peers to think critically as they read their copies.

4. Allow for *silence* after the oral reading. Give people time to formulate their response. Expect to feel awkward and tense at first.

If students don't expect the inevitable awkward silence, they rush to fill the void with a comment meant to put the writer at ease. A comment such as "Well, this is good—I like this" freezes further discussion. Few students have the courage to follow with "Well, I don't think it's so good."

5. Begin discussion by *encouraging the writer to ask questions* about what he or she wants responses to—what she is not satisfied with or where he had difficulty.

Students who lack experience with group process are too likely to throw their draft into the arena and then retreat into silence while the others sniff it over or tear it apart. Having the writer begin discussion reduces the initial tension, provides a focus for discussion, and allows the writer to maintain a sense of authorship. Because the writer is asking for help, the group members feel more comfortable in the role of critics.

6. Ask the writer questions. These questions may be to elicit more detail, to see why the author made certain choices, to clarify the writer's intention.

Because students have for years been trained to answer rather than ask questions, they need to work hard to perform the question-asking roles. Teachers can conduct practice sessions using anonymous student papers to help students identify the difference between strong and mediocre writing. They can remind students of central concerns appropriate to specific kinds of writing. If, for example, a paper is expository, then early concerns are maybe these: Has the writer found a thesis yet? Does she stay with each point long enough? Are the examples convincing? Students may also need guidance in translating these concerns into questions to the writer. If there is no thesis, then they might ask, "What is it you really want to say about _____?" They may also need criteria for deciding whether the thesis is adequate: does it go beyond stating the obvious? is it narrow enough for the paper?

7. Elaborate on one another's responses and encourage everyone to contribute. Discussion should be between three or four people, not two people.

Teachers who take time to introduce the concept of group roles will find that once students are aware that there are certain positions to be played, they work more effectively within the groups. Two roles especially important to the functioning of a writing response group are elaborator ("I noticed that too, but my reaction was . . .") and gatekeeper ("What do the rest of you think?"). When one student picks up on another's comment and extends it further, she is heightening the credibility of peer response, deepening the group's understanding of what is often undeveloped, and also insuring that one student doesn't dominate the discussion. Learning to elaborate on one another's responses also helps students narrow the focus of discussion. Gatekeeping, opening the door to quiet or blocked-out members, contributes to the likelihood of elaboration as well as helping the group achieve strong consensus.

8. Keep in mind the assignment and the audience for the piece. *Make suggestions* for strengthening it.

Students often serve as surrogate critics. If someone else will be evaluating the final piece, they must role-play that audience and they must know the criteria that will be used for evaluation.

9. *Point to specific parts* that seemed strong to you. Ask others how they reacted to the same parts.

I used to ask students to begin discussion by commenting on strengths in the piece, but I found that focusing too early on strengths created a false sense of euphoria, making the group reluctant to move on to criticism. Saving specific compliments for later gives the writer confidence about the work ahead and ends discussion on a positive note.

10. Conclude by naming one or two things the writer should concentrate on revising. Encourage everyone to contribute.

This summing up provides the opportunity to rank concerns. Someone's early comment about wordiness may now be seen as a problem with the development of a paper rather than with style.

The guidelines conclude with three reminders *to the writers* about how to use responses:

1. *Take notes* on your peers' responses. Don't trust your memory. Fresh insight is too easily lost.
2. Decide what changes *you* want to make. Check with me when you would like another opinion.
3. Restrict your defense. Listen to your readers and ask them questions. If you find yourself talking more about your *subject* than your paper, it often means you have more writing to do.

Conclusion

Even with carefully written guidelines, students' ability to function well in small groups improves slowly. Teachers can hasten improvement by showing as well as telling students about the behavior expected of them. Teachers can provide live models of the unfamiliar process by holding small group conferences instead of individual ones. For me, this means periodically canceling regular classes and having individual groups meet with me for forty- to sixty-minute seminar-style discussions of their drafts. Students frequently tell me that they have a better idea of what is expected of them after having watched a "pro" do it. They recognize with relief that the typical group session starts awkwardly, gropes for direction, ends on uncertainty, but still motivates and guides revision. Students need to see the teacher coping with issues (or the lack thereof) in the papers and asking the writer questions that demonstrate a lack of understanding but still show an interest and concern. They need to see that it isn't cheating to tell the writer what she might be saying even though she hasn't said it yet. Most importantly, they need to see that no one expects them to offer magic formulas or prescriptions. What we *do* expect is the expressive talk, the hard questions that ask the writer to re-think his subject, forcing him back to the chaos of an early draft but eventually bringing a sharper focus and more thorough development.

Reading and responding thoughtfully and sensitively to draft essays is demanding work. When my students know that their effort "counts," they work harder to develop the ability to judge and to communicate their responses. My students know at the beginning of the term that members of their group will evaluate them on four points: (1) their oral and written response, (2) their contribution to group functioning, (3) their attitude toward group work, and (4) their attendance. This peer evaluation accounts for twenty percent of their final grade and has the paradoxical effect of making them more comfortable accepting the responsibility of group work.

Lessons from the Computer Writing Problems of Professionals

Gerald Grow
Florida A & M University

In the past seven years, I have edited the writing of a number of professionals—including instructional designers, engineers, management consultants, environmental planners, biologists, psychologists, Army officers,

Grow, Gerald. "Lessons from the Computer Writing Problems of Professionals." *College Composition and Communication* 39, no. 2 (May 1988): 217–220. Copyright 1988 by the National Council of Teachers of English. Reprinted with permission.

and journalists—who write with computers. Like most users of word processing, these are not "writers"—they are professionals whose work requires them to write. Few of these have ever heard of "the writing process," and few have had any formal training since freshman English 20 years ago. For them, like millions of others, writing by computer is largely a self-taught enterprise.

Although most of these professionals share the belief that computers help them write, they display specific writing problems that may actually be caused, or accentuated, by the fact that they write on computers.

There are two reasons why the writing problems of professionals may be important to teachers of writing. First, students that I have taught (graduate students in instructional development and education; juniors and seniors majoring in communication and journalism) show similar tendencies when they write on computers. Though student writers may not have enough experience to demonstrate all of them, they distinctly gravitate toward the writing problems described here.

Second, many students from writing classes will soon be surrounded by people who have largely taught themselves writing and word processing. These self-taught professionals will become your graduates' next writing instructors—and their bosses. Unless students bring with them enough experience to maintain and defend good writing habits—the kind that make them effective, productive writers—they may be swamped by the kind of writing habits and writing problems common among self-taught professionals.

I will describe the problems I have observed among "real world" users of word processing and suggest some strategies for working with future professionals while they are still your students. What I have to say will apply best to nonfiction writing that is amenable to strong focus and clear organization—functional writing of the kind required of professionals in many fields.

The Editing Trap

Computers seem to tempt people to substitute writing for thinking. When they write with a computer, instead of rethinking their drafts for purpose, audience, content, strategy, and effectiveness, most untrained writers just keep editing the words they first wrote down. I have seen reports go through as many as six versions without one important improvement in the thought. In such writing, I find sentences that have had their various parts revised four or five times on four or five different days. Instead of focusing, simplifying, and enlivening the prose, these writers tend to graft on additional phrases, till even the qualifiers are qualified and the whole, lengthening mess slows to a crawl.

Drawn in by the word processor's ability to facilitate small changes, such writers neglect the larger steps in writing. They compose when they need to be planning, edit when they need to be revising.

Problems in Collaboration by Computer

Computers encourage more collaborative writing, and they encourage the collaboration to be far more intense. Before computers, the usual form of collaboration consisted of dividing up the work so that different authors wrote different chapters; then they reviewed one another's work. Writing with computers, though, collaborators can enter into one another's work so readily and revise it so easily that, in effect, co-authors can mutually co-write each sentence.

This kind of collaborative writing can be difficult to read. No two writers have quite the same sense about punctuation, tone, rhythm, headings, sentence variation, and the like. In collaborative works, I sometimes find grammatical conventions changing from the beginning to the end of the same sentence—because one author started the sentence and the other finished it.

In the worst cases, collaborative writing becomes a colloid of conflicting styles. In a document I recently edited, one section was written by a psychologist with a propensity for theoretical language, another by a computer programmer concerned mainly with the technical characteristics of machinery, another by a manager recording the history of the project. To complicate things, each author had inserted a few sentences (in his own style) in the midst of the other sections. Every time I reached a new major heading, the narrator changed voice—and the voices occasionally jumped around from sentence to sentence. It was schizophrenic prose, with faults that had been amplified by the easy editing made possible by the word processor.

Reused Prose

Writers easily become attached to what they have written, even when it serves the purpose badly. The computer frees many writers from this attachment by making the text fluid and continuously editable; for some writers, though, computers make this attachment harder to break. Typewriters challenge this attachment; in writing with a typewriter, writers typically retype each passage several times, which forces them to reread word for word and presents an excellent occasion to hear the passage and make changes. By contrast, a word processor enables writers to reuse passages from the developing piece so easily that reuse becomes a universal, invisible step in writing.

Being pragmatic, professionals often reuse blocks of material from previous reports. A good writer can do this well; but a less accomplished writer easily succumbs to a clumsy kind of self-plagiarism. Most of the adult writers I have worked with reuse "boilerplate" materials in a simple, modular fashion, stacking blocks of self-contained material in the midst of new passages, having little sense of how to combine the different parts. Most of them are tone deaf to the lurches, shifts in convention, and changes in tone between new and old writings.

I often advise authors to throw out these drafts and rewrite from scratch, but no one ever has. In part, they are always too busy; but more important, they are not writers. They are unaccustomed to taking responsibility for a piece of writing, devising an effective strategy, and seeing it through. Few of them have developed an effective writing process, and their approaches to writing lack flexibility. Such people do not need an editor; they need a writing instructor—something they lack but your students are fortunate enough to have.

Distortions of Length: Prolix and Telegraphic

The ease of writing on a microcomputer liberates many writers. And though this liberation helps reticent students, aids brainstorming, and makes many professionals more productive, the very ease of writing can lead to problems. People who have little to say suddenly take a long time to say it. Word-inflation multiplies. Instead of saying it well

one time, unfocused writers devise dozens of ways of coming close to saying what they mean. They continue writing. The words pile up. The results look impressive, but I never know quite what the writers meant to say.

Computers have the opposite effect on their writers. Normally intelligible, they become cryptic. Each mysterious word stands for phrases, sentences, even whole pages of unwritten intensions. I have to pry the words apart to uncover the thoughts concealed between them.

Small Screens and Clean Printouts

It is hard enough for writers to get an overview of their work when they write on a typewriter; when they use a screen that shows, at most, 24 lines at a time, an overview is almost impossible. Without a clear sense of direction, prose wanders, partially expressed ideas recur, and the point gets muddied.

Faced with this problem, I advocate various strategies, including extensive use of an outlining program. Among the professionals I have assisted, however, few use outlining programs (something available to them all), and only a few others outline effectively in any form. Generally, they try to write as they talk—offhand, or from a few casual notes.

Even one of the nicest features of word processing can cause an unexpected problem. At any stage, writers can always get a beautiful print-out of perfectly-formatted clean copy. Unfortunately, some writers confuse clean copy with completed copy. When they see their words on nice-looking pages, they begin to feel finished long before they have a reason to. When I think this is happening, I deliberately mess up their printouts with markings—to shock them into distrusting the finished appearance of their working drafts.

Computer Self-Defense for Future Professionals

You may find the problems I have described worth discussing in technical and professional writing classes where students are writing with computers. I raise them in courses on editing and magazine article writing. The recommendations that follow may not seem new, but they offer a different perspective from which to justify, explain, and reinforce certain teaching methods.

Over-Editing. Emphasize that writers should think through the topic until they develop a strong focus. Reinforce the distinction between large-scale revision and small-scale editing. Give them exercises in revising passages that need, not just editing, but rethinking.

Collaboration. Show students ways to divide responsibilities among the writers preparing a collaborative writing project. (For excellent charts to track the various roles of collaborators, see Dan Poynter and Mindy Bingham, *Is There A Book Inside You?* 1985, 154–56.) Have them write in collaboration with at least two other students. Give the class practice editing a work of multiple authorship so that it has a single, consistent style.

Boilerplate. Show students how to reuse blocks of prose from previous work, and show them the problems that may come up. Give an assignment that requires such reuse. For example, give them an article on disk to rewrite for a different audience, reusing large portions of the original text. Or have them design form letters that use boilerplate passages, then add appropriate transitions and edit the results for continuity.

Problems of Length. Experience is probably the best cure. Make sure your students write enough to become comfortable with a word processor. Show them the problems. Teach them to edit out prolixity, to identify and expand cryptic passages. By reading aloud examples of good writing, train their minds to hear the rhythm of prose that speaks and breathes. Interview cryptic writers as a way of drawing out their intentions and helping them move past a draft filled with enigmas.

Small-Screen Thinking. Help students clarify the topics they are writing about. Show them ways to keep the focus of the piece in mind while writing. For a simple method, tape a one-sentence summary to the side of the screen. For a more complex method, demonstrate how a project-management chart can be adapted to track a complex article as it develops. Study examples of how good writing is organized. Practice devising alternate plans for organizing and developing a given topic, preferably with the help of an outline processor like *ThinkTank.* Ask students to be ready at any time with a clear, short answer to the question, "What are you writing about?"

So they won't confuse clean copy with finished copy, encourage students to leave in extra headings, embed working questions, etc., so their working drafts look distinctly different from their final drafts.

Remind students that you are helping prepare them for a world after college—a world in which many of them will write with computers. By helping your students develop ways of writing that make them effective and productive, you give them the best general defense against the pressures facing them when they are required to write on the job.

Counter-Statement: Using Written Dialogue to Develop Critical Thinking and Writing

Stephen Hahn
William Paterson College

Many students "get stuck" in the development of critical thinking skills and in writing because they underconceptualize the context in which controversy occurs, and because they have not been shown that the monologues of written discourse are actually fragments of ongoing debates. Until they have been socialized to the conventions of discourse in the academy and the professions, they are unaware that significant written discourse almost always arises to counter some previous statement or assumption. The academic writing and lecturing students are exposed to often seems like the monological exposition of truth, remote from their own concerns. No wonder so many students come up empty when set to compose a critical or argumentative essay. Hearing or reading what probably seem like unmotivated monologues, students must suspect that some similar monologue should be occurring in-

Hahn, Stephen. "Counter-Statement: Using Written Dialogue to Develop Critical Thinking and Writing." *College Composition and Communication* 38 (February 1987): 97–100. Copyright 1987 by the National Council of Teachers of English. Reprinted with permission.

side their heads, just waiting for transcription. Thinking and writing therefore seem more solitary than they are. While successful writers in the professions see themselves actively engaged in continuing debates, novice writers do not see themselves as participants in such debates. Yet they have the potential for developing this sense of themselves before they achieve professional certification in a field.

One way to raise the student's awareness of being implicated in a continuing debate is to use the written dialogue as a basis for an extended writing assignment that combines exposition, analysis, and argumentation. James Moffett has outlined some useful strategies for using dialogue as a tool for writing and learning across the curriculum in *Active Voice* (Boynton/Cook, 1981, 49–71); the strategy described here, however, differs from his in using an explicit counter-statement as a prompt.

A counter-statement focuses the dialogue on an issue to be resolved. For example, in response to the prompt "But that's not music"—a counter-statement that might be uttered by one of the participants in a debate—students typically discover the roots of an argument. These are excerpts from two dialogues written by college freshmen:

"That's not music. That's junk."
"What do you mean, 'junk'?"
"There's not one song today where I can hear the words clearly and there's only one kind of beat to it. Boom, boom, boom, boom."

"That's not music."
"Why isn't it?"
"Because nobody in thirty years will even remember it."
"How do you know what people will remember in thirty years?"

After they have written their dialogues, I ask the students to form groups to share what they have written and to comment on the relationship between the characters or voices in the dialogues, implicit assumptions that the interlocutors make, and the expressed or implied attitudes of the speakers about the things to which they refer. The end-in-view is to generate thinking and writing, to show how arguments may begin with counter-statements, and to locate concern with argumentation in the student's own context. By beginning with a counterstatement, one is encouraged to develop an awareness of the social context of thinking and writing, and to see the difference between isolated "reasons" and reasoned arguments. It helps teachers avoid highly formalistic or moralistic approaches to teaching logic and argumentation.

In the examples above, more is at stake than an apparent difference in taste or opinion. In order to answer the counterstatement "But that's not music," the writer must imagine a complex response, or the conflict will not be resolved. In the dialogue in which one character states that "nobody in thirty years will even remember it," the writer may—and implicitly does—question the authority of this assertion (who knows?), and its relevance to the case (is memorableness a criteria for musicality? who remembers yesterday's songs?). In the other case, the writer may also ask if "understanding the words" is an essential part of music, cite unambiguous counter-examples (opera, scat singing, and the obvious cases of classical and jazz instrumental music) and question if it is true that there is only one beat to the music in question. From this point the writer may develop her own argument, having found and analyzed an example of "what is said," in her own context, to argue against.

As one might expect, the dialogues written by students in late adolescence focus on controversies involving parents or peers and tap into expressive resources stemming from actual situations. The most common scenario for the dialogue on the music prompt is a drive in the car with Mom or Dad. Older, "returning" students generally respond well to the prompt, "But that's not really work." Issues that come to the surface here often have to do with home or family conflicts regarding the legitimacy or importance or difficulty of academic study. Sometimes, too, students become sensitized to their own internal resistance to learning or their lack of esteem for themselves as learners (which often stems from the home conflicts). The content of many of these dialogues raises issues about our complex ideas regarding remuneration, pleasure, suffering, autonomy, and the flexibility of "time" (i.e., in contrast to the rigidity of hourly work schedules) as they inform casual definitions of "work." For example, a student writes:

"I'm too exhausted to do anything."
"What's the matter?"

"I had so much work to do for school."
"What work? I thought you liked school."
"I do, but it *is* work."
"But it's not really *work*. I mean, you don't get paid for it."

Having produced this text, the student can be led to identify the unstated assumptions and suppressed premises of the counter-statement, for example:

Work is effort expended for pay.
You do not get paid for school work.
Therefore, school work is not *work*.

Or:

Work is something you do because you have to, not because you enjoy it.
You enjoy school.
Therefore, school work is not work.

Once they are explicated in this way, statements that seem to be "logical," probable, or to carry the force of definite pronouncement, can be analyzed effectively. Argument becomes more than an abstract method used by philosophers; it becomes a practical means of thinking and deriving one's own conclusions.

If there is one problem that presents itself in using the counter-statement prompt, it is that dialogues often end with a hostile, "That's just your opinion." This rather abrupt ending, probably quite true-to-life, can be used as a key to enhance the purpose of exploring argumentation. Since students are only just becoming socialized to the academic and professional conduct of "arguments," and since, as Perry argues in *Forms of Intellectual and Ethical Development in the College Years* (Holt, Rinehart and Winston: New York, 1968), students *have* become socialized to atomistic and dualistic distinctions between "opinion" and "fact," this provides another occasion for learning. What does it mean to say "That's just your opinion"? What is the purpose of reasoned argument? What makes a strong or a weak argument, if, to argue at all, we have to overcome this impasse? When dialogues end with the discounting of one point of view as mere opinion, we can deal with the question we have to deal with

anyway: How does one frame an assertion to answer alternate claims without succumbing to the idea that we can all think what we want in our own small worlds?

Whatever the key prompt-word, two possibilities for extended writing arise from the use of the dialogue exercise. One is to write an essay that analyzes how the dialogue poses a problem or conflict and how the problem/conflict can be resolved or mediated. The following assignments may help the student to shape an essay from his or her text, depending on the kind of controversy implicit in it:

1. People often get confused about the difference between a word and the thing it stands for, or about the use of a word to mean different things. Using your dialogue as a case study, explain how such a confusion may be involved in an "argument" and how the confusion can be clarified by analysis.
2. Using your dialogue as a case study, explain the role of assumptions in guiding our thinking and in creating controversy, and explain how what is unstated in the dialogue affects what is actually stated.
3. Write a definition of _____ in which you show the need to arrive at common definitions by using your dialogue as an example of the problems created by mismatched definitions. State the context in which your definition holds and demonstrate your awareness of contexts in which it might be questionable.

These assignments increase students' self-awareness, because the contexts supplied by the dialogue often express conflicts and questions that the students experience *actually* and not merely verbally or formally in the academic environment.

As I have suggested above, the dialogue may also provide material for an argumentative essay in which the writer argues for the validity of a certain kind of music or for the validity of academic study, or child-rearing, or housekeeping as a kind of work. The exercise can be adapted to deal with any number of issues (sexism, racial or ethnic prejudice, cheating, or other relevant concerns) by changing the prompt.

Perhaps because the dialogue is not a traditional academic exercise, students feel free to experiment and to take risks in writing dialogues. Where a typical freshman English assignment often forces students to "drag in" examples to support a thesis or conclusion they care very little about, this assignment fosters discovery. It helps the student to discover a thesis *through* examples drawn from experience. It elicits considerable experiential and expressive content, enhancing the student's commitment to his or her eventual thesis. It also enables us to intervene before the commitment stage and before a student has relegated ideas to the dustbin of "mere opinion." By drawing on the student's resources to imagine controversies in his or her own context, the exercise demonstrates the background of thinking in experience and places discourse in a social context.

Responding to Student Writing: Goals, Methods, Alternatives

Joseph M. Moxley
University of South Florida

Thanks to substantial research about ways teachers respond to student writing and the effect of teachers' commentaries on writing improvement, composition teachers are well informed about evaluation procedures and goals. Below are some of the most important recommendations that have emerged from composition research:

- **Provide "formative" as opposed to "summative" evaluations.** In other words, our purpose is not to justify a grade or do what Mallone and Breihan call "character assassination." Instead, we need to decipher the writer's intentions and propose, when pertinent, several alternatives to realizing these intentions. Ultimately, our

Moxley, Joseph M. "Responding to Student Writing: Goals, Methods, Alternatives. *Freshman English News* 17, no. 2 (1989): 3–11. Copyright 1989 by *Freshman English News*. Reprinted with permission.

concern should be more with teaching a student to ask the critical questions that writers ask when revising than with the quality of any one particular manuscript.

- **Require multiple drafting.** According to over thirty years of research, students benefit from our responses to their writing only when we respond to several drafts. To transform grading papers into a learning process, we must allow students to revise their work in light of our criticism. Otherwise, they tend to ignore our commentaries, no matter how wise our responses may be. (Bamberg; Burton and Arnold; Buxton; Fellows; Marzano; Knoblauch and Brannon; Sommers).

- **Place students in small groups and teach them to evaluate each other's work.** Allowing students to evaluate each other's work in small groups promotes critical thinking and leads to the development of essential editorial skills. Peer reviews also help students to better understand the needs, interests and expectations of audiences other than the teacher. Discussing various topics and treatments helps students better understand assignments and alternatives, and shy students can ask questions that they might not otherwise ask.

- **Avoid "appropriating" students' texts and simplifying students' roles to that of army privates following orders:**

In other words, the teacher's proper role is not to tell the student explicitly what to do but rather to serve as a sounding board enabling the writer to see confusions in the text and encouraging the writer to explore alternatives that he or she may not have considered. The teacher's role is to attract a writer's attention to the relationship between intention and effect, enabling a recognition of discrepancies, but finally leaving decisions about alternative choices to the writer, not the teacher (Brannon and Knoblauch 162).

Most composition scholars (such as Erika Lindemann and David Fuller) argue that we need to provide written commentaries that outline alternative ways to improve student writing.

- **Play the role of the students' intended audience.** By role playing our students' intended audiences, we teach that writers compose for an audience, instead of "performing for a verdict" (Elbow 225).

- **Encourage students to view revision to be an opportunity to clarify and discover one's meaning.** We must encourage students to perceive revision to be an inevitable and important aspect of composing, not punishment for not getting it right the first time (Murray).
- **Avoid overburdening students with advice by identifying only one or two patterns of error at a time.** We need to teach students that writing well means more than forming grammatically correct sentences. We can teach students that what they say is more important—or at least as important—as how they say it, if we primarily respond to the substance and significance of their topics.
- **Praise positive attributes in each paper.** Like everyone else, students respond to encouragement and positive reinforcement. When papers are smeared with red ink, even the hardiest ego can be slow to recover. Sam Dragga and Frances Zak have argued that we should provide solely positive responses, and the work of learning theorists supports the assumption that we are most inclined to learn more when we are rewarded for positive behaviors as opposed to being punished for negative behaviors. Indeed, some preliminary research has suggested that praiseworthy grading improves students' attitudes about writing and results in more writing on the students' part than traditional faultfinding grading (Dragga; Zak).
- **Avoid excessive abstract, formulaic textbook language, such as "edit for efficiency!"; "transition?"; "v/ag"; "p/ag"; etc.** Students are not professional copy editors, and past research indicates that they don't appear to understand or respond to our abbreviations.
- **Omit grades on individual papers.** Grades often transform the effective coaching role of a teacher to that of a judge and gatekeeper. As an alternative to grading individual papers, we can have students keep a portfolio of papers, and select a few papers to grade at the end of each semester (Murray; Burkland and Grimm).

Although well-intentioned and theoretically convincing, some of the above recommendations are difficult to put into practice. For example, in my experience teaching undergraduate writing courses, I have found that not providing grades can be incredibly stressful for students, and colleagues have reported similar results. As much as we may hope that students truly want to improve their writing, grades are their lives and without them—in the unfamiliar territory of an English classroom—a few students may become fearful, anxious about plunging grade point averages. And even though I wholeheartedly agree with being positive when responding to students' work, I believe we need to show students why some passages are weaker than others.

Although most of us are quick to support many (if not all) of the recommendations outlined above, past research suggests that we still emphasize editorial concerns and simplify the demands of global revision (Gee; Connors; Lindemann). In fact, studies of teachers' commentaries indicate that many of us fail to respond in meaningful ways to student papers. For example, in Nancy Sommers' examination of 35 teachers at the University of Oklahoma and New York University, she discovered that "most teachers' comments are not text-specific and could be interchanged, rubber-stamped, from text to text" (152). Arthur Applebee's relatively large scale study of writing in the secondary schools drew similar conclusions. Of the 138 high-school teachers he surveyed, 87.7% routinely "indicate mechanical errors." After examining the response patterns of 454 teachers in the content areas—foreign language, math, science, social science, and business—Applebee concluded:

> Techniques that directly engaged the ideas that the student was expressing—posing counterarguments, responding to the teacher's own views, or suggesting related topics the student might explore—were used routinely by no more than a fifth of the teachers; those techniques were also rated as least important among possible reactions (85).

Sommers' and Applebee's findings are obviously a powerful indictment. Despite all of the errors we identify, as Barbara Mallone and John Breihan have found in their research, students tend to ignore our error hunts (particularly when revisions are not assigned):

> Furthermore, overly-specific faculty action on the text, especially the conscientious red-flagging of every error, did not encourage students in their own problem-solving or teach them to be their own editors.

Mere retyping of faculty corrections was more often than not a mindless task for the student (215).

Robert J. Connors has traced our concern for correctness to the late 19th century when writing teachers were expected to teach 140 to 200 students and grade 216 themes a week: "At some point between 1870 and 1900 the teacher as commentator on the general communicative success of a piece of student writing—form and content—was succeeded by a simplified concept: the teacher as spotter and corrector of formal errors" (67). Our passion for editing and grammar is particularly intriguing in light of Patrick Hartwell's comprehensive review of over fifty years of research that demonstrates that teaching grammar does not affect the quality of student writing or the frequency of error.

Given the overwhelming evidence that red-flagging mechanical errors fails to improve student writing, our tendency to respond in this way is perplexing. Perhaps we highlight errors because we love language and expect students to craft their ideas with care. As writers and critics, we believe sentence-level errors are important and cannot be dismissed, that our students must be competent editors if they are to succeed as business people, and that they will learn from our copy-editing. Also, we're good at editing and can easily justify grades by marking errors.

Like the overworked composition teachers of the 1870's, many of us are faced with unrealistic case loads. Obviously, time restraints play an important role in the way we respond to student writing. With standard courseloads of two or three writing courses a semester (and community college teachers often teach four or five writing courses a semester) and with an average of twenty-five students per course, most writing teachers spend thirty to forty hours each week evaluating papers. Accustomed to marking errors and faced with mountains of student papers, many of us grow weary and understandably have some difficulty embracing the above recommendations or experimenting with new techniques, no matter how well grounded they are in theory and research. In some ways we are like the weary miner who has been digging in the same cave for as long as we can remember. Finding occasional gold nuggets convinces us that the next shovelful will hit the motherlode.

Nevertheless, in light of the process movement, we are more aware than ever before of the importance of responding to students' ideas. And yet how can we reasonably be expected to provide formative commentaries on seventy-five papers a week? How can we integrate the recommendations of recent research, and still deal effectively with the paper load?

Naturally, the single most reasonable action would be a reduction in class sizes for writing teachers and for teachers who assign extensive writing assignments. Expecting teachers to review more than forty essays a week is unrealistic and counterproductive. Also, teachers must experiment with the portfolio method discussed above.

However, faced with a society that demands grades—that produces bumper stickers that exclaim "My son is an honor student at such and such elementary school"—teachers are typically expected to evaluate all writing by assigning grades, even though the best pedagogical response is positive feedback and suggestions for revision.

Fortunately, we have a powerful alternative way to respond to student writing that allows us to achieve most of the goals outlined above in roughly the same amount of time as it takes to flag errors: taped evaluations. Like the many other teachers and scholars who have experimented with taped, oral commentaries, I have found that students prefer taped responses because they are more comprehensive. For many reasons, as analyzed below, taped evaluations can enable us to provide more formative responses—that is, the kinds of commentaries that show students alternative ways to develop their material.

Because speaking is approximately five hundred times faster than writing (Klammer), we can more effectively use the same time dedicated to writing marginal comments to discussing content-level issues, showing syntactical options, suggesting stylistic alternatives, and altering our intonation to personalize the evaluation (Petite; Rubens; Olsen; Stratton; Klammer; Carson and McTasney). The extra time provided by this method provides us with the opportunity to show students how their intended audiences would respond. We can also demonstrate the flexibility of language—the wide variety of stylistic and syntactical options. Instead of concentrating on superficial surface problems—places where students have made grammatical errors or deviated from expectations and standard techniques or formats—we can discuss content-level issues. In short, we can transcend our role as gatekeepers and police of English. We can *show* students where meaning breaks down and why by mentioning *how* we are confused or misled by vague statements and confusing passages.

When I have asked students to respond anonymously to questions about whether they prefer oral or written evaluations, many of them have indicated that they prefer taped commentaries because of their more comprehensive nature:

> I think the oral evaluations are much more personal and give a better relationship between teacher and student. It is an excellent idea.

> I like oral evaluations because they help me understand, not only where my mistakes are, but possibly why I made them and definitely help me understand how to change them and possibly avoid making them again.

> This is the first class I have ever taken where oral evaluations were provided. While written evaluations may pinpoint errors and provide visible corrections, they are time consuming and as a result may end up being cryptic requiring "translation" for the students. Oral evaluations provide an indication that a professor has taken time and interest in your work; also in a subtle way she/he has read the whole paper. Oral responses enable the exchange of more information. I like a combination of both.

Some additional advantages to taped responses have been identified since they were introduced in the 1960's. For example, because taped responses are multisensory, they broaden the learning experience. Education specialists have frequently pointed out that we all learn best in response to a variety of approaches (Petite; Coleman; Volger; Olsen).

In addition, many of my students have mentioned that they prefer taped commentaries because they provide a permanent, cumulative record of each writer's strengths and weaknesses, a record that each writer can refer to at any time:

Oral evaluations are much better because I am able to go back to old essays and listen to the tape again. With written evaluations the special marks are often forgotten and I can't be sure what mistakes were made.

Students also have mentioned to me that they appreciate being spared the embarrassment and shock of receiving marked papers back in front of their peers.

Other teachers who have experimented with taped responses report positive results: In independent investigations, Robert Lumsden, Bernard Tanner, Philip Rubens, Jean McGrew, and John Harris have reported that oral evaluations quickly establish a rapport, are appreciated by students, and prompt more frequent exchanges between teachers and students than traditional grading practices. David L. Carson and John B. McTasney, both from the United States Air Force Academy, examined students' perceptions of taped responses. They found 94% of the 367 students who participated in a semester-long study reported at the end of the semester that they preferred cassette grading to conventional, written grading. My informal polls have netted similar results:

> Although oral evaluations take more time, it seems like private tutoring.

> When the professor is talking to me I pay attention. If I don't understand a problem, at least I have his words to consider—problems are basically easier to attack when I can hear what he's saying. I guess this boils down to a PERSONAL, INDIVIDUAL approach which motivates me more than scratches in a margin.

> I prefer oral evaluations because I can hear the voice and I usually don't have to guess what was meant by a statement. Oral evaluations are nice because hearing the inflection in the voice adds meaning to the words.

Of course, not focusing, for example, on grammatical errors does not mean that we totally ignore such problems. When students are polishing completed drafts, rather than writing "review subject-verb agreement," we can explain what the problems are, and we can refer students to specific reference books and previous class discussions. Responding orally to grammatical errors gives us the flexibility to reduce the confusion that can occur when students fail to understand our abbreviations: awk; agr; ref?; frag; run-on. Also, by checking the aural against the written word, our students learn the helpfulness of a powerful editing technique: hearing writing spoken out loud helps writing (Bartholomae; Petite). By reading their sentence fragments aloud we can help students identify the dangling incompleteness of their thoughts, as well as demonstrate by example the value of reading work out loud. Moreover, Joseph Petite has discovered in his research that taped commentaries significantly reduce the number of fragments that his developmental students write.

While the vast majority of students prefer the more comprehensive and personal nature of taped commentaries, a few students are uncomfortable with this approach. For example, one student writes:

> Oral evaluations are more personal although at times they can be more threatening as the human element of voice is involved. All-and-all, I think both approaches should be used.

Certainly, the above student raises some important concerns. Evaluation, no matter how it's presented, can be difficult for all of us to digest. Inexperienced writers, who have so much to learn about the craft, can become stymied in response to evaluation. We clearly need to be sensitive to how much each of our students is open to criticism. Thus, we must remember Donald Murray's advice to establish a positive ambience in the classroom, one that promotes constructive criticism by establishing it as an inevitable and important aspect of composing. Taped responses do not necessarily preclude written comments. In fact, I often write a sentence or two at the end of the student's manuscript.

Ultimately, to best tailor our responses to our students, we should allow them to decide whether their writing is evaluated in writing or orally. It is interesting to note that although I only require students to try taped responses *once* at the beginning of the semester, very few students—say four or five out of sixty each semester—elect not to use them.

Clearly, taped evaluations provide a useful alternative to traditional written evaluations. Ultimately, however,

we must remember that the message is more important than the medium. Poor advice on our part is still poor advice, whether it is conveyed in written or oral form. Also, we need to recognize that taped evaluations could be more pernicious than written ones because of their more personal and comprehensive nature if we don't follow the guidelines discussed above for responding to student writing.

For those teachers who wish to try responding to student writing via tape recorders, I make the following recommendations:

1) To avoid confusion, have students write their names on their work folders, cassettes, and cassette boxes. 2) Read the entire piece and briefly review the student's rough drafts before making any comments. (Asking students to submit their drafts with the final product helps you see the amount of work a student has completed, and it also serves as a useful check against plagiarized papers.) 3) Identify the major strengths and weaknesses of the piece before tape recording. I typically greet the student, mention which report I'm reviewing, respond to the significance of the subject matter, refer to the audience the student has identified in an audience profile, and then respond to some of the interesting ideas he or she has raised. Whenever possible, I praise strong writing. As I mentioned earlier, I also think it's important to mention one or two major problems with the manuscript. After this summary, I clarify, when pertinent, how a particular word, sentence or paragraph exhibits these weaknesses and/or strengths.

Near the beginning of each writing course, I read through each student's manuscript line-by-line and try to provide what Peter Elbow calls "movies of the mind." In other words, I try to model the questions the student's intended audience would ask *while they read the piece.* As the course progresses, though, I make fewer of these line-by-line reviews. 4) Somewhere near the end of each commentary, I assign the paper a grade, *if it is a final draft.* Many students have mentioned that they would prefer that I place their grades on their papers, but I think it is more effective to mention the grade somewhere on the tape because it ensures that they listen to the tape. In a sense, this is a tricky way to ensure that they listen to my response. 5) Then, at the end of commentary, I summarize the goals that the student should address when rewriting the current piece or when writing his or her next paper. Also, I always encourage each student to come see me if he or she has any questions.

Ultimately, taped responses cannot replace the immediacy of the one-to-one teacher-conference. Yet, it is an extremely valuable technique, one worth trying.

Works Cited

Applebee, Arthur N. *Writing in the Secondary School.* Urbana, Illinois: NCTE, 1981.

Bamberg, Betty. "Composition Instruction Does Make a Difference: A Comparison of the High School Preparation of College Freshmen in Regular and Remedial English Classes." *Research in the Teaching of English* 12 (1978): 47–59.

Bartholomae, David. "The Study of Error." *Rhetoric and Composition.* Ed. Richard L. Graves. Upper Montclair, New Jersey: Boynton/Cook. 1984.311–327.

Brannon, Lil and C. H. Knoblauch. "On Students' Rights to Their Own Texts: A Model of Teacher Response." *College Composition and Communication* 33 (1972): 157–166.

Burkland, Jim and Nancy Grimm. "Motivating Through Responding." *Journal of Teaching Writing* 5 (1986): 237–247.

Burton, D. and L. Arnold. "Effects of Frequency of Writing Intensity of Teacher Evaluation Upon High School Students' Performance in Written Composition." Florida State University, 1963.

Buxton, E. "An Experiment to Test the Effects of Writing Frequency and Guided Practice Upon Students' Skill in Written Expression." Diss. Stanford University, 1958.

Carson, David L. and John B. McTasney. "Grading Technical Reports with the Cassette Tape Recorder: The Results of a Test Program at the United States Air Force Academy. *Journal of Technical Writing and Communication* 3 (1973): 131–144.

Connors, Robert J. "Mechanical Correctness as a Focus in Composition Instruction." *College Composition and Communication* 36 (1983): 61–72.

Dragga, Sam. "Praiseworthy Grading." *Journal of Teaching Writing* 4 (1985): 61–72.

Elbow, Peter. *Embracing Contraries.* New York and Oxford: Oxford University Press, 1986.

Fellows, John E. "The Influence of Theme Readings and Theme Correction on Eliminating Technical Errors in the Written Composition of Ninth Grade Pupils." Diss. University of Iowa, 1936.

Fuller, David C. "Teacher Commentary that Communicates: Practicing What We Preach in Writing Class." *Journal of Teaching Writing* 6 (1987): 307–317.

Gee, Thomas C. "Students' Responses to Teacher Comments." *Research in the Teaching of English* 6 (1972): 212–221.

Harris, John S. "The Use of the Tape Recorder in Grading." *Teaching Technical Writing: Cassette Anthology*, Ed. Charles R. Stratton. Idaho: Association of Teachers of Technical Writing, 1979.

Hartwell, Patrick. "Grammar, Grammars, and the Teaching of Grammar." *College English* 47 (1985): 105–127.

King, Jean A. "Teachers' Comments on Students' Writing: A Conceptual Analysis and Empirical Study." Diss. Cornell University, 1979.

Klammer, Enno. "Cassettes in the Classroom." *College English* 35 (1973): 179–189.

Knoblauch, C. H. and Lil Brannon. "Teacher Commentary on Student Writing: The State of the Art." *Freshman English News* 10 (1981): 1–4.

Lindemann, Erika. Teaching as a Rhetorical Art. *The CEA Forum*, Vol. 15: 2 (1985): 9–12.

Lumsden, Robert. "Evanston, lllinois, Township High Adds to Its Program." *The Bulletin of the National Association* (1962): 223–226.

Mallone, Barbara C. and John R. Breihan. "Responding to Students' Drafts: Interdisciplinary Consensus." *College Composition and Communication* 36 (1985): 213–231.

McColly, W. and Remstad, R. *Comparative Effectiveness of Composition Skills Learning Activities in the Secondary School.* Madison, Wisconsin: University of Wisconsin,1963.

Marzano, R. J. and S. Arthur. "Teacher Comments on Student Essays: It Doesn't Matter What You Say." Study at University of Colorado, Denver, ERIC, 1977. ED 142864.

McGrew, Jean B. "An Experiment to Assess the Effectiveness of the Dictation Machine as an Aid to Teachers in the Evaluations and Improvement of Student Compositions." Report to Lincoln Public Schools, Lincoln, Nebraska. ERIC, 1969. ED 034776.

Murray, Donald. *Learning by Teaching.* New Jersey: Boynton/Cook, 1982.

Olsen, Gary A. "Beyond Evaluation: The Recorded Response to Essays." *Teaching English in the Two-Year College,* 8 (1982): 121–122.

Petite, Joseph. "Tape Recorders and Tutoring." *Teaching English in the Two-Year College* 9 (1983): 123–125.

Robertson, Michael. "Is Anybody Listening?': Responding to Student Writing." *College Composition and Communication* 37 (1986): 87–91.

Rubens, Philip M. "Oral Grading Techniques: An Interactive System for the Technical Writing Classroom." *Technical Writing Teacher* 10 (1982): 41–44.

Sommers, Nancy I. "Responding to Student Writing." *College Composition and Communication* 33 (1982): 148–156.

Stratton, Charles R. "The Electric Report Card: A Follow-up on Cassette Grading." *Journal of Technical Writing and Communication* (1975): 17–22.

Tanner, Bernard. "Teacher to Disc to Student." *English Journal* (1964): 362–363.

Volger, Stephen H. "Grading Themes: A New Approach; A New Dimension." *English Journal* (1971): 70–74.

Zak, Frances. "Between the Red Pencil and the Smiley Face: More Ways to Respond to Student Writing." Conference on College Composition and Communication. St. Louis, Missouri, 18 March 1988.

Freshmen, "Focus," and Writing Across the Curriculum

Barbara Walvoord
Loyola College in Maryland

Teaching students to "focus" an essay—to limit and define a controlling main idea (or "thesis") and to marshall reasoning, illustration, and evidence around that main idea—is a central goal in most freshman composition courses. We like to tell ourselves that focusing is one skill that students can carry with them to their writing in other disciplines and in the professions. That assumption, however, is more complicated than it seems, and teaching students how to transfer their focusing skills to writing outside the composition class requires deliberate and thoughtful pedagogical strategies.

The difficulty is illustrated by a typical freshman composition student of mine I'll call Tami—a bright, responsible student who paid attention when we studied the importance of focusing around a limited, specific thesis, and of organizing evidence logically to support the thesis, and when we discussed the uses of writing to discover a focus or thesis. She wrote several nicely focused essays. l told the class that these same skills should be applied to all their writing. The same semester, however, Tami wrote a history paper that appeared as though it had been produced by someone who had never heard of the need to focus around a specific, limited thesis. It was a cut-and-paste pastiche of encyclopedia facts about the assigned topic. Tami was not synthesizing her freshman writing skills with her history class, nor was she using writing to synthesize information in history. When I asked her what her paper's thesis was, she looked at me in shock.

The history professor obviously had a part in what happened—more about that later. Nevertheless, I couldn't

Walvoord, Barbara. "Freshmen, 'Focus,' and Writing Across the Curriculum." *Freshman English News* 14, no. 2 (1985): 13–17. Copyright 1985 by Freshman English News. Reprinted with permission.

help feeling that if Tami was shocked at the very idea that "thesis" was relevant in the history class, I had not done all I could for her. And not I alone. Preliminary analysis of data from a longitudinal study of three college students by Lucille McCarthy indicates an ambiguous and problematic relationship between their freshman writing course and their writing as sophomores in other courses.

Before we can determine how best to help our students acquire and use focusing skills, we need some idea of how "focus" appears in a variety of disciplines. The record of Tami's writing process in history, as well as data colleagues and I have collected from college students writing in biology and business classes, will offer a "bore hole" sample of the shapes "focus" takes, and the skills needed to achieve it, in courses at various levels and in various disciplines across the curriculum.

Tami's assignment in her freshman core history class was delivered orally. The instructor, an inexperienced adjunct, asked students to write a paper on the colonization of a third world country. According to Tami's notes, the instructor suggested students would need information about three aspects—the situation before colonization, the events of colonization, and the results of colonization. Tami's log and her notes and drafts reveal several important features about her concept of focus and her attempts to achieve it in that environment:

1. She spent an inordinate amount of time and thought on the choice of a country.
2. She appears never to consider that she might use or need a thesis for her paper.
3. She seems to have assumed without question that the instructor's three aspects were to be the three sections of the paper.
4. She used the topic and the three sections as a guide to information-gathering and to focusing and organizing her paper. After collecting a book and two encyclopedias about Algeria, she went through them, taking notes first on aspect 1, then 2, then 3. She wrote her paper the same way—putting down some facts that would fall under aspect 1, then 2, then 3. The day before the paper was due, she went through her draft, and, as she states in her log, "Cut out detail that didn't seem connected. Made sure that all included information was directly related to the guidelines for the paper."

She had the concept of checking and revising a draft for focus, but her guideline seems to have been only "Is there anything in any section of the paper that is not related to the teacher-suggested topic for that section?" She relied on a single revision strategy—deletion.

In a large state university upperclass biology class, biologist Virginia Johnson Gazzam and I have collected notes, drafts, interviews, classroom records, and think-aloud tapes produced in the course of students' natural fulfillment of their assignments. Their instructor was skilled and experienced, with extensive training in a writing across the curriculum workshop. There was extensive use of writing as a tool for learning, and careful guidance of the writing and thinking process, as students developed their assignment—to conduct original research and write a scientific report. Also, they were asked to read a section from a composition text that suggested formation of a thesis as one tool for achieving focus. Yet one of the most difficult problems these students reveal in their drafts and in their think-aloud tapes while composing, was the problem of focusing. None of them appeared to use the composition text's discussion of thesis, though one conscientious student actually reads it aloud on her tape.

In writing a report on original scientific research, the focus problem for students assumed two facets. First, they had to know exactly what they were trying to discover, and to shape a method that controlled all relevant variables and that addressed their specific hypothesis or question, while keeping extraneous material out of the experiment and out of the written report.

Second, students had to deal with the conventional scientific report format. Here their think-aloud tapes reveal two problems—first, reconciling the order of the format with the order of their experimental activities and with the order that might naturally suggest itself while beginning to write about the experiment, and, second, deciding how to apportion the information appropriately among the various sections. For example, while composing the initial draft of the materials and methods section, a student would muse aloud, "maybe this should go in the results section." The train of thought in drafting was interrupted by quandaries about placement of material in the final report format. Despite the difficulties in drafting, some students eventually sorted it all out appropriately; some did not.

The scientific report format may discourage the kind of focus one finds in an essay—arguments marshalled in logical order to support a thesis. Bazerman has pointed out that in the social sciences, professional writing has moved from argumentative essays, where points followed the author's choice of sequence, to a rigid format in which each section is labelled ("Codifying"). Today's format suggests, perhaps especially to inexperienced student writers, that all one need do is chuck into each section whatever material has something to do with the topic of that section—a temptation much like that faced by Tami with her instructor-given aspects, which she took as sections of the paper.

At a selective, mid-size private college, Daniel Singer, of its management/marketing department, and I studied three sections of an upperclass business management class taught by the case method. In their case analyses, students were instructed first to compose a "strategic statement" about the firm, summarizing its present situation, then to examine the aspects of that situation (resources, environment, etc.), then to analyze the firm's central problems and make recommendations. In tandem with this upperclass capstone course, I also examined drafts and revisions of essays written in response to a short case, by introductory economics students, with a skilled senior instructor, at another, smaller and less selective private liberal arts college, in a different area of the country. In both situations, a number of students correctly identified the firm's problems, and made good suggestions for remedies, but were not able to focus their papers around the major problems and recommendations. One important key to success, as we examined the notes and drafts of the upperclass management students, seemed to be the student's ability to break away very early from the case's organization and begin to organize even notes and early drafts around key ideas that would become the central problems and recommendations in the final written analysis. Unsuccessful students were marked by their adherence in notes and drafts to the case's own, deliberately mixed organization. For example, an unsuccessful student retained throughout her drafting the case's division into the steel division and the aluminum division of the firm, making separate, isolated recommendations to each division, while more successful students organized the paper around the central problem—the firm was not sure which

business it was in—and interpreted the interrelationship between the steel and aluminum divisions as symptomatic of that larger problem. Their instructor's suggestion that they work out the strategic statement and the discussion of environment and resources first, was partly a guide to a working method, and partly a guide to organizing the paper, but if used too narrowly as a guide to organizing, the suggested order was incomplete, since the best papers were clearly organized from the beginning as arguments to support the writer's final problem analysis and recommendations.

In the introductory economics class, students were given no suggested order of sections. The less successful students merely gave a list of recommendations, with no priority order and little internal coherence. A number of students' drafts contained key insights—for example, the suggestion that the restaurant owner had not specifically enough identified his market—but some students buried this key insight among other recommendations such as firing some waitresses or changing the menu. Obviously, they could read the case and state the key problem, but in that environment they could not organize a focused case analysis around the key problem.

This sampling is not broad enough to represent every major demand or problem students face in focusing their writing across the curriculum, but it does help a freshman composition instructor to understand what his or her students will be asked to do after the composition class is over:

1. To apply the concept that readers need and want focus, even when the instructor assumes rather than states the need for focus, or when the instructor's emphasis upon topic and upon parts or aspects masks the need for focus in the final paper.
2. To expect that a major goal of the student in the course is to come to understand what constitutes focus in a particular genre, discipline, and classroom, and for a particular audience.
3. To correctly interpret the assignment: is the student being given a focus, or merely a broad topic area? Are parts in the assignment a selection of get-started questions, a guide for the order of thinking or investigation, an inclusive list of material to be covered, or an outline of required sections for the final paper?

4. To be able to depart from the order of written text and/or the instructor's suggested aspects, in order to find a focus and to organize material in support of that focus.
5. When working within a required format, to impose whatever larger focus is needed, transcending individual sections, and to choose and order material within the parts, consistent with the larger focus.

What do these insights mean for the freshman composition class? Joseph Williams, in a recent address, made the point that some aspects of increased maturation as a writer are not inbuilt developmental stages to be mastered once forever, but rather stages in the development of skill in a particular type of writing—stages, then, that must be mastered anew by each person learning that type of writing, no matter how sophisticated in other types of writing. He and others have suggested that we look upon each classroom as a separate "culture" or "speech community," whose singular expectations and constraints shape the writing that takes place among its students. A discipline, likewise, may be seen as a "culture." It would follow, then, that a person who can function comfortably and effectively in one culture will not function effectively in another culture until that person has learned *its* particular expectations and constraints. Thus the composition class does not simply push students up a few rungs in a single human developmental ladder, but also creates its own culture, which students master in order to write successfully in it. But if the composition class is to fulfill a goal of preparing students for writing in other disciplines, it must equip students with the skills and expectations that will help them adapt and learn as quickly as possible, once they step into the new culture represented by a course in a different discipline, with a different instructor and set of constraints and influences.

Our success in teaching student skills that are applicable to other "cultures" depends, then, upon our teaching them not just to focus the essays we assign in composition class, but our equipping them with the expectations and skills that will help them learn to focus their writing appropriately within the "culture" of other disciplines and classrooms. l will suggest several pedagogical strategies I believe could help achieve these ends. Needed in the future are studies that will trace the effects of such strategies upon student attitudes and behaviors.

First, I would suggest that students can profit from graphic illustration that certain attitudes about the writer and about the nature of writing are common to nearly every writing culture. Tami would have profited from the knowledge that, in most academic classrooms, writers must be makers of meaning, not merely shovelers of information into a row of ready bins. In addition, she needed to know firmly that readers, because they are human, need to perceive a focus in what they read. An instructor can present focus as a human need. Ask students to read a focused piece of writing and an unfocused one, then test their recall. Bring into the freshman classroom a variety of writings, from various fields and situations, and analyze what is the writer's main idea. If your instruction for freshman essays uses thesis formation as a tool, point out that in some types of writing, the main idea cannot readily be expressed as a "thesis," but rather can be understood as a dominating purpose (to teach someone to assemble a bicycle; to present information someone will need to conduct further research). Students must see that, in some form or another, the concept of focus applies to most writing situations.

Second, one can specifically teach the forms of writing students will likely meet immediately after their composition class. Discuss the academic paper in its several variations, telling students that sometimes an instructor will have them write a paper that is little more than a transcription of library information, arranged by categories, copied down for the purpose of learning it, but that *most* academic papers are intended as a preparation for the papers academics write to each other, to present information for a specific purpose, and defend positions. Discuss with students how academic papers function, why they are written, how a discipline builds upon the past work of its scholars, and what is the rhetorical stance of the scholar who interprets a phenomenon and then proposes and defends that interpretation before other scholars (Bazerman, "What").

One can make the lesson about readers and about the academic paper more graphic by bringing into class writers from other fields. A psychologist has had my class on the edge of their seats as she described how she got an idea for a particular paper, researched work already done, discovered someone had already written on her idea, saw a related area still unexplored, shaped a hypothesis, designed an investigation, wrote her report, and got it published.

In addition to knowing that focus, in some form, is necessary for readers and will be crucial to the success of most writing in the disciplines, students should also be led to see that audience analysis, in some form, will be universally useful. Again, in addition to talking broadly about analyzing the education, expectations, and needs of various segments of the public or of business audiences, encourage immediate application by discussing the next audience students will face after composition class—the college instructors who assign them their history and chemistry papers. (Specific advice to students about their instructors as audiences is found in Walvoord, *Writing,* pp. 13–14.) You can bring to your class instructors from other disciplines to explain what they look for when they read student papers. To avoid vague disquisitions, ask each instructor to bring sample student writing and actually to demonstrate how the paper meets or does not meet the instructor's needs and expectations. Students see that instructors are real people, with specific expectations, to whom an academic paper is expected to communicate a real meaning. In my experience, the common refrain that will ring again and again in these talks is that instructors want interpretations, focus, coherence, not just recital of random information. My students are always impressed by the strength of this common expectation.

My final suggestions involve the student's ability to handle assignments and conventional formats. Students can be taught to analyze an instructor's assignment, just as they can be taught close reading and interpretation of any other text. Particularly, one can give them practice in determining whether their instructor is giving them a focus, or merely a broad topic area, within which they must narrow and focus. This is the advice that would have warned Tami not to accept her instructor's three aspects as the parts of the paper, and not to assume that those three aspects would provide her only needed focus. Teach students to ask themselves, "Within the expected number of pages, can I say everything there is to say about the assigned subject? If not, does the assignment make clear how I should limit and define? If not, how

shall I focus?" Teach them to ask questions of their instructor, such as "Should we try to cover everything about this topic, or should we find an angle or define a narrower aspect?" In doing so, you help them to overcome the disadvantage of an instructor who may not be elucidating expectations clearly, or who has not carefully thought through the assignment. (Two texts that teach assignment analysis are Walvoord, *Writing,* and Hashimoto, Kroll, and Schafer).

Since so much of students' writing outside the composition class must adhere to conventional or instructor-given formats, the composition class can help students make that significant transition from the essay to the format-driven piece of writing. Give students direct practice in writing within a required format. Formats are all around us; we need merely to adapt one to our uses. For example, an art instructor at my former college asked her students to write student-teaching reports that had the following sections, in this order:

1. Description of the community
2. Description of the school
3. Description of the art department
4. Recommendations for improvement in the art department

As samples of her students' writing show (Walvoord, *Writing,* p. 44), a major problem for students is to create a unifying focus, so that the descriptions of community, school, and art department include not just a random selection of facts, but the information that prepares the reader for the recommendations. l show my students a sample of art student reports, and we discuss how the first 3 sections should be planned and shaped to lead the reader logically to the recommendations. I also show them a well written real estate assessor's report (Walvoord, *Writing,* pp. 42–44), which I got from the president of the local real estate assessor's organization. Within its traditional format, a real estate assessment, like the art students' reports, also has a "neighborhood description" for the property whose value is being assessed. Students are asked to justify every piece of information in the neighborhood description. They quickly see that only that information is given that helps establish the thesis of the assessment—"This property is worth $250,000."

More than showing students that material in every section must fit the focus and purpose of the whole, l help them see that required formats, such as are used for real estate assessment and thousands of other situations in college and in the professions, have arisen because they are logical ways to arrange information to support a particular point or focus. In some cases, such formats also embody the assumptions and methodologies of the discipline (see Bazerman's articles), but you can treat only as much of that aspect as your students are ready to understand. Before my students see the real estate assessment, l ask them to plan how they would support the thesis, "This property is worth $250,000," writing for an audience of the property owner, potential buyers, and a governmental or legal audience. They usually come up with the suggestion that one would first establish a definition of "value" and how it would be determined. Then one would describe aspects of the property such as its neighborhood, its history, its improvements, its current condition. One would also have to establish the qualifications and defend the procedures of the assessor. I now show my students that they have composed a rough approximation of the conventional sections of a real estate assessment. One can do the same for the conventional scientific report format, or for the diagnostic evaluation or client progress report that, in one form or another, is used in many agencies that deal with clients seeking therapy, advice, or services. l got such a format from my college's speech therapy clinic (Walvoord, *Helping*, pp. 103–106).

Once you have some formats, adapt them for yourself and your freshmen. For example, have them assess the value of anything they own, and make them follow a simple version of the real estate assessment format. Or have them assess the needs of someone they are helping or could help, using a variation of the diagnostic evaluation forms. In doing so, they must tackle the same problems the biology majors in our study tackled—what goes where? Should I draft material in the same order as the paper's final required sections? Need I draft the whole paper at once? If not, which sections should be drafted first? Can I change the format? If so, in what situations? What should be the relation between one section and another? How shall I mark the passage between one section and another? Do I need formal transitions as in an argumentative essay, or is a subheading sufficient? It doesn't matter whether the answers for your adapted format are different from the answers students will work out in science class or speech therapy clinic—what matters is that they get a first pass at some of the strategies involved in writing to a specified format.

Obviously, integrating the kinds of instruction I have suggested with the composition class could be implemented gradually over several semesters, learning as one went along. A necessary component is the study of the types of writing students will be asked to produce in other disciplines and professions. One can approach colleagues and simply ask them to address one's class, or to bring a sample piece of writing from their field and talk to you about it over lunch, or to share a batch of student papers and discuss with you what makes them succeed or fail. Another excellent way of learning about writing is to offer your composition students the option, as one of their composition papers, to work with you on an assignment in one of their other classes. The paper they write is for that other class. To get credit in composition, the student must keep a careful log, detailing how he or she worked on the paper, must bring you all notes and drafts, and must meet with you periodically to show you the progress on their papers. These sessions are enjoyable to me because, while I am helping the student to improve the paper, the student also gets to be the "instructor," introducing me to the constraints and traditions of a discipline new to me. The cooperative effort also gives me a chance to communicate with the professor, to clarify what is expected, and to compare our reactions to the student's work—usually a mutually pleasurable and beneficial exchange.

Once we begin to know what our students will actually be asked to do out there, and what are the common problems they face, we can use these or other methods to prepare them more effectively to use their focusing skills across the curriculum.

Works Cited

Bazerman, Charles. "Codifying the Social Scientific Style: The APA Publication Manual as a Behaviorist Rhetoric." Unpublished manuscript, 1985.

————. "Modern Evolution of the Experimental Report in Physics: Spectroscopic Articles in *Physical Review*, 1893–1980." *Social Studies of Science* 14 (1984): 163–96.

————. "What Written Knowledge Does: Three Examples of Academic Discourse." *Philosophy of Social Science* 11 (1981): 361–87.

Hashimoto, Irvin, Barry M. Kroll, and John C. Schafer. *Strategies for Academic Writing: A Guide for College Students.* Ann Arbor: University of Michigan Press, 1982.

McCarthy, Lucille P. "A Stranger in Strange Lands: A College Student Writing Across the Curriculum." Diss. U. of Pennsylvania, 1985.

Meyer, Bonnie J. F. "Reading Teaches Plans for Writing." *Forum* 3 (1982): 67–70.

Walvoord, Barbara Fassler. *Helping Students Write Well: A Guide for Teachers in All Disciplines.* New York: MLA, 1982.

————. *Writing: Strategies for All Disciplines.* Englewood Cliffs, NJ: Prentice Hall, 1985.

Walvoord, Barbara, and Daniel Singer. "Process-Oriented Writing in a Case Method Class." *Proceedings of the Academy of Management.* Ed. John A. Pearce 11 and Richard B. Robinson, Jr. Boston: Academy of Management, 1984. ERIC ED 249 500.

Williams, Joesph. Keynote Address for "Writing in the Humanities and Sciences: Assignments that Teach." Conference of the Delaware Valley Writing Council and PATHS. Philadelphia, 2 March 1985.

Recommended Reading: Recent Articles on Composition

Adams, Katherine H., and John L. Adams. "The Paradox Within: Origins of the Current-Traditional Paradigm." *Rhetoric Society Quarterly* 17 (1987): 421–431.

Anderson, Worth, Cynthia Best, Alycia Black, John Hurst, Brandt Miller, and Susan Miller. "Cross-Curricular Underlife: A Collaborative Report On Ways With Academic Words." *College Composition and Communication* 41 (1990): 11–36.

Berlin, James A. "Contemporary Composition: The Major Pedagogical Theories." *College English* 44 (1982): 765–777.

———. "Rhetoric and Ideology in the Writing Class." *College English* 50 (1988): 477–494.

Bizzell, Patricia. "What Happens When Basic Writers Come to College?" *College Composition and Communication* 37 (1986): 294–301.

Bloom, Lynn Z. "Teaching College English as a Woman." *College English* Vol. 54 no. 7 (Nov. 1992): 818–825.

Bruffee, Kenneth. "Collaborative Learning and the 'Conversation of Mankind.'" *College English* 46 (1984): 635–642.

———. "Social Construction, Language, and the Authority of Knowledge: A Bibliographical Essay." *College English* 28 (1986): 773–790.

Brooke, Robert. "Underlife and Writing Instruction." *College Composition and Communication* 38 (1987): 141–153.

Chase, Geoffrey. "Accommodation, Resistance, and the Politics of Student Writing." *College Composition and Communication* 39 (1988): 13–22.

Connors, Robert J. "Personal Writing Assignments." *College Composition and Communication* 38 (1987): 166–183.

———. "Textbooks and the Evolution of the Discipline." *College Composition and Communication* 37 (1986): 178–194.

Dillon, George L. "My Words of an Other." *College English* 50 (1988): 63–73.

Eden, Rick, and Ruth Mitchell. "Paragraphing for the Reader." *College Composition and Communication* 37 (1986): 416–430.

Elbow, Peter. "Ranking, Evaluating, and Liking: Three Forms of Judgment." *College English* Vol. 55 no. 2 (Feb. 1993): 187–206.

Faigley, Lester. "Competing Theories of Process: A Critique and a Proposal." *College English* 48 (1986): 527–542.

———. "Judging Writing, Judging Selves." *College Composition and Communication* 40 (1989): 395–412.

Flynn, Elizabeth A. "Composing as a Woman." *College Composition and Communication* 39 (1988): 423–435.

Fontaine, Sheryl 1. "The Unfinished Story of the Interpretive Community." *Rhetoric Review* 7 (1988): 86–96.

Hamilton-Wieler, Sharon. "Empty Echoes of Dartmouth: Dissonance Between the Rhetoric and the Reality." *The Writing Instructor* 8 (1988): 29–41.

Harris, Joseph. "The Idea of Community in the Study of Writing." *College Composition and Communication* 40 (1989): 11–22.

———. "Rethinking the Pedagogy of Problem Solving." *Journal of Teaching Writing* 7 (1988): 157–165.

Hashimoto, 1. "Voice as Juice: Some Reservations about Evangelic Composition." *College Composition and Communication* 38 (1987): 70–80.

Holt, Mara. "Towards a Democratic Rhetoric: Self and Society in Collaborative Theory and Practice." *Journal of Teaching Writing* 8 (1989): 99–112.

Juncker, Clara. "Writing (with) Cixous." *College English* 50 (1988): 424–436.

Karis, Bill. "Conflict in Collaboration: A Burkean Perspective." *Rhetoric Review* 8 (1989): 113–126.

Lu, Min-zhan. "From Silence to Words: Writing as Struggle." *College English* 49 (1987): 437–448.

MacDonald, Susan Peck. "Specificity in Context: Some Problems for the Inexperienced Writer." *College Composition and Communication* 37 (1986): 195–203.

Mack, Nancy. "The Social Nature of Words: Voices, Dialogues, Quarrels." *The Writing Instructor* 8 (1989): 157–165.

Perelman, Les. "The Context of Classroom Writing." *College English* 48 (1986): 471–479.

Porter, James E. "Intertextuality and the Discourse Community." *Rhetoric Review* 5 (1986): 34–47.

Purdy, Dwight. "A Polemical History of Freshman Composition in Our Time." *College English* 48 (1986): 791–796.

Reither, James A., and Douglas Vipond. "Writing as Collaboration." *College English* 51 (1989): 855–867.

Ritchie, Joy S. "Beginning Writers: Diverse Voices and Individual Identity." *College Composition and Communication* 40 (1989): 152–174.

Rose, Shirley K. "Reading Representative Anecdotes of Literacy Practice; or 'See Dick and Jane read and write!'" *Rhetoric Review* 8 (1990): 244–259.

Russell, David R. "Romantics on Writing: Liberal Culture and the Abolition of Composition Courses." *Rhetoric Review* 6 (1988): 132–148.

Shen, Fan. "Classroom and the Wider Culture: Identity as a Key to Learning English Composition." *College Composition and Communication* 40 (1989): 459–466.

Sirc, Geoffrey. "Gender and 'Writing Formations' in First-Year Narratives." *Freshman English News* 18 (1989): 4–11.

Stewart, Donald C. "Collaborative Learning and Composition: Boon or Bane?" *Rhetoric Review* 7 (1988): 58–83.

Tebo-Messina, Margaret. "Authority and Models of the Writing Workshop: All Collaborative Learning is Not Equal." *The Writing Instructor* 8 (1989): 86–92.

Tobin, Lad. "Bridging Gaps: Analyzing Our Students' Metaphors of Composing." *College Composition and Communication* 40 (1989): 444–458.

Trimbur, John. "Beyond Cognition: The Voices in Inner Speech." *Rhetoric Review* 5 (1987): 211–221.

———. "Consensus and Difference in Collaborative Learning." *College English* 51 (1989): 602–616.

Welch, Kathleen. "Ideology and Freshman Textbook Production: The Place of Theory in Writing Pedagogy." *College Composition and Communication* 38 (1987): 269–282.

Zuber, Sharon and Ann M. Reed. "The Politics of Grammar Handbooks: Geueric *He* and Singular *They*." *College English* vol. 55 no. 5 (Sept. 1993): 515–530.

The
Holt Handbook

FOURTH EDITION

The
Holt Handbook

FOURTH EDITION

Laurie G. Kirszner

Philadelphia College of Pharmacy and Science

Stephen R. Mandell

Drexel University

Harcourt Brace College Publishers

Fort Worth   Philadelphia   San Diego   New York   Orlando   Austin   San Antonio
Toronto   Montreal   London   Sydney   Tokyo

Publisher	Ted Buchholz
Acquisitions Editor	Michael Rosenberg
Developmental Editor	Camille Adkins
Project Editor	Barbara Moreland
Production Manager	Tad Gaither
Art Director	Nick Welch
Permissions Editor	Sheila Shutter

ISBN: 0-15-501226-6

Library of Congress Catalog Card Number: 93-81249

Address for Editorial Correspondence: Harcourt Brace College Publishers, 301 Commerce Street, Suite 3700, Fort Worth, TX 76102.

Address for Orders: Harcourt Brace & Company, 6277 Sea Harbor Drive, Orlando, FL 32887-6777. 1-800-782-4479, or 1-800-433-0001 (in Florida).

Preface

Our goal for the fourth edition of *The Holt Handbook* is the same as it was for the first—to create a writer's handbook that serves as a classroom text, as a comprehensive reference, and as a writer's companion. In preparing the first edition we concentrated on making the book inviting, accessible, useful, and interesting for both teachers and students. In the fourth edition, we have kept this goal in mind, adding distinctive new design features that make information even easier to locate than before. **Close-up boxes,** which focus on special problems related to the topic under discussion, are light yellow and identified by a magnifying glass. **Writing** and **Revision Checklists,** which provide added guidance for writers, are also light yellow; these are distinguished by a large check mark. **Summary Boxes,** which highlight the information users consult most often, are surrounded by a blue ruled line. **Marginal cross-references,** which direct users to related discussions in other parts of the book, are indicated by small blue arrows. In addition, we have taken special care to make headings clear and descriptive and to position them logically in the text. We believe the result is a comprehensive reference work that not only guides writing and revision but also enables writers to find and apply information quickly and easily.

Although *The Holt Handbook*, Fourth Edition, is grounded in the most up-to-date research in composition, it is also informed by our many years of classroom experience. As teachers, we continue to search for what works for our students, trying to give them what they need to succeed in college. Our hope is that this book will continue to reflect our commitment to our teaching and to our students—some of whose writing appears on its pages. With its logical organization, its process approach, its emphasis on revision, and its focus on student writing, *The Holt Handbook* is truly a writing-centered text. And, of course, its descriptive approach to grammar and nonthreatening tone also make it a student-centered book.

As we began this revision of *The Holt Handbook,* our goal was to retain the features that have made the book so satisfying to users, while adding new material to make it a more valuable reference text and writing guide. Thoughtful and incisive comments from users of the first three editions and our own classroom experience with the book led us to make a number of changes in the fourth edition.

The Fourth Edition at a Glance

- **A new English as a Second Language (ESL) chapter: Chapter 26, "Language Issues for International Students"** Unlike ESL chapters in other books, this chapter provides students with a context for English grammar by contrasting its rules with those that govern their own languages. Its unique approach gives this chapter appeal for native as well as non-native English speakers.

- **Collaborative exercises** Throughout *The Holt Handbook* asterisks identify exercises suitable for collaborative work.

- **New "Student Writer at Work" exercises** These new essays-in-progress, like those retained from the third edition, reinforce the connection between the material discussed in the chapters and student writing.

- **Extensive revision of Chapters 1–3** Three drafts of a new student essay, "My Problem: Escaping the Stereotype of the 'Model Minority,'" have been added to Section 1. New material on thesis and support has also been added.

- **Expanded discussion of reading critically in Chapter 5, "Reading Critically and Writing Critical Responses"**

- **Toulmin Model added to Chapter 6, "Thinking Logically"** Instructors can now supplement their discussion of traditional logic with the Toulmin model of critical thinking.

- **New student paper in Chapter 7, "Writing an Argumentative Essay"** The new student paper, "The Returning Student: Older Is Definitely Better," uses sources to support an argumentative thesis.

- **Material on computer-based research highlighted in Chapter 38, "Research for Writing"**

- **Expanded discussions of summary, paraphrase, and plagiarism in Chapter 39, "Working With Sources"**

- **Extensive revision of Chapter 40** Expanded discussions of the most current MLA, Chicago, APA, and CBE documentation styles make this chapter more accessible and more comprehensive.

- **New research paper, "Athletic Scholarships: Who Wins?" in Chapter 41, "Writing a Research Paper"** This paper is supplemented by a new case study tracing the student's research.

- **New literary papers added to Chapter 43, "Writing in the Humanities," and Chapter 47, "Writing about Literature"** A new literary paper with sources and MLA documentation has been added to Chapter 43. A second literary paper without sources has been added to Chapter 47. The fourth edition of *The Holt Handbook* now has **three** papers on literary subjects.

- **New design** Writing checklists, summary boxes, and close-up boxes have distinct and readily identifiable designs that make them easy to locate.

In Part 1, we have expanded and clarified the discussion of thesis and support. In response to suggestions from readers, we now show how the thesis can help writers organize their essays by suggesting a rhetorical pattern. We have also added three drafts of a new student essay, "My Problem: Escaping the Stereotype of the 'Model Minority.'" This essay, along with first-person excerpts from a student writer's journal, provides a thorough, personal account of the writing process. This treatment, enhanced by a new design, makes the discussion segments and the Student Case Study segments genuinely complementary and easy to follow. We have also streamlined and condensed the treatment of paragraphing in Chapter 4 while substituting many new sample paragraphs on varied and interesting topics. Finally, in this section and throughout the book, we have given additional emphasis to collaborative learning by designating certain exercises as suitable for collaborative activity. Instructors can use these exercises, which are identified with asterisks, for group work. (Additional suggestions for collaborative activities appear in the annotated teacher's edition of *The Holt Handbook*.)

Because critical thinking is such an important part of the writing process, we devote all of Part 2 to this subject. Chapter 5, "Reading Critically and Writing Critical Responses," now contains a discussion of active reading strategies as well as a discussion of critical reading. Chapter 6, "Thinking Logically," which explains the principles of inductive and deductive reasoning, now contains discussions of the Toulmin model and Rogerian argument, enabling instructors to supplement their discussion of traditional logic with these approaches. In addition, this chapter contains a helpful new summary box that compares deductive and inductive reasoning. Chapter 7, "Writing an Argumentative Essay," includes a new student paper, "The Returning Student: Older Is Definitely Better." Because this essay contains material derived from sources, instructors can now illustrate how information from articles and interviews can supplement a student's own ideas to support an argumentative thesis.

Throughout Parts 3 through 7 we have edited and redesigned material on style, grammar, punctuation, and mechanics so that definitions, guidelines, and key concepts are emphasized visually as well as stylistically. We have carefully scrutinized every example and exercise in these sections and have edited, revised, eliminated, or replaced material when necessary. In addition, headings have been reworded, redesigned, or relocated to make information more accessible. Whenever possible, we have reformatted important information into summary boxes or checklists to make it easy to identify. Chapter 17, now titled "Awkward or Confusing Sentences" to better reflect its content, has been refocused and streamlined. The most noticeable addition to Part 6 is Chapter 26, "Issues for International Students." Unlike ESL chapters in other books, this chapter offers students a context for understanding English grammar rules by contrasting them to those of their own language. Far from being a watered-down version of information presented elsewhere in the book, Chapter 26 uses contrastive grammar and rhetoric to present a unique view of writing in English.

Part 8, "Writing with Sources," has received especially favorable attention in past editions. Now we have fine-tuned this section to make it even more useful. In Chapter 38, "Research for Writing," a new chart presents a clear picture of the research process. In addition, material reflecting recent developments in computer-based research has been highlighted. The discussions of writing summaries and paraphrases in Chapter 39, "Working With Sources," have also been expanded, as has the treatment of plagiarism. Detailed instructions, checklists, and side-by-side comparisons of a paraphrase and summary help students with these sometimes troublesome assignments. Chapter 40, "Documentation," has also been expanded; it now includes the formats presented in the fourteenth edition of the *Chicago Manual of Style* and the fourth edition of the *Publication Manual of the American Psychological Association*. New summary boxes provide quick guides to the typing conventions for MLA, CMS, APA, and CBE documentation styles. The most dramatic change in this section is the addition of a new research paper—"Athletic Scholarships: Who Wins?"—to Chapter 41. A new case study that traces the student's research process has also been added.

In Part 9, "Understanding the Disciplines," discussions of each of the disciplines have been updated to include the most current print and database resources. Sample research papers in each discipline have been edited to conform to the latest conventions of styles and documentation. In addition, a new literature paper with sources and MLA documentation ("Assertive Men and Passive Women: A Comparison of Adrienne Rich's 'Aunt Jennifer's Tigers' and 'Mathilde in Normandy'") has been added to Chapter 43, "Writing in the Humanities," and a new literature paper without sources ("Irony in Delmore Schwartz's 'The True-Blue American'") has been added to Chapter 47, "Writing about Literature." *The Holt Handbook* now has *three* papers on literary subjects— one in Chapter 43 and two in Chapter 47.

In its fourth edition, *The Holt Handbook* continues to approach writing as a recursive process, giving students the opportunity to practice planning, shaping, writing, and revising. This approach encourages students to become involved with every stage of the process and to view revision as a natural and ongoing part of writing. The style, grammar, and mechanics and punctuation chapters present clear, concise definitions of key concepts followed by examples and exercises that gradually increase in difficulty and sophistication. (Whenever possible, sentence-level skills are reinforced in groups of related sentences that focus on a single high-interest topic instead of in isolated sentences.) Thus students learn incrementally, practicing each skill as it is introduced. In this way they begin to recognize and revise sentence-level problems within longer units of discourse, duplicating the way they must actually interact with their own writing. This approach has been useful to hundreds of thousands of students who have used the first three editions, and we continue to believe strongly in its effectiveness.

The Holt Handbook, Fourth Edition, is a classroom text, a reference book, and—above all—a writing companion that students can turn to again and again for advice and guidance as they write in college and beyond. Our goal for each edition has remained the same: to combine the best of current composition research with our own instincts as experienced teachers. We continue to believe that we have the obligation to give not just the rule but the rationale behind it. Accordingly, we are careful to explain the principles that writers must understand if they are to make informed choices about grammar, usage, rhetoric, and style. The result is a book that students and instructors can continue to use with ease, confidence, and we hope, with pleasure.

With this edition, an even more comprehensive ancillary package is available for students and instructors.

Preface

For Students

The Holt Composition Workbook. This workbook, which follows the organization of *The Holt Handbook,* Fourth Edition, offers practice in grammar, mechanics, punctuation, spelling, editing, and revision. A special section on research is included.

The Research Sourcebook, Second Edition. This combination guide and workbook addresses common problems students encounter in research assignments and clarifies the process of gathering and integrating source material.

The Harcourt Brace Guide to Documentation and *Writing in the Disciplines.* These useful resources introduce students to the types of writing they will encounter in the humanities, social sciences, and sciences.

Preparing for the TASP Using The Holt Handbook. This guide enables students to connect sample test material to instruction and exercises available in the handbook.

Supplementary Exercises. These additional grammar and composition exercises are provided for students who need reinforcement of basic skills.

The Holt Workbook for International Students. This collection of exercises is designed to supplement material that appears in the ESL section (Chapter 26) of the handbook. Cross references also lead students into the main body of the text.

For Instructors

Diagnostic Test Package. This complete testing program, cross-referenced to *The Holt Handbook,* Fourth Edition, includes general grammar proficiency and diagnostic tests, as well as Florida CLAST-based tests, Texas TASP-based tests, and a Tennessee Proficiency Examination-based test.

The Holt Guide to Teaching Composition. In this collection of essays, composition directors and course coordinators from across the country discuss the practical aspects of teaching in their particular English programs.

The Holt Guide to Using Daedalus. This brief guide offers class-tested suggestions for setting up and conducting a composition course using Daedalus software along with *The Holt Handbook*, Fourth Edition.

Writing Tutor IV. These self-paced tutorial programs offer students practice in grammar, punctuation, mechanics, parts of speech, sentence errors, editing, and revision.

Holt On-Line 2. This version of the handbook works within a word processor to provide access to handbook topics for students as they write their papers.

Exam-Master. This testbank contains 800 questions for skills testing, diagnostic evaluation, and state test preparation. Chapter references relate each fill-in, multiple-choice, true/false, matching, essay, or short answer question directly to a specific section in *The Holt Handbook*.

Acknowledgments

We thank the following reviewers for their sound advice on the development of the fourth edition.

Henry Castillo, Temple Junior College
Laurie Chesley, Grand Valley State University
Scott Douglass, Chattanooga State Technical Community College
Maurice R. Duperre, Midlands Technical College
Nancy Ellis, Mississippi State University
Jane Frick, Missouri Western State University
G. Dale Gleason, Hutchinson Community College
Maureen Hoag, Wichita State University
Susan B. Jackson, Spartanburg Technical College
Anne Maxham-Kastrinos, Washington State University
J. L. McClure, Kirkwood Community College
John Pennington, St. Norbert College
Robert Perry, Lock Haven University of Pennsylvania
Robbie C. Pinter, Belmont University
Mary Sue Ply, Southeastern Louisiana University
Linda Rollins, Motlow State Community College
Laura Ross, Seminole Community College
Anne B. Slater, Frederick Community College
Connie White, Salisbury State University
Karen W. Willingham, Pensacola Junior College

Special thanks to the following colleagues for their expert contributions: Clyde Moneyhun, University of Arizona, Language Issues for International Students; Carol Lea Clark, University of Texas, El Paso, and Deborah Barberousse, Coker College, Exercises; Richard Louth and Carole McAllister, Southeastern Louisiana University, Collaborative Activities for Instructor's Edition; Linda Daigle, Houston Community College, Annotations for Instructor's Edition; and David Roberts, Samford University, Guide to Writing with Computers.

We also thank the following colleagues for their valuable comments on the development of the first and second editions: Chris Abbott, University of Pittsburgh; Virginia Allen, Iowa State University; Stanley Archer, Texas A & M University; Lois Avery, Houston Community College; Rance G. Baker, Alamo Community College; Julia Bates, St. Mary's College of Maryland; John G. Bayer, St. Louis Community College/Meramec; Larry Beason, Texas A & M University; Al Bell, St. Louis Community College at Florissant Valley; Debra Boyd, Winthrop College; Margaret A. Bretschneider, Lakeland Community College; Pat Bridges, Grand Valley State College; Alma Bryant, University of South Florida; Wayne Buchman, Rose State College; David Carlson, Springfield College; Patricia Carter, George Washington University; Faye Chandler, Pasadena City College; Peggy Cole, Arapahoe Community College; Sarah H. Collins, Rochester Institute of Technology; Charles Dodson, University of North Carolina/Wilmington; Margaret Gage, Northern Illinois University; Sharon Gibson, University of Louisville; Owen Gilman, St. Joseph's University; Margaret Goddin, Davis and Elkins College; Ruth Greenberg, University of Louisville; George Haich, Georgia State University; Robert E. Haines, Hillsborough Community College; Ruth Hamilton, Northern Illinois University; Iris Hart, Santa Fe Community College; John Harwood, Penn State University; Michael Herzog, Gonzaga University; Clela Hoggatt, Los Angeles Mission College; Keith N. Hull, University of Wyoming/Laramie; Anne Jackets, Everett Community College; Zena Jacobs, Polytechnic Institute of New York; LaVinia Jennings, University of North Carolina/Chapel Hill; D. G. Kehl, Arizona State University; Philip Keith, St. Cloud State University; George Kennedy, Washington State University; William King, Bethel College; Edward Kline, University of Notre Dame; Susan Landstrom, University of North Carolina/Chapel Hill; Marie Logye, Rutgers University; Helen Marlborough, DePaul University; Nancy Martinez, University of New Mexico/Valencia; Marsha McDonald, Belmont College; Vivien Minshull-Ford, Wichita State University; Robert Moore, SUNY/Oswego;

George Murphy, Villanova University; Robert Noreen, California State University/Northridge; L. Sam Phillips, Gaston College; William Pierce, Prince George's Community College; Robbie Pinter, Belmont College; Nancy Posselt, Midlands Technical College; Robert Post, Kalamazoo Valley Community College; Richard N. Ramsey, Indiana University/Purdue University; Mike Riherd, Pasadena City College; Emily Seelbinder, Wake Forest University; Charles Staats, Broward Community College/North; Frank Steele, Western Kentucky University; Barbara Stevenson, Kennesaw College; Jim Stick, Des Moines Area Community College; James Sodon, St. Louis Community College at Florissant Valley; Josephine K. Tarvers, Rutgers University; Kathleen Tickner, Brevard Community College/Melbourne; George Trail, University of Houston; Daryl Troyer, El Paso Community College; Ben Vasta, Camden County Community College; Connie White, Salisbury State College; Joyce Williams, Jefferson State Junior College; Branson Woodard, Liberty University; and Peter Zoller, Wichita State University.

We express our appreciation to our colleagues who offered suggestions and advice for revisions to the third edition of *The Holt Handbook:* Lynne Diane Beene, University of New Mexico, Albuquerque; Elizabeth Bell, University of South Carolina; Jon Bentley, Albuquerque Technical-Vocational Institute; Debra Boyd, Winthrop College; Judith Burdan, University of North Carolina/Chapel Hill; Phyllis Burke, Hartnell College; Sandra Frisch, Mira Costa College; Gerald Gordon, Black Hills State University; Mamie Hixson, University of West Florida; Sue Ellen Holbrook, Southern Connecticut State University; Linda Hunt, Whitworth College; Rebecca Innocent, Southern Methodist University; Gloria John, Catonsville Community College; Gloria Johnson, Tennessee State University; Suzanne Liggett, Montgomery College; Richard Pepp, Massasoit Community College; Nancy Posselt, Midlands Technical College; Robert Peterson, Middle Tennessee State University; Randy Popkin, Tarleton State University; George Redmond, Benedict College; Linda Rollins, Motlow State Community College; Gary Sattelmeyer, Trident Technical College; Father Joseph Scallon, Creighton University; Emily Seelbinder, Queens College; Cynthia Smith, University of West Florida; Bill Stiffler, Harford Community College; Nancy Thompson, University of South Carolina, Columbia; Warren Westcott, Frances Marion College; Connie White, Salisbury State University; and Helen Yanko, California State University, Fullerton; Terry Roberts, University of North Carolina/Chapel Hill; Gary Sattelmeyer, Trident Technical College; and Kathleen Tickner, Brevard Community College.

Preface

Among the many people at Harcourt Brace who contributed to this project, we would like to single out Michael Rosenberg for his energy and vision.

In addition, we want to acknowledge the very vital day-to-day contributions of two people in particular: Camille Adkins and Barbara Moreland. Camille, our developmental editor, brought to the project a winning combination: a delightfully droll sense of humor and a pragmatic yet creative editorial approach. Barbara, our project editor, contributed high standards, incredible patience, and a constant willingness to put in extra time (even when she didn't have it). Both also brought to the project their years of experience in the classroom, and for this, and many other things, we are grateful.

Once again, we would like to thank our families—Mark, Adam, and Rebecca Kirszner and Demi, David, and Sarah Mandell—who gave us no editorial assistance, did not type the manuscript, and offered no helpful suggestions, but who managed to put up with the general chaos for the duration of another edition.

Finally, we would like to thank each other for making this book a collaboration in the truest sense.

Contents

Contents

Contents

Contents

Contents

Contents

Contents

Contents

Contents

Contents

Contents

Contents

Contents

Contents

Contents

Composing an Essay

CHAPTER 1

Planning an Essay

1a   Understanding the Writing Process

Writing is no easy task. Almost everyone who has labored over an assignment or stared at an empty page hoping for inspiration has wondered whether there is a better way to communicate. But the truth is, there is not. Despite their ease of use and mass appeal, the sounds and images conveyed by telephone, tape recorder, and video camera are imperfect substitutes for the written word.

Writing is a process, and the process of writing enables writers to discover ideas, make connections, and see with new perspectives. When you write, you can reread, reconsider, and revise until you discover exactly what you want to say and how you want to say it. In this sense, writing is not just a physical activity but also a demanding, creative process of thinking and learning—about yourself, about others, and about your world. In another sense, writing is a tool that empowers you: It enables you to participate in the ongoing dialogue among all kinds of educated people who "talk" to each other in letters, memos, petitions, reports, articles, editorials, and books. In short, writing is worth the trouble. If you can write, you can communicate.

Writing is a constant process of decision making, of selecting, deleting, and rearranging material.

THE WRITING PROCESS
Planning: Consider purpose, audience, and tone; choose topic; discover ideas.

continued on the following page

2

continued from the previous page

Shaping: Decide how to organize your material.

Writing: Draft your essay.

Revising: "Re-see" what you have written; write additional drafts.

Editing: Check grammar, spelling, punctuation, and mechanics.

Proofreading: Check for typographical errors.

The neatly defined stages listed above communicate neither the complexity nor the flexibility of the writing process. Although we will examine these stages separately, they actually overlap: As you seek ideas you begin to shape your material; as you shape your material you begin to write; as you write a first draft you change your organization; as you revise you continue to discover more material. Moreover, various stages are repeated again and again throughout the writing process.

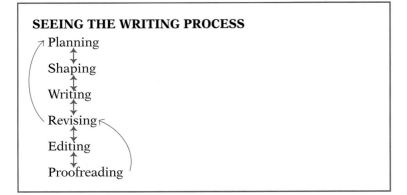

SEEING THE WRITING PROCESS

Planning

Shaping

Writing

Revising

Editing

Proofreading

During your college years and in the years that follow, you will develop your own version of the writing process and use it whenever you write, adapting it to the audience, purpose, and writing situation at hand.

1b Thinking about Writing

Writing presents many situations in which you must think critically—make judgments, weigh alternatives, analyze, compare, question, evaluate, and engage in other decision-making activities. Virtually all writing demands that you make informed

QUOTATIONS

"The essay is a literary device for saying almost everything about almost anything."

–Aldous Huxley

"*Essay.* A funny word, when you think of it, like a bit of literary Pig Latin. In French, the word is *essai*. It means "a try, an attempt." Loosely translated, it might mean something like 'giving it a whirl' or 'taking a whack at it.' The French writer Montaigne pinned the word to a collection of short prose pieces written in 1580. Montaigne's *Essais* didn't claim to be the last word on the subject. They were not research papers, but attempts to get at the truth."

–Jan Farrington, quoted in *Writing!* magazine

Ask the students to find and bring to class examples of newspaper articles written for each of four major purposes: (1) to reflect on emotions or experiences, (2) to report information, (3) to explain a concept or process, and (4) to persuade. Form small groups of four or five students to read and discuss the examples. Ask the students to find and list the common characteristics of articles in each of the four groups. Use those lists as a point of departure for discussing specific techniques that help writers achieve different purposes. Then have each student write a short newspaper article to fit one of the four purposes, using the techniques identified.

choices about your subject matter and about the manner in which you present your ideas.

Planning your essay—thinking about what you want to say and how you want to say it—begins well before you actually put your thoughts on paper in any organized way. This planning is as important a part of the writing process as the writing itself.

(1) Determining your purpose

We write for a variety of **purposes:** *to reflect* on emotions or experiences, exploring ideas and feelings in a diary, an autobiographical memoir, or a personal letter; *to report* information, perhaps in a news article, an encyclopedia entry, or an essay that focuses on observations of people, communities, or phenomena; *to explain* a concept or process—for example, in a textbook, an instruction manual, or an essay that defines or that examines causal connections; or *to persuade*, trying to convince an audience to agree with a position on an issue in a proposal, an editorial, or an essay that takes a stand on a moral, ethical, political, or social issue. Whenever you write, you may have any of these purposes—or you may have other, more specific aims or a combination of purposes.

Your purpose determines the material you choose and the way you arrange and express it. For instance, a memoir written by an adult remembering summer camp might *reflect* on the emotional upheavals experienced during a month away from home—or explore memories of mosquitoes, poison ivy, institutional food, shaving cream fights, and so on. A magazine article about summer camps could *report* information, explaining how camping has changed in the past twenty years, for example. Such an article would present facts and statistics straightforwardly, focusing on the big picture rather than on individual details. To *explain* the concept of summer camp to those unfamiliar with it, a writer would thoroughly discuss such features as recreational programs and sports facilities. An advertising brochure designed to recruit potential campers could *persuade* by enumerating the benefits of the camping experience. Such a brochure would stress positive details—the opportunity to meet new friends, for example—and deemphasize the possibilities of homesickness and rainy weather. In each case, purpose determines what material is selected and how it is presented.

College writing may call for any of a wide variety of approaches, but your general purpose is often to report information or to explain a concept.

✔ WRITING CHECKLIST: PURPOSE

Is your purpose

- to reflect on emotions or experiences?
- to report information?
- to explain?
- to persuade?
- to amuse or entertain?
- to evaluate?
- to discover?
- to analyze?
- to debunk?

- to criticize?
- to satirize?
- to speculate?
- to warn?
- to reassure?
- to take a stand?
- to identify problems?
- to suggest solutions?
- to define causes?
- to predict effects?

(2) Identifying your audience

Writing is often such a solitary activity that it is easy to forget about your audience. But except for personal diaries or journals, everything you write addresses an **audience**—a particular set of readers.

At different times, in different roles, you address different kinds of audiences. As a citizen, consumer, or member of a community, civic, political, or religious group, you may respond to pressing social, economic, or political issues by writing letters to a newspaper editor; a public official; a representative of a special interest group, business, or corporation; or another recipient you do not know well or at all. In your personal life, you may write notes and letters to friends and family and perhaps diary or journal entries to yourself. As an employee, you may write letters, memos, and reports to your superiors, to workers you supervise, or to workers on your level; you may also be called on to address customers or critics, board members or stockholders, funding agencies or the general public. As a student, you write essays, reports, and other papers addressed to one or more instructors and sometimes to other students or to outside evaluators.

As you progress through the stages of the writing process, you shape your paper increasingly in terms of what you believe your audience needs and expects. Your assessment of your readers' interests, educational level, biases, and expectations determines not only the information you include but also the emphasis, arrangement of material, and style or tone you choose.

FURTHER READING

Patricia Bizzell. "Cognition, Convention, and Certainty: What We Need to Know about Writing." *PRE/TEXT* 3.3 (Fall 1982): 213–43. This article is frequently cited in discussions of writing programs and research into the composing process(es). The author distinguishes between "inner-directed" and "outer-directed" theorists, favoring a balance between the two, yet finally emphasizing the value of the latter over the former. The author is particularly critical of Linda Flower and John Hayes's work in the composing process and in fostering a view of writing as primarily "problem-solving."

The inner-directed theorists (such as Flower and Hayes) are searching for a *universal* view of the composing process, or universal, innate cognitive thought patterns that can be taught. Once this process is identified, a model of the composing process can be constructed that can be taught to all gender, ethnic, and disciplinary groups. The author argues that such inner-directed theorists are biased in favor of standard norms of cognition and language use.

The outer-directed theorists, in contrast, believe that even *if* universal cognitive structures can be found, they cannot be taught, and that thinking and language use should be considered in social contexts. Individuals are within specific discourse communities that help shape their style of thinking, speaking, and writing. What must be discovered, therefore, are "local" rather than universal structures of cognition.

In concern for the local, the author relies on Richard Rorty (*Philosophy*
continued on following page

continued from previous page

and the Mirror of Nature), Stanley Fish (*Is There a Text in This Class?*), and Thomas Kuhn (*The Structure of Scientific Revolutions*). Stanley Fish has since responded to this article describing Bizzell's assertions as "crypto-foundationalist." See his "Anti-Foundationalism, Theory Hope, and the Teaching of Composition" in *The Current in Criticism,* Ed. Clayton Koelb and Virgil Lokke. West Lafayette, Indiana: Purdue University Press, 1987: 65–79.

QUOTATION

"Your audience is one single reader. I have found that sometimes it helps to pick out one person—a real person you know, or an imagined person—and write to that one."

–John Steinbeck

COMPUTER TIP

Analyzing Audiences
A worksheet template file can help students analyze their audience. A typical audience-analysis worksheet might look like the one below.

Worksheet for Audience Analysis
RHETORICAL PURPOSE:
The effect I want to have on my readers is to cause them to
THINKING ABOUT WHO MY READERS ARE:
If I were to describe my readers, I would say the following things:
THINKING ABOUT MY READERS' POSITION:
What are the attitudes of my readers toward my subject or problem already?

continued on following page

The Academic Audience   As a student, you most often write for an audience of one: the instructor who assigns the paper. When addressing an instructor, you assume he or she represents a class of readers who apply accepted standards of academic writing.

What are these standards? Basically, instructors expect correct information, standard grammar and spelling, logical presentation of ideas, and some stylistic fluency. Instructors further ask that you define your terms and support your generalizations with specifics. They want to know what you know and whether you can express what you know clearly and accurately. Instructors assign written work to encourage you to use critical thinking skills, so the way you organize and express your ideas can be as important as the ideas themselves.

You can assume that all your instructors are specialists in their fields, so you can generally omit long overviews and basic definitions. But outside their areas of expertise, most instructors are simply general readers. If you think you may know more about a subject than your instructor does, be sure to provide ample background, supplying the definitions, examples, and analogies that will make your ideas clear.

Of course, each of your instructors is also a unique person with special knowledge and interests. This means that even though they may well have similar general requirements concerning the quality of your work, individual instructors will look for particular emphases or approaches. Consequently, your assessment of a particular teacher's special requirements also influences what and how you write.

The course for which you are writing is another factor that should influence your writing choices. If, for example, you decide to write about the underground mine fires that for years have

TOPIC: UNDERGROUND FIRES

Course	Emphasis
Chemistry	Chemical analysis of fumes
Sociology	Relocation patterns of residents
Economics	Effects on local businesses and real estate market
Psychology	Emotional impact of fires on children
Political Science	Role of federal or state government

been burning out of control near your hometown of Centralia, Pennsylvania, you would focus on different aspects of the topic for different courses.

Finally, keep in mind that different academic disciplines have their own formats, documentation styles, methods of collecting and reporting data, systems of formulas and symbols, technical vocabularies, and stylistic conventions. Instructors in different disciplines may therefore have different expectations, and you should take this into account as you write.

▶ See Pt.9

✔ ### WRITING CHECKLIST: AUDIENCE

Is your audience

- an individual?
- a member of a group?
- specialized?
- general?

Can you identify your audience's

- needs?
- expectations?
- educational level?
- biases?
- interests?

Do you need to supply your audience with

- definitions?
- overviews?
- examples?
- analogies?

What special conventions does your audience expect concerning

- format?
- documentation style?
- methods of collecting and reporting data?
- systems of formulas and symbols?
- specialized vocabulary?
- writing style?

---

(3) Setting your tone

Your **tone** is the attitude you adopt as you write. This attitude, or mood, may be serious or frivolous, respectful or condescending, intimate or detached. Tone gives your readers clues about your feelings toward your material and helps them understand what you have to say. Therefore, your tone must remain consistent with your purpose and your audience as you shape, write, and revise your material.

---

continued from previous page

How much interest in my subject or problem do my readers have?

What interests or relevant experiences do they have that I can appeal to?

PLANNING MY STRATEGY:

How can I appeal to uninterested or antagonistic readers?

What information can I include about the subject or problem to make it clear and understandable to my readers?

What facts or explanations should I furnish?

What specialized terminology or concepts should I explain?

STUDENTS WORKING TOGETHER

Place students in groups of four or five. Their task is to rewrite a fairy tale for a particular audience (e.g., lawyers, rappers, etc.). When each group has finished, it will exchange its story with another group. Its task now is to guess the audience and explain what led the group to its conclusion.

IN THE CLASSROOM

Use different types of music to make the point that not all music is appropriate for all audiences (and that musicians play to certain audiences). For example, people who enjoy something like Billy Idol's "Aftershock" (loud, heavy metal) probably won't care for something by Al Jarreau (jazz). Play a few songs (or parts of songs) for the class and ask students to describe the audience each artist wants to reach (and to give reasons for their choices).

Tone/Voice

Because student writers deal with a variety of writing tasks, they are unlikely to settle on one distinctive tone or "voice" in their own writing. Many professional writers, however, develop a tone or voice that becomes their literary stock in trade. To illustrate a variety of writing tones and voices, collect samples of the work of newspaper or magazine columnists and editorial writers. How does Mike Royko's tone differ from Judith ("Miss Manners") Martin's? Russell Baker's from Erma Bombeck's? William Buckley's from Dave Barry's? These writers succeed, in part, because readers come to enjoy the tone of their writing—the attitude from which they view the world.

Conversely, editorial writers sometimes work to eliminate a sense of tone or voice from their writing in order to focus attention on the opinion offered, not the writer. Discuss the samples and the effect of various tones (or lack of tones) on the reader.

Play a tape of someone reading an essay, poem, or story. Have students describe the reader's tone of voice and discuss why they (students) described it as they did. Was it inflection, pitch, word choice? Then explore how to create the same tone in writing, perhaps by having students work in groups to produce a particular tone (serious, ironic, humorous, etc.).

How you feel about your readers—sympathetic or superior, concerned or indifferent, friendly or critical—is also revealed by your tone. For instance, if you identify with your readers or feel close to them, you use a personal and conversational tone. When you address a generalized, distant reader indirectly or anonymously, you use a more formal, impersonal tone.

When your audience is an instructor and your purpose is to report information or to explain (as is often the case in college writing situations), you should use an objective tone, neither too personal and informal nor too detached and formal—unless you are told otherwise. This paragraph from a student paper on the resistance to various drugs of a specific group of microorganisms achieves an appropriate tone for its audience (an instructor—and perhaps students—in a medical technology lab) and purpose (to report information).

> One of the major characteristics of streptococci is that they are gram-positive. This means that after a series of dyes and rinses they take on a violet color. (Gram-negative organisms take on a red color.) Streptococci are also non-spore forming and non-motile. Most strains produce a protective shield called a capsule. They use organic substances instead of oxygen for their metabolism. This process is called fermentation.

An English composition assignment asking students to write a short, informal essay expressing their feelings about the worst job they ever had calls for an entirely different tone. In the next paragraph, the student's sarcastic tone effectively conveys his attitude toward his job, and his use of the first person encourages audience identification. Ironic comments ("good little laborer," "Now here comes the excitement!") contribute to the informal effect.

> Every day I followed the same boring, monotonous routine. After clocking in like a good little laborer, I proceeded over to a grey file cabinet, forced open the half-caved-in doors, and removed a staple gun, various packs of size cards, and a blue ballpoint pen. Now here comes the excitement! Each farmer had a specific number assigned to his name. As his cucumbers were being sorted into their particular size, they were loaded into two-hundred-pound bins which I had to label with a stapled size card with the farmer's number on it. I had to complete a specific size card for every bin containing that size cucumber. Doesn't it sound wonderful? Any second grader could have handled it. And all the time I worked, the machinery moaned and rattled and the odor of cucumbers filled the air.

In a letter applying for a job, however, the same student would have different objectives. In this situation, his distance from his

Plan 1c

audience and his purpose (to impress readers with his qualifications) would call for a much more objective and straightforward tone.

> My primary duty at Germaine Produce was to label cucumbers as they were sorted into bins. I was responsible for making sure each 200-pound bin bore the name of the farmer who had grown those cucumbers and also for keeping track of the cucumbers' sizes. Accuracy was extremely important in this task.

*EXERCISE 1

Take as your general subject a book that you liked or disliked very much.

1. How would each of the following writing situations affect the content, style, organization, tone, and emphasis of an essay about this book?

   * A journal entry recording your informal impressions of the book
   * An examination question that asks you to summarize the book's main idea
   * A book review for a composition class in which you evaluate both the book's strengths and its weaknesses
   * A letter to your local school board in which you try to convince your readers that, regardless of the book's style or content, it should not be banned from the local public library
   * An editorial for your school newspaper in which you try to persuade other students that the book is not worth reading

2. Choose one of the writing situations listed in number 1, and write an opening paragraph for the specified assignment.

1c   Getting Started

Before you begin any writing task, be sure you understand the exact requirements of your assignment. It is very important to keep these guidelines in mind as you write and revise. Do not make any guesses—and do not assume anything. Ask questions, and be sure you understand the answers.

GUIDELINES FOR GETTING STARTED

* Has your instructor assigned a specific topic? A general subject area? A question to answer? Or are you free to choose your own topic?

continued on the following page

9

ANSWERS TO EXERCISE 1

Your students' answers will vary. The following considerations should be raised in discussion:

Journal entry: content likely to be a personal response to the book rather than an analysis of content; style very informal; organization probably not evident; tone judgmental, depending on student's reaction to book; emphasis on personal response and specific details.

Examination question: content concerned mainly with summarizing material in book; style formal; careful organization, probably following that of book; tone neutral, nonjudgmental; emphasis on providing information about book's content.

Book review: content an evaluation of book's material and presentation; style formal; organization formal, probably presenting book's strengths and weaknesses in an orderly manner; tone neutral; emphasis on balance of evaluation.

Letter to school board: content likely to be persuasive reasons not to ban books in general, with or without acknowledging faults of this particular book; style formal, given large, mostly unknown audience; careful organization, probably with a hierarchy of reasons not to ban book; tone committed, but not overly emotional; emphasis probably on importance of free access to books.

Editorial for newspaper: content probably reasons not to read the book; style somewhat less formal, given audience of student peers; organization probably less formal, but orderly; tone likely to be committed and, because of shared assumptions of audience, probably more emotional; emphasis on book's lack of value.

continued from the previous page

- What is the word, paragraph, or page limit?
- How much time do you have to complete your assignment?
- Is the assignment to be done in class or at home?
- Can you take notes or do research?
- Will you have an opportunity to get feedback on your ideas for your paper in class discussion? In collaborative activities? In a conference with your instructor?
- If the assignment requires a specific format, do you know what its conventions are?

(1) Choosing a topic

If your instructor permits you to choose your own topic, be sure to choose one you know something about—or, at least, one you want to learn about. Perhaps a class discussion or assigned reading has suggested a topic; maybe you have seen a movie or television special or had a conversation (or even an argument) about an interesting subject. If so, you are off to a good start. If not, your instructor may be able to help you develop an unfocused idea into a workable topic—or, you may be able to use a collaborative brainstorming session to help you find a promising topic.

See ◄ 1c2

Although you are sometimes able to choose your topic freely, more often your assignment limits your options somewhat. For example, your instructor may pose a specific question for you to answer—"How did the boundaries of Europe change following World War I?" "What are the advantages and disadvantages of the Federal Guaranteed Student Loan Program?"—or give you a specific length, format, and subject area or a list of subjects from which to choose.

- Write a two-page critical analysis of a film. (Specifies length, format, and subject area)
- Write an essay explaining the significance of one of these court decisions: *Marbury* v. *Madison, Baker* v. *Carr, Brown* v. *Board of Education, Roe* v. *Wade.* (Gives list of specific subjects from which to choose)

Even in such cases, you cannot start to write immediately. First, you must narrow the assignment to a workable topic that suits your purpose, audience, and assignment.

Tracking the Writing Process

1. Make a process-of-writing assignment. Assign a topic or allow the students to pick their own. Ask them to compile a folder that represents all stages of the writing process. You might ask for specific strategies such as brainstorming, topic tree, informal outline, rough draft, peer or self-criticism, first revision with editing symbols, and a polished final draft.
2. As a continuing exercise in tracking the writing process, suggest that students keep a writing journal or scrapbook including outlines, drafts, written peer and teacher criticism (or notes of oral criticism), and the student's own notes on problems, solutions, writing strategies, etc.

Have students write a short in-class essay in which they describe their own writing process, perhaps how they normally get started, where they work best, where their best ideas come from. They could also compare their own processes with those suggested in *The Holt Handbook*. Providing examples of your own drafts (or something from a professional writer like "Writing" by William Stafford—a wonderfully positive essay) may help them think seriously and positively about writing and the process they normally follow. Knowing how students actually write will also help instructors understand why students may or may not have trouble following the "normal" process of prewriting, drafting, revising, and editing.

10

NARROWING AN ASSIGNMENT

Course	Assignment	Topic
Public Health	Analyze one effect of AIDS on American society.	How has the spread of AIDS affected metropolitan-area blood banks?
Sociology	Identify and evaluate the success of one resource available to the homeless population of one major American city.	The role of the Salvation Army in providing for Chicago's homeless
Composition	Describe a place that is very important to you.	Cape May: A town that never changes
Psychology	Write a three- to five-page paper assessing one method of treating depression.	How dogs interact with severely depressed patients

---

As you go through the process of deciding on a topic, keep your assignment, your audience, and your purpose in mind; also consider the things you know best and like best. If your assignment is to write about a place that is important to you, do not write about your dorm room simply because it is the first thing you think of or about the treasures of the Metropolitan Museum of Art just because you think the topic will impress your instructor. If, however, you see your dorm room as representing your independence from your family and your essay is supposed to focus on the idea of making a fresh start, your room could be an ideal topic. Similarly, if you know the collections at the Metropolitan well and plan to study art, you may legitimately use this knowledge in a paper about how your career goals developed. The best advice about choosing a topic is not to settle automatically for a routine glance at the Mississippi River when you can describe Catfish Creek vividly.

Make certain that you understand the difference between what you *can* write about and what you *want* to write about. You may

STUDENTS WORKING TOGETHER

Students work with partners. Each takes a turn interviewing the other for information about significant memories stemming from childhood, background, interests, and so on.

STUDENTS WORKING TOGETHER

Following the interview, have students brainstorm for 5-10 minutes on promising story topics generated from the interview. Then choose one topic and tell the story to a partner. The partner listens, responds to the story (e.g., questions about missing details, etc.), and then tells the author what he or she found significant in the story.

STUDENTS WORKING TOGETHER

Students tell a story about their childhood in a group. Then group members individually compose five questions raised by the story. As a group, students discuss their questions to find similarities and differences; instructor leads class discussion about the nature and kinds of questions listeners/readers ask.

Have students brainstorm in groups to narrow and direct one topic (homelessness, perhaps) for a particular discipline. Instead of using different topics and showing how they can be narrowed for a particular assignment, ask students to think about how different disciplines treat a topic differently. For example, a public health course might require a paper examining the sorts of diseases spread among the homeless; a sociology course might discuss the social groups and community feeling among the homeless—how they take care of each other; a composition course might focus on describing an encounter with a homeless person; a psychology paper might focus on the role that mental illness plays in homelessness. Students could also be asked to come up with a pattern of development (comparison/contrast, process, narration, description, cause and effect, etc.) for each topic as a way to show how writing patterns come naturally from certain topics. Once their brainstorming session is over, each group could present the results to the entire class.

be very much interested in the Yukon but have nothing interesting to say about it; you may know more than you care to admit about McDonald's but not have the faintest desire to write about it.

Throughout the first three chapters of this text, we will be following the writing process of Nguyen Dao, a first-year composition student. Nguyen's instructor gave the class the following assignment.

> Write a short essay about a problem you face that you believe is unique to you. Be sure your essay has a clearly stated thesis and helps readers to understand your problem and why it troubles you.

The class had two weeks to complete the assignment, and library research was not permitted. The instructor explained that she was going to require some collaborative work, so Nguyen knew his classmates would read and react to his paper.

Because this paper was the class's first full-length assignment, the instructor asked them to take a very thorough, systematic approach to the writing process, engaging in many different kinds of activities designed to help them plan, shape, write, and revise their essays. She pointed out that not all of these strategies would work equally well for each student or each topic. Still, she wanted students to discover which activities worked best for them. For this reason, she asked them to write a brief assessment of each

NGUYEN DAO'S WRITING PROCESS: CHOOSING A TOPIC

Thinking up possible topics wasn't too hard. I've had plenty of problems lately. Some topics I thought about were specific things like the trouble I had convincing my parents to let me live at school, and my decision to study political science and give up the idea of being a doctor. I also thought about more general, long-standing problems: dealing with my parents' strict rules and their limited English skills. I saw all these different problems as related to my being Asian American. Then this made me think about the problem of how Asian Americans are always seen as family oriented, science oriented, or success oriented, which is not always true. Dealing with what people expect of me--because I'm supposed to be a "typical Asian" or "model minority"--is a problem I've always faced.

activity after they completed it. Nguyen Dao's writing process, which appears in shaded boxes throughout Chapters 1–3, includes his comments about how each activity worked for him. (Keep in mind that your own experience with the writing process may be quite different from Nguyen's.)

*EXERCISE 2

Read the following excerpt from Ron Kovic's autobiographical *Born on the Fourth of July.* Then list ten possible essay topics about your own childhood suggested by Kovic's memories of his. Each topic should be suitable for a short essay directed to your composition instructor. Your purpose is to give your audience a vivid sense of what some aspect of your childhood was like. Choose the one topic you feel best qualified to write about, and explain the reasons for your choice.

When we weren't down at the field or watching the Yankees on TV, we were playing whiffle ball and climbing trees checking out birds' nests, going down to Fly Beach in Mrs. Zimmer's old car that honked the horn every time it turned the corner, diving underwater with our masks, kicking with our rubber frog's feet, then running in and out of our sprinklers when we got home, waiting for our turn in the shower. And during the summer nights we were all over the neighborhood, from Bobby's house to Kenny's, throwing gliders, doing handstands and backflips off fences, riding to the woods at the end of the block on our bikes, making rafts, building tree forts, jumping across the streams with tree branches, walking and balancing along the back fence like Houdini, hopping along the slate path all around the back yard seeing how far we could go on one foot.

And I ran wherever I went. Down to school, the candy store, to the deli, buying baseball cards and Bazooka bubblegum that had the little fortunes at the bottom of the cartoons.

When the Fourth of July came, there were fireworks going off all over the neighborhood. It was the most exciting time of year for me next to Christmas. Being born on the exact same day as my country I thought was really great. I was so proud. And every Fourth of July, I had a birthday party and all my friends would come over with birthday presents and we'd put on silly hats and blow these horns my dad brought home from the A&P. We'd eat lots of ice cream and watermelon and I'd open up all the presents and blow out the candles on the big red, white, and blue birthday cake and then we'd all sing "Happy Birthday" and "I'm a Yankee Doodle Dandy." At night everyone would pile into Bobby's mother's old car and we'd go down to the drive-in, where we'd watch the fireworks display. Before the movie started, we'd all get out and sit up on the roof of the car with our blankets wrapped around us watching the rockets and Roman candles

Your students' answers will vary. While students should be encouraged to think of topics for themselves, some that can be suggested include a typical summer or summer day; family vacations; birthdays, Christmas, the Fourth of July, other holidays with family traditions and significance; special childhood friends or experiences; movies or television programs; a memorable historical event.

Unlike the necessarily private world of freewriting, brainstorming lends itself to collaborative activity. Divide the class into groups of four or five students. Write a short list of topics on the board (perhaps relating to a current or future class assignment). Ask each group to choose one topic and brainstorm for ideas. Suggest a time limit (five to ten minutes should suffice), and ask each group to appoint one member to write down brainstorming notes along the lines of Nguyen Dao's list on page 19.

Ask students to evaluate their brainstorming sessions. Did everyone participate? What was the atmosphere within the group? Was there a time during the session when ideas really begin to take off? Could they have produced more if they had more time? Remind students that writing flows from talk—both from overt, externalized conversations with other people and from internal, ruminative conversations with themselves. Eventually, good writers learn how to initiate the kind of internal conversations (brainstorming) that lead to the writing ideas they need.

going up and exploding into fountains of rainbow colors, and later after Mrs. Zimmer dropped me off, I'd lie on my bed feeling a little sad that it all had to end so soon. As I closed my eyes I could still hear strings of firecrackers and cherry bombs going off all over the neighborhood. . . .

The whole block grew up watching television. There was Howdy Doody and Rootie Kazootie, Cisco Kid and Gabby Hayes, Roy Rogers and Dale Evans. The Lone Ranger was on Channel 7. We watched cartoons for hours on Saturdays—Beanie and Cecil, Crusader Rabbit, Woody Woodpecker—and a show with puppets called Kukla, Fran, and Ollie. I sat on the rug in the living room watching Captain Video take off in his spaceship and saw thousands of savages killed by Ramar of the Jungle.

I remember Elvis Presley on the Ed Sullivan Show and my sister Sue going crazy in the living room jumping up and down. He kept twanging this big guitar and wiggling his hips, but for some reason they were mostly showing just the top of him. My mother was sitting on the couch with her hands folded in her lap like she was praying, and my dad was in the other room talking about how the Church had advised us all that Sunday that watching Elvis Presley could lead to sin.

(2)   Finding something to say

Once you have a topic, you can begin to discover ideas for your paper, using one (or several) of the following strategies.

STRATEGIES FOR FINDING SOMETHING TO SAY	
Reading and observing	Clustering
Keeping a journal	Asking journalistic questions
Freewriting	Asking in-depth questions
Brainstorming	

Reading and Observing   A good way to start looking for material to write about is by opening your mind to new ideas. It is unlikely that all the information you need for your essay will already be in your mind, just waiting to be discovered. For this reason, it is important that you keep your eyes and ears open from the time you receive your assignment until you turn it in. As you read textbooks on various subjects or look through magazines and newspapers, be on the lookout for new ideas that pertain to your topic. Also, make a point of talking informally with friends or family about your topic.

CLOSE-UP

FINDING SOMETHING TO SAY

Many of the activities for finding something to say that are discussed in this chapter can be done collaboratively. With your instructor's permission, you can **collaborate** (work in pairs or small groups) on brainstorming, clustering, and applying specific questions to a topic. Not only is such group activity often more enjoyable than solitary work, but it also gives you the added benefit of doubling (or even tripling or quadrupling) the size of your pool of ideas.

▶ See 3e2

Collaborative work can be useful at other stages of the writing process as well—for example, when you work on revising your essay.

If your instructor encourages you to do formal research, you can use material from nonprint sources such as films and television programs as well as material from books and articles. Interviews, telephone calls, letters, and questionnaires can be as fruitful as library research. But remember to document ideas that are not your own.

▶ See Pt.8

NGUYEN DAO'S WRITING PROCESS: READING AND OBSERVING

My first reaction was that writing about a problem I faced that was unique to me was not a topic I needed to do any reading about. I know my problem. I've had it for almost 19 years. I figured it would be a good idea to listen more closely to the kind of things people say about Asians all the time (the things I usually tune out) on TV or even right to my face. One thing I heard was the idea that Asians come to the U. S. and take unskilled entry-level jobs away from American citizens. (Someone on talk radio was really steamed about this.) Maybe I can use this idea--maybe not.

Keeping a Journal Professional writers sometimes keep **journals**, writing in them regularly whether or not they have a specific project in mind. Such a record of thoughts and ideas can be a valuable resource when you run short of material. Journals,

Brainstorming and Discovering Meaning

Use the computer to encourage students to carry out the "Strategies for Finding Something to Say" (p. 14 in *The Holt Handbook*). Show them that brainstorming and freewriting are easily done on a computer; that journal entries can be easily incorporated into subsequent drafts; and that material generated when asking questions about a subject can be copied directly into formal writing assignments.

Lab Activity

Ask students to come to class with a subject for an essay already chosen. Have them begin by creating a file for brainstorming.

Tell them that they have a limited period of time (perhaps fifteen or twenty minutes) to generate as much material as possible on the subject they've chosen. (An option, of course, is to have them do focused brainstorming on a narrowed topic.)

Emphasize that, since brainstorming is for the writer's use only, they shouldn't pay attention to mistakes in punctuation, spelling, or typing; nor should they worry about writing in grammatically correct or complete sentences. Finally, make sure they know that they won't be evaluated on what they produce.

"The journal is a tool for recording the process of our lives. . . . This is not to remove ourselves from involvement with our lives, but is an additional function, a special vantage point. . . . [The journal] makes it possible for one part of ourselves to write and one to read, so that one part may ask and one may answer, one may act and one may reflect, one may explore and one may comprehend the exploration."

–Christina Baldwin, *One to One: Self-Understanding through Journal Writing*

NGUYEN DAO'S WRITING PROCESS: KEEPING A JOURNAL

Journal Entry

I'm not really comfortable writing about being Asian American, but I have to admit it's a good topic for a paper about a problem I have. A lot of my problems seem to come from the ideas about what other people think Asians are supposed to do or be. But I don't want to get too personal because other people are going to read what I write and it could get embarrassing. I don't want them to know about fights I have with my parents over what they think a good Asian son should be. It's like in that story where the Chinese mother tells her daughter there are two kinds of daughters--obedient ones and the other kind: "Americans." I don't feel like analyzing my whole family in public. What I want to write about is just the pressures on Asians in general, and mention a few things about my own life to support these general ideas.

I really hate the idea of writing in a journal, but I guess it works. When I reread this entry, it helped to remind me that my paper didn't have to be a confession or an argument about what's wrong with my life.

---

unlike diaries, do more than record personal experiences and reactions. In a journal you explore ideas as well as events and emotions, thinking on paper and drawing conclusions. You might, for example, explore the evolution of your position on a political issue or solve on paper a problem you find difficult to work through in your mind. A journal is an excellent place to record the evolution of your ideas about an assignment. You can also record quotations that mean something special to you or make notes about important news events, films, or conversations. A good journal is a scrapbook of ideas that you can leaf through in search of new material and new ways of looking at old material. The important thing is to *write regularly*—every day if possible—so that when an idea comes along, you won't miss the opportunity to record it.

Freewriting Another strategy that can help you to discover ideas is **freewriting**. Freewriting is comparable to the stretching exercises athletes do to warm up—it is a relatively formless, low-key activity, but it is also serious preparation for the highly focused, sometimes strenuous, work that lies ahead. When you freewrite, you try to forget you have a particular assignment or deadline. You let yourself go and write *nonstop*, about anything that comes to mind, moving as quickly as you can. Give yourself a set period of time—say, five minutes—and do not stop to worry about punctuation, spelling, or grammar, or about where your mind is wandering. This strategy encourages your mind to make free associations; it unearths ideas you probably are not aware

NGUYEN DAO'S WRITING PROCESS: FREEWRITING

Freewriting (excerpt)

I really don't want to do this freewriting but I have to--I'm being forced--I have no choice but it seems stupid. If I have ideas they'll come and if not they won't. I don't see why I need an idea anyway--I wish she'd just say write about your summer vacation like they did in high school and I'd write about Colorado last summer and the mountains and that lake I can't remember the name but we had a boat and I could see--This is too hard. I need to write about Asians but there weren't any Asians in that town on that lake. People looked at me like someone from outer space half the time. I wonder what they thought, or who they thought I was.

Focused Freewriting (excerpt)

Being Asian in the Colorado mountains--wondering whether people thought I was a Japanese tourist--Asian cowboy--Asian hiker/athlete/mountain man. People never expect Asians to be athletes. Just engineers or violinists. Or maybe own fruit stands or be kung fu teachers. Being a teacher could be good for me--being a role model for kids, showing them other things to be.

I liked the idea about not having to worry about grammar, spelling, etc. Still, I felt pretty self-conscious doing freewriting, but I actually got one idea I might be able to use in my paper—the fact that Asians are expected to be certain things and not others.

you have. When your time is up, look over what you have written and see if you have anything you can use. If you do, be sure to underline, bracket, or star this material so you will be able to find it later on. Sometimes you will have nothing at all; if you are lucky, you may find one good idea about your topic, and you can use this idea as the center of a focused freewriting exercise.

When you do **focused freewriting**, you zero in on your topic instead of trying to forget about it. Here too you write without stopping to reconsider or reread, so you have no time for counterproductive reactions—no time to be self-conscious about style or form, to worry about the relevance of your ideas, or to count how many words you have and panic about how many more you think you need. At its best, focused freewriting can suggest new details, a new approach to your topic, or even a more interesting topic.

Brainstorming   One of the most useful ways to accumulate ideas is **brainstorming.** This strategy encourages you to recall pieces of information and to see connections among the pieces.

When you brainstorm, you list all the points you can think of that seem pertinent to your topic. Keeping your topic in mind, write down all the ideas that surface—comments, questions, single words, symbols, or diagrams—as quickly as you can. As the ideas flow, write them down without pausing to consider their relevance or explain their significance. Your goal is to let one idea suggest another, so do nothing to slow down your momentum. Only when you finish your brainstorming should you think about starting to group, sort, and classify the ideas on your list.

You can also do **collaborative brainstorming**—that is, you can brainstorm with your classmates in small groups that your instructor sets up or enlist some friends to help you work through some ideas. Sometimes you can meet individually with your instructor to brainstorm about your paper. This kind of brainstorming—invariably part discussion, part argument, and part dictation—can be very productive and enlightening. As ideas are suggested, write them down quickly and uncritically, asking for clarification only if it is absolutely necessary. Do not stop to worry about where in your paper you will use a particular idea. Even if an idea seems irrelevant, make a note of it. It may come in handy later in your writing process.

Clustering   **Clustering**—sometimes called *webbing* or *mapping*—is similar to brainstorming in some respects. As with brainstorming, you do not worry about writing complete sentences,

Plan 1c

NGUYEN DAO'S WRITING PROCESS:
BRAINSTORMING

Brainstorming List

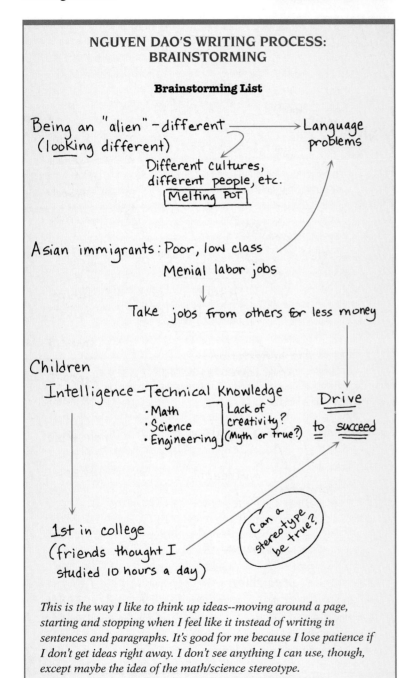

Being an "alien" – different ────────→ Language
(looking different)   problems

                Different cultures,
                different people, etc.
                Melting POT

Asian immigrants: Poor, low class

              Menial labor jobs

                 ↓

        Take jobs from others for less money

Children

   Intelligence – Technical Knowledge   Drive
          · Math   Lack of
          · Science   creativity?   to succeed
          · Engineering (Myth or true?) =

  1st in college
  (friends thought I   Can a stereotype be true?
   studied 10 hours a day)

*This is the way I like to think up ideas--moving around a page,
starting and stopping when I feel like it instead of writing in
sentences and paragraphs. It's good for me because I lose patience if
I don't get ideas right away. I don't see anything I can use, though,
except maybe the idea of the math/science stereotype.*

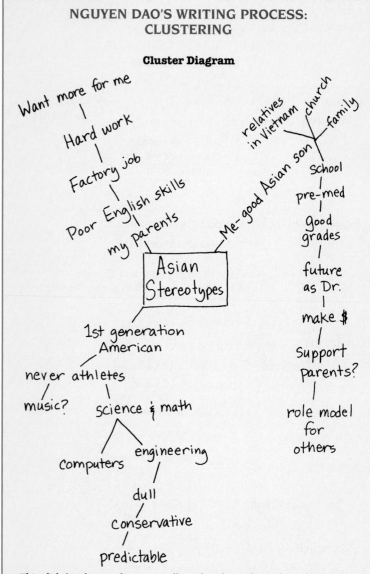

NGUYEN DAO'S WRITING PROCESS: CLUSTERING

Cluster Diagram

This didn't take much time at all, and it showed me that I have basically three parts to my topic--stereotypes about Asian immigrants, stereotypes about their children, and ideas about myself. I'm not sure yet how this all ties together.

and you jot ideas down quickly, without pausing to evaluate their usefulness or analyze their logical relationships to other ideas. However, clustering encourages you to explore your topic in a somewhat more systematic (and more visual) manner.

You begin by writing your topic in the center of a sheet of paper. Then, you surround your topic with details as they occur to you, moving outward from the general topic in the center and writing down increasingly specific ideas and details as you move toward the outer edges of the page. Eventually, following the path of one idea at a.time, you create a diagram (often lopsided rather than symmetrical) that arranges ideas on spokes of branches radiating out from a central core (your topic). If your clustering is a collaborative activity, you can even work on a single large sheet of paper along with one or more other students, with each of you moving in a different direction, working toward a different corner of the page.

Asking Journalistic Questions   A more structured way of finding something to say about your topic is to ask questions. Your answers to these questions will enable you to explore your topic in an orderly and systematic fashion. One strategy involves asking six simple questions: *Who? What? Why? Where? When? How?* Journalists often use these questions to assure themselves that they have explored all angles of a story, and you can use them to see whether you have considered all aspects of your topic.

JOURNALISTIC QUESTIONS

Who?	Where?
What?	When?
Why?	How?

Asking In-Depth Questions   If you have time, and if you want to generate ideas systematically, you can ask questions that suggest familiar ways of organizing material. These questions can not only give you a great deal of information about your topic but can also suggest ways you can eventually shape your ideas into paragraphs and essays. (As you review these questions, keep in mind that not every question will apply to every topic.)

▶ See 2b5

STUDENTS WORKING TOGETHER

Use a get-acquainted activity to introduce students to the use of journalistic questions. Divide the class into pairs or small groups and have them interview each other: *Who* are you? *Where* do you come from? *What* are you studying? *When* will you graduate? *Why* are you in school? *How* do you study? Ask the students to write a brief biography of the classmate interviewed. After students have interviewed each other, they could introduce group members to the class (as a way to become comfortable with oral presentation early in the term).

COMPUTER TIP

Lab Activity

If students have the possibility of working in a networked lab with some sort of interactive conferencing software (such as Daedalus Interchange or Pacerforum), create a prompt with a few questions about a reading or writing assignment and ask students to brainstorm with other students. It's important to give clear directions for an interchange since students may veer to personal concerns if they are just asked what they think about something. Often, these class brainstorming sessions (or reading responses) produce some excellent writing—writing that can be included in an essay. Students also have the opportunity to find out what other students think and how they write. (Conferences on general topics like childhood, masculinity and femininity, and the media work very well.)

QUOTATION

"I am an author because I want to ask questions. If I had answers I'd be a politician."

–Eugene Ionesco

IN-DEPTH QUESTIONS

What happened? When did it happen? Where did it happen?	Suggest narration (account of your first day of school; a summary of Emily Dickinson's life)
What does it look like? What does it sound like, smell like, taste like, or feel like?	Suggest description (of the Louvre; of the electron microscope)
What are some typical cases or examples of it?	Suggests exemplification (three infant day-care settings; four popular fad diets)
How did it happen? What makes it work? How is it made?	Suggest process (how to apply for financial aid; how a bill becomes a law)
Why did it happen? What caused it? What does it cause? What are its effects?	Suggest cause and effect (events leading to the Korean War; the results of Prohibition; the impact of a new math curriculum on slow learners)
How is it like other things? How is it different from other things?	Suggest comparison and contrast (of the popular music of the 1950s and the 1960s; of two paintings)
What are its parts or types? Can they be separated or grouped? Do they fall into a logical order? Can they be categorized?	Suggest division and classification (components of the catalytic converter; kinds of occupational therapy; kinds of dietary supplements)
What is it? How does it resemble other members of its class? How does it differ from other members of its class?	Suggest definition (What is Marxism? What is photosynthesis? What is schizophrenia?)

Plan 1c

Creating and Using Heuristic Files

Suggest to students that they create permanent files containing the journalistic questions and the in-depth questions. Whenever they brainstorm for a paper, they can call up one or both of these files.

Your students' answers will vary. You might wish to do your own list to stimulate discussion and to provide another perspective as you talk about using the material generated in brainstorming to discover a topic and plan a paper.

NGUYEN DAO'S WRITING PROCESS: ASKING QUESTIONS

Journalistic Questions

Who stereotypes Asians? Who suffers from this?

What is a stereotype? What exactly is an Asian (Vietnamese, Chinese, Korean, Japanese, Indian?) What is an Asian American? What is an American?

Why do people stereotype others? Why do they stereotype Asians? Why do people expect so much of Asians? Why do we ourselves accept these stereotypes? Why do we use them?

Where does most stereotyping occur? In places where a lot of Asians live? In places where hardly any live? Where do stereotypes appear? In newspapers? On TV? In casual conversation?

How has the stereotype of the "typical Asian" changed over the years? How are immigrants seen? How are their children seen? How do people see me? How are various kinds of Asians alike? How are they different?

In-Depth Questions (excerpt)

What are some typical cases or examples of it? (suggests exemplification)

Asians are seen as good in math, science, engineering, and computers.

Everyone thinks I study all day.

People think my parents make me work hard.

What are its causes? (suggests cause and effect)

People don't understand other cultures. There are a lot of different kinds of people in the U. S., so we have a lot of confusion and conflict.

This took a lot of time, but it did help me to see a possible shape for my essay. I thought the Why? *and* What are its causes? *questions were the most interesting (although probably the hardest to answer).*

EXERCISE 3

List all the sources you encounter in one day (people, books, magazines, observations, and so on) that could provide you with useful information for the essay you are writing.

Your students' answers will vary.

Your students' answers will vary.

EXERCISE 4

Make a cluster diagram and a brainstorming list for the topic you selected in Exercise 2. If you have trouble thinking of material to write about, try freewriting. Then write a journal entry assessing your progress and evaluating the different strategies for finding something to say. Which approach worked best for you? Why?

EXERCISE 5

Using the question strategies described on pages 21–22 to supplement the work you did in Exercises 3 and 4, continue generating material for a short essay on your topic from Exercise 2.

CHAPTER 2

Shaping Your Material

2a Grouping Ideas: Making a Topic Tree

As you begin to see the direction your ideas are taking, you will start to sort your notes, to sift through your ideas and choose those you can use to build the most effective essay. At this point you may find it useful to make a **topic tree**, a diagram that enables you to arrange material logically and see relationships among ideas.

Begin by reviewing all your notes—from freewriting, journal entries, brainstorming, and so on—and selecting the three or four basic categories of information that best represent your material. Write these categories across the top of a piece of paper. Then go through your notes again. This time, identify specific ideas or details related to each category and write each detail under the appropriate heading, drawing lines to connect each bit of specific information to the more general heading above. Continue going through your notes and adding related information, skipping those details that do not seem to fit into any of your categories. As you move down the page, move from general information to increasingly specific details.

Of course you will add to, delete from, rearrange, and revise your topic tree as you accumulate additional material and as you continue to review the material you have. When you complete your topic tree, you will be able to see how your ideas are related. Understanding the nature of these relationships will help you develop a tentative thesis and organize supporting information in your essay. If you go on to prepare an informal outline, each branch of your topic tree will very likely correspond to one of your outline's major divisions.

▶ See 2c

If your students have trouble converting subject ideas into complete sentences, use the list below as a reinforcement exercise. Distribute copies of the list (or write it on the board) and ask students to turn each subject into a complete statement—a thesis that would be appropriate for a short paper.

Censorship
One-year "national service"
  requirement
The fall of communism
Junior-year-abroad programs
The NC-17 movie rating
Children
The VCR
Baby boomers

Note: Remind students that their thesis statements do not have to include the actual wording of the subjects above. For example, *Junior-year-abroad programs* might evolve into the statement: *The year I spent studying in Italy changed my life.*

COMPUTER TIP

Lab Activity

When students finish brainstorming on the computer, ask them to look for similarities among ideas and group ideas that go together. (In carrying out this part of the exercise, they can work individually or in pairs.) Demonstrate how to form idea groups by "cutting and pasting"—that is, by deleting text from one location and inserting it in another.

Developing a Thesis

After students have grouped items, ask them to develop a tentative thesis

continued on following page

NGUYEN DAO'S WRITING PROCESS: MAKING A TOPIC TREE

Topic Tree

This diagram seemed at first to make things more complicated, but after I looked at it for a while I realized it showed some true ideas as well as some inaccurate stereotypes. Obviously, there are *some things Asian Americans have in common. In my paper I'll have to be careful to distinguish between reality and stereotype.*

Shape 2b

MAKING A TOPIC TREE

- Review all your notes carefully.
- Identify the three or four general categories of information that best represent your material.
- Write these categories across the top of a piece of paper.
- Review your notes again to select ideas and details related to each category.
- Arrange these ideas under relevant headings, moving from general information to increasingly specific details as you move down the page.
- Draw lines to connect ideas within each category.

EXERCISE 1

Reread your notes from the work you did for Exercises 2, 4, and 5 in Chapter 1. Use this material to help you construct a topic tree.

2b Developing a Thesis

Your **thesis** is the main idea of your essay, the central point your essay supports.

(1) Understanding thesis and support

Your essay will eventually be composed of several paragraphs—an **introductory paragraph (see 4g2)**, which opens your essay and introduces your thesis; a **concluding paragraph (see 4g3)**, which closes your essay and gives it a sense of completion, perhaps restating your thesis; and a number of **body paragraphs**, which provide the support for your essay's thesis. The concept of **thesis and support**—stating the thesis, or main idea, and supplying information that explains and develops the thesis—is central to a good deal of the writing you will do in college.

---

continued from previous page

statement that the newly organized material supports. Once they have decided which material supports their thesis, tell them to delete the extraneous material.

ANSWERS TO EXERCISE 1

Your students' answers will vary. A student's first attempts at a topic tree will probably not be as well developed as Nguyen's, but with practice, your students should improve.

COMPUTER TIP

Defining an Effective Thesis
To help students develop an effective thesis statement and to help them understand the differences among a general topic, a statement of fact, an announcement of intent, and a thesis, create a template such as the following.

Moving toward a Thesis
Subject:
Statement of fact:
Announcement:
Thesis:

Have students spend fifteen or twenty minutes working through this heuristic. At the end of this period, ask volunteers to submit their results for class scrutiny and discussion.

You can also turn the "Checklist for Identifying an Effective Thesis" (p. 32 in *The Holt Handbook*) into a word-processing file.

27

John Trimbur. "Consensus and Difference in Collaborative Learning." *College English* 51.6 (October 1989): 602–16.

As the title implies, the author raises questions about "consensus" in relation to Kenneth Bruffee's views on collaborative learning. The aim of collaborative learning is to reach consensus between and among students and faculty in small discussion groups and then between the class and the teacher. The author identifies two notions of consensus, both of which see shortcomings with the concept and approach, but one of which modifies the concept for correct political ends. The first view, totally against Bruffee, sees consensus as stifling individuality by suppressing differences and enforcing conformity. It is—as one critic labels consensus—"group think." The second group is informed by a leftist critique. Although it agrees that the first view is correct in its suspicions about consensus, it disagrees over the reasons. The second group sees the concept of (romantic) individualism, which is associated with the first group, as untenable. Human beings, after all, are social. What consensus should do is allow individuals to function productively in social life. This second view of consensus is realized through differences in commonality.

See ◄ 7a2

Trimbur questions Richard Rorty's concept of "conversation," which Bruffee borrows. Rorty allows for difference through what he calls "abnormal discourse." When, for example, new speakers enter the conversation,

continued on following page

CLOSE-UP

UNDERSTANDING THESIS AND SUPPORT

A general thesis-and-support structure links the paragraphs in an essay.

(2) Defining an effective thesis

An effective thesis clearly communicates your essay's main idea. It tells your readers not only what your essay's topic is, but also how you will approach that topic and what you will say about it. Thus, the thesis reflects your essay's purpose. If your aim is to persuade, your thesis will take a strong stand. If your purpose is to convey information, your thesis can present the specific points you will discuss or give an overview that suggests how the essay will be organized. A laboratory report, for instance, simply explains a process and does not take a position. To impose an argumentative thesis on such a report would be confusing or distracting. A descriptive or narrative essay may be written to express feelings or convey impressions; its purpose may be simply to describe a scene or recount an experience, not to argue a point about that scene or experience. In such an essay, the thesis might reflect the organizing principle or communicate the dominant impression.

An effective thesis is more than a general subject, a statement of fact, or an announcement of your intent. Consider the differences among the following statements.

Shape 2b

Subject	Statement of Fact	Announcement
The Draft	The United States currently has no peacetime draft.	In this essay I will reconsider our country's need for a draft.

Thesis While once the military draft may have been necessary to keep the armed forces strong, today's all-volunteer force has eliminated the need for a draft.

Subject	Statement of Fact	Announcement
Intelligence Tests	Intelligence tests are used extensively in some schools.	The paragraphs that follow will advance the idea that intelligence tests may be inaccurate.

Thesis Although intelligence tests are widely used for placement in many schools, they are not always the best measure of a student's academic performance.

Subject	Statement of Fact	Announcement
Music Videos	Music videos can enhance record sales.	As I will argue in this paper, music videos are an important part of our culture.

Thesis It may be true that music videos present stale images in place of listeners' original interpretations, but the shared images these videos show us play an important role in establishing a common culture.

Subject	Statement of Fact	Announcement
Math Anxiety	Math anxiety is a problem for many young girls.	My paper will attempt to show why young girls have problems with mathematics.

Thesis Young boys tend to outperform young girls in math classes not because of their superior ability but because girls are afraid of being seen as unfeminine if they excel in math.

continued from previous page

they introduce a form of abnormal discourse that helps change, for the better, the ongoing conversation. The second, leftist group also critiques this view as romantic individualism. Instead, abnormal discourse should be seen as dissensus in the face of consensus, or difference in the face of sameness. (For this view of dissensus, the author relies on Greg Myers and Victor Vitanza.) Specifically, what students need to learn is the activity of dissensus that actually pervades professional writing, which manifests itself as "systematic misunderstanding and concealment." Such a manifestation often occurs, as Susan Wells has pointed out, in professional-business writing, in which particular secrets are not to be shared, for personal or corporate reasons, with the collaborators.

The author continues by distinguishing between "consensus as an acculturative practice" and "consensus as an oppositional one." The first produces business as usual; the second is a challenge to prevailing means of production. The first, it may be discovered, is "'spurious' consensus" because it only creates the illusion of free deliberation.

The author concludes with examples of how the second, leftist view of consensus can be applied in the study of literature and writing.

(The work by Kenneth A. Bruffee referred to is "Collaborative Learning: Some Practical Models." *College English* 34 (1973): 634–43.)

CLOSE-UP

DEFINING AN EFFECTIVE THESIS

Effective thesis statements should not include phrases like "As I will show," "In my paper, I plan to demonstrate," and "It seems to me." These expressions tend to distract readers, calling attention to you rather than to your subject. Moreover, such meaningless phrases weaken your credibility by suggesting that your conclusions are based on opinion rather than on reading, observation, and experience.

An effective thesis is carefully worded. In order to communicate your main idea, an effective thesis should be clearly and accurately worded, with careful phrasing that makes your meaning apparent to your readers. It is usually expressed in a single, concise sentence. Your thesis should be direct and straightforward, avoiding vague, abstract language and overly complex terminology. It should include no unnecessary details that might confuse or mislead readers and make no promises that your essay will not fulfill. Although your thesis cannot enumerate every idea your essay will develop, it may list your most important points; in any case, it should be specific enough to give readers a good idea of how you will develop your essay.

CLOSE-UP

DEFINING AN EFFECTIVE THESIS

Thesis statements are weakened by vague phrases such as *centers on, deals with, involves, revolves around, has to be, has a lot to do with,* and *is primarily concerned with.*

INEFFECTIVE THESIS STATEMENT: The real problem in our schools does not *revolve around* the absence of nationwide goals and standards; the problem *is primarily concerned with* the absence of resources with which to implement them.

EFFECTIVE THESIS STATEMENT: The real problem in our schools *is* not the absence of nationwide goals and standards; the problem *is* the absence of resources with which to implement them.

Finally, an effective thesis suggests your essay's direction, emphasis, and scope. Your thesis should help you visualize the most effective way to arrange your material and connect your thesis with your support. Ideally, it suggests how your ideas are related, in what order your major points should be introduced, and where you should place your emphasis.

The following thesis statement conveys a good deal of information:

> Widely ridiculed as escape reading, romance novels are becoming increasingly important as a proving ground for many never-before-published writers and, more significantly, as a showcase for strong heroines.

The main thing this thesis tells is you that this essay focuses primarily on what the writer considers to be the two major new roles of the romance novel: providing markets for new writers and (more importantly) presenting strong female characters. The thesis also tells you that the role of the romance as escapist fiction may also be treated. The thesis statement even suggests a possible order for the various ideas discussed.

Introduction

Discussion of romance formulas; general settings; plots and characters Thesis: Widely ridiculed <u>as escape reading</u>, romance novels are becoming increasingly important as a proving ground for many <u>never-before-published writers</u>, and, more significantly, as a showcase for <u>strong heroines</u>.

Romance novels as escape reading

Romance novels as outlet for unpublished writers

Romance novels as showcase for strong heroines

Body

Restatement of thesis. Review of major points; significance of recent trends in romance novels

Conclusion

Beginning writers often choose thesis statements that are simply "too big" for the length of the assignment, for the available research time, or perhaps for the level of audience (and teacher) interest. Constructing ladders is an effective way to help your students learn how to "downsize" a thesis that is too broad or general for the writing task at hand.

FRIENDS
↓
CHILDHOOD FRIENDS
↓
JAMES AND DAVID
↓
A DISAGREEMENT THAT LED TO A FIGHT

BASEBALL
↓
YANKEES
↓
CASEY STENGEL
↓
CASEY STENGEL'S HUMOR

ADVERTISING
↓
TELEVISION ADVERTISING
↓
BEER COMMERCIALS
↓
STEREOTYPES IN BEER COMMERCIALS

MOVIES
↓
OLD MOVIES ON T.V.
↓
MUSICALS FROM THE 50's
↓
PRODUCTION NUMBERS IN 50's MUSICALS

Divide the class into small groups, providing each group with a list of ten one-word general subjects. Ask the groups to construct one five-rung ladder for each subject. Allow students to choose their own working strategy. The group may want to brainstorm together, with all members allowed complete freedom to suggest "rungs" and voice support/opposition to final choices. Or, as an alternative, group members could sit in a fixed circle. One member will be given a topic and allowed to choose the first rung: then, working clockwise around the circle, each member in turn will add the next rung. (Be sure you let each group member have the chance to start at least two rungs, so that no one is limited to making either the most general or most specific choices.)

WRITING CHECKLIST: IDENTIFYING AN EFFECTIVE THESIS

- Does your thesis clearly communicate your essay's main idea? Does it suggest the approach you will take toward your material? Does it reflect your essay's purpose?
- Is your thesis more than a subject, a statement of fact, or an announcement of your intent?
- Is your thesis carefully worded?
- Does your thesis suggest your essay's direction, emphasis, and scope?

(3)   Deciding on a thesis

Occasionally—especially if you know a lot about your topic—you may begin writing with a thesis in mind. Most often, however, your thesis evolves out of the reading, questioning, and grouping of ideas you do during the writing process. If you have difficulty finding a thesis, try reviewing your notes. Some writers find that doing focused freewriting at this stage helps: By concentrating on a key idea about your topic and then freewriting about it, you may be able to discover a possible thesis.

See ◀
1c2

The thesis you develop as you plan your essay is only a tentative one. It gives you sufficient focus to guide you through your first draft, but you should expect to modify it in subsequent drafts. As you write and rewrite, you often think of new ideas and see new connections; as a result, you may change your essay's direction, emphasis, and scope several times. It stands to reason, then, that you will constantly change and sharpen your thesis to keep it consistent with your paper's changing goals. Notice how the following pairs of thesis statements changed as the writers moved through successive drafts of their essays.

DECIDING ON A THESIS

Tentative Thesis (rough draft)	**Revised Thesis (final draft)**
Professional sports can easily be corrupted by organized crime.	Although proponents of legalized gambling argue that organized crime cannot make inroads into professional

continued on the following page

continued from the previous page

	sports, the way in which underworld figures compromised the 1919 World Series suggests the opposite.
Laboratory courses provide valuable educational experiences.	By providing students with the actual experience of doing scientific work, laboratory courses encourage precise thinking, careful observation, and creativity.
It is difficult to understand Henry James's short novel *The Turn of the Screw* without examining the personality of the governess.	A careful reading of Henry James's *The Turn of the Screw* suggests that the governess is an unreliable narrator, incapable of distinguishing appearance from reality.

(4) Stating your thesis

As a beginning writer, you may find it helpful to state your thesis early in your paper. Not only will this placement immediately signal the focus of your discussion to your readers, but it will also serve as a constant reminder to you of your essay's direction, emphasis, and scope. Your thesis can appear anywhere in your essay, however, as long as it makes your essay's main idea clear to your readers. Where you state your thesis largely depends on the effect you wish your essay to have on a particular audience. In an argument, you may have to lead your readers gradually to your controversial thesis instead of stating it at once. To do otherwise could alienate a segment of your audience. In a research paper, you may have to present several paragraphs of background material before your audience is able to understand your thesis.

Although many of the essays you write in college will include a thesis, you will not always want to state it explicitly. Sometimes your thesis can be *implied* through the arrangement of the points in your essay. Like an explicit thesis, an implied thesis must clearly and specifically convey your essay's main idea to your readers. Professional writers often use this technique, preferring to make their points with some subtlety, thus allowing readers to arrive at their own conclusions.

▶ See 7b

▶ See 41d, 41f

STUDENTS WORKING TOGETHER

Give groups of three students a broad subject, such as "School," or "Ethnic Stereotypes." Each group must generate three different thesis statements from this single subject and then rank order the statements from best to worst. Afterwards, the group exchanges its thesis statements (unranked) with another group for this group to rank order. Collect all the "first choice" thesis statements and inform the class of the selections. Discussion follows from there.

STUDENTS WORKING TOGETHER

In groups of four or five, students create their own body of information, writing six related statements on transparencies. On a separate transparency, they then write the thesis statement generated from this body of information. Then they exchange the six statements with another group, which reads them and itself devises a thesis. Collect all transparencies and show them to the class for discussion.

(5)   Using a thesis statement to shape your essay

As you have seen, the wording of your thesis statement can suggest a possible order and emphasis for your essay's ideas. More specifically, the way the thesis is worded can suggest the paper's specific pattern of development—narration, description, exemplification, process, cause and effect, comparison and contrast,

See ◄ 4e

division and classification, or definition. These familiar patterns may also shape individual paragraphs of your essay.

In each of the following examples, the thesis statement suggests not only the order in which ideas will be presented, but also a specific pattern of development. Remember, though, that a pattern of development is never imposed on your essay; it emerges naturally in the course of the writing process.

Thesis Statement	Pattern of Development
As the months went by and I grew more and more involved with the developmentally delayed children at the Learning Center, I came to see how important it is to treat every child as an individual.	Narration
Looking around the room where I had spent my childhood, I realized that every object I saw told me I was now an adult.	Description
The risk-taking behavior that characterizes the 1990s can be illustrated by the increasing interest and involvement in such high-risk sports as mountain biking, ice climbing, sky diving, and bungee jumping.	Exemplification

continued on the following page

continued from the previous page

Armed forces basic training programs take recruits through a series of tasks designed to build skills, confidence, and camaraderie.	Process
The exceptionally high birthrate of the post-World War II years had many significant social, economic, and political consequences.	Cause and Effect
Although people who live in cities and people who live in small towns have some obvious basic similarities, their views on issues like crime, waste disposal, farm subsidies, and educational vouchers tend to be very different.	Comparison and Contrast
The section of the proposal that recommends establishing satellite health centers is quite promising; unfortunately, however, the sections that call for the creation of alternative educational programs, job training, and low-income housing are seriously flawed.	Division and Classification
Until quite recently, most people assumed rape was an act perpetrated by a stranger, but today's wider definition encompasses acquaintance rape as well.	Definition

Divide the class into groups of four or five students. Give each group fifteen minutes to come up with a workable thesis statement for a short essay on one of the following subjects and to prepare an informal outline for the essay.

Bicycle helmets
Year-round school schedules
Professional athletes
The environment
MTV

1. An announcement, not a thesis.
2. A subject, not a thesis. Gives no indication of essay's focus or direction, let alone writer's position.
3. A subject, not a thesis. Why should it be avoided? What coast? What kind of development? What constitutes overdevelopment?
4. No position indicated. What aspects will be considered? What pattern of development might be used? What standards of judgment will be used?
5. A good start, but "but it has some disadvantages" is not specific enough.
6. A subject, not a thesis. What position? Do all environmentalists share a single position?
7. States a fact.
8. A good start, but needs clearer focus. Thesis is too general.
9. States a fact.
10. States a personal opinion. Question of taste. Hard to develop objectively or concretely.

ANSWERS TO EXERCISE 3

1. Henry David Thoreau's "Civil Disobedience" changed my life by introducing me to the idea of passive resistance.
2. Colleges should do more to discourage cheating.
3. The SAT is not a valid measurement of every student's potential for college success.
4. If women are equal to men, then they should be permitted to serve in military combat as men do.

continued on following page

NGUYEN DAO'S WRITING PROCESS: DEVELOPING A THESIS

Tentative Thesis

As an Asian American, I am frequently a victim of ethnic stereotyping, and this has been a serious problem for me.

I tried some other ways to word my tentative thesis, but I wanted to make sure it included the most important ideas I was going to talk about--that I'm Asian American, that people stereotype me, and that this is a problem. This thesis seemed to do all that. It seemed to suggest I could include some examples of things that happened to me or things other people assume about me. And I could also dig into some causes and effects, explaining why this stereotyping has been a problem. This thesis was pretty specific and to the point, not too wordy or general. So, I thought I'd stick with it for the time being.

*EXERCISE 2

Analyze the following statements or topics and explain why none of them qualifies as an effective thesis. Be prepared to explain how each could be improved.

1. In this essay, I will examine the environmental effects of residential and commercial development on the coastal regions of the United States.
2. Residential and commercial development in the coastal regions of the United States
3. How to avoid coastal overdevelopment
4. Coastal Development: Pro and Con
5. Residential and commercial development of America's coastal regions benefits some people, but it has some disadvantages.
6. The environmentalists' position on coastal development
7. More and more coastal regions in the United States are being overdeveloped.
8. Residential and commercial development guidelines need to be developed for coastal regions of the United States.
9. Coastal development is causing beach erosion.
10. At one time I enjoyed walking on the beach, but commercial and residential development has ruined the experience for me.

Shape 2b

*EXERCISE 3

For three of the following topics, formulate a clearly worded thesis statement.

1. A literary work that has influenced your thinking
2. Cheating in college
3. The validity of SAT scores as the basis for college admissions
4. Should women in the military serve in combat?
5. Private vs. public education
6. Should college health clinics provide birth control services?
7. Is governmental censorship of art ever justified?
8. The role of the individual in saving the earth
9. The portrayal of an ethnic group in film and television
10. Should smoking be banned from all public places?

*EXERCISE 4

Read the following sentences excerpted from the essay "Territoriality and Dominance" by René Dubos. Then develop a thesis that will tie together the information in the sentences. Make sure your thesis is well constructed and follows the requirements outlined in this chapter.

- Whenever the population density of a group increases beyond a safe limit, many of the low-ranking animals in the social hierarchy are removed from the reproductive pool.
- The remarkable outcome of these automatic mechanisms is that, in the case of many animal species, animal populations in the wild remain on the average much more stable than would be expected from the maximum reproductive potential.
- When males fight, the combat is rarely to the death.
- The losing animal in a struggle saves itself . . . by an act of submission, an act usually recognized and accepted by the winner.
- The view that destructive combat is rare among wild animals . . . is at variance with the "Nature, red in tooth and claw" legend.
- Since ritualization of behavior is widespread among the higher apes, it is surprising that humans differ from them, as well as from most other animals, in practicing warfare extensively with the intent to kill.

EXERCISE 5

Review the topic tree you made in Exercise 1. Use it to help you to develop a thesis for an essay on the topic you chose in Exercise 2 in Chapter 1.

continued from previous page

5. Although private schools typically offer smaller classs and more individual attention, public schools can expose children to wider educational and cultural opportunities.
6. College health centers should provide birth control services because such services promote safe sex.
7. If the first amendment is valid, governmental censorship of art is never justified.
8. Every individual can help save the earth by taking a few simple measures.
9. Despite the progress made in the last few decades, Arabs are still underrepresented and misrepresented in American films and television.
10. To protect the rights and health of nonsmokers, smoking should be banned in all public places.

ANSWERS TO EXERCISE 4

Although students' answers will vary, the following thesis statements might reasonably be derived from the material:

Whereas humans practice warfare extensively with the intent to kill, most other animal species engage in destructive combat as a means to maintain stable population densities in the wild.

Although male animals in the wild sometimes engage in battle to establish dominance, leading to the "Nature, red in tooth and claw" legend, most ritualistic destructive combat with the intent to kill is reserved for maintaining stable population densities.

continued on following page

continued from previous page

The "Nature, red in tooth and claw" legend is played out in most animal species through automatic and ritualistic mechanisms that help to maintain stable population densities in the wild and to establish dominance.

ANSWERS TO EXERCISE 5

Your students' answers will vary.

COMPUTER TIP

Organizing Material

A worksheet template file can be helpful to students at the stage between an outline and a first draft. The worksheet can help students begin to address the problem of finding sufficient and relevant support for their theses.

A typical worksheet for an essay might look like the one below.

Worksheet for an Essay

Write down your SUBJECT. It can be a thing (the island of Antigua), a person (your grandfather), or a process (founding an organization). Finish the sentence below in fewer than ten words.

My subject is

Make a STATEMENT about your subject. Write *one* sentence that tells what you want your readers to know.

I want my readers to know that

SUPPORT your statement. Finish the sentences below by giving the **MAIN SUPPORTING POINTS** that your readers need to know in order for them to understand and accept your statement. Next, briefly give the

continued on following page

NGUYEN DAO'S WRITING PROCESS: PREPARING AN INFORMAL OUTLINE

Informal Outline: Asian-American Stereotypes

Thesis: As an Asian American, I am frequently a victim of ethnic stereotyping, and this has been a serious problem for me.

Stereotypes of Asian immigrants
    —Poor English
    —Can't use skills and education here
    —Low-paid jobs
        —Sacrifice for children
        —Take from U.S. citizens
Stereotypes of children of Asian immigrants
    —Hard workers
    —Study hard
        —Pushed by parents
    —Focus on math and science
        —Pre-med
Stereotypes applied to me
    —Science major
    —Forced to study all day
        —No social life
Reality
    —I have outside interests
    —I'm not pre-med
    —My parents don't push me

This outline seemed kind of short, even after I added a category (Reality) that wasn't on my topic tree, but I thought I could follow it pretty easily as I did my rough draft, maybe getting a paragraph out of each section.

2c   Preparing an Informal Outline

An **informal outline** is a blueprint for an essay, a plan that gives you more detailed, specific guidance than a thesis statement. You need not always prepare such an outline; a short essay on a topic with which you are familiar may require nothing beyond a thesis and a list of major divisions or main supporting points. More often, however, you will find that you need additional help. An informal outline arranges your essay's main points and supporting ideas and details in an informal but orderly way to guide you as you write.

continued from previous page

ideas or details that support your main supporting points.

> One main supporting point for my statement is that
>> Supporting ideas or details
>> Supporting ideas or details
>> Supporting ideas or details

Using the Copy and Insert commands, make as many box sets as you need to support your statement about your subject.

You'll find that your **statement**, your **main supporting points**, and your **supporting ideas or details**, taken all together, form a rough outline that can help you write a well-organized essay.

PREPARING AN INFORMAL OUTLINE

- Write down the categories and subcategories from your topic tree.
- Arrange the categories and the subcategories within them in the order in which you will discuss them.
- Expand the outline with additional material from your notes, adding any new points that come to mind.

For a short paper, an informal outline is usually sufficient. Sometimes, however—particularly when you are writing a long or complex essay—you will need to construct a **formal outline**.

▶ See 41g

EXERCISE 6

Read the paragraphs below, which are excerpted from a newspaper editorial. Prepare an informal outline that includes all the main ideas and supporting points.

The college football season is starting with a familiar flurry of resignations and probations.

At the University of Washington, Don James resigned as head coach after failing to notice that his quarterback owned three cars. Of course, the athletic staff was not part of the money-lending scheme that ensnared the player and brought down the coach. It was one of those pesky "boosters" who always seem handy when blame is assigned.

So among big-time teams, Washington joined Auburn University on two-year probation. The N.C.A.A. is expected to add Texas A and M to the list as footballs are being teed up for the television cameras that provide the money that seduces universities into being willing co-conspirators in exploiting young athletes.

College football *is* a great show, and at its best, it still provides some of the emotion and spectacle missing in the cogwheel perfection of pro football. But big-time college football has a corruption at its center that can and must be cured. Why the nation's college presidents and boards of trustees, acting through the N.C.A.A., have not taken the obvious steps is a mystery.

Informal Outlines
In high school, some of your students may have been forced to write detailed outlines and then follow them to the letter (or Roman numeral). Make sure students understand that an informal outline is a guide, not a straitjacket. The outline **can** and **should** be changed at any point during the research and writing process, as the writer discovers new information or thinks of a different approach to the topic. Emphasize the notion that *the writing process itself* often shapes the final form of an essay or paper as much as the informal outline does.

ANSWERS TO EXERCISE 7

Your students' answers will vary. Small group discussions of informal outlines may be a good classroom exercise.

One such step is to limit participation rights from teams that fail to honor Principle VII of the Knight Foundation Commission on Intercollegiate Athletics. That principle says that athletes will graduate in the same proportion as non-athletes. The national graduation rate is around 50 percent. Five of the top 20 teams in The Times preseason poll—including reigning national champion Alabama—have graduation rates below 40 percent.

Starting immediately, the N.C.A.A. should set up standards so that private and public schools can be measured fairly against one another and then begin to punish schools that fail to graduate a proportionate number of athletes on time.

It should then do something about the money problem. Athletes make millions for their schools and receive piddling amounts for tuition and board. Their poverty in comparison to their market value as college players and potential earning power as pros make corruption inevitable.

A sensible first step is to lift student athletes out of poverty by giving them realistic scholarships based on the *entire* cost of keeping a student in school: tuition, housing, food, plus the allowances a parent would provide for clothes, transportation, social activities and spending money.

With tuition alone at many N.C.A.A. schools running from $10,000 to $20,000, many middle-income families spend a total of $30,000 or more a year in after-tax dollars. And it is the total cost of keeping a student in comfortable style that an athletic scholarship should cover. Presently N.C.A.A. scholarships cover only tuition, room and board, books and up to $2,400 in cash.

Nothing can bring back the era of pure amateurism in college football. But simple reforms could make it financially clean and academically respectable as well as entertaining.

EXERCISE 7

After reviewing your notes, prepare an informal outline for a paper on the thesis you developed in Exercise 5.

CHAPTER 3

Writing and Revising

QUOTATIONS

"Just get it down on paper, then we'll see what to do with it."

–Maxwell Perkins

"I believe in miracles in every area of life *except* writing. Experience has shown me that there are no miracles in writing: the only thing that produces good writing is hard work."

–Isaac Bashevis Singer

"Writing is one of the few areas in life where 'Do it right the first time or don't do it at all' doesn't apply. If we tried to do it right the first time, most of us would never finish a sentence."

–Ronald John Donovan

3a Writing a Rough Draft

Because its purpose is to get your emerging ideas down on paper so you can react to them, a rough draft is often messy, full of false starts and information that seems to go nowhere. Try not to see this disorder as a problem; at this stage, it is not. Experienced writers know—and so should you—that an essay will generally be rewritten several times. You should expect to cross out words and sentences and to rewrite and reorder paragraphs that seem choppy or disconnected. You will have plenty of time to make changes when you revise, so do not waste time now worrying about things that seem less than perfect.

As you write your draft, you will probably discover some new ideas. In fact, you may even find your paper taking an unexpected detour. If so, do not panic. Your new perspective may lead you to a more interesting paper than the one you have been planning.

When you write a rough draft, it is usually simplest to focus on the body of your essay, writing only sketchy—or even nonexistent—introductory and concluding paragraphs. If ideas for your opening and closing paragraphs are elusive, do not waste time struggling to discover them. The effort will slow you down; besides, these paragraphs are likely to change in subsequent drafts anyway. For now, concentrate on drafting the support paragraphs of your essay.

You will probably be revising much of what you write, and careful preparation of your first draft will make these revisions less painful.

COMPUTER TIP

Using Temporary Headings to Organize Ideas

To help students organize material, encourage them to use temporary headings and format them differently from the rest of the paper (perhaps as bold uppercase italics). They can delete these headings before printing the final draft. Here is an example:

Differences Between High School and College Field Hockey

PRACTICES

When I tried out for the Babson women's field hockey team this fall, I discovered some major differences from playing in high school. First of all, I received a complete conditioning package from the Babson coach at the beginning of the summer and was expected to be in shape by the first day of double sessions. In high school, on the other hand, a great deal of time was taken out of practice for conditioning. Unlike high school, practices in college are quick moving and intense. There is not much fooling around.

TALENT

Another significant difference was the talent of my teammates. In high school anyone can play hockey. There are no cuts, so there is minimal skill. When I came to Babson, however, I learned that most of the players were recruits and there was an incredible amount of skill on the team.

PLAYING GAMES

The competition between teams is tough. There are no weak teams as there are in high school, and the

continued on following page

See ◄
1c2

STRATEGIES FOR WRITING A ROUGH DRAFT

- **Prepare your work area.** Once you begin to write, you should not have to stop because you need a sharp pencil, better lighting, important notes, or anything else. Unscheduled breaks can ruin your concentration.

- **Fight writer's block. Writer's block**—an inability to start (or continue) writing—may be caused by fear that you will not write well or that you have nothing to say. If you really do not feel ready to write, taking a short break may give your ideas time to incubate, which, in turn, may release new ideas. If you decide that you really do not have enough material to get you started, return to one of the strategies for finding something to say. But remember, the least productive response to writer's block is procrastination.

- **Get your ideas down on paper as quickly as you can.** Do not worry about sentence structure, spelling and punctuation, or finding exactly the right word. Concentrate on recording your ideas. Writing quickly helps you uncover new ideas or new connections between ideas. If you have made an informal outline, you may find that following it enables you to move smoothly from one point to the next. If, however, you find this structure too inhibiting or confining, feel free to write without consulting your outline.

- **Take regular breaks as you write.** If you are finding it difficult to stay focused on your topic, try writing one section of your essay at a time. When you have completed a section or paragraph, take a break. Your unconscious mind may continue to focus on your assignment while you do other things. When you return to your essay, writing may be easier.

- **Leave yourself enough time to revise.** All writing benefits from revision, so be sure you have time to reconsider your work and to write as many drafts as necessary.

DRAFTING WITH REVISION IN MIND

- **Write on every other line.** (If you type your draft, triple-space.) This makes errors more obvious and also gives

continued on the following page

continued from the previous page

you plenty of room to add new material or to try out new versions of sentences.

- **Develop a system of symbols,** each indicating a different type of revision. For instance, you can circle individual words or box longer groups of words (or even entire paragraphs) that you want to relocate. You can use an arrow to indicate the new location, or you can use asterisks or matching numbers or letters to indicate how you want to rearrange ideas. When you want to add words, use a caret like ^this. You can also use proofreader's marks.

- **Write on only one side of a sheet of paper.** This technique enables you to reread your pages side by side. Writing on only one side also permits you to cut and paste without destroying material on the other side of the page. This strategy gives you the flexibility to keep reorganizing the sections or paragraphs of your paper until you find their most effective arrangement.

- **Print out every draft** if you are composing on a computer, so you can revise on paper rather than on the screen, adding handwritten notations and symbols to your typed draft and then returning to the computer to type in your changes. In addition, be very careful not to erase or discard any information you may need later.

continued from previous page

games are more controlled and quicker moving. The whistles are minimal in college, where in high school there is a whistle every fifteen seconds.

▶ See App. A

▶ See App. B

TEACHING STRATEGY

Proofreaders' Marks

If the situation seems appropriate, this may be a good time briefly to review the common proofreaders' marks found in Appendix A.

These symbols simplify the writer's revision and editing of his or her own work and allow other readers (teachers, typesetters, etc.) to easily understand any alterations. If you have a set policy on the use of these marks in final drafts, make it clear to the students.

NGUYEN DAO'S WRITING PROCESS: WRITING A ROUGH DRAFT

ROUGH DRAFT

Asian-American Stereotypes

1   The United States prides itself on being the "melting pot" of the world. However, in reality, the abundance of different cultures in America often causes misunderstandings and even conflicts within the society. These misunderstandings and conflicts result from the society's lack of knowledge

continued on the following page

continued from the previous page

COMPUTER TIP

Lab Activity

The following techniques work only if you have a means of projecting your computer screen onto a screen at the front of the classroom or if students' computers are networked so that you can project your screen onto their screens.

If you are brave enough to take chances, try composing in front of your students. If your word processor has macro capabilities, record all the keystrokes you make while composing a short piece and replay them in class while commenting on what you did, why you did it, and what went through your mind while doing it.

Just as you can write publicly for your students, you can also revise publicly for them. Bring in an example of your own work and talk about (and execute) revisions in front of them. An alternative method is to revise a paper from another class while asking for suggestions from students.

STUDENTS WORKING TOGETHER

Divide the class into small discussion groups. Ask all students to read Nguyen Dao's first draft; then give the groups ten minutes for a discussion of the essay's successes and shortcomings. (Someone in each of the groups might take notes on the discussion.) Where does the draft seem to tell too little—or too much? Are there points of confusion? Does the essay hold reader interest or lose it at one or more points? Encourage students to focus on general issues, not hunt for grammatical or mechanical errors.

about other cultures. As an Asian American, I am frequently a victim of ethnic stereotyping, and this has been a serious problem for me.

2   It has been within the last twenty years or so that the United States has seen a large rise in the number of Asian immigrants. First-generation immigrants are seen as an underclass of poor who struggle in low paying jobs so their children will have a better future. Many accuse immigrants of accepting less pay for their work than the established majority is willing to accept, thus putting the established majority out of work. While it is true that most newly arrived Asians do seek low paying, low skill jobs, they are just following the same trend that other immigrant groups followed when they first arrived in the United States. Because the first generation of Asian Americans have poorly developed skills in English, they are forced into jobs which do not require those skills. Many Asians received degrees from institutions in their native countries or received advanced training of some kind but cannot use those skills in the United States.

3   Asian-American children are seen as hard workers who are pushed by their families to succeed. Asian children are

continued on the following page

continued from the previous page

seen as intelligent but only in scientific and technical knowledge. The media likes to point to the facts that most Asians succeed only in the math, science, and engineering fields and that there is an inordinate number of Asian college students who identify themselves as "pre-med."

4   In my personal experience, in college, many of my friends assume that I am either a science or an engineering major and that my parents force me to study five to ten hours a day. They believe that I sacrifice all my free time and social life in pursuit of a high grade point average.

5   In fact, I am a political science major. I also like to go to basketball games, listen to music, and read automotive magazines, just like other college students I know. My parents do encourage me to do well in school because they see that education is a stepping stone to social class mobility; however, I am lucky because my parents do not push me in one direction or another as some of my Asian friends' parents do. Many of my friends do not realize that just two or three generations ago, their parents and grandparents were going through the same process of social adjustment that all immigrants endure.

6   It is important to remember that not all Asians fit into the overachieving, success-oriented stereotype.

continued on the following page

continued from the previous page

> *This first draft really only got me started writing. I wrote a short introduction to get into the subject and identify my problem; however, at this point I really didn't have a clear vision of where my paper was headed or how I was going to support my points. I thought I might like to compare my experiences with the experiences of other ethnic groups, particularly other minorities that are like Asians. I also saw that I had to revise my thesis to make it more focused on the problems stereotyping causes. Knowing I'd have to do a lot of revision, I triple-spaced so I could write in my changes.*

EXERCISE 1

Write a rough draft of the essay you began planning in Chapter 1.

3b Understanding Revision

Revision does not simply follow planning, shaping, and writing as the next step in a sequence. Rather, it is a process you engage in from the moment you begin to discover ideas for your essay. As you work you are constantly rethinking your ideas and reconsidering their relevance, their relative importance, their logical and sequential relationships, and the pattern in which you arrange them. Revision is a creative and individual aspect of the writing process, and everyone does it somewhat differently. You will have to do a lot of experimenting before you find the particular techniques that work best for you.

Inexperienced writers sometimes believe they have failed if their first drafts are not perfect, but more experienced writers expect to revise. They also differ from inexperienced writers in *how* they revise. Inexperienced writers are often afraid to make significant changes. In fact, they may do little more than refine word choices, correct grammatical or mechanical errors, or recopy their papers to make them neater. Experienced writers, however, see revision as more than surface editing; they expect it to involve a major upheaval. They are therefore willing to rethink a thesis and to disassemble and reassemble an entire essay—and you too should be willing, even eager, to do this kind of revision.

You may have written your rough draft without thinking much about your audience. When you revise, however, you should begin to shape your writing for others, to "re-see" what you have written

Your students' answers will vary. Encourage students to use the prewriting exercises from Chapter 1 and the revision checklists in this chapter to improve their work. Some of the checking (for logical outline, e.g.) might be done in pairs or in groups.

Collaborative Revision
Collaborative revision is an increasingly popular technique in writing classes, but teachers should provide students with suggestions and techniques and be sensitive to any developing problems. Students may be embarrassed to share their writing with peers or be unskilled in giving—or accepting—constructive criticism. Before students tackle one another's work, distribute copies of an anonymous "paper with problems" and lead the class in a freewheeling oral critique of the paper. Guide the discussion by inserting your own comments and questions and help students learn to be frank and honest without sounding scornful, sarcastic, or aggressive. At some point during the session, it may be valuable to ask the question: "If the author were sitting here, would you be asking the same questions, using the same tone of voice and word choice? If not, why would you change?"

and make the changes necessary to enable your readers to understand and appreciate your ideas. Thus, revision should reflect not only your private criticism of your first draft but also your anticipation of your readers' reactions.

WRITING WITH YOUR READER IN MIND

• **Present one idea at a time and summarize when necessary.** When you overload your paper with more information than readers can take in, you lose them. Readers should not have to backtrack constantly to understand your message.

• **Organize your ideas clearly.** Readers expect your paper to do what your thesis says it will do, with major points introduced in a logical order and supported in your body paragraphs.

• **Provide clear signals to establish coherence.** Repeating key points, constructing clear topic sentences, using transitional words and phrases to link ideas logically, and providing verbal cues that indicate the precise relationships among ideas all help provide continuity and coherence.

3c Applying Strategies for Revision

Everyone revises differently, and every writing task demands a slightly different kind of revision. Four strategies in particular can help you revise.

(1) Using a formal outline

Making a **formal outline** of a draft helps you to check the structure of your paper. A formal outline reveals at once whether points are irrelevant or poorly placed—or, worse, missing. It also reveals the hierarchy of your ideas—which points are dominant and which are subordinate. This strategy is especially helpful early in revision, when you are reworking the larger structural elements of your essay.

▶ See 41g

COMPUTER TIP

Collaborative Revision
If your school has a computer lab, exchanging monitors or computers is an effective way to encourage students to respond critically to each other's work.

After students have completed a first or rough draft, have them exchange seats so that they are sitting at someone else's computer. (You can achieve the same effect by having students exchange disks while they remain at their own computers.) As they read the drafts on the screens in front of them, ask them to enter comments in bold, underlined, or uppercase characters within the text itself or to enclose their comments in brackets.

If students' software supports *hidden text* or *invisible writing*, they can elect to hide their comments until the writers choose to view them; in this way, criticism can be completely unobtrusive.

During the prewriting stage (or even later, when writers need to evaluate their arguments and give support evidence), have students engage in computerized dialogues by swapping monitors. Designate one student to be the "writer" and the other to be the "critic." Ask the writer to enter the thesis statement or one of the main supporting arguments; the writer's entry will appear only on the monitor in front of the critic. Then ask the critic to enter a response, either a refutation backed by evidence or a question that the writer needs to answer; this response will appear only on the monitor in front of the writer.

After the students have drafted a paper, divide the class into teams of two or three for collaborative revision. Have the students read each other's papers and write responses to the "Questions for Collaborative Revision" on page 46. Then ask them to discuss the comments. You may have the students work with the same peer critics for each assignment or change the teams each time.

(2) Doing collaborative revision

Instead of trying to imagine an audience, you can address a real audience by asking a friend, classmate, or family member to read your draft and comment on it (with your instructor's permission, of course). The response should tell you whether or not your essay has the effect you intended and perhaps why it succeeds or fails. **Collaborative revision** can also be more formal. Your instructor may conduct the class as a workshop, assigning students to work in groups to critique other students' essays. Or, he or she

WRITING CHECKLIST: QUESTIONS FOR COLLABORATIVE REVISION

- What is the essay about? Is the topic suitable for this assignment?
- What is the main point of the essay? Is the thesis stated? If so, is it clearly worded? If not, how can the wording be improved? Is the thesis stated in an appropriate place?
- Is the essay arranged logically? Do the body paragraphs appear in an appropriate order?
- What ideas support the thesis? Does each body paragraph develop one of these ideas?
- Does each body paragraph have a unifying idea? Are topic sentences needed to summarize the information in the paragraphs? Are the topic sentences clearly related to the thesis?
- Is any necessary information missing? Identify any areas that seem to need further development. Is any information irrelevant? If so, indicate possible deletions.
- Can you contribute anything to the essay? Can you think of any ideas or examples from your own reading, experience, or observations that would strengthen the writer's points?
- Can you follow the writer's ideas? If not, would clearer connections between sentences or paragraphs be helpful? If so, where are such connections needed?
- Is the introductory paragraph interesting to you? Would another kind of introduction work better?
- Does the conclusion leave you with a sense of completion? Would another kind of conclusion be more appropriate?
- Is anything unclear or confusing?
- What is the essay's greatest strength?
- What is the essay's greatest weakness?

may ask you to exchange essays with another student and write an evaluation of the paper.

In any case, approach a fellow student's draft responsibly, and take the analysis of your own essay seriously. Collaborative revision, like all collaborative work, is a dialogue. You should always ask yourself what you can contribute—and what advice someone else can offer you.

(3) Using instructors' comments

Instructors' comments can also help you revise. In general, these comments fall into three categories.

Correction Symbols Frequently an instructor will indicate an area of concern about style, grammar, mechanics, or punctuation with one of the correction symbols listed on the inside back cover of this book. Instead of correcting a problem, the instructor will simply identify it and supply the number of the section in the handbook that deals with the error. After reading the appropriate section in the handbook, you should be able to make the necessary changes. The symbol and number beside the following sentence referred a student to 18f2, the section in the handbook that discusses sexist usage.

Equal access to jobs is a desirable goal for all mankind. *Sexist Usage see 18f*

After reading section 18f2, the student made this change.

Equal access to jobs is a desirable goal for everyone.

(For examples of an instructor's use of correction symbols, **see 3d2**).

Marginal Comments Instructors frequently make marginal comments when they read a paper. These comments may suggest changes in content or structure, such as additional supporting information or a different arrangement of paragraphs within the essay, or they may recommend stylistic changes, such as more varied sentences. Instructors may also use marginal comments to question the logic of a line of thought, suggest a more explicit thesis statement, ask for clearer transitions, or propose a new direction for a discussion. In some cases, you can consider these comments to be suggestions rather than corrections. You may decide to incorporate these ideas into a revised draft of your essay, and then again, you may not. In all instances, however, you should consider your instructor's comments seriously. After all,

Commenting on Student Papers
If you use the same hardware and software as your students, you can have students hand in their floppy disks along with (or instead of) printed versions of their essays. You'll then have the option of loading their essay files and entering comments directly into the texts themselves. Format your comments in bold.

If you load students' files and enter comments directly into their texts, and if your software program has a glossary file, you can create a series of text entries representing comments that you make dozens of times in a set of papers and hundreds of times during a term. Such boilerplate entries can be made up of personal comments and/or references to sections of *The Holt Handbook*.

Here is an example of a student's text in which a teacher has injected a boilerplate comment:

Though hunting is a popular sport that often requires a gun, there are alternatives. For instance, hunting with a bow and arrow. <<**Check to see if the previous sentence has both a subject and a finite verb. Read about sentence fragments (Chapter 13) in <u>The Holt Handbook</u>.**>> This method is fairer to the animal being hunted. An exciting element of challenge would also be presented in becoming skillful. Surely this alternative to gun hunting is not only fair, but challenging for the hunter as well.

FURTHER READING

Lil Brannon and C. H. Knoblauch. "On Students' Rights to Their Own Texts: A Model of Teacher Response." *College Composition and Communication* 33.2 (May 1982): 157–66.

The authors begin by saying that readers grant "authority" to writers by virtue of an author's name, a journal's reputation, or the reasonableness of an author's prose. In other words, readers begin in favor of the writer until he or she gives the reader a reason to think otherwise. In reading student papers, however, teachers seldom grant authority to the author who is, after all, *a student writer*. The teachers assume authority over what is to be written. Often, the teacher will have some "Ideal-Platonic Discourse" in mind by which to measure student writing. What is taught, therefore, is that the teacher's agenda is more important than the student's. What is equally taught, or untaught, is that what the student has to say—and any possible authority the student might muster to support it—is of little value. Such teaching is "paternalistic."

The authors' task was to determine whether "adherence to an Ideal Text interferes with the ability to read student writing in ways that help writers to achieve their goals." The students were asked to write a paper on the Lindbergh kidnapping trial. Forty teachers assessed the writing *in light of what the writer was attempting.* The teachers responded in one of two ways, neither of which considered student-writers' intentions. The two groups broke down roughly into "conservative" and "liberal."

continued on following page

they are the product of years of experience—and they represent a level of objectivity that a beginning writer is seldom able to bring to his or her own draft. (For an example of an instructor's use of marginal comments, **see 3d2**.)

Conferences Your instructor may require a conference. If not, try to arrange one, especially if you have questions about the written comments on your draft. During a conference, you can respond to the instructor's questions and ask for clarification of marginal comments. Sometimes your instructor will work along with you to revise a thesis so it says what you want it to say and to help you decide how to support it effectively. If you see a certain concept or section of your paper as a problem, use your conference time to focus on it, perhaps asking for help in sharpening a sentence or choosing a more accurate word. In short, your conference time should be tailored to your individual needs and your paper's particular problems.

GETTING THE MOST OUT OF AN INSTRUCTOR CONFERENCE

- **Make an appointment.** Make an appointment with your instructor before coming to his or her office, and arrive on time. Do not barge in and expect him or her to drop everything to help you. (If you find you are unable to keep your appointment, be sure to call your instructor to reschedule it.)

- **Read your paper carefully.** Before coming to the conference, review your notes and drafts, and go over all the instructor's comments and suggestions. Look up any correction symbols and read the sections of the handbook to which each refers you. Make all the changes you can on your draft. If you are coming to a conference to discuss ideas for a paper in progress, be sure your instructor is aware of this fact.

- **Have a list of questions.** Do not expect the instructor to anticipate your questions. Preparing a list in advance will enable you to get the most out of the conference in the allotted time.

- **Bring your draft.** Come to the conference with a draft of your paper. Without it, both you and the instructor can talk only in generalities about your writing. If you have several drafts, you may want to bring them all, but be

continued on the following page

---

continued from the previous page

sure you bring the draft that has the instructor's comments on it.

- **Take notes.** As you discuss your paper, write down any suggestions that you think will be helpful. Chances are that if you do not write down important suggestions at the time you hear them, you will forget them when you revise.
- **Participate actively.** Be prepared to discuss your draft. Your instructor is there to help you clarify your thoughts and to answer your questions, but he or she does not expect to deliver a monologue. A successful conference is not one-sided; it should be an open exchange of ideas between you and your instructor.

(4) Using checklists

A **revision checklist**—one your instructor prepares or one you develop yourself—enables you to examine your writing systematically. A checklist helps you focus on revising one element at a time. Depending on the problems you have and the time you have to deal with them, you can survey your paper using all the questions on the checklist or only some of them.

The four checklists that follow are keyed to sections of this text. They parallel the normal revision process, moving in stages from

REVISION CHECKLIST: THE WHOLE ESSAY

- Is your tone consistent with your purpose? **(See 1a3.)**
- Have you maintained an appropriate distance from your readers? **(See 1a3.)**
- Are thesis and support logically related, with each body paragraph supporting your thesis? **(See 2b1.)**
- Is your thesis clearly and specifically worded? **(See 2b2.)**
- Have you discussed everything promised in your thesis? **(See 2b2.)**
- Have you included any irrelevant points? **(See 2b2.)**
- Did you present your ideas in a logical sequence? Can you think of a different arrangement that might be more appropriate for your purpose? **(See 2b3.)**
- Do clear transitions between paragraphs allow your readers to follow your essay's structure? **(See 4d6.)**

continued from previous page

Brannon and Knoblauch recommend, as the result of their research, that teachers of writing "relinquish control" by allowing students to write multiple drafts, by emphasizing revision, by expecting student writers to determine their intentions, and by supporting those intentions. This "process of negotiation" allows both writer and peers or writer and teacher/tutor to collaborate in making the relationship between intention and effect successful.

Duplicate two or three student essays (from another class or year). Then use the revision checklists to evaluate and discuss the essays.

"The mere act of reading aloud puts his work before him in a new light and, by constraining his attention to every line, made him judge it more rigorously. I always intend to read, and generally do read, what I write aloud to some one; any one almost will do, but he should not be so clever that I am afraid of him. I feel weak places at once when I read aloud where I thought, as long as I read to myself, that the passage was alright."

–Samuel Butler

Teachers and Revision
Resist the temptation to be your students' editor. Marking each problem

continued on following page

continued from previous page

on students' drafts and asking them to revise following your directions not only overburdens you but also makes your students dependent. Training your students to be their own editors will enable them to become independent and to perform more competently in classes outside the English discipline.

As a less intrusive alternative, you might use the revision checklists on pages 50–51 in *The Holt Handbook* to create a progress chart. Give each student a copy of the chart with checks beside problem areas when the class turns in the first assignment. Then ask for the chart to be returned with each essay assignment. The pattern of checks from one assignment to the next will help you identify areas where additional work may be needed.

FURTHER READING

Nancy Sommers. "Revision Strategies of Student Writers and Experienced Adult Writers." *College Composition and Communication* **31** (December 1980): 378–88. The author claims the predominant model of the composing process is the *linear stage model*, as espoused, e.g., by Gordon Rohman or James Britton. The linear stage model is based on a model that does not allow for revision. In place of this model, the author proposes the *recursive model* and says there is only rewriting, or revision.

The author conducted a study questioning the value of the linear model. The case-study approach was employed, with writers reporting ret-

continued on following page

 REVISION CHECKLIST: PARAGRAPHS

- Does each body paragraph have one unifying idea? **(See 4c.)**
- Are topic sentences clearly worded and logically related to the thesis? **(See 4c1.)**
- Are your body paragraphs developed fully enough to support your points? **(See 4e.)**
- Does your introductory paragraph arouse reader interest and prepare readers for what is to come? **(See 4g2.)**
- Do your paragraphs have a clear organizing principle? **(See 4d1.)**
- Are the relationships of sentences within paragraphs clear? **(See 4d2–5.)**
- Are your paragraphs arranged according to familiar patterns of development? **(See 4e.)**
- Does your concluding paragraph sum up your main points? **(See 4g3.)**
- Have you provided transitional paragraphs where necessary? **(See 4g1.)**

REVISION CHECKLIST: SENTENCE STRUCTURE

- Have you strengthened sentences with repetition, balance, and parallelism? **(See 10c–d, 16a.)**
- Have you avoided overloading sentences with too many clauses? **(See 11c.)**
- Have you used correct sentence structure? **(See Chs. 13 and 14.)**
- Have you placed modifiers clearly and logically? **(See Ch. 15.)**
- Have you avoided potentially confusing shifts in tense, voice, mood, person, or number? **(See 17a–d.)**
- Are your sentences constructed logically? **(See 17f–h.)**
- Have you used emphatic word order? **(See 10a.)**
- Have you used sentence structure to signal the relative importance of clauses in a sentence and their logical relationship to one another? **(See 10b.)**
- Have you eliminated nonessential words and needless repetition? **(See 11a–b.)**
- Have you varied your sentence structure? **(See Ch. 12.)**
- Have you combined sentences where ideas are closely related? **(See 12b.)**

continued from previous page

✔ REVISION CHECKLIST: WORD CHOICE

- Is your level of diction appropriate for your audience and your purpose? **(See 18a–b.)**
- Have you selected words that accurately reflect your intentions? **(See 18c1.)**
- Have you chosen words that are specific, concrete, and unambiguous? **(See 18c3–4.)**
- Have you enriched your writing with figurative language? **(See 18e.)** '
- Have you eliminated jargon, neologisms, pretentious diction, clichés, ineffective figures of speech, and offensive language from your writing? **(See 18d,f,g.)**

the most global to the most specific concerns. These checklists may be more helpful when you have become familiar with concepts discussed later in the book. For now, you can begin with the questions on the first revision checklist, which focuses on the essay as a whole. As your understanding of the writing process increases and you are better able to identify the strengths and weaknesses of your writing, you can narrow the focus of your revision. Perhaps you will even add points to the checklists. You can also use your instructors' comments to tailor these checklists to your own needs.

 3d Revising Your Drafts

(1) Revising the rough draft

After you finish your rough draft, set it aside for a day if possible. When you look it over, you will probably notice problems that need attention. You cannot solve every problem at once, however. It makes more sense to focus on only a few areas at a time and to rework your essay in several drafts. As you review your first draft, focus on your essay's *content, organization,* and *thesis.* Once you have given attention to these areas, you can attend to other problems more easily.

As you reread your draft, you may want to consider most carefully the items in the first revision checklist on page 51. As you review your first draft, you may also have the benefit of a

rospectively about their approaches to writing. The study included twenty freshmen and twenty experienced writers (journalists, editors, and academics). Each writer wrote three essays (expressive, explanatory, and persuasive), produced three drafts, was interviewed three times after writing, and suggested revisions for an anonymous paper.

The author counted and categorized the revision strategies made by the writers for their own papers and the anonymous papers. Four strategies were identified: deletion, substitution, addition, and reordering. The revisions were at the level of word, phrase, sentence, and theme. Next she identified the revision strategies of the student and experienced writers. The *student writers* approached revision by searching for a better word or phrase, the "thesaurus philosophy of writing." Some of the students knew there were larger problems, but did not know or have strategies available to reconceive them. The author says these inadequate strategies are "teacher-based" and are "directed towards a teacher-reader who expects compliance with rules."

Experienced writers reconceive their writing at every level. When revising, they search for "the form or shape of their argument" they have "a concern for their readership" and are constantly in a "process of discovering meaning altogether." What is most characteristic of experienced writers is their view of writing as recursive. The author concludes that writing, or rewriting, must be the approach employed by teachers and students.

Revision Strategy
To teach formatting skills while teaching revision, ask students to put material they might want to delete in **bold** characters, material they might want to rearrange in UPPERCASE characters, and material they might want to add in underlined characters. After discussing possible changes as a class or in small groups, students can carry out the ones they decide to make.

Donald M. Murray. "Teaching the Other Self: The Writer's First Reader." *College Composition and Communication* 33.2 (May 1982): 140–47.
The author posits a model of writing that fosters a view of teaching writing that allows a student writer to write first for him or her "self" instead of for a wider audience. This model is not limited to expressive writing, nor should it produce what Linda Flowers would negatively call "writer-based" prose. Instead, student writers are encouraged to think of themselves as writing for a first reader, who is "the other self." In other words, they are taught to understand that there are, at least, two writers in each writer: one who initially writes, and one who observes, interprets, and critiques what has been written. Murray concludes that it is the writing teacher's responsibility to recruit and teach this other self so that student writers can learn to be their own teacher-critics.

continued on following page

collaborative revision session or a conference with your instructor. If you do, consider your readers' comments carefully, focusing for now on their suggestions about content, organization, and thesis and support.

NGUYEN DAO'S WRITING PROCESS: REVISING THE ROUGH DRAFT

When I met with my group to discuss my draft, people said I had a good topic. They liked the idea that I was going to talk about stereotypes instead of more common problems like grades, money, or family. Also, they don't know much about Asian Americans (except the stereotypes I mention in my paper), so they thought my paper could be pretty interesting. But they said what I had so far wasn't specific enough, and it wasn't really focused on my problem. I mostly talk about society in general or Asians in general. They said I should put in examples from my own experiences (which I was planning on doing anyway, but I forgot). Then people in the group talked about their own problems with being stereotyped, and I wrote those down. It's not just Asians--someone in my group said people always assume she's on a scholarship because she's black, but her parents own a successful business. A football player said his professors always assume he's dumb (which I admit I assumed too). I'm sure if I ask around, I'll find other examples, and I think I can use them in my next draft.

ROUGH DRAFT WITH HANDWRITTEN REVISIONS

The unique
(Characteristic of American culture is its genuine desire to understand & embrace the wide range of traditions and values of its people.
The United States prides itself on being the "melting
-- a nation where diverse cultures intermingle to form a unique and enlightened society.
pot" of the world. However, in reality, the abundance of different cultures in America often causes misunderstandings and even conflicts within the society. These misunderstandings and conflicts result from the society's lack of knowledge about other cultures.
Still, as
edge about other cultures. As an Asian American, I am

continued on the following page

continued from the previous page

frequently a victim of ethnic stereotyping, ~~and this has~~
been a serious problem for me. People keep trying to make
me something I'm not, and this is a

It has been within the last twenty years or so that the

United States has seen a large rise in the number of Asian

immigrants. First-generation immigrants are seen as an un-

derclass of poor people who struggle in low paying jobs so that their

children will have a better future. Many accuse immigrants

of accepting ~~less~~ lower pay for their work than other American citizens are ~~the established ma-~~

~~jority is~~ willing to accept, thus putting the established major-

ity out of work. ¶ While it is true that most newly arrived

Asians do seek low paying, low-skill jobs, they are just fol-

lowing the same ~~trend~~ road that other immigrant groups followed

when they first arrived in the ~~U.S.~~ United States. Because the first genera-
The Irish who escaped the potato famines came to the U.S.
without many advanced skills. To this day, many Latinos come to
tion of Asian Americans, have poorly developed skills in Eng-
my parents included the U.S. in search of a
better life with little
lish, they are forced into jobs which do not require those more
than
the shirts on their
skills. Many Asians received degrees from institutions in backs.

their native countries or received advanced training of some

kind but cannot use those skills in the United States. Therefore, they
have no
choice but to
accept whatever low-paying job they can get--
not to steal jobs from others, but to survive.
Asian-American children are ~~seen as~~ hard workers who
Along with the view of the first-generation Asian Americans
as low skilled workers comes the notion that
are pushed by their families to succeed. Asian children are

seen as intelligent, but only in terms of scientific and technical knowl-

edge. The media likes to point out ~~to the facts~~ that most Asians

continued on the following page

continued from previous page

This method of teaching the other self is done, Murray claims, in conference. Murray says that he—as a teacher—will teach less and allow the students to learn more. The student is always expected to teach first in the conference. During the conference, the student responds to his or her writing and then Murray responds to the student's response. In this way, Murray "teaches the other self." The writing teacher's responsibility in the conference, therefore, is to address and awaken a self that the student writer did not know existed. Most importantly, the teacher's response should focus on the student's successes rather than on correcting errors. To be sure, the student is aware of another self that is overly negative and speaks only in counterproductive ways. It is, therefore, the positive, supportive other self that the writing teacher must appeal to and the student must come to rely on.

continued from the previous page

succeed only in ~~the~~ math, science, and engineering ~~fields~~

and that ~~there is an inordinate~~ a great number of Asian college

students ~~who~~ identify themselves as "pre-med. what the media seems to forget is that other immigrant groups also seem to prize success above all else.

In my personal experience, in college, many of my

friends assume that I am either a science or an engineering

major and that my parents force me to study five to ten

hours a day. They believe that I sacrifice all my free time

and social life in pursuit of a high grade point average. my friends are quite surprised when I tell them that I am taking a drawing class and that I am majoring in political science--not as a

~~In fact, I am a political science major. I also like to go~~ stepping stone into law school, but as a study of man & society.

~~to basketball games, listen to music, and read automotive~~

~~magazines, just like other college students I know. My par-~~

~~ents do encourage me to do well in school because they see~~

~~that education is a stepping stone to social class mobility;~~

Add new ¶

however, I am lucky because my parents do not push me in

one direction or another as many of my Asian friends' par-

ents do. Many of my friends do not realize that just two or

three generations ago, their parents and grandparents were

going through the same process of social adjustment that all

immigrants endure. these cultural stereotypes. For example,

It is important to remember that ~~not all Asians~~ most people don't fit into fit not all Asians

the over-achieving, success-oriented stereotype. When any child comes from an economically disadvantaged background, many times they must sacrifice their academic pursuits in order to support their families. Also, as Asians, particularly

continued on the following page

continued from the previous page

the children, become more integrated into the society, the traditional Asian values of hard work & familial obligations will certainly clash with the American pursuits of recreation & individualism. The resolution of that conflict will add yet another facet to the complexity of America's society.

New ¶ (add after "personal experience" ¶)

This practice of assuming people of similar ethnic backgrounds share certain traits is certainly not limited to Asian Americans. African-American students complain people expect them to be athletes, to like rap music, to be on scholarship, to be from single-parent families -- even to be gang members. Athletes say people expect them to be dumb jock, to drink a lot, and to mistreat their girlfriends. Latinos say people assume their parents are immigrants and that they speak Spanish better than English. Business majors say people think they're politically conservative and not creative. Engineering students are expected to be dull & wear pocket protectors. Women are supposed to be weak in math & science. Overweight people are expected to be class clowns. In fact, my friends (of all ethnic groups) buy their clothes where I do, & we listen to the same music & laugh at the same jokes. But outsiders don't know this. They have different expectations for each of us, & these expectations are based on culture, not ability.

Your students' answers will vary.

EXERCISE 2

Revise your rough draft, using one or more of the strategies for revision discussed in 3c. At this point, it is probably most sensible to focus on the general concept of thesis and support and on paragraph development, unity, and coherence. Try not to worry at this stage about narrower stylistic issues, such as sentence variety and word choice.

(2) Writing a second draft

When you have read over your rough draft several times, marking it up with notes and outlining plans for revision, you are ready to write a second draft.

NGUYEN DAO'S WRITING PROCESS: WRITING A SECOND DRAFT

SECOND DRAFT WITH INSTRUCTOR'S COMMENTS

The Danger of Stereotypes

1   The United States prides itself on being the "melting pot"

of the world--a nation where diverse cultures intermingle to

form a unique and enlightened society. However, in reality,

the abundance of different cultures in America often causes

misunderstandings and even conflicts within the society.

These misunderstandings result from the society's lack of

knowledge about other cultures. The unique characteristic of

American culture is its genuine desire to understand and
Can you sharpen the thesis so it takes a stand?
embrace the wide range of traditions and values of its peo-
Why is stereotyping a problem?
ple. Still, as an Asian American, I am a victim of ethnic

continued on the following page

continued from the previous page

stereotyping. People keep trying to make me something I'm

not, and this is a serious problem for me.

Wordy — see 11a

2   It has been within the last twenty years or so that the

United States has seen a large rise in the number of Asian

immigrants. First-generation immigrants are seen as an

underclass of poor people who struggle in low-paying jobs

so that their children will have a better future. Many accuse

immigrants of accepting lower pay for their work than

other American citizens are willing to accept, thus putting

Good background. But you might condense

the established majority out of work. *¶s 2 & 3 a bit.*

You're wandering from your topic.

3   While it is true that most newly arrived Asians do seek

low-paid, low-skill jobs, they are just following the same

road that other immigrant groups followed when they first

arrived in the United States. The Irish who escaped the

potato famines came to the United States without many ad-

vanced skills. To this day, many Latinos come to the United

Cliché (See 18d4). Also — you

States in search of a better life with little more than the

may be guilty of stereotyping here.

shirts on their backs. Because the first generation of Asian

Americans, my parents included, have poorly developed

skills in English, they are forced into jobs which do not re-

quire those skills. Many Asians received degrees from insti-

tutions in their native countries or received advanced

continued on the following page

continued from the previous page

training of some kind but cannot use those skills in the

pronoun ref— see 22d 1

United States. Therefore, they have no choice but to accept

whatever low-paying job they can get--not to steal jobs from

others, but to survive.

4　Along with the view of the first-generation Asian

Americans as low-skilled workers comes the notion that

Asian-American children are hard workers who are pushed

by their families to succeed. Asian children are seen as

intelligent, but only in terms of scientific and technical

knowledge. The media likes to point to the facts that most

Asians succeed only in math, science, and engineering and

that a great number of Asian college students identify them-

agreement— see 24 a 6

selves as "pre-med." What the media seems to forget is that

(media = plural; medium = singular)

other immigrant groups also seem to prize success above

all else.

Wordy— see 11a

5　In my personal experience, in college, many of my

friends assume that I am either a science or an engineering

major and that my parents force me to study five to ten

hours a day. They believe that I sacrifice all my free time

and social life in pursuit of a high grade point average.

My friends are quite surprised when I tell them that I am

taking a drawing class and that I am majoring in political

continued on the following page

continued from the previous page

science--not as a stepping stone into law school, but as a

study of (man) and society. *sexist language*
See 18F2

6   This practice of assuming people of similar ethnic back-

grounds share certain traits is certainly not limited to Asian

Americans. African-American students complain people ex-

are you sure you need all this? Your focus in
pect them to be athletes, to like rap music, to be on scholar-
this paper is on ethnic (specifically Asian)
ship, to be from single-parent families--even to be gang
stereotypes, remember?
members. Athletes say people expect them to be dumb

jocks, to drink a lot, and to mistreat their girlfriends. Lati-

nos say people assume their parents are immigrants and

that they speak Spanish better than English. Business ma-

jors say people think they're politically conservative and

not creative. Engineering students are expected to be dull

and wear pocket protectors. Women are supposed to be

weak in math and science. Overweight people are expected

to be class clowns. In fact, my friends (of all ethnic groups)

buy their clothes where I do, and we listen to the same

music and laugh at the same jokes. But outsiders don't

know this. They have different expectations for each of us,

and these expectations are based on culture, not ability.
good point – but wordy (see 11a)
7   It is important to remember that most people do not fit

these cultural stereotypes. For example, not all Asians fit

continued on the following page

continued from the previous page

into the overachieving, success-oriented stereotype. When

any child comes from an economically disadvantaged back-

agreement

ground, many times they must sacrifice their academic pur-

Are you sure this is   *see 24b*

suits in order to support their families. Also, as Asians,

what you want to leave

particularly the children, become more integrated into the

your readers with? It

society, the traditional Asian values of hard work and famil-

doesn't really address your

ial obligations will certainly clash with the American pur-

essay's main point.

suits of recreation and individualism. The resolution of that

conflict will add yet another facet to the complexity of

America's society.

I like what you've done here, but something important is still missing. You really do need more examples from your own experience. Also, think about this question before our conference on Tuesday: Exactly why are the stereotypes you enumerate so harmful, so damaging? This idea needs to be developed in some detail (it's really the heart of your paper), and it should certainly be addressed in your thesis and conclusion as well.

　　When I reread my second draft before I handed it in, I liked it better than the rough draft--especially the material in paragraphs 5 and 6, which my writing group wanted me to add. Now I thought I might like to add something about how I have been limited by my

continued on the following page

continued from the previous page

Your students' answers will vary.

own preconceptions. For example, I did start out as a pre-med student, although I am now considering political science as a major. Another idea I might want to look into is that many Asians, including some of my friends, are now reconsidering whether the drive for success is worth the price. I saw this as an interesting possibility, but I felt as if I was trying to do too much. Was I writing about a problem in U.S. society or my own personal problem (the conflict between what people expect a good Asian boy to do vs. what I want to do)? Was I going to deal with all *stereotypes? All ethnic stereotypes? Or only Asian-American stereotypes?*

My conference with Professor Cross, and her comments on my draft, helped me solve some of my problems and get my assignment into focus. It also brought up some new problems. First, she reminded me that my paper was only supposed to be two to three pages long, something I was starting to forget. So I should stop looking for new material and start sorting through what I had. She said my knowledge of the Asian-American experience was my paper's biggest strength, but that I was going in too many directions at once. She thought all the stuff in paragraph 6 was interesting, but she said it didn't really fit if my paper was going to be on Asians and on the problems I *face. I hate to take this paragraph out, but I guess she's right. She wants me to revise my thesis so it takes a stand about* why *and* how *stereotyping has been damaging to me (and, for the same reasons, to other people). She says this kind of thesis will make everything easier, and then all I'll have to do is add support based on what I know firsthand.*

This all makes sense, but it means I'll have to take out some of the background on immigrants in paragraphs 2 and 3 and cut most of paragraph 6. I won't have anything left! Now I'll have to list and explain the problems I have, telling why they're problems, *and I'll also have to redo my introduction and conclusion so they fit with the new ideas. It sounds like a lot of work.*

EXERCISE 3

Review the second draft of your paper, this time paying attention not only to the way your thesis is worded and supported and to paragraphing, topic sentences, and transitions, but also to the way you structure your sentences and select your words. If possible, ask a friend to read your draft and respond to the collaborative revision questions in 3c2. Then revise your draft, incorporating any suggestions you find helpful.

TEACHING STRATEGY

Editing and Proofreading

Use a "problem paper" (from another class or year) to demonstrate the use of *The Holt Handbook's* grammar, punctuation, and mechanics sections in the editing process. Go through the paper (on the overhead or using individual student copies) and identify problems. Then move to the revision checklist (pp. 64–65) for a page number or chapter reference. After reading the suggested material, return to the paper and let students suggest revisions. (If you are using individual progress charts, you may want to make general class assignments in Parts 3 and 4 of *The Holt Handbook* whenever the charts indicate shared problems or a common lack of progress.)

(3) Preparing a final draft: Editing and proofreading

After you have revised your drafts to your satisfaction, two final steps remain: editing and proofreading your paper.

Editing   When you edit, you concentrate on grammar and spelling, punctuation and mechanics. You will have done some of this work as you revised previous drafts of your paper, but now your *focus* is on editing. Approach your work as a critical reader would, reading each sentence carefully. As you proceed, you may find it helpful to review the items on the editing checklist that follows. Keep your preliminary notes and drafts and reference books (such as this handbook and a current dictionary) nearby as you edit. If you are typing your paper on a computer, now is the time to run a spell check—and (with your instructor's permission) perhaps a grammar and usage check as well.

See ◄
App.
B

> ✔ **REVISION CHECKLIST: EDITING FOR GRAMMAR, PUNCTUATION, MECHANICS, AND SPELLING**
>
> **GRAMMAR**
>
> - Have you used the appropriate case for pronouns? **(See 22a–b.)**
> - Are pronoun references clear and unambiguous? **(See 22c–d.)**
> - Are verb forms correct? **(See 23a.)**
> - Are tense, mood, and voice of verbs logical and appropriate? **(See 23b–k.)**
> - Do subjects and verbs agree? **(See 24a.)**
> - Do pronouns and antecedents agree? **(See 24b.)**
> - Are adjectives and adverbs used correctly? **(See 25a–d.)**
>
> **PUNCTUATION**
>
> - Is end punctuation used correctly? **(See 27a–c.)**
> - Are commas used correctly? **(See Ch. 28.)**
> - Are semicolons used correctly? **(See Ch. 29.)**
> - Are apostrophes used correctly? **(See Ch. 30.)**
> - Are quotation marks used where they are required? **(See 31a–e.)**
> - Are quotation marks used correctly with other punctuation marks? **(See 31f.)**
> - Are other punctuation marks—colons, dashes, parentheses, brackets, slashes, and ellipses—used correctly? **(See Ch. 32.)**

continued on the following page

continued from the previous page

MECHANICS

- Is capitalization consistent with standard English usage? **(See Ch. 33.)**
- Are italics used correctly? **(See Ch. 34.)**
- Are hyphens used where required and placed correctly within and between words? **(See Ch. 35.)**
- Are abbreviations used where convention calls for their use? **(See Ch. 36.)**
- Are numerals and spelled-out numbers used appropriately? **(See Ch. 37.)**

SPELLING

- Are all words spelled correctly? (Run a spell check or check a dictionary if necessary.)

Proofreading After you have completed your editing, print a final draft. Even when it is complete, however, you are still not finished. Now you must proofread, rereading every word carefully to make sure you did not make any errors as you typed. You must also make sure the final typed copy of your paper conforms to your instructor's format requirements.

▶ See App. A

CLOSE-UP

CHOOSING A TITLE

When you decide on a title for the final draft of your essay, keep these criteria in mind.

- In all cases, a title should be descriptive, giving an accurate sense of your essay's focus. Whenever possible, it should include one or more of the key words and phrases that are central to your paper.
- A title's wording can echo the wording of your assignment, reminding you (and your instructor) that you have not lost sight of your goal.
- Ideally, a title should arouse interest, perhaps by using a provocative question or an apt quotation (or, if appropriate, by introducing a note of controversy).

continued on the following page

Lab Activity

Isolating One Sentence at a Time
Students may find it easier to spot surface errors if they can see only a small portion of text. Ask students to split their screens so that they create a window small enough to accommodate only one or two lines of text. Have them proofread their own drafts or someone else's.

Suggest, also, that they read the essay "backwards"—last sentence first and first sentence last.

Using the Search Command
Students can check for some of the more common errors in grammar, punctuation, and usage by using the search command. Ask students to search their own or someone else's essay for commonly confused words *(lay/lie, effect/affect)*, homophones *(their/there, its/it's, too/to)*, problem punctuation (semicolons, colons), or sexist language *(he, his, him, man)*.

(Oral or Written) In groups, students read each other's work. Then they respond to **content only** in a nonevaluative way.

continued from the previous page

Assignment: Write about a problem faced on college campuses today.
Topic: Free speech on campus
Possible titles:
Free Speech: A Problem for Today's Colleges (descriptive; echoes wording of assignment and includes key words of essay)
How Free Should Free Speech on Campus Be? (provocative question)
The Right to "Shout 'Fire' in a Crowded Theater" (quotation)
Hate Speech: A Dangerous Abuse of Free Speech on Campus (controversial position)

*EXERCISE 4

Read Nguyen Dao's final draft, which appears on the following pages. What exactly has Nguyen Dao changed between his second and third drafts? List each of these changes and evaluate them as any careful reader would. Are all his changes for the better? Are any other changes called for?

EXERCISE 5

Using the revision checklists in 3d4 as a guide, create a customized checklist—one that reflects the specific concerns you need to consider to revise your essay. Then revise your essay according to this checklist.

EXERCISE 6

Edit your essay and then prepare a final draft, being sure to proofread it carefully.

NGUYEN DAO'S WRITING PROCESS: PREPARING A FINAL DRAFT

FINAL DRAFT

Nguyen Dao   Dao 1
Professor Cross
English 101
10 October 1994

My Problem:
Escaping the Stereotype of the "Model Minority"

The United States prides itself on being a nation where diverse cultures intermingle to form a unique and enlightened society. However, in reality, the existence of so many different cultures in America often causes misunderstandings within the society. These misunderstandings result from most people's lack of knowledge about other cultures. The unique characteristic of American culture is its genuine desire to understand and embrace the wide range of traditions and values of its people. Still, as an Asian American, I am frequently confronted with other people's ideas about who I am and how I should behave. Such ideas are not just limiting to me, but also dangerous to the nation because they challenge the image of the United States as a place where people can be whatever they want to be.

Within the last twenty years, the United States has experienced a sharp rise in the number of Asian immigrants-- and, consequently, of Asian stereotypes. First-generation immigrants are seen as an underclass of poor people who struggle in low-paying jobs, working long hours so that their children will have a better future. Along with the view of the first-generation Asian Americans as driven, low-skilled workers comes the notion that all Asian-American children are hard workers who are pushed by their families to succeed. Asian children are seen as intelligent, but only in terms of scientific and technical knowledge. The media like to point out that most Asians succeed only in math, science,

The major change Nguyen Dao makes between the second draft and the third is to narrow his focus. He may have read his second draft and said to himself, "I have been affected by ethnic stereotyping, and other people have too. So what?" At the end of paragraph 1 in the third draft, he introduces a thesis to answer his question about the consequences of ethnic stereotypes: "Such ideas are not just limiting to me, but also dangerous to the nation." This statement suggests that the paper will concentrate on how ethnic stereotypes have affected Nguyen Dao personally and how they may potentially affect the whole nation. Past experiences of various immigrant groups described in paragraph 3 and common expectations concerning these groups as described in paragraph 6 of the second draft are not exactly within the limits Nguyen Dao sets in his revised thesis statement. So he deletes paragraphs 3 and 6 from the second draft, but he retains a one-sentence reference to stereotypical expectations for various ethnic groups in paragraph 5 of the third draft.

The conclusion in the second draft is not very satisfactory because it focuses on information that is already common knowledge. The closing of the third draft consists of mostly new material, arranged to fit the revised thesis. Paragraph 4 considers the effects of stereotypes on Nguyen Dao; paragraph 5, the effects on various individuals including Nguyen Dao; and paragraph 6, the consequences for the nation. The revised conclusion is more satisfying because it describes the "complexity" referred to in the second draft's closing in more specific

continued on following page

continued from previous page

terms. Asian children, Nguyen Dao concludes, are free to choose what balance they will maintain between traditional Asian values and American individualism. "Freedom to choose" is the idea that makes the essay close rather than just stop.

Nguyen Dao also makes sentence- and word-level changes. He deletes the cliche of the melting pot from the first sentence of the second draft. He eliminates generalities that may be unfair or exaggerated. "Society's lack of knowledge about other cultures" becomes "most people's lack of knowledge"; "I am a victim of ethnic stereotyping" becomes "I am frequently confronted with ethnic stereotyping." He deletes unnecessary words. Nguyen Dao also revises to eliminate sexist language. "A study of man and society" become "a foundation for a liberal arts education." And he makes one adjustment in subject-verb agreement, changing "media likes" and "media seems" to "media like" and "media seem."

Dao 2

and engineering and that a disproportionately large number of Asian college students identify themselves as pre-med. What the media seem to forget is that many other immigrant groups also value success. In a larger sense, America has always been seen as the land of opportunity, where everyone is in search of the American dream.

Many of my college friends assume that I am either a science or an engineering major and that my parents force me to study many hours each day. They believe that I sacrifice all my free time and social life in pursuit of a high grade point average. My friends are quite surprised when I tell them that I am taking a drawing class and that I am interested in political science--not as a stepping stone into law school, but as a foundation for a liberal arts education. They are also surprised to find that I am not particularly quiet or shy, that I do not play a musical instrument, and that I do not live in Chinatown. (I don't know why this surprises people; I'm not even Chinese.)

I try to see these stereotypes as harmless, but I have learned that even neutral or positive stereotypes can have negative consequences. I have seen how teachers have unreasonable expectations for me, and how those expectations create pressure for academic success. Teachers expect me to do well, but only in certain areas. They have encouraged me to take AP math and science classes, try out for band, sign up for an advanced computer seminar, and join the chess club. No one has ever suggested that I (or any other Asian American I know) pursue athletics, creative writing, or debating. I spent my high school years trying to be what other people wanted me to be, and I got to be pretty good at it.

I realize now, however, that I have been limited, and that similar stereotypes also limit the options other groups have. The law says we can choose our activities and choose our careers, but it does not always work out that way.

Dao 3

Often, because of long-held stereotypes, we are gently steered (by peers, teachers, bosses, parents, and even by ourselves) in a certain direction, toward some options and away from others. We may have come a long way from the time when African Americans were expected to be domestics or blue-collar workers, Latinos to be migrant farmers or gardeners, and Asians to be restaurant workers. But at the college, high school, and even elementary school level, students are expected to follow certain predetermined paths, and too often these expectations are based on culture, not interests or abilities.

Most people do not fit these cultural stereotypes. For example, not all Asians fit into the overachieving, success-oriented mold. When children come from an economically disadvantaged background, as the children of some recent Asian immigrants do, they must work hard and study hard. But this situation is only temporary. As Asian children become more integrated into American society, they retain the traditional Asian values of hard work and family obligations, but they also acquire the American drive for individualism. We are Americans, and therefore we are individuals. Ethnic stereotypes limit individual choice. And freedom to choose our futures, to be whoever we want to be, is what the United States is supposed to be all about.

ANSWERS TO EXERCISES 5 AND 6

Your students' answers will vary. Encourage students to use the comments and corrections you've made on previous work as they prepare this list so that their revising checklists are tailored to their particular problems.

Paragraphs

A dip-in-at-need approach to this chapter might work best. If your students are writing essays, you might begin at the end of the chapter with introductory paragraphs, then study general paragraph structure and topic sentences at the beginning of the chapter, and then move back to concluding paragraphs. Methods of development can be introduced when appropriate as you assign essay topics. The material on unity and coherence can be discussed as part of class sessions on organization and on revision.

QUOTATION

"One cannot be arbitrary about paragraphing. It is a means of grouping thoughts, but much more it is a visual device. Much depends on the subject, the typography, the purpose of what is being written, the readers to whom it is addressed and the conditions under which they are likely to read it."

–Theodore M. Bernstein,
Miss Thistlebottom's Hobgoblins

CHAPTER 4

Writing Paragraphs

A **paragraph** is a group of related sentences, which may be complete in itself or part of a longer piece of writing. Paragraphs serve three important purposes:

1. They join sentences together so each group of sentences works as a unit to support an essay's main idea.
2. They provide visual breaks in the text that give readers a chance to pause and assimilate ideas.
3. They signal the movement of ideas in an essay. In most short essays, each paragraph presents a specific point that supports a more general thesis. In a longer essay, a group of related paragraphs, called a **paragraph cluster**, may support each point.

CLOSE-UP

WRITING PARAGRAPHS

Different kinds of paragraphs have different specialized functions in an essay.

Body paragraphs support an essay's thesis. **(See 2b1; 4a–f.)**

Transitional paragraphs connect body paragraphs within an essay. **(See 4g1.)**

Introductory paragraphs introduce the essay's subject and give an overview of its scope. **(See 4g2.)**

Concluding paragraphs reinforce the essay's key points. **(See 4g3.)**

4a Determining When to Paragraph

Paragraphs enable you to control the organization of your paper and convey your ideas clearly to your readers.

WHEN TO PARAGRAPH

- **To signal a shift in focus.** Whenever you move from one major point to another, you begin a new paragraph.
- **To signal a shift in time or place.** It is often a good idea to begin a new paragraph when you move your readers from one time period or location to another.
- **To clarify sequence.** In enumerating steps or tasks, you may begin a new paragraph every time you begin discussing a new step in a sequence or a new stage in a process.
- **To make ideas more emphatic.** You can underscore important ideas by isolating them in separate paragraphs.
- **To set off dialogue.** When recording dialogue, convention requires that you begin a new paragraph every time a new person speaks.
- **To set off introductions and conclusions.** Begin a new paragraph to signal the end of your introduction, and begin your conclusion with a new paragraph.

4b Charting Paragraph Structure

Charting the ideas in a paragraph helps you recognize its underlying structure. Begin by assigning the sentence that expresses the main or **unifying idea** of the paragraph to level 1. If no sentence in the paragraph expresses the unifying idea, compose a sentence that does. Then, read each sentence of the paragraph. Assign to level 1 sentences as important as the one containing the unifying idea. Indent and assign to level 2 more specific sentences, those which qualify or limit the unifying idea. Indent again and assign to level 3 any sentences that support level-2 sentences. Do this for every sentence in the paragraph, assigning increasingly higher numbers to more specific sentences.

Notice how charting reveals the structure of the following paragraph.

Taking a Break

Student writers generally want cut-and-dried rules on where to break for paragraphs—but such rules simply don't exist. Writers, teachers, and grammarians continue to wrangle and disagree about the rights and wrongs of paragraphing. You can tell students, however, that if they state and develop one idea in each paragraph, they are probably on the right track.

As a possible strategy, try reading aloud a section from a recent magazine article. As you read, ask students to raise their hands each time they sense a shift that would (on paper) mean a new paragraph.

Charting the Structure of a Paragraph

Help your students write developed paragraphs by using their word processors to represent visually the *levels of generality* among sentences in a paragraph taken from a published work—and, later, from their own or their classmates' work.

First, prepare an example to show students. "Explode" a paragraph (turn each sentence into a separate paragraph) by inserting a return between each sentence. Indent each of these sentence–paragraphs a distance from the left margin appropriate to its degree of generality. (For example, the topic sentence will be on the margin; first-level-support sentences will be indented five spaces; second-level-support sentences will be indented ten spaces, etc.)

continued on following page

continued from previous page

Here is an example of how "exploding" a student paragraph and using the Tab key to indicate levels of generality clearly demonstrate a lack of development:

I found my first year at Babson College to be difficult and challenging.

- First of all, most of my classes went into subjects deeper than my classes in high school did.

- In addition, I was constantly being pushed to do better by others who were a lot smarter than the kids I had been going to school with for the previous four years.

- Finally, I found it really difficult to budget my time and get all my work done while still having a social life.

STUDENTS WORKING TOGETHER

After discussing section 4c, divide the class into pairs. Distribute a copy of a recent magazine or journal article to each pair of students. Ask them to identify and underline the topic sentence of each paragraph in the piece and to place a star beside any paragraph that has an implied topic sentence. In addition, you might tell them to place a check at any trouble spots—for example, beside paragraphs where the topic sentence was hard to identify or where the two students disagreed on *which* sentence was the topic sentence.

Finally, ask the pairs to make marginal notes indicating where the topic sentence appears in each paragraph: B (beginning), M (middle), E (end), or I (implied). Do the results show a variety of placements? Discuss any problems experienced during the exercise.

1 My grandmother told me that fifty years ago life was not easy for a girl in rural Italy.
  2 At the age of six a girl was expected to help her mother with household chores.
    3 Girls of this age were no longer permitted to play games or to indulge in childish activities.
  2 At the age of twelve, a girl assumed most of the responsibilities of an adult.
    3 She worked in the fields, prepared meals, carried water, and took care of the younger children.
    3 Education was usually out of the question; it was an unusual family that allowed a girl to enroll in one of the few convent schools that took peasant children.

The first sentence (level 1) of this paragraph introduces the unifying idea, and each level-2 sentence gives an illustration of that idea. The level-3 sentences support these examples with specifics.

CLOSE-UP   **CHARTING PARAGRAPH STRUCTURE**

Note that a properly constructed paragraph has only one level-1 sentence. If charting reveals more than one level-1 sentence, you will need to revise your paragraph, perhaps dividing it into two paragraphs.

By charting your paragraphs, you can make certain that each one is unified, coherent, and well developed.

4c   Writing Unified Paragraphs

A paragraph is **unified** when it focuses on a single idea and develops it. You can create unified paragraphs by making sure that each paragraph has a **topic sentence** and that all the sentences in the paragraph support the unifying idea the topic sentence expresses.

(1)   Using topic sentences

Although many experienced writers do not use topic sentences in all their paragraphs, they do organize each of their paragraphs

See ◄
4c2

around a single unifying idea. In such cases, the topic sentence is **implied**. However, if you are a beginning writer, it makes good sense for you to use a topic sentence—at the beginning, in the middle, or at the end of a paragraph—to make your unifying idea clear both to you and to your readers.

Topic Sentence at the Beginning   Placing a topic sentence first and supporting sentences afterward in a paragraph is effective when you want your readers to understand your paragraph's unifying idea immediately. Beginning with the topic sentence also helps you to stay focused on your subject.

> I was a listening child, careful to hear the very different sounds of Spanish and English. Wide-eyed with hearing, I'd listen to sounds more than words. First, there were English (*gringo*) sounds. So many words were still unknown that when the butcher or the lady at the drugstore said something to me, exotic polysyllabic sounds would bloom in the midst of their sentences. Often the speech of people in public seemed to me very loud, booming with confidence. The man behind the counter would literally ask, "What can I do for you?" But by being so firm and so clear, the sound of his voice said that he was a *gringo*; he belonged in public society. (Richard Rodriguez, "Aria: A Memoir of a Bilingual Childhood")

Topic Sentence in the Middle   Placing a topic sentence in the middle of a paragraph enables you to build up to a point gradually or give background information before you state and support your unifying idea. This strategy is especially effective if you are refuting opposing points of view or presenting unfamiliar or unexpected information.

> African-American servicemen have played a role in the U.S. military since revolutionary times. In the years before World War II, however, they were employed chiefly as truck drivers, quartermasters, bakers, and cooks. Then, in July 1941, a program was set up at Alabama's Tuskegee Institute to train black fighter pilots. Eventually, nearly one thousand flyers—about half of whom fought overseas—were trained there; sixty-six of these men were killed in action. Ironically, even as African-American servicemen were fighting valiantly against fascism in Europe, they continued to experience discrimination in the U.S. military. Black officers encountered hostility and even violence at officers' clubs. Enlisted men and women were frequently the target of bigoted remarks. Throughout the war, in fact, African-American servicemen were placed in separate, all-black units. This segregation was official army policy until 1948, when President Harry S. Truman signed an executive order to desegregate the military. (Student Writer)

Lab Activity
Have students split their computer screens so that the top window is wide enough to show only their topic sentence. Suggest that they keep their topic sentence in view as they compose to ensure that everything written contributes to supporting it. (They can, of course, use the same technique with paragraphs and thesis statements.)

Placing the Topic Sentence
Moving topic sentences to different locations within paragraphs introduces students to some of the strategies writers employ and makes them self-conscious about their composing habits.

Have students compose a "typical" paragraph in which the topic sentence occurs at the beginning. Ask them to copy it twice so that they have a total of three.

Tell them to leave the topic sentence in the first paragraph at the beginning; tell them to move the topic sentence in the second paragraph to a point near the middle; and finally, tell them to move the topic sentence in the third paragraph to the end. Ask them to address the following questions:

- In each case, what are the advantages or disadvantages in having the topic sentence located where it is?

- When they place the topic sentence in a new location, do they have to make other changes in the rest of the paragraph? What changes? Why?

Lab Activity
Tell students to identify the thesis statement and all topic sentences in their essays by specially formatting them—perhaps by underlining thesis statements and putting topic sentences in bold. Next, ask them to move the thesis statement and all topic sentences (in the order in which they appear in the essay) into another file.

continued on following page

Topic Sentence at the End   Placing a topic sentence at the end of a paragraph enables you to present a controversial idea effectively. If you open a paragraph with an unusual, surprising, or hard-to-accept idea, you risk alienating your audience. However, if you lead your readers through a logical and carefully thought out chain of reasoning and *then* present your conclusion in the topic sentence, you are more likely to convince readers that your conclusion is reasonable.

> These sprays, dusts and aerosols are now applied almost universally to farms, gardens, forests, and homes—nonselective chemicals that have the power to kill every insect, the "good" and the "bad," to still the song of birds and the leaping of fish in the streams, to coat the leaves with a deadly film, and to linger on in soil—all this though the intended target may be only a few weeds or insects. Can anyone believe it is possible to lay down such a barrage of poisons on the surface without making it unfit for life? <u>They should not be called "insecticides," but "biocides."</u> (Rachel Carson, "The Obligation to Endure," *Silent Spring*)

Unifying Idea Implied   In some situations—especially in narrative or descriptive paragraphs—a topic sentence can seem forced or artificial. In such cases, your unifying idea should be implied instead of stated in a topic sentence. In the following paragraph, for example, the author wants readers to come gradually to the conclusion (as she herself did) that because she was female, she was considered inferior. Explicitly stating her unifying idea would defeat that purpose.

> I am eight years old and a tomboy. I have a cowboy hat, cowboy boots, checkered shirt and pants, all red. My playmates are my brothers, two and four years older than I. Their colors are black and green, the only difference in the way we are dressed. On Saturday nights we all go to the picture show, even my mother; Westerns are her favorite kind of movie. Back home, "on the ranch," we pretend we are Tom Mix, Hopalong Cassidy, Lash LaRue (we've even named one of our dogs Lash LaRue); we chase each other for hours rustling cattle, being outlaws, delivering damsels from distress. Then my parents decide to buy my brothers guns. These are not "real" guns. They shoot "BBs," copper pellets my brothers say will kill birds. Because I am a girl, I do not get a gun. Instantly I am relegated to the position of Indian. Now there appears a great distance between us. They shoot and shoot at everything with their new guns. I try to keep up with my bow and arrows. (Alice Walker, "Beauty: When the Other Dancer Is the Self," *In Search of Our Mothers' Gardens*)

(2)   Testing for unity

Each sentence in a paragraph should support its unifying idea, whether that idea is stated or implied. When you reread your paragraphs, look carefully for sentences that do not support the unifying idea and revise or delete them to bring your paragraph into focus. The following paragraph contains sentences that wander from its subject.

> One of the first problems students have is learning to use a computer. All students were required to buy a computer before school started. Throughout the first semester we took a special course to teach us to use a computer. The Macintosh Classic II has a large memory and can do word processing and spreadsheets. It has an eighty-character screen and a hard drive. My parents were happy that I had a computer, but they were concerned about the price. Tuition was high, and when they added in the price of the computer, it was almost out of reach. To offset expenses, I arranged for a part-time job in the school library. Now I am determined to overcome "computer anxiety" and to master my Macintosh by the end of the semester. (Student Writer)

The lack of unity in the preceding paragraph becomes obvious when you chart its sentences.

▶ See 4b

1 One of the first problems students have is learning to use a computer.
  2 All students were required to buy a computer before school started.
    3 Throughout the first semester we took a special course to teach us to use a computer.
1 The Macintosh Classic II has a large memory and can do word processing and spreadsheets.
  2 It has an eighty-character screen and a hard drive.
1 My parents were happy that I had a computer, but they were concerned about the price.
  2 Tuition was high, and when they added in the price of the computer, it was almost out of reach.
    3 To offset expenses, I arranged for a part-time job in the school library.
1 Now I am determined to overcome "computer anxiety" and to master my Macintosh by the end of the semester.

Each level-1 sentence represents a topic that should be developed in its own paragraph; in other words, this paragraph has not one but four topic sentences. Instead of writing one unified paragraph, the writer has made a series of false starts.

To unify this paragraph around the idea of the problems he faced in learning to use his computer, the writer took out the sentences about his parents' financial situation and the computer's

continued from previous page

Students will end up with topic-sentence outlines of their essays that will reveal strengths and weaknesses in unity, coherence, logic, and support—as in the following example:

The Advantages of Advertising

Although advertising has been the target of a lot of criticism, it provides many advantages that may not be apparent to consumers.
    Through the mass media, advertising has helped the consumer enormously.
    Advertising is responsible for an increase in competition among manufacturers.
    By increasing competition, advertising increases the quality and efficiency of production.
    Consumers get the advantage of reduced prices due to a cut in production costs for the manufacturer.
    Advertising helps the economy by increasing the overall demand for the product.
    Finally, advertising helps the economy by properly distributing information about products.

If students notice that the order of the supporting ideas seems to be weak or illogical, suggest that they use the delete, move, and insert commands to reorder the topic sentences; if material seems to be missing, tell them to suggest suitable topic sentences; if extraneous material is present, tell them to delete the gratuitous topic sentences.

characteristics, keeping only those details related to the unifying idea.

> One of the first problems I had as a college student was learning to use my computer. All first-year students were required to buy a Macintosh Classic II computer before school started. Throughout the first semester, we took a special course to teach us to use the computer. In theory this system sounded fine, but in my case it was a disaster. In the first place, the closest I had ever come to a computer was the hand-held calculator I used in math class. In the second place, I could not type. And to make matters worse, many of the people in my computer orientation course already knew how to operate a computer. By the end of the first week I was convinced that I would never be able to work with my Macintosh.

*EXERCISE 1

Each of the following paragraphs is unified by a central idea, but that idea is not explicitly stated. Identify the unifying idea of each paragraph, write a topic sentence that expresses it, and decide where in the paragraph to place it.

A. The narrator in Ellison's novel leaves an all-black college in the South to seek his fortune—and his identity—in the North. Throughout the story he experiences bigotry in all forms. Blacks as well as whites, friends as well as enemies, treat him according to their preconceived notions of what he should be, or how he can help to advance their causes. Clearly this is a book about racial prejudice. However, on another level, *Invisible Man* is more than the account of a young African-American's initiation into the harsh realities of life in the United States before the civil rights movement. The narrator calls himself invisible because others refuse to see him. He becomes so alienated from society—black and white—that he chooses to live in isolation. But, when he has learned to see himself clearly, he will emerge demanding that others see him too.

B. "Lite" can mean a product has fewer calories, or less fat, or less sodium, or it can simply mean the product has a "light" color, texture, or taste. It may mean none of these. Food can be advertised as 86 percent fat free when it is actually 50 percent fat, because the term "fat free" is based on weight and fat is extremely light. Another misleading term is "no cholesterol," which is found on some products that never had any cholesterol in the first place. Peanut butter, for example, contains no cholesterol—a fact that manufacturers have recently made an issue—but it is very high in fat and so would not be a very good food for most dieters. Sodium labeling presents still another problem. The terms, "sodium free," "very low sodium," "low sodium," "reduced sodium," and "no salt added" have very specific meanings, frequently not explained on the packages on which they appear.

A. Unifying idea: The message of Ralph Ellison's novel *Invisible Man*

Topic sentence: In *Invisible Man* Ralph Ellison is concerned at least as much with self-discovery as he is with race relations. (This sentence could be placed at either the beginning or the end of the paragraph.)

B. Unifying idea: Problems in food labeling

Topic sentence: Food labels have recently begun to give consumers a great deal of nutritional information, but too often this information is difficult to decipher. (Make this the first sentence in the paragraph.)

STUDENTS WORKING TOGETHER

In groups of four or five, students scramble a paragraph (e.g., an unread selection from their text). On the top portion of a transparency they write the scrambled paragraph as a list of numbered sentences; on the bottom they write the correct version of the paragraph. Place the top half of the transparency on screen. The group task is to determine the topic sentence and justify that choice.

STUDENTS WORKING TOGETHER

Follow the same procedure as in the previous exercise. The group task is to determine what links sentences together (organizing principle, transitions, parallelism, etc).

Para 4d

4d Writing Coherent Paragraphs

A paragraph is **coherent** if all its sentences are logically related to one another.

> **TECHNIQUES FOR ACHIEVING PARAGRAPH COHERENCE**
> • Arrange details according to an organizing principle.
> • Use transitional words and phrases.
> • Use pronouns.
> • Use parallel constructions.
> • Repeat key words and phrases.

(1) Arranging details

Even if a paragraph's sentences are all about the same subject, they lack coherence until they are arranged according to an organizing principle—*spatial, chronological,* or *logical.*

Spatial Order Paragraphs arranged in **spatial** order establish the perspective from which readers will view details. For example, an object or scene can be viewed from top to bottom or from near to far. Spatial order is central to **descriptive paragraphs**. Notice how the following descriptive paragraph begins on top of a hill, moves down to a valley, follows a river through the valley into the distance, and then moves to a point behind the speaker, where Mt. Adams stands.

▶ See 4e2

> East of us rose another hill like ours. Between the hills, far below, was the highway which threaded south into the valley. This was the Yakima valley; I had never seen it before. It is justly famous for its beauty, like every planted valley. It extended south into the horizon, a distant dream of a valley, a Shangri-la. All its hundreds of low, golden slopes bore orchards. Among the orchards were towns, and roads, and plowed and fallow fields. Through the valley wandered a thin, shining river; from the river extended fine, frozen irrigation ditches. Distance blurred and blued the sight, so that the whole valley looked like a thickness or sediment at the bottom of the sky. Directly behind us was more sky, and empty lowlands blued by distance, and Mount Adams. Mount Adams was an enormous, snow-covered volcanic cone rising flat, like so much scenery. (Annie Dillard, "Total Eclipse")

Chronological Order   Paragraphs arranged in **chronological** order present details in sequence, using transitional phrases that establish the sequence of events—*at first*, *yesterday*, *later*, and so on. This type of organization is central to **narrative paragraphs**. The following narrative paragraph is unified by the orderly sequence of events.

See ◀ 4e1

> They married in February, 1921, and began farming. Their first baby, a daughter, was born in January, 1922, when my mother was 26 years old. The second baby, a son, was born in March, 1923. They were renting farms; my father, besides working his own fields, also was a hired man for two other farmers. They had no capital initially, and had to gain it slowly, working from dawn until midnight everyday. My town-bred mother learned to set hens and raise chickens, feed pigs, milk cows, plant and harvest a garden, and can every fruit and vegetable she could scrounge. She carried water nearly a quarter of a mile from the well to fill her wash boilers in order to do her laundry on a scrub board. She learned to shuck grain, feed threshers, shuck and husk corn, feed corn pickers. In September, 1925, the third baby came, and in June, 1927, the fourth child—both daughters. In 1930, my parents had enough money to buy their own farm, and that March they moved all their livestock and belongings themselves, 55 miles over rutted, muddy roads. (Donna Smith-Yackel, "My Mother Never Worked")

Chronological order is also used to arrange details in **process** paragraphs, which explain how something works or how to carry out a procedure.

See ◀ 4e4

Logical Order   Paragraphs arranged in **logical** order present details or ideas in terms of their relative emphasis. For example, the ideas in a paragraph may move from *general to specific*, as in the conventional topic-sentence-at-the-beginning paragraph, where a relatively general topic sentence is supported by specific details. Conversely, a paragraph's ideas may progress from *specific to general*, as they do when the topic sentence appears at the end of the paragraph. A writer may also choose to move from *most familiar to least familiar* idea—or from *least familiar to most familiar*. Alternatively, a paragraph can begin with the *least important* idea and move to the *most important*. In some cases—particularly in technical or business writing—a writer might do just the opposite: begin with the *most important* idea and progress to the *least important*.

The following paragraph moves from a *general* statement about the need to address the problem of the injury rate in boxing to *specific* proposals for solutions.

Several reforms would help solve the problem of the high injury rate in boxing. First, all boxers should wear protective equipment—head gear and kidney protectors, for example. This equipment is required in amateur boxing and should be required in professional boxing. Second, the object of boxing should be to score points, not to knock out opponents. An increased glove weight would make knockouts almost impossible. And finally, all fights should be limited to ten rounds. Studies show that most serious injuries occur in boxing between the eleventh and fifteenth rounds—when the boxers are tired and vulnerable. By limiting the number of rounds a boxer could fight, officials could substantially reduce the number of serious injuries. (Student Writer)

(2)   Using transitional words and phrases

Transitional words and phrases aid coherence by indicating the relationships among sentences, establishing spatial, chronological, and logical connections among the ideas in a paragraph. The following paragraph shows how the omission of transitional words and phrases can make a passage difficult to understand.

Without transitional words and phrases

Napoleon certainly made a change for the worse by leaving his small kingdom of Elba. He went back to Paris, and he abdicated for a second time. He fled to Rochfort in hope of escaping to America. He gave himself up to the English captain of the ship *Bellerophon*. He suggested that the Prince Regent should grant him asylum, and he was refused. All he saw of England was the Devon coast and Plymouth Sound as he passed on to the remote island of St. Helena. He died on May 5, 1821, at the age of fifty-two.

Although the unifying idea of this paragraph is stated in the first sentence, the exact chronological relationships among events are not clear. With no transitional words or phrases, the paragraph is just a list of unconnected events. In the following paragraph, words and phrases such as *after*, *finally*, *once again*, and *in the end* provide the links that clarify the chronological order of the events enumerated in the passage.

Napoleon certainly made a change for the worse by leaving his small kingdom of Elba. After Waterloo, he went back to Paris, and he abdicated for a second time. A hundred days after his return from Elba, he fled to Rochfort in hope of escaping to America. Finally, he gave himself up to the English captain of the ship *Bellerophon*. Once again, he suggested that the Prince Regent grant him asylum, and once again, he was refused. In the end, all he saw of England was the

79

Devon coast and Plymouth Sound as he passed on to the remote island of St. Helena. <u>After six years of exile</u>, he died on May 5, 1821, at
the age of fifty-two. (Norman Mackenzie, *The Escape from Elba*)

USING TRANSITIONAL EXPRESSIONS
To Signal Sequence or Addition

again	furthermore
also	in addition
and	last
besides	next
finally	one . . . another
first . . . second . . . third	still
	too

To Signal Time

afterward	immediately
as soon as	in the meantime
at first	later
at length	meanwhile
at the same time	next
before	now
earlier	soon
eventually	subsequently
finally	then
	until

To Signal Comparison

also	likewise
by the same token	similarly
in comparison	

To Signal Contrast

although	nevertheless
but	nonetheless
despite	on the contrary
even though	on the one hand . . .
however	on the other hand
in contrast	still
instead	whereas
meanwhile	yet

To Signal Examples

for example	specifically
for instance	thus
namely	

continued on the following page

continued from the previous page

To Signal Narrowing of Focus

after all	in particular
indeed	specifically
in fact	that is
in other words	

To Signal Conclusions or Summaries

as a result	in summary
consequently	therefore
in conclusion	thus
in other words	to conclude

To Signal Concession

admittedly	naturally
certainly	of course
granted	

To Signal Causes or Effects

accordingly	since
as a result	so
because	then
consequently	therefore
hence	

(3) Using pronouns

Because **pronouns** refer to nouns or other pronouns, they establish connections among sentences. By drawing ideas together and establishing coherence, clear well-placed pronoun references can lead readers through a paragraph. Unclear pronoun references, like those in the paragraph that follows, can make a writer's ideas difficult to understand.

Unclear pronoun reference

Like Martin Luther, John Calvin wanted to return to principles of early Christianity described in the New Testament. Martin Luther founded the evangelical churches in Germany and Scandinavia, and he founded a number of reformed churches in other countries. A third Protestant branch, episcopacy, developed in England. They rejected the word *Protestant* because they agreed with Roman Catholicism on most points. They rejected the primacy of the Pope. They accepted the Bible as the only source of revealed truth, and they held that faith, not good works, defined a person's relationship to God. (Student Writer)

1
2
3
4
5
6

81

Parallel Constructions

Parallel construction and repetition of key words can add great power to the written or spoken word. Look for examples of parallelism in some of the great speeches of the twentieth century and read excerpts aloud to the class. As an example, here is a little-quoted but powerful excerpt from Dr. Martin Luther King, Jr.'s, speech, given on the steps of the Lincoln Memorial in August 1963.

"In a sense we've come to our nation's Capital to cash a check. When the architects of our republic wrote the magnificent words of the Constitution and the Declaration of Independence, they were signing a . . . promise that all men—yes, black men as well as white men—would be guaranteed the unalienable rights of life, liberty, and the pursuit of happiness.

"It is obvious today that America has defaulted on this promissory note. . . . America has given the Negro people a bad check; a check which has come back marked 'insufficient funds.' But we refuse to believe that the bank of justice is bankrupt. We refuse to believe that there are insufficient funds in the great vaults of opportunity of this nation. So we've come to cash this check—a check that will give us upon demand the riches of freedom and the security of justice. . . ."

Unclear pronoun references make the preceding paragraph quite confusing. Does *he* in sentence 2 refer to Luther or Calvin? Does *they* in sentence 4 refer to Luther and Calvin or to the three Protestant branches?

In this revision, unclear pronoun references have been eliminated, and the paragraph's ideas are much easier to follow.

Like Martin Luther, John Calvin wanted to return to the principles of early Christianity described in the New Testament. Martin Luther founded the evangelical churches in Germany and Scandinavia, and John Calvin founded a number of reformed churches in other countries. A third Protestant branch, episcopacy, developed in England. Its members rejected the word *Protestant* because they agreed with Roman Catholicism on most points. All these sects rejected the primacy of the Pope. They accepted the Bible as the only source of revealed truth, and they held that faith, not good works, defined a person's relationship to God.

The writer has replaced *he* in sentence 2 with *John Calvin* and changed *they* in sentence 4 to *its members*, to which the sentence's *they* now clearly refers. By adding the phrase *all these sects*, the writer makes clear that the two uses of *they* in sentence 6 refer to *sects*, not to *episcopacy*.

(4) Using parallel constructions

Parallelism—the repeated use of similar grammatical constructions—can help to increase the coherence of a paragraph. Absence of parallelism can blur the relationships among ideas. The following paragraph does not use parallel constructions.

Without parallel constructions

Thomas Jefferson was born in 1743 and died at Monticello, Virginia, on July 4, 1826. During his eighty-four years he accomplished a number of things. Although best known for his draft of the Declaration of Independence, Jefferson was a delegate to the Continental Congress. Not only was Jefferson a patriot, he was also a profound thinker. During the Revolution he drafted the Statute for Religious Freedom. He drafted an ordinance for governing the West, and he formulated the first decimal monetary system. After being elected president, he abolished internal taxes, reduced the national debt, and made the Louisiana Purchase. Jefferson also designed Monticello and the University of Virginia. (Student Writer)

Note in the following paragraph how now the basic sentence structure is used to introduce each of Thomas Jefferson's accomplishments. This use of parallelism helps the reader comprehend the material and at the same time adds emphasis.

▶ **See 10c**

> Thomas Jefferson was born in 1743 and died at Monticello, Virginia, on July 4, 1826. During his eighty-four years he accomplished a number of things. Although best known for his draft of the Declaration of Independence, Jefferson was a man of many talents who had a wide intellectual range. He was a patriot who was one of the revolutionary founders of the United States. He was a reformer who, when he was governor of Virginia, drafted the Statute for Religious Freedom. He was an innovator who drafted an ordinance for governing the West and devised the first decimal monetary system. He was a president who abolished internal taxes, reduced the national debt, and made the Louisiana Purchase. And, finally, he was an architect who designed Monticello and the University of Virginia.

(5) Repeating key words and phrases

Repeating **key words and phrases**—those essential to meaning—throughout a paragraph aids coherence by reminding readers how the sentences relate to one another and to the paragraph's unifying idea.

You should not repeat words and phrases monotonously—a well-written paragraph has variety. But you have to balance the need to vary your vocabulary against your audience's need to understand what you have written. In the following paragraph, the absence of repeated key words and phrases that point to the paragraph's subject makes the discussion difficult to understand.

Without repeated key words or phrases

> Mercury poisoning is a problem that has long been recognized. "Mad as a hatter" refers to the condition prevalent among nineteenth-century workers who manufactured felt hats. Workers in many other industries, such as mining, chemicals, and dentistry, were also affected. In the 1950s and 1960s there were cases of poisoning in Minamata, Japan. Research showed that there were high levels of pollution in streams and lakes surrounding the village. In the United States in 1969 a New Mexico family got sick from eating tainted food. Since then certain pesticides have been withdrawn from the market, and chemical wastes can no longer be dumped into the ocean. (Student Writer)

1
2
3
4
5
6
7

The preceding paragraph demands a lot from readers. Sentence 1 introduces mercury poisoning as the subject of the paragraph, but sentences 2 through 7 never mention it. Readers must decide for themselves how the examples relate to the topic sentence. The following revision uses repetition of key words to help readers focus on the subject.

> Mercury poisoning is a problem that has long been recognized. "Mad as a hatter" refers to the condition prevalent among nineteenth-century workers who were exposed to mercury during the manufacturing of felt hats. Workers in many other industries, such as mining, chemicals, and dentistry, were similarly affected. In the 1950s and 1960s there were cases of mercury poisoning in Minamata, Japan. Research showed that there were high levels of mercury pollution in streams and lakes surrounding the village. In the United States this problem came to light in 1969 when a New Mexico family got sick from eating food tainted with mercury. Since then pesticides containing mercury have been withdrawn from the market, and chemical wastes can no longer be dumped into the ocean.

1
2
3
4
5
6
7

The use of the words *mercury* and *mercury poisoning* throughout the paragraph now reminds readers of the subject. Notice that to avoid monotony the writer sometimes refers indirectly to this subject with phrases such as *similarly affected* (sentence 3) and *this problem came to light* (sentence 6).

(6)   Achieving coherence among paragraphs

The same methods you use to establish coherence within paragraphs may also be used to link paragraphs in an essay. In addition, you can use topic sentences to connect paragraphs, or you can use a transitional paragraph as a bridge between two paragraphs.

ACHIEVING COHERENCE AMONG PARAGRAPHS

- Arrange paragraphs within an essay according to an organizing principle. **(See 4dl.)**
- Use transitional words and phrases to connect paragraphs. **(See 4d2.)**
- Use pronouns to connect paragraphs. **(See 4d3.)**
- Use parallel constructions to connect paragraphs. **(See 4d4.)**
- Connect paragraphs by repeating key words and phrases. **(See 4d5.)**
- Use topic sentences to connect paragraphs. **(See 4c1.)**
- Use transitional paragraphs to connect paragraphs. **(See 4g1.)**

The following paragraph cluster shows how some of these strategies work.

> A language may borrow a word directly or indirectly. A direct borrowing means that the borrowed item is a native word in the language it is borrowed from. *Festa* was borrowed directly from French and can be traced back to Latin *festa*. On the other hand, the word *algebra* was borrowed from Spanish, which in turn borrowed it from Arabic. Thus *algebra* was indirectly borrowed from Arabic, with Spanish as an intermediary.
>
> Some languages are heavy borrowers. Albanian has borrowed so heavily that few native words are retained. On the other hand, most Native American languages have borrowed little from their neighbors.
>
> English has borrowed extensively. Of the 20,000 or so words in common use, about three-fifths are borrowed. Of the 500 most frequently used words, however, only two-sevenths are borrowed, and because these "common" words are used over and over again in sentences, the actual frequency of appearance of native words is about 80 percent. Morphemes such as *and, be, have, it, of, the, to, will, you, on, that,* and *is* are all native to English. (Victoria Fromkin and Robert Rodman, *An Introduction to Language,* 4th ed.)

The three paragraphs above form a tightly knit unit. They are arranged according to a logical organizing principle, moving from general to specific ideas: from the general concept of borrowing words to a specific discussion of English, which borrows extensively. Although the paragraphs are not linked to one another by transitional words and phrases or by pronouns, they are explicitly connected by other strategies. For example, each topic sentence contains a variation of the word group *A language may borrow*, repeating or echoing the key words *borrow* and *language*. Throughout the three paragraphs, some form of these key words (as well as *word* and the names of various languages) appears in almost every sentence, further reinforcing the connections among the three paragraphs.

*EXERCISE 2

A. Read the following paragraph and determine how the author achieves coherence. Identify parallel elements, pronouns, repeated words, and transitional words and phrases that link sentences.

> Some years ago the old elevated railway in Philadelphia was torn down and replaced by the subway system. This ancient El with its barnlike stations containing nut-vending machines and scattered food scraps had, for generations, been the favorite feeding ground of flocks of pigeons, generally one flock to a station along the route of the El. Hundreds of pigeons were dependent upon the system. They flapped

A. Your students' answers will vary, as individual perceptions of coherence will lead students to different conclusions. Certainly they should notice the obvious links, such as parallel elements in the first sentence, the repetition of the words *pigeons* and *El,* and the repetition of the pronoun *they.*

B. The theory of continental drift was first put forward by Alfred Wegener in 1912. *He believed that* the continents fit together like a giant jigsaw puzzle. *Because* the opposing Atlantic coasts, especially those of South America and Africa, seem to have been attached, *he also believed that* at one time, probably 225 million years ago, there was one supercontinent *which* broke into parts that eventually drifted into their present positions. The theory, *which* stirred controversy during the 1920s, eventually was ridiculed by the scientific community. *However,* the *theory of continental drift* was revived in 1954 and is *now* accepted as a reasonable geological explanation of the continental system.

Your students' answers will vary. The simplest way to improve the coherence of these paragraphs is probably to use transitional expressions at the beginning of each new paragraph: "In the 1970s *The Brady Bunch* featured six children. . . " and "More recently, *The Cosby Show* was extremely popular. . . ."

in and out of its stanchions and steel work or gathered in watchful little audiences about the feet of anyone who rattled the peanut-vending machines. They even watched people who jingled change in their hands, and prospected for food under the feet of the crowds who gathered between trains. Probably very few among the waiting people who tossed a crumb to an eager pigeon realized that this El was like a food-bearing river, and that the life which haunted its banks was dependent upon the running of the trains with their human freight. (Loren Eiseley, *The Night Country*)

B. Supplying the missing transitional words and phrases, revise the following paragraph to make it coherent.

The theory of continental drift was first put forward by Alfred Wegener in 1912. The continents fit together like a gigantic jigsaw puzzle. The opposing Atlantic coasts, especially South America and Africa, seem to have been attached. He believed that at one time, probably 225 million years ago, there was one supercontinent. This continent broke into parts that drifted into their present positions. The theory stirred controversy during the 1920s and eventually was ridiculed by the scientific community. In 1954 the theory was revived. The theory of continental drift is accepted as a reasonable geological explanation of the continental system. (Student Writer)

*EXERCISE 3

Read the following paragraph cluster. Then revise as necessary to increase coherence among paragraphs.

Leave It to Beaver and *Father Knows Best* were typical of the late 1950s and early 1960s. Both were popular during a time when middle-class mothers stayed home to raise their children while fathers went to "the office." The Beaver's mother, June Cleaver, always wore a dress and high heels, even when she vacuumed. So did Margaret Anderson, the mother on *Father Knows Best*. Wally and the Beaver lived a picture-perfect small-town life, and Betty, Bud, and Kathy never had a problem that father Jim Anderson couldn't solve.

The Brady Bunch featured six children and the typical Mom-at-home and Dad-at-work combination. Of course, Florence Brady did wear pants, and the Bradys were what today would be called a "blended family." Nevertheless, *The Brady Bunch* presented a hopelessly idealized picture of upper-middle-class suburban life. The Brady kids lived in a large split level house, went on vacations, had two loving parents, and even had a live-in maid, the ever-faithful, wisecracking Alice. Everyone in town was heterosexual, employed, able-bodied, and white.

The Cosby Show was extremely popular. It featured two professional parents, a doctor and a lawyer. They lived in a townhouse with original art on the walls, and money never seemed to be a problem. In addition to warm relationships with their siblings, the Huxtable children also had close ties to their grandparents. *The Cosby Show* did introduce problems, such as son Theo's dyslexia, but in many ways it replicated the 1950s formula. Even in the 1990s, it seems, father still knows best.

4e   Patterns of Paragraph Development

The pattern of a paragraph, like the pattern of an entire essay, reflects the way the writer thinks. Most of the time writers do not consciously decide in advance on a particular pattern of development and then write their paragraphs accordingly. Rather, after drafting, when they see the patterns into which their thoughts fall, they are able to rearrange information to support the unifying idea more effectively.

Of course paragraphs, like essays, can combine more than one pattern of development. After you have developed your paragraph skills, you may want to try combining various strategies in a single paragraph.

(1)   Narration (What happened?)

Narrative paragraphs tell a story, but they do not necessarily arrange events in strict chronological order. Sometimes a narrative begins in the middle of a story, or even at the end, and then moves back to the beginning.

The following paragraph opens with a topic sentence that introduces the narrative. The movement from one event to the next is signaled by transitional words like *by midterms, by the end of the semester,* and *at the beginning of my second semester,* which keep the chronology clear.

> My academic career almost ended as soon as it began when, three weeks after I arrived at college, I decided to pledge a fraternity. By midterms I was wearing a straw hat and saying "Yes sir" to every fraternity brother I met. When classes were over I ran errands for the fraternity members, and after dinner I socialized and worked on projects with the other people in my pledge class. In between these activities I tried to study. Somehow I managed to write papers, take tests, and attend lectures. By the end of the semester, though, my grades had slipped and I was exhausted. It was then that I began to ask myself some important questions. I realized that I wanted to be popular, but not at the expense of my grades and my future career. At the beginning of my second semester I dropped out of the fraternity and got a job in the biology lab. Looking back, I realize that it was then that I actually began to grow up. (Student Writer)

IN THE CLASSROOM

Compose a paragraph that is poorly developed, or adapt a paragraph from a magazine article or other source, deleting the support material (examples, statistics, opinions, etc.). Write the paragraph on the board or use the overhead. Ask students to analyze the paragraph along the lines of the example in 4f, using the level 1-2-3 method. At what points in the paragraph is additional development and support needed? What questions are left unanswered? If possible, see whether students can come up with some of the support information the paragraph needs.

IN THE CLASSROOM

After covering section 4e, show students a photograph of an event, such as a holiday celebration or a store's grand opening. Ask them to write several paragraphs about the event, each using a different pattern of development. For example, you might ask first for description, then narration, then process or cause and effect (explaining how or why the event happened).

Patterns Within Paragraphs
Paragraphs can be organized according to *chronological order* (time), *spatial order* (physical space), or *logical order* (focusing on relationships of ideas or details). Descriptive paragraphs are generally arranged in spatial order. As an example, read the following excerpt from Boris Pasternak's *Doctor Zhivago* and ask students to note how the description moves in physical space, out from the doorway of the house to the snow and the wolves, then drawing back into the house and, finally, to Yuri's lamplit desk.

He was dazzled by the white glow playing on the shadowless, moonlit snow and could at first see nothing. Then the long, whimpering, deep-bellied howl sounded around, muffled by the distance, and he noticed four long shadows, no thicker than pencil strokes, at the edge of the clearing just beyond the gully.

The wolves stood in a row, their heads raised and their muzzles pointing at the house, baying at the moon or at its silver reflection on the windows. . . . Going back, he shut all the doors between the cold rooms and the heated part of the house, pushed rugs and clothes against the cracks to keep out the draft, and went back to his desk. The lamplight was bright and welcoming as before. But he was no longer in the mood to write. He couldn't settle down. He could think of nothing but wolves and of looming dangers and complications of every kind. . . .

(2) Description (What does it look like?)

In **descriptive** paragraphs the most natural arrangement of details reflects the way you actually look at a person, scene, or object: near to far, top to bottom, side to side, or front to back. (Sometimes senses other than sight come into play in a description. As you focus on what something looks like, you may also consider how it sounds, smells, tastes, or feels.) The arrangement of details is made clear by transitions that identify the spatial relationships.

CLOSE-UP

WRITING PARAGRAPHS: DESCRIPTION

Although many descriptive paragraphs do not have topic sentences, they must be unified by a **dominant impression,** a descriptive unifying idea to which all the details contribute.

The following descriptive paragraph begins with an exterior view of the Jemez Day School; it then narrows its focus to specific parts of the school and finally moves to the interior of the living quarters.

The day school was a large stucco building in the pueblo style, not unlike the Pueblo Church in certain respects, especially that part of it which was officially the "school"; it had vigas, a dozen large windows along the south wall admitting light into the classrooms, and a belfry in front. The two classrooms, situated end to end, were of about the same size; each could accommodate about thirty pupils comfortably. The highest enrollment during my parents' tenure here was sixty-eight. The front classroom was my mother's; she taught the beginners (who comprised a kind of kindergarten class, and most of whom could not speak English when they came), the first, the second, and the third graders. In the other classroom my father, who was the principal, taught the fourth, fifth, and sixth graders. My parents were assisted by a "housekeeper," a Jemez woman whose job it was to clean the classrooms, supervise the playground, and prepare the noon meal for the children. Frequently she assisted as an interpreter as well. On the opposite side of the building were the teachers' quarters, roughly equivalent in size to one of the classrooms. There were a living room, a kitchen, two bedrooms, and a bath. There was also a screened porch in front, where it was comfortable to sit in the good weather, and another, much smaller, in back; the latter we used largely for storage;

there was our woodbox, convenient to the living room, in which there was a fireplace, and to the kitchen, in which, when we moved there, there was an old wood-burning range. In the living room there was a door which opened upon my mother's classroom, and beside this door a wall telephone with two bells and a crank. We shared a party line with the San Diego Mission and the Jemez Trading Post; there were no other telephones in the pueblo; we answered to two long rings and one short. (N. Scott Momaday, *The Names*)

(3) Exemplification (What are some typical cases or examples of it?)

Exemplification paragraphs use specific illustrations to clarify a general statement. In the following paragraph a series of well-chosen examples supports the topic sentence, with the movement from one example to the next effectively signaled by the repetition of the key words *cannot* and *can't*.

> Illiterates cannot travel freely. When they attempt to do so, they encounter risks that few of us can dream of. They cannot read traffic signs and, while they often learn to recognize and to decipher symbols, they cannot manage street names which they haven't seen before. The same is true for bus and subway stops. While ingenuity can sometimes help a man or woman to discern directions from familiar landmarks, buildings, cemeteries, churches, and the like, most illiterates are virtually immobilized. They seldom wander past the streets and neighborhoods they know. Geographical paralysis becomes a bitter metaphor for their entire existence. They are immobilized in almost every sense we can imagine. They can't move up. They can't move out. They cannot see beyond. Illiterates may take an oral test for drivers' permits in most sections of America. It is a questionable concession. Where will they go? How will they get there? How will they get home? Could it be that some of us might like it better if they stayed where they belong? (Jonathan Kozol, *Illiterate America*)

In the next paragraph, a single extended example gives readers enough detail to help them imagine the world the writer creates.

> Let us imagine a country in which reading is a popular voluntary activity. There, parents read books for their own edification and pleasure, and are seen by their children at this silent and mysterious pastime. These parents also read to their children, give them books for presents, talk to them about books and underwrite, with their taxes, a public library system that is open all day, every day. In school—where an attractive library is invariably to be found—the children study certain books together but also have an active reading life of their own. Years later it may even be hard for them to remember if they read *Jane Eyre* at home and Judy Blume in class, or the other way around.

Ask each student to write either a) a one-paragraph set of directions from the classroom to another well-known point on campus or in town, or b) a recipe they know well. (Remind students that these are to be written in the form of instructions.) Then divide the class into pairs and ask each pair to exchange their "process" paragraphs, read them carefully, and discuss what works and what does not work—focusing primarily on problems of clarity, chronological order, and so on. Ask some pairs to share their results with the class or have students actually try out the process rather than just imagine whether or not it is effectively described. You could expand the response between groups by making it written: Have each group either a) suggest revisions for the other or b) evaluate each other's process in a brief memo or report. The process should be revised in either case and tried out with another group to see if it now gets a more positive response. It also could be useful for a few process papers to be photocopied and distributed to groups so the groups can rank them. This could then lead to useful class discussion.

In college young people continue to be assigned certain books, but far more important are the books they discover for themselves—browsing in the library, in bookstores, on the shelves of friends, one book leading to another, back and forth in history and across languages and cultures. After graduation they continue to read, and in the fullness of time produce a new generation of readers. Oh happy land! I wish we all lived there. (Katha Pollitt, "Canon to the Right of Me . . .")

(4) Process (How did—or does—it happen?)

Process paragraphs describe how something works, presenting a series of steps in strict chronological order. The topic sentence (when there is one) identifies the process, and the rest of the paragraph presents the steps involved. Transitional words such as *first, next, then,* and *finally* link the steps in the process. The paragraph that follows explains the process by which members of the Supreme Court decide whether or not to grant an appeal.

Members of the court have disclosed, however, the general way the conference is conducted. It begins at ten A.M. and usually runs on until late afternoon. At the start each justice, when he enters the room, shakes hands with all others there (thirty-six handshakes altogether). The custom, dating back generations, is evidently designed to begin the meeting at a friendly level, no matter how heated the intellectual differences may be. The conference takes up, first, the applications for review—a few appeals, many more petitions for certiorari. Those on the Appellate Docket, the regular paid cases, are considered first, then the pauper's applications on the Miscellaneous Docket. (If any of these are granted, they are then transferred to the Appellate Docket.) After this the justices consider, and vote on, all the cases argued during the preceding Monday through Thursday. These are tentative votes, which may be and quite often are changed as the opinion is written and the problem thought through more deeply. There may be further discussion at later conferences before the opinion is handed down. (Anthony Lewis, *Gideon's Trumpet*)

When a process paragraph presents instructions to enable readers actually to perform the process, it is written in the present tense and in the imperative mood: "*Remove* the cover . . . and *check* the valve." This directness of both tense and mood helps readers follow the directions more easily. The following paragraph presents a set of instructions for making a quilt.

Take a variety of fabrics: velvet, satin, silk, cotton, muslin, linen, tweed, men's shirting; mix with a variety of notions: buttons, lace, grosgrain, or thick silk ribbon lithographed with city scenes, bits of drapery, appliqués of flora and fauna, honeymoon cottages, and clouds. Puff them up with: down, kapok, soft cotton, foam, old stockings. Lay between the back cloth a large expanse of cotton batting; stitch it all together with silk thread, embroidery thread, nylon thread. The stitches must be small, consistent, and reflect a design of their own. (Whitney Otto, *How to Make an American Quilt*)

(5) Cause and effect (What caused it? What are its effects?)

Cause and effect paragraphs explore why events occur and what happens as a result of them.

CLOSE-UP

WRITING PARAGRAPHS: CAUSE AND EFFECT

Because cause-and-effect relationships are often complicated, topic sentences and transitional words and phrases (*one cause, another cause, a more important result, because, as a result*) are especially important. Choose them carefully to help clarify causal connections.

▶ **See 4d2**

In the following paragraph the writer suggests a *cause* of thumb sucking and then summarizes a study to support his statement.

The main reason that a young baby sucks his thumb seems to be that he hasn't had enough sucking at the breast or bottle to satisfy his sucking needs. Dr. David Levy pointed out that babies who are fed every 3 hours don't suck their thumbs as much as babies fed every 4 hours, and that babies who have cut down on nursing time from 20 minutes to 10 minutes . . . are more likely to suck their thumbs than babies who still have to work for 20 minutes. Dr. Levy fed a litter of puppies with a medicine dropper so that they had no chance to suck during their feedings. They acted just the same as babies who don't get enough chance to suck at feeding time. They sucked their own and each other's paws and skin so hard that the fur came off. (Benjamin Spock, *Baby and Child Care*)

In the next paragraph, the writer focuses on the *effects* of threatened violence on his grandfather. He begins by identifying his grandfather's reactions and then steps back to examine the causes for his grandfather's behavior.

Causal Relationships
Remind your students that, when investigating causal relationships, writers should note the danger of assuming that cause/effect is a simple, neat pattern. These four variations suggest the complexity of this structure.

Single cause ⟶ **Single effect**
    Forget to remove cake from oven ⟶ burnt cake

Single cause ⟶ **Multiple effects**
    Telephone out of order ⟶
        Inability to call
        Inability to use answering machine
        Need to arrange repair

Multiple Causes ⟶ **Single Effect**
    Taking careful notes
    Reading assignments ⟶
        Earning high grades
        Studying effectively

Multiple Causes ⟶
Multiple Effects
    Local school condemned
    Town refuses to appropriate building funds ⟶
        State freezes building aid funds
        No available local school
        Children bused to school in another town
        Recall effort launched against local school committee

Give a piece of writing to the whole class. In groups of four or five, students determine how many methods of development are used in the essay. Are there any "pure" (single method of development) paragraphs?

On December 8, 1941, the day after the Japanese attack on Pearl Harbor in Hawaii, my grandfather barricaded himself with his family—my grandmother, my teenage mother, her two sisters and two brothers—inside of his home in La'ie, a sugar plantation village on Oahu's North Shore. This was my maternal grandfather, a man most villagers called by his last name, Kubota. It could mean either "Wayside Field" or else "Broken Dreams," depending on which ideograms he used. Kubota ran La'ie's general store, and the previous night, after a long day of bad news on the radio, some locals had come by, pounded on the front door, and made threats. One was said to have brandished a machete. They were angry and shocked, as the whole nation was in the aftermath of the surprise attack. Kubota was one of the few Japanese Americans in the village and president of the local Japanese language school. He had become a target for their rage and suspicion. A wise man, he locked all his doors and windows and did not open his store the next day, but stayed closed and waited for news from some official. (Garrett Hongo, "Kubota")

(6) Comparison and contrast (How is it like other things? How is it different?)

Comparison-and-contrast paragraphs examine the similarities and differences between two subjects. Comparison emphasizes similarities, while contrast stresses differences.

CLOSE-UP

WRITING PARAGRAPHS: COMPARISON AND CONTRAST

When using comparison and contrast, be sure that the subjects you compare have elements in common and that you compare the same or similar qualities of both. Also be sure that the likenesses or differences are of some significance, not so obvious that a discussion of them would be pointless. Do not forget to use transitional words and phrases (*similarly, likewise, however, but, on the contrary, nevertheless*) to signal comparison or contrast and to indicate movement from one subject to another.

See ◄
4d2

Comparison and contrast can be organized in one of two ways. One is to compare and contrast the subjects point by point. This organization is especially useful in a complex paragraph in which your readers may have trouble keeping track of your points. The following paragraph uses **point-by-point comparison.**

> There are two Americas. One is the America of Lincoln and Adlai Stevenson; the other is the America of Teddy Roosevelt and the modern superpatriots. One is generous and humane, the other narrowly egotistical; one is self-critical, the other self-righteous; one is sensible, the other romantic; one is good-humored, the other solemn; one is inquiring, the other pontificating; one is moderate, the other filled with passionate intensity; one is judicious and the other arrogant in the use of great power. (J. William Fulbright, *The Arrogance of Power*)

A second way to organize a comparison-and-contrast paragraph is subject by subject. Treat one subject in its entirety at the beginning of your paragraph and the other subject in its entirety at the end. Use this organization when your readers will easily remember details about the first subject while they read about the second. In the following **subject-by-subject comparison,** notice how the writer shifts from one subject to the other with the transitional word *now*.

> This seems to be an era of gratuitous inventions and negative improvements. Consider the beer can. It was beautiful—as beautiful as the clothespin, as inevitable as the wine bottle, as dignified and reassuring as the fire hydrant. A tranquil cylinder of delightfully resonant metal, it could be opened in an instant, requiring only the application of a handy gadget freely dispensed by every grocer. Who can forget the small, symmetrical thrill of those two triangular punctures, the dainty *pffff,* the little crest of suds that foamed eagerly in the exultation of release? Now we are given, instead, a top beetling with an ugly, shmoo-shaped "tab," which, after fiercely resisting the tugging, bleeding fingers of the thirsty man, threatens his lips with a dangerous and hideous hole. However, we have discovered a way to thwart Progress, usually so unthwartable. *Turn the beer can upside down and open the bottom.* The bottom is still the way the top used to be. True, this operation gives the beer an unsettling jolt, and the sight of a consistently inverted beer can might make people edgy, not to say queasy. But the latter difficulty could be eliminated if manufacturers would design cans that looked the same whichever end was up, like playing cards. What we need is Progress with an escape hatch. (John Updike, *Assorted Prose*)

An **analogy** is a special kind of comparison that explains an unfamiliar concept or object by likening it to a familiar one. Extended analogies resemble comparison-and-contrast paragraphs, with one important difference: Whereas comparisons give equal weight to both things being compared, extended analogies use one part of the comparison for the *sole* purpose of shedding light on the other. Here an author uses the behavior of people to explain the behavior of ants.

> Ants are so much like human beings as to be an embarrassment. They farm fungi, raise aphids as livestock, launch armies into wars, use chemical sprays to alarm and confuse enemies, capture slaves. The families of weaver ants engage in child labor, holding their larvae like shuttles to spin out the thread that sews the leaves together for their fungus gardens. They exchange information ceaselessly. They do everything but watch television. (Lewis Thomas, "On Societies as Organisms")

(7) Division (What are its parts?) and classification (Into what categories can its parts be arranged?)

In **division**, you take a single item and break it into its components. In **classification**, you take many separate items and group them into categories according to qualities or characteristics they have in common. In the following paragraph, a student *divides* blood into several components. The opening sentence identifies the subject the paragraph will analyze, and subsequent sentences identify the components of blood, moving from the most frequently to the least frequently found elements.

> The blood can be divided into four distinct components: plasma, red cells, white cells, and platelets. Plasma is 90 percent water and holds a great number of substances in suspension. It contains proteins, sugars, fat, and inorganic salts. Plasma also contains urea and other by-products from the breaking down of proteins, hormones, enzymes, and dissolved gasses. In addition, plasma contains the red blood cells that give it color, the white cells, and the platelets. The red cells are most numerous; they get oxygen from the lungs and release it in the tissues. The less numerous white cells are part of the body's defense against invading organisms. The platelets, which occur in almost the same number as white cells, are responsible for clotting. (Student Writer)

The following paragraph establishes its subject, scientific frauds, and then goes on to *classify* frauds into three categories.

> Charles Babbage, an English mathematician, reflecting in 1830 on what he saw as the decline of science at the time, distinguished among three major kinds of scientific fraud. He called the first "forging," by which he meant complete fabrication—the recording of observations that were never made. The second category he called "trimming"; this consists of manipulating the data to make them look

better, or, as Babbage wrote, "in clipping off little bits here and there from those observations which differ most in *excess* from the mean and in sticking them on to those which are too small." His third category was data selection, which he called "cooking"—the choosing of those data that fitted the researcher's hypothesis and the discarding of those that did not. To this day, the serious discussion of scientific fraud has not improved on Babbage's typology. (Morton Hunt, *New York Times Magazine*)

(8)   Definition

A **formal definition** includes the term you are defining, the class to which it belongs, and the details that distinguish it from other members of its class.

Carbon is a nonmetallic element
(term)   (class to which it belongs)

occurring as diamond, graphite, and charcoal.
                (distinguishing details)

A paragraph of **extended definition** develops the formal definition with other patterns, defining *happiness*, for instance, by telling a story (narration), or defining a diesel engine by telling how it works (process). Extended definitions may also include the background or origins of a term and may define terms by telling what they are like (using synonyms) or what they are not (using negation).

The following paragraph begins with a straightforward definition of *gadget* and then cites an example.

A gadget is nearly always novel in design or concept and it often has no proper name. For example, the semaphore which signals the arrival of the mail in our rural mailbox certainly has no proper name. It is a contrivance consisting of a piece of shingle. Call it what you like, it saves us frequent frustrating trips to the mailbox in winter when you have to dress up and wade through snow to get there. That's a gadget! (*Smithsonian*)

EXERCISE 4

Determine one possible method of development for a paragraph on each of these topics. Then write a paragraph on one of the topics.

1. What success is (or is not)
2. How to write a résumé
3. The kinds of people who listen to radio talk shows
4. My worst date

5. American vs. Japanese education
6. The connection between stress and the immune system
7. Budgeting money wisely
8. Self-confidence
9. Preparing for the perfect vacation
10. Drinking and driving

 ## 4f Writing Well-Developed Paragraphs

See ◄
5b2   A paragraph is **well developed** when it contains the supporting evidence—examples, statistics, expert opinion, and so on—that readers need to understand the unifying idea.

(1) Charting paragraph development

See ◄
4c2   Just as charting structure can help you see whether a paragraph is unified, it can also help you determine whether it is well developed.

The following paragraph contains a number of interesting ideas. At first glance it may seem quite well developed.

> From Thanksgiving until Christmas, children are saturated with ads for violent toys. Advertisers persist in thinking that only toys that appeal to children's aggressiveness will sell. Far from improving the situation, video games have escalated the arms race. The real question is why toy manufacturers continue to pour millions of dollars into violent toys, especially in light of the success of toys that promote learning and cooperation. (Student Writer)

Charting the underlying structure of the paragraph, however, reveals a problem.

1 From Thanksgiving until Christmas, children are saturated with ads for violent toys.
  2 Advertisers persist in thinking that only toys that appeal to children's aggressiveness will sell.
  2 Far from improving the situation, video games have escalated the arms race.
  2 The real question is why toy manufacturers continue to pour millions of dollars into violent toys, especially in light of the success of toys that promote learning and cooperation.

The first sentence of this paragraph is a level-1 sentence. The level-2 sentences expand the discussion, but the paragraph offers no level-3 examples. What kinds of toys appeal to a child's aggressive tendencies? Which particular video games does the writer object to? The following revision includes specific examples that convincingly support the topic sentence (sentence 1).

> From Thanksgiving until Christmas, children are saturated with ads for violent toys. Advertisers persist in thinking that only toys that appeal to children's aggressiveness will sell. <u>One television commercial praises the merits of a commando team that attacks and captures a miniature enemy base. Toy soldiers wear realistic uniforms and carry automatic rifles, pistols, knives, grenades, and ammunition. Another commercial shows laughing children shooting one another with plastic rocket fighters and tanklike vehicles.</u> Far from improving the situation, video games have escalated the arms race. <u>The most popular video games involve children in realistic combat situations. One game lets children search out and destroy enemy rocket fighters in outer space. Other best-selling games simulate attacks on enemy fortresses or raids by hostile creatures.</u> The real question is why toy manufacturers continue to pour millions of dollars into violent toys, especially in light of the success of toys that promote learning and co-operation.

(2) Revising for paragraph development

Adequate length—the number of sentences or number of examples in a paragraph, for instance—does not in itself constitute adequate development. The *amount* and *kind* of support you need in a paragraph depends not only on your purpose, your audience, and the scope of your unifying idea, but also on your paragraph's pattern of development. If, for example, your purpose is to explain a complicated process, describe an unfamiliar or unusual place, draw a comparison between two unlikely subjects, or trace a series of causal relationships, your paragraph will need to provide a good deal of support. If, however, your unifying idea is one your readers will have little difficulty understanding, you will not need to supply as many supporting details.

REVISION CHECKLIST:
PARAGRAPH DEVELOPMENT

1. Chart your paragraph **(see 4f1)** to gauge the level of its support. How specific does the paragraph get?
2. Identify your purpose. Will your paragraph give readers a general overview of a topic, or will it present complex information? Given your purpose, is the paragraph adequately developed?

continued on the following page

continued from the previous page

3. Identify your audience. Will readers be familiar with your subject? Are they likely to be receptive to the point you are making about the subject? Given the needs of your audience, is the paragraph adequately developed?
4. Identify your paragraph's pattern of development **(see 4e)**.

Narration Do you present enough events to enable readers to understand what occurred? Do you support your unifying idea with descriptive details and dialogue?

Description Do you supply enough detail about what things look like, sound like, smell like, taste like, and feel like? Will your readers be able to visualize the person, object, or setting your paragraph describes?

Exemplification Do you present a sufficient number of individual examples to support your paragraph's unifying idea? Is a single extended example developed in enough detail to enable readers to understand how it supports the paragraph's unifying idea?

Process Do you present enough steps to enable readers to understand how the process is performed? If you are writing instructions, do you include enough explanation—including reminders and warnings—to enable readers to perform the process?

Cause and Effect Do you identify enough causes (subtle as well as obvious, minor as well as major) to enable readers to understand why something occurred? Do you identify enough effects to show the significance of the causes and the impact they had?

Comparison and Contrast Do you supply a sufficient number of details to illustrate and characterize each of the subjects in the comparison? Do you present a comparable number of details for each subject? Do you discuss the same or similar details for each subject?

Division and Classification Do you present enough information to enable readers to identify each category and distinguish one from another?

Definition Do you present enough support (examples, analogies, descriptive details, and so on) to enable readers to understand the term you are defining and distinguish it from others in its class?

*EXERCISE 5

a. Read each of the following paragraphs and then answer these questions. In general terms, how could each paragraph be developed further? What pattern of development might be used in each case?

b. Choose one paragraph, and rewrite it to develop it further. To assess the development of your revised paragraph, consult the checklist on pp. 97–98. Revise again if necessary.

1. Many new words and expressions have entered the English language in the last ten years or so. Some of them come from the world of computers. Others come from popular music. Still others have politics as their source. There are even some expressions that have their origins in films or television shows.

2. Making a good spaghetti sauce is not a particularly challenging task. First, assemble the basic ingredients: garlic, onion, mushrooms, green pepper, and ground beef. Cook these ingredients in a large saucepan. Then, add canned tomatoes, tomato paste, and water, and stir. At this point, you are ready to add the spices: oregano, parsley, basil, and salt and pepper. Don't forget a bay leaf! Cook for about two hours, and serve over spaghetti.

3. High school and college are not at all alike. Courses are a lot easier in high school, and the course load is lighter. In college, teachers expect more from students; they expect higher quality work, and they assign more of it. Assignments tend to be more difficult and more comprehensive, and deadlines are usually shorter. Finally, college students tend to be more focused on a particular course of study—even a particular career—than high school students are.

4g   Writing Special Kinds of Paragraphs

So far we have been talking only about **body paragraphs**, the paragraphs that carry the weight of your essay's discussion. Other kinds of paragraphs, however, have specialized functions.

CLOSE-UP

WRITING PARAGRAPHS

Although most paragraphs you write will be part of an essay, some will stand alone. For example, see Writing a Summary **(39b1),** Writing a Paraphrase **(39b2),** Writing a Response Statement **(43b1),** the Abstract **(45b1),** and Writing Paragraph-length Exam Answers **(46d).**

ANSWERS TO EXERCISE 5

1. The paragraph could be developed further by exemplification—that is, by giving examples of words that came into the English language from computer terminology, from popular music, from politics, and from films or TV. If enough examples are given, the paragraph can be expanded into an essay.

2. This process would be easier to replicate if certain steps in the process were described more exactly. For example, the garlic, onions, mushrooms, and peppers should be chopped; the ingredients should be cooked in olive oil; and so forth. The writer also needs to specify the amount of each ingredient required for the recipe.

3. This comparison/contrast paragraph could be more fully developed by adding specific details to support the generalizations. For example, high school calculus is easier than college calculus. Anecdotes could demonstrate how one teacher expects more than another; easy and difficult assignments could be described; and so on. Depending on the details that are added, this paragraph might be expanded into an essay.

Students sometimes have difficulty envisioning a way to begin a paper other than with a bold statement of the thesis. To help students identify and analyze the various techniques for introductory paragraphs, bring a variety of magazines and journals to class or ask students to bring their own. Once the class has covered the text of 4g2, distribute copies of magazines to individual students and ask them to identify the opening strategies used in the introductions of at least three articles. Share the results with the class as time permits. Do any of the introductions seem to use more than one of the techniques listed in the text?

(1) Transitional paragraphs

Longer essays frequently include one or more **transitional paragraphs** whose function is to signal a change in subject while providing a bridge between one section of an essay and another. At their simplest, transitional paragraphs can be single sentences that move readers from one point to the next.

Let us examine this point further.

This idea works better in theory than in practice.

Of course there are other avenues we can explore.

Let us begin with a few estimates.

More often, writers use a transitional paragraph to present a concise summary of what they have already said, reinforcing important concepts by encouraging readers to pause to consider what they have read before moving on to a new point. The following transitional paragraph uses a series of questions to restate some of the alarming ideas the writer has been advancing about the threat of overpopulation. In the next part of his essay, he goes on to answer these questions.

> Can we bleed off the mass of humanity to other worlds? Right now the number of human beings on Earth is increasing by 80 million per year, and each year that number goes up by 1 and a fraction percent. Can we really suppose that we can send 80 million people per year to the Moon, Mars, and elsewhere, and engineer those worlds to support those people? And even so, nearly remain in the same place ourselves? (Isaac Asimov, "The Case Against Man")

(2) Introductory paragraphs

Writing Introductions   Some introductions are straightforward, concerned primarily with presenting information. They move from general to specific information, introducing the subject, narrowing it down, and then stating the essay's thesis.

> Although modern architecture is usually not intricate in design, it often involves remarkable engineering accomplishments. Most people do not realize the difficulties an architect encounters when designing a "great" modern structure. The new wing of the Smithsonian in Washington, in its simplicity, is such a masterpiece of engineering and design. (Student Writer)

Not all subjects appeal to all readers, however, so at times you must find a way to arouse your audience's interest. Often writers do so by using one of the strategies for effective introductions listed below.

STRATEGIES FOR EFFECTIVE INTRODUCTIONS

Quotation or Series of Quotations

When Mary Cassatt's father was told of her decision to become a painter, he said: "I would rather see you dead." When Edgar Degas saw a show of Cassatt's etchings, his response was: "I am not willing to admit that a woman can draw that well." When she returned to Philadelphia after twenty-eight years abroad, having achieved renown as an Impressionist painter and the esteem of Degas, Huysmans, Pissarro, and Berthe Morisot, the *Philadelphia Ledger* reported: "Mary Cassatt, sister of Mr. Cassatt, president of the Pennsylvania Railroad, returned from Europe yesterday. She has been studying painting in France and owns the smallest Pekingese dog in the world." (Mary Gordon, "Mary Cassatt," *Good Boys and Dead Girls*)

Question or Series of Questions

Of all the disputes agitating the American campus, the one that seems to me especially significant is that over "the canon." What should be taught in the humanities and social sciences, especially in introductory courses? What is the place of the classics? How shall we respond to those professors who attack "Eurocentrism" and advocate "multiculturalism"? This is not the sort of tedious quarrel that now and then flutters through the academy; it involves matters of public urgency. I propose to see this dispute, at first, through a narrow, even sectarian lens, with the hope that you will come to accept my reasons for doing so. (Irving Howe, "The Value of the Canon")

Definition

Moles are collections of cells that can appear on any part of the body. With occasional exceptions, moles are absent at birth. They first appear in the early years of life, between ages two and six. Frequently moles appear at puberty. New moles, however, can continue to appear throughout life. During pregnancy new moles may appear and old ones darken. There are three major designations of moles, each with its own characteristics. (Student Writer)

Unusual Comparison

Once a long time ago, people had special little boxes called refrigerators in which milk, meat, and eggs could be

continued on the following page

COMPUTER TIP

Beginning at the End and Ending at the Beginning

Students who have not been trained in the writing process will likely be trapped in a linear composing mode. If so, they will insist upon beginning at the beginning and ending at the end.

Various techniques are helpful in breaking students of old, nonproductive composing habits concerning beginnings and endings. These can be implemented easily using computers.

Collaborative Revision

- After students have produced a first draft, have them write two or three different introductions and test each one out on a peer critic or the entire class.

- Then have them turn one of the introductions into a conclusion.

- After students have produced a first draft, have them write two or three different conclusions and test each one on a peer critic or the entire class.

- Then have them turn one of the conclusions into an introduction.

- After students have produced a first draft, have them switch their introduction with their conclusion, rewriting as necessary.

continued from the previous page

kept cool. The grandchildren of these simple devices are large enough to store whole cows, and they reach temperatures comparable to those at the South Pole. Their operating costs increase each year, and they are so complicated that few home handymen attempt to repair them on their own. Why has this change in size and complexity occurred in America? It has not taken place in many areas of the technologically advanced world (the average West German refrigerator is about a yard high and less than a yard wide, yet refrigeration technology in Germany is quite advanced). Do we really need (or even want) all that space and cold? (Appletree Rodden, "Why Smaller Refrigerators Can Preserve the Human Race")

Controversial Statement

Something had to replace the threat of communism, and at last a workable substitute is at hand. "Multiculturalism," as the new menace is known, has been denounced in the media recently as the new McCarthyism, the new fundamentalism, even the new totalitarianism—take your choice. According to its critics, who include a flock of tenured conservative scholars, multiculturalism aims to toss out what it sees as the Eurocentric bias in education and replace Plato with Ntozake Shange and traditional math with the Yoruba number system. And that's just the beginning. The Jacobins of the multiculturalist movement, who are described derisively as P.C., or politically correct, are said to have launched a campus reign of terror against those who slip and innocently say "freshman" instead of "freshperson," "Indian" instead of "Native American" or, may the Goddess forgive them, "disabled" instead of "differently abled." (Barbara Ehrenreich, "Teach Diversity—with a Smile")

Revising Introductions   Whether or not it includes a thesis, your introduction should lead naturally into the body of your paper. It should not be at odds with your subject or seem imposed on it. It must also be consistent with the purpose, tone, and style of the rest of your essay. A serious, formal discussion should have the same kind of introduction. If your discussion is relaxed and informal, your introduction should also be. Finally, you should

REVISION CHECKLIST: INTRODUCTIONS

- Does your introduction arouse your readers' interest?
- Does it contain the thesis of your essay?
- Does it lead naturally into the body of your essay?
- Is it consistent with the purpose, tone, and style of the rest of your essay?
- Does it avoid statements that simply announce your subject or that undercut your credibility?

avoid opening statements that do no more than announce your subject ("In my paper I will talk about Lady Macbeth") or that undercut your credibility ("I don't know much about alternative energy sources, but I would like to present my opinion about the subject").

In the following draft of an introduction, the writer simply announces the subject of the paper and does little to create interest or to clarify the paper's focus.

INEFFECTIVE INTRODUCTION:

Dimensioned drawings are a standard method of pictorially communicating any object. In this paper I will discuss dimensioned drawings and show how they are used.

The revised introduction begins with familiar examples that immediately draw readers into the essay. It also defines the term *dimensioned drawing* and presents an accurate and engaging picture of what the essay will be about.

EFFECTIVE INTRODUCTION:

Every manufactured object, from a paper clip to a skyscraper, begins as a dimensioned drawing. Dimensioning, the process of drawing an object so it appears to have depth, is involved in at least one step of the production process. Because dimensioned drawings are a standard method of pictorially representing objects, the American National Standard Institute (ANSI) has established rules for placing dimensions on drawings. Despite these rules, experienced draftspersons have adopted their own methods of dimensioning objects, and for this reason, the art of dimensioning can be quite creative.

(3)   Concluding paragraphs

A **conclusion** should reinforce an essay's major ideas and give readers a sense of completion. In short essays, the conclusion is

Photocopy a variety of recent news-paper editorials. After covering the various concluding-paragraph tech-niques mentioned in the text, pass out copies of the editorials to individ-ual students or small groups. Ask stu-dents to identify the technique used in the conclusion of the editorial and to analyze the relationship between the introduction and conclusion. Share the results as time permits. (As an alternative, you might place the editorials on the overhead and let the class work as a group to identify the technique used.)

usually only one paragraph; in longer essays, it can be two or more.

Writing Conclusions   Many conclusions begin with specifics —restating the thesis or reviewing the main points—and then move to more general statements.

> As an Arab-American, I feel I have the best of two worlds. I'm proud to be part of the melting pot, proud to contribute to the tremendous diversity of cultures, customs and traditions that makes this country unique. But Arab-bashing—public acceptance of hatred and bigotry—is something no American can be proud of. (Ellen Man-soor Collier, "I Am Not a Terrorist")

Several strategies for effective conclusions are presented below.

Revising Conclusions   A conclusion should give readers a sense of completion. It should not introduce new points or go off in new directions. Because your conclusion is your last word, a weak or uninteresting one detracts from an otherwise strong essay. Do not simply repeat your introduction in different words or apologize or in any way cast doubt on your concluding points

---

STRATEGIES FOR EFFECTIVE CONCLUSIONS

Prediction

Looking ahead, [we see that] prospects may not be quite as dismal as they seem. As a matter of fact, we are not doing so badly. It is something of a miracle that creatures who evolved as nomads in an intimate, small-band, wide-open-spaces context manage to get along at all as villagers or surrounded by strangers in cubicle apartments. Consid-ering that our genius as a species is adaptability, we may yet learn to live closer and closer to one another, if not in utter peace, then far more peacefully than we do today. (John Pheiffer, "Seeking Peace, Making War")

Opinion

A piece of writing is never finished. It is delivered to a deadline, torn out of the typewriter on demand, sent off with a sense of accomplishment and shame and pride and frustration. If only there were a couple more days, time for just another run at it, perhaps then . . . (Donald Murray, "The Maker's Eye: Revising Your Own Manuscripts")

continued on the following page

continued from the previous page

Quotation

When we let freedom ring, when we let it ring from every village and every hamlet, from every state and every city, we will be able to speed up that day when all of God's children, black men and white men, Jews and Gentiles, Protestants and Catholics, will be able to join hands and sing in the words of the old Negro spiritual, "Free at last! Free at last! Thank God almighty, we are free at last!" (Martin Luther King, Jr., "I Have a Dream")

("I may not be an expert" or "At least this is my opinion"). If possible, try to end with a statement that readers will remember.

The following conclusion lists points instead of reinforcing the focus of the essay. Readers are left with no idea of the essay's thesis and get no sense of closure.

INEFFECTIVE CONCLUSION:

A draftsperson must know how to apply the rules of dimensioning. The draftsperson must also know how to use the two types of dimension. In addition, he or she must avoid inconsistencies. When a draftsperson changes the codes of a drawing, workers will waste a lot of time trying to decipher the illustration.

The revised conclusion reinforces the focus of the essay and re-states its major points. It ends by emphasizing the major point of the essay—that dimensioned drawing is a creative activity that requires skill and concentration.

EFFECTIVE CONCLUSION:

In spite of the rules that govern the placement of dimensions on drawings, draftspersons have a great deal of flexibility in interpreting information: They can use different methods of dimensioning and labeling, and they can align objects differently and modify the way relationships are shown. As long as workers are able to interpret a drawing quickly and easily, many variations are possible. Over time, a draftsperson comes to realize that dimensioned drawing requires not only precision and concentration but also creativity, intelligence, and artistic talent.

Ineffective Conclusions
Ineffective conclusions can mar even the best student writing. Briefly discuss the following "don'ts" with your students.

1. Don't use cliché closers such as "In conclusion," "To summarize," or "In closing. . . ."
2. Don't merely restate the thesis. The writer can return to the thesis, but the restatement should be an amplified, expanded version of the thesis.
3. Don't apologize or qualify your ideas in a way that dissolves reader confidence: "Of course, further research may prove me wrong. . . ."
4. Don't make a "big claim" for your thesis at the last minute, especially if the claim is unsupported by the body of the paper: "Furthermore, if this plan isn't adopted, I would guess that America as we know it will not exist by the year 2000."

Writing Paragraphs

Your students' answers will vary. (You might encourage them to try using an anecdote or quotation in either the introduction or conclusion.) Students should be reminded to provide transitional devices, if they are necessary, between the introduction and the body and between the body and the conclusion.

Sample Introduction

High school is an exciting time for most young people, and I was no exception. I look back on those years as a time of growth and achievement, grateful for the education I received and the opportunities set before me. My curriculum was challenging, and I am fortunate that so many of my courses were stimulating. Although I learned a great deal in all of my classes, my chemistry class taught me some lasting lessons, not only about chemistry, but about life as well.

Sample Conclusion

Now that I look back on my high school years, I recognize that each of my classes prepared me for the challenges to come. However, my chemistry class in particular gave me knowledge and experience that extended far beyond the walls of the classroom. The lessons learned there laid the cornerstone upon which I have built my education and planned my future.

✔ **REVISION CHECKLIST: CONCLUSIONS**

- Does your conclusion provide a sense of completion?
- Does it remind readers of the primary focus of your essay?
- Does it avoid digressions and new directions?
- Does it avoid repeating the introduction's points?
- Does it avoid apologies?
- Does it end memorably?

STUDENT WRITER AT WORK

WRITING PARAGRAPHS

The following draft of a student essay has a weak introduction and conclusion. Rewrite the opening and closing paragraphs to increase their effectiveness.

Without a doubt, the best class I had during my time in high school was chemistry.

One thing that made the class great was the subject matter and the way it was presented. I can still remember the first day. We just dove right in, with no introduction at all! Maybe that was why I liked the class so much: There was no fooling around. Everything was presented clearly, and I ate it up. It was one of those classes that you take home with you and apply to everyday life. I used to tell my friends what I learned each day in chemistry. I also used to whine to them about how difficult the class was. Still, the challenge was one of the reasons I enjoyed the class. Chemistry was one of the first classes I worked diligently at, and it set the standard for the rest of my high school career.

Another reason I liked the class was, simply enough, the teacher. Mr. Karoulis was a large, bold, and outspoken man. At first, my classmates and I feared his boisterousness. Over time, though, we grew fond of the man. Since the class was "advanced," Mr. Karoulis gave us freedom. We were allowed to leave the class when we finished exams, and he let us voice our opinions and ask questions. Also, Mr. Karoulis did not just teach chemistry; he taught me many valuable lessons about life in general. If Mr. Karoulis thought something in the news was important, he would stop class and discuss the topic with us. Along the way, he taught me the importance of my heritage and the value of being an individual--a leader, and not a follower.

continued on the following page

continued from the previous page

Of course, I also learned a lot about chemistry and about science in general. The things I learned in that class established the foundation for my later physics and chemistry classes. I still use the knowledge I gained in Mr. Karoulis's class and apply it to new concepts with which I am presented. Most important of all, Mr. Karoulis's class taught me that I could enjoy science, and that I could do well in what I had always before seen as an impossibly difficult field.

That class laid the cornerstone upon which I have built my education and planned my future.

Thinking Critically

FURTHER READING

Ann E. Berthoff. "Is Teaching Still Possible? Writing, Meaning, and Higher Order Reasoning." *College English* 46.8 (December 1984): 743–55.

In this article, Berthoff continues her critique of problem-solving approaches to teaching and then offers alternative views of language and learning that will foster thinking about thinking.

She constructs a parallel distinction between "a pedagogy of exhortation" and "a pedagogy of knowing." With these distinctions, the author critiques empiricism and cognitive psychology, neither of which are aware of the basic concept of interpreting interpretations. Berthoff refers to a psycholinguist who explained that to determine the "cohesive devices" (e.g., such connectives as "however" and "although") in a piece of discourse, it was necessary to diminish the role of meaning. She claims that this kind of empirical research institutionalizes a pedagogy of exhortation. A writer should not, as this view of cohesive devices encourages, simply insert such clues for cohesion by adding them mechanically. Done mechanically, via exhortation, there has been no thinking about thinking in the process of making meaning.

While the first section of the paper is written in the negative, the second is in the affirmative. The author distinguishes between "abstraction" and "generalization." Abstraction does not have to be taught; however, if we wish to move our students from a pedagogy of exhortation to one of

continued on following page

CHAPTER 5

Reading Critically and Writing Critical Responses

The fact that something is printed in a book does not automatically establish it as free of bias or contradictions, let alone meaningful, insightful, or even accurate. For this reason, you should read with a critical eye, approaching any text with a healthy skepticism. When you **read critically**, your goal should be not simply to understand the literal meaning of what you read, but also to consider its subtleties—and, eventually, to assess the credibility of the writer and evaluate the soundness of his or her ideas. Keep in mind, though, that reading critically does not mean searching for flaws and inadequacies in a text. Rather, it means remaining open to new ideas while challenging and questioning what you read and how you react.

As you read, remember that questioning a writer's ideas is not heresy; in fact, scholars realize that probing other writers' ideas helps to create new interpretations, new perspectives, and new theories. The questioning process contributes to the ongoing intellectual debates in various fields, dialogues that help to keep ideas current and fresh.

Approaching a text with a critical eye is no easy task. Before you even begin to read, you must agree to certain conditions.

APPROACHING A TEXT

- Keep an open mind.
- Withhold judgment.
- Acknowledge your own limitations and biases as a reader.
- Consider the possible reactions and questions of other readers—both to the text and to your ideas.

In addition, you must begin to develop a set of standards by which you will judge the text. For instance, will you be judging it in terms of its usefulness for a particular writing assignment? In terms of its consistency with the theories of a particular academic discipline or a particular school of criticism? On the basis of its logic? Its historical accuracy? Its willingness to break new ground? Any of these criteria is perfectly legitimate; which you select depends on your own critical stance.

Once you begin to read, you must be willing to read the text several times. At first, you will simply *analyze* it, identifying its subject and emphasis; then you will *make connections*, comparing ideas so you can identify relationships and understand what the writer is saying. Eventually you will *evaluate* the text, making judgments about the writer's ideas so you can begin to formulate a critical response.

STEPS IN CRITICAL READING AND WRITING

1. Read actively.
2. Read the text with a critical eye.
3. Record your reactions.
4. Write a critical response.

 5a | **Reading Actively**

Part of the process of becoming a critical reader is reading actively, with pen in hand, physically marking the text to help you distinguish important points from not-so-important ones; identify interesting parallels and significant relationships; and connect

continued from previous page

knowing, we must show students "how to move from abstraction in the nondiscursive mode to discursive abstraction, to generalization." In other words, we must get students to thinking about their thinking and to take charge of the criticism of their assumptions and thoughts.

The author argues against the communications triangle (writer, reader, message), which leaves out purpose, meaning, and intention. In its place she favors C. S. Pierce's triad, which includes "representation or symbol," "object or referent," and "interpretant or reference." The author concludes by explaining how this triad contributes to a reclamation of the imagination and an establishment of a pedagogy of knowing.

TEACHING STRATEGY

Have students bring articles from newspapers or magazines that represent points of view different from their own and discuss the articles with the class. They might need to summarize the content of the articles briefly. This should work especially well with controversial topics such as abortion, gun control, sexual harassment, and the like. Have a list of questions ready for them to apply (to avoid nonproductive argument): What are the possible positions on the subject? What view do you support? What view does the author support? What is his or her rationale? Has he or she considered opposing points of view?

FURTHER READING

Jerry J. Murphy. "Rhetorical History as a Guide to the Salvation of American Reading and Writing: A Plea for Curricular Courage." In *The Rhetorical Tradition and Modern Writing.* Ed. James J. Murphy. New York: The Modern Language Association, 1982. 3–12.

The author begins with a complaint made by Cicero that one set of teachers taught students how to think, while another taught them how to use language. Similarly, today one set of teachers teaches students how to read, and another teaches students how to write. Educators must come to understand this division of labor historically. The author gives a brief account of how this division occurred during the nineteenth century at Harvard and Johns Hopkins Universities.

More important, the author claims, is an understanding of the history of rhetoric, basically developed for oral communication by the ancient Greeks and then later systematized by the Romans. Quintilian's *Institutes of Oratory* (A.D. 95) is "perhaps the most influential single book on education yet published." This book influenced the approach of teaching language by integrating reading, speaking, listening, and writing, and by integrating sight, sound, and sense. The book was influential in the Middle Ages and the Renaissance. The influence continued until the past century.

It is time to reexamine this separation of reading from writing in light of concern about the "failure" of education. "Curricular courage" is required for reintegration to be realized. What is necessary is first a

continued on following page

cause with effect, generalization with specific example. This process of **active reading** will help you understand a text's ideas and prepare you to evaluate its arguments.

(1) Previewing

The first time you approach a text, you should **preview** it—that is, skim it to gain a sense of the author's subject and emphasis. When you preview a book, begin by looking at its table of contents, especially at the sections that pertain to your topic. A quick glance at the index will reveal the kind and amount of coverage the book gives to subjects that may be important to you. As you leaf through the chapters, take note of any pictures, graphs, or tables, reading the captions that appear under them. When previewing magazine articles, look at headings or boxed excerpts that may appear throughout the text. Also scan the introductory and concluding paragraphs for summaries of the author's main points. Journal articles in the sciences and social sciences often begin with summaries called **abstracts**. Read these as part of your previewing process.

See ◄
44b1

In addition, look for the visual cues that writers use to emphasize ideas.

VISUAL CUES

Headings	Color
Capital letters	Italics
Underlining	Lists—with items numbered
Boxes	or set off with bullets (•)
Boldface	

Thesis statements, topic sentences, repeated key terms, transitional words and phrases, and transitional paragraphs can also give you insight into a text's meaning and emphasis, thereby preparing you to react critically to its ideas.

(2) Highlighting

You **highlight** a text to identify the writer's points and their relationships to one another. As you highlight, use symbols and underlining to identify important ideas. (If you are working with library material, photocopy the pages you need and then highlight them.) Be sure to use symbols that you will be able to understand when you reread your text at a later time.

HIGHLIGHTING SYMBOLS
- Underline to indicate information you should read again.
- Box or circle key words or important phrases.
- Put a question mark next to confusing passages, unclear points, or words you have to look up.
- Draw lines or arrows to show connections between ideas.
- Number points that appear in sequence.
- Draw a vertical line in the margin to set off an important section of text.
- Place an asterisk (∗) next to an especially important idea.

The student who highlighted the following passage used various symbols to help her isolate the author's key ideas and to clarify the progression of ideas in the passage.

Public zoos came into existence at the beginning of the period which was to see the disappearance of animals from daily life. The zoo to which people go to meet animals, to observe them, to see them, is, in fact, a monument to the impossibility of such encounters. Modern zoos are an epitaph to a relationship which was as old as man. They are not seen as such because the wrong questions have been addressed to zoos.

When they were founded—the London Zoo in 1828, the Jardin des Plantes in 1793, the Berlin Zoo in 1844—they brought considerable prestige to the national capitals. The prestige was not so different from that which had accrued to the private royal menageries. These menageries, along with gold plate, architecture,

---

continued from previous page

historical understanding of the separation, and second an acknowledgment of "our rhetorical sins." One rhetorical sin is that educators "have allowed writing to become isolated from the writer." Speaking out against recent research the author writes: "No human being operates in the Flower-Hayes model." The second sin is despairing that education is hopeless. The author says we must think in positive terms and specifically adapt Quintilian's approach to education in our times.

FURTHER READING

Janet Emig. "Writing as a Mode of Learning," *College Composition and Communication* **28 (May 1977): 122–28.**
This article is packed with information and conclusions that have spawned numerous other articles by leading theorists/pedagogues in the profession. Emig's thesis is simple: "Writing represents a unique mode of learning." In fact, writing is the approach *par excellence* for learning.

The author first distinguishes writing from other "languaging processes," specifically from listening, talking, and reading. She surveys some of the literature that posits different distinctions. Some linguists distinguish between first-order and second-order processes, with the first (talking and listening) passively learned and the second (reading and writing) actively learned. Other theorists argue that listening is not passive but equally active. The argument for active learning in relation to listening, reading, writing, and talking is that each involves creativity and

continued on following page

continued from previous page

origination. Though all three appear to be equal in that they are all active, writing is, nonetheless, different from and superior to the other three when it comes to learning. (The author briefly mentions that writing is not written speech, as some textbooks claim.)

The author then continues her argument in favor of writing over the other three languaging processes. Moreover, she argues that writing involves the fullest possible functioning of the brain; writing is bispheral, that is, uses both left and right brains. Lastly, writing is connective as process; it creates, as Vygotsky says, "a web of meaning" and at all levels of representation.

As a final argument, the author compares (in a chart) the parallels between successful learning strategies and selected attributes of writing.

orchestras, players, furnishings, dwarfs, acrobats, uniforms, horses, art and food, had been demonstrations of an emperor's or king's power and wealth. Likewise in the 19th century, public zoos were an endorsement of modern colonial power. The capturing of the animals was a symbolic representation of the conquest of all distant and exotic lands. "Explorers" proved their patriotism by sending home a tiger or an elephant. The gift of an exotic animal to the metropolitan zoo became a token in subservient diplomatic relations.

Yet, like every other 19th century public institution, the zoo, however supportive of the ideology of imperialism, had to claim an independent and civic function. The claim was that it was another kind of museum, whose purpose was to further knowledge and public enlightenment. And so the first questions asked of zoos belonged to natural history; it was then thought possible to study the natural life of animals even in such unnatural conditions. A century later, more sophisticated zoologists such as Konrad Lorenz asked behavioristic and ethological questions, the claimed purpose of which was to discover more about the springs of human action through the study of animals under experimental conditions. (John Berger, *About Looking*)

Notice how symbols helped the student understand the passage. For example, she underlined and starred the main idea of the passage and used arrows to show the relationship of one point to another. In addition, she circled and put a question mark next to unfamiliar words, phrases, and names—*ideology of imperialism, ethological, behavioristic,* and *Konrad Lorenz*—that she will have to look up. Finally, she numbered the two reasons why imperial governments established public zoos. Once this student had highlighted the passage, she could go on to record her reactions to its ideas—to ask questions, make connections, and draw tentative conclusions—in the form of marginal annotations.

▶ See 5c

EXERCISE 1

Preview the following passage and then read it more carefully, highlighting it to help you understand the writer's ideas. Then, answer the following questions:

What is the writer's subject?

What is the writer's most important point?

Which points are related?

What is their relationship to one another?

How does the writer make connections among related ideas clear?

My father loved to tell the story of how he got into college. It was 1947 and my father, poor, black and brilliant, was a 15-year-old high school senior in rural Sylvester, Ga. One day he was called into the principal's office to meet a visiting state education official, a white bureaucrat who had learned of my father's academic prowess. The state, the official said, had decided that it wanted to send "a nigra" to college. "You can go to any college in the state," the official told my father. "Except . . . for the University of Georgia, Georgia Tech, Georgia A & M, Emory . . ."

My father was happy and proud to attend Atlanta's Morehouse, perhaps the finest black college in America. But there was always a trace of bitterness when he told this story since his choice of college had, in effect, been made for him.

Times had changed when I applied to college in 1978. Thanks to my father's success as a financial consultant, I grew up in a solidly middle-class home in the Bronx, and, thanks largely to the social advances wrought by the civil rights movement, I was able to attend private school and get into Harvard.

I arrived in Cambridge just as the national backlash against affirmative action was gaining momentum. Many critics were suggesting that African Americans were inherently inferior students, below the standards of the great universities. I found this argument fatuous,

Your students' answers will vary. The writer's subject is affirmative action. The most important point is that students who benefit from affirmative action are stigmatized whereas students who benefit from other forms of preferential treatment are not. The writer makes the relationships between his points clear by applying the term "legacy" to himself and his classmate "Ted." The writer's father, who was denied the chance to enter any predominantly white university, attended Morehouse—"perhaps the finest black college in America." Ted's father attended Harvard. Both sons entered Harvard at a time when "critics were suggesting that African Americans were inherently inferior students" but when wealthy white students, especially the sons of alumni, were "allegedly . . . accepted on 'merit.'" The writer points out that individuals from all backgrounds have succeeded through talent and perseverance, but believes that historical advantages and legislated advantages are not viewed equally in American society.

FURTHER READING

Linda Flower and John R. Hayes. "The Cognition of Discovery: Defining a Rhetorical Problem." *College Composition and Communication* 31 (February 1980): 21–32.
The authors maintain two basic premises: that writers do not find meaning, but make it, and that the notion of discovery is crippled if the teacher or student of writing believes in and acts on the myth of romantic inspiration. The authors define the making of meaning as the defining of a rhetorical problem. Flower and Hayes's research indicates that expert writers know and practice the two premises while novice writers do not.

The authors' method of studying the process of constructing a rhetorical problem is that of prospective protocols, which demand that the research subjects speak their thoughts out loud while attempting to define and solve a rhetorical problem. From the protocols of two groups—which included novice student writers and NEH fellows—Flower and Hayes construct what they call a composite of the cognition of discovery, or the defining of a rhetorical problem. The composite has two sections: the *rhetorical situation*, which includes the audience and the assignment and the *set of goals* that the writer sets.

Flower and Hayes point out that their findings are similar to the communications model of reader, writer, world, and word. They also point out that the concept of rhetorical situation is identical to Lloyd Bitzer's components of exigency, audience, and constraints. (See Bitzer, "The Rhetorical Situation." *Philosophy and Rhetoric* 1 [January 1968]: 1–14.)

continued on following page

particularly when I encountered some of the less illustrious white students who had allegedly been accepted on "merit." There was, for example, the charming, wealthy young man I'll call "Ted." Intellectually incurious, struggling in most of his courses, Ted said he had been rejected by every college to which he had applied, except Harvard, the alma mater of his father and grandfather.

Ted was what is known as a "legacy." According to Harvard's dean of admissions, William Fitzsimmons, approximately 40 percent of alumni children who apply are admitted each year as against 14 percent of nonalumni applicants.

So Ted and I were beneficiaries of two different forms of affirmative action. He was accepted largely because his forebears had attended Harvard. I was accepted largely because my father had been denied the chance to apply to any predominantly white universities. Yet the type of affirmative action that benefited me is relentlessly assailed while the more venerable form of preferential treatment that Ted enjoyed goes virtually unchallenged.

Whether one is listening to Clarence Thomas's tortuous rationalizations about how he didn't really benefit from affirmative action, or George Bush's railing against racial quotas, which have never been widely supported by Americans, white or black, the underlying message is the same: were it not for affirmative action, America would function as a perfect meritocracy.

Of course there are people of all backgrounds who have succeeded solely through talent and perseverance. But at least as many have been assisted by personal connections, old-boy networks, family ties and the benefits traditionally accorded certain, primarily white, primarily male segments of the American population. Why, in the interminable debate over affirmative action, have these historic advantages generally been brushed aside?

I am not suggesting that most "legacies" or other beneficiaries of long-established de facto affirmative action programs are unqualified for the placements and positions they get. Most of the "legacies" I met at Harvard did just fine there, but so did the great majority of African Americans. The difference was that the "legacies" were not stigmatized by their extra edge and the black students were. (Jake Lamar, "Whose Legacy Is It, Anyway?")

5b   Reading Critically

Once you have a sense of what a writer is saying, you can begin to evaluate what you have read. In other words, you can begin to think critically about your text—to distinguish fact from opinion, assess the writer's support, and detect bias. In addition, you can look for faulty reasoning, logical fallacies, and unfair appeals. As you turn a critical eye to your text, you can begin to express your reactions in the form of written annotations. ▶ **See 5c**

Thinking critically about what you read means more than just forming opinions about a writer's ideas. You must also be prepared to allow what you read to call your own ideas into question, and you must be prepared to seriously consider the merits of opposing points of view. As you read, remain open-minded; realize that a text may test your own accepted beliefs and expose your own biases. Be willing to listen to a particularly strong argument even if it goes against your own views.

(1)   Distinguishing fact from opinion

As you read and react critically to a text, you should be evaluating how effectively the writer supports his or her points. This supporting evidence may be in the form of *fact* or *opinion*.

A **fact** is a verifiable statement that something is true or that something happened. Because our individual experience is limited, we have to accept many facts we are not able to verify personally. We accept that the planet Jupiter is 483 million miles from the sun because we trust the reference sources that give us this information. Of course, facts may change as new information is uncovered and new discoveries are made. To most Europeans who lived during the late fifteenth century, it was a *fact* that there was no sea route to India. Thirty years ago it was a *fact* that a person could not be given an artificial heart.

An **opinion** is a conclusion or belief that is not substantiated by proof and is, therefore, debatable.

FACT: Measles is a potentially deadly disease.

OPINION: All children should be vaccinated against measles.

An opinion may be *supported* or *unsupported*.

UNSUPPORTED OPINION: All children should be vaccinated against potentially deadly diseases, even if their parents' religious beliefs oppose this practice.

SUPPORTED OPINION: Several years ago, in Philadelphia, a number of children enrolled in a church-related school whose parents' religious doctrines oppose vaccination died of measles, a preventable disease. The risks are such that all children should be vaccinated against potentially deadly diseases, even if their parents' religious beliefs oppose this practice.

As you read, be very careful to distinguish between fact and opinion as well as between opinion that is supported by evidence

continued from previous page

Flower and Hayes conclude by reporting the differences between the novice writers and NEH writers. The NEH writers on more occasions explicitly represented the aspects of the rhetorical situation and the set of goals, whereas the novice writers less frequently represented their thoughts across the model. Expert writers respond to all aspects of the rhetorical problem, create a rich network of goals for influencing their readers, and finally, represent the problem in more breadth and depth.

and opinion that is unsupported. Remember, too, that supporting evidence can only make a statement more convincing; it cannot turn an opinion into a fact.

EXERCISE 2

Some of the following statements are facts; others are opinions. Identify each fact with the letter *F* and each opinion with the letter *O*. Then consider what kind of evidence, if any, could support each opinion.

1. Snickers bars contain peanuts.
2. Snickers bars taste better than Milky Way bars.
3. Some people are allergic to chocolate and peanuts.
4. Young people watch too much television.
5. Television has changed many people's lives.
6. Lucretia Mott was active in the nineteenth-century abolitionist and women's rights movements.
7. The NBA drafts college basketball players every summer.
8. The spotted owl is on the endangered species list.
9. Preserving the jobs of loggers in the Pacific Northwest is more important than saving the spotted owl.
10. Skiing is a great sport.

(2)   Evaluating supporting evidence

The more reliable the supporting evidence, the more convincing a statement will be—and the more willing readers will be to accept it. Statements may be supported by *examples,* by *statistics,* or by *expert testimony.*

KINDS OF EVIDENCE

Statement Supported By Examples

The American Civil Liberties Union is an organization that has been unfairly characterized as left wing. It is true that it has opposed prayer in the public schools, defended conscientious objectors, and challenged police methods of conducting questioning and searches of suspects. However, it has also backed the antiabortion group Operation Rescue in a police brutality suit and presented a brief in support of a Republican politician accused of violating an ethics law.

continued on the following page

ANSWERS TO EXERCISE 2

1. F	6. F
2. O	7. F
3. F	8. F
4. O	9. O
5. F	10. O

TEACHING STRATEGY

Have students bring short editorials (the type written by the editors of a newspaper will probably make the best examples) and examine support. Many of these rely on one or two instances and draw a conclusion, often advising readers to think one way or the other about a topic. Even a brief examination of the evidence presented should show students how little support these pieces often have. An alternative would be to have them look at letters to the editor and notice that people often draw conclusions with little support.

continued from the previous page

Statement Supported By Statistics

A recent National Institute of Mental Health study concludes that mentally ill people account for more than 30 percent of the homeless population. Because so many homeless people have psychiatric disabilities, the federal government should seriously consider expanding the state mental hospital system.

Statement Supported By Expert Testimony

Clearly no young soldier ever really escapes the emotional consequences of war. As William Manchester, noted historian and World War II combat veteran, observes in his essay "Okinawa: The Bloodiest Battle of All," "the invisible wounds remain" (72).

No matter what kind of supporting evidence writers use, however, it must be *accurate, sufficient, representative,* and *relevant.*

Evidence must be accurate. Evidence is most likely to be accurate if it comes from a trustworthy source. Such a source will, for example, quote *exactly* and not present remarks out of context. It will also present examples, statistics, and expert testimony fairly and without bias, drawing them from similarly reliable sources. For this reason, statistics published in a journal known for its balanced treatment of the issues should carry more weight than those published in a journal known for its support of a particular cause or political position. Similarly, a newspaper poll of voters is likely to be more accurate than one commissioned by a candidate, who may wish to use the results to enhance his or her image.

Evidence must be sufficient. A writer must present an adequate amount of evidence. It is not enough, for instance, for a writer to cite just one example in an attempt to demonstrate that impoverished women are receiving high-quality prenatal care. Even though a single detailed case study might be quite convincing, this one example is not enough to support such a sweeping statement. In addition, the statistical sample must be large enough to be meaningful. An article in the *Journal of the American Medical Association* recently noted that some researchers were publishing studies that included as few as ten participants. In many cases, the article pointed out, the results of such studies are misleading because a change in response by as few as two people can alter the results by 20 percent.

FURTHER READING

John Schilb. "On Personally Constructing 'Social Construction': A Response to Richard Rorty." *Journal of Advanced Composition* 10.1 (1990): 146–49. (The interview with Rorty appeared in *JAC* 9.1-2 [1989]: 1–9.)

Schilb laments the fact that since Rorty's and Bruffee's uses of it, the phrase "social construction of reality" has become almost meaningless or so full of different meanings that it is useless. He cites Knoblauch's similar sentiments.

Schilb says that the interview has "two major surprises." The first of these is that Rorty is "unfamiliar" with Bruffee's "social constructionisms" and that, therefore, Bruffee *can be* "accused of misleading us" when he says he is working from identical or similar principles. We should not assume that Bruffee and Rorty are in line with each other. Schilb points to E. Said's notion (in *The World, The Text, The Critic*) that theories "travel," or, as a case in point, that Rorty's theory has been appropriated by Bruffee, and that Bruffee's appropriation has been appropriated by others in our field.

The second surprise is Rorty's view that the freshman composition class should restrict itself to "basic skills." Schilb says that Rorty refuses to see and reject the very foundationalist habits that he himself theoretically scorns. Rorty's identification with E. D. Hirsch's view of literacy and his rejection of R. Ohmann's is another example of his *de facto* foundationalism.

continued on following page

continued from previous page

Rorty's, as well as Bruffee's, rejection of the importance of theorists and critics of class, race, and gender only deflects and obscures "processes of domination and struggle." Rorty in the interview seems to favor a view of the individual ("I write for myself") over community and thereby undercuts his previous position in favor of the social.

Similarly, a single expert, no matter how stellar his or her reputation or how compelling his or her testimony, may be disputed by other authorities. Look for the testimony of several experts, combined with other supporting evidence, in evaluating a writer's evidence.

Evidence must be representative. To present a balanced picture of an issue, writers must select evidence that is representative of a fair range of facts and viewpoints; they should not permit their biases to influence their choice of examples or expert testimony. See ◀ 5b3 For example, if a writer is trying to convince readers that Asian immigrants as a group have had great success in achieving professional status in the United States, he or she must draw evidence from the experiences of a range of Asian immigrant groups—Vietnamese, Chinese, Japanese, Korean—and a representative sample of professions—law, medicine, teaching, and so on. No matter how accurate the information or how numerous the examples, a writer cannot draw a general conclusion about *all* Asian immigrants to the United States or *all* professions by citing a limited range of examples—Chinese in San Francisco becoming accountants or Koreans in Philadelphia becoming pharmacists, See ◀ 6c2 for example—or by providing statistics or expert testimony that applies to a single group, region, or profession.

Examples must be relevant. If a writer is arguing for U.S. medical aid to developing nations, it is not relevant to present examples or statistics that support—however convincingly—U.S. efforts to reform its own health-care system. Moreover, expert testimony is only as convincing as the expert. Not only should the authority a writer cites be a recognized expert in his or her field of study, but that field of study should be relevant to the point the expert is being called upon to support. Julia Child may be widely recognized as an authority on cooking and Spike Lee as a talented director; Lee's pronouncements on food or Child's on film, however, should carry no more weight than those of any other individual.

CLOSE-UP

EVALUATING SOURCES

See ◀ 39a1 Before you evaluate a source, you should know whether it is a **primary source**—an original document—or a **secondary source**—an interpretation of that original document.

EXERCISE 3

Read the following student paragraph and evaluate its supporting evidence.

> The United States is becoming more and more violent every day. I was talking to my friend Gayle, and she mentioned that a guy her roommate knows was attacked at dusk and had his skull crushed by the barrel of a gun. Later she heard that he was in the hospital with a blood clot in his brain. Two friends of mine were walking home from a party when they were attacked by armed men right outside the A-Plus Mini Market. These two examples make it very clear to me how violent our nation is becoming. My English professor, who is in his fifties, remembers a few similar violent incidents occurring when he was growing up, and he was even mugged in London last year. He believes that if London police carried guns, the city would be safer. Only two of the twenty-five people in our class have been the victims of violent crime, and I feel lucky that I am not one of them.

(3) Detecting bias

As you read a text and form critical judgments about it, you should be alert for signs of the writer's bias. A **bias** is a predisposition to think a certain way. A writer is biased when he or she bases conclusions on preconceived ideas rather than on evidence. Recognizing that a writer may be influenced by certain political beliefs or cultural or class biases can help you evaluate his or her statements for accuracy and fairness and assess the soundness of his or her position.

Some bias that you will encounter in your reading will be obvious and easy to detect.

- **Sexist or racist statements.** A writer who assumes that all doctors are male and that all nurses are female reflects a clear bias. A researcher who states that certain racial groups are intellectually superior to others is also presenting a biased view. In both cases, readers should be aware that bias may have led the researcher to see only what he or she wants to see and to select data that support one outcome over all others.

- **A writer's stated beliefs.** In a recent article about the Middle East, a writer declared herself a strong supporter of the Palestinian position. This statement should tell readers that it is unlikely she will present an unbiased view of Israel's policies in the West Bank. Such a bias does not automatically invalidate the writer's points. On the contrary, she may offer an interesting perspective. What she will *not* do, however, is present a *balanced* view of the subject. For this reason it is essential that you

The first example is based on hearsay, what the writer's friend Gayle heard about "a guy her roommate knows." The violent incidents reported in this paragraph do not prove the violence of an entire nation; the mugging in London is irrelevant. The one statistic the writer presents is not very convincing: only two of the twenty-five people in class have been victims of violence. And the paragraph contains no expert testimony. (The writer's English teacher is not an expert on the subject at hand.)

explore other points of view before drawing your own conclusions.

- **Slanted language.** Slanted language can reveal a writer's bias. Saying "the politician presented an *impassioned* speech," gives one impression. Saying "the politician delivered a *diatribe*" gives another. Similarly, describing a person as "a Wall-Street type" is quite different from saying that he or she is successful in business. In either case, language used to describe something determines the way a reader will perceive it.

- **Tone.** The tone of a piece of writing indicates a writer's attitude toward readers or toward his or her subject. As you read, ask yourself if the writer is being matter-of-fact, ironic, bitter, sarcastic, playful, tentative, angry, apologetic, or self-confident. In many cases, the tone of an essay can alert you to the possibility that the writer is slanting his or her case. An angry writer, for example, might not be able or willing to present an accurate account of an opponent's position, and an apologetic writer might inadvertently dilute the strength of his or her case in an attempt to avoid offending readers.

See ◄
5b2

- **Choice of examples.** As you read, try to evaluate a writer's use of examples. Frequently, the examples selected reveal the writer's biases—that is, a writer may include only examples that support a point and leave out examples that might contradict it.

See ◄
5b2

- **Choice of experts.** In order to support a point effectively, a writer must cite experts who represent a fair range of opinion. If, for instance, a writer discussing the president's economic policies toward Japan includes only statements by economists who advocate protectionism, he or she is presenting a biased case. The absence of statements by economists who advocate free trade should alert you to this problem.

Some bias is so subtle that it is extremely difficult to detect. Cultural biases creep into a text when a writer accepts certain ideas as universal, not realizing that they are limited to the culture in which he or she lives. Assumptions about concepts such as material success, technology, progress, personal freedom, and family values, for instance, are often cultural. A farmer in one South American country was shocked to hear a worker from the World Health Organization suggest that he should limit the size of his family. To him, a large number of male children meant prosperity and assured care in his old age. Similarly, the American emphasis on individual achievement is not universally shared. A Japanese business executive was recently quoted as saying that the problem with American workers is that they are more interested in themselves than in the welfare of the companies for which they work. Other biases can also affect a writer's thinking. Gender or social class, for instance, can determine how a writer sees the world and therefore influence his or her views about the status of women or about the welfare system.

CLOSE-UP

DETECTING BIAS

Remember that your *own* biases can also affect your reaction to a text—how you interpret a writer's ideas, whether you are convinced by what you read and whether you react with sympathy or anger, for example. When you read, then, it is important to remain aware of your own values and beliefs and alert to how they may affect your reactions.

(4)   Recognizing faulty reasoning

As a critical reader, you should carefully scrutinize a writer's reasoning. The relationship between a writer's evidence and conclusions should be clear, and the writer's inferences should be based on a logical chain of reasoning, with no missing links or unwarranted conclusions. Be sure to look for **faulty reasoning** when you read and to avoid it in your own writing.

▶ See 6a,b

(5)   Recognizing logical fallacies

Writers who use **logical fallacies**—flawed arguments—cannot be trusted. A writer who uses these fallacies inadvertently is not thinking clearly or logically; a writer who uses them intentionally is trying to deceive readers. In either case, your identification of logical fallacies should lead you to challenge a writer's credibility.

▶ See 6c

(6)   Recognizing unfair appeals

Another clue that will help you spot a weak or illogical argument is its use of **unfair appeals**—for instance, appeals to the reader's prejudices or fears. These appeals, like faulty reasoning and logical fallacies, should also be avoided in your own writing.

▶ See 7a8

Kenneth A. Bruffee. "Response to the *JAC* Interview with Richard Rorty." *Journal of Advanced Composition* 10.1 (1990): 145–46. (The interview appeared in *JAC* 9. 1-2 [1989]: 1–9.)

Bruffee says that he is in sympathy with Rorty's view of America and finds that Rorty's language has said "things that I have been unsuccessful trying to say myself." He also says— in his introduction—that he is thankful for an opportunity to respond.

Bruffee says, however, that the interviewer (Gary Olson) did both him and Rorty a "disservice in the way [the interviewer] framed . . . questions and summarized the interview." Bruffee says that neither Rorty nor he is a "theorist." He says, "social construction is not a theory"; "it is a way of talking, a language, a vernacular." It is a concept used to describe collaborative learning. That is all; to attempt to make more of it is to misrepresent. He agrees with Rorty that it is advisable to "play back and forth" with an idea to see whether it works. He says that this description sums up his twenty-year experience with collaborative learning. He claims that such "playing" finally taught him that "foundations" were not possible, and this conclusion led him to embrace the antifoundationalist positions of Rorty as well as those of Kuhn and Geertz. Bruffee finally says that the interview and summary were couched in foundationalist terms and that he laments that the field of composition is moving more and more in that direction.

QUESTIONS FOR READING CRITICALLY

- What points is the writer making? What is stated? What is suggested? **(See 5a.)**
- Do you agree with the writer's ideas? **(See 5a.)**
- Are the writer's points supported primarily by fact or by opinion? Does the writer present opinion as fact? **(See 5b1.)**
- Does the writer offer supporting evidence for his or her statements? What kind of evidence is provided? How convincing is it? **(See 5b2.)**
- Is the evidence accurate? Is enough evidence provided? Is the evidence representative? Is the evidence relevant? **(See 5b2.)**
- Does the writer display any bias? If so, is the bias revealed through language, tone, or choice of evidence? **(See 5b3.)**
- Does the writer present a balanced picture of the issue? **(See 5b3.)**
- Are any alternative perspectives omitted? **(See 5b3.)**
- Does the writer omit pertinent examples? **(See 5b3.)**
- Do your reactions reveal biases in your own thinking? **(See 5b3.)**
- Does the writer challenge your own values, beliefs, and assumptions? **(See 5b3.)**
- Does the writer use valid reasoning? **(See 6a,b.)**
- Does the writer use logical fallacies? **(See 6c.)**
- Does the writer use unfair persuasion tactics such as appeals to prejudice or fear? **(See 7a8.)**
- Does the writer oversimplify complex ideas?
- Does the writer make unsupported generalizations? **(See 6c1, 6c2.)**
- Does the writer make reasonable inferences? **(See 6a2.)**
- Does the writer represent the ideas of others accurately? Fairly? **(See 7a8.)**
- Does the writer distort the ideas of others or present them out of context? **(See 7a8.)**

Questions such as the preceding ones can help you move beyond a text's literal meaning to consider its subtleties. Using such questions as a guide, you can begin to analyze, interpret, and evaluate a text and to consider its implications. In short, such questions can help you develop your critical reading skills.

EXERCISE 4

Read the following excerpt from a statement on comparable worth, a method by which some people seek to balance inequities in jobs occupied primarily by women. First, identify the facts and opinions in the excerpt. Then, evaluate the quantity and quality of the writer's supporting evidence and try to determine what biases, if any, she has. Finally, evaluate the writer's reasoning, identifying logical fallacies and unfair appeals. Use the Questions for Reading Critically on page 124 as a guide.

My name is Phyllis Schlafly, president of Eagle Forum, a national profamily organization. I am a lawyer, writer, and homemaker.

We oppose the concept called *comparable worth* for two principal reasons: (*a*) it's unfair to men and (*b*) it's unfair to women.

The comparable worth advocates are trying to freeze the wages of blue-collar men while forcing employers to raise the wages of *some* white- and pink-collar women above marketplace rates. According to the comparable worth rationale, blue-collar men are overpaid and their wages should be frozen until white- and pink-collar women have their wages artificially raised to the same level. The proof that this is really what the comparable worth debate is all about is in both their rhetoric and their statistics.

I've been debating feminists and listening to their arguments for more than a decade. It is impossible to overlook their rhetoric of envy. I've heard feminist leaders say hundreds of times, "It isn't fair that the man with a high school education earns more money than the women who graduated from college or nursing or secretarial school." That complaint means that the feminists believe that truck drivers, electricians, plumbers, mechanics, highway workers, maintenance men, policemen, and firemen earn more money than feminists think they are worth. And how do the feminists judge "worth"? By paper credentials instead of by apprenticeship and hard work and by ignoring physical risk and unpleasant working conditions.

So the feminists have devised the slogan *comparable worth* to make the blue-collar man feel guilty for earning more money than women with paper credentials and to trick him into accepting a government-enforced wage freeze while all available funds are used to raise the wages of *some* women.

Statistical proof that the aim of comparable worth is to reduce the relative earning power of blue-collar men is abundantly available in the job evaluations commissioned and approved by the comparable worth advocates. You can prove this to yourself by making a job-by-job

ANSWERS TO EXERCISE 4

After analyzing this selection, students will find it is mostly unsupported opinion. Although they may agree with Schlafly's position, they should acknowledge the weakness of her case. You may want to point out that Schlafly seems to make little effort to reach those who either disagree with her points or who are skeptical.

1. The facts are: a. "I've been debating feminists and listening to their arguments for more than a decade." (paragraph 3), and b. the discussion of *AFSCME v. State of Washington* (paragraphs 8–9). Most of the discussion of this case, however, is opinion.

   The opinions are: a. The thesis—the assertion that comparable worth is unfair to men and unfair to women (paragraph 2), b. "The comparable worth advocates are trying to freeze the wages of the blue-collar men . . ." (paragraph 3), c. "It's impossible to overlook their rhetoric of envy." (paragraph 4), d. "How do feminists judge 'worth'? By paper credentials . . ." (paragraph 5), e. "So the feminists have devised the slogan *comparable worth* . . ." (paragraph 6), f. "You can prove this to yourself by making a job-by-job examination of *any* study . . ." (paragraph 7), g. "The federal court accepted the Willis evaluation as though it were some kind of divine law . . ." (paragraph 9), h. "The federal court decision (unless it is overturned on appeal) means . . ." (paragraph 9).

continued on following page

continued from previous page

2. Schlafly's support consists mainly of opinion. In paragraph 1, she attempts to establish that she is an authority on her subject: She is president of Eagle Forum, a national profamily organization, a lawyer, a writer, and a homemaker. (You might want to ask students why they think she includes "homemaker" in her list and what they make of Schlafly's Eagle Forum affiliation.) In addition, she says that she has debated feminists for more than a decade. Although her credentials are impressive, Schlafly does not establish that she is an expert on comparative worth and does little to support her claims. Finally, her remarks about feminists indicate that she is both hostile and biased. In fact, her sarcasm and unsupported allegations could turn off many readers.

In addition to her opinion, Schlafly presents one example in paragraphs 8 and 9. Although she says *AFSCME v. State of Washington* is a famous case, she does not establish that it is representative of the cases that the courts have heard about comparative worth. As a result, this single example is not enough to support Schlafly's point that comparative worth is "an elaborate scheme to devalue the blue-collar man."

3. Schlafly seems to have a hostile and possibly prejudiced attitude toward feminists. A number of her statements reveal this bias. In addition, a case could be made that Schlafly's remarks indicate cultural as well as class biases.

examination of *any* study or evaluation made with the approval of comparable worth advocates; it is always an elaborate scheme to devalue the blue-collar man.

For example, look at the Willis evaluation used in the famous case called *AFSCME v. State of Washington.* Willis determined that the electricians and truck drivers were overvalued by the state and that their "worth" was really far less than the "worth" of a registered nurse. More precisely, Willis produced an evaluation chart on which the registered nurse was worth 573 points, whereas the electrician was worth only 193 points (one-third of the nurse), while the truck driver was only worth 97 points (one-sixth of the nurse).

The federal court accepted the Willis evaluation as though it were some kind of divine law (refusing to listen to the Richard Jeanneret "PAQ" evaluation which produced very different estimates of "worth"). The federal court decision (unless it is overturned on appeal) means that the electricians and the truck drivers will probably have their salaries frozen until the state finds a way to pay the registered nurse three times and six times as much, respectively. [In a September 4, 1985, decision, the Ninth U.S. Circuit Court of Appeals overturned the decision.] (Phyllis Schlafly, "Comparable Worth: Unfair to Men and Women")

5c Recording Your Reactions

As you read more critically, you should begin to **annotate** the text—to record your reactions to what you read in the form of notes in the margins or between the lines. This activity can help you focus on your own evaluation of a text's ideas. At first you may write down some relatively uncritical responses—for example, define new words, identify unfamiliar references, or jot down brief summaries. Eventually, however, you will write down more critical responses that may identify points that confirm (or challenge) your own thinking, question the appropriateness or accuracy of the writer's support, uncover the writer's biases, identify slanted language or faulty reasoning, or even question the writer's conclusions. Even if you are not an expert in a particular field, you may still be qualified to challenge a writer's ideas—perhaps because you have different kinds of knowledge, have a different perspective on an issue, or are able to remain emotionally detached.

The following passage illustrates a student's annotations of a section of an article by Joseph Nocera about the decline of American public schools.

One of the most compelling arguments about the Vietnam War is that it lasted as long as it did because of its "classist" nature. The central thesis is that because neither the decision makers in the government *nor anyone they knew* had children fighting and dying in Vietnam, they had no personal incentive to bring the war to a halt. The government's generous college-deferment system, steeped as it was in class distinctions, allowed the white middle class to avoid the tragic consequences of the war. And the people who did the fighting and dying in place of the college-deferred were those whose voices were least heard in Washington: the poor and the disenfranchised.

I bring this up because I believe that the decline of the public schools is rooted in the same cause. Just as with the Vietnam War, as soon as the white middle class no longer had a stake in the public schools, the surest pressure on school systems to provide a decent education instantly disappeared. Once the middle class was gone, no mayor was going to get booted out of office because the schools were bad. No incompetent teacher had to worry about angry parents calling for his or her head "downtown." No third-rate educationalist at the local teachers college had to fear having his or her methods criticized by anyone that mattered.

Is this comparison valid? (seems forced)

bias

127

The analogy to the Vietnam War can be extended even to the extent of the denial. It amuses me sometimes to hear people like myself decry the state of the public schools. We bemoan the lack of money, the decaying facilities, the absurd credentialism, the high foolishness of the school boards. We applaud the burgeoning reform movement. And everything we say is deeply, undeniably true. We can see every problem with the schools clearly except one: the fact that our decision to abandon the schools has helped create all the other problems. One small example: In the early 1980s, Massachusetts passed one of those tax cap measures, called Proposition 2 1/2, which has turned out to be a force for genuine evil in the public schools. Would Proposition 2 1/2 have passed had the middle class still had a stake in the schools? I wonder. I also wonder whether 20 years from now, in the next round of breast-beating memoirs, the exodus of the white middle class from the public schools will finally be seen for what it was. Individually, every parent's rationale made impeccable sense—"I can't deprive my children of a decent education"—but collectively, it was deeply destructive act.

The main reason the white middle class fled, of course, is race, or more precisely, the complicated admixture of race and class and

Who are these people? Does he really represent them?

Is this "one small example" enough to support his claim?

Oversimplification— Do all parents have same motives?

Is this a valid assumption?

good intentions gone awry. The (fundamental) (good intention)—which even today strikes one as both moral and right—was to integrate the public classroom, and in so doing, to equalize the resources available to all school children. In Boston, this was done through enforced busing. In Washington, it was done through a series of judicial edicts that attempted to spread the good teachers and resources throughout the system. In other big city districts, judges weren't involved; school committees, seeing the handwriting on the wall, tried to do it themselves.

Why does he assume intent was "good" & "moral"? Is he correct?

However moral the intent, the result almost (always) was the same. The white middle class left. The historic parental vigilance I mentioned earlier had had a lot to do with creating the two-tiered system—one in which schools attended by the kids of the white middle class had better teachers, better equipment, better everything than those attended by the kids of the poor. This did not happen because the white middle-class parents were racists, necessarily; it happened because they knew how to manipulate the system and were willing to do so on behalf of their kids. Their neighborhood schools became little havens of decent education, and they didn't much care what happened in the other public schools.

Interesting point — but is it true?

In retrospect, this behavior, though perfectly understandable, was (tragically) short-sighted. When the judicial fiats made those safe havens untenable, the white middle class quickly discovered what the poor had always known: There weren't enough good teachers, decent equipment, and so forth to go around. For that matter, there weren't even enough good students to go around; along with everything else, middle-class parents had to start worrying about whether their kids were going to be mugged in school.

slanted language (over emotional)

generalization?

Faced with the (grim) fact that their children's education was quickly deteriorating, middle-class parents essentially had two choices: They could stay and pour the energy that had once gone into improving the neighborhood school into improving the entire school system—a frightening task, to be sure. Or they could leave. (Invariably,) they chose the latter.

Either/or fallacy? Were there other choices?

oversimplification Are there no exceptions?

And it wasn't just the white middle class that fled. The black middle class, and even the black poor who were especially ambitious for their children, were getting out as fast as they could too, though not to the suburbs. They headed mainly for the parochial schools, which subsequently became integration's great success story, even as the public schools became integration's great failure. (Joseph Nocera, "How the Middle Class Has Helped Ruin the Public Schools")

As she read and reread the article, the student referred to the Questions for Reading Critically on page 124 and expressed her reactions in the form of annotations. Among other things, she questioned some of Nocera's assumptions, pointing out examples of generalization, oversimplification, and bias (in the form of slanted language). She also challenged a number of his conclusions and identified at least one logical fallacy. As she made her annotations, this student also thought about her own educational experiences and reconsidered her view of the importance of the public schools.

When she finished her annotations, the student was ready to draw her ideas together and write a critical response to the text. ▶ See 5d

EXERCISE 5

Read the following short article, highlighting it as you read. Then, read the questions that follow the article. Reread the article, recording your reactions in the form of annotations. Finally, answer the questions.

"Go to Wall Street," my classmates said.
"Go to Wall Street," my professor advised.
"Go to Wall Street," my father threatened.

Whenever I tell people about my career indecisiveness, their answer is always the same: Get a blueprint for life and get one fast. Perhaps I'm simply too immature, but I think 20 is far too young to set my life in stone.

Nobody mentioned any award for being the first to have a white picket fence, 2.4 screaming kids and a spanking new Ford station wagon.

What's wrong with uncertainty, with exploring multiple options in multiple fields? What's wrong with writing, "Heck, I don't know" under the "objective" section of my résumé?

Parents, professors, recruiters and even other students seem to think there's a lot wrong with it. And they are all pressuring me to launch a career prematurely.

My sociology professor warns that my generation will be the first in American history not to be more successful than our parents' generation. This depressing thought drives college students to think of success as something that must be achieved at all costs as soon as possible.

My father wants me to emulate his success: Every family wants its children to improve the family fortune. I feel that desire myself, but I realize I don't need to do it by age 25.

Students' answers will vary. The writer's main point is that young people should have the chance to explore different fields before deciding on a lifelong career. To support his point, he cites friends recently graduated who are already miserable because of their jobs; professors who emphasize gaining knowledge as a way to earn income rather than for enjoyment; and a Wall Street interviewer who suggests that having a high GPA means more than learning. His support is drawn from his personal experience with family, friends, teachers, and job recruiters.

The writer uses the expert opinions of professors and job recruiters, whose arguments are not convincing given the nature of the writer's argument; their opinions would better support an argument about the competitive atmosphere of Wall Street. The writer might have used statistics to improve the argument—perhaps about the number of people who burn out quickly after entering a career early, or the percentage of highly successful people in a field who had previously held a variety of positions in other fields.

The examples the writer uses are not representative, since it is doubtful that most college graduates now working are miserable, that most professors would emphasize earning over learning, and that interviewers outside highly competitive Wall Street firms would offer the same advice.

This pressure to do better, to compete with the achievements of our parents in a rapidly changing world, has forced my generation to pursue definitive, lifelong career paths at far too young an age. Many of my friends who have graduated in recent years are already miserably unhappy.

My professors encourage such pre-professionalism. In upper level finance classes, the discussion is extremely career-oriented. "Learn to do this and you'll be paid more" is the theme of many a lecture. Never is there any talk of actually enjoying the exercise.

Nationwide, universities are finally taking steps in the right direction by re-emphasizing the study of liberal arts and a return to the classics. If only job recruiters for Wall Street firms would do the same.

"Get your M.B.A. as soon as possible and you'll have a jump on the competition," said one overly zealous recruiter from Goldman Sachs. Learning for learning's sake was completely forgotten: Goldman Sachs refused to interview anybody without a high grade-point average, regardless of the courses composing that average.

In other interviews, it is expected that you know exactly what you want to do or you won't be hired. "Finance?" they say, "What kind of finance?"

A recruiter at Dean Witter Reynolds said investment banking demands 80 to 100 hours of work per week. I don't see how anyone will ever find time to enjoy the gobs of money they'll be making.

The worst news came from a partner at Salomon Brothers. He told me no one was happy there, and if they said they were, they're lying. He said you come in, make a lot of money and leave as fast as you can.

Two recent Wharton alumni, scarcely two years older than I, spoke at Donaldson, Lufkin & Jenrette's presentation. Their jokes about not having a life outside the office were only partially in jest.

Yet, students can't wait to play this corporate charade. They don ties and jackets and tote briefcases to class.

It is not just business students who are obsessed with their careers. The five other people who live in my house are not undergraduate business majors, but all five plan to attend graduate school next year. How is it possible that, without one iota of real work experience, these people are willing to commit themselves to years of intensive study in one narrow field?

Mom, dad, grandpa, recruiters, professors, fellow students: I implore you to leave me alone.

Now is my chance to explore, to spend time pursuing interests simply because they make me happy and not because they fill my wallet. I don't want to waste my youth toiling at a miserable job. I want to make the right decisions about my future.

Who knows, I may even end up on Wall Street. (Michael Finkel, "Undecided—and Proud of It")

1. What is the writer's main point? Do you agree with him?
2. How does he support this point?

3. Is his supporting evidence primarily fact or opinion? Does he support his opinions? Is the support convincing?
4. Where does the writer use expert testimony? How convincing are the experts he cites?
5. Should the writer have used other kinds of support? For example, should he have used statistics? If so, where? What kind of statistical evidence might have made his case more convincing?
6. Does the writer's choice of examples reveal any biases? What leads you to your conclusion?
7. The writer is a student at an Ivy League university. Do you think this status might give him a limited or unrealistic view of college students' professional options?
8. Do you have any biases against the writer based on your assessment of his economic status, social class, or educational level?
9. The writer is a college senior. Do his age and his lack of experience in the working world make his article less credible to you?

EXERCISE 6

Read, highlight, and annotate the short essay that follows. Then, draw a vertical line down the middle of a piece of paper. In the left-hand column, list the essay's main points. In the right-hand column, write down your reactions to the writer's ideas in the form of complete sentences (statements or questions). You may explain or clarify, question or contradict, probe for further information, or relate the writer's ideas to ideas of your own or to ideas from other sources. Use the Questions for Reading Critically on page 124 as a guide.

It happens in public, not behind a closed office door. There is no "he said/she said" dispute about the facts. Everybody can see what's going on. Friends, classmates, teachers.

A boy backs a girl up against her junior high locker. Day after day. A high school junior in the hallway grabs a boy's butt. A sophomore in the playground grabs a girl's blouse. An eighth-grade girl gets up to speak in class and the boys begin to "moo" at her. A ninth-grader finds out that her name and her "hot number" are posted in the boy's bathroom.

It's all quite normal, or at least it's become the norm. This aberrant behavior is now as much a part of the daily curriculum, the things children learn, as math or social studies. Or their worth in the world.

This is the searing message of another survey that came spilling out of the schoolhouse door last week. This one, commissioned by the American Association of University Women, confirmed the grim fact that four out of five public school students between grades 8 and 11—85 percent of the girls and 76 percent of the boys—have experienced sexual harassment.

That's if sexual harassment means—and it does—"unwanted and unwelcome sexual behavior which interferes with your life." That's if sexual harassment includes—and it does—sexual comments, touching, pinching, grabbing, and worse.

The girls in schools are the more frequent targets of the more serious verbal and physical assaults. They suffer more painful repercussions in their lives, their grades, their sense of well-being.

But the notion that "everybody does it" is not far off the mark. If some 81 percent of the students in the AAUW survey were targets, here's another figure to remember. Some 59 percent—66 percent of the boys and 52 percent of the girls—admitted that they had done unto others what was done to them.

In public spaces in public schools, nearly every student is then a target or a perpetrator or a bystander—or all three in turn. The vast majority have been up close and too personal with sexual harassment. Yet we are still grappling with how it happened and how to change the schoolhouse and hallway.

In Minnesota, the agent of change has been a fistful of lawsuits. In California, a new law was passed that allows expulsions. Elsewhere, schools are looking for a magic bullet, a one-day workshop, a 10-point program.

But cultural change requires more than a crash curriculum; there is no quick fix in the creeping court system. Indeed Mary Rowe of MIT, who has studied harassment for over a decade, has learned that the vast majority of students won't bring their stories to any formal grievance procedure, let alone a courtroom. They won't tattle tale.

For a host of reasons, she and others, like Nan Stein of Wellesley College, have come to believe that the schools need a wider range of choices to fill the space between doing nothing and suing. They need teachers who see and say no to harassment in class. They need designated adults in schools who can listen and help. They need to help students address each other directly and honestly. Indeed in one tactic, a student is encouraged to write a personal letter to the classmate who hurt her . . . maybe unwittingly.

A school culture of sexual harassment exists in a wide and troubling social context, but change ultimately rests in the hands of the students themselves. After all, not all boys will be boys. Not all girls follow the leader.

So, these days, when Nan Stein goes into a school, she says, "I talk a lot about courage." She thinks the role that everybody plays, the bystander, is pivotal. "Kids have to learn to speak out, to make moral judgments. I tell them not to be moral spectators."

Sexual harassment is, as Stein says, an older cousin to bullying. Students who understand the dividing line between teasing and bullying can learn the line between sexual play and harassment. They can draw that line.

The most powerful tool for the everyday garden-variety misery of name-calling, body-pinching and sexual bullying that turns a school hallway into a gauntlet may not be a lawsuit. It may be one high school senior walking by who says, "Don't do that, it's gross." It may be one group of buddies who don't laugh at the joke.

In our society, the courts are the last-ditch place for resolving conflicts. The schools must become the place for teaching basics. Like respect and courage. (Ellen Goodman, "Sexual Harassment in Schools: Nearly Everybody 'Is Doing It'")

5d Writing a Critical Response

After you have read critically and evaluated the ideas in the text, you may want to draw together your reactions by writing a critical response.

In a **critical response** you express the result of your interaction with the text. As in any convincing article or essay, the judgments and conclusions included in your response must be supported with specific examples. In addition, you must supply the logical and sequential links (transitions, topic sentences, and so on) that will help your readers follow the progression of your ideas.

Before you begin writing your critical response, review your annotations. Carefully reconsider your judgments and reactions

WRITING CHECKLIST: WRITING A CRITICAL RESPONSE

Before writing:

- Review your annotations.
- Reconsider your judgments, taking your own biases into account.
- Formulate a statement that summarizes your critical reaction.

In your critical response:

- Identify your text and summarize the writer's position.
- State your position about the writer's stance.
- Summarize and evaluate the writer's key points, using paraphrase and quotation as needed.
- Restate your critical reaction to the text.
- Reread your response, checking for accuracy, clarity, and fair-mindedness.

to the text in light of any biases you may have uncovered in your own thinking. Then, formulate a statement that summarizes your critical reactions to the text.

As you write, begin by identifying your text and summarizing its position. Next, state your own position about the writer's stance and present your support: Summarize and respond to the writer's key points one by one, carefully paraphrasing ideas and quoting key words and phrases where appropriate, supporting your judgments with the ideas you wrote down as you annotated. At every stage, continue to test and reexamine your ideas and to remain as open-minded as possible to the ideas in the text—no matter how they may challenge (or even contradict) your own values or beliefs. Conclude by restating your critical response to the text. When you have finished, reread what you have written, making sure that it is as accurate, clear, and fair-minded as possible.

See ◀
39b,c

Following is a student's critical response to Joseph Nocera's "How the Middle Class Has Helped Ruin the Public Schools," an annotated section of which is reproduced on pages 127–131. In her remarks, the student uses her own knowledge, observations, and experiences to question and challenge Nocera's points.

Identification of text

Summary of writer's position

Statement of position about writer's stance

Summary and evaluation of writer's key points (¶2–4)

In his article "How the Middle Class Has Helped Ruin the Public Schools," first published in the February 1989 issue of *The Washington Monthly* and later reprinted in the September/October 1990 *Utne Reader*, Joseph Nocera tries to have it both ways. He is confessing the guilt he feels for contributing to the decline of public education and attacking those middle-class parents who have made the same choices he has made. What he seems to be asking his audience to do is to feel both sympathy and outrage towards parents in this situation. This is asking a lot.

Early in his essay Nocera tells readers that he moved to a small town because of its good public schools--which he attributes to "a large group of white middle-class parents deeply involved in the public school system" (67). He cites the "outrages" of public schools in Boston, Washington, and New York and concludes that "The destruction of the large public school systems in America is one of the great tragedies of our time" (67). Then, he apologetically explains that he has chosen, for the sake of his children's education, "not to stand and fight"

(68). After all, he argues, "Parents aren't willing to sacrifice their children on the altar of their social principles" (71). Throughout the essay, he seems to assume his readers will agree with him simply because he is an ordinary middle-class parent.

Admitting that, as this typical middle-class parent, he has options poor parents do not have, Nocera tries to justify his private decision despite the widespread public disaster he admits it has helped to create. In doing so, he reveals his biases. At various points in his discussion he attacks the courts, unions, bureaucrats, and school committees. He talks about the nation's problems, but all his examples are drawn from the urban northeast, particularly Boston. He also suggests that any middle-class parents who, unlike himself, keep their children in public schools are sacrificing their children's education for some abstract social principles. This conclusion ignores the fact that some parents may consider the understanding of such social principles to be a valuable part of their children's education. In addition, not all urban public schools provide an inferior education. Finally, he reveals a racial bias when he makes the assumption that white middle-class students (and parents) are the most valuable in a school system and that without these ingredients the system is doomed.

Nocera's reasoning is sometimes faulty. For instance, when he says "Since the white middle class left, the system [in Boston] has simply fallen apart. Can this be sheer coincidence? I think not" (71) he is making a sweeping (and unsupported) generalization. He is also assuming that either white flight or coincidence--and no other factor--must have caused the schools' decline. Similarly, when he says the middle class could either try to improve the whole system or abandon it, he ignores the possibility of other options--such as working to improve one particular school in a system.

Generalizations and oversimplifications like these reduce a complex issue to a simplistic, either-or situation. Given this limited perspective, it is not surprising that Nocera can offer no solution. Predictably, he believes change should come not from

Restatement of critical reaction to the text

137

Students' answers will vary.

people like him but from those outside the system. In his conclusion Nocera reveals that what he is really looking for is not a platform from which to effect change but an opportunity for confession and a plea for forgiveness.

As you write and revise your critical response, you should scrutinize your work as carefully as you would any other text. Remember, you are the writer now, and all the criteria listed in the Questions for Critical Reading on page 124 apply to your own work. Like any other writer, you should take care to distinguish fact from opinion; present evidence that is accurate, sufficient, representative, and relevant; and avoid bias, faulty reasoning, logical fallacies, and unfair appeals.

CLOSE-UP

WRITING A CRITICAL RESPONSE

See ◄ 40a

If you use the ideas or words of a source in your critical response, remember to document them.

EXERCISE 7

The following letter (*Utne Reader,* November/December 1990) was written by a Denver parent in response to Nocera's essay. Putting yourself in Nocera's place, write a brief critical response to this parent's ideas.

Multicultural Education

I believe Joseph Nocera's article is entirely accurate except for one major flaw. That flaw is the belief that sending your children to urban public schools is "sacrificing them." I believe it is in their best interest to do so.

I am a parent of four children in the Denver public school system. I have observed all of the rationalizations described in Nocera's article as dozens of good white liberals in our integrated neighborhood have fled to the suburbs and private schools. For years I tried to appeal to their altruism to support the schools. Then I realized that the reason I was sending my kids to the public schools was not "a sacrifice for my social principles" as Nocera would have us believe, but was actually because I knew it was in their own best interest, to help them learn how to relate and function within a multicultural, multiracial environment. Let me illustrate this point with personal anecdotes.

My son, who is a good fourth-grade student, asked why our friend's son was leaving our public school. I told him it was because "his parents didn't want him to be the only white boy in his class." Then I asked him, "Has that ever happened to you?" He pondered a moment and then said, "I don't think so. I can't remember." Actually, only the year before, he was the only white boy in his third-grade class, but he obviously had overcome any anxiety or prejudice or even awareness of this recent experience.

My daughter, who is a straight-A seventh grader, asked me, "What is an ethnic minority?" I explained that it was a small minority group like blacks. She said with a totally straight face, "Come on, Dad, blacks aren't a minority." In her world they aren't.

Duane Gall
Denver, CO

EXERCISE 8

Read this newspaper column and, after highlighting and annotating it carefully, write a brief critical response to the ideas it discusses. Use the Questions for Critical Reading on page 124 to guide you.

The thought police are busy this year. With all the best intentions in the world—because aren't their intentions always honorable, from preserving the sanctity of the family to promoting racial equality?—they have proferred petitions, organized boycotts, brought lawsuits. Last week they even had the capacity to surprise us. They came out of left field, and they were represented by those people who traditionally have been the quirky, the avant-garde, the antithesis of the thought police type.

That is, actors.

In Central Park this summer, Morgan Freeman, a gifted actor who is black, has played Petruchio in "The Taming of the Shrew," and Denzel Washington, a gifted actor who is black, is portraying Richard III.

This is the way I choose to describe them. I use sentence structure to make their race, in this context, descriptive, not definitive. The definitive reasons they were chosen are clear: craft, and celebrity.

Actors' Equity, the union that represents them and others far less successful, believes that this is insufficient.

Last week Equity torpedoed a musical called "Miss Saigon," which includes parts for dozens of Asian actors. The celebrated actor Jonathan Pryce plays the Engineer in the London production, and was to do so on Broadway. Equity said it could not approve of Mr. Pryce playing the role, which he apparently does brilliantly, because the Engineer is Eurasian and Mr. Pryce is Caucasian.

That is, race as definition, not as description.

Let's be clear: this wish for politically correct casting goes only one way, the way designed to redress the injuries of centuries. When Pat

Carroll, who is a woman, plays Falstaff, who is not, the casting is considered a stroke of brilliance. When Josette Simon, who is black, plays Maggie in "After the Fall" a part Arthur Miller patterned after Marilyn Monroe and which has traditionally been played, not by white women, but by blonde white women, it is hailed as a breakthrough.

But when the pendulum moves the other way, the actors' union balks. It is noted, quite correctly, that it is insufferable that roadshow companies of "The King and I" habitually use Caucasian men wearing eyeliner to play the King, rather than searching for suitable Asian actors. But the conclusion drawn from this is that a white man should never be permitted inside the skin of an Asian one, although all of acting is about getting inside someone else's skin, someone different, someone somehow foreign.

Anyone who has ever faced discrimination knows that bigotry begins when race, when gender, when ethnicity are applied as sole definition. The black man doesn't get into the country club; the savvy banker who is black might. The woman may not be a prime candidate for chief financial officer; the terrific numbers-cruncher who is a woman may get the job.

Actors are accustomed to this, too. How many of them have slunk off to coffee shops after a casting call, defeated and despondent after hearing "We're looking for a tall redhead."

Definitive can be limiting, reductive, making of a diverse group a collection of sameness.

Descriptive tries to force people to see us as individuals, with certain attributes, chief among them the proud one of race, the cherished one of gender, and the ineffable, unmistakable one of talent.

We women are familiar with what has become a tradition in the corporate culture. A woman is given an executive job, say vice president for human resources, and we all cheer. Except that, as the years go by, a woman is always vice president for human resources, but never anything else. Never controller. Never president. Vice president for human resources has become The Woman's Job. There is no need to worry about gender equity anymore.

Perhaps "Miss Saigon," if it ever comes to this country, will become the Asian show. An Asian actor will play the Engineer, and if he is seen as an Asian actor, he may never be offered Othello, or the Stage Manager in "Our Town," or the lead in a new John Guare play.

If he is seen as a brilliant actor who is Asian, it may be a different story.

It is dangerous to form a thought police force. Other people have been at it much longer; they have had months and months of fighting the ghost of Robert Mapplethorpe and trashing the righteous rage of Karen Finley. Who knows? One day one of them might run Actors' Equity.

And then perhaps Denzel Washington will not be playing Richard III. Perhaps there will be an open call for white male actors. Once you open certain doors, they swing both ways. (Anna Quindlen, "Error, Stage Left")

CHAPTER 6

Thinking Logically

Before you can think critically about a text you read or an essay you write, you should know something about the process of reasoning. Such knowledge is important because it enables you to evaluate the arguments you read and shape convincing arguments of your own.

▶ See 5b

▶ See 7b

The two most common ways of reasoning are *induction* and *deduction*. **Inductive reasoning** moves from specific observations or experiences to a general conclusion. **Deductive reasoning** moves from generalizations believed to be true or self-evident to more specific conclusions. Quite often writers use a combination of inductive and deductive reasoning—relying both on conclusions drawn from evidence and on well-established beliefs. Here, however, we consider induction and deduction separately.

6a Reasoning Inductively

(1) Moving from observations to conclusion

Writers use inductive reasoning whenever they draw conclusions based on specific observations. You can see how inductive reasoning operates by studying the following statements.

- At Princeton University 50 percent of the admission decision is based on academic credentials and 50 percent on nonacademic factors.
- Academic factors include grades and rank in high school, SAT scores, achievement-test scores, and recommendations from teachers and counselors.
- Nonacademic factors include demonstration of leadership, extracurricular activities, sports, and a personal interview.

141

TEACHING STRATEGY

Although it's true that we all rely a lot on inductive reasoning, it's also true that sometimes we think we're reasoning from evidence when in fact we're interpreting evidence according to ideas we already accept (because of our upbringing, religious beliefs, hidden prejudices). Having students examine their attitudes toward certain groups (college instructors or athletes, for example) might be a good starting point. Attitudes toward controversial topics (abortion, gun control, freedom of speech, capital punishment, etc.) often come from preconceived "truths."

- Special attention may be given to athletes, minorities, and alumni and faculty children.
- The SAT is an admission requirement for all applicants.
- Fewer than 52 percent of applicants for a recent class with SAT verbal scores between 750 and 800 were accepted.
- Fewer than 39 percent of those with similar SAT math scores were offered admission.
- Approximately 18 percent of those with SAT verbal scores between 550 and 599 and about 19 percent of those with similar SAT math scores were admitted.

This list of statements focuses on the relationship between SAT scores and college admissions. Using inductive reasoning, you can draw a general conclusion from the specific points on the list: Although important, SAT scores are not always the determining factor in the college admission process.

(2)   Making inferences

No matter how much evidence is presented, inductive conclusions are never certain, only probable. They are arrived at by what is called an **inductive leap**. When we make an inductive leap, we are actually making an **inference**, a conclusion about the unknown based on the known. For example, we commonly infer the size of people's incomes from the kind of cars they drive. Similarly, an engineer infers the cause of bridge collapse after finding signs of metal fatigue on a supporting girder, and a physician infers a diagnosis by observing a patient's symptoms.

Naturally, the more observations we make, the narrower the gap between our observations and our conclusion and the better our chance of drawing an accurate conclusion. Even so, absolute certainty is not possible. At some point, writers must decide they have enough evidence to present a convincing conclusion to their readers.

For example, suppose a student in a summer internship program is told to find a way of keeping her state free of the many soda bottles and cans that litter its towns and roadways. Her reading suggests a number of possible actions. The state could hire unemployed teenagers to pick up the litter. It could also place brightly colored refuse containers around the state to encourage people to dispose of bottles and cans properly. Finally, the state could require a deposit from all those who buy beverages in bottles or cans. Reviewing her reading, the student finds that the first two solutions have had no long-term effect on litter in states that tried them. However, mandatory deposit regulations, along with a prohibition of plastic beverage containers, have significantly decreased the number of bottles and cans in the two states that instituted such measures. Still, because the conditions in the student's state are different from those in the states she studied, she must make a leap from the known—the states she studied—to the unknown—the situation in her state. By means of inductive reasoning, the student is able to infer that a mandatory deposit law could be a good solution to the problem.

EXERCISE 1

Read the paragraphs below and answer the questions that follow them.

A. The role—and reputation—of pawn shops seems to be changing. Once considered by many to be somewhat seedy, disreputable establishments with an equally suspect clientele, pawn shops are now becoming more socially acceptable. As many as one out of every ten adult Americans borrows money from a pawn shop each year. In the past five years, the number of pawn-shop licenses has increased by more than a third in some states. Many individuals who cannot get credit from banks or other lending institutions are able to do so at pawn shops. They can borrow modest amounts of cash at no interest simply by presenting identification, offering a piece of collateral, and signing an agreement. Although often criticized for being legalized usurers or receivers of stolen goods, pawn-shop owners insist that they legitimately fill a gap in the loan business. Pawn-shop owners are working diligently to improve their image and have begun to attract middle-class and even upper-middle-class borrowers.

Which of the following statements can be inferred from the paragraph?

1. More and more Americans are borrowing money.
2. Pawn shops are trying to improve their image by refusing to buy stolen goods.
3. Pawn shops are still not socially acceptable to some people.
4. Some states are trying to ban pawn shops because they are legalized usurers and receivers of stolen goods.
5. Pawn shops offer loans at no interest to qualified borrowers for any amount of money.

B. Americans are becoming more ecologically aware with each passing year, but their awareness may be limited. Most people know about the destruction of rain forests in South America, for example, or the vanishing African elephant, but few realize what is going on in their own backyards in the name of progress. Even people who are knowledgeable about such topics as the plight of the wild mustang, the dangers of toxic waste disposal, and acid rain frequently fail to realize either the existence or the importance of "smaller" ecological issues. The wetlands are a good case in point. In recent decades, more than 500,000 acres of wetlands a year

TEACHING STRATEGY

Students might test their "prejudices" by sitting in a shopping mall or other public place such as a restaurant or sporting event, possibly somewhere on campus, observing the people who pass by and writing down the inferences they make about specific shoppers or other groups. Ask students to describe the person and try to determine which details about their clothing, age, general appearance, hairstyle, actions, speech, or gestures they (the students) use to classify these people, what inferences they make based on dress, age, hairstyle, gait, and so on. Although this activity would probably be an out-of-class assignment, students could practice oral skills by presenting their findings to the class. Make sure that the discussion stays focused on the inferences students made as they watched these people. Students might then discuss whether they used inductive reasoning or deductive reasoning to reach their conclusions.

ANSWERS TO EXERCISE 1

Your students' answers will vary.

A. 3.   B. 1, 3, 5.

STUDENTS WORKING TOGETHER

The whole class sits in a circle. Together, they will compose a narrative. Beginning with one sentence (either contributed by instructor or first person in circle), each person contributes a sentence to the story, building logically upon what has been said previously. When the story no longer seems logical, a gong (hit by the instructor or by a student) sounds. Discussion centers on the necessary logic in the relationship between pieces of content. Try the same exercise to create an expository essay.

have been filled, and it seems unlikely that the future will see any great change. What has happened in recent times is that U.S. wetlands are filled in in one area and "restored" in another location, a practice that is legal according to Section 404 of the Clean Water Act, and one that does in fact result in "no net loss" of wetlands. Few see the problem with this. To most, wetlands are mere swamps, and getting rid of swamps is viewed as something positive. In addition, the wetlands typically contain few spectacular species—the sort of glamour animals, such as condors and grizzlies, that easily attract publicity and sympathy. Instead, they contain boring specimens of flora and fauna unlikely to generate great concern among the masses. Yet the delicate balance of the ecosystem *is* upset by the elimination or "rearrangement" of such marshy areas. True, cosmically speaking, it matters little if one organism (or many) is wiped out. But even obscure subspecies might provide some much-needed product or information in the future. We should not forget that penicillin was made from a lowly mold.

Which of the following statements can be inferred from the paragraph?

1. The loss of even a single species may be disastrous to the ecosystem of the wetlands.
2. Even though the wetlands are considered swamps, most people are very concerned about their fate.
3. Section 404 of the Clean Water Act is not sufficient to protect the wetlands.
4. Few Americans are concerned about environmental issues.
5. Most people would agree that the destruction of rain forests is worse than the destruction of the wetlands.

EXERCISE 2

Read this essay carefully, and answer the questions that follow.

As Deerfield Academy, following Lawrenceville, departs from its traditional mission of single-sex education for boys, I, like others involved in girls' schools, am saddened that few voices were apparently raised to defend a mission I cannot help comparing to our own.

Advocates of single-sex education for girls are enthusiastic defenders of the cause. We feel validated every day in our classrooms, dormitories, councils of student government, science and computer laboratories, and yearbook editing offices. We see growth, developing self-esteem, individuality, and leadership all around us.

It is not that good co-educational schools cannot offer girls these things; it is that girls' schools do so consistently. We are a fail-safe producer of first-class citizenship for girls in a world in which they are not guaranteed this opportunity elsewhere. We, like the women's colleges, provide not only "equal opportunity, but every opportunity," to quote Dr. Nannerl Keohane, the president of Wellesley College.

A colleague of mine described a vignette in her all-girls kindergarten class: A small girl surveyed the room, arms akimbo, and sized up the situation. "Thank heavens," she said. "No boys in the block corner." No, there aren't. She won't have to establish her right to build with blocks, just as later on she won't have to elbow her way to the computer terminals or perhaps feel out of place spending extra time in the physics lab.

Her voice will be heard in class, her opinion sought—on every topic—and taken seriously. Whatever the athletic facilities, they are for her alone. Moreover, leadership roles are more available: Girls get experience in managing radio stations, editing student newspapers and literary magazines, heading the debate and mathematics teams—all without having to fight for a place in the sun, because sex stereotyping does not complicate life in the school.

Since failure is less threatening, risk-taking becomes more bearable. For example, like many girls' schools we have a wonderful dance program, but our dancers don't have to care if they don't have figures like the models in *Seventeen* magazine—and few teen-agers do. Like most people, they come in various shapes and sizes, yet they know no one will laugh at them or make disparaging remarks. So they learn to carry themselves with poise and grace, to be proud of their bodies—and stand a chance of becoming good dancers besides.

Relationships can flourish, both among peers and between students and adults. Communication with teachers and other adults is open and warm. Friendships grow strong and last long into adult life, as I have observed time and again by watching the alumnae of women's schools and colleges network and support each other in myriad ways. All this creates a learning and teaching atmosphere that is almost tangible: Our classrooms are lively, exciting places.

A graduate, finishing her sophomore year at a major New England (formerly all-male) college, visited Miss Porter's School last summer. She and I had known each other somewhat, had talked during her years with us but had never discussed the roles and expectations of women as such.

She and her friends were fighting for better health services for women at her college and were frustrated with their lack of progress. They were not being heard, she said. People were not taking them seriously. She talked about the climate of the classrooms, which she found alienating, and the effort she felt she must continually make to claim her equal place. The intensity and warmth of the conversation surprised me. "I wanted to talk to you," she said. "I knew you'd understand." That solidarity and that strength is what a women's single-sex school or college can provide.

We are sorry that Lawrenceville, and now Deerfield, did not feel that they could raise their voices in support of their historic educational environments. This is not a judgment—it is a sentiment—because women's schools, like women's colleges, find their mission constantly validated.

Ours is a co-educational world—no doubt about it—and single-sex education needs its defenders, promoters, believers and proselytizers. Yes, some people think girls' schools are anachronisms, but they suc-

1. Belash makes the following specific observations in her article.

- Single-sex schools produce first-class citizenship for girls. (¶3)
- A female's voice is heard in class. (¶5)
- Risk taking is more bearable. (¶6)
- Relationships can flourish. (¶7)
- Women's single-sex schools can provide solidarity and strength. (¶9)

2. Paragraph 5 is an example of inductive reasoning. It presents specific examples that lead to the conclusion that sex stereotypes do not complicate life in the school.

3. In paragraph 10 Belash makes an inductive leap from the examples above to infer that women's colleges find their mission constantly validated. She makes another inductive leap from this contention to the more general conclusion she states in paragraph 11.

4. Belash concludes that single-sex education is necessary "in a world where men and women still do not work together equally as professionals." Some students will say she has inadequately supported a relationship between receiving a single-sex education and working in a "co-educational world."

FURTHER READING

Frank J. D'Angelo. "The Evolution of the Analytic Topoi: A Speculative Inquiry." *Essays on Classical Rhetoric and Modern Discourse.* Ed. Robert J. Connors, Lisa S. Ede, and Andrea A. Lunsford. Carbondale: Southern Illinois University Press, 1984. 50–68.

This article is of importance to anyone designing a writing program with the design based on a theory of the development of human consciousness. (The article is a further development of a previous article, "An Ontological Basis for a Modern Theory of the Composing Process." *Quarterly Journal of Speech* 64 [February 1978]: 79–85.) The author argues that the development of human consciousness parallels the development of the "analytic *topoi*" as realized in Aristotle, Cicero, and contemporary rhetoricians. The development of these *topoi*, according to the author, moves in three stages.

In the first stage (the Global Stage), the analytic *topoi* are implicitly present in oral poetic discourse and then in narratives, from which they are eventually abstracted. Agreeing with Eric Havelock, the author maintains that "proto-thinkers" used rhythm and oral formulas to arrange experience and only later used narratives. A cataloguing technique—as found in Hesiod—was employed and comprised cumulative commonplaces, such as proverbs and parables, which eventually helped develop epic narratives. Throughout this stage, thought is in an undifferentiated matrix.

continued

ceed, better than most people realize, and remain necessary in a world where men and women still do not work equally together as professionals. (Rachel Phillips Belash, "Why Girls' Schools Remain Necessary")

1. What specific observations does Belash include in her article?
2. Identify two examples of inductive reasoning.
3. At what point does Belash make an inductive leap? What does she infer?
4. What conclusion does Belash reach? Does this conclusion make sense in light of the observations she presents?

6b   Reasoning Deductively

Writers use deductive reasoning whenever they begin with a general or self-evident proposition and establish a chain of reasoning that leads to a necessary and specific conclusion. The process of deduction has traditionally been illustrated with a **syllogism**, a three-part set of statements or propositions that contains a *major premise*, a *minor premise*, and a *conclusion*.

> **MAJOR PREMISE:** All books from that store are new.
>
> **MINOR PREMISE:** These books are from that store.
>
> **CONCLUSION:** Therefore, these books are new.

In a syllogism, the premises contain all the information expressed in the conclusion; that is, no terms are introduced in the conclusion that have not already appeared in the major and minor premises.

The major premise of a syllogism makes a general statement that the writer believes to be true. The minor premise presents a specific instance of the belief stated in the major premise. If the reasoning is sound, the conclusion should follow from the premises. For example, suppose you wish to argue that the government should take steps to protect people who live along polluted streams. You begin with the general statement that the government is obliged to protect its citizens. You then say that people who live along polluted streams are citizens. Your conclusion—that the safety of these residents should therefore be ensured—follows from these assumptions. Stated as a syllogism, your argument looks like this.

> **MAJOR PREMISE:** All citizens should be protected by the government.
>
> **MINOR PREMISE:** People who live along polluted streams are citizens.

CONCLUSION: Therefore, people who live along polluted streams should be protected by the government.

The strength of a deductive argument is that if readers accept the premises, they usually grant the conclusion. Because the major premise is so important, it is a good idea for writers to choose an idea that their readers already accept as self-evident.

(1)   Constructing valid syllogisms

A syllogism is **valid** (or logical) when its conclusion follows from its premises. In other words, *validity* depends on the form of a syllogism. A syllogism is **true** when it makes accurate claims—that is, when the information it contains is consistent with the facts. To be **sound**, a syllogism must be both valid and true. However, a syllogism may be valid without being true or true without being valid. The following syllogism, for example, is valid but not true.

MAJOR PREMISE: All politicians are male.

MINOR PREMISE: Patricia Schroeder is a politician.

CONCLUSION: Therefore, Patricia Schroeder is male.

As odd as it may seem, this syllogism is valid. In the major premise, the phrase *all politicians* establishes that the entire class *politicians* is male. Once Patricia Schroeder is identified as a politician, the conclusion that she is male automatically follows. Common sense tells us, however, that Patricia Schroeder is female. Because the major premise of this syllogism is not true, no conclusion based upon it can be true. For this reason, even though the logic of the syllogism is correct, its conclusion is not.

Just as a syllogism can be valid but not true, it can also be true but not valid. In each of the following situations, the structure of the syllogism undercuts its logic.

Syllogism with an Illogical Middle Term   The **middle term** of a syllogism is the term that appears in both premises but not in the conclusion. In the following invalid syllogism the middle term is *male*.

INVALID SYLLOGISM

MAJOR PREMISE: All fathers are *male*.

MINOR PREMISE: Bill Cosby is *male*.

CONCLUSION: Therefore, Bill Cosby is a father.

Even though the premises of this syllogism are true, the conclusion—*Bill Cosby is a father*—does not follow logically. As it now

continued from previous page

In the second stage (the Analytic Stage), there is increasing complexity as found in the thought of the pre-Socratics. There is a reliance on binary oppositions, which may reflect the lateral specialization of the brain. (The author relies on Julian Jaynes's *The Origin of Consciousness in the Breakdown of the Bicameral Mind*.) What is reflected in early arguments and explanations is a penchant for polarity and analogy. There were, however, according to the author, no clear-cut distinctions between contradictions and contraries, which were to be developed later by Plato and Aristotle. What happens, then, between the first stage and this second stage is that the cumulative *topoi* begin to increase in levels of abstractions. A high level of abstraction is eventually reached when Aristotle in the *Rhetoric* lists twenty-eight topoi with exemplary enthymemes. For example, the first Aristotelian *topos* is "opposites," from which a rhetor can construct the enthymeme "Temperance is good/intemperance is bad"; or "If temperance is good, then intemperance is bad."

The final stage (the Stage of Synthesis) parallels the shift in the history of rhetoric from tradition to theory. During this period—say, beginning in the eighteenth and nineteenth centuries—the analytic *topoi* are eclipsed by an emphasis on style in place of invention. The eclipse is only superficial, for the *topoi* are still to be found in style (schemes and figures) and in paragraph development. That *invention* is recapitulated in *style, memory, arrangement*, and per

continued

continued from previous page

haps *delivery* only illustrates the movement toward integration. The author examines developments in the present century, focusing on the work done by William F. Nelson; W. Ross Winterowd; Kenneth Burke; Young, Becker, and Pike, etc. The author concludes by saying also that the evolution of the analytic topoi parallels the composition process itself.

stands, the syllogism asserts that because Bill Cosby is a male, he must also necessarily be a father. This, of course, makes no sense. In order for the syllogism to be valid, the middle term must be the one preceded by *all* in the major premise.

Consider the following version of the preceding syllogism.

VALID SYLLOGISM

MAJOR PREMISE: All *fathers* are male.

MINOR PREMISE: Bill Cosby is a *father*.

CONCLUSION: Therefore, Bill Cosby is male.

The middle term in this syllogism is *father*, the term that is preceded by *all* in the major premise. This syllogism now says that because Bill Cosby is a father, he must also be male. This assertion seems reasonable even if you are not acquainted with the rules of formal logic. Since the major premise establishes that *all* fathers are male and the minor premise says that Bill Cosby is a father, the conclusion—Bill Cosby is male—logically follows.

See ◄
6c3 ***Syllogism with a Term Whose Meaning Shifts*** A syllogism in which the meaning of a key term shifts cannot have a valid conclusion.

INVALID SYLLOGISM

MAJOR PREMISE: Only man contemplates the future.

MINOR PREMISE: No woman is a man.

CONCLUSION: Therefore, no woman contemplates the future.

In the major premise, *man* is used to denote all human beings. In the minor premise, however, *man* refers to gender. You can avoid this problem by making certain that the meaning of each key term in the major premise remains the same throughout the syllogism.

VALID SYLLOGISM

MAJOR PREMISE: Only human beings contemplate the future.

MINOR PREMISE: No dog is a human being.

CONCLUSION: Therefore, no dog contemplates the future.

See ◄
18f2 (Note that *human beings* is also preferable to *men* because it conforms to nonsexist usage.)

Syllogism with Negative Premises A syllogism in which one of the premises is negative cannot have a valid affirmative conclusion.

Logic 6b

INVALID SYLLOGISM

MAJOR PREMISE: No person may be denied employment because of a physical disability.

MINOR PREMISE: Deaf persons have a physical disability.

CONCLUSION: Therefore, a deaf person may be denied employment because of his or her disability.

In most cases, syllogisms that have this problem are obviously invalid. For example, in the preceding syllogism, the only conclusion possible is, a negative one. ("Therefore, *no* deaf person may be denied employment because of his or her disability.")

A syllogism in which both premises are negative cannot have a valid conclusion.

INVALID SYLLOGISM

MAJOR PREMISE: Injured workers may not be denied workers' compensation.

MINOR PREMISE: Frank is not an injured worker.

CONCLUSION: Therefore, Frank may not be denied workers' compensation.

In the syllogism above, one of the premises must be positive if it is to yield a valid conclusion. As they now stand, the two negative premises do not establish a chain of reasoning. Indeed, they cannot tell us anything positive about the terms in the syllogism. (How, for example, can Frank get workers' compensation if he is *not* an injured worker?)

VALID SYLLOGISM

MAJOR PREMISE: Injured workers may not be denied workers' compensation.

MINOR PREMISE: Frank is an injured worker.

CONCLUSION: Therefore, Frank may not be denied workers' compensation.

(2)   Recognizing enthymemes

As intellectually challenging as syllogisms can be, they do not present a realistic picture of how people actually form arguments. Many examples of deductive reasoning take the form of statements in which assumptions are implied rather than stated. An **enthymeme** is a syllogism in which one of the premises—often the major premise—is unstated.

Melissa is on the Dean's List.

She is a good student.

Enthymemes

In section 6b2 your students will discover the *enthymeme*, a syllogism in which the major premise is not stated but implied—often because it is a widely accepted idea or belief. Just as a correctly structured syllogism may be logically valid but not true, so an enthymeme may be based on a false premise—a premise which, if stated, might make it much harder for the writer to persuade readers. Prejudiced or bigoted arguments are often based on an unstated premise: For example, the argument "Don't pay any attention to Father X's opinion on abortion—he's a Catholic and a man" is based on the unstated premise "No Catholics or men can have a valid opinion on abortion."

Can your students come up with other situations in which the major premise is left unstated in order to deflect objections and controversy?

These statements contain the minor premise and the conclusion of a syllogism. The reader must fill in the missing premise (in this case, the major premise) in order to complete the syllogism.

MAJOR PREMISE: All those on the Dean's List are good students.

MINOR PREMISE: Melissa is on the Dean's List.

CONCLUSION: Therefore, Melissa is a good student.

Enthymemes often occur as sentences which contain words that signal conclusions—*therefore, consequently, for this reason, for, so, since,* or *because.* Notice how *so* in the following sentence indicates the presence of an enthymeme.

Deer are intelligent creatures, *so* they should not be hunted.

After the major premise is supplied, the structure of the syllogism becomes clear.

MAJOR PREMISE: Intelligent creatures should not be hunted.

MINOR PREMISE: Deer are intelligent creatures.

CONCLUSION: Therefore, deer should not be hunted.

Some writers deliberately use unstated premises in an attempt to influence an audience unfairly. By keeping their basic assumptions ambiguous or by pretending that assumptions are so self-evident that they need not be stated, these writers hope to influence an audience unfairly. For example, the mayor of a large city recently said that the city would not provide aid to the homeless because doing so would require an increase in taxes. In this case, the underlying assumption is that a city should not be required to do anything that would cause it to raise taxes. Although some people would agree with the sentiment the mayor expressed, others would disagree.

Whenever you encounter an enthymeme, whether in your own writing or in the writing of others, supply the missing premise to determine whether the enthymeme is sound.

REVIEW: INDUCTIVE AND DEDUCTIVE REASONING

Inductive	Deductive
1. Begins with specific observations.	1. Begins with a general statement or proposition.
2. Moves from the specific to the general.	2. Moves from the general to the specific.
3. Conclusion is probable, never certain.	3. Conclusion can be sound or unsound.

continued on the following page

continued from the previous page

4. Progresses by means of inference.
5. Draws a conclusion about the unknown based on what is known.

4. Progresses by means of the syllogism.
5. Draws a necessary conclusion about the known based on what is known.

CLOSE-UP

USING TOULMIN LOGIC

The philosopher Stephen Toulmin has introduced another method for structuring arguments. **Toulmin logic** divides arguments into three parts: *the claim, the grounds,* and *the warrant.*

The claim is the main point the writer makes in the essay. Usually the writer states the claim, but in some arguments it may be implied.

The grounds are the evidence and reasons on which the claim is based. They are the support the writer uses to bolster the claim.

The warrant is the assumption that links the claim to the data. It shows how the evidence supports the claim.

In the simplest terms, an argument following Toulmin's pattern would look like this.

Jane graduated from medical school. ——— Jane is a doctor.
(Grounds) | (Claim)
A person who graduates from medical
school is a doctor.
(Warrant)

Notice that the claim presents a specific situation, whereas the warrant is a general principle that can apply to a number of situations. In this sense the warrant is similar to the major premise of a syllogism and the claim to the conclusion. (The grounds are the premises from which the claim is derived or the premises that make the claim probable or possible.)

Although Toulmin logic is an alternative method of constructing arguments, it still relies on inductive and deductive reasoning. You arrive at your claim by moving inductively from your reading, and the relationship of your grounds and warrant to your claim is deductive.

Inductive and Deductive Reasoning

Are your students having trouble differentiating between inductive and deductive arguments? A class discussion or debate may help these concepts take deeper root. One option: Fictional detective Sherlock Holmes was fond of saying he had *deduced,* for example, that a man came from East Sussex from the color and texture of the dirt on his boots. But was this the result of *deductive* reasoning, or—as some writers have noted—actually a form of *inductive* argument? A case can be made for the following deductive line of thought:

1. Dirt like this comes from East Sussex.
2. This man's boots carry dirt of this type.
3. Therefore, this man has recently come from East Sussex.

But some critics of Sir Arthur Conan Doyle's skill as a literary logician argue that Holmes's ability to state the first premise above ("Dirt like this . . .") was the result of a lifetime of the kind of *inductive* research that is central to scientific inquiry. In other words, they argue, Holmes was able to "deduce" only because he had, at some earlier date, made an extensive "inductive" study of English soil types—first making observations and collecting evidence, and then drawing conclusions based on that evidence.

You may want to present both sides of the case to the class and let them argue it out. Try to guide the discussion so that the differences between induction and deduction become increasingly clear.

ANSWERS TO EXERCISE 3

A. 1. The article's unstated premises include the following.

- That the audience accepts Lamm's terminology and definitions

- That they see a common language as a unifier

- That they agree in principle that citizenship and English as a national language will lessen alienation and fragmentation

- That they believe too many non-English-speaking immigrants are a danger to the cohesion of the country

- That they see naturalization, citizenship, and "being American" as synonymous

- That they think immigrants want to be U.S. citizens

2. Lamm supports his position with deductive reasoning (¶1); inductive reasoning (¶s 7, 8, and 9); statistics (¶4); expert opinion (¶s 3 and 6); examples (¶5).

3. Students will find that Lamm begins his conclusion as early as paragraph 8. Lamm states that if the U.S. does not move toward greater integration, it will move toward fragmentation and alienation. In paragraphs 11 and 12 he states language should be one of the unifiers of our nation's people and "we should be color blind but linguistically cohesive."

4. Lamm's argument can be stated as the following syllogism:

continued

EXERCISE 3

Read this essay carefully.

A nation succeeds only if the vast majority of its citizens succeed. It therefore stands to reason that with immigrants accounting for about 40 percent of our population growth, the future economic and social success of the United States is bound up with the success of these new Americans. Demography, in a word, is destiny.

This is an important principle to keep in mind as we try to come to grips with the problems and opportunities presented by the flood of legal and illegal immigrants from Mexico and other parts of South and Central America, who now constitute by far our largest immigrant group.

How are we doing in our efforts to assimilate these largely Hispanic newcomers and provide them with a bright future? Some signs are disturbing.

John Garcia, associate professor of political science at the University of Arizona, writing in International Migration Review, finds that the average rate of naturalization of Mexican immigrants is one-tenth that of other immigrant naturalization rates. The Select Commission on Immigration and Refugee Policy made a similar finding. Increasingly, immigrants are separated from everyone else by language, geography, ethnicity and class.

The future success of this country is closely linked to the ability of our immigrants to succeed. Yet 50 percent of our children of Hispanic background do not graduate from high school. Hispanic students score 100 points under the average student on Scholastic Aptitude Test scores. Hispanics have much higher rates of poverty, illiteracy and need for welfare than the national average. This engenders social crisis.

Not all the indicators of assimilation are pessimistic: the success of many Indochinese immigrants has been gratifying. But the warning signs of nonassimilation are increasing and ominous.

America must make sure the melting pot continues to melt: immigrants must become Americans. Seymour Martin Lipset, professor of political science and sociology at the Hoover Institution, Stanford University, observes: "The history of bilingual and bicultural societies that do not assimilate are histories of turmoil, tension and tragedy. Canada, Belgium, Malaysia, Lebanon—all face crises of national existence in which minorities press for autonomy, if not independence. Pakistan and Cyprus have divided. Nigeria suppressed an ethnic rebellion. France faces difficulties with its Basques, Bretons and Corsicans."

The United States is at a crossroads. If it does not consciously move toward greater integration, it will inevitably drift toward more fragmentation. It will either have to do better in assimilating all of the other peoples in its boundaries or it will witness increasing alienation and fragmentation. Cultural divisiveness is not a bedrock upon which a nation can be built. It is inherently unstable.

The nation faces a staggering social agenda. We have not adequately integrated blacks into our economy and society. Our education system is rightly described as "a rising tide of mediocrity." We have the most violent society in the industrial world; we have startlingly high rates of illiteracy, illegitimacy and welfare recipients.

It bespeaks a hubris to madly rush, with these unfinished social agendas, into accepting more immigrants and refugees than all of the rest of the world and then to still hope to keep a common agenda.

America can accept additional immigrants, but we must be sure that they become American. We can be a Joseph's coat of many nations, but we must be unified. One of the common glues that hold us together is language—the English language.

We should be color-blind but linguistically cohesive. We should be a rainbow but not a cacophony. We should welcome different peoples but not adopt different languages. We can teach English through bilingual education, but we should take great care not to become a bilingual society. (Richard D. Lamm, "English Comes First")

A. Answer the following questions about the essay.

1. In developing the preceding essay, Richard D. Lamm relies on a number of unstated premises about his subject that he expects his audience to share. What are some of these premises?
2. What kinds of information does Lamm use to support his position?
3. Where does Lamm state his conclusion? Restate the conclusion in your own words.
4. In paragraph 1 Lamm uses deductive reasoning. Express this reasoning as a syllogism.

B. Evaluate the reasoning in the following statements. (If the statement is in the form of an enthymeme, supply the missing term before evaluating it.)

1. All immigrants should speak English. If they do not, they are not real Americans.
2. Richard D. Lamm was born in the United States and grew up in an English-speaking household. Therefore, he has no credibility on the subject of bilingualism.
3. Spanish-speaking immigrants should be required by law to learn English. After all, most eastern European immigrants who came to this country early in the twentieth century learned English.
4. If immigrants do not care enough about our country to learn English, we should not allow them to become citizens.
5. Some immigrants have become financially successful even though they did not learn English. Obviously, then, learning English does not increase an immigrant's chances for success.
6. All Cuban immigrants speak Spanish. Secretary of Housing and Urban Development Henry Cisneros speaks Spanish, so he must be a Cuban immigrant.
7. As sociologist Seymour Martin Lipset points out, bilingual societies can be threatened by tension and political unrest. Therefore, it is important that immigrants not be bilingual.

continued from previous page

MAJOR PREMISE: A nation succeeds only if the vast majority of its citizens succeed.

MINOR PREMISE: America is a nation in which immigrants account for about 40 percent of its population growth.

CONCLUSION: Therefore, America's success depends on the success of its new citizens.

B. All of the arguments are unsound.
1. Missing premise, that only *real Americans* speak English, is faulty. The phrase is also ambiguous.
2. Missing premise, that one must be bilingual to understand the ramifications of bilingualism, is faulty. Argument concludes that native speakers cannot have authoritative knowledge of the subject.
3. Argument is faulty. Conclusion assumes that migrant groups are alike and, as such, that they should be treated alike.
4. Argument is faulty. An untrue assumption is made that learning the language is the only way in which immigrants can show that they care about the country.
5. Argument is a hasty generalization.
6. Argument is faulty because it contains an illogical middle term. Therefore, it erroneously concludes that only Cuban immigrants speak Spanish.
7. Argument is faulty. There is no necessary causal connection between the two statements.

TEACHING STRATEGY

Critical Thinking

Lively discussion usually results when students are asked to consider the ethical and practical implications of using logical fallacies. Some pragmatists will point out that using these fallacies obviously works (and they can cite innumerable advertisements and political speeches to support their point). If the fallacies help writers accomplish their purpose, why shouldn't such strategies be used? More idealistic students will usually respond that such manipulation is unethical and that the effectiveness of shifted issues does not justify its use. The question becomes more complicated, however, if someone introduces the idea of a "worthy" cause being supported by appeals to the emotions. Most students will readily condemn advertisers or unnamed politicians for playing on their audience's fear, hopes, and prejudices, but they are not so certain when they consider, for example, a flyer from an environmental group that gets readers' attention by emphasizing a horrible future possibility.

Students might be asked whether shifting the issue to an audience's prejudices is ever justified. Suggest that each student bring to class at least two examples of advertisements, editorials, or summaries of television programs in which writers or speakers use some of the tactics described in section 6c. One of the examples should support a point of view the student favors; the other should support a point of view with which the student disagrees. (Advertising by anti-gun organizations, by the NRA,

continued

6c   Recognizing Logical Fallacies

Fallacies are arguments so weak that the premises do not support the conclusion. On the surface the premises may seem relevant to the conclusion, but they are not. Because fallacies are so common, many of us do not recognize them when we encounter them in our reading or in our writing. This failure to recognize fallacies can make us gullible readers and unconvincing writers. Unscrupulous writers often use logical fallacies to manipulate readers, appealing to prejudices and fears, for example, instead of to reason. Even when you write with the best of intentions and the greatest care, you may accidentally include fallacies. When readers detect these fallacies, they may conclude you are illogical—or even dishonest. It makes sense, then, to learn to recognize fallacies—to challenge them when you read and to avoid them when you write.

(1)   Hasty generalization

A **hasty generalization** occurs when a conclusion is based on too little evidence. For example, a single bad experience with an elected official is not enough to warrant the statement that you will never vote again, and one stimulating class taught by a certain instructor does not mean that all classes she teaches will be interesting. The number of examples you need depends on the statement you are supporting. A few examples could be enough to make the point that you and your former boyfriend or girlfriend were incompatible. Many more would be needed to support the statement that writers of color have a difficult time getting their work published by mainstream American publishers.

See ◄
5b2

(2)   Sweeping generalization

Sometimes confused with the hasty generalization, the **sweeping generalization** occurs when a statement cannot be adequately supported no matter how much evidence is supplied. **Absolute statements**, for example, are so sweeping they allow for no exceptions.

> Everyone should exercise.

Certainly, most people would agree that regular exercise promotes good health. This does not mean, however, that *all* people should exercise. What if, for example, a person has a heart condition? To avoid making statements that cannot be supported, you should get used to qualifying your statements. For example, instead of saying, "Everyone should exercise," you might say, "Most people benefit from regular exercise." The easiest way to identify absolute statements is to examine each use of words such as *always, all, never,* and *everyone.* In many cases, a word such as *often, seldom, some,* or *most* will be more accurate.

Stereotypes are sweeping generalizations about the members of a race, religion, gender, nationality, or other group. Although such generalizations may apply to some individuals in a group, they do not pertain to all members.

▶ **See 18f**

(3)   Equivocation

Equivocation occurs when the meaning of a key word or phrase shifts so that the conclusion seems to follow logically from the premises—when, in fact, it does not.

Equivocation can be subtle.

> It is in the public interest for the government to provide for the welfare of those who cannot help themselves. The public's interest becomes aroused, however, when it hears of welfare recipients getting thousands of dollars by cheating or by fraud.

In the first sentence, *public interest* refers to social good and *welfare* to well-being. In the second sentence, *the public's interest* refers to self-interest and *welfare* to financial assistance provided by the government.

(4)   The either/or fallacy

The **either/or fallacy** occurs when a complex situation is presented as if it has only two sides when actually it has more. If you ask whether the policies of the United States toward Central America are beneficial or harmful, you acknowledge only two possibilities, ruling out all others. In fact, the policies of the United States toward some Central American countries may be beneficial, but toward others they may be harmful. Or in any given country they may be *both* beneficial and harmful. Avoid the either/or fallacy by acknowledging the complexity of an issue. Do not misrepresent issues by limiting them.

Of course, *some* either/or situations lead to valid conclusions. In biology lab, a test either will or will not indicate the presence of a certain enzyme. An either/or statement is valid as long as it encompasses *every* possible alternative. Thus, the statement "Either Kim took the test or she did not" is valid because there are no other possibilities. But the statement "Either Kim took the test or she went to the Student Health Center" is an example of the

continued from previous page

by anti-abortion and by pro-choice groups provide useful samples.) After identifying the tactics, the class may discuss either as a whole or in small groups their responses to what they have seen and read. How convincing do they find the approaches? Do they feel any of the tactics are justified?

STUDENTS WORKING TOGETHER

Put students in small groups and give each group a different essay containing an argument. Using the "Guide to Logical Fallacies" on page 160, each group inserts as many fallacies as possible to ruin the argument. Then each group exchanges its reconstructed fallacy-filled argument with another in order to identify and remove the fallacies used by the other group. Then both groups meet to compare their findings.

QUOTATION

"You don't tell deliberate lies; but sometimes you have to be evasive."

–Margaret Thatcher

either/or fallacy. To disprove the statement, all someone has to do is point out that Kim went somewhere else.

(5) Post hoc, ergo propter hoc

Post hoc, ergo propter hoc is Latin for "after this, therefore because of this." The **post hoc** fallacy occurs when you mistakenly infer that because one event follows another in time, the first event *caused* the second.

Many arguments depend on establishing cause-and-effect relationships, but the link between the causes and effects presented must actually exist. In some arguments this is not the case. For example, after the United States sold wheat to Russia, the price of wheat and wheat products rose dramatically. Many people blamed the wheat sale for this rapid increase. One event followed the other closely in time, so people falsely assumed that the first event caused the second. In fact, a complicated series of farm price controls that had been in effect for years caused the increases in wheat prices.

Make certain that you identify the actual causes and effects of the events you discuss. Cause-and-effect relationships are difficult to prove, so you may have to rely on expert testimony to support your claim.

(6) Begging the question

A writer **begs the question** when the premise he or she intends to prove is restated as if it were a conclusion that has already been proved or disproved. Often this fallacy occurs when a person incorrectly assumes that a proposition is so obvious it needs no proof. Consider this exchange.

"I believe in Darwin's theory of evolution."
"Why?"
"Because Darwin says it is so."
"But why should we believe Darwin?"
"Because he formulated the theory of evolution."

The initial statement assumes the truth of Darwin's theory of evolution, which is exactly what is at issue. In other words, the argument goes in a circle.

Begging the question is not always as obvious as it is in the preceding example. In some cases it can be subtle.

Sadistic experiments on animals should be stopped because they clearly constitute cruel and unusual punishment.

Logic 6c

Certainly sadistic experimentation is cruel. What has to be proven, however, is that the experiments actually are sadistic. By simply saying the same thing twice, the writer sidesteps this issue entirely.

(7) False analogy

Analogies—extended comparisons—enable a writer to explain something unfamiliar by comparing it to something familiar. ▶ See 4e6 Skillfully used, an analogy can be quite convincing, as when a student illustrates his frustration with the registration process at his college by comparing students to rats in a maze. By itself, however, an analogy establishes nothing; it is no substitute for supporting evidence. The student's analogy between rats and people is certainly effective: Both are rewarded if they succeed and punished if they do not. But people are not rats, and the student would still have to provide support if his intent was to show the shortcomings of the registration process.

A **false analogy** (or faulty analogy) assumes that because issues or concepts are similar in some ways, they are similar in other ways. On a television talk show recently, a psychiatrist who was asked to explain why people commit crimes gave the following response.

> People commit crimes because they are weak and selfish. They are like pregnant women who know they shouldn't smoke but do anyway. They have a craving that they have to give in to. The answer is not to punish criminals, but to understand their behavior and to try to change it.

Admittedly, the analogy between criminals and pregnant women who smoke is convincing. However, it oversimplifies the issue. A pregnant woman does not intend to harm her unborn child by smoking; many criminals do intend to harm their victims. To undercut the doctor's argument, you need only point out the shortcomings of his analogy.

(8) Red herring

The **red herring** fallacy occurs when a writer changes the subject to distract the audience from the actual issue. Consider, for example, the statement "This company may charge high prices, but it gives a great deal of money to charity each year." The latter observation has nothing to do with the former; still, it somehow manages to obscure it.

QUOTATION

"The great enemy of truth is very often not the lie—deliberate, contrived, and dishonest—but the myth—persistent, persuasive, and unrealistic."

–John F. Kennedy

Many people use a red herring when they are backed into a corner. By switching the subject, they hope to change direction and begin on safer ground. Here is an example from a student essay.

> The appeals court should overturn the lower court's decision to ban prayer at high school graduations. The school board members who support this decision should have more pressing things on their minds. Maybe they should spend more time wondering how they will finance public education in this city next year.

This student avoids discussing the point she sets out to support. Instead of explaining why prayer should be allowed at high school graduations, she introduces an irrelevant point about financing public education.

(9) Argument to ignorance (*argumentum ad ignorantiam*)

The **argument to ignorance** fallacy occurs when a writer says that something is true because it cannot be proved false or that something is false because it cannot be proved true. This fallacy occurred recently during a debate about a policy allowing children who have AIDS to attend public school. A parent asked an AIDS researcher, "How can you tell me to send my child to school where there are children with AIDS? After all, you doctors can't say for sure that my child won't catch AIDS from these children." In other words, the parent was saying, "My children are likely to contract AIDS from other children in school because it has never been proven that they cannot." As persuasive as this line of reasoning can sometimes be, it is logically flawed. The fact remains that no evidence has been presented to support the speaker's conclusion.

(10) Bandwagon

The **bandwagon** fallacy occurs when a writer tries to establish that something is true or worthwhile because everyone believes it is. For example, a recent newspaper editorial makes the point that a state should raise the speed limit on its highways to sixty-five miles an hour because nearly everyone exceeds the present fifty-five mile an hour limit. Instead of focusing on the drawbacks of the present speed limit, the editorial relies on an appeal to numbers. Certainly, the fact that many people ignore the speed limit is important, but this fact does not in itself establish that the speed limit should be raised.

(11) Skewed sample

The **skewed sample** is a problem that can occur during the collection of statistical evidence. To present accurate results, a statistical sample should be *representative*; that is, it should be typical of the broader population it represents. When a statistical sample is collected so that it favors one segment of the population over others, it is said to be *skewed*. For example, a study of the spending habits of Americans would most likely be skewed in favor of relatively affluent individuals if it were based on respondents chosen from lists of luxury-car owners. Similarly, census questions asked only in English would disproportionately skew results in favor of English-speaking respondents.

(12) You also (*tu quoque*)

The **you also** fallacy occurs when a writer says a point has no merit because the person making it does not follow his or her own advice. Such an argument is irrelevant because it focuses attention on the person rather than on the issue being debated.

> If you think exercise is beneficial, why don't you exercise?

> You're telling me to save? Look at how much money you spent last year.

(13) Argument to the person (*ad hominem*)

Arguments ***ad hominem*** attack a person rather than an issue. By casting aspersions on an opponent, they turn attention away from the facts of the case.

> That woman has criticized the president's commitment to women's rights. But she believes in parapsychology. She thinks that she can communicate with the dead.

> Congressman Rodriguez supports increases in the defense budget. What do you expect from a man who worked for a defense contractor before he ran for public office?

Although this tactic may persuade some people, others will recognize the fallacy and question the writer's credibility.

(14) Argument to the people (*ad populum*)

Arguments ***ad populum*** appeal to people's prejudices. A senatorial candidate seeking support in a state whose textile industry has been hurt by foreign competition may allude to "foreigners who are attempting to overrun our shores." By exploiting prejudices of the audience, the candidate is able to avoid the concrete issues of the campaign. Of course, many political speeches contain emotional appeals—to patriotism, for instance—and this is appropriate when used to enlist support, but should not be accepted as a substitute for specific evidence.

GUIDE TO LOGICAL FALLACIES

- **Hasty Generalization**   Drawing a conclusion on the basis of too little evidence

- **Sweeping Generalization**   Making a generalization that cannot be supported no matter how much evidence is supplied

- **Equivocation**   Shifting the meaning of a key word during an argument

- **Either/Or Fallacy**   Treating a complex issue as if it has only two sides

- *Post Hoc*   Establishing an unjustified link between cause and effect

- **Begging the Question**   Stating a debatable premise as if it were true

- **False Analogy**   Assuming that because things are similar in some ways they are similar in other ways

- **Red Herring**   Changing the subject to distract an audience from the issue

- **Argument to Ignorance**   Saying that something is true because it cannot be proved false, or vice versa

- **Bandwagon**   Trying to establish that something is true because everyone believes it is true

- **Skewed Sample**   Collecting a statistical sample so that it favors one population over another

- **You Also**   Accusing a person of not upholding the position that he or she advocates

- **Argument to the Person**   Attacking the person and not the issue

- **Argument to the People**   Appealing to the prejudices of the people

---

EXERCISE 4

Identify the fallacies in the following statements. In each case, name the fallacy and rewrite the statement to correct the problem.

1. Robert Redford is an excellent actor. He should concentrate on making movies instead of criticizing the nation's environmental policy.
2. My opponent says that he wants to be mayor. He has been divorced twice. Obviously, he should get his own life in order before he thinks of running for public office.
3. How can we not support railroads? Railroads are the arteries of our nation, and the trains are the lifeblood that bring sustenance to all parts of the country.
4. The school's mail-in registration program will either make things easier for students or result in total chaos.
5. During the last flight of the space shuttle, there was heavy rainfall throughout the entire Northeast. Therefore, the launch must have disturbed the weather patterns for that region of the country.
6. I just received a pamphlet that urges people to buy savings bonds. How can the government talk about saving? Look at the size of the national debt.
7. What the mayor did was wrong, but many people in public office have done a lot worse.
8. No truly intelligent person would deny that the government's welfare programs are shamefully mismanaged.
9. The public should be responsible for the indigent and the elderly. Therefore, public utilities must make certain that the indigent and the elderly have heat this winter.
10. No responsible scientist has been able to establish that second-hand smoke will definitely cause lung disease in a particular individual. Therefore, the surgeon general should withdraw her warnings.

*EXERCISE 5

Read the following excerpt. Identify as many logical fallacies as you can. Then, write a letter to the author, pointing out the fallacies and explaining how they weaken his argument.

Hunting and eating a free-roaming wild deer is one thing; slaughtering and eating a [wounded] deer is another.

The point . . . is that—despite what our enemies are saying—hunters are just as compassionate as the next fellow. It hurts us to see an animal suffer, and when we can help an animal in need, we go out of our way to do whatever we can.

A case in point is the story . . . about SCI Alaska vice president Dave Campbell's efforts to help a cow moose. That animal had carried a poorly shot arrow in its body for weeks until Campbell saw it and made certain it got help.

1. red herring and argument to the person. Robert Redford is an excellent actor. Although he has the right to criticize the nation's environmental policy, his credentials do not qualify him as an expert in that area.
2. argument to the person. My opponent says that he wants to be mayor. However, he has no managerial experience.
3. false analogy. Railroads should be supported. They provide a way to transport goods and supplies across the country.
4. either/or fallacy. The school's mail-in registration program will probably make registering easier for students, though some confusion may occur initially.
5. *post hoc, ergo propter hoc.* During the last flight of the space shuttle, there was heavy rainfall throughout the entire Northeast. However, no causal relationship has been established between the launch and weather patterns.
6. *tu quoque.* I just received a pamphlet that urges people to buy savings bonds. The government urges citizens to buy bonds because purchasing them strengthens the U.S. economy.
7. Bandwagon. What the mayor did was wrong. Even though others have done worse, he should be held accountable for his actions.
8. begging the question. Many intelligent people believe that the government's welfare programs are mismanaged.

continued

continued from previous page

9. equivocation. Public utilities should be responsible for the indigent and the elderly. Therefore, they should make certain that the indigent and the elderly have heat this winter.

10. argument to ignorance. Scientists have demonstrated that smoking may cause lung disease. Thus, smokers should not ignore their warnings.

ANSWERS TO EXERCISE 5

The letters the students will write will vary. Even so, they should mention some of the following fallacies.

- Straw man: "Despite what our enemies are saying. . . ." (¶2)

- Straw man: "Despite how some media have handled the story. . . ." (¶4)

- Argument to ignorance: "We do it all the time. . . ." (¶5)

- Hasty generalization: The writer bases his argument on just two sources.

- Argument to the Person: "The vegetarian/animal rights herd is a wacko bunch. . . ." (¶10)

You might also point out that the writer uses unrepresentative examples—sources that represent an extreme point of view—to support his points. Show students that he makes no effort to present a cross-section of opinion or to represent opposing points of view fairly.

You may want to ask students to try to set up the argument in this essay as a syllogism. It will prove difficult and should illustrate to them the oversimplification inherent in the author's position—that hunters "are just as compassionate as the next fellow."

Despite how some media handled that story, there is no irony in hunters coming to the rescue of the same species we hunt.

We do it all the time.

A story of hunters showing compassion for an animal is something you'll never see in *The Bunny Huggers' Gazette* (yes, there *is* such a publication. It's a bimonthly magazine produced on newsprint. According to the publisher's statement, it provides information about vegetarianism, and "organizations, protests, boycotts or legislation on behalf of animal liberation . . .").

Among the protests announced in the June issue of BHG are boycotts against the countries of Ireland and Spain, the states and provinces of the Yukon Territory, Alberta, British Columbia, Pennsylvania and Alaska, the companies of American Express, Anheuser-Busch, Bausch & Lomb, Bloomingdale's, Coca-Cola Products, Coors, Gillette, Hartz, L'Oreal, McDonald's, Mellon Bank, Northwest Airlines, Pocono Mountain resorts and a host of others.

Interestingly, BHG tells how a subscribing group, Life Net of Montezuma, New Mexico, has petitioned the U.S. Forest Service to close portions of the San Juan and Rio Grande National Forests between April and November to all entry "to provide as much protection as possible" for grizzly bears that may still exist there. Another subscriber, Predator Project of Bozeman, Montana, is asking that the entire North Cascades region be closed to coyote hunting because gray wolves might be killed by "sportsmen (who) may not be able to tell the difference between a coyote and a wolf."

Although it's not a new idea, another subscriber, Prairie Dog Rescue, is urging persons who are opposed to hunting to apply for limited quota hunting permits because "one permit in peaceful hands means one less opportunity for a hunter to kill."

And if you ever doubted that the vegetarian/animal rights herd is a wacko bunch, then consider the magazine's review on *Human Tissue, A Neglected Experimental Resource*. According to the review, the 24-page essay encourages using human tissues to test "medicines and other substances, any of which would save animals' lives." (Bill Roberts, "The World of Hunting")

CHAPTER 7

Writing an Argumentative Essay

For many people, the true test of their critical thinking skills comes when they write an argumentative essay. When you write an argumentative essay, you follow the same process you use when you write any essay. The special demands of argument, though, require you to employ some additional strategies to make your ideas convincing to readers.

▶ See Chs. 1–3

WRITING ARGUMENTATIVE ESSAYS
- Choose a debatable topic.
- Formulate an argumentative thesis.
- Define your terms.
- Accommodate your audience.
- Consider opposing arguments.
- Gather evidence.
- Establish your credibility.
- Present your points fairly.

7a Planning an Argumentative Essay

(1) Choosing a topic

An argumentative essay must focus on a **debatable topic**, one about which reasonable people disagree. Factual statements—

about which people do not disagree—are, therefore, not suitable for argument. Neither are statements that simply convey information. Facts *become* debatable only if you voice an opinion about them.

> **FACT:** Many countries hold political prisoners.

> **DEBATABLE TOPIC:** The United States *should not* trade with countries that hold political prisoners.

> **FACT:** First-year students are not required to buy a meal plan from the university.

> **DEBATABLE TOPIC:** First-year students *should* be required to buy a meal plan from the university.

In addition to being debatable, your topic should be one about which you know something. The more evidence you can provide, the more likely you are to sway your audience. General knowledge is seldom convincing by itself, so you will probably have to do some research.

See ◄ Ch. 38

CLOSE-UP

ARGUMENT AND PERSUASION

In general conversation, many people use the terms *argument* and *persuasion* as if they meant the same thing. Students of writing draw a distinction between the two, however. *Persuasion* is a general term that refers to the various ways a writer can encourage readers to accept his or her position—appealing to their emotions or to their reason, for example. *Argument* usually refers to formal written compositions that use logic and evidence to convince readers.

Before you select a topic, you should understand what you want to accomplish in an argumentative essay. Your purpose is to change or clarify your reader's view of an issue. To accomplish this end, you must be able to define various sides of an issue, isolate crucial points, and state your own ideas.

It helps if you care about your topic, but that is not an absolute requirement. In fact, when you feel very strongly about an issue, you may not be able to view it clearly. If this is the case, consider another topic or consider writing an argument for the other side. (Dr. Samuel Johnson, the eighteenth-century lexicographer and critic, said that he preferred to argue on the wrong side of an issue because all the interesting things were to be said there.)

Finally, you should make sure your topic is narrow enough to write about in the page limit you have been given. Remember, in an argumentative essay you will have to present not only your own ideas but also the evidence to support these ideas. In addition, you will have to point out the strengths and weaknesses of opposing arguments. If your topic is too broad, you will not be able to cover it in enough detail.

CLOSE-UP

PLANNING AN ARGUMENTATIVE ESSAY

Some topics—no matter how controversial—have been discussed and written about so often that you will not be able to say anything new or interesting about them. They usually inspire tired, uninteresting essays that add little or nothing to a reader's understanding of an issue. For this reason you should avoid general topics such as "The Need for Gun Control" or "The Effectiveness of the Death Penalty" unless you are sure you can add something to the debate.

(2) Formulating an argumentative thesis

Once you have chosen a topic, your next step is to state your position in an argumentative thesis that asserts or denies something about that topic. Properly worded, this thesis lays the foundation for the rest of your argument.

Because the purpose of an argumentative essay is to convince readers to accept your position, your thesis must take a stand. One way to make sure your thesis actually does take a stand is to formulate an **antithesis**, a statement that takes an arguable position opposite from yours. If you can create an antithesis, your main idea takes a stand. If you cannot, your statement needs further revision to make it argumentative.

> **THESIS:** Put the blame [for the prize-fighter Benny Paret's death] where it belongs—on the prevailing mores that regard prize-fighting as a perfectly proper enterprise and vehicle of entertainment. (Norman Cousins, "Who Killed Benny Paret?")

> **ANTITHESIS:** Paret's death cannot be blamed on the prevailing mores that regard prize-fighting as a perfectly proper enterprise and vehicle of entertainment.

Putting a controversial topic (question) at the beginning of a conferencing session and asking students to write their own views (anonymously), along with their reasons for thinking as they do, would give you a good overview of your class. Then insert an opposing view or two to see how students will respond (especially if everyone seems to have the same view). This sort of interchange could also help students produce a collaborative paper on a particular topic. Asking students to play "devil's advocate" as they examine their classmates' views would help them delve more deeply into a topic. They might also look for specific fallacies in their classmates' reasons (support). (Note: Most conferencing software allows several conferences at the same time.)

Critical Thinking
To help students understand the concept of audience analysis, bring to class several advertisements from magazines or newspapers aimed at very different (and very specific) groups of readers. *People, Family Circle, Esquire, The New Yorker,* and *Newsweek* offer widely varying advertisements. Ask students to work in groups analyzing the audiences for whom they believe the advertising was created. Each group should choose a recorder who will later report the group's evaluation to the class. If time allows, distribute photocopies of brief excerpts from articles in each of the magazines from which

continued on following page

Keep in mind that you do not decide on your thesis arbitrarily. Whenever possible, you should test a tentative thesis on classmates—either informally in classroom conversations or formally in collaborative work. You may also want to talk to your instructor or do some reading about your topic. Your goal should be to get a grasp of your topic so you can make an informed statement about it. Before you feel ready to do this, however, you may want to review your annotations, your writing journal, and any notes you made when you were considering your topic. If you still have trouble formulating your thesis, you may want to try some of the techniques for getting started discussed in Chapter 1.

See ◀ 1c

(3) Defining your terms

You and your readers must agree about the terms you use in your argument. An argument that one rock group is *superior* to another means nothing unless your audience knows how you define *superior.* Never assume that everybody will know exactly what you mean.

You should also be careful to use precise language in your thesis, avoiding vague words such as *wrong, bad, good, right,* and *immoral,* which convey different meanings to different people.

VAGUE: Censorship of pay TV would be wrong.

CLEARER: Censorship of pay TV would unfairly limit free trade.

As a rule, you should define any potentially ambiguous terms you use in your argument. Quite often the soundness of an entire argument may hinge on the definition of a word that may mean one thing to one person and another to someone else. In the United States, *democratic* elections involve the selection of government officials by popular vote. In other countries rulers have used the same term to describe elections in which only one candidate has run or in which several candidates—all from the same party—have run. For this reason, when you use a term such as *democratic,* you should make sure your readers know exactly what you mean. The same is true for other potentially slippery terms, such as *freedom of speech, cruel and unusual punishment,* and *political correctness.*

In some cases you may want to use a *formal definition* in your essay. This is especially useful when you want a concise, exact definition of a term. Formal dictionary definitions, however, are not always the best. Often they do not apply to the situation you are discussing, and they are frequently overused by beginning writers. Instead, you may want to use an *extended definition.* This type of definition can often be a paragraph or more in length and contain examples from your own experience or reading. Not only can it be

See ◀ 4e8

tailored to your specific situation, but it can also provide much more insight than a dictionary definition. In fact, some terms—such as the ones mentioned in the preceding paragraph—are so abstract that they can be defined only in this way.

(4) Accommodating your audience

Plan your strategy with a specific audience in mind. Who are your readers? Are they unbiased observers or people deeply concerned about the issue you plan to discuss? Can they be cast in a specific role—concerned parents, victims of discrimination, irate consumers—or are they so diverse that they cannot be categorized? If you cannot be certain who your readers are, you will have to direct your arguments to a general audience.

In an argument you nearly always assume a skeptical audience. Even if they are sympathetic to your position, you cannot assume they will accept your ideas without question. At times—especially if your topic is highly controversial or emotionally charged—you may even have to assume your readers are hostile. The strategies you use to convince your readers will vary according to your relationship with them. Skeptical readers may only need to see that your argument is logical and that your evidence is solid. Hostile readers may need a good deal of reassurance that you understand their concerns and that you concede some of their points. However, you will never be able to convince some readers that your conclusion is valid. The best you can hope for is that these readers will grant the strength of your argument.

(5) Considering opposing arguments

To argue effectively, you must know how to refute opposing arguments. You can do this by showing that opposing views are untrue, unfair, illogical, unimportant, or irrelevant. For example, in the following paragraph from an essay criticizing the practices associated with whaling, a student refutes an argument against her position.

> Of course some will say Sea World only wants to capture a few whales. George Will makes this point in his commentary in *Newsweek,* pointing out how valuable the research on whales would be. Unfortunately, Will downplays the fact that Sea World wants to capture a hundred whales, not just "a few." And after releasing ninety whales, Sea World intends to keep ten for "further work." At hearings in Seattle last week, several noted marine biologists went on record condemning Sea World's research program. We must wonder, as they do, why Sea World needs such a large number of whales to carry out its project.

continued from previous page

you selected advertising and ask students, on the basis of their audience evaluation, to match their advertisement with the article from the same source.

IN THE CLASSROOM

Students frequently shy away from an in-depth look at opposing arguments—perhaps fearing they might find they've taken the "losing" side of an issue. As an in-class exercise, write a list of currently controversial issues on the board or overhead. A short list might include the following:

Legalization of drugs

Establishment of a smoke-free campus

Video stores renting X-rated movies to teenagers

Ask for volunteers to argue the "pro" and "con" side of these or other issues and then turn the tables by asking all volunteers to switch and argue the *other* side. The exercise may produce a lively discussion—and convince students that they *can* deal with opposing arguments.

167

Lab Activity

Opposing Arguments

Have students switch computers or floppy disks and respond to one another's arguments by typing comments directly into the writer's file. Peer critics can keep their responses distinct from writers' texts by formatting their comments in bold, underlined, or uppercase characters or by enclosing them within brackets. After fifteen or twenty minutes, students can reclaim their own computers or disks and see what other students have written. To carry the procedure one step further, after students have had a chance to consider comments made by peer critics, have writers and critics consult with each other; in this way, writers will have an opportunity to defend their original versions, ask for additional comments or clarification, and try out suggested revisions.

When an opponent's position is so strong that it cannot be dismissed, concede that the point is well taken, and then, if possible, discuss its limitations. Martin Luther King, Jr., uses this tactic in his "Letter from Birmingham Jail."

> You express a great deal of anxiety over our willingness to break laws. This is certainly a legitimate concern. Since we so diligently urge people to obey the Supreme Court's decision of 1954 outlawing segregation in the public schools, at first glance it may seem rather paradoxical for us consciously to break laws. One may well ask: "How can you advocate breaking some laws and obeying others?" The answer lies in the fact that there are two types of laws: just and unjust. I would be the first to advocate obeying just laws. One has not only a legal but a moral responsibility to obey just laws. Conversely, one has a moral responsibility to disobey unjust laws. I would agree with St. Augustine that "an unjust law is no law at all."

CLOSE-UP

CONSIDERING OPPOSING ARGUMENTS

When you acknowledge an opposing view, do not distort it or present it as ridiculously weak. This tactic, called creating a **straw man**, could seriously undermine your credibility.

(6) Gathering evidence

Most arguments are built on **assertions**—claims you make about a debatable topic—backed by **evidence**—supporting information, in the form of examples, statistics, or expert opinion, which reinforces your argument. You could, for instance, assert that law-enforcement officials are losing the war against violent crime. You could then support this assertion by referring to a government report stating that violent crime in the ten largest U.S. cities has increased during the last five years. This report would be one piece of persuasive evidence.

Certain assertions need no proof: statements that are *self-evident* ("All human beings are mortal"), statements that are true by *definition* (2 + 2 = 4), and *statements of fact* that you can expect the average person to know ("The Atlantic Ocean separates England and the United States"). All other kinds of assertions need supporting evidence.

See ◀
5b2

CLOSE-UP

GATHERING EVIDENCE

Remember that an argumentative essay never proves a thesis conclusively—if it did, there would be no argument. The best you can do is to establish a high probability that your thesis is correct or establish that it is reasonable.

(7)   Establishing your credibility

Clear reasoning, compelling evidence, and pointed refutations go a long way toward making an argument solid. But these elements are not sufficient to create a convincing argument. In order to sway readers, you have to satisfy them that you are someone they should listen to—in other words, that you have credibility.

Certain individuals, of course, bring credibility with them every time they speak. When a Nobel Prize winner in physics makes a speech about the need to control proliferation of nuclear weapons, we automatically assume that he or she knows the subject. But most people do not have this kind of authority and so must work to establish it—by *finding common ground, demonstrating knowledge,* and *maintaining a reasonable tone.*

Finding Common Ground   When you write an argument, it is tempting to go on the attack, emphasizing the differences between your position and those of your opponents. However, this tactic can antagonize not only readers who do not agree with you but also those who might be sympathetic. In argument, as in labor negotiations and foreign policy, lack of agreement can cause animosity, mistrust, and eventually a total breakdown of communications. Writers of effective arguments know they must avoid this deterioration at all costs.

Thomas Jefferson establishes common ground at the beginning of the Declaration of Independence when he presents the self-evident truths that he feels all people will accept. In "Letter from Birmingham Jail," Martin Luther King, Jr., also employs this strategy when he takes great pains to demonstrate to his audience of white clergy that they share the same goal—to fight injustice—and differ only on the methods that should be used to eliminate the problems. Of course, Thomas Jefferson could have verbally attacked England, and Martin Luther King, Jr., could

Give your students a quick lesson in writing for what writer Robert Graves called "the reader over your shoulder." Ask each of your students to write a short letter to a close friend: one or two paragraphs that deal with a public issue about which the student holds strong opinions. (This might be anything from a national social issue to a change in campus policy.) Then ask the students to write one or two paragraphs—on the same issue—as a "public" document arguing the position: a letter to the editor, to the college president, to a politician, and so on.

Discuss the results. Did the students instinctively change their tone, style, choice of words, facts, and arguments to fit the changed audience? Were some students unable to change their tone for the second assignment? Which of their two letters do the students think is the stronger piece of writing? Chances are that the first (private) letter is stronger, more vibrant writing. Why? Too often, beginning writers can write strong informal arguments but are unable to keep the same life, color, and strength in their writing in more formal situations.

have called the white clergy bigots, but they knew these comments would have alienated their audiences and undermined their own credibility.

One way to avoid a confrontational stance, and thereby increase your credibility, is to use the techniques of **Rogerian argument**, based on the work of the psychologist Carl Rogers. According to Rogers, you should think of the members of your audience as colleagues with whom you must collaborate to find solutions to problems. Instead of verbally assaulting them, you try to arrive at points of agreement. In this way, you establish common ground and work toward a resolution of the issue you are discussing. For example, you could begin an essay in which you argue against the imposition of speech codes on your campus by asserting that everyone is interested in protecting first amendment rights. Once you and your readers occupy this common ground, you will more easily convince them of the dangers of speech codes.

Demonstrating Knowledge   Including pertinent personal experiences in your argumentative essay can show readers that you know a lot about your subject and thus give you some authority. Your attendance at a National Rifle Association conference can give you authority when you write an essay arguing for (or against) gun control. Similarly, the fact that you worked for a landscaping company can give you credibility when you argue for (or against) banning certain lawn-care products.

You can also establish credibility by showing you have done research into a subject. By mentioning important research sources you have consulted and documenting the information you got from them, you show readers that you have laid the necessary groundwork. Including references to several sources—not just one—suggests to readers that you have a balanced knowledge of your subject.

CLOSE-
UP   **ESTABLISHING CREDIBILITY**

See ◀
Ch. 39
Make certain that you document source material very carefully. Questionable sources, inaccurate documentation, and factual errors can undermine an argument. For many readers, an undocumented quotation or even an incorrect date can call an entire paper into question.

Maintaining a Reasonable Tone   The tone you adopt is almost as important as the information you convey. Avoid sounding high-handed or pedantic. Talk *to* your readers, not *at* them. If you lecture your readers or appear to talk down to them, you will only succeed in alienating them. Remember, readers are more likely to respond to a writer who is conciliatory than one who is insulting. At the end of "Letter from Birmingham Jail," for example, Martin Luther King, Jr., deliberately understates his position, saying he hopes he did not go so far that he insulted his readers. In this way, King not only reassures his readers but also strengthens his argument.

In addition, use moderate language. Words and phrases such as *never, all,* and *in every case* can make your claims seem exaggerated and unrealistic. Learn to qualify your statements so they seem reasonable. The statement "Mercy killing is never acceptable," for example, leaves no room for debate or differing points of view. A more conciliatory statement might be "In cases of extreme suffering one can understand a patient's desire for death, but in most cases the moral and social implications of mercy killing make it unacceptable." By qualifying your position, you demonstrate you are making every effort to adopt a reasonable point of view.

▶ See 6c2

GUIDELINES FOR ESTABLISHING YOUR CREDIBILITY

Finding Common Ground
- Identify the various sides of the issue.
- Identify the points on which you and your readers are in agreement.
- Work the areas of agreement into your argument.

Demonstrating Knowledge
- Include relevant personal experiences.
- Include relevant special knowledge of your subject.
- Include the results of any relevant research you have done.

Maintaining a Reasonable Tone
- Avoid sounding high-handed or pedantic.
- Use moderate language and qualify your statements.

---

(8) Being fair

The line between being persuasive and being unfair is a fine one, and no clear-cut rules exist to help you make this distinction.

Writers of effective and sometimes brilliant argumentative essays are often less than fair to their opponents. We could hardly call Jonathan Swift "fair" when, in "A Modest Proposal," he implies that the English are cannibals. A supporter of George III would argue that Thomas Jefferson and the other writers of the Declaration of Independence were less than fair when they criticized British policy in America.

Of course, "A Modest Proposal" is bitter satire, and Swift employs overstatement to express his rage at social conditions. In justifying their break with England, the writers of the Declaration of Independence did not intend to be fair to the king. Argument promotes one point of view, so it is seldom objective.

For better or for worse, however, college writing requires that you stay within the bounds of fairness. To be sure the support for your argument is not misleading or distorted, you should take the following steps.

Avoid Distorting Evidence   Distortion is misrepresentation. Writers sometimes intentionally misrepresent their opponents' views by exaggerating them and then attacking this extreme position. For example, a Democratic governor of a northeastern state proposed requiring mothers receiving welfare to name their children's fathers and supply information about them. His Republican challenger attacked him with the following statements.

> What is the governor's next idea in his headlong rush to embrace extreme right-wing radicalism? A program of tattoos for welfare mothers? A badge sewn on to their clothing identifying them as welfare recipients? Creation of colonies, like leper colonies, where welfare recipients would be forced to live? How about an involuntary relocation program into camps?

The governor made a controversial proposal, and his opponent could certainly have challenged it on its own merits. Instead, by distorting his position, she attacked it unfairly.

Avoid Quoting Out of Context   A writer or speaker quotes out of context by taking someone's words from their original setting and using them in another. When you select certain words from a statement and ignore others, you can change the meaning of what someone has said or implied. Consider the following example.

MR. N, TOWNSHIP RESIDENT: I don't know why you are opposing the new highway. According to your own statements the highway will increase land value and bring more business into the area.

MS. L, TOWNSHIP SUPERVISOR: I think you should look at my statements more carefully. I have a copy of the paper that printed my interview and what I said was [*reading*]: "The highway will increase land values a bit and bring some business to the area. But at what cost? One hundred and fifty families will be displaced, and the highway will divide our township in half." My comments were not meant to support the new highway but to underscore the problems that its construction will cause.

By repeating only some of Ms. L's remarks, Mr. N alters her meaning to suit his purpose. In context, Ms. L's words indicate that although she concedes the highway's few benefits, she believes its drawbacks outweigh them.

Avoid Slanting Support When you select information that supports your case and ignore information that does not, you are slanting. For example, if you support your position that smoking should not be prohibited in public places by choosing only evidence provided by the American Tobacco Institute, you are guilty of slanting supporting information. Inflammatory language also creates biases in your writing. A national magazine slanted its information, to say the least, when it described a reputed criminal as "a hulk of a man who looks as if he could burn out somebody's eyes with a propane torch." Although one-sided presentations frequently appear in newspapers and magazines, you should avoid such distortions when you seek to present a rational argument.

Avoid Using Unfair Appeals Traditionally, writers of arguments have used three types of appeals to influence readers. The **logical appeal** relies on an audience's logical faculties. Using the principles of inductive and deductive reasoning discussed in Chapter 6, the logical appeal moves from evidence to conclusions. The **emotional appeal** addresses the emotions of a reader, and the **ethical appeal** calls the reader's attention to the credibility of the writer.

All three of these appeals are acceptable as long as they are used fairly. Problems arise, however, when these appeals are used unfairly. For example, a writer can use fallacies to fool readers ▶ See 6c into thinking a conclusion is logical when it is not. A writer can also employ inappropriate emotional appeals—to prejudice or fear, for example—to influence readers. And finally, a writer can use qualifications in one area of expertise to bolster his or her ▶ See 5b2 stature in another area that he or she is not qualified to discuss. Not only are these techniques misleading, but they are also dishonest and frequently backfire. Even the most compelling argument will be dismissed if readers begin to question a writer's integrity.

Worksheet for an Argumentative Essay

Students can assess the support for their argument by filling out the worksheet for an argumentative essay.

Here is a sample worksheet for an argumentative essay.

Worksheet for an Argumentative Essay

Write down your SUBJECT. Finish the sentence below in fewer than ten words.

My subject is

Make an ASSERTION about your subject. Finish the sentence below in order to state what you have discovered (or what you think you will discover) about your subject.

I believe that

SUPPORT your assertion about your subject. Finish the sentences below by supplying **MAIN REASONS** for believing the assertion you have made about your subject. Next, briefly give the specific **EVIDENCE** you found (or that you will find) that supports your main reasons.

One reason the assertion is true is that
    supporting evidence:
    supporting evidence:
    supporting evidence:
      counter evidence:

Once you have carried out these steps, you will find that your **assertion**, your **main reasons**, and the **supporting evidence** will together form an INFORMAL OUTLINE that can help you write a well-organized essay.

continued on following page

7b   Shaping an Argumentative Essay

In its simplest form, an argument consists of a thesis and supporting evidence. However, argumentative essays contain additional elements calculated to win audience approval and to overcome potential opposition. Depending on your purpose and audience, you may arrange these elements in various ways; in some cases, you may omit one or more elements entirely.

See ◀
4g2

ELEMENTS OF AN ARGUMENTATIVE ESSAY

The Introduction

The introduction of your argumentative essay orients your audience to your subject. Here you can show how your subject concerns the reader, note why it is interesting, or explain how it has been misunderstood.

The Background Statement

In this section you present a brief overview of the subject you will discuss. This section may include a narrative of past events, a summary of others' opinions on your subject, or a summary of basic facts. Keep your background statement short; long, drawn-out discussions at this point will distract your readers from the focus of your argument.

The Thesis Statement

Your thesis statement can appear anywhere in your argumentative essay. Frequently, you present your thesis after you have given your readers an overview of your subject. However, in highly controversial arguments—those to which you believe your audience might react negatively—you may postpone stating your thesis until later in your essay.

The Arguments in Support of Your Thesis

See ◀
6a,
6b

This section is the center of your essay; it contains the arguments that support your thesis. Here you present your points and the evidence you have gathered to support them.

One problem you may have at this stage is deciding on the arrangement of your points. Most often, you begin with your weakest argument and work up to your strongest. If you move in the opposite direction, your essay will be anticlimactic. Not only will you emphasize your weakest argu-

continued on the following page

continued from the previous page

ments by putting them last, but you will also make these arguments appear to be afterthoughts. If all your arguments are equally strong, you might want to begin with points with which your readers are already familiar (and which they are therefore likely to accept) and then move on to relatively unfamiliar points.

The Refutation of Opposing Arguments

In a face-to-face debate, you are confronted by the arguments against your thesis, and you have the opportunity to refute them. In an argumentative essay, however, you must bring these points up yourself: You must anticipate these arguments and assess their soundness. If you do not confront these opposing arguments, doubts about your case will remain in the minds of your readers. If strong opposing arguments exist, admit their strengths and then refute the arguments early in your paper, before you present your own points. If the opposing arguments are relatively weak, refute them after you have made your case.

The Conclusion

Your conclusion can summarize key points, restate your thesis, reinforce the weaknesses of opposing arguments, or underscore the logic of your position. Most often, the conclusion restates in general terms the major arguments you have marshaled in support of your thesis. Many writers like to end their arguments with a strong last line, one calculated to stay in the minds of their readers. An apt quotation or a statement that crystallizes the sentiments or captures the intensity of your argument works well.

▶ See 4g3

continued from previous page

Students can use this worksheet in at least two ways:

1. They can begin the writing process by *first* completing the worksheet. Then, when they begin writing the first draft of their essay, they can use the **assertion** as a tentative thesis sentence, their **main reasons** as tentative topic sentences, and the **supporting evidence** as material for supporting sentences within paragraphs.
2. They can complete the worksheet *after* they have written a rough draft, using the rough draft to help them fill in the worksheet. The filled-in worksheet will be a way of checking whether they're on their way to producing an essay that's really going to convince readers.

7c   Writing and Revising an Argumentative Essay

(1)   Writing an argumentative essay

The following essay contains many of the elements listed in the box on pages 174–175. The student, Samantha Masterton, was asked by her instructor to write an argument, drawing her supporting evidence both from her own experience and from library and nonlibrary research.

▶ See 38b, 38c

Writing an Argumentative Essay
As a final exercise for Chapter 7, you may want to have your students plan and write an argumentative essay on an issue of their choice, following the instructions given to Samantha Masterton for the sample essay beginning on the previous page. Suggest that students look for an issue that will allow them to draw on their personal experiences and observations. (If time allows, you might let students work in pairs, asking each pair to read each other's essays and then revise them together, following the revision checklist on page 181.)

Samantha Masterton   Masterton 1
Professor Wade
English 102
15 March 1994

The Returning Student:
Older Is Definitely Better

Introduction   After graduating from high school, young people must decide what they want to do with the rest of their lives. Many graduates (often without much thought) decide to continue their education uninterrupted, and they go on to college. This group of teenagers makes up what many see as the typical first-year college student. Recently, however, this stereotype is being challenged by an influx of older students into American colleges and universities. Not only do these students make a valuable contribution to the schools they attend, but they also present an alternative to young people who go to college simply because it is the thing to do. A few years off between high school and col-

Thesis   lege can give many--perhaps most--students the life experience they need to appreciate the value of higher education.

Background statement   The college experience of an eighteen-year-old is quite different from that of an older student. The typical teenager is often concerned with things other than cracking books-- going to parties, dating, and testing personal limits, for example. Although the maturation process from teenager to adult is something we must all go through, college is not necessarily the appropriate place for this to occur. My experience as an adult enrolled in a university has convinced me that many students would benefit from delaying entry into college. I almost never see older students cutting lectures or not studying. Most have saved for tuition and want to get their money's worth, just as I do. Many are also balancing the demands of home and work to attend classes, so they know how important it is to do well.

Masterton 2

Generally, young people just out of high school have not been challenged by real-world situations that include meeting deadlines and setting priorities. Younger college students often find themselves hopelessly behind or scrambling at the last minute simply because they have not learned how to budget their time. Although success in college depends on the ability to set realistic goals and organize time and materials, college itself does little to help students develop these skills. On the contrary, the workplace--where reward and punishment are usually immediate and tangible--is the best place to learn such lessons. Working teaches the basics that college takes for granted: the value of punctuality and attendance, the importance of respect for superiors and colleagues, and the need for establishing priorities and meeting deadlines.

Argument in support of thesis

The adult student who has gained experience in the workplace has advantages over the teenaged freshman. In general, the older student enrolls in college with a definite course of study in mind. As Laura Mansnerus reports in her article "A Milieu Apart," for the older student, "college is no longer a stage of life but a place to do work" (17). For the adult student, then, college becomes an extension of work rather than a place to discover what work will be. This greater sense of purpose is not lost on college instructors. Dr. Laurin Porter, Assistant Professor of English at the University of Texas at Arlington, echoes the sentiments of many of her colleagues when she says, "Returning older students, by and large, seem more focused, more sure of their goals, and more highly motivated."

Argument in support of thesis

Given his or her age and greater experience, the older student brings more into the classroom than does the younger student. Eighteen-year-olds have only been driving for a year or two; they have just earned the right to vote; and they usually have not lived on their own. They cannot be expected to have formulated definite goals or developed firm ideas about themselves or about the world in which

Argument in support of thesis

Masterton 3

they live. In contrast, the older student has generally had
a variety of real-life experiences. Most have worked for
several years; many have started families. Their years in
the "real world" have helped them to become more focused
and more responsible than they were when they graduated
from high school. As a result, they are better prepared for
college. Thus, they not only bring more into the classroom,
but take more out of it.

Refutation
of oppos-
ing argu-
ment

Of course, postponing college for a few years is not for
everyone. There are certainly some teenagers who have a
definite sense of purpose and a maturity well beyond their
years, and these individuals might benefit from an early
college experience, so they can get a head start on their
careers. Charles Woodward, a law librarian, went to college directly after high school, and for him the experience
was positive. "I was serious about learning, and I loved my
subject," he said. "I felt fortunate that I knew what I
wanted from college and from life." For the most part,
though, students are not like Woodward; they graduate
from high school without any clear sense of purpose. For
this reason, it makes sense for most students to stay away
from college until they are mature enough to benefit from
the experience.

Refutation
of oppos-
ing argu-
ment

Granted, some older students do have difficulties when
they return to college. Because these students have been
out of school so long, they may have difficulty studying
and adapting to the routines of academic life. Some older
students may even feel ill at ease because they are in class
with students who are many years younger than they are
and because they are too busy to participate in extracurricular activities. As I have seen, though, these problems
soon disappear. After a few weeks, older students get into
the swing of things and adapt to college. They make
friends, get used to studying, and even begin to participate
in campus life.

Masterton 4

Conclu-
sion

All things considered, higher education is wasted on the young, who are either too immature or too unfocused to take advantage of it. Taking a few years off between college and high school would give these students the breathing room they need to make the most of a college education. The increasing numbers of older students returning to college would seem to indicate that many students are taking this path. According to one study, 45 percent of the students enrolled in American colleges in 1987 were twenty-five years of age or older (Aslanian 57). These older students have taken time off to serve in the military, to get a job, or to raise a family. Many have traveled, engaged in informal study, and taken the time to grow up. By the time they get to college they have defined their goals and made a commitment to achieve them. It is clear that postponing college for a few years can result in a better educational experience for both students and teachers. As Dr. Porter says, when the older student brings more life experience into the classroom, "everyone benefits."

Masterson 5

Works Cited

Aslanian, Carol B. "The Changing Face of American Campuses." USA Today Magazine May 1991: 57–59.

Mansnerus, Laura. "A Milieu Apart." New York Times 4 Aug. 1991, late ed.: A7.

Porter, Laurin. Personal interview. 23 Feb. 1994.

Woodward, Charles B. Personal interview. 25 Feb. 1994.

Your students' answers will vary. The paragraph should be deleted. It is irrelevant and does not arrive at a logical conclusion. Moreover, the information it presents overlaps with information presented in paragraph 5.

STUDENT WRITER AT WORK

Writing an Argumentative Essay
The thesis is the sweeping generalization that parental supervision will solve the problem of violence on television. The paper does not deal with the issue of parental supervision and, in fact, only mentions it again in the conclusion.

Commentary on Original
Paragraph 1: referring to one group of protesters as "the weirdest" is an *ad hominem* argument. Suggesting that "these people" find better things to do with their time is a *red herring*.
Paragraph 2: uses a *false analogy* by specifically defining the broadcasting industry as a business. Also, the argument is based on the *either/or fallacy* that *either* television programming will give people what they want (evidently, in this case, violence) *or* the industry will be out of business within a year. Also, referring to the protestors as so-called experts is an *ad hominem* argument. The writer also undercuts his credibility by introducing a point with "I am not an expert on the subject, but. . . ."
Paragraph 3: "I am a citizen, so I should be able to watch whatever I want" falsely concludes that free speech is the same as freedom to watch television violence. By using *equivocation*—shifting the meaning of a key term—the writer constructs an invalid deductive argument. The reference to the Founding Fathers is a legitimate appeal to authority although

continued on following page

(2) Revising an argumentative essay

See ◄
3c

You revise your argumentative essay using the same strategies you use for any essay. In addition, you concentrate on some specific concerns which are listed in the Revision Checklist on p. 181.

CLOSE-UP

USING TRANSITIONS IN ARGUMENTATIVE ESSAYS

Argumentative essays should include transitional words and phrases to indicate which paragraphs are *arguments in support of the thesis*, which are *refutations* of evidence that opposes the thesis, and which are *conclusions*. These transitional words and phrases send important signals that help readers follow the progression of the argument.

Arguments in support of thesis	Because, for example, for instance, in general, given, generally, since
Refutations	Although, admittedly, certainly, granted, naturally, of course
Conclusions	All things considered, as a result, in conclusion, in summary, therefore, thus

EXERCISE

The following paragraph was deleted by Samantha Masterton from her essay "The Returning Student: Older Is Definitely Better." Was Samantha right to delete it? If it belongs in the essay, where would it go? Is it relevant? Logical? Does it need any revision?

The dedication of the adult student is evident in the varied roles he or she must play. Many of the adults who return to school are seeking to increase their earning power. They have established themselves in the working world, only to find they cannot advance without more education or a graduate degree. The dual-income family structure enables many of these adults to return to school, but it is unrealistic for them to put their well-established lives on hold while they pursue their education. In addition to the rigors of college, the older student is often juggling a home, a family, and a job. However, the adult student makes up in determination what he or she lacks in time. In contrast, the younger student often lacks the essential motivation to succeed in school. Teenagers in college often have no clear idea of why they are there and, lacking this sense of purpose, may do poorly even though they have comparatively few outside distractions.

continued from previous page

✔	**REVISION CHECKLIST: ARGUMENTATIVE ESSAYS**

- Is your topic debatable?
- Does your essay include an argumentative thesis?
- Have you adequately defined the terms you use in your argument?
- Have your considered the opinions, attitudes, and values of your audience?
- Have you identified and refuted opposing arguments?
- Have you supported your assertions with evidence?
- Have you established your credibility?
- Have you been fair?
- Are your arguments logically constructed?
- Have you avoided logical fallacies?
- Does the structure of your essay suit your material and your audience?
- Have you provided your readers with enough background information?
- Do you present your points clearly and organize them logically?
- Do you have an interesting introduction and a strong conclusion?

some students might mistakenly call it an *ad populum* argument. Citing two countries such as Iran and China to support a conclusion is a *hasty generalization* that sets up an *either/or fallacy*: *Either* we allow violence on television *or* we will end up with complete loss of freedom. Comparing Iran and China to the United States is also a *false analogy* that assumes because certain conditions exist in China and Iran as well as in the United States, other conditions will too.

Paragraph 4: uses the *red herring* tactic of distracting readers by mentioning all the other ills of society. Stating that "there is absolutely no proof that the violence people see on television causes them to act violently" is an *argument to ignorance*. Also, by leaning on a catchy phrase (which is actually an enthymeme), "movies don't kill people; people kill people," the writer implies a quick fix.

Paragraph 5: based on *sweeping generalizations*. Also, the reference to the majority of "law-abiding" people is an *ad populum* argument that is also based on the *bandwagon* fallacy. The essay ends with an *either/or fallacy* that *either* violence be allowed on television *or* all programming will be dull.

Possible Revision

Television Violence: Let Us
Exercise Our Choice

Television began as what many people thought was a fad. Now, over fifty years later, it is the subject of much controversy. As a result, numerous activist groups have formed with the intent to examine television's role in society. Some such groups are devoted entirely to protesting against violence on television. However, what these groups fail to recognize is that banning television violence is a form of censorship that does more harm than good.

In examining censorship of violence on television, it is important to remember that the broadcast industry is a business. Success in business

---

STUDENT WRITER AT WORK

WRITING AN ARGUMENTATIVE ESSAY

Revise the following draft of an argumentative essay, paying particular attention to the essay's logic, its use of support, and the writer's efforts to establish credibility. Be prepared to identify the changes you made and to explain how they make the essay more convincing. If necessary, revise further to strengthen coherence, unity, and style.

Television Violence: Let Us Exercise Our Choice

Television began as what many people thought was a fad. Now, over fifty years later, it is the subject of arguments and controversy. There are even some activist groups who spend all their time protesting television's role in society. The weirdest of these groups is definitely the one that attacks television for being too violent. As far as I am concerned, these people should find better things to do with their time. There is nothing wrong with American television that a little bit of parental supervision wouldn't fix.

continued on following page

continued from previous page

is most obviously achieved by meeting the demands of the customer. In the case of television, the viewers are the customers, and many express the desire for realism in television programming.

Additionally, the first amendment of the United States Constitution guarantees all citizens the right of free speech. In part, the first amendment is a reflection of the concept that an informed citizenry is the best defense against tyranny. An examination of other countries such as Iran and China, where the people see only what the government wants them to see, provides examples of censorship at its most extreme. Although censoring violence on television would not necessarily result in a complete loss of our freedoms, it might well open the door to more far-reaching forms of censorship in the future.

Although television reflects the violence in American society, it does not create the violence. Obviously, other factors in society such as drugs, the proliferation of guns, and increased unemployment contribute to the increase in violence in America. Only by being aware of these problems, in part through the realism of television, can we acknowledge them and begin to work toward resolving the issues.

Even though there are no simple solutions to the problem of violence in America, television censorship of violence is not the answer. Such censorship would only result in the absence of realism and would blind Americans to the ills in our society that have led to violence in the first place. Since television is a powerful medium, perhaps we should examine ways to expand its influence to help us recognize our problems and find solutions to them rather than shackling programming with censorship.

The best argument against these protest groups is that television gives people what they want. I am not an expert on the subject, but I do know that the broadcasting industry is a business, a very serious business. Television programming must give people what they want or they won't watch it. This is a fact that many of the so-called experts forget. If the television networks followed the advice of the protestors, they would be out of business within a year.

Another argument against the protest groups is that the first amendment of the U.S. Constitution guarantees all citizens the right of free speech. I am a citizen, so I should be able to watch whatever I want to. If these protestors do not want to watch violent programs, let them change the channel or turn off their sets. The Founding Fathers realized that an informed citizenry is the best defense against tyranny. Look at some of the countries that control the programs citizens are able to watch. In Iran and in China, for example, people only see what the government wants them to see. A citizen can be put in prison if he or she is caught watching an illegal program. Is this where our country is heading?

Certainly, American society is too violent. No one can deny this fact, but you cannot blame all the problems of American society on television violence. As far as I know, there is absolutely no proof that the violence people see on television causes them to act violently. Violence in society is probably caused by a number of things--drugs, the proliferation of guns, and increased unemployment, for example. Before focusing on violence on television, the protestors should address these things. Protestors should also remember that movies don't kill people; people kill people. It would seem that the protestors are forgetting this fact.

All things considered, the solution to violence on television is simple: Parents should monitor what their children see. If they don't like what their children are watching, they should change the channel or turn off the television. There is no reason why the majority of television watchers--who are for the most part law-abiding people--should have to stop watching programs they like. I for one do not want some protestor telling me what I can or cannot watch, and if, by chance, these protestors do succeed in eliminating all violence, the result will be television programming that is boring: Television, the vast wasteland, will suddenly be turned into the dull wasteland.

Composing Sentences

Commentary on Possible Revision
Even after correcting many of the original essay's problems with logic and use of evidence, the writer still had more work to do. His arguments were still based on premises that many people would find debatable. In addition, specific examples of possible positive results of violence on television, analogies with censorship in other media, or instances of illogical or overzealous censorship of television violence might strengthen the writer's argument. Statistical evidence might also be useful.

continued on following page

CHAPTER 8

Building Simple Sentences

8a   Identifying the Basic Sentence Elements

A **sentence** is an independent grammatical unit that contains a subject and a predicate and expresses a complete thought.

The quick brown fox jumped over the lazy dog.

It came from outer space.

A **simple subject** is a noun or noun substitute (*fox, it*) that tells who or what the sentence is about. A **simple predicate** is a verb or verb phrase (*jumped, came*) that tells or asks something about the subject. The **complete subject** of a sentence, however, includes the simple subject plus all its modifiers (*the quick brown fox*). The **complete predicate** includes not only the verb or verb phrase but all the words associated with it—such as modifiers, objects, and complements (*jumped over the lazy dog, came from outer space*).

8b   Constructing Basic Sentence Patterns

A **simple sentence** consists of at least one subject and one predicate. The most basic English sentences may be built from one of the following five patterns.

(1)   Subject + intransitive verb (s + v)

The simplest sentence pattern consists of only a subject and a verb or **verb phrase** (the main verb plus all its auxiliaries).

$$\overset{s}{\underline{\text{The price of gold}}} \; \overset{v}{\underline{\text{rose}}}.$$

$$\overset{s}{\underline{\text{Stock prices}}} \; \overset{v}{\underline{\text{may fall}}}.$$

In both sentences the verbs (*rose, may fall*) are **intransitive**—that is, they do not need an object to complete their meaning. (A dictionary can tell you which verbs are transitive, which are intransitive, and which may be either, depending on context.)

▶ See 19b4

(2) Subject + transitive verb + direct object (s + v + do)

In another pattern, the sentence consists of the subject, a transitive verb, and a direct object. A **transitive** verb is one that requires an object to complete its meaning in the sentence. A **direct object** indicates where the verb's action is directed and who or what is affected by it.

$$\overset{s}{\underline{\text{Van Gogh}}} \; \overset{v}{\underline{\text{created}}} \; \overset{do}{\textit{The Starry Night}}.$$

$$\overset{s}{\underline{\text{Caroline}}} \; \overset{v}{\underline{\text{saved}}} \; \overset{do}{\text{Jake}}.$$

In each sentence the direct object tells *who* or *what* received the verb's action.

(3) Subject + transitive verb + direct object + object complement (s + v + do + oc)

This pattern, similar to the preceding one, includes an **object complement** that renames or describes the direct object.

$$\overset{s}{\underline{\text{The class}}} \; \overset{v}{\underline{\text{elected}}} \; \overset{do}{\text{Bridget}} \; \overset{oc}{\text{treasurer}}.$$ (Object complement *treasurer* renames direct object *Bridget*.)

$$\overset{s}{\underline{\text{I}}} \; \overset{v}{\underline{\text{found}}} \; \overset{do}{\text{the exam}} \; \overset{oc}{\text{easy}}.$$ (Object complement *easy* describes direct object *exam*.)

(4) Subject + linking verb + subject complement (s + v + sc)

This kind of sentence consists of a subject, a **linking verb** (a verb that connects a subject to its complement) and the **subject complement** (the word or phrase that describes or renames the subject).

▶ See 21c1

$$\overset{s}{\underline{\text{The injection}}} \; \overset{v}{\underline{\text{was}}} \; \overset{sc}{\text{painless}}.$$

$$\overset{s}{\underline{\text{John Major}}} \; \overset{v}{\underline{\text{became}}} \; \overset{sc}{\text{prime minister}}.$$

continued from previous page

if both the subject and verb are compounded. The following examples may be useful.

> *The factory closed.* (Simple sentence, basic subject-verb structure.)
>
> *Both the canning and meat-packing factories reportedly will close as soon as this week's orders are filled.* (Still a simple sentence, but with compound subject and modifying adverb clause.)

QUOTATION

"Understanding the distinction between transitive and intransitive can prevent frustration in selecting the correct verb from troublesome pairs such as *lay/lie, set/sit,* and *raise/rise.* The first verb in these combinations is transitive: it requires an object that answers the question *what.*

> Lay the <u>book</u> on the table.
> Have you set your <u>clocks</u> back yet?
> Let's raise the <u>roof</u> tonight.
> (objects underlined)

On the other hand, verbs such as *lie, sit* and *rise* are intransitive.

> Would you please *lie* down?
> *Sit* in this chair, please.
> I will *rise* early tomorrow."

–Lauren Kessler and Duncan McDonald,
When Words Collide

In the first sentence the subject complement is an adjective, called a **predicate adjective**, that describes the subject; in the second sentence the subject complement is a noun, called a **predicate nominative**, that renames the subject. In each case the linking verb is like an equal sign, equating the subject with its complement (*John Major = prime minister*).

(5)   Subject + transitive verb + indirect object + direct object (s + v + io + do)

In this sentence pattern the **indirect object** tells to whom or for whom the verb's action was done.

<u>Cyrano</u> <u>wrote</u> Roxanne a poem. (Cyrano wrote a poem for Roxanne.)
<small>s   v   io   do</small>

<u>The officer</u> <u>handed</u> Frank a ticket. (The officer handed a ticket to Frank.)
<small>s   v   io   do</small>

BASIC SENTENCE PATTERNS

Subject + intransitive verb	<u>The bell</u> <u>rang</u>.
Subject + transitive verb + direct object	<u>Toni Morrison</u> <u>won</u> the 1993 Nobel Prize.
Subject + transitive verb + direct object + object complement	<u>Artists and writers</u> <u>found</u> Paris exciting.
Subject + linking verb + subject complement	<u>The cowardly lion</u> <u>looked</u> frightened.
Subject + transitive verb + indirect object + direct object	<u>The 1882 Chinese Exclusion Act</u> <u>denied</u> a specific ethnic group citizenship. (The act denied citizenship to a specific ethnic group.)

EXERCISE 1

In each of the following sentences underline the subject once and the verb twice. Then label direct objects, indirect objects, subject complements, and object complements.

       EXAMPLE: Isaac Asimov was a science fiction writer. [sc]

1. Isaac Asimov first saw science fiction stories in the newsstand of his parents' Brooklyn candy store.
2. He practiced writing by telling his schoolmates stories.
3. Asimov published his first story in *Astounding Science Fiction*.
4. The magazine's editor, John W. Campbell, encouraged Asimov to continue writing.
5. The young writer researched scientific principles to make his stories better.
6. Asimov's "Foundation" series of novels is a "future history."
7. The World Science Fiction Convention awarded the series a Hugo Award.
8. Asimov used Paul French as a pseudonym for the publication of his series of David Starr space ranger novels.
9. *Biochemistry and Human Metabolism* was Asimov's first nonfiction book.
10. Asimov coined the term *robotics*.

8c   Identifying Phrases and Clauses

Individual words may be joined to build *phrases* and *clauses*.

(1)   Identifying phrases

    A **phrase** is a grammatically ordered group of related words that lacks a subject or predicate or both and functions as a single part of speech. It cannot stand alone as an independent sentence.

    A **verb phrase** consists of a main verb and all its auxiliary verbs.

    Time is flying.

    A **noun phrase** includes a noun or pronoun plus all related modifiers.

    I'll climb the highest mountain.

    A **prepositional phrase** consists of a preposition, its object, and any modifiers of that object.

▶ See 8e1

1. Isaac Asimov first saw science [do] fiction stories in the newsstand of his parent's Brooklyn candy store.
2. He practiced [do] writing by telling his schoolmates stories.
3. Asimov published [do] his first story in *Astounding Science Fiction*.
4. The magazine's editor, John W. Campbell, encouraged [do] Asimov to continue writing.
5. The young writer researched [do] scientific principles to make his stories better.
6. Asimov's "Foundation" series of novels is [sc] a "future history."
7. The World Science Fiction Convention awarded [io] the series a [do] Hugo Award.
8. Asimov used [do] Paul French as a pseudonym for the publication of his series of David Starr space ranger novels.
9. *Biochemistry and Human Metabolism* was Asimov's first [sc] nonfiction book.
10. Asimov coined [do] the term *robotics*. [oc]

They discussed the ethical implications of the animal studies.

He was last seen heading into the sunset.

See ◀
8e2
A **verbal phrase** consists of a verbal and its related objects, modifiers, or complements. A verbal phrase may be a **participial phrase**, a **gerund phrase**, or an **infinitive phrase**.

> Encouraged by the voter turnout, the candidate predicted a victory. (participial phrase)
>
> Taking it easy always makes sense. (gerund phrase)
>
> The jury recessed to evaluate the evidence. (infinitive phrase)

See ◀
8e3
An **absolute phrase** usually consists of a noun or pronoun and a participle, accompanied by modifiers. It modifies an entire independent clause rather than a particular word or phrase.

> Their toes tapping, they watched the auditions.

(2) Identifying clauses

A **clause** is a group of related words that includes a subject and a predicate. An **independent** (main) **clause** may stand alone as a sentence, but a **dependent** (subordinate) **clause** must always be accompanied by an independent clause.

> [Lucretia Mott was an abolitionist.] [She was also a pioneer for women's rights.] (two independent clauses)
>
> [Lucretia Mott was an abolitionist] [who was also a pioneer for women's rights.] (independent clause, dependent clause)
>
> [Although Lucretia Mott was most widely known for her support of women's rights,] [she was also a prominent abolitionist.] (dependent clause, independent clause)

Depending on how they function in a sentence, dependent clauses may be classified as *adjective*, *adverb*, or *noun* clauses.

Adjective clauses, sometimes called **relative clauses**, modify nouns or pronouns and always follow the nouns or pronouns they modify. They are introduced by relative pronouns—*that, what, whatever, which, who, whose, whom, whoever,* or *whomever.* The adverbs *where* and *when* can be used as relative pronouns when the adjective clause modifies a place or time.

> The television series *M*A*S*H,* which depicted life in an army hospital in Korea during the Korean War, ran for eleven years. (Adjective clause modifies the noun *M*A*S*H.*)

> Celeste's grandparents, who were born in Romania, speak little English. (Adjective clause modifies the noun *grandparents.*)

The Pulitzer Prizes for journalism are prestigious awards that are presented in areas such as editorial writing, photography, editorial cartooning, and national and international reporting. (Adjective clause modifies the noun *awards*.)

William Styron's novel *Sophie's Choice* is set in Brooklyn, where the narrator lives in a house painted pink. (Adjective clause modifies the noun *Brooklyn*.)

Adverb clauses modify single words (verbs, adjectives, or adverbs), entire phrases, or independent clauses. They are always introduced by subordinating conjunctions. Adverb clauses provide information to answer the questions *how? where? when? why?* and *to what extent?*

▶ See 9b

Exhausted after the match was over, Kim decided to take a long nap. (Adverb clause modifies *exhausted*, telling *when* Kim was exhausted.)

Mark will go wherever there's a party. (Adverb clause modifies *will go*, telling *where* Mark will go.)

Because 75 percent of its exports are fish products, Iceland's economy is heavily dependent on the fishing industry. (Adverb clause modifies independent clause, telling *why* the fishing industry is so important.)

Life would be easier if weekends were longer. (Adverb clause modifies independent clause, telling *under what conditions* life would be easier.)

Noun clauses do not act as modifiers; they function in a sentence as nouns (as subjects, direct objects, indirect objects, or complements). A noun clause may be introduced by a relative pronoun or by *whether, when, where, why,* or *how.*

Whatever happens to us will be for the best. (Noun clause serves as subject of sentence.)

They finally decided which candidate was most qualified. (Noun clause serves as direct object of verb *decided.*)

What you see is what you get. (First noun clause serves as subject; second noun clause serves as subject complement.)

Elliptical clauses are grammatically incomplete—that is, a part of the subject or predicate or the entire subject or predicate is missing. If the missing part can be easily understood from the context of the sentence, such constructions are acceptable.

▶ See 17h1

Although [they were] full, they could not resist dessert.

He has never been able to read maps as well as his brother [can read maps].

EXERCISE 2

Which of the following groups of words are independent clauses? Which are dependent clauses? Which are phrases? Mark each word group IC, DC, or P.

EXAMPLE: Coming through the rye. (P)

1. Beauty is truth.
2. When knights were bold.
3. In a galaxy far away.
4. He saw stars.
5. I hear a symphony.
6. Whenever you are near.
7. The clock struck ten.
8. The red planet.
9. Slowly I turned.
10. For the longest time.

8d Building Simple Sentences with Individual Words

A simple sentence can consist of just a subject and predicate.

Jessica fell.

Simple sentences can also be considerably more elaborate.

1. Jessica fell in love with Henry Goodyear.

2. Jessica and her sister fell in love with Henry Goodyear.

3. Jessica and her younger sister Victoria almost immediately fell hopelessly in love with the very mysterious Henry Goodyear.

4. Rebounding from unhappy love affairs, Jessica and her younger sister Victoria almost immediately fell hopelessly in love with the very mysterious Henry Goodyear.

5. Rebounding from unhappy love affairs, Jessica and her younger sister Victoria almost immediately fell hopelessly in love with the very mysterious Henry Goodyear, an unfortunate condition for both.

The addition of prepositional phrases (sentence 1), compounds (sentence 2), modifying words (sentence 3), verbal phrases (sentence 4), and absolute phrases (sentence 5) can change the substance and meaning of *Jessica fell* quite substantially. Yet all of these examples are simple sentences because each includes only a single subject and predicate.

(1) Building simple sentences with adjectives and adverbs

Descriptive adjectives and adverbs enrich the meaning of a sentence. Reread sentence 3 above.

▶ See
25a

> Jessica and her younger sister Victoria almost immediately fell hopelessly in love with the very mysterious Henry Goodyear.

In this sentence, two adjectives describe nouns.

Adjective	*Noun*
younger	sister
mysterious	Henry Goodyear

Four adverbs describe the action of verbs or modify adjectives or other adverbs.

Adverb	
almost	immediately (adverb)
immediately	fell (verb)
hopelessly	fell (verb)
very	mysterious (adjective)

▶ See
25a

EXERCISE 3

Label all descriptive adjectives and adverbs in the following sentences.

adj   adv   adj
EXAMPLE: The new freshmen waited patiently in the registration line.

1. He chewed tobacco and spat in the black water beneath the pier. He chewed like a farmer. His pickup, though of recent model, was sufficiently weathered and had a dusty-road look about it. (John Grisham, *The Pelican Brief*)
2. Now when I had mastered the language of this water and had come to know every trifling feature that bordered the great river as familiarly as I knew the letters of the alphabet, I had made a valuable acquisition. (Mark Twain, "Two Ways of Viewing the River")
3. Presently he proceeded again on his forward way. The battle was like the grinding of an immense and terrible machine to him. Its complexities and powers, its grim processes, fascinated him. He must go close and see it produce corpses. (Stephen Crane, *The Red Badge of Courage*)
4. After the hurried good-bys the door had closed and they sat at the table with the tragic wreck of the Thanksgiving turkey before them, their heads turned regretfully toward the young folks' laughter in the hall. (Ralph Ellison, "Did You Ever Dream Lucky")
5. His face was bony and sweaty and bright, with a little pointed nose in the center of it, and his look was different from what it had been at the dinner table. (Flannery O'Connor, "Good Country People")

ANSWERS TO EXERCISE 4

Your students' answers will vary. Some samples follow.

1. Charlotte drove her car <u>quickly</u> (adv)
   along, singing <u>softly</u>. (adv)

2. Upstairs, the <u>sky-blue</u>, <u>feather-</u> (adv) (adj) (adj)
   <u>soft</u>, <u>lightly</u> <u>quilted</u> canopy fluttered (adv) (adj)
   <u>softly</u> over the <u>baby's</u> bassinet. (adv) (adj)

3. The sky was <u>dark</u>, <u>heavy</u> with (adj) (adj)
   clouds.

4. The methods taught for taking tests have been challenged
   <u>frequently</u> by <u>many</u> educators. (adv) (adj)

5. There was a <u>tall</u>, <u>dark</u>, <u>handsome</u>, (adj) (adj) (adj)
   <u>mysterious</u> stranger from town (adj)
   with him.

ANSWERS TO EXERCISE 5

Your students' answers will vary. Some samples follow.

1. garbage (truck)
   computer (program)
   TV (dinner)
   football (player)
   record (player)
   bookstore
   table (lamp)
   sofa (bed)
   shoe (horn)
   light (switch)

2. folded   running
   dried   burning
   chopped   singing
   parked   playing
   irritated   rocking

continued on following page

EXERCISE 4

Using these five sentences as models, write five original simple sentences. Use adverbs and adjectives where the model sentences use them and then underline and label these modifiers.

> **EXAMPLE:** Cathy <u>carefully</u> put the <u>baby</u> bird in the nest. (adv) (adj)
>
> He <u>angrily</u> called the <u>old</u> man into the kitchen. (adv) (adj)

1. Nick turned his head carefully away smiling sweatily. (Ernest Hemingway, *In Our Time*)
2. Outside, the fire-red, gas-blue, ghost-green signs shone smokily through the tranquil rain. (F. Scott Fitzgerald, "Babylon Revisited")
3. The pavement was wet, glassy with water. (Willa Cather, "The Old Beauty")
4. The therapy used for treating burns has been improved considerably in recent years. (Lewis Thomas, "On Medicine and the Bomb")
5. There was a strange, inflamed, flurried, flighty recklessness of activity about him. (Herman Melville, *Bartleby the Scrivener*)

(2) Building simple sentences with nouns and verbals

Nouns and verbals can also help you build richer simple sentences.

Nouns   Nouns can sometimes act as adjectives modifying other nouns.

He needed two cake pans for the layer cake.

See ◄ 21c2   *Verbals*   **Verbals** include present and past participles, infinitives, and gerunds.

Verbals may act as modifiers.

All of the living former presidents attended the funeral. (Present participle serves as adjective.)

The Grand Canyon is the attraction to visit. (Infinitive serves as adjective.)

The puzzle was impossible to solve. (Infinitive serves as adverb.)

Or verbals may act as nouns.

When the going gets tough, the tough get going. (Gerund serves as noun.)

To err is human. (Infinitive serves as noun.)

It took me an entire three-hour lab period to identify my <u>unknown</u>. (Past participle serves as noun.)

*EXERCISE 5

1. List ten nouns that can be used as modifiers.

    EXAMPLES: <u>word</u> processor, <u>truck</u> stop, <u>peanut</u> butter

2. List ten participles that can be used as modifiers.

    EXAMPLES: crushed, ringing

3. Choosing words from your lists, write five original sentences, each of which includes both a noun and a participle used as modifiers.

    EXAMPLE: The <u>word</u> processor was a <u>crushed</u> mass of metal and plastic.

4. Then add adjectives and adverbs to enrich the sentence further.

    EXAMPLE: The <u>new</u> word processor was a <u>gruesomely</u> crushed mass of metal and plastic.

EXERCISE 6

For additional practice in building simple sentences using individual words, combine each of the following groups of sentences into one simple sentence that contains several modifiers. You will have to add, delete, or reorder words.

> **EXAMPLE:** The night was cold. The night was wet. The night scared them. They were terribly scared.
>
> **REVISED:** The cold, wet night scared them terribly.

1. The ship landed. The ship was from space. The ship was tremendous. It landed silently.
2. It landed in a field. The field was grassy. The field was deserted.
3. A dog appeared. The dog was tiny. The dog was abandoned. The dog was a stray.
4. The dog was brave. The dog was curious. He approached the spacecraft. The spacecraft was burning. He approached it carefully.
5. A creature emerged from the spaceship. The creature was smiling. He was purple. He emerged slowly.
6. The dog and the alien stared at each other. The dog was little. The alien was purple. They stared meaningfully.
7. The dog and the alien walked. They walked silently. They walked carefully. They walked toward each other.
8. The dog barked. He barked tentatively. He barked questioningly. The dog was uneasy.
9. The alien extended his hand. The alien was grinning. He extended it slowly. The hand was hairy.

continued from previous page

3. The garbage truck backed into the parked car.

    After the short circuit, the table lamp was a burning tangle of wood and wire.

    A TV dinner of chopped liver was not my idea of gourmet food.

    The bookstore hired a singing clown to celebrate its grand opening.

    The new suntan lotion soothed his irritated skin.

4. The overloaded garbage truck backed into the illegally parked car.

    After the short circuit, the designer table lamp was a fitfully burning tangle of wood and wire.

    A cold TV dinner of tough chopped liver was not my idea of gourmet food.

    The new bookstore hired a ridiculous singing clown to celebrate its grand opening.

    The newly marketed suntan lotion immediately soothed his red, irritated skin.

ANSWERS TO EXERCISE 6

1. The tremendous spaceship landed silently.
2. It landed in a deserted, grassy field.
3. A tiny, stray, abandoned dog appeared.
4. The brave, curious dog carefully approached the burning spacecraft.
5. A smiling purple creature slowly emerged from the spaceship.
6. The little dog and the purple alien stared meaningfully at each other.

continued on following page

continued from previous page

7. The dog and the alien walked silently and carefully toward each other.
8. The uneasy dog barked questioningly and tentatively.
9. The grinning alien slowly extended his hairy hand.
10. In his hand was a green canvas laundry bag.

STUDENTS WORKING TOGETHER

Put students in small groups and give all groups the same paragraph containing numerous modifiers. Groups locate all modifiers in the paragraph and change each modifier so that the paragraph is still logical and coherent but is totally different in meaning. Groups may then compare different new versions of the paragraph with the instructor leading a discussion on modifiers.

STUDENTS WORKING TOGETHER

Small groups each create a short story of about two paragraphs in length, then break that story down into its kernel (or most basic) sentences. Their list of sentences based on the story should resemble the list in Exercise 6. They exchange this list with another group, and then each group combines the kernel sentences it has received into a story. Groups get together and compare their combined versions to the original versions. Together, the groups pick one sentence from each story that combines the same kernels but has different emphases. Using the board or a projector, compare the two sentences for the class. You may also use this as an opportunity to discuss grammar rules in context.

10. In his hand was a bag. The bag was made of canvas. The bag was green. The bag was for laundry.

8e Building Simple Sentences with Phrases

See ◀ 8c1
You can also enrich your sentences with phrases. Because a phrase lacks a subject or predicate (or both), it cannot stand alone as a sentence. Within a sentence, however, phrases add information and provide connections between ideas.

(1) Building simple sentences with prepositional phrases

A **preposition** indicates a relationship between a noun or noun substitute and other words in a sentence. A **prepositional phrase** See ◀ 21f consists of the preposition, its object (the noun or noun substitute), and any modifiers of that object.

> prep modifiers obj
>
> John is the father of two small children. (Prepositional phrase functions as adjective modifying the noun *father.*)

In this example the preposition *precedes* its object. In informal speech, however, the preposition often appears *after* the object.

> obj
>
> I see the person you are looking for.
>
> prep

Prepositional phrases can function in a sentence as adjectives or as adverbs.

> Carry Nation was a crusader for temperance. (Prepositional phrase functions as adjective modifying the noun *crusader.*)

> The Madeira River flows into the Amazon. (Prepositional phrase functions as adverb modifying the verb *flows.*)

EXERCISE 7

Read the following sentences. Underline each prepositional phrase, and then connect it with an arrow to the word it modifies. Finally, tell whether each phrase functions as an adjective or an adverb.

EXAMPLE: During the afternoon the school teacher had been to see Doctor Welling concerning her health. (Sherwood Anderson, *Winesburg, Ohio*) (Prepositional phrase functions as an adverb.)

1. She spread her skirts on the bank around her and folded her hands over her knees. (Eudora Welty, "A Worn Path")
2. A letter doesn't take us by surprise in the middle of dinner, or intrude when we are with other people, or ambush us in the midst of other thoughts. (Ellen Goodman, "Life in a Bundle of Letters")
3. He passed the old shack on the corner—the wooden fire-trap where S. Goldberg ran his wiener stand. Then he passed the Singer place next door, with its gleaming display of new machines. (Thomas Wolfe, "The Lost Boy")
4. The practical thing was to find rooms in the city, but it was a warm season, and I had just left a country of wide lawns and friendly trees, so when a young man at the office suggested that we take a house together in a commuting town, it sounded like a great idea. (F. Scott Fitzgerald, *The Great Gatsby*)
5. The front yard was parted in the middle by a sidewalk from gate to door-step, a sidewalk edged on either side by quart bottles driven neck down into the ground on a slant. (Zora Neale Hurston, "The Gilded Six-Bits")

EXERCISE 8

For additional practice in using prepositional phrases, combine each pair of sentences to create one simple sentence that includes a prepositional phrase. You may add, delete, or reorder words. Some sentences may have more than one correct version.

EXAMPLE: America's drinking water is being contaminated. Toxic substances are contaminating it.

REVISED: America's drinking water is being contaminated by toxic substances.

1. Toxic waste disposal presents a serious problem. Americans have this problem.
2. Hazardous chemicals pose a threat. People are threatened.
3. Some towns, like Times Beach, Missouri, have been completely abandoned. Their residents have abandoned them.
4. Dioxin is one chemical. It has serious toxic effects.
5. Dioxin is highly toxic. The toxicity affects animals and humans.
6. Toxic chemical wastes like dioxin may be found. Over fifty thousand dumps have them.
7. Industrial parks contain toxic wastes. Open pits, ponds, and lagoons are where the toxic substances are.
8. Toxic wastes pose dangers. The land, water, and air are endangered.
9. In addition, toxic substances are a threat. They threaten our public health and our economy.
10. Immediate toxic waste cleanup would be a tremendous benefit. Americans are the ones who would benefit.

ANSWERS TO EXERCISE 7

1. on the bank→spread (adverb)
   around her→spread (adverb) or→bank (adjective)
   over her knees→folded (adverb)
2. by surprise→take (adverb)
   in the middle→take (adverb)
   of dinner→middle (adjective)
   with other people→are (adverb)
   in the midst→ambush (adverb)
   of other thoughts→midst (adjective)
3. on the corner→shack (adjective)
   with its gleaming display→place (adjective)
   of new machines→display (adjective)
4. in the city→rooms (adjective) or→find (adverb)
   of wide lawns and friendly trees→country (adjective)
   at the office→man (adjective)
   in a commuting town→house (adjective) or→take (adverb)
5. in the middle→parted (adverb)
   by a sidewalk→parted (adverb)
   from gate to doorstep→sidewalk (adjective) or→parted (adverb)
   on either side→edged (adverb)
   by quart bottles driven neck down→edged (adverb)
   into the ground→driven (adverb)
   on a slant→driven (adverb)

ANSWERS TO EXERCISE 8

1. Toxic waste disposal presents a serious problem for Americans.
2. People are threatened by hazardous chemicals.
3. Some towns, like Times Beach, Missouri, have been completely abandoned by their residents.
4. Dioxin is one chemical with serious toxic effects.

continued on following page

continued from previous page

5. Dioxin is highly toxic to animals and humans.
6. Toxic chemical wastes like dioxin may be found in over fifty thousand dumps.
7. Industrial parks contain toxic wastes in open pits, ponds, and lagoons.
8. Toxic wastes pose dangers to the land, water, and air.
9. In addition, toxic substances are a threat to our public health and economy.
10. Americans would benefit from immediate toxic waste cleanup.

STUDENTS WORKING TOGETHER

Assign each student a partner. Both students compose sentences using verbals and then exchange sentences for the partner to identify the verbals.

(2) Building simple sentences with verbal phrases

A **verbal phrase** consists of a verbal (participle, gerund, or infinitive) and its related objects, modifiers, or complements. Some verbal phrases act as nouns. **Gerund phrases**, for example, like gerunds themselves, are always used as nouns. **Infinitive phrases** may also be used as nouns.

<u>Making a living</u> isn't always easy. (Gerund phrase serves as sentence's subject.)

Wendy appreciated <u>Tom's being honest</u>. (Gerund phrase serves as object of verb *appreciated*.)

The entire town was shocked by <u>their breaking up</u>. (Gerund phrase is object of preposition *by*.)

<u>To know him</u> is <u>to love him</u>. (Infinitive phrase *to know him* serves as sentence's subject; infinitive phrase *to love him* is subject complement.)

Other verbal phrases act as modifiers. **Participial phrases** are always used to modify nouns or pronouns, and **infinitive phrases** may function as adjectives or as adverbs.

<u>Fascinated by Scheherazade's story</u>, they waited anxiously for the next installment. (Participial phrase modifies pronoun *they*.)

The next morning young Goodman Brown came slowly into the street of Salem Village, <u>staring around him like a bewildered man</u>. (Nathaniel Hawthorne, "Young Goodman Brown") (Participial phrase modifies noun *Goodman Brown*.)

Henry M. Stanley went to Africa <u>to find Dr. Livingstone</u>. (Infinitive phrase modifies verb *went*.)

It wasn't the ideal time <u>to do homework</u>. (Infinitive phrase modifies noun *time*.)

See ◄
15a2
15b1

CLOSE-UP

BUILDING SIMPLE SENTENCES

When you use verbal phrases as modifiers, be especially careful not to create misplaced or dangling modifiers.

EXERCISE 9

For practice in using verbal phrases, combine each of these sentence pairs to create one simple sentence that contains a participial phrase, a gerund phrase, or an infinitive phrase. Underline and label the verbal phrase in your sentence. You will have to add, delete, or reorder words, and you may find more than one way to combine each pair.

EXAMPLE: The American labor movement has helped millions of workers. It has won them higher wages and better working conditions.

REVISED: The American labor movement has helped millions of
participial phrase
workers, <u>winning them higher wages and better working conditions</u>.

1. In 1912 the textile workers of Lawrence, Massachusetts, went on strike. They were demonstrating for "Bread and Roses, too."
2. The workers wanted higher wages and better working conditions. They felt trapped in their miserable jobs.
3. Mill workers toiled six days a week. They earned about $1.50 for this.
4. Most of the workers were women and children. They worked up to sixteen hours a day.
5. The mills were dangerous. They were filled with hazards.
6. Many mill workers joined unions. They did this to fight exploitation by their employers.
7. They wanted to improve their lives. This was their goal.
8. Finally, twenty-five thousand workers walked off their jobs. They knew they were risking everything.
9. The police and the state militia were called in. Attacking the strikers was their mission.
10. After 63 days, the American Woolen Company surrendered. This ended the strike with a victory for the workers.

(Adapted from William Cahn, *Lawrence 1912: The Bread and Roses Strike*)

(3)　Building simple sentences with absolute phrases

An **absolute phrase** is usually composed of a noun or pronoun and a past or present participle, along with its modifiers.

<u>All things considered</u>, I prefer Maine's cold winters to California's smog.

Sometimes, however, an infinitive phrase functions as an absolute phrase.

<u>To make a long story short</u>, our team lost.

Absolute phrases act as modifiers, but they are not connected grammatically to any particular word or phrase in a sentence. Instead, an absolute phrase modifies the whole independent clause to which it is linked.

1. In 1912 the textile workers of Lawrence, Massachusetts, went on strike, <u>demonstrating for "Bread and Roses, too."</u> (participial phrase)
2. The workers, <u>feeling trapped in their miserable jobs</u>, wanted higher wages and better working conditions. (participial phrase)
3. Mill workers toiled six days a week, <u>earning about $1.50</u>. (participial phrase)
4. Most of the workers were women and children, <u>working up to sixteen hours a day</u>. (participial phrase)
5. The mills were dangerous, <u>filled with hazards</u>. (participial phrase)
6. <u>Many mill workers joined unions to fight exploitation by their employers</u>. (infinitive phrase)
7. Their goal was <u>to improve their lives</u>. (infinitive phrase)
8. Finally, <u>risking everything</u>, twenty-five thousand workers walked off their jobs. (participial phrase)
9. The police and the state militia were called in <u>to attack the strikers</u>. (infinitive phrase)
10. After 63 days, the American Woolen Company surrendered, <u>ending the strike with a victory for the workers</u>. (participial phrase)

1. Notre Dame stood majestically, its rose window glowing in the darkness.
2. Our feet [being] tired, we took a boat ride down the Seine.
3. Its doors closed only on Tuesdays, the Louvre is open six days a week.
4. The Jeu de Paume displays Impressionist paintings, its exhibits showcasing Manet, Degas, Renoir, and van Gogh.
5. Our francs spent, we were forced to cut our vacation short.

1. *I Know Why the Caged Bird Sings,* the first book in Maya Angelou's autobiography, deals primarily with her life as a young girl in Stamps, Arkansas.
2. Catgut, a tough cord generally made from the intestines of sheep, is used for tennis rackets, violin strings, and for surgical stitching.
3. Hermes, the messenger of the Greek gods, is usually portrayed as an athletic youth wearing a cap and winged sandals.
4. Emiliano Zapata, a hero of the Mexican Revolution, is credited with effecting land reform in his home state of Morelos.
5. Pulsars, celestial objects that emit regular pulses of radiation, were first discovered in 1967.

Demonstrate to the class how a simple sentence may be enriched by each kind of addition discussed in this

continued on following page

*EXERCISE 10

For practice in using absolute phrases, combine each group of sentences to create one simple sentence that includes an absolute phrase. You will have to change, add, delete, or reorder some words.

EXAMPLE: Paris was beautiful. Its streets were exceptionally clean.

REVISED: Paris was beautiful, its streets [being] exceptionally clean.

1. Notre Dame stood majestically.
   Its rose window glowed in the darkness.
2. We took a boat ride down the Seine.
   Our feet were tired.
3. The Louvre is open six days a week.
   Its doors are closed on Tuesdays.
4. The Jeu de Paume displays Impressionist paintings. Its exhibits showcase Manet, Degas, Renoir, and van Gogh.
5. We were forced to cut our vacation short.
   Our francs were spent.

(4) Building simple sentences with appositives

An **appositive** is a noun or a noun phrase that identifies, in different words, the noun or pronoun it follows.

Roy's horse Trigger was a golden palomino. (Appositive *Trigger* identifies noun *horse.*)

Farrington hated his boss, a real tyrant. (Appositive *real tyrant* identifies noun *boss.*)

Appositives expand a sentence by defining the nouns or pronouns they modify, giving them new names or adding identifying details. Appositives can substitute for the nouns or pronouns to which they refer.

Francie Nolan, the protagonist of Betty Smith's novel *A Tree Grows in Brooklyn,* is determined to finish high school and make a better life for herself. (Francie Nolan = the protagonist of the novel)

As in the examples here, appositives are frequently used without special introductory phrases. They may also be introduced by *such as, or, that is, for example,* or *in other words.*

A regional airline, such as Southwest, frequently accounts for more than half the departures at so-called second-tier airports.

Rabies, or hydrophobia, was nearly always fatal until Pasteur's work.

EXERCISE 11

For practice in using appositives when you write, build ten new simple sentences by combining each of the following pairs. Make one sentence in each pair an appositive. You may need to delete or reorder words in some cases.

EXAMPLE: René Descartes was a noted French philosopher. Descartes is best known for his famous declaration, "I think, therefore I am."

REVISED: René Descartes, a noted French philosopher, is best known for his famous declaration, "I think, therefore I am."

1. *I Know Why the Caged Bird Sings* is the first book in Maya Angelou's autobiography. It deals primarily with her life as a young girl in Stamps, Arkansas.
2. Catgut is a tough cord generally made from the intestines of sheep. Catgut is used for tennis rackets, for violin strings, and for surgical stitching.
3. Hermes was the messenger of the Greek gods. He is usually portrayed as an athletic youth wearing a cap and winged sandals.
4. Emiliano Zapata was a hero of the Mexican Revolution. He is credited with effecting land reform in his home state of Morelos.
5. Pulsars are celestial objects that emit regular pulses of radiation. Pulsars were first discovered in 1967.

(5) Building simple sentences with compound constructions

Compound constructions consist of two or more grammatically equivalent items, parallel in importance. Within simple sentences, compound words or phrases—subjects, predicates, complements, or modifiers—are joined in one of three ways.

• With commas:

He took one <u>long</u>, <u>loving</u> look at his '57 Chevy.

• With a coordinating conjunction:

They <u>reeled</u>, <u>whirled</u>, <u>flounced</u>, <u>capered</u>, <u>gamboled</u>, and <u>spun</u>. (Kurt Vonnegut, Jr., "Harrison Bergeson")

• With a pair of correlative conjunctions (*both/and, not only/but also, either/or, neither/nor, whether/or*):

<u>Both milk and carrots</u> contain Vitamin A.

<u>Neither the twentieth-century poet Sylvia Plath nor the nineteenth-century poet Emily Dickinson</u> achieved recognition during her lifetime.

continued from previous page

chapter. Begin with a two-word sentence—for example, "Maureen laughed." With the help of the students, write additional simple sentences that enrich the original using prepositional phrases, adjectives, and adverbs, and so forth. Then, write several basic sentences on the board. Ask students to enrich those sentences by adding compound constructions. Call upon volunteers to share their enriched sentences and discuss the methods they used.

ANSWERS TO EXERCISE 12

Your students' answers will vary. Some samples follow.

A.
1. Cortés and DeSoto explored the New World.
2. Virginia Woolf and Gertrude Stein wrote novels.
3. Edison invented the phonograph and devised a complete system of electric power distribution for potential customers.
4. PBS television and radio produce and air educational programming.
5. Thomas Jefferson wrote and signed the Declaration of Independence.

B.
1. Hernán Cortés and Hernando De Soto explored the New World for Spain in the 1500s.
2. Two of the most influential literary figures of the early twentieth century, Virginia Woolf and Gertrude Stein wrote novels, essays, and autobiographies.
3. After he moved to his laboratory in Menlo Park, New Jersey, in

continued on following page

continued from previous page

1876, the famed U.S. inventor Thomas Alva Edison invented the phonograph and devised a complete system of electrical power distribution for potential customers.

See ◄
16a

4. In addition to providing music, entertainment specials, and news, PBS airs educational television and radio programming.

5. Before he became George Washington's Secretary of State and the third American president, Thomas Jefferson wrote and signed the Declaration of Independence, a seminal document in the history of the United States.

ANSWERS TO EXERCISE 13

1. Robert Ludlum, Tom Clancy, and John le Carré write best-selling spy thrillers.
2. Smoking, a high-fat, high-cholesterol diet, and stress can cause heart disease.
3. Sylvia Plath and Adrienne Rich were important twentieth-century poets who frequently wrote about depression and despair.
4. Successful rock bands give concerts, record albums, make videos, and license merchandise bearing their names and likenesses.
5. Sports superstars such as Michael Chang and Michael Jordan earn additional income by making personal appearances and endorsing products.

CLOSE-UP 🔍

BUILDING SIMPLE SENTENCES

Be sure to use parallel structure when you join two or more grammatically equivalent words or phrases in a compound construction.

EXERCISE 12

A. Expand each of the following sentences by using compound subjects and/or predicates.

EXAMPLE: Bill played guitar.

REVISED: Bill and Juan played guitar and sang.

B. Then expand your simple sentence with modifying words and phrases, using compound constructions whenever possible.

EXAMPLE: Despite butterflies in their stomachs and a restless audience, Bill and Juan played guitar and sang.

1. Cortés explored the New World.
2. Virginia Woolf wrote novels.
3. Edison invented the phonograph.
4. PBS airs educational television programming.
5. Thomas Jefferson signed the Declaration of Independence.

EXERCISE 13

To practice building sentences with compound subjects, predicates, and modifiers, combine the following groups of sentences into one.

EXAMPLE: Marion studied. Frank studied. They studied quietly. They studied diligently.

REVISED: Marion and Frank studied quietly and diligently.

1. Robert Ludlum writes best-selling spy thrillers. Tom Clancy writes best-selling spy thrillers. John le Carré writes best-selling spy thrillers.
2. Smoking can cause heart disease. A high-fat, high-cholesterol diet can cause heart disease. Stress can cause heart disease.
3. Sylvia Plath was an important twentieth-century poet. Adrienne Rich was an important twentieth-century poet. They both frequently wrote about depression and despair.
4. Successful rock bands give concerts. They record albums. They make videos. They license merchandise bearing their names and likenesses.
5. Sports superstars like Michael Chang and Michael Jordan earn additional income by making personal appearances. They earn money by endorsing products.

CHAPTER 9

Building Compound and Complex Sentences

9a   Building Compound Sentences

The pairing of similar elements—words, phrases, or clauses—to give equal weight to each is called **coordination**. Coordination can be used in simple sentences to join similar elements into compound subjects, predicates, complements, or modifiers. It can also join two independent clauses to form a compound sentence.

▶ See
8e5

> **USE COMPOUND SENTENCES**
> - to show addition (*and, in addition to, not only . . . but also,* semicolon)
> - to show contrast (*but, however*)
> - to show cause and effect (*so, therefore, consequently*)
> - to present a choice of alternatives (*or, either . . . or*)

A **compound sentence** is formed when two or more independent clauses are connected with *coordinating conjunctions, conjunctive adverbs, correlative conjunctions, semicolons,* or *colons*.

(1)   Using coordinating conjunctions

You can join two independent clauses with a coordinating conjunction.

[The cowboy is a workingman], yet [he has little in common with the urban blue-collar worker]. (John R. Erickson, *The Modern Cowboy*)

Complex Sentences
You may want to teach this chapter along with the sections on punctuating compound and complex sentences in the chapters on commas and semicolons. For advanced students, you might also include some work on balanced sentences in the chapter on parallel structures.

Students may find it helpful to review the concept of the *phrase* (presented in section 8c). A *phrase* is a group of related words without a subject or predicate. The phrase can expand and add to the information conveyed in a sentence *but does not express a complete thought.* Clear recognition of prepositional, infinitive, participial, and gerund phrases will prevent confusion as students begin to study clause patterns.

STUDENTS WORKING TOGETHER

Have students analyze one of their own papers, determining which sentences are simple, compound, and complex, and which sentences need to be enriched. Encourage the students to read their work out loud to determine whether they are using too many sentences of any one type. For example, too many simple sentences will sound choppy, and too many compound sentences will produce a singsong rhythm. It might also be helpful to devote part of a class session to enriching compound and complex sentences (perhaps using sentences "volunteered" by students from their own work). Suggest some ways to help students distinguish between enriched and overblown sentences.

IN THE CLASSROOM

Increase student awareness of how frequently and easily we use adverb clauses in spoken and written English. Working with the class or group as a whole, introduce subordinating conjunctions one at a time—*because, when, now that, if, after, as soon as,* and so on—and ask students to create sentences that begin with an adverb clause. Write some of the sentences on the board; then, ask students to determine which relationship (time, cause, purpose, etc.) is expressed.

STUDENTS WORKING TOGETHER

Students work with partners. Each student takes one paragraph from a partner's piece of writing and changes each type of sentence to another type—for example, a compound sentence to a complex sentence. Together, partners examine both versions of each paragraph and answer these questions: Which is better? How have meanings been altered on the sentence or paragraph level?

[In the fall the war was always there], but [we did not go to it any more]. (Ernest Hemingway, "In Another Country")

[She carried a thin, small cane made from an umbrella], and [with this she kept tapping the frozen earth in front of her]. (Eudora Welty, "A Worn Path")

CLOSE-UP

BUILDING COMPOUND SENTENCES

See ◄ 28a

Use a comma before a coordinating conjunction—*and, or, nor, but, for, so,* or *yet*—that joins two independent clauses.

(2) Using conjunctive adverbs and transitional expressions

You can join two independent clauses with a **conjunctive adverb** or a **transitional expression**.

[Peter dropped History 102]; instead, [he took Education 242].

[The saxophone does not belong to the brass family]; in fact, [it is a member of the woodwind family].

[Aerobic exercise can help lower blood pressure]; however, [those with high blood pressure should still limit salt intake].

See ◄ 21e

See ◄ 4d2

Commonly used conjunctive adverbs include *consequently, finally, still,* and *thus.* Commonly used transitional expressions include *for example, in fact, on the other hand,* and *for instance.*

CLOSE-UP

BUILDING COMPOUND SENTENCES

Use a semicolon—not a comma—before a conjunctive adverb that joins two independent clauses.

(3) Using correlative conjunctions

See ◄ 21g

You can use **correlative conjunctions** to join two independent clauses into a compound sentence.

Sharon not only passed the exam, but she also received the highest grade in the class.

Either he left his coat in his locker, or he left it on the bus.

(4)   Using semicolons

A **semicolon** can link two closely related independent clauses.

▶ See 29a

[Alaska is the largest state]; [Rhode Island is the smallest].

[Theodore Roosevelt was president after the Spanish American War]; [Andrew Johnson was president after the Civil War].

(5)   Using colons

A **colon** can link two independent clauses.

▶ See 32a

He got his orders: He was to leave for France on Sunday.

They thought they knew the outcome: Truman would lose to Dewey.

EXERCISE 1

Bracket the independent clauses in these compound sentences.

EXAMPLE: [He saw them and admired them], but [he felt no joy]. (Thomas Wolfe, "The Lost Boy")

1. It was dusk; the whippoorwills had already begun. (William Faulkner, "Barn Burning")
2. There were a couple of old women ahead of her, and then a miserable-looking poor devil came and wedged me in at the other side, so that I couldn't escape even if I had the courage. (Frank O'Connor, "First Confession")
3. Lights shone from the scattered houses, and a gang of labourers who stood beside the track waved their lanterns. (Willa Cather, "Paul's Case")
4. He'd been a waiter at the best hotel in Tangier just before he'd come to the United States, and a prince, a duke, and two princesses had been among his patrons. (Prudencio de Pereda, "Conquistador")
5. It was such a very uproarious joke that her face turned slightly purple, and she screamed with laughter. (Katherine Anne Porter, "Rope")

*EXERCISE 2

After reading the following paragraph, use coordination to build as many compound sentences as you think your readers need to understand the links between ideas. When you have finished, bracket the independent clauses and underline the coordinating conjunctions, correlative conjunctions, or punctuation marks that link clauses.

ANSWERS TO EXERCISE 1

1. [It was dusk]; [the whippoorwills had already begun].
2. [There were a couple of old women ahead of her], and [then a miserable-looking poor devil came and wedged me in at the other side, so that I couldn't escape even if I had the courage].
3. [Lights shone from the scattered houses], and [a gang of labourers who stood beside the track waved their lanterns].
4. [He'd been a waiter at the best hotel in Tangier just before he'd come to the United States], and [a prince, a duke, and two princesses had been among his patrons].
5. [It was such a very uproarious joke that her face turned slightly purple], and [she screamed with laughter].

ANSWERS TO EXERCISE 2

Your students' answers will vary.

[Alan Bakke applied to medical school at the University of California at Davis], but [he was rejected in 1973 and 1974.] [Bakke's grades were good.] [He said he was the victim of reverse discrimination], for [the medical school had designated sixteen out of every hundred slots for minority students,] and [Bakke was white.] [Bakke said some minority students, less qualified than he, had been admitted,] so [he sued the University of California.] [The case went to the Supreme Court,] and [in 1978, Bakke won his suit.] [Today, Bakke is a doctor.]

ANSWERS TO EXERCISE 3

1. The average American consumes 128 pounds of sugar each year; *therefore*, most Americans eat much more sugar than any other food additive, including salt.
2. Many of us are determined to reduce our sugar intake; *consequently*, we have consciously eliminated sweets from our diets.
3. Unfortunately, sugar is found *not only* in sweets, *but also* in many processed foods.
4. Processed foods like puddings and cake contain sugar, *and* foods like ketchup and spaghetti sauce do, too.
5. We are trying to cut down on sugar, *but* we find limiting sugar intake extremely difficult.
6. Processors may *not only* use sugar in foods for taste, *but* they may *also* use it to help prevent foods from spoiling and to help improve their texture and appearance.
7. Sugar comes in many different forms, *so* it is easy to overlook it on a package label.
8. Sugar may be called sucrose or fructose, *and* it may also be called corn syrup, corn sugar, brown sugar, honey, or molasses.
9. No sugar is more nourishing than the others; *therefore*, it really doesn't matter which is consumed.
10. Sugars contain empty calories; thus, whenever possible, they should be avoided.

The case of Alan Bakke presents an interesting footnote to the history of affirmative action legislation. Bakke applied to medical school at the University of California at Davis. He was rejected in 1973 and 1974. Bakke's grades were good. He said he was the victim of reverse discrimination. The medical school had designated sixteen out of every hundred slots for minority students. Bakke is white. Bakke said some minority students, less qualified than he, had been admitted. Bakke sued the University of California. The case went to the Supreme Court. In 1978, Bakke won his suit. Today, Bakke is a doctor.

EXERCISE 3

Add appropriate coordinating conjunctions, conjunctive adverbs, or correlative conjunctions as indicated to combine each pair of sentences into one well-constructed compound sentence that retains the meaning of the original pair. Be sure to use correct punctuation.

EXAMPLE: The American population is aging. People seem to be increasingly concerned about what they eat. (coordinating conjunction)

REVISED: The American population is aging, so people seem to be increasingly concerned about what they eat.

1. The average American consumes 128 pounds of sugar each year. Most of us eat much more sugar than any other food additive, including salt. (conjunctive adverb)
2. Many of us are determined to reduce our sugar intake. We have consciously eliminated sweets from our diets. (conjunctive adverb)
3. Unfortunately, sugar is not found only in sweets. It is also found in many processed foods. (correlative conjunction)
4. Processed foods like puddings and cake contain sugar. Foods like ketchup and spaghetti sauce do, too. (coordinating conjunction)
5. We are trying to cut down on sugar. We find limiting sugar intake extremely difficult. (coordinating conjunction)
6. Processors may use sugar in foods for taste. They may also use it to help prevent foods from spoiling and to improve their texture and appearance. (coorelative conjunction)
7. Sugar comes in many different forms. It is easy to overlook it on a package label. (coordinating conjunction)
8. Sugar may be called sucrose or fructose. It may also be called corn syrup, corn sugar, brown sugar, honey, or molasses. (coordinating conjunction)
9. No sugar is more nourishing than the others. It really doesn't matter which is consumed. (conjunctive adverb)
10. Sugars contain empty calories. Whenever possible, they should be avoided. (conjunctive adverb) (Adapted from *Jane Brody's Nutrition Book*)

9b Building Complex Sentences

You use **subordination** when you want to indicate that one idea is less important than another. You place the more important idea in the independent clause and the less important idea in the dependent clause. The result is a complex sentence.

A **complex sentence** consists of one independent or main clause and at least one dependent or subordinate clause. **Independent clauses** can stand alone as sentences.

The hurricane began.

The town was evacuated.

Dependent clauses, introduced by subordinating conjunctions or relative pronouns, cannot stand alone.

After the town was evacuated

Which threatened to destroy the town

Dependent clauses must be combined with independent clauses to form sentences. The subordinating conjunction or relative pronoun links the independent and dependent clauses and shows the relationship between them.

      dependent clause   independent clause
[After the town was evacuated], [the hurricane began].

    independent clause   dependent clause
[Officials watched the storm], [which threatened to destroy the town].

Sometimes a dependent clause may be placed within an independent clause.

          dependent clause
Town officials, [who were very concerned], watched the storm.

Depending on their function in a sentence, dependent clauses may be **adverb** or **adjective clauses**. ▶ See 8c2

When the school board voted to ban Judy Blume's books, parents protested. (Adverb clause tells when parents protested.)

The Graduate was the film that launched Dustin Hoffman's career. (Adjective clause modifies *film*.)

Adverb clauses are introduced by **subordinating conjunctions**.

Independent clauses are designated IC; dependent clauses are designated DC.

1. [The people were clustered thickly about the old man,] all of them intermittently flicking glances toward me [*as* they talked animatedly in their Mandinka tongue.] (Dependent clause serves as an adverb.)

2. [The first thing we saw [when we climbed the riverbank to the village of Providence] was the deer.] (dependent clause, embedded within independent clause, serves as an adverb)

3. [We cannot help regarding a camel as aloof and unfriendly] [*because* it mimics, quite unwittingly and for other reasons, the "gesture of haughty rejection" common to so many human cultures.] (Dependent clause serves as an adverb.)

4. [These are both hopeful and frustrating times for those] [*who* want to improve the nation's science and math education.] (Dependent clause serves as an adjective.)

5. [*Although* there was always generosity in the Negro neighborhood,] [it was indulged on pain of sacrifice.] (Dependent clause serves as an adverb.)

ANSWERS TO EXERCISE 5

Your students' imitations will vary.

1. [I said] [what I meant.]
   [He realized] [what he needed.]
2. [Isadora Duncan is the dancer]
   [who best exemplifies the phrase
   "poetry in motion."]
   [Roger Marris is the baseball
   player] [who broke Babe Ruth's
   home run record.]
3. [Because she was considered a
   heretic,] [Joan of Arc was burned
   at the stake.]
   [Because he believed in nonvio-
   lence,] [Martin Luther King, Jr.,
   advocated the ideal of reform
   through passive resistance.]
4. [The oracle of Delphi predicted]
   [that Oedipus would murder his
   father and marry his mother.]
   [The leaders of the French Revolution believed] [that they could
   bring about liberty, equality, and
   fraternity.]
5. [The ghost vanished] [before
   Hamlet could question him further.]
   [The flour should be sifted] [be-
   fore it is added to the batter.]

ANSWERS TO EXERCISE 6

1. Some twelve million people who
   are currently out of work need
   new skills for new careers.
2. Although talented high school
   graduates are usually encouraged to go to college, some high
   school graduates are now starting
   to see that a college education
   may not guarantee them a job.
3. Since a college education can
   cost a student more than $80,000,
   vocational education is becoming
   increasingly important.

continued on following page

COMMONLY USED SUBORDINATING CONJUNCTIONS

after	in order that	unless
although	now that	until
as	once	when
as if	rather than	whenever
as though	since	where
because	so that	whereas
before	that	wherever
even though	though	while
if		

Adjective clauses are introduced by **relative pronouns**.

RELATIVE PRONOUNS

that	whatever	who (whose, whom)
what	which	whoever (whomever)

EXERCISE 4

Bracket and label the independent and dependent clauses in these sentences, and then underline the subordinating conjunctions or relative
pronouns. Finally, indicate the function of each dependent clause.

                IC   DC

  EXAMPLE: ["Jet-stream art" is created] when paint thrown into the
exhaust of a jet engine is spattered onto a giant canvas]. (Dependent
clause serves as an adverb.)

1. The people were clustered thickly about the old man, all of them intermittently flicking glances toward me as they talked animatedly in their
   Mandinka tongue. (Alex Haley, *Roots*)
2. The first thing we saw when we climbed the riverbank to the village
   Providencia was the deer. (Annie Dillard, "The Deer at Providencia")
3. We cannot help regarding a camel as aloof and unfriendly because it
   mimics, quite unwittingly and for other reasons, the "gesture of
   haughty rejection" common to so many human cultures. (Stephen Jay
   Gould, *The Panda's Thumb*)
4. These are both hopeful and frustrating times for those who want to
   improve the nation's science and math education. (Arlen J. Large, *Wall
   Street Journal*)

5. Although there was always generosity in the Negro neighborhood, it was indulged on pain of sacrifice. (Maya Angelou, *I Know Why the Caged Bird Sings*)

EXERCISE 5

Bracket the independent and dependent clauses in the following complex sentences. Then, using these sentences as models, create two new complex sentences in imitation of each. For each set of new sentences, use the same subordinating conjunction or relative pronoun that appears in the original.

1. I said what I meant.
2. Isadora Duncan is the dancer who best exemplifies the phrase "poetry in motion."
3. Because she was considered a heretic, Joan of Arc was burned at the stake.
4. The oracle at Delphi predicted that Oedipus would murder his father and marry his mother.
5. The ghost vanished before Hamlet could question him further.

EXERCISE 6

Use a subordinating conjunction or relative pronoun to combine each of the following pairs of sentences into one well-constructed sentence. The conjunction or pronoun you select must clarify the relationship between the two sentences. You will have to change or reorder words, and in most cases you have a choice of connecting words.

EXAMPLE: Some colleges are tightening admissions requirements. The pool of students is growing smaller.

REVISED: Although the pool of students is growing smaller, some colleges are tightening admissions requirements.

1. Some twelve million people are currently out of work. They need new skills for new careers.
2. Talented high school students are usually encouraged to go to college. Some high school graduates are now starting to see that a college education may not guarantee them a job.
3. A college education can cost a student more than $80,000. Vocational education is becoming increasingly important.
4. Vocational students complete their work in less than four years. They can enter the job market more quickly.
5. Nurses' aides, paralegals, travel agents, and computer technicians do not need college degrees. They have little trouble finding work.
6. Some four-year colleges are experiencing growth. Public community colleges and private trade schools are growing much more rapidly.

continued from previous page

4. Because vocational students complete their work in less than four years, they can enter the job market more quickly.
5. Nurses' aides, paralegals, travel agents, and computer technicians, who do not need college degrees, have little trouble finding work.
6. Some four-year colleges are experiencing growth, though public community colleges and private trade schools are growing much more rapidly.
7. The best vocational schools, which are responsive to the needs of local businesses, train students for jobs that actually exist.
8. For instance, a school in Detroit might offer advanced auto design, while one in New York City might focus on fashion design.
9. Other schools, which offer courses in horticulture, respiratory therapy, and computer programming, are able to place their graduates easily.
10. Laid-off workers, returning housewives, recent high school graduates, and even college graduates, who all hope to find rewarding careers, are reexamining vocational education.

ANSWERS TO EXERCISE 7

1. compound-complex
2. compound
3. complex
4. complex
5. simple

[ ] = independent clauses
( ) = dependent clauses
＿ = whole phrase as subject

continued on following page

continued from previous page

1. [Use of the telephone involves

   personal risk (because it involves
   exposure)]; for some, [to be

   "hung up on" is among the worst

   fears]; [others dream of a ring-

   ing telephone and wake up with a
   pounding heart.]

2. [This nation is even more litigious
   than religious], and [the

   school prayer issue has

   prompted more, and more sophisticated, arguments about constitutional law than about the nature
   of prayer.]

3. [The first time (I ever went

   naked in mixed company) was at

   the house of a girl (whose father

   had a bad back and had built himself a sauna in the corner of the
   basement).]

4. [I am the son of Mexican-

   American parents,] (who speak a
   blend of Spanish and English,

   but who read neither language
   easily.)

5. [The most alarming of all man's

   assaults upon the environment is
   the contamination of air, earth,
   rivers, and the sea with dangerous
   and even lethal materials.]

ANSWERS TO EXERCISE 8

Your students' answers will vary.

7. The best vocational schools are responsive to the needs of local businesses. They train students for jobs that actually exist.

8. For instance, a school in Detroit might offer advanced automotive design. A school in New York City might focus on fashion design.

9. Other schools offer courses in horticulture, respiratory therapy, and
computer programming. They are able to place their graduates easily.

10. Laid-off workers, returning housewives, recent high school graduates, and even college graduates are reexamining vocational education. They all hope to find rewarding careers.

9c **Building Compound-Complex Sentences**

A **compound-complex sentence** consists of two or more independent clauses and at least one dependent clause.

dependent clause
[When small foreign imports began dominating the U.S. automobile

independent clause   independent clause
industry], [consumers were very responsive], but [American auto workers were dismayed].

EXERCISE 7

In each of these sentences, identify subjects and verbs; bracket and label
dependent and independent clauses; and identify each sentence as simple, compound, complex, or compound-complex.

1. Use of the telephone involves personal risk because it involves exposure; for some, to be "hung up on" is among the worst fears; others
dream of a ringing telephone and wake up with a pounding heart.
(John Brooks, *Telephone: The First Hundred Years*)

2. This nation is even more litigious than religious, and the school prayer
issue has prompted more, and more sophisticated, arguments about
constitutional law than about the nature of prayer. (George F. Will,
Newsweek)

3. The first time I ever went naked in mixed company was at the house of
a girl whose father had a bad back and had built himself a sauna in the
corner of the basement. (Garrison Keillor, *New Yorker*)

4. I am the son of Mexican-American parents, who speak a blend of
Spanish and English, but who read neither language easily. (Richard
Rodriguez, *Aria: A Memoir of a Bilingual Childhood*)

5. The most alarming of all man's assaults upon the environment is the
contamination of air, earth, rivers, and sea with dangerous and even
lethal materials. (Rachel Carson, *Silent Spring*)

*EXERCISE 8

Write an original sentence in imitation of each sentence in Exercise 7.

STUDENT WRITER AT WORK

BUILDING SENTENCES

A student in a freshman composition class was assigned to interview a grandparent and write a short paper about his or her life. When she set out to turn her grandmother's words into a paper, the student faced a set of choppy notes—words, phrases, and simple sentences—that she had jotted down as her grandmother spoke. She needed to fill out and combine these fragments and short sentences to produce varied, interesting sentences that would establish the relationships among her ideas.

Read the notes, turn them into complete sentences when necessary, and combine sentences wherever it seems appropriate. Your goal is to build simple, compound, and complex sentences enriched by modifiers—without adding any information. When you have finished, revise further to strengthen coherence, unity, and style.

Notes

67 years old. Born in Lykens, PA (old coal-mining town). Got her first paying job at 13. Her parents lied about her age. Working age was 14. Parents couldn't afford all the mouths they had to feed. Before that, she helped with the housework. At work, she was a maid. Got paid only about a dollar a week. Most of that went to her parents. Ate her meals on job. Worked in house where 3 generations of men lived. They all worked in the mines. Had to get up at 4 A.M. First chore was to make lunch for the men. She'd scrub the metal canteens. Then she'd fill them with water. Then she'd make biscuits and broth. Then she'd start breakfast. Mrs. Muller would help. Cooking for 6 hungry men was a real job. Then she did the breakfast dishes. Then she did the chores. The house had 3 stories. She had to scrub floors, dust, and sweep. It wasn't easy. Then Mrs. Muller would need help patching and darning. She had just enough time to get dinner started. Grabbed her meals after the family finished eating. Had no spare time. When not working she had chores to do at home. In spring and summer she would grow vegetables. Canned vegetables for her family. What was left over, she sold. Got married at 16.

STUDENT WRITER AT WORK

Building Sentences

Possible Revision

My grandmother is sixty-seven years old. Born in Lykens, Pennsylvania, an old coal-mining town, she got her first paying job at thirteen. Her parents lied about her age. The working age was fourteen, but her parents could not afford all the mouths they had to feed. Before that, she helped with the housework. At work, she was a maid and got paid only about a dollar a week. Most of that went to her parents, and she ate her meals on the job. She worked in a house where three generations of men, who all worked in the mines, lived. She had to get up at 4 A.M. Her first chore was to make lunch for the men. She would scrub the metal canteens, then she would fill them with water, and then she would make biscuits and broth. Then she'd start breakfast. Mrs. Muller would help because cooking for six hungry men was a real job. Then my grandmother did the breakfast dishes. Then she did the chores. The house had three stories, and she had to scrub floors, dust, and sweep. It wasn't easy. Then Mrs. Muller would need help patching and darning. My grandmother had just enough time to get dinner started. She grabbed her meals after the family finished eating. She had no spare time. When she wasn't working, she had chores to do at home. In spring and summer she would grow vegetables, which she canned for her family. What was left over, she sold. When she was sixteen, she got married.

Commentary on Possible Revision

The writer attempts to build compound and complex sentences; however, there is still little sentence variety and *then* is overused. In addition, the material will have to be divided into paragraphs, and transitions between sentences and paragraphs will have to be supplied.

Writing Emphatic Sentences
This chapter may be most appropriate for writers who have progressed well beyond the basics of grammar and punctuation. The concept of emphasis is a difficult one to teach—in part because (as you may want to remind students) professional writers tend to make instinctive, rather than analytical, choices about word order, sentence structure, repetition, and so on. As your students experiment with these choices in their own writing, conscious use should (over time) give way to a less calculating, more instinctive use of these strategies. Remind students that the too-frequent use of any strategy may result in monotony and "de-emphasis" —in short, in the exact opposite of the intended effect.

Suggest that students go back to some favorite piece of writing and analyze how the writer uses word order, sentence structure, and other strategies to keep the reader's focus exactly where the writer wants it to be. Students may want to bring and share some examples with the class. The following excerpt may be a useful place to start.

"The hotel is empty; the other guests have risked a punishing journey to Swat in hopes of being received by His Highness the Wali. You sleep soundly under a tent of mosquito net and are awakened by the fluting of birds for an English breakfast that begins with porridge and ends with a kidney."

–Paul Theroux,
The Great Railway Bazaar

[Emphasis is on both the beginnings and endings of sentences. But

continued on following page

CHAPTER 10

Writing Emphatic Sentences

When speaking, you add emphasis to your ideas with facial expressions, with gestures, and by raising or lowering your voice. When writing, you use other techniques to highlight important points.

STRATEGIES FOR WRITING EMPHATIC SENTENCES
- Use emphatic word order.
- Use emphatic sentence structure.
- Use parallelism and balance.
- Use repetition.
- Use active voice.

10a   Achieving Emphasis through Word Order

Where you place words, phrases, and clauses within a sentence emphasizes or de-emphasizes their importance. For example, readers focus on the *beginning* and the *end* of a sentence, expecting the most important information to appear in these places.

(1)   Beginning with important ideas

One way of conveying emphasis clearly and forcefully is to place key ideas at the beginning. Look at the following sentence.

In a landmark study of alcoholism, Dr. George Vaillant of Harvard followed 200 Harvard graduates and 400 inner-city working-class men from the Boston area.

Here, greatest emphasis is placed on the importance of the study, not on those who conducted it or who participated in it. Rephrasing changes the emphasis by focusing attention on the researcher and de-emphasizing the information about the study.

> Dr. George Vaillant of Harvard, in a landmark study of alcoholism, followed 200 Harvard graduates and 400 inner-city working-class men from the Boston area.

Situations that demand a straightforward presentation—laboratory reports, memos, technical papers, business correspondence, and the like—call for sentences that present vital information first and qualify ideas later.

> Treating cancer with interferon has been the subject of a good deal of research. (emphasizes the treatment, not the research)
>
> Dividends will be paid if the stockholders agree. (emphasizes the dividends, not the stockholders)

CLOSE-UP

WRITING EMPHATIC SENTENCES

Because sentence beginnings are so strategic, the use of unemphatic, empty phrases like *there is* or *there are* in this position weakens the sentence.

UNEMPHATIC: There is heavy emphasis placed on the development of computational skills at MIT.

EMPHATIC: Heavy emphasis is placed on the development of computational skills at MIT.

or

MIT places heavy emphasis on the development of computational skills.

(2) Ending with important ideas

The ending of a sentence can also be an emphatic position for important ideas. Key elements may be placed at the end of a sentence in a number of conventional ways.

With a Colon or Dash A colon or a dash can add emphasis by isolating an important word or phrase at the end of a sentence.

continued from previous page

while the sentence openings are purely informational, endings take on an ironic, amused tone: ". . . His Highness the Wali." ". . . ends with a kidney."]

TEACHING STRATEGY

Novice writers tend to "bury the lead" in too many of their sentences by placing less important information ahead of the main idea. For example, offer your students this newspaper-style lead sentence.

Though neighbors say he was good to his mother and school records reveal he was an A student at Morningside High School, Albert Bunker was arrested yesterday for the murder of three local women.

Ask students to revise an essay, paper, or business letter to make sure the main idea isn't camouflaged by the less-important information that precedes it. (Note, of course, that writers CAN save the main idea for the end of a sentence IF the sentence is structured to place emphasis on the closing words.)

Beth had always dreamed of owning one special car: a 1953 Corvette.

The elderly need a good deal of special attention—and they deserve that attention.

With Subordinate Elements at the Beginning Putting modifiers or other subordinate elements at the beginning of a sentence allows you to lead readers to the more important elements at the end of the sentence.

UNEMPHATIC: The Philadelphia Eagles and the Pittsburgh Steelers became one professional football team, nicknamed the Steagles, <u>because of the manpower shortage during World War II</u>. (Modifying phrases at end detract from main idea and weaken sentence.)

EMPHATIC: <u>Because of the manpower shortage during World War II</u>, the Philadelphia Eagles and the Pittsburgh Steelers became one professional football team, nicknamed the Steagles. (correctly emphasizes the newly created team)

CLOSE-UP

WRITING EMPHATIC SENTENCES

Unless you have a good reason to do so, do not waste the end of a sentence on qualifiers such as conjunctive adverbs. In that position, a qualifier loses its power as a linking expression that indicates the relationship between ideas. Put transitional words earlier, where they can fulfill their functions and add emphasis.

LESS EMPHATIC: Smokers do have rights; they should not try to impose their habit on others, however. (conjunctive adverb at end of clause)

MORE EMPHATIC: Smokers do have rights; however, they should not try to impose their habit on others. (conjunctive adverb at beginning of clause)

With Climactic Word Order **Climactic word order**, the arrangement of items in a series from the least to the most important, stresses the key idea—the last point—while building suspense and heightening interest.

When the key idea is buried in the middle of a sentence, the sentence's emphasis will not be clear.

Duties of a member of Congress include serving as a district's representative in Washington, making speeches, and answering mail. (Which duty is most important?)

When you use climactic word order, placing the key idea at the end, the momentum of the sentence gives this important idea added emphasis.

The nation's most prominent orchestras all boast large annual budgets, locations in important cities, and the most talented musicians and conductors. (Talent is the key idea.)

EXERCISE 1

Underline the most important word group in each sentence of the following paragraph. Then identify the device the writer used to emphasize those key words. Are the key ideas placed at the beginning or end of a sentence? Does the writer use climactic order?

Buried in the basement of the computer center, often for hours on end, the campus computer hackers work. Day after day they sit at their terminals, working on games, class assignments, research projects, or schemes to conquer the world. Pausing only for occasional meals or classes, some computer hackers are totally absorbed in their terminals. A few hackers become almost reclusive, spending little time on recreational activities or social relationships. Hackers structure their lives around their computers, sacrificing grades, exercise, dating, and contact with the outdoors. With their own slang, their own habits, and their own hangouts, computer hackers tend to set themselves apart from their fellow students. But these computer addicts feel that the computer experience is worth the sacrifices they must make: Computers have opened up a whole new world for them.

(3)   Inverting word order

The word order of most sentences is subject-verb-object (or complement). When you depart from expected word order, you call attention to the word, phrase, or clause that you have inverted. You may even call attention to the entire sentence.

More modest and less inventive than Turner's paintings are John Constable's landscapes.

Here the writer calls special attention to the modifying phrase *more modest and less inventive than Turner's paintings* by turning the sentence around, thereby stressing the comparison with Turner's work.

EXERCISE 2

Revise the following sentences to make them more emphatic. For each, decide which ideas should be highlighted and group key phrases at sentence beginnings or endings, using climactic order or inverted order where appropriate.

213

Your students' answers will vary. A possible version appears below. In-class discussion of this exercise could be useful in helping students clarify the uses of emphasis.

Buried in the basement of the computer center, often for hours on end, the computer hackers work. [end] Day after day they sit at their terminals, working on games, class assignments, research projects, or schemes to conquer the world [climactic order]. Pausing only for occasional meals or classes, some computer hackers are totally absorbed in their terminal keyboards. [end] A few hackers become almost reclusive, spending little time on recreational activities or social relationships. [beginning] Hackers structure their lives around their computers, sacrificing grades, exercise, dating, and contact with the outdoors. [beginning] With their own slang, their own habits, and their own hangouts, computer hackers tend to set themselves apart from their fellow students. [end] But these computer addicts feel that the computer experience is worth the sacrifices they must make: Computers have opened up a whole new world for them. [end]

1. Because criminals are better armed than ever before, police want to upgrade their firepower.
2. A few years ago felons used small-caliber six-shot revolvers—so-called Saturday night specials.
3. Now these weapons have been replaced by semiautomatic pistols

continued on following page

continued from previous page

capable of firing fifteen to twenty rounds, along with paramilitary weapons like the AK-47.

4. In order to gain equal footing with their adversaries, police are adopting new fast-firing shotguns and 9mm automatic pistols.
5. Automatic pistols, the weapon of choice among law enforcement officers, have numerous advantages over the traditional .38-caliber police revolver, including faster reloading and a hair trigger.

STUDENTS WORKING TOGETHER

Two pairs of partners work together. Each pair takes a paragraph from one of the students' works and rewrites it, increasing parallelism for forcefulness. Both the original paragraph and the rewrite are exchanged with another pair's original and rewrite. Each pair must now figure out which of the two samples they received is more forceful and why. Take a few recommended samples to review with the whole class.

ANSWERS TO EXERCISE 3

1. A. <u>However different in their educational opportunities,</u> [both Jefferson and Lincoln as young men became known to their contemporaries as "hard students"]. (Periodic)
B. Both Lincoln and Jefferson as young men became known to their contemporaries as "hard students," however different their educational opportunities.

continued on following page

1. Police want to upgrade their firepower because criminals are better armed than ever before.
2. A few years ago felons used so-called Saturday night specials, small-caliber six-shot revolvers.
3. Now semiautomatic pistols capable of firing fifteen to twenty rounds, along with paramilitary weapons like the AK-47, have replaced these weapons.
4. Police are adopting such weapons as new fast-firing shotguns and 9mm automatic pistols in order to gain an equal footing with their adversaries.
5. Faster reloading and a hair trigger are among the numerous advantages that automatic pistols, the weapons of choice among law enforcement officers, have over the traditional .38-caliber police revolver.

10b Achieving Emphasis through Sentence Structure

Skillful use of subordination can sharpen a sentence's emphasis by de-emphasizing less important ideas and emphasizing more important ones.

See ◀ 9b

(1) Using cumulative sentences

Most English sentences are cumulative. A **cumulative sentence** begins with an independent clause that is followed by additional words, phrases, or clauses that expand or develop it.

> She holds me in strong arms, arms that have chopped cotton, dismembered trees, scattered corn for chickens, cradled infants, shaken the daylights out of half-grown upstart teenagers. (Rebecca Hill, *Blue Rise*)

Because it presents its main idea first, a cumulative sentence tends to be clear and straightforward. When you want to communicate an idea in a direct manner, a cumulative sentence is the appropriate choice.

(2) Using periodic sentences

A **periodic sentence** places the main idea at the end of the sentence. It moves from supporting details, expressed in modifying phrases and dependent clauses, to the main idea, which is placed in the independent clause.

> Unlike World Wars I and II, which ended decisively with the unconditional surrender of the United States's enemies, the war in Vietnam did not end when American troops withdrew.

In the preceding sentence, the writer adds emphasis to his main idea not only by placing it in the independent clause but also by keeping readers waiting for it.

In some periodic sentences the modifying phrase or dependent clause comes between subject and predicate.

> Columbus, after several discouraging and unsuccessful voyages, finally reached America.

Longer, more complex periodic sentences can be even more forceful. Piling up phrases and clauses, such sentences can gradually build in intensity, and sometimes in suspense, until a climax is reached in the independent clause.

> The problems of soiled artificial flowers, soggy undercrust, leaky milk cartons, sour dishrags, girdle stays jabbing, meringue weeping, soda straws sticking out of bag lunches, shower curtains flapping out of the tub, creases in the middle of the tablecloth sticking up, wet boxes in the laundry room, roach eggs in the refrigerator motor, shiny seam marks on the front of recently ironed ties, flyspecks on chandeliers, film on bathroom tiles, steam on bathroom mirrors, rust in Formica drain-boards, road film on windshields—all were acknowledged and certified, probably for the first time ever, in "Hints from Heloise." (Ian Frazier, *New Yorker*)

Periodic sentences are generally more emphatic than cumulative sentences, but the most emphatic sentence is not always the best choice. Because the periodic structure forces readers to wait—or even to search—for the delayed main idea, periodic sentences tend not to be as straightforward as cumulative ones.

EXERCISE 3

A. Bracket the independent clause(s) in each sentence and underline each modifying phrase and dependent clause. Label each sentence cumulative or periodic.

B. Relocate the supporting details to make cumulative sentences periodic and periodic sentences cumulative, adding words or rephrasing to make your meaning clear.

C. Be prepared to explain how your revision changes the emphasis of the original sentence.

> **EXAMPLE:** Feeling isolated, sad, and frightened, [the small child sat alone in the train depot.] (cumulative)
>
> **REVISED:** The small child sat alone in the train depot, feeling isolated, sad, and frightened. (periodic)

1. However different in their educational opportunities, both Jefferson and Lincoln as young men became known to their contemporaries as

continued from previous page

2. A. [The road came into being slowly], league by league, river crossing by river crossing. (Cumulative)
   B. League by league, river by river, the road came into being slowly.

3. A. Without willing it, [I had gone from being ignorant of being ignorant to being aware of being aware]. (Periodic)
   B. I had gone from being ignorant of being ignorant to being aware of being aware, without willing it.

4. A. To those of us who remain committed mainly to the exploration of moral distinctions and ambiguities, [the feminist analysis may have seemed a particularly narrow and cracked determinism]. (Periodic)
   B. The feminist analysis may have seemed a particularly narrow and cracked determinism to those of us who remain committed mainly to the exploration of moral distinctions and ambiguities.

5. A. [(Henry) Moore's personal history is as familiar in outline as are his sculptures]: his birth in 1898 as the seventh child of a Yorkshire coal-mining family; his early skill at carving; a conservative artistic education at the Royal College of Art in London. (Cumulative)
   B. His birth in 1898 as the seventh child of a Yorkshire coal-mining family; his early skill at carving; a conservative artistic education at the Royal College of Art in London—[Henry] Moore's personal history is as familiar in outline as are his sculptures.

215

ANSWERS TO EXERCISE 4

Your students' preferences about the effectiveness of the cumulative and periodic examples that follow may vary. Generally cumulative sentences tend to be clear and straightforward, while periodic sentences build to a climax.

1. a. Many politicians oppose the MX missile, believing it is too expensive and that a smaller, single-warhead missile is preferable.
   b. Believing it is too expensive and that a smaller, single-warhead missile is preferable, many politicians oppose the MX missile.
2. a. Smoking poses a real danger because it is associated with various cancers, linked to heart disease and stroke, and may even threaten nonsmokers.
   b. Because it is associated with various cancers, linked to heart disease and stroke, and may even threaten nonsmokers, smoking poses a real danger.
3. a. Infertile couples who want children sometimes go through a series of difficult processes, trying adoption, artificial insemination or in vitro fertilization, or even seeking out surrogate mothers.
   b. Trying adoption, artificial insemination, or in vitro fertilization, or even seeking out surrogate mothers, infertile couples who want children sometimes go through a series of difficult processes.
4. a. The Thames is a river that meanders through southern England, the inspiration for literary

continued on following page

"hard students." (Douglas L. Wilson, "What Jefferson and Lincoln Read," *Atlantic Monthly*)
2. The road came into being slowly, league by league, river crossing by river crossing. (Stephen Harrigan, "Highway 1," *Texas Monthly*)
3. Without willing it, I had gone from being ignorant of being ignorant to being aware of being aware. (Maya Angelou, *I Know Why the Caged Bird Sings*)
4. To those of us who remain committed mainly to the exploration of moral distinctions and ambiguities, the feminist analysis may have seemed a particularly narrow and cracked determinism. (Joan Didion, "The Women's Movement")
5. [Henry] Moore's personal history is as familiar in outline as are his sculptures: his birth in 1898 as the seventh child of a Yorkshire coal-mining family; his early skill at carving; a conservative artistic education at the Royal College of Art, in London. (Kay Larson, *New York Magazine*)

*EXERCISE 4

A. Combine each of the following sentence groups into one cumulative sentence, subordinating supporting details to main ideas.
B. Then combine each group into one periodic sentence. Each group can be combined in a variety of ways, and you may have to add, delete, change, or reorder words.
C. Be prepared to explain how the two versions of the sentence differ in emphasis.

EXAMPLE: More people of color than ever before are running for office. They are encouraged by the success of minority candidates.

CUMULATIVE: More people of color than ever before are running for office, encouraged by the success of minority candidates.

PERIODIC: Encouraged by the success of minority candidates, more people of color than ever before are running for office.

1. Many politicians opposed the MX missile. They believed it was too expensive. They felt that a smaller, single-warhead missile was preferable.
2. Smoking poses a real danger. It is associated with various cancers. It is linked to heart disease and stroke. It even threatens nonsmokers.
3. Infertile couples who want children sometimes go through a series of difficult processes. They may try adoption. They may also try artificial insemination or in vitro fertilization. They may even seek out surrogate mothers.
4. The Thames is a river that meanders through southern England. It has been the inspiration for literary works such as *Alice's Adventures in Wonderland* and *The Wind in the Willows*. It was also captured in paintings by Constable, Turner, and Whistler.
5. Black-footed ferrets are rare North American mammals. They prey on

prairie dogs. They are primarily nocturnal. They have black feet and black-tipped tails. Their faces have raccoonlike masks.

*EXERCISE 5

Combine each of the following sentence groups into one sentence in which you subordinate supporting details to main ideas. In each case, create either a periodic or a cumulative sentence, depending on which structure you think will best convey the sentence's emphasis. Add, delete, change, or reorder words when necessary.

> **EXAMPLE:** The fears of today's college students are based on reality. They are afraid there are too many students and too few jobs.

> **REVISED:** The fears of today's college students—that there are too many students and too few jobs—are based on reality. (periodic)

1. Today's college students are under a good deal of stress. Job prospects are not very good. Financial aid is not as easy to come by as it was in the past.
2. Education has grown very expensive. The job market has become tighter. Pressure to get into graduate and professional schools has increased.
3. Family ties seem to be weakening. Students aren't always able to count on family support.
4. College students have always had problems. Now college counseling centers report more—and more serious—problems.
5. The term *student shock* was coined several years ago. This term describes a syndrome that may include depression, anxiety, headaches, and eating and sleeping disorders.
6. Many students are overwhelmed by the vast array of courses and majors offered at their colleges. They tend to be less decisive. They take longer to choose a major and to complete school.
7. Many drop out of school for brief (or extended) periods or switch majors several times. Many take five years or longer to complete their college education.
8. Some colleges are responding to the pressures students feel. They hold stress-management workshops and suicide-prevention courses. They advertise the services of their counseling centers. They train students as peer counselors. They improve their vocational counseling services.

10c Achieving Emphasis through Parallelism and Balance

By highlighting corresponding grammatical elements, parallelism helps writers convey information clearly, quickly, and emphatically.

▶ See
16a

217

continued from previous page

works such as *Alice's Adventures in Wonderland* and *The Wind in the Willows* and paintings by Constable, Turner, and Whistler.

b. The inspiration for literary works such as *Alice's Adventures in Wonderland* and *The Wind in the Willows* and paintings by Constable, Turner, and Whistler, the Thames is a river that meanders through southern England.

5. a. Black-footed ferrets are rare North American mammals that prey on prairie dogs, are primarily nocturnal, and have black feet, black-tipped tails, and raccoon-like masks on their faces.

b. Primarily nocturnal, with black feet, black-tipped tails, and raccoonlike masks on their faces, black-footed ferrets are rare North American mammals that prey on prairie dogs.

ANSWERS TO EXERCISE 5

Your students' answers will vary; possible answers appear below.

1. Today's college students are under a good deal of stress because financial aid is not as easy to come by as it was in the past, and job prospects are not very good. (cumulative)
2. As the job market has become tighter, education has grown very expensive, and pressure to get into graduate and professional schools has increased. (periodic)
3. With family ties apparently weakening, students aren't always able to count on family support. (periodic)

continued on following page

continued from previous page

4. Though college students have always had problems, college counseling centers now report more—and more serious—problems. (periodic)

5. The term *student shock* was coined recently to describe a syndrome that may include depression, anxiety, headaches, and eating and sleeping disorders. (cumulative)

6. Overwhelmed by the vast array of courses and majors offered by their colleges, students tend to be less decisive and take longer to choose a major and complete school. (periodic)

7. Many drop out of school for brief (or extended) periods or switch majors several times, often taking five years or longer to complete their college education. (cumulative)

8. Some colleges are responding to the pressures students feel, advertising the services of their counseling centers, training students as peer counselors, improving vocational counseling services, and holding stress management workshops and suicide prevention courses. (cumulative)

STUDENTS WORKING TOGETHER

In groups of three, students create a checklist for judging emphasis on the sentence level. This checklist might take the form of a holistic rating scale with accompanying descriptions (1 Unemphatic . . . 4 Emphatic) or some other form altogether. Whatever form it takes, it should be a useful tool for judging sentences. Using a paper from each student in the

continued on following page

We seek an individual who is a self-starter, who owns a late-model automobile, and who is willing to work evenings. (classified advertisement)

Do not pass go; do not collect $200. (instructions)

Discuss the role of women in the short stories of Ernest Hemingway and F. Scott Fitzgerald, paying special attention to their relationships with men, to their relationships with other women, and to their roles in their jobs and/or marriages. (examination question)

The Faust legend is central in Benét's *The Devil and Daniel Webster,* in Goethe's *Faust,* and in Marlowe's *Dr. Faustus.* (examination answer)

A **balanced sentence** is neatly divided between two parallel structures. Although balanced sentences are typically compound sentences made up of two parallel clauses, they can also be complex sentences. The symmetrical structure of a balanced sentence highlights correspondences or contrasts between clauses.

In the 1950s, the electronic miracle was the television; in the 1980s, the electronic miracle was the computer.

Alive, the elephant was worth at least a hundred pounds; dead, he would only be worth the value of his tusks, five pounds, possibly. (George Orwell, "Shooting an Elephant")

10d **Achieving Emphasis through Repetition**

Ineffective repetition makes sentences dull and monotonous as well as wordy.

He had a good arm and also could field well, and he was also a fast runner.

We got three estimates, and the one we got from the Johnson Brothers seemed more reasonable than the one we got from County Carpenters.

Effective repetition, however, can place emphasis on key words or ideas. For example, repeating a word or word group in a parallel series can add emphasis.

They decided to begin again: to begin hoping, to begin trying to change, to begin working toward a goal.

Repeating a key word or phrase just once can also add emphasis.

During those years when I was just learning to speak, my mother and father addressed me only in Spanish; in Spanish I learned to reply. (Richard Rodriguez, *Aria: A Memoir of a Bilingual Childhood*)

If ever two groups were opposed, surely those two groups are runners and smokers. (Joseph Epstein, *Familiar Territory*)

Words may be repeated within a sentence or throughout a paragraph—or even a paragraph cluster. In the following group of sentences, the parallel structure and repetition of *still* add emphasis, stressing how hard the author's mother worked.

▶ See 4d6

> Still she sewed—dresses and jackets for the children, housedresses and aprons for herself, weekly patching of jeans, overalls, and denim shirts. She still made pillows, using the feathers she had plucked, and quilts every year—intricate patterns as well as patchwork, stitched as well as tied—all necessary bedding for her family. Every scrap of cloth too small to be used in quilts was carefully saved and painstakingly sewed together in strips to make rugs. She still went out in the fields to help with the haying whenever there was a threat of rain. (Donna Smith-Yackel, "My Mother Never Worked")

*EXERCISE 6

Revise the sentences in this paragraph, using parallelism and balance whenever possible to highlight corresponding elements and using repetition of key words and phrases to add emphasis. (To achieve repetition, you must change some synonyms.) You may combine sentences and add, delete, or reorder words.

> Many readers distrust newspapers. They also distrust what they read in magazines. They do not trust what they hear on the radio and what television shows them either. Of these media, newspapers have been the most responsive to audience criticism. Some newspapers even have ombudsmen. They are supposed to listen to reader complaints. They are also charged with acting on these grievances. One complaint many people have is that newspapers are inaccurate. Newspapers' disregard for people's privacy is another of many readers' criticisms. Reporters are seen as arrogant, and readers feel that journalists can be unfair. They feel reporters tend to glorify criminals, and they believe there is a tendency to place too much emphasis on bizarre or offbeat stories. Finally, readers complain about poor writing and editing. Polls show that despite its efforts to respond to reader criticism, the press continues to face hostility. (Adapted from *Newsweek*)

10e Achieving Emphasis through Active Voice

The active voice is generally more emphatic—and more concise—than the passive voice.

continued from previous page

group, the group tests its tool, ranking the papers from best to worst on sentence emphasis. The group exchanges its papers with another group, judges their papers and tries to reach consensus, then meets with the other groups to compare tools and findings.

ANSWERS TO EXERCISE 6

Your students' answers will vary; a possible version appears below.

Many readers distrust newspapers and magazines; they also distrust what they hear on radio and television. Of these, newspapers have been the most responsive to audience criticism. Some newspapers even have ombudsmen, who listen to reader complaints and act on these grievances. Many people complain that newspapers are inaccurate, that they disregard people's privacy, and that reporters are arrogant and unfair. Reporters, they say, tend to glorify criminals and to place too much emphasis on bizarre and offbeat stories. Readers also complain about poor writing and editing. Polls show there is more hostility toward the press, as well as other media, now than in years past.

TEACHING STRATEGY

Active vs. Passive Voice
Student writers tend to use passive voice far too frequently, and some keep the habit throughout their later writing careers. Suggest that students focus on *avoiding* passive voice in their current or upcoming writing assignment—except when they can justify its use.

continued on following page

continued from previous page

If students seem to be having trouble rewriting passive-voice sentences as they work from a rough to final draft, suggest the following step-by-step procedure.

1. Find the action being performed (the *verb*) in the sentence.
2. Decide *who* or *what* is performing the action. Remember, some passive-voice sentences don't directly identify the person/thing who performs the action.
3. Rewrite the sentence so that the person/thing performs the action.

For example:

PASSIVE

The winning home run was <u>hit</u> by the team's recently acquired Italian <u>outfielder</u>, Renato Sforzi.

ACTIVE

The team's recently acquired Italian outfielder, Renato Sforzi, hit the winning home run.

PASSIVE

A plan to open a new north-south runway at the regional airport was <u>revealed</u> today. (The *who* is omitted.)

ACTIVE

The City Council today revealed a plan to open a new north-south runway at the regional airport.

ANSWERS TO EXERCISE 7

Jack Dempsey, the heavyweight champion between 1919 and 1926, had an interesting but uneven career. Many considered him one of the greatest boxers of all time. Dempsey began fighting as "Kid Blackie," but his career didn't take off until 1919, when

continued on following page

PASSIVE: The prediction that oil prices will rise is being made by economists.

ACTIVE: Economists are now predicting that oil prices will rise.

The passive voice tends to focus your readers' attention on the action or on its receiver rather than on who is performing it. The receiver of the action is the subject of a passive sentence, so the actor fades into the background *(by economists)* or is omitted *(the prediction is now being made).*

See ◀ 231 Sometimes, of course, you *want* to stress the action rather than the actor. If so, it makes sense to use the passive voice. To stress the exploration of the West, you would write the following.

PASSIVE: The West was explored by Lewis and Clark. (*or* The West was explored.)

To stress the contribution of Lewis and Clark, however, you would use the active voice.

ACTIVE: Lewis and Clark explored the West.

The passive is also used when the identity of the actor is irrelevant or unknown.

The course was canceled.

Littering is prohibited.

See ◀ 45c1 The passive voice is frequently used in scientific and technical writing.

The beaker was filled with a saline solution.

EXERCISE 7

Revise this paragraph to eliminate awkward or excessive use of passive constructions.

Jack Dempsey, the heavyweight champion between 1919 and 1926, had an interesting but uneven career. He was considered one of the greatest boxers of all time. Dempsey began fighting as "Kid Blackie," but his career didn't take off until 1919, when Jack "Doc" Kearns became his manager. Dempsey won the championship when Jess Willard was defeated by him in Toledo, Ohio, in 1919. Dempsey immediately became a popular sports figure; Franklin Delano Roosevelt was one of his biggest fans. Influential friends were made by Jack Dempsey. Boxing lessons were given by him to the actor Rudolph Valentino. He made friends with Douglas Fairbanks, Sr., Damon Runyon, and J. Paul Getty. Hollywood serials were made by Dempsey, but the title was lost by him to Gene Tunney, and Dempsey failed to regain it the following year. Meanwhile, his life was marred by unpleasant developments such as a bitter legal battle with his manager and his 1920 indictment for draft evasion. In subsequent years, after his boxing career declined, a restaurant was opened by Dempsey, and many major sporting events were attended by him. This exposure kept him in the public eye until he lost his restaurant. Jack Dempsey died in 1983.

STUDENT WRITER AT WORK

WRITING EMPHATIC SENTENCES

Identify the strategies a freshman composition student has used in this draft to add emphasis. Revise the draft to make sentences more emphatic and then revise again if necessary to strengthen coherence, unity, and style.

"Fight, Fight, Fight for the Home Team"

It is pathetic for an athlete being paid millions of dollars to get into a fight for something so ridiculous as a bad call by a referee or an accidental push by a player. Players have to be tough in order to win. This shouldn't mean they have to be violent, however. The amount of violence increases every day in the world of sports. It is a national embarrassment that this situation exists.

It is a disgrace to see professional athletes getting into fights. This situation occurs in nearly every game. If a fight breaks out at an event, the game is disrupted, the players involved in the altercation are fined, and spectators are disappointed or angry. What does a fight solve, therefore? Nothing. Things are only made worse by fights.

In hockey, there are always fights that break out. These fights aren't even stopped by the referees. The players are allowed to continue beating each other to a pulp instead. The crowd gets into the hitting and checking, and a fight erupts in the stands before you know it. A family can seldom go to a peaceful game without witnessing a fight between players or a riot in the stands.

America's pastime, baseball, can be a beautiful sport to watch. Imagine the crack of the bat, the smell (in some ballparks) of fresh-cut grass, the cheers for a home run, and the sounds of both teams rushing out of the dugouts to the mound to demolish the other team. Fighting shouldn't be part of the game, obviously, but often it seems it is. A batter should just walk it off on the way to first base if he gets hit by a pitch. Usually, he runs after the pitcher, though. Within ten seconds, the dugouts are empty and fists are flying on the mound.

continued from previous page

Jack "Doc" Kearns became his manager. Dempsey won the championship when he defeated Jess Willard in Toledo, Ohio, in 1919. Dempsey immediately became a popular sports figure; Franklin Delano Roosevelt was one of his biggest fans. Jack Dempsey made influential friends, gave boxing lessons to the actor Rudolph Valentino, made friends with Douglas Fairbanks, Sr., Damon Runyon, and J. Paul Getty, and made Hollywood serials. But he lost the title to Gene Tunney and failed to regain it in the following year. Meanwhile, unpleasant developments such as a bitter legal battle with his manager and his 1920 indictment for draft evasion marred his life. In subsequent years, after his boxing career declined, Dempsey opened a restaurant and attended many major sporting events. This exposure kept him in the public eye until he lost his restaurant. Jack Dempsey died in 1983.

STUDENT WRITER AT WORK

Possible Revision

"Fight, Fight, Fight
for the Home Team"

In the world of professional sports, it is not unusual for athletes to be paid millions of dollars, and each new contract negotiation seems to result in even higher salaries. Unfortunately, the increase in salaries seems to be accompanied by an increase in violence in professional sports. It is not unusual for athletes to engage in fights over such minor incidents as a bad call by a referee or an accidental push by another player. While it is true that athletes must be tough in order to win, it is not true that they must be violent. The increasing violence in the world of sports has become a national embarrassment.

Violence occurs in almost every professional game, and it is a disgrace to all involved. When a fight breaks out at an event, the game is disrupted, the players involved are fined, and the spectators

continued on following page

continued from previous page

are disappointed or angry. Worst of all, a fight solves nothing. In fact, it actually makes things worse.

In professional hockey, fights break out regularly. They are so common that the referees do not stop them. Instead, the players are allowed to continue beating each other to a pulp. Soon the crowd gets into the hitting and checking, and, before long, fights erupt in the stands. A family can seldom go to a peaceful hockey game without witnessing a fight between players or a riot in the stands.

Baseball, America's premier pastime, can be a beautiful sport to watch. However, the crack of the bat, the smell of fresh-cut grass, and the cheers for a home run are often overpowered by the sounds of both teams rushing out of the dugouts to the mound to demolish the other team. If a batter gets hit by a pitch, he should just walk it off on the way to first; usually, though, he runs after the pitcher, and within ten seconds, the dugouts are empty and fists are flying on the mound.

Although violence is found in hockey and baseball, it is even more obvious in contact sports such as football, lacrosse, boxing, and rugby. In fact, violence, whether on the field or the court, is making it harder and harder for fans to go to a sporting event and have a plain old-fashioned good time. Certainly, a player can have a "bad" or a "hungry" attitude, but that attitude should not mean getting into violent fights. Whatever happened to the spirit of "Buy me some peanuts and Cracker Jack, I don't care if I never get back"?

Commentary on Possible Revision

Although the writer has revised to make sentences more emphatic and has made a number of other stylistic revisions, he needs to look again at his thesis statement. Some reorganization of the main points and the addition of some transitional phrases would also improve the essay's unity and effectiveness.

Violence is even more obvious in sports like football, lacrosse, boxing, and rugby. It is becoming harder and harder to go to a sporting event and have a plain old-fashioned good time because of all the violence on the field or court. A player can be "bad" or "hungry," but this shouldn't mean getting into violent fights. What ever happened to the spirit of "Buy me some peanuts and Cracker Jack, I don't care if I never get back"?

CHAPTER 11

Writing Concise Sentences

A concise sentence comes directly to the point. It contains only the number of words necessary to achieve its effect or to convey its message. But a sentence is not concise simply because it is short. Conciseness is always related to content, to how much you have to say. If you can eliminate words without reducing the amount of information you present, you should do so.

Every word serves a purpose in a concise sentence. Because they are free of unnecessary words and convoluted constructions that come between writer and reader, concise sentences are also clear and emphatic.

STRATEGIES FOR WRITING CONCISE SENTENCES
- Eliminate nonessential words.
- Eliminate needless repetition.
- Tighten rambling sentences.

11a Eliminating Nonessential Words

One way to find out which words are essential to the meaning of a sentence is to underline the key words. Then, look carefully at the remaining words so you can see which are unnecessary and delete them.

It seems to me that it doesn't make sense to allow any <u>bail</u> to be <u>granted</u> to <u>anyone</u> who has ever been <u>convicted</u> of a <u>violent crime</u>.

TEACHING STRATEGY

Conciseness
It might be interesting to discuss the implications of the kind of "wordy" or "imprecise" language that has become all too common in government, political and business circles—language as camouflage rather than communication. Ask students to read George Orwell's famous essay "Politics and the English Language" and to look for quotations or excerpts from recent sources that seem to connect with his concern for concise, clear writing. (A look at a recent issue of the *Congressional Record* could be a good place to start; see also Philip Howard's book *Weasel Words.*)

QUOTATION

"A sentence should contain no unnecessary words, a paragraph no unnecessary sentences, for the same reason that a drawing should have no unnecessary lines and a machine no unnecessary parts. This requires not that the writer make all his sentences short or that he avoid all detail . . . but that every word tell."

–William Strunk, Jr., and E.B. White,
The Elements of Style

223

Lab Activity

Eliminating Deadwood

The computer is a powerful tool for helping students gain practice in eliminating verbal deadwood. Turn the exercises in Chapter 11 of *The Holt Handbook* (or examples from student essays) into word-processing documents. Have students, working individually or in pairs, eliminate deadwood, circumlocution, and needless repetition and rewrite where necessary.

The underlining shows you immediately that none of the words in the long introductory phrase are essential. A revision of this sentence includes all the key words and the minimum number of other words needed to give the key ideas coherence.

> Bail should not be granted to anyone who has ever been convicted of a violent crime.

Nonessential words fall into three general categories: *deadwood, utility words,* and *circumlocution.*

(1) Delete deadwood

Deadwood denotes unnecessary phrases that take up space and add nothing to meaning.

WORDY	CONCISE
There were many factors that influenced his decision to become a priest.	Many factors influenced his decision to become a priest.
Kareem Abdul-Jabbar was considered to be a great center.	Kareem Abdul-Jabbar was considered a great center.
Shoppers who are looking for bargains often patronize outlets.	Shoppers looking for bargains often patronize outlets.
They played a racquetball game which was exhausting.	They played an exhausting racquetball game.
The box that was in the middle contained a surprise.	The box in the middle contained a surprise.

Many familiar expressions—such as *as the case may be, I feel, it seems to me, all things considered, without a doubt, in conclusion,* and *by way of explanation*—are simply padding. You may think they balance or fill out a sentence or make your writing sound more authoritative, but the reverse is true.

WORDY	CONCISE
With reference to your memo, the points you make are worth considering.	The points in your memo are worth considering.
In my opinion, the characters seem undeveloped.	The characters seem undeveloped.
As far as this course is concerned, it looks interesting.	This course looks interesting.
For all intents and purposes, the two brands are alike.	The two brands are essentially alike.

It is important to note that the results were identical in both clinical trials.	The results were identical in both clinical trials.

(2) Delete or replace utility words

Utility words are vague all-purpose words that act as fillers and contribute nothing to a sentence. They may be nouns, usually those with imprecise meanings (*factor, kind, type, quality, aspect, thing, sort, field, area, situation,* and so on); adjectives, usually those so general as to be almost meaningless (*good, nice, bad, fine, important, significant*); or adverbs, usually common ones concerning degree (*basically, completely, actually, very, definitely, quite*).

WORDY	**CONCISE**
The registration situation was disorganized.	Registration was disorganized.
The scholarship offered Fran a good opportunity to study Spanish.	The scholarship offered Fran an opportunity to study Spanish.
It was actually a worthwhile book, but I didn't completely finish it.	It was a worthwhile book, but I didn't finish it.

When you find yourself using a utility word, delete it or replace it with a more specific word. The result will be a more economical sentence.

(3) Avoid circumlocution

Taking a roundabout way to say something (using ten words when five will do) is called **circumlocution**. When you use long words, complicated phrases, and rambling constructions instead of short, concrete, specific words and phrases, you cannot write concise sentences. Notice how the revised versions of these sentences use fewer words and simpler constructions to say the same thing.

WORDY	**CONCISE**
The experience that changed Andy most would have to be the hunting trip.	The experience that changed Andy most was the hunting trip.
The curriculum was of a unique nature.	The curriculum was unique.

Suggest that students read their words out loud when revising an essay to help them eliminate wordy phrases and other deadwood. Student writers frequently *write* words and phrases they would never dream of *saying* aloud—whether in casual speech to a friend or in a formal presentation. Ask the students to underline possible cuts as they read, going along as much as possible with their instinctive reaction to the words, their feeling that somehow, for some reason, this word or phrase isn't "it." Discuss the results of the exercise in class.

Sally Ride, an astrophysicist, was the first American woman astronaut. NASA selected Ride as an astronaut in 1978 because she was a first-rate athlete and had done graduate work in X-ray astronomy and free-electron lasers. Earlier, while she was a graduate student at Stanford, Ride knew she could become a theoretical physicist. At NASA she helped to design the space shuttle's remote manipulator arm, and later she relayed flight instructions to astronauts until she was assigned to a flight crew. Now, Ride teaches at the University of California. Although Ride is no longer a NASA employee, she remains something special: America's first woman in space.

It is not unlikely that the trend toward smaller cars will continue.	The trend toward smaller cars will probably continue.
Joel was in the army during the same time that I was in college.	Joel was in the army while I was in college.
It is entirely possible that the lake is frozen.	The lake may be frozen.

Wordy phrases can almost always be controlled or avoided. Always choose simple, easily understood terms; when you revise, strike out wordy, convoluted constructions.

WORDY	CONCISE
at the present time	now
at this point in time	now
for the purpose of	for
due to the fact that	because
on account of the fact that	because
until such time as	until
in the event that	if
by means of	by
in the vicinity of	near
have the ability to	be able to

*EXERCISE 1

Revise the following paragraph to eliminate deadwood, utility words, and circumlocution. When a word or phrase seems superfluous, delete it or replace it with a more concise expression.

Sally Ride is an astrophysicist who was selected to be the first American woman astronaut. It seems that there were many good reasons why she was chosen. She is a first-rate athlete, and she did graduate work in X-ray astronomy and free-electron lasers. As a result of these and other factors, NASA accepted Ride as a "mission specialist" astronaut in the year 1978. Prior to that time, Ride had been a graduate student at Stanford who knew she had the capability of becoming a specialist in the area of theoretical physics. At NASA she helped to design the remote manipulator arm of the space shuttle, and at a later point she relayed flight instructions to astronauts until such time as she was assigned to a flight crew. Now, Ride teaches at the University of California. Although she is no longer employed by NASA, at this point in time she remains something quite definitely special: America's very first woman in space.

11b   Eliminating Needless Repetition

Repetition of words or concepts can add emphasis to your writing, but unnecessary repetition annoys readers and obscures your meaning. Repeated words and **redundant** word groups (words or phrases that say the same thing) are the chief problems. Consider the following sentences.

▶ **See 10d**

> **WORDY:** Ernest Hemingway, one of the most <u>famous</u> and <u>well-known</u> <u>authors</u> in American literary history, is the <u>author</u> of <u>novels</u> like *The Sun Also Rises* and other <u>novels</u>. (*Famous* and *well-known* are redundant, while *author* and *novels* are repeated needlessly.)

> **CONCISE:** Ernest Hemingway, one of the most famous writers in American literary history, is the author of *The Sun Also Rises* and other novels.

You can correct needless repetition in a number of different ways.

(1)   Delete repeated words

> **WORDY:** The childhood <u>disease</u> chicken pox occasionally leads to dangerous complications, such as the <u>disease</u> known as Reye's Syndrome.

> **CONCISE:** The childhood disease chicken pox occasionally leads to dangerous complications, such as Reye's Syndrome. (Repeated words are deleted.)

(2)   Substitute a pronoun

> **WORDY:** Agatha Christie's <u>Hercule Poirot</u> solves many difficult cases. *The Murder of Roger Ackroyd* was one of <u>Hercule Poirot's</u> most challenging cases.

> **CONCISE:** Agatha Christie's Hercule Poirot solves many difficult cases. *The Murder of Roger Ackroyd* was one of his most challenging cases. (The pronoun *his* is substituted for *Hercule Poirot*.)

(3)   Use elliptical clauses

> **WORDY:** The Quincy Market <u>is a popular tourist attraction</u> in Boston; the White House <u>is a popular tourist attraction</u> in Washington, D.C.; and the Statue of Liberty <u>is a popular tourist attraction</u> in New York City.

▶ **See 28f1**

▶ **See 10d**

▶ **See 28f1**

CONCISE: The Quincy Market is a popular tourist attraction in Boston; the White House, in Washington, D.C.; and the Statue of Liberty, in New York City. (Elliptical clauses eliminate unnecessary repetition.)

(4) Use appositives

WORDY: Red Barber <u>was</u> a sportscaster. He <u>was</u> known for his colorful expressions.

CONCISE: Red Barber, a sportscaster, was known for his colorful expressions. (Appositive eliminates unnecessary repetition.)

(5) Create compounds

WORDY: Wendy <u>found the exam difficult</u>, and Karen <u>also found it hard</u>. Ken <u>thought it was tough, too</u>.

CONCISE: Wendy, Karen, and Ken all found the exam difficult. (Compound subject eliminates unnecessary repetition.)

WORDY: *Huckleberry Finn* <u>is</u> an adventure story. <u>It is also</u> a sad account of an abused, neglected child.

CONCISE: *Huckleberry Finn* is both an adventure story and a sad account of an abused, neglected child. (Compound complement eliminates unnecessary repetition.)

WORDY: In 1964 Ted Briggs was discharged from the Air Force. <u>He</u> then got a job with Maxwell Data Processing. <u>He</u> married Susan Thompson that same year.

CONCISE: In 1964 Ted Briggs was discharged from the Air Force, got a job with Maxwell Data Processing, and married Susan Thompson. (Compound predicate eliminates unnecessary repetition.)

(6) Use subordination

WORDY: Americans value <u>freedom of speech</u>. <u>Freedom of speech</u> is guaranteed by the First Amendment.

CONCISE: Americans value freedom of speech, which is guaranteed by the First Amendment. (Subordinating one clause to the other eliminates needless repetition.)

EXERCISE 2

Eliminate any unnecessary repetition of words or ideas in this paragraph. Also revise to eliminate deadwood, utility words, or circumlocution.

For a wide variety of different reasons, more and more people today are choosing a vegetarian diet. There are three kinds of vegetarians: strict vegetarians eat no animal foods at all; lactovegetarians eat dairy products, but they do not eat meat, fish, poultry, or eggs; and ovolactovegetarians eat eggs and dairy products, but they do not eat meat, fish, or poultry. Famous vegetarians include such well-known people as George Bernard Shaw, Leonardo da Vinci, Ralph Waldo Emerson, Henry David Thoreau, and Mahatma Gandhi. Like these well-known vegetarians, the vegetarians of today have good reasons for becoming vegetarians. For instance, some religions recommend a vegetarian diet. Some of these religions are Buddhism, Brahmanism, and Hinduism. Other people turn to vegetarianism for reasons of health or for reasons of hygiene. These people believe that meat is a source of potentially harmful chemicals, and they believe meat contains infectious organisms. Some people feel meat may cause digestive problems and may lead to other difficulties as well. Other vegetarians adhere to a vegetarian diet because they feel it is ecologically wasteful to kill animals after we feed plants to them. These vegetarians believe *we* should eat the plants. Finally, there are facts and evidence to suggest that a vegetarian diet may possibly help people live longer lives. A vegetarian diet may do this by reducing the incidence of heart disease and lessening the incidence of some cancers. (Adapted from *Jane Brody's Nutrition Book*)

11c   Tightening Rambling Sentences

Rambling, out-of-control sentences are the inevitable result of using nonessential words, unnecessary repetition, and complicated syntax. Making such sentences concise involves more than crossing out a word or two. In fact, revising rambling sentences can require ruthless deletion. As you write and revise, the following techniques can help you keep your sentences under control.

(1)   Eliminate excessive coordination

Stringing a series of clauses together with coordinating conjunctions often creates a rambling sentence. Such excessive coordination is not only wordy but also misleading: It presents all your ideas as if they have equal weight when they do not. To revise such sentences, first identify the main idea and then subordinate the supporting details.

▶ See 12c

TEACHING STRATEGY

Eliminating Passive Constructions

In passive voice constructions, the verb is usually flanked by both an auxiliary verb and a preposition. These words clutter the sentence and tend to bury its strong verb—thereby sapping the life from the writing. All around us, officials and academics who believe it is the voice of authority use passive voice.

On the board or on an overhead projector, write a short paragraph entirely in the passive voice. Then ask students to rewrite it in active voice. How many words were eliminated?

WORDY: Puerto Rico is the fourth largest island in the Caribbean, and it is predominantly mountainous, and it has steep slopes, and they fall to gentle coastal plains.

CONCISE: Fourth largest island in the Caribbean, Puerto Rico is predominantly mountainous, with steep slopes falling to gentle coastal plains. (*National Geographic*)

(2) Eliminate excessive subordination

A series of adjective clauses is likely to produce a rambling sentence. To correct this problem, substitute concise modifying words or phrases for the adjective clauses.

WORDY: *Moby-Dick*, which is a book about a whale, was written by Herman Melville, who was friendly with Nathaniel Hawthorne, who encouraged him to revise the first draft of his novel.

CONCISE: *Moby-Dick*, a book about a whale, was written by Herman Melville, who revised the first draft of his novel at the urging of his friend Nathaniel Hawthorne.

(3) Eliminate passive constructions

The active voice is usually more concise than the passive voice.

WORDY: "Buy American" rallies are being organized by concerned Americans who hope jobs can be saved by such gatherings.

CONCISE: Concerned Americans are organizing "Buy American" rallies, hoping such gatherings can save jobs.

WORDY: Water rights are being fought for in court by Indian tribes like the Papago in Arizona and the Pyramid Lake Paiute in Nevada.

CONCISE: Indian tribes like the Papago in Arizona and the Pyramid Lake Paiute in Nevada are fighting in court for water rights.

(4) Eliminate wordy prepositional phrases

Often you can tighten a rambling sentence by substituting single adjectives or adverbs for wordy prepositional phrases used as modifiers.

WORDY: The trip was one of danger but also one of excitement.

CONCISE: The trip was dangerous but exciting. (Adjectives replace prepositional phrases.)

WORDY: He spoke in a confident manner.

CONCISE: He spoke confidently. (Adverb replaces prepositional phrase.)

Concise 11c

(5) Eliminate wordy noun constructions

You can also tighten a rambling sentence by substituting strong verbs for wordy noun phrases that express the action of the sentence.

WORDY: The normalization of commercial relations between the United States and China in 1979 led to an increase in trade between the two countries.

CONCISE: When the United States and China normalized commercial relations in 1979, trade between the two countries increased.

WORDY: We have made the decision to postpone the meeting until after the appearance of all the board members.

CONCISE: We have decided to postpone the meeting until all the board members appear.

EXERCISE 3

Revise the rambling sentences in this paragraph by eliminating excessive coordination and subordination, unnecessary use of the passive voice, and overuse of wordy prepositional phrases and noun constructions. As you revise, make your sentences more concise by deleting nonessential words and superfluous repetition.

Some colleges that have been in support of fraternities for a number of years are at this time in the process of conducting a reevaluation of the position of those fraternities on campus. In opposition to the fraternities are a fair number of students, faculty members, and administrators who claim fraternities are inherently sexist, which they say makes it impossible for the groups to exist in a coeducational institution, which is supposed to offer equal opportunities for members of both sexes. And, more and more members of the college community see fraternities as elitist as well as sexist and favor their abolition. The situation has already begun to be dealt with at some colleges. For instance, Williams College made a decision in favor of the abolition of fraternities. At Wesleyan University a decision was made by officials to sever formal ties to all-male fraternities. This was done because of the university's inability to persuade the fraternities to consider the acceptance of women. Some fraternities became coed "literary societies." At Middlebury College in Vermont, the fraternities (which are now called "houses") now offer membership to women as well as men. In some cases, however, students, faculty, and administration remain wholeheartedly in support of traditional fraternities, which they believe are responsible for helping students make the acquaintance of people and learn the leadership skills which they believe will be of assistance to them in their future lives as adults. Supporters of fraternities believe

Some colleges that have supported fraternities for many years are reevaluating the fraternities' positions on campus. Opposing the fraternities are students, faculty, and administrators, who claim that fraternities are inherently sexist and, therefore, are unacceptable in coed institutions which offer equal opportunities. Many members of the college community see fraternities as elitist as well as sexist and favor their abolition. Some colleges have already begun to deal with these problems: Williams College, for instance, has abolished fraternities, and Wesleyan University severed formal ties to all-male fraternities when it failed to persuade them to accept women. Some fraternities became coed "literacy societies." At Vermont's Middlebury College, fraternities are now coed "houses." In some cases, however, students, faculty, and administrators wholeheartedly support traditional fraternities, believing they help students meet people and acquire leadership skills that will help them as adults. Supporters of fraternities believe that students should retain the right to make their own social decisions, including joining a fraternity, and that fraternities provide valuable services, such as tutoring, raising money for charity, and running campus escort services. Therefore, they oppose abolishing fraternities.

STUDENT WRITER AT WORK

Writing Concise Sentences

Possible Revision

Often in a literary work, characters may not be a part of their society. This problem often leads to conflicts either between a misfit and his or her own identity or between a misfit and society.

In "The Minister's Black Veil," Reverend Hooper surprises the townspeople and his parishioners by wrapping a piece of black material over his face. This move confuses readers as thoroughly as it does the townspeople, until we learn, in his sermon, that he is covering his face to atone for the sins of mankind. We are never sure why he feels he must do this, nor whether Reverend Hooper feels this guilt for some sin in his past or for the sins of everyone. Still he clearly believes it is his duty to isolate himself from the world. To the Reverend, the problem is insoluble, and he wears the veil all his life. Even after his death he insists that the veil remain, for he says his face must not be revealed on earth.

A terrible conflict exists within Reverend Hooper, and he voluntarily makes himself a misfit at the expense of losing everything, even his true love Elizabeth.

Commentary on Possible Revision

The deletion of unnecessary words and phrases has strengthened the original considerably. It is now more focused and coherent. Note, however, the first-person references ("we"), which are usually avoided in such an analysis. In a subsequent draft, the student might try to analyze Hooper's motivation more closely, adding specific examples and quotations from the story to support the idea of Hooper as a self-created misfit.

students should retain the right to make their own social decisions and that joining a fraternity is one of those decisions, and they also believe fraternities are responsible for providing valuable services and some of these are tutoring, raising money for charity, and running campus escort services. Therefore, they are not of the opinion that the abolition of fraternities makes sense.

STUDENT WRITER AT WORK

WRITING CONCISE SENTENCES

Revise this excerpt from an essay examination in American literature to make it more concise. After you have done so, revise further if necessary to strengthen coherence, unity, and style.

Oftentimes in the course of a literary work, characters may find themselves misfits in the sense that they do not seem to be a real part of the society in which they find themselves. This problem often leads to a series of genuinely serious and severe problems, conflicts either between the misfits and their own identities or possibly between them and that society into which they so poorly fit.

In "The Minister's Black Veil" Reverend Hooper all of a sudden gives to the townspeople and members of his parish a surprise: a piece of black material that he has wrapped over his face, which causes readers to be as completely and thoroughly confused as the townspeople about the possible reason for the minister's decision to hide his face, until readers learn, in his sermon, that he is covering his face (from God, his fellow man, and himself) to atone for the sins of mankind. As far as readers can tell, they are never quite sure exactly why he is in possession of the notion that this act must be carried out by him, and they are never completely sure whether Reverend Hooper feels this guilt for some sin that may exist in his own past or for those sins that may have been committed by mankind in general, but in any case it is clear that he feels it is his duty to place himself in isolation from the world at large around him. To the Reverend, there is no solution to his problem, and he lives his whole entire life wearing the veil. Even after his death he insists that the veil remain covering his features, for it is said by the Reverend that his face must not be revealed on earth.

For Reverend Hooper, a terrible conflict exists within himself, and so Reverend Hooper voluntarily makes himself a misfit even at the expense of losing everything, even his true love Elizabeth.

CHAPTER 12

Writing Varied Sentences

Varying your sentences will help you convey emphasis accurately and hold reader interest.

> ### STRATEGIES FOR WRITING VARIED SENTENCES
> - Vary sentence length.
> - Combine choppy simple sentences.
> - Break up strings of compounds.
> - Vary sentence types.
> - Vary sentence openings.
> - Vary standard word order.

12a Varying Sentence Length

A mixture of long and short sentences not only gives a pleasing texture to your writing but also keeps readers interested.

(1) Mix long and short sentences

A paragraph consisting entirely of short sentences (or entirely of long ones) can be dull.

> Drag racing began in California in the 1940s. It was an alternative to street racing, which was illegal and dangerous. It flourished in the 1950s and 1960s. Eventually, it became almost a rite of passage. Then, during the 1970s, almost one-third of America's racetracks closed. Today, however, drag racing is making a modest comeback.

TEACHING STRATEGY

Writing Varied Sentences
Distribute copies of the following three passages and then read each aloud to the class. Ask students to talk about the rhythm, pace, and overall effect of each passage. How do sentence length and structure work to help create the final impact on the reader?

"We were all indestructible. I remember speeding down a highway through a snowstorm with some friends, late at night; the car skidded and swerved, and suddenly we were crashing through a guardrail and the car was tumbling down a gully. When it came to rest, we looked around at one another; there we were, in our high school letter jackets, unhurt, and the radio was still blaring and there were the Beatles singing, 'I Want to Hold Your Hand.' And we stood in the snow, laughing; young and new and invincible, laughing."

–Bob Greene,
American Beat

"I unlocked my door and went in. There were three men in there, all in high-domed peaked caps and white overalls, unrolling a large Indian-type rug in the cleared free center [of the hotel room].

'What . . .' I began. And I thought: it's Sunday.

I spun on my heel to retreat, but it was already too late.

A fourth man was blocking the doorway; advancing, stretching out his arms, thrusting me forcefully backwards into the room.

continued on following page

continued from previous page

I looked into his eyes . . . and
knew him.
I thought in lightning flashes.
I thought: I've lost.
I thought: I'm dead . . .
I thought: I'm a fool. . . ."

–Dick Francis,
The Danger

"There is a lovely road that runs
from Ixopo into the hills. These
hills are grass-covered and rolling,
and they are lovely beyond any
singing of it. The road climbs
seven miles into them, to
Carlsbrooke; and from there, if
there is no mist, you look down on
one of the fairest valleys of Africa.
About you there is grass and
bracken and you may hear the for-
lorn crying of the titihoya, one of
the birds of the veld. Below you is
the valley of the Umzimkulu, on
its journey from the Drakensberg
to the sea; and beyond and behind
the river, great hill after great hill;
and beyond and behind them, the
mountains of Ingeli and East
Griqualand."

–Alan Paton,
Cry, the Beloved Country

ANSWERS TO EXERCISE 1

1. a. Chocolate is composed of more
than 300 compounds, including
phenylethylamine, the chemical whose presence in the brain
may be linked to falling in love.
   b. Maybe that's the reason lovers
give gifts of chocolates.
2. a. Americans consume a good deal
of chocolate—on average, over
nine pounds per year apiece;

continued on following page

The following revision combines some sentences to create units
of various lengths.

Drag racing began in California in the 1940s as an alternative to
street racing, which was illegal and dangerous. It flourished in the
1950s and 1960s, eventually becoming almost a rite of passage. Then,
during the 1970s, almost one-third of America's racetracks closed.
Today, however, drag racing is making a modest comeback.

(2) Follow a long sentence with a short one

Using a short sentence after one or more long ones immediately attracts reader attention. The following passages illustrate
how this shifting of gears emphasizes the short sentence and its
content while adding variety.

There are two social purposes for family dinners—the regular exchange of news and ideas and the opportunity to teach small children not to eat like pigs. These are by no means mutually exclusive.
(Judith Martin, "Miss Manners")

In arguing the need for [vitamin] supplements, doctors like to
point out that the normal diet supplies the RDA (Recommended
Dietary Allowance) minimums. Nutritionists counter that the RDA,
as established by the National Academy of Sciences, is only the minimum daily dose necessary to prevent the diseases associated with
particular vitamin deficiencies. Over the years, vitamin boosters say,
a misconception has grown that as long as there are no signs or
symptoms of say, scurvy, then we have all of the vitamin C we need.
Although we know how much of a particular vitamin or mineral will
prevent clinical disease, we have practically no information on how
much is necessary for peak health. In short, we know how sick is
sick, but we don't know how well is well. (*Philadelphia Magazine*)

EXERCISE 1

A. Combine each of the following sentence groups into one long sentence.
B. Then, compose a relatively short sentence to follow each long one.
C. Finally, combine all the sentences into a paragraph, adding a topic
sentence and any transitions necessary for coherence. Proofread your
paragraph to be sure the sentences are varied in length.

1. Chocolate is composed of more than 300 compounds. Phenylethylamine is one such compound. Its presence in the brain may be linked
to the emotion of falling in love.
2. Americans now consume a good deal of chocolate. On average, they
eat more than nine pounds of chocolate per person per year. The typical Belgian, however, consumes almost fifteen pounds per year.

Varied 12b

continued from previous page

3. In recent years, Americans have begun a serious love affair with chocolate. Elegant chocolate boutiques sell exquisite bonbons by the piece. At least one hotel offers a "chocolate binge" vacation. The bimonthly *Chocolate News* for connoisseurs is flourishing. (Adapted from *Newsweek*)

12b Combining Choppy Simple Sentences

Strings of short simple sentences can be tedious—and sometimes hard to follow. Revise such sentences by combining them with adjacent sentences, using *coordination*, *subordination*, or *embedding*.

(1) Use coordination

Coordination is one way to revise choppy simple sentences, such as these notes for part of a short paper on freedom of the press in America.

▶ See 9a1

> John Peter Zenger was a newspaper editor. He waged and won an important battle for freedom of the press in America. He criticized the policies of the British governor. He was charged with criminal libel as a result. Zenger's lawyers were disbarred by the governor. Andrew Hamilton defended him. Hamilton convinced the jury that Zenger's criticisms were true. Therefore, the statements were not libelous.

The information is here, but the presentation is flat and lacks coherence. Notice how coordination adds interest and clarity.

> John Peter Zenger was a newspaper editor. He waged and won an important battle for freedom of the press in America. He criticized the policies of the British governor, and as a result, he was charged with criminal libel. Zenger's lawyers were disbarred by the governor. Andrew Hamilton defended him. Hamilton convinced the jury that Zenger's criticisms were true. Therefore, the statements were not libelous.

This revision links two of the choppy simple sentences with *and* to create a compound sentence. The result is a smoother paragraph.

(2) Use subordination

Subordination clarifies the relationships among ideas. The following revision uses subordination to change two simple sentences into dependent clauses, creating two complex sentences.

▶ See 9b

the typical Belgian, however, consumes almost fifteen pounds per year.

b. Anyone who has ever tasted Belgian chocolate knows why.

3. a. In recent years, Americans have begun a serious love affair with chocolate: Elegant chocolate boutiques sell chocolate by the piece; at least one hotel offers a "chocolate binge" vacation; and the bimonthly *Chocolate News* for connoisseurs is flourishing.

b. Chocolate has become big business.

c. Chocolate is composed of more than 300 compounds, including phenylethylamine, the chemical whose presence in the brain is linked with falling in love. Maybe that's the reason lovers give gifts of chocolates. Whether they receive chocolates as gifts or purchase them, Americans now consume a good deal of chocolate—on average, over nine pounds of chocolate per year apiece. The typical Belgian, however, consumes almost fifteen pounds a year. Anyone who has ever tasted Belgian chocolate knows why.) The fact that Americans are having a serious love affair with chocolate has become evident in recent years: Elegant chocolate boutiques sell chocolate by the piece; at least one hotel offers a "chocolate binge" vacation; and the bimonthly *Chocolate News* for connoisseurs is flourishing. Clearly, chocolate has become big business.

The first modern miniature golf course, built in New York in 1925, was an indoor course with 18 holes. As the game caught on, entrepreneurs Drake Delanoy and John Ledbetter built 150 more indoor and outdoor courses; Garnet Carter, who made miniature golf a worldwide fad with his elaborate miniature courses, later joined with Delanoy and Ledbetter to build more courses. They abbreviated playing distances and highlighted the game's hazards at the expense of skill, making miniature golf much more popular. By 1930 there were 25,000 miniature golf courses in the United States, with courses growing more elaborate and hazards becoming more bizarre. The craze spread to London and Hong Kong as the expansion of miniature golf grew out of control. Then, interest in the game declined, and by 1931 most miniature golf courses were out of business. The game was revived in the early 1950s, and today there are between eight and ten thousand miniature golf courses. The architecture of miniature golf remains an enduring form of American folk art.

Demonstrate and discuss the effect of strings of compounds, too many long sentences, complex sentences, complicated parallels, or periodic sentences upon reader comprehension. Select examples of the prose styles of past centuries to illustrate your point—for example, a paragraph from Milton's *Areopagitica* or from one of the works of John Lyly.

John Peter Zenger was a newspaper editor who waged and won an important battle for freedom of the press in America. He criticized the policies of the British governor, and as a result, he was charged with criminal libel. When Zenger's lawyers were disbarred by the governor, Andrew Hamilton defended him. Hamilton convinced the jury that Zenger's criticisms were true. Therefore, the statements were not libelous.

The revised paragraph now includes two complex sentences: One links ideas with a relative pronoun (*who*) and one with a subordinating conjunction (*when*).

(3) Use embedding

See ◀ **Embedding**—working additional words and phrases into sentences—is another strategy for varying sentence structure. The following revision illustrates this technique.

> John Peter Zenger was a newspaper editor who waged and won an important battle for freedom of the press in America. He criticized the policies of the British governor, and as a result, he was charged with criminal libel. When Zenger's lawyers were disbarred by the governor, Andrew Hamilton defended him, convincing the jury that Zenger's criticisms were true. Therefore, the statements were not libelous.

In this revision the sentence *Hamilton convinced the jury . . .* has been reworded to create a phrase (*convincing the jury*) that modifies the independent clause *Andrew Hamilton defended him.*

The final revision of the original string of choppy sentences is a varied, readable paragraph that uses coordination, subordination, and embedding to vary sentence length but retains the final short simple sentence for emphasis. This revision, of course, represents only one of the many possible ways to achieve sentence variety.

EXERCISE 2

Using coordination, subordination, and embedding, revise this string of choppy simple sentences into a more varied and interesting paragraph.

The first modern miniature golf course was built in New York in 1925. It was an indoor course with 18 holes. Entrepreneurs Drake Delanoy and John Ledbetter built 150 more indoor and outdoor courses. Garnet Carter made miniature golf a worldwide fad. Carter built an elaborate miniature golf course. He later joined with Delanoy and Ledbetter. Together they built more miniature golf courses. They abbreviated playing distances. They highlighted the game's hazards at the expense of skill. This made the game much more popular. By 1930 there were 25,000 miniature golf courses in the United States. Courses grew more elaborate. Hazards grew more bizarre. The craze spread to London and Hong Kong. The expansion of miniature golf grew out of control. Then, interest in the game declined. By 1931 most miniature golf courses were out of business. The game was revived in the early 1950s. Today there are between eight and ten thousand miniature golf courses. The architecture of miniature golf remains an enduring form of American folk art. (Adapted from *Games*)

12c Breaking Up Strings of Compounds

An unbroken series of compound sentences can be dull—and unemphatic. When you connect clauses with coordinating conjunctions only, you may fail to indicate emphasis or relationships accurately.

▶ See 11c1

UNEMPHATIC: A volcano that is erupting is considered *active,* but one that may erupt is designated *dormant,* and one that has not erupted for a long time is called *extinct.* Most active volcanoes are located in "The Ring of Fire," a belt that circles the Pacific Ocean, and they can be extremely destructive. Italy's Vesuvius erupted in A.D. 79, and it destroyed the town of Pompeii. In 1883 Krakatoa, located between the Indonesian islands of Java and Sumatra, erupted, and it caused a tidal wave, and more than 36,000 people were killed. Martinique's Mont Pelée erupted in 1902, and its lava and ash killed 30,000 people, and this completely wiped out the town of St. Pierre.

EMPHATIC: A volcano that is erupting is considered *active;* one that may erupt is designated *dormant;* and one that has not erupted for a long time is called *extinct.* **[compound sentence]** Most active volcanoes are located in "The Ring of Fire," a belt that circles the Pacific Ocean. **[simple sentence with modifier]** Active volcanoes can be extremely destructive. **[simple sentence]** Erupting in A.D. 79, Italy's Vesuvius destroyed the town of Pompeii. **[simple sentence with modifier]** When Krakatoa, located between the Indonesian islands of Java and Sumatra, erupted in 1883, it caused a tidal wave that killed 36,000 people. **[compound-complex sentence with modifier]** The eruption of Martinique's Mont Pelée in 1902 produced lava and ash that killed 30,000 people, completely wiping out the town of St. Pierre. **[complex sentence with modifier]**

*EXERCISE 3

Revise the compound sentences in this passage so the sentence structure is varied and the writer's meaning and emphasis are clear.

Dr. Alice I. Baumgartner and her colleagues at the Institute for Equality in Education at the University of Colorado surveyed 2,000 Colorado schoolchildren, and they found some startling results. They

ANSWERS TO EXERCISE 3

In surveying 2,000 Colorado schoolchildren, Dr. Alice I. Baumgartner and her colleagues at the Institute for Equality in Education found some startling results. They asked, "If you woke up tomorrow morning and discovered that you were a (girl) (boy), how would your life be different?" The answers were sad and shocking. Although the researchers assumed they would find that boys and girls would see advantages in being either male or female, the survey found instead that both sexes had a fundamental contempt for females. Many elementary school boys titled their answers "The Disaster" or "Doomsday" and described the terrible life they would lead as girls. The girls, however, seemed to think they would be better off as boys and would be able to do more and have easier lives.

Boys and girls alike realized that girls, being judged by their appearance, have to pay more attention to their looks, so both groups perceived boys as having an advantage. In addition, boys' activities were valued more highly than girls' by both groups, and both agreed that "women's work" is less valuable and less valued than "men's work." Both boys and girls, feeling that boys are expected to behave differently, thought that boys could get away with more and be more active. However, they felt that girls have one big advantage: the ability to express their emotions openly.

Finally, both boys and girls agreed that boys are treated better and respected more than girls. This prejudice against females among boys and girls, coupled with sex stereotyping, is a psychological handicap for both men and women.

Varying Sentence Types

Oral work is especially helpful here. Many students will hear a problem before they see it. Devise oral exercises to illustrate the monotony of using too many sentences of the same type. Written examples might include a paragraph of choppy, short sentences, a paragraph of strung-together compounds, a paragraph with repeated use of participial phrases as sentence openers, a paragraph in which every sentence begins with a subject. Have students read the examples out loud so that they can hear as well as see the monotony. Ask them to revise one or more of the monotonous paragraphs. Suggest they test their own writing for monotonous style by first reading their papers out loud and listening for repeated use of patterns.

asked, "If you woke up tomorrow and discovered that you were a (boy) (girl), how would your life be different?" and the answers were sad and shocking. The researchers assumed they would find that boys and girls would see advantages in being either male or female, but instead they found that both boys and girls had a fundamental contempt for females. Many elementary schoolboys titled their answers "The Disaster" or "Doomsday," and they described the terrible lives they would lead as girls, but the girls seemed to feel they would be better off as boys, and they expressed feelings that they would be able to do more and have easier lives.

Boys and girls alike realized that girls are judged by their looks more than boys, and both felt girls had to pay more attention to their looks, so all children perceived boys as having an advantage. In addition, boys and girls both valued boys' activities more highly, and boys and girls agreed that "women's work" is less valuable and less valued than "men's work." Both boys and girls also felt that boys are expected to behave differently, and they felt that boys could get away with more and be more active, but girls did have one advantage and that was that they could express their feelings openly.

Finally, both boys and girls agreed that boys are treated better and respected more than girls, so in other words there is a prejudice against females among both boys and girls, and this sex stereotyping is a psychological handicap for both men and women. (Adapted from *Redbook*)

12d Varying Sentence Types

You can vary sentence types by mixing **declarative sentences** (statements) with occasional **imperative sentences** (commands or requests), **exclamations**, and **rhetorical questions** (questions that the reader is not expected to answer). The following paragraph does just this.

Local television newscasts seem to be delivering less and less news. Although we stay awake for the late news hoping to be updated on local, national, and world events, only about 30 percent of most newscasts is devoted to news. Up to 25 percent of the typical program—even more during "sweeps weeks"—can be devoted to feature stories, with another 25 percent reserved for advertising. The remaining time is spent on weather, sports, and casual conversation between anchors. Given this focus on "soft" material, what options do those of us wishing to find out what happened in the world have? **[rhetorical question]** Critics of local television have a few suggestions. First, write to your local station's management voicing your concern and threatening to boycott the news if changes are not made; then, try to get others who feel the way you do to sign a petition. **[imperatives]** If changes are not made, try turning off your television and reading the newspaper! **[exclamation]**

Other options for varying sentence types include mixing simple, compound, and complex sentences **(see 12b and c)**; mixing cumulative and periodic sentences **(see 10b)**; and using balanced sentences where appropriate **(see 10c)**.

EXERCISE 4

The following paragraph is composed entirely of declarative sentences. To make it more varied, add three sentences—one exclamation, one rhetorical question, and one command—anywhere in the paragraph. Be sure the sentences you create are consistent with the paragraph's purpose and tone.

> When the Fourth of July comes around, the nation explodes with patriotism. Everywhere we look we see parades and picnics, firecrackers and fireworks. An outsider might wonder what all the fuss is about. We could explain that this is America's birthday party, and all the candles are being lit at once. There is no reason for us to hold back our enthusiasm—or to limit the noise that celebrates it. The Fourth of July is watermelon and corn on the cob, American flags and sparklers, brass bands and more. Everyone looks forward to this celebration, and everyone has a good time.

12e Varying Sentence Openings

In addition to varying sentence length and type, you can vary the openings of your sentences. Rather than resigning yourself to beginning every sentence with the subject, strengthen your emphasis and clarify the relationship of a sentence to those that surround it by opening some sentences with modifying words, phrases, or clauses.

(1) Adjectives, adverbs, and adverb clauses

<u>Proud and relieved</u>, they watched their daughter receive her diploma. (adjectives)

<u>Hungrily</u>, he devoured his lunch. (adverb)

ANSWERS TO EXERCISE 4

What a day of excitement! When the Fourth of July comes around, the nation explodes with patriotism. Everywhere we look, we see parades and picnics, firecrackers and fireworks. An outsider might wonder what all the fuss is about. *What could we tell this person?* We could explain that this is America's birthday party, and all the candles are being lit at once. There is no reason for us to hold back our enthusiasm—or to limit the noise that celebrates it. *So join in.* The Fourth of July is watermelon and corn on the cob, American flags and sparklers, brass bands and more. Everyone looks forward to this celebration, and everyone has a good time.

Your students' answers will vary.

1. When he was a very young child, Momaday was taken to Devil's Tower, the geological formation in Wyoming that is called Tsoai (Bear Tree) in Kiowa, and given the name Tsoai-talee (Bear Tree Boy). (adverb clause)

2. In the Kiowa myth of the origin of Tsoai, a boy playfully chases his seven sisters up a tree, which rises into the air as the boy is transformed into a bear. (prepositional phrase)

3. Becoming increasingly ferocious, the bear-boy claws the bark of the tree, which becomes a great rock with a flat top and deeply scored sides. (participial phrase)

4. Eventually, after climbing higher and higher to escape their brother's wrath, the sisters become the seven stars of the Big Dipper. (adverb)

5. A constant in Momaday's works, this story from which he received one of his names appears in *The Way to Rainy Mountain, House Made of Dawn,* and *The Ancient Child.* (appositive)

While Woodrow Wilson was incapacitated by a stroke, his wife unofficially performed many presidential duties. (adverb clause)

(2) Prepositional phrases, participial phrases

For better or worse, alcohol has been a part of human culture through the ages. (*Consumer Reports*) (prepositional phrase)

Located on the west coast of Great Britain, Wales is part of the United Kingdom. (participial phrase)

(3) Coordinating conjunctions, conjunctive adverbs

The Big Bang may be the beginning of the universe, or it may be a discontinuity in which information about the earlier history of the universe was destroyed. But it is certainly the earliest event about which we have any record. (Carl Sagan, *The Dragons of Eden*) (coordinating conjunction)

Pantomime was first performed in ancient Rome. However, it remains a popular dramatic form today. (conjunctive adverb)

(4) Absolute phrases, restrictive appositives

His interests widening, Picasso designed ballet sets and illustrated books. (absolute phrase)

British scientist Alexander Fleming is famous for having discovered penicillin. (restrictive appositive)

No comma is required to set a restrictive appositive off from the rest of the sentence.

EXERCISE 5

Each of these sentences begins with the subject. Rewrite each so that it has a different opening and then identify the opening strategy you used.

EXAMPLE: N. Scott Momaday, the prominent Native American writer, tells the story of his first fourteen years in *The Names.*

REVISED: Prominent Native American writer N. Scott Momaday tells the story of his first fourteen years in *The Names.* (appositive)

1. Momaday was taken as a very young child to Devil's Tower, the geological formation in Wyoming that is called Tsoai (Bear Tree) in Kiowa, and there he was given the name Tsoai-talee (Bear Tree Boy).

2. The Kiowa myth of the origin of Tsoai is about a boy who playfully chases his seven sisters up a tree, which rises into the air as the boy is transformed into a bear.

3. The boy-bear becomes increasingly ferocious and claws the bark of the tree, which becomes a great rock with a flat top and deeply scored sides.

4. The sisters climb higher and higher to escape their brother's wrath, and eventually they become the seven stars of the Big Dipper.

5. This story, from which Momaday received one of his names, appears as a constant in his works—*The Way to Rainy Mountain, House Made of Dawn,* and *The Ancient Child*

12f Varying Standard Word Order

You can vary standard subject-verb-object (or complement) word order in two ways: by intentionally inverting this usual order or by placing words between subject and verb.

(1) Invert word order

You can invert conventional word order by placing the complement or direct object *before* the verb instead of in its conventional position. Or, you can place the verb *before* the subject instead of after it.

In each of the following sentences, the unusual word order draws attention to what has been inverted.

(subject)
Nature I loved and, next to Nature, Art. (Walter Savage Landor)
(object)   (verb)

(complement)
The book was extremely helpful; especially useful was its index.
(verb)   (subject)

CLOSE-UP

WRITING VARIED SENTENCES

Because it is an unexpected departure from natural order, inverted word order can be distracting. Therefore, you should use it in moderation; when overused, inversion loses its force and can sound pretentious.

ANSWERS TO EXERCISE 6

1. A band called the Beach Boys, consisting of Brian Wilson, his brothers Carl and Dennis, their cousin Mike Love, and Alan Jardine, a friend, was formed in 1961.
2. Attracting national attention was the group's first single, "Surfin.'"
3. Capitol Records, feeling the band had potential, signed the group to record "Surfin' Safari."
4. Writing, arranging, and producing most of the other top-twenty singles during the next five years was Brian Wilson.
5. Their songs, focusing on California sun and good times, included "I Get Around," "Be True to Your School," "Fun, Fun, Fun," and "Good Vibrations."

Sample Paragraph

A band called the Beach Boys, consisting of Brian Wilson, his brothers Carl and Dennis, their cousin Mike Love, and Alan Jardine, a friend, was formed in 1961. Attracting national attention was the group's first single, "Surfin.'" Capitol Records, feeling the band had potential, signed the band to record "Surfin' Safari." Writing, arranging, and producing most of the other top-twenty singles during the next five years was Brian Wilson. Their songs, focusing on California sun and good times, included "I Get Around," "Be True to Your School," "Fun, Fun, Fun," and "Good Vibrations."

Ask students how this paragraph could be further edited to achieve greater sentence variety.

(2) Separate subject from verb

Placing words or phrases between subject and verb is another way to vary standard word order.

> Many <u>states</u> <u>require</u> that infants and young children ride in government-approved car seats because they hope this will reduce needless fatalities. (subject and verb together)

> Many <u>states</u>, hoping to reduce needless fatalities, <u>require</u> that infants and young children ride in government-approved car seats. (subject and verb separated)

CLOSE-UP

WRITING VARIED SENTENCES

See ◄ 15a4

Be sure the word or phrase you place between subject and verb does not obscure their connection.

EXERCISE 6

The following sentences use conventional word order. Revise each in one of two ways: Either invert the sentence, or vary the word order by placing words between subject and verb. After you have completed your revisions, create a varied five-sentence paragraph by linking the sentences together.

> **EXAMPLE:** Exam week is invariably hectic and not much fun.
>
> Invariably hectic and not much fun is exam week.
>
> or
>
> Exam week, invariably hectic, is not much fun.

1. The Beach Boys formed a band in 1961, and the group consisted of Brian Wilson, his brothers Carl and Dennis, their cousin Mike Love, and Alan Jardine, a friend.
2. The group's first single was "Surfin'," which attracted national attention.
3. Capitol Records signed the band to record "Surfin' Safari" because the company felt the group had potential.
4. The Beach Boys had many other top-twenty singles during the next five years, and most of these hits were written, arranged, and produced by Brian Wilson.
5. Their songs focused on California sun and good times and included "I Get Around," "Be True to Your School," "Fun, Fun, Fun," and "Good Vibrations."

WRITING VARIED SENTENCES

Read this draft of a student essay carefully. Then, revise it to achieve greater sentence variety by varying the length, type, openings, and word order of the sentences. After you have done so, revise further if necessary to strengthen coherence, unity, and style.

The American Dream

Although I was only seven at the time, I can vividly recall my family's escape from Vietnam to America. My mother, brother, and I came to America together. My father had fought against the communists during the war. As a result, my brother and I were forbidden from obtaining any education beyond the high school level. Also, Vietnam was an impoverished nation. It seemed as if we had no choice but to leave.

The boat we left in was small. It held only fifty-two people. We left at two o'clock in the morning. At that time passage by boat cost a sum equal to fifteen hundred American dollars for each child and three thousand for each adult. This was a great deal of money, and very few families could afford to escape. My family was fairly well off in Vietnam. Still, this was a lot of money for us.

Unlike other "boat people," we were fortunate to be rescued by a Norwegian liner only three days after we left Vietnam. When we saw this ship, all the men in our boat hid. They told the women and children to sit up and stay in the ship's view. They hoped the crew would feel sorry for the women and children and rescue us. This is exactly what happened. When we first boarded the ship, the crew asked us if we wanted anything. We cupped our hands and brought them to our lips. This showed them we were thirsty.

The Norwegian ship left us in Japan at a United Nations refugee camp. We stayed there nearly a year. Then, United Nations staff came to pick up people who had sponsors in other countries. My father had come to the United States two years earlier, so he was our sponsor.

When my mother, brother, and I arrived in America, we stayed for two months in a boardinghouse with my father. Then we all moved into an apartment of our own. My mother found a job as a housekeeper in a hospital. This was hard for her at first because she could not speak or understand English. It was frustrating to her when she was accused of not doing her work. She couldn't defend herself. My brother and I were more fortunate. At

continued on the following page

Writing Varied Sentences

Possible Revision

The American Dream
Although I was only seven at the time, I can vividly recall when my mother, brother, and I escaped from Vietnam and came to America together. Because my father had fought against the communists during the war, my brother and I were forbidden from obtaining any education beyond the high school level. In addition, Vietnam had become an impoverished nation, so it seemed as if we had no choice but to leave.

We left Vietnam at two o'clock in the morning on a small boat which held only fifty-two people. At that time, passage by boat cost a great deal of money, a sum equal to fifteen hundred American dollars for each child and three thousand dollars for each adult. As a result, very few families could afford to escape. Although my family was fairly well off in Vietnam, this was still a lot of money for us as well.

We were fortunate, unlike some other "boat people," to be rescued by a Norwegian liner only three days after we left Vietnam. When we saw the ship, all the men in our boat hid, telling the women and children to sit up and stay in the ship's view, hoping that the crew would feel sorry for the women and children and rescue us. This is exactly what happened. When we first boarded the ship, the crew asked us if we wanted anything, and we cupped our hands and brought them to our lips to indicate that we were thirsty.

The Norwegian ship left us at a United Nations refugee camp in Japan, where we stayed for nearly a year. Then the United Nations staff came to pick up people who had

continued on following page

continued from previous page

sponsors in other countries. Since my father had come to the United States two years earlier, he was our sponsor.

When we arrived in America, my mother, brother, and I stayed for two months in a boardinghouse with my father before we all moved into an apartment of our own. My mother found a job as a housekeeper in a hospital, but the job was hard and frustrating for her at first. When she was accused of not doing her work, she could not defend herself because she could not speak or understand English. My brother and I were more fortunate, however, because at our school there were many people to help us adjust to American life. Even so, we had to fend for ourselves when we came home from school since both of our parents worked until 6 p.m.

Before they arrived, my parents had expected life to be much easier in America than it had been in Vietnam, but they soon found out just how hard life could be here. Because of their limited education, my parents cannot expect a life much better than the one they currently have; still, they do expect more for their children. We left Vietnam so my brother and I could have a better education and a better life. Now we are in America; the rest is up to us.

Commentary on Possible Revision

In this revision, the writer has achieved greater variety in sentence structure. However, the writer needs to revise this paper further to cut out deadwood and to improve style. Additional details about the family's problems at home, at school, and at work would give the essay greater meaning for readers unfamiliar with the struggles faced by immigrants.

continued from the previous page

our school there were many people to help us adjust to American life. When we came home from school, however, we had to fend for ourselves. Both of our parents worked until 6 P.M.

Before they arrived in America, my parents expected life to be much easier in America than it was in Vietnam. They were surprised to find out just how hard life could be here. Because of the limited education my parents had, they cannot expect a life much better than what they currently have. However, they do expect more from their children. We left Vietnam so my brother and I could have a better education and a better life. Now we are in America. The rest is up to us.

Solving Common Sentence Problems

Sentence Fragments

continued on following page

A **sentence fragment** is an incomplete sentence, a phrase or clause punctuated as if it were a complete sentence.

A sentence may be incomplete because it lacks a subject or a verb.

Many astrophysicists now believe that galaxies are distributed in clusters. <u>And even form super cluster complexes.</u> (subject missing)

Researchers are engaged in a variety of studies. <u>All suggesting a link between alcoholism and heredity.</u> (verb missing)

The streets of many large cities are home to increasing numbers of mentally ill homeless people. <u>Unable to find shelter.</u> (both subject and verb missing)

The second and third of the underlined fragments in the examples above may appear to contain verbs, but they do not. The participle *suggesting* and the infinitive *to find* are verb forms called **verbals.** Verbals cannot serve as main verbs in a sentence.

See ◄
21c2

A sentence may also be incomplete because it is actually a dependent clause, introduced by a subordinating conjunction or by a relative pronoun.

Bishop Desmond Tutu was awarded the Nobel Peace Prize. <u>Because he struggled to end apartheid.</u> (introduced by subordinating conjunction)

The pH meter and the spectrophotometer are two scientific instruments. <u>That changed the chemistry laboratory dramatically.</u> (introduced by relative pronoun)

When readers cannot see where sentences begin and end, they have difficulty understanding what you have written. For instance, it is impossible to tell to which independent clause the fragment in each of the following sequences belongs.

The course requirements were changed last year. Because a new professor was hired at the very end of the spring semester. I was unable to find out about this change until after preregistration.

In *The Ox-Bow Incident* the crowd is convinced that the men are guilty. Even though the men insist they are innocent and Davies pleads for their lives. They are hanged.

TESTS FOR SENTENCE COMPLETENESS
1. A sentence must have a subject.
2. A sentence must have a verb.
3. A sentence cannot consist of a dependent clause alone. (A sentence cannot consist of a single clause that begins with a subordinating conjunction; unless it is a question, it cannot consist of a single clause beginning with *how, who, which, where, when,* or *why.*)

If your sentence does not pass these three tests, it is a fragment and should be revised. Once you determine a sentence is incomplete, you can use one or more of the following strategies to revise the fragment.

STRATEGIES FOR REVISING SENTENCE FRAGMENTS
- Attach the fragment to an adjacent complete sentence.

   Fragment: According to German legend, Lohengrin is the son of Parzival. And a knight of the Holy Grail.

   Revised: According to German legend, Lohengrin is the son of Parzival and a knight of the Holy Grail.

- Supply the missing subject and/or verb.

   Fragment: Lancaster County, Pennsylvania, is home to many Pennsylvania Dutch. Descended from eighteenth-century settlers from southwest Germany.

   Revised: Lancaster County, Pennsylvania, is home to many Pennsylvania Dutch. They are descended from eighteenth-century settlers from southwest Germany.

- Delete the subordinating conjunction or relative pronoun.

   Fragment: Property taxes rose sharply. Although city services showed no improvement.

   Revised: Property taxes rose sharply. City services showed no improvement.

---

continued from previous page

sentence will automatically make it incomplete. "Melinda learns easily" is a complete sentence. But "If/Because/Although Melinda learns easily . . ." is a fragment.

TEST 2: HAVE YOU ALLOWED A RELATIVE PRONOUN TO COME BETWEEN YOUR SUBJECT AND ITS VERB?

When *who, what,* or *which* comes between a subject and its complete verb, it grabs the verb and creates a fragment. "College football draws more fans every year" is a sentence; "College football, *which draws* more fans every year" is not.

TEST 3: DOES YOUR WORD GROUP LACK A NOUN OR A SUBSTITUTE FOR A SUBJECT?

When writing quickly, students sometimes leave out the noun, pronoun, noun clause, or verbal a sentence needs for its subject. Without a functioning subject, what is left is a *phrase fragment.*

TEST 4: DOES YOUR WORD GROUP LACK A COMPLETE VERB?

Verbals (infinitives, gerunds, past participles) cannot complete sentences because none of them can stand alone as the verb of a predicate. "Hershel *to run* the 100-yard dash" is a fragment; to create a complete sentence, you must add a finite verb: "Hershel *will run* the 100-yard dash."

Revising for Fragments
Some students find it easier to find fragments in their writing using the time-honored proofreader's "trick" of starting at the end of an essay—reading it in reverse order, one punctuated sentence at a time. Reading in this way will help them forget about content and concentrate on sentence structure.

The following sections identify the grammatical structures most likely to appear as fragments and illustrate the most effective ways of revising each.

13a Revising Dependent Clauses

A **dependent clause** contains a subject and a verb, but it cannot stand alone as a sentence. Because it needs an independent clause to complete its meaning, a dependent clause (also called a subordinate clause) must always be attached to at least one independent clause. You can recognize a dependent clause because it is always introduced by a subordinating conjunction or a relative pronoun.

See ◄ 9b

To correct dependent clause fragments, either join the dependent clause to a neighboring independent clause, or delete the subordinating conjunction or relative pronoun, creating a complete sentence with a subject and a verb. (In many cases, you will have to replace the relative pronoun with another word that can serve as the clause's subject.)

> **FRAGMENT:** The United States declared war. <u>Because the Japanese bombed Pearl Harbor.</u> (Dependent clause is punctuated as a sentence.)
>
> **REVISED:** The United States declared war because the Japanese bombed Pearl Harbor. (Dependent clause has been attached to an independent clause to create a complete sentence.)
>
> **REVISED:** The Japanese bombed Pearl Harbor. The United States declared war. (Subordinating conjunction *because* has been deleted; the result is a complete sentence.)
>
> **FRAGMENT:** The battery is dead. <u>Which means the car won't start.</u> (Dependent clause is punctuated as a sentence.)
>
> **REVISED:** The battery is dead, which means the car won't start. (Dependent clause has been attached to an independent clause to create a complete sentence.)
>
> **REVISED:** The battery is dead. This means the car won't start. (Relative pronoun *which* has been replaced by *this*, an acceptable subject, to create a complete sentence.)

EXERCISE 1

Identify the sentence fragments in the following paragraph and correct each, either by attaching the fragment to an independent clause or by deleting the subordinating conjunction or relative pronoun to create a sentence that can stand alone. In some cases you will have to replace a relative pronoun with another word that can serve as the subject of an independent clause.

The drive-in movie came into being just after World War II. When both movies and cars were central to the lives of many Americans. Drive-ins were especially popular with teenagers and young families during the 1950s. When cars and gas were relatively inexpensive. Theaters charged by the carload. Which meant that a group of teenagers or a family with several children could spend an evening at the movies for a few dollars. In 1958, when the fad peaked, there were over 4,000 drive-ins in the United States. While today there are fewer than 3,000. Many of these are in the sunbelt, with most in California. Although many sunbelt drive-ins continue to thrive because of the year-round warm weather. Many northern drive-ins are in financial trouble. Because land is so expensive. Some drive-in owners break even only by operating flea markets or swap meets in daylight hours. While others, unable to attract customers, are selling their theaters to land developers. Soon drive-ins may be a part of our nostalgic past. Which will be a great loss for many who enjoy them.

13b Revising Prepositional Phrases

A **prepositional phrase** consists of a preposition, its object, and any modifiers of the object. It cannot stand alone as a sentence.

▶ See
8e1

To correct this kind of fragment, attach it to the independent clause that contains the word or word group modified by the prepositional phrase.

FRAGMENT: President Lyndon Johnson decided not to seek reelection. For a number of reasons. (Prepositional phrase is punctuated as a sentence.)

REVISED: President Lyndon Johnson decided not to seek reelection for a number of reasons. (Prepositional phrase has been attached to an independent clause.)

FRAGMENT: He ran sixty yards for a touchdown. In the final minutes of the game. (Prepositional phrase is punctuated as a sentence.)

REVISED: He ran sixty yards for a touchdown in the final minutes of the game. (Prepositional phrase has been attached to an independent clause.)

Most college athletes are caught in a conflict between their athletic and academic careers. Sometimes college athletes' responsibilities on the playing field make it hard for them to be good students. Often athletes must make a choice between sports and a degree. Some athletes would not be able to afford college without athletic scholarships. But, ironically, their commitments to sports (training, exercise, practice, and travel to out-of-town games, for example) deprive athletes of valuable classroom time. The role of college athletes is constantly being questioned. Critics suggest they exist only to participate in and promote college athletics. Because of the importance of this role to academic institutions, scandals occasionally develop, with coaches and even faculty members arranging to inflate athletes' grades to help them remain eligible for participation in sports. Some universities even lower admissions standards. To help remedy this and other inequities, the controversial Proposition 48, passed at the NCAA convention in 1982, established minimum scores on aptitude tests. But many people feel the NCAA remains overly concerned with profits rather than education. As a result, college athletic competition is increasingly coming to resemble pro sports, from the coaches' pressure on the players to win to the network television exposure to the wagers on the games' outcomes.

Many food products have well-known trademarks, identified by familiar
continued on following page

EXERCISE 2

Read the following passage and identify the sentence fragments. Then correct each one by attaching it to the independent clause that contains the word or word group it modifies.

Most college athletes are caught in a conflict. Between their athletic and academic careers. Sometimes college athletes' responsibilities on the playing field make it hard for them to be good students. Often athletes must make a choice. Between sports and a degree. Some athletes would not be able to afford college. Without athletic scholarships. But, ironically, their commitments to sports (training, exercise, practice, and travel to out-of-town games, for example) deprive athletes. Of valuable classroom time. The role of college athletes is constantly being questioned. Critics suggest athletes exist only to participate in and promote college athletics. Because of the importance of this role to academic institutions, scandals occasionally develop. With coaches and even faculty members arranging to inflate athletes' grades to help them remain eligible. For participation in sports. Some universities even lower admissions standards. To help remedy this and other inequities. The controversial Proposition 48, passed at the NCAA convention in 1982, established minimum College Board scores and grade standards for college students. But many people feel that the NCAA remains overly concerned. With profits rather than with education. As a result, college athletic competition is increasingly coming to resemble pro sports. From the coaches' pressure on the players to win to the network television exposure to the wagers on the games' outcomes.

13c Revising Verbal Phrases

See ◀ 8e2

A **verbal phrase** consists of a present participle *(walking)*, past participle *(walked)*, infinitive *(to walk)*, or gerund plus related objects and modifiers *(walking along the lonely beach)*. Because a verbal cannot serve as a sentence's main verb, a verbal phrase is not a complete sentence and should not be punctuated as one.

To correct a verbal phrase fragment, either attach the verbal phrase to a related independent clause or change the verbal to a verb and add a subject.

FRAGMENT: In 1948 India became independent. <u>Divided into the nations of India and Pakistan.</u> (Verbal phrase is punctuated as a sentence.)

REVISED: Divided into the nations of India and Pakistan, India became independent in 1948. (Verbal phrase has been attached to related independent clause to create a complete sentence.)

REVISED: In 1948 India became independent. It was divided into the nations of India and Pakistan. (Verbal *divided* has been changed to verb *was divided*, and subject *it* has been added; the result is a separate independent clause.)

FRAGMENT: The pilot changed course. <u>Realizing the weather was worsening.</u> (Verbal phrase is punctuated as a sentence.)

REVISED: The pilot changed course, realizing the weather was worsening. (Verbal phrase has been attached to the related independent clause to create a complete sentence.)

REVISED: The pilot changed course. She realized the weather was worsening. (Verb *realized* has been substituted for verbal *realizing*, and subject *she* has been added; the result is a separate independent clause.)

EXERCISE 3

Identify the sentence fragments in the following paragraph and correct each. Either attach the fragment to a related independent clause or add a subject and a verb to create a new independent clause.

> Many food products have well-known trademarks. Identified by familiar faces on product labels. Some of these symbols have remained the same, while others have changed considerably. Products like Sun-Maid Raisin, Betty Crocker potato mixes, Quaker Oats, and Uncle Ben's Rice use faces. To create a sense of quality and tradition and to encourage shopper recognition of the products. Many of the portraits have been updated several times. To reflect changes in society. Betty Crocker's portrait, for instance, has changed five times since its creation in 1936. Symbolizing women's changing roles. The original Chef Boy-ar-dee has also changed. Turning from the young Italian chef Hector Boiardi into a white-haired senior citizen. Miss Sunbeam, trademark of Sunbeam Bread, has had her hairdo modified several times since her first appearance in 1942; the Blue Bonnet girl, also created in 1942, now has a more modern look, and Aunt Jemima has also been changed. Slimmed down a bit in 1965. Similarly, the Campbell's Soup kids are less chubby now than in the 1920s when they first appeared. But the Quaker on Quaker Oats remains as round as he was when he first adorned the product label in 1877. The Morton Salt girl has evolved gradually. Changing several times from blonde to brunette and from straight- to curly-haired. But manufacturers are very careful about selecting a trademark or modifying an existing one. Typically spending a good deal of time and money on research before a change is made. After all, a trademark of long standing can help a product's sales. Giving shoppers the sense that they are using products purchased and preferred by their parents and grandparents.

continued from previous page

faces on product labels. Some of these symbols have remained the same, while others have changed considerably. Products like Sun-Maid Raisins, Betty Crocker potato mixes, Quaker Oats, and Uncle Ben's Rice use faces to create a sense of quality and tradition and to encourage shopper recognition of the products. Many of the portraits have been updated several times to reflect changes in society. Betty Crocker's portrait, for instance, has changed five times since its creation in 1936, symbolizing women's changing roles. The original Chef Boy-ar-dee has also changed, turning from the young Italian chef Hector Boiardi into a white-haired senior citizen. Miss Sunbeam, trademark of Sunbeam Bread, has had her hairdo modified several times since her first appearance in 1942; the Blue Bonnet girl, also created in 1942, now has a more modern look, and Aunt Jemima has also been changed, slimmed down a bit in 1965. Similarly, the Campbell's Soup kids are less chubby now than in the 1920s when they first appeared. But the Quaker on Quaker Oats remains as round as he was when he first adorned the product label in 1877. The Morton Salt girl has evolved gradually, changing several times from blonde to brunette and from straight- to curly-haired. But manufacturers are very careful about selecting a trademark or modifying an existing one, typically spending a good deal of time and money on research before a change is made. After all, a trademark of long standing can help a product's sales, giving shoppers the sense that they are using products purchased and preferred by their parents and grandparents.

The domestic responsibilities of Colonial women were many, their fates sealed by the absence of mechanical devices that have eased the burdens of women in recent years. Washing clothes, for instance, was a complicated procedure, the primary problem being the moving of some 50 gallons of water from a pump or well to the stove (for heating) and washtub (for soaking and scrubbing). Home cooking also presented difficulties, the main challenges for the housewife being the danger of inadvertently poisoning her family and the rarity of ovens. Even much later, housework was extremely time-consuming, especially for rural and low-income families, their access to labor-saving devices remaining relatively limited. (Just before World War II, for instance, only 35 percent of farm residences in the United States had electricity.)

13d   Revising Absolute Phrases

An **absolute phrase**—a modifying phrase that is not connected grammatically to any one word in a sentence—usually consists of a noun or pronoun plus a participle and any related modifiers. (Infinitive phrases sometimes function as absolute phrases; **see 8e3**.) An absolute phrase always contains a subject, but it lacks a verb and, therefore, cannot stand alone as a sentence.

To correct this kind of fragment, attach the absolute phrase to the clause it modifies or substitute a verb for the participle or infinitive in the absolute phrase.

FRAGMENT: India has over 16 million child laborers. <u>Their education cut short by the need to earn money.</u> (Absolute phrase is punctuated as a sentence.)

REVISED: India has over 16 million child laborers, their education cut short by the need to earn money. (Absolute phrase has been attached to the clause it modifies.)

FRAGMENT: The Vietnam War memorial is a striking landmark. <u>Its design featuring two black marble slabs meeting in a V.</u> (Absolute phrase is punctuated as a sentence.)

REVISED: The Vietnam War memorial is a striking landmark. Its design features two black marble slabs meeting in a V. (Verb *features* has been substituted for participle *featuring*.)

EXERCISE 4

Identify the sentence fragments in the following passage and correct each by attaching the absolute phrase to the clause it modifies or by substituting a verb for the participle or infinitive in the absolute phrase.

The domestic responsibilities of colonial women were many. Their fates sealed by the absence of the mechanical devices that have eased the burdens of women in recent years. Washing clothes, for instance, was a complicated procedure. The primary problem being the moving of some 50 gallons of water from a pump or well to the stove (for heating) and washtub (for soaking and scrubbing). Home cooking also presented difficulties. The main challenges for the housewife being the danger of inadvertently poisoning her family and the rarity of ovens. Even much later, housework was extremely time-consuming, especially for rural and low-income families. Their access to labor-saving devices remaining relatively limited. (Just before World War II, for instance, only 35 percent of farm residences in the United States had electricity.) (Adapted from Susan Strasser, *Never Done: A History of American Housework*)

13e Revising Appositives

An **appositive**—a noun or noun phrase that identifies or renames the person or thing it follows—cannot stand alone as a
sentence. An appositive must directly follow the person or thing it
renames, and it must do so within the same sentence.

▶ See
8e4

To correct an appositive fragment, attach the appositive to the
independent clause that contains the word or word group the appositive renames.

> **FRAGMENT:** Piero della Francesca was a leader of the Umbrian
> school. A school that remained close to the traditions of Gothic art.
> (Appositive, a fragment that renames *the Umbrian school*, is punctu-
> ated as a complete sentence.)

> **REVISED:** Piero della Francesca was a leader of the Umbrian school,
> a school that remained close to the traditions of Gothic art. (Apposi-
> tive has been attached to the noun it renames.)

Appositives included for clarification are sometimes introduced by a word or phrase like *or, that is, for example, for instance,
namely,* or *such as.* These additions do not change anything: Appositives still cannot stand alone as sentences. Once again, the
easiest way to correct such fragments is to attach the appositive
to the preceding independent clause.

> **FRAGMENT:** Fairy tales are full of damsels in distress. Such as Snow
> White, Cinderella, and Rapunzel. (Appositive, a phrase that identifies
> *damsels in distress*, is punctuated as a sentence.)

> **REVISED:** Fairy tales are full of damsels in distress, such as Snow
> White, Cinderella, and Rapunzel. (Appositive has been attached to the
> noun it renames.)

You can also correct an appositive fragment by embedding the
appositive within the related independent clause.

> **FRAGMENT:** Some popular novelists are highly respected by later
> generations. For example, Mark Twain and Charles Dickens. (Apposi-
> tive, a phrase that identifies *some popular novelists*, is punctuated as a
> sentence.)

> **REVISED:** Some popular novelists—for example, Mark Twain and
> Charles Dickens—are highly respected by later generations. (Apposi-
> tive has been embedded within the preceding independent clause,
> directly following the noun it renames.)

Note that a nonrestrictive appositive is set off by commas, but
a restrictive appositive takes no commas.

▶ See
27d1

TEACHING STRATEGY

Revising Appositives

Fragments created when appositives
are punctuated as a sentence sometimes go unrecognized simply because they are *long*—and some
students believe all sentence fragments are short, two- or three-word
affairs. To illustrate the inaccuracy of
this idea, give examples of short sentences and long fragments of the various types already mentioned in the
text. Section 13g will provide further
examples.

EXERCISE 5

Identify the fragments in this paragraph and correct them by attaching each to the independent clause containing the word or word group the appositive modifies.

The Smithsonian Institution in Washington, D.C., includes fourteen different buildings. All museums or art galleries. These include well-known landmarks. Such as the National Air and Space Museum and the National Gallery of Art. The Smithsonian also includes the National Zoo. Home of the giant pandas Ling-Ling and Hsing-Hsing. The Smithsonian contains many entertaining and educational exhibits. For example, the Wright Brothers' plane, Lindbergh's *Spirit of St. Louis*, and a moon rock. The Smithsonian also includes eleven other museums. The National Museum of American History, the National Museum of Natural History, the Hirshhorn Museum, the Freer Gallery, the Arts and Industries Building, the National Museum of American Art and National Portrait Gallery, the Renwick Gallery, the National Museum of African Art, The Castle, the Arthur M. Sackler Gallery, and the Anacostia Neighborhood Museum. The National Museum of American History contains one especially historic item. Edison's first light bulb. The National Museum of Natural History also includes a spectacular exhibit. A giant squid that washed ashore in Massachusetts in 1980. These and other entertaining and educational exhibits make up the Smithsonian. An attraction that should not be missed.

The Smithsonian Institution in Washington, D.C., includes fourteen different buildings, all museums or art galleries. These include well-known landmarks such as the National Air and Space Museum and the National Gallery of Art. The Smithsonian also includes the National Zoo, home of the giant pandas Ling-Ling and Hsing-Hsing. The Smithsonian contains many entertaining and educational exhibits--for example, the Wright Brothers' plane, Lindbergh's *Spirit of St. Louis*, and a moon rock. The Smithsonian also includes eleven other museums: the National Museum of American History, the National Museum of Natural History, the Hirshhorn Museum, the Freer Gallery, the Arts and Industries Building, the National Museum of American Art and National Portrait Gallery, the Renwick Gallery, the National Museum of African Art, The Castle, the Arthur M. Sackler Gallery, and the Anacostia Neighborhood Museum. The National Museum of American History contains one especially historic item, Edison's first light bulb. The National Museum of Natural History also includes a spectacular exhibit, a giant squid that washed ashore in Massachusetts in 1980. These and other entertaining and educational exhibits make up the Smithsonian, an attraction that should not be missed.

13f Revising Compounds

When detached from its subject, the last part of a **compound predicate** cannot stand alone as a sentence.

To correct this kind of fragment, connect the detached part of the compound predicate to the rest of the sentence.

FRAGMENT: People with dyslexia have trouble reading. And may also find it difficult to write. (Fragment, part of the compound predicate *have . . . and may also find,* is punctuated as a sentence.)

REVISED: People with dyslexia have trouble reading and may also find it difficult to write. (Detached part of the compound predicate has been connected to the rest of the sentence.)

The last part of a **compound object** or **compound complement** cannot stand alone as a sentence either. Be sure to attach such compounds to the rest of the sentence.

FRAGMENT: They took only a compass and a canteen of water. And some trail mix. (Fragment, part of the compound object *compass . . . canteen . . . trail mix,* is punctuated as a sentence.)

REVISED: They took only a compass, a canteen of water, and some trail mix. (Detached part of the compound object has been connected to the rest of the sentence.)

FRAGMENT: When their supplies ran out, they were surprised. And hungry. (Fragment, part of the compound complement *surprised and hungry*, is punctuated as a sentence.)

REVISED: When their supplies ran out, they were surprised and hungry. (Detached part of the compound complement has been connected to the rest of the sentence.)

EXERCISE 6

Identify the sentence fragments in this passage and correct them by attaching each detached compound to the rest of the sentence.

One of the phenomena of the 1990s is the number of parents determined to raise "superbabies." Many affluent parents, professionals themselves, seem driven to raise children who are mentally superior. And physically fit as well. To this end, they enroll babies as young as a few weeks old in baby gyms. And sign up slightly older preschool children for classes that teach computer skills or violin. Or swimming or Japanese. Such classes are important. But are not the only source of formal education for very young children. Parents themselves try to raise their babies' IQs. Or learn to teach toddlers to read or to do simple math. Some parents begin teaching with flash cards when their babies are only a few months—or days—old. Others wait until their children are a bit older. And enroll them in day-care programs designed to sharpen their skills. Or spend thousands of dollars on "educational toys." Some psychologists and child-care professionals are favorably impressed by this trend toward earlier and earlier education. But most have serious reservations, feeling the emphasis on academics and pressure to achieve may stunt children's social and emotional growth.

13g Revising Incomplete Clauses

Not all sentence fragments are short. In a long sentence with several modifiers, you can easily lose track of the direction of a sentence and leave it unfinished.

To correct such a fragment, you must add, delete, or change words to create a complete independent clause.

FRAGMENT: *Ancient Evenings*, Norman Mailer's 1983 novel, more than ten years in the making and considered by critics to be a major work, which is set in Egypt before the birth of Christ. (Subject *Ancient Evenings* has no verb.)

One of the phenomena of the 1990s is the number of parents determined to raise "superbabies." Many affluent parents, professionals themselves, seem driven to raise children who are mentally superior and physically fit as well. To this end, they enroll babies as young as a few weeks old in baby gyms and sign up slightly older preschool children for classes that teach computer skills or violin or swimming or Japanese. Such classes are important, but are not the only source of formal education for very young children. Parents themselves try to raise their babies' IQs or learn to teach toddlers to read or to do simple math. Some parents begin teaching with flash cards when their babies are only a few months—or days—old. Others wait until their children are a bit older and enroll them in day-care programs designed to sharpen their skills or spend thousands of dollars on "educational toys." Some psychologists and child-care professionals are favorably impressed by this trend toward earlier and earlier education. But most have serious reservations, feeling the emphasis on academics and pressure to achieve may stunt children's social and emotional growth.

The Brooklyn Bridge, completed in 1883, has been the subject of poems, paintings, and films for many years, which helped it to capture the imagination of the public as well as the artist as few other structures have. Artists like George Bellows, Georgia O'Keeffe, Andy Warhol, and Joseph Stella have used the bridge as a subject, and it has appeared in the novels of John Dos Passos and Thomas Wolfe and the poems of Hart Crane and Marianne Moore, who have all seen it as a major symbol of America. The Brooklyn Bridge has also made an appearance in essays by writers like Henry James and Lewis Mumford and has been seen in films from *Tarzan's New York Adventure* and Laurel and Hardy's *Way Out West* to *Annie Hall* and *Sophie's Choice*, which ensured its visibility to the public. Over the years, the bridge has also turned up in songs, in Bugs Bunny cartoons, and on product labels, record jackets, and T-shirts, making it one of the most recognizable structures in America.

REVISED: *Ancient Evenings,* Norman Mailer's 1983 novel, more than ten years in the making and considered by critics to be a major work, is set in Egypt before the birth of Christ. (Relative pronoun *which* has been deleted; *Ancient Evenings* is now the subject of the verb *is set.*)

FRAGMENT: Because of Wright Morris's ambition to become a writer, which led him to travel to Paris just as Ernest Hemingway and Gertrude Stein had. (Two dependent clauses; no independent clause.)

REVISED: Wright Morris's ambition to become a writer led him to travel to Paris just as Ernest Hemingway and Gertrude Stein had. (Subordinating conjunction *because* and relative pronoun *which* have been deleted, leaving one complete independent clause.)

REVISED: Because of his ambition to become a writer, Wright Morris traveled to Paris just as Ernest Hemingway and Gertrude Stein had. (Relative pronoun *which* has been deleted and some words have been changed, leaving a complete independent clause preceded by a dependent clause.)

CLOSE-UP

SENTENCE FRAGMENTS

In certain limited contexts, sentence fragments may be acceptable. For example, we commonly use fragments in speech and in informal writing.

See you later. No sweat.

Back soon. Could be trouble.

Advertising copywriters often use fragments.

Finally. Vegetables with no salt added.

Journalists and creative writers frequently use fragments to achieve special effects—for instance, to represent casual conversation or to convey disconnected thinking.

They tell me that apathy is in this year. Very chic. (Ellen Goodman, *Close to Home*)

Then the curtains breathing out of the dark upon my face, leaving the breathing upon my face. A quarter hour yet. And then I'll not be. The peacefullest words. (William Faulkner, *The Sound and the Fury*)

Keep in mind, however, that in most college writing situations, sentence fragments are not acceptable. Do not use incomplete sentences without carefully considering their suitability for your audience and purpose.

EXERCISE 7

Identify the fragments in the following paragraph and correct each by adding, deleting, or changing words to create a complete independent clause.

The Brooklyn Bridge, completed in 1883 and the subject of poems, paintings, and films for many years, which helped it to capture the imagination of the public as well as the artist as few other structures have. Because artists like George Bellows, Georgia O'Keeffe, Andy Warhol, and Joseph Stella have used the bridge as a subject, and because it has appeared in the novels of John Dos Passos and Thomas Wolfe and the poems of Hart Crane and Marianne Moore, who have all seen it as a major symbol of America. The Brooklyn Bridge, also making an appearance in essays by writers like Henry James and Lewis Mumford and seen in films from *Tarzan's New York Adventure* and Laurel and Hardy's *Way Out West* to *Annie Hall* and *Sophie's Choice*, which ensured its visibility to the public. Over the years, the bridge has also turned up in songs, in Bugs Bunny cartoons, and on product labels, record jackets, and T-shirts, making it one of the most recognizable structures in America.

STUDENT WRITER AT WORK

SENTENCE FRAGMENTS

Carefully read this excerpt from a draft of a student essay. Identify all the sentence fragments and determine why each is a fragment. Correct each sentence fragment by adding, deleting, or modifying words to create a sentence or by attaching the fragment to a neighboring independent clause. Finally, go over the draft again and, if necessary, revise further to strengthen coherence, unity, and style.

Ab Snopes: A Trapped Man

Abner (Ab) Snopes, the father in William Faulkner's story "Barn Burning," is trapped in a hopeless situation. Disgusted with his lack of status yet unable to do much to remedy his dissatisfaction. He has little control over his life, but he still struggles. Fighting his useless battle as best he can.

Ab is a family man. Responsible for a wife, children, and his wife's sister. Unfortunately, he is unable to meet his responsibilities. Such as providing a stable home for his family. Evicted because of Ab's "barn burnings," the family constantly moves from town to town. All its belongings piled on a wagon. But Ab continues to burn barns. Because he hopes that these acts will give him power as well as revenge.

continued on the following page

STUDENT WRITER AT WORK

Sentence Fragments

Possible Revision

Abner (Ab) Snopes, the father in William Faulkner's story "Barn Burning," is trapped in a hopeless situation, disgusted with his lack of status yet unable to do much to remedy his dissatisfaction. [verbal phrase] He has little control over his life, but he still struggles, fighting his useless battle as best he can. [verbal phrase]

Ab is a family man, responsible for a wife, children, and his wife's sister. [appositive] Unfortunately, he is unable to meet his responsibilities, such as providing a stable home for his family. [appositive] Evicted because of Ab's "barn burnings," the family constantly moves from town to town, all its belongings piled on a wagon. [absolute phrase] But Ab continues to burn barns because he hopes that these acts will give him power as well as revenge. [dependent clause]

To the rich landowners he works for, Ab is of little significance, poor, uneducated, uncultured. [appositive] There are many men just like him who can work the land. [dependent clause] Ab understands this situation but is unwilling to accept his inferior status. [compound predicate] Consequently, he approaches new employers with arrogance, his actions and manner soon causing trouble. [absolute phrase] This behavior, of course, ensures his eventual dismissal. Ab feels that since he can never gain their respect, he should not even bother behaving in a civilized manner. [dependent clause] So he insists on playing the role of a belligerent, raging man. [prepositional phrase]

This behavior sets in motion a self-fulfilling prophecy. Each time Ab's behavior causes an employer to ask

continued on following page

continued from previous page

him to leave, his prophecy that he will be mistreated is fulfilled. He pretends that the failure is his employer's, not his own, and vents his frustration [compound predicate] by destroying their property with fire. [prepositional phrase] He also feels that such actions will earn him respect. People will be frightened of him, and he will create a name for himself. Only Ab's son, Sarty, sees the truth, that Ab is to his employers "no more . . . than a buzzing wasp." [appositive]

Commentary on Possible Revision

Note that in this revision, the writer has revised carefully to eliminate sentence fragments. In a subsequent draft, she might want to support her statements with quotations from the story and references to specific events.

continued from the previous page

To the rich landowners he works for, Ab is of little significance. Poor, uneducated, uncultured. There are many men just like him. Who can work the land. Ab understands this situation. But is unwilling to accept his inferior status. Consequently, he approaches new employers with arrogance. His actions and manner soon causing trouble. This behavior, of course, ensures his eventual dismissal. Ab feels that since he can never gain their respect. He should not even bother behaving in a civilized manner. So he insists on playing the role. Of a belligerent, raging man.

Ab's behavior sets in motion a self-fulfilling prophecy. Each time Ab's actions cause an employer to ask him to leave, his prophecy that he will be mistreated is fulfilled. He pretends that the failure is his employer's, not his own. And vents his frustration. By destroying their property with fire. He also feels that such actions will earn him respect. People will be frightened of him, and he will create a name for himself. Only Ab's son, Sarty, sees the truth. That Ab is to his employers "no more . . . than a buzzing wasp."

CHAPTER 14

Comma Splices and Fused Sentences

A **comma splice** occurs when two independent clauses are joined only by a comma. A **fused sentence** occurs when two independent clauses are joined with no punctuation.

COMMA SPLICE: Charles Dickens created the character of Mr. Micawber, he also created Uriah Heep.

FUSED SENTENCE: Charles Dickens created the character of Mr. Micawber he also created Uriah Heep.

REVISED: Charles Dickens created the character of Mr. Micawber. He also created Uriah Heep.

You can revise comma splices and fused sentences in one of four ways.

REVISING COMMA SPLICES AND FUSED SENTENCES

Comma Splice

1. Substitute a period for the comma.
2. Substitute a semicolon for the comma.
3. Add an appropriate co-ordinating conjunction.

4. Subordinate one clause to the other.

Fused Sentence

1. Add a period between clauses.
2. Add a semicolon between clauses.
3. Add a comma and an appropriate coordinating conjunction.
4. Subordinate one clause to the other.

TEACHING STRATEGY

Fused Sentences
Some students, in their zeal to avoid fragments, may write fused sentences; others, in their zeal to avoid fused sentences, may write fragments. Integrating work on the two problem areas may help students to eliminate both kinds of errors. Devise exercises that include both fragments and fused sentences. Ask the students to revise and correct the errors. Then discuss their answers in class.

Just as some students confuse sentence fragments with short sentences, students often confuse fused sentences with long sentences. Providing examples of long sentences that are not fused and short ones that are may help clarify the difference.

"If you must use . . . the semicolon to connect two independent clauses . . . be sure that the two clauses actually need some connection and that they wouldn't be better off as separate sentences. Don't write:

The car slid off the roadway into the muddy embankment; the police arrived later to find that no one had survived.

when you could write:

The car slid off the roadway into the muddy embankment. When police arrived later, they found three bodies crushed by the overturned car."

–Lauren Kessler and
Duncan McDonald,
When Words Collide

14a Revising with Periods

Use a period to separate independent clauses and create two separate sentences. A comma splice or fused sentence can be revised in this way when the clauses are of equal importance but are not related closely enough to be joined in one sentence.

COMMA SPLICE: In the late nineteenth century Alfred Dreyfus, a Jewish captain in the French army, was falsely convicted of treason, his struggle for justice pitted the army and the Catholic establishment against the civil libertarians.

FUSED SENTENCE: In the late nineteenth century Alfred Dreyfus, a Jewish captain in the French army, was falsely convicted of treason his struggle for justice pitted the army and the Catholic establishment against the civil libertarians.

REVISED: In the late nineteenth century Alfred Dreyfus, a Jewish captain in the French army, was falsely convicted of treason. His struggle for justice pitted the army and the Catholic establishment against the civil libertarians.

CLOSE-UP

COMMA SPLICES AND FUSED SENTENCES

Substituting a period is the best way to revise a comma splice created by the incorrect punctuation of a broken quotation.

COMMA SPLICE: "This is a good course," Eric said, "in fact, I wish I'd taken it sooner."

REVISED: "This is a good course," Eric said. "In fact, I wish I'd taken it sooner."

14b Revising with Semicolons

If two clauses of equal importance are closely related, and if you want to underscore that relationship, use a semicolon.

COMMA SPLICE: In pre-World War II Western Europe only a small elite had access to a university education, this situation changed dramatically after the war.

FUSED SENTENCE: In pre-World War II Western Europe only a small elite had access to a university education this situation changed dramatically after the war.

REVISED: In pre-World War II Western Europe only a small elite had access to a university education; this situation changed dramatically after the war.

You can also use a semicolon to revise a comma splice or fused sentence when the ideas in the joined clauses are presented in parallel terms. In such cases, the semicolon emphasizes the parallelism between the two clauses.

COMMA SPLICE: Chippendale chairs have straight legs, Queen Anne chairs have curved legs.

FUSED SENTENCE: Chippendale chairs have straight legs Queen Anne chairs have curved legs.

REVISED: Chippendale chairs have straight legs; Queen Anne chairs have curved legs.

CLOSE-UP

COMMA SPLICES AND FUSED SENTENCES

Remember that you cannot correct a comma splice or fused sentence simply by adding a conjunctive adverb *(however, nevertheless, therefore,* and so on) or a transitional expression *(for example, in fact, on the other hand)* between the independent clauses. If you do, you will still have a comma splice or fused sentence.

▶ **See 29c**

COMMA SPLICE: The International Date Line is drawn north and south through the Pacific Ocean, largely at the 180th meridian, thus, it separates Wake and Midway Islands.

FUSED SENTENCE: The International Date Line is drawn north and south through the Pacific Ocean, largely at the 180th meridian thus, it separates Wake and Midway Islands.

REVISED: The International Date Line is drawn north and south through the Pacific Ocean, largely at the 180th meridian; thus, it separates Wake and Midway Islands.

REVISED: The International Date Line is drawn north and south through the Pacific Ocean, largely at the 180th meridian. Thus, it separates Wake and Midway Islands.

Your students' answers will vary. To illustrate the various responses, each sentence below is followed with the four possible types of corrections. Class discussion should consider the need to balance the types of choices in a piece of writing rather than adhering to a single method of correction.

Entrepreneurship is the study of small businesses, college students are embracing it enthusiastically.

1. businesses. College students
2. businesses; college students
3. businesses, and college students
4. Entrepreneurship, which is the study of small businesses, is being embraced enthusiastically by college students.

Many schools offer one or more courses in entrepreneurship, these courses teach the theory and practice of starting a small business.

1. entrepreneurship. These courses
2. entrepreneurship; these courses
3. entrepreneurship, and these courses
4. entrepreneurship, which teach the theory and practice of starting a small business.

Students are signing up for courses, moreover they are starting their own businesses.

1. courses. Moreover,
2. courses; moreover,
3. courses, and, moreover,
4. Students who sign up for courses are even starting their own businesses.

One student started with a car-waxing business, now he sells condominiums.

continued on following page

continued from previous page

1. business. Now
2. business; now
3. business, and now
4. One student, who started with a car-waxing business, now sells condominiums.

Other students are setting up catering services, they supply everything from waiters to bartenders.

1. services. They
2. services; they
3. services, and they
4. services that supply everything from waiters to bartenders.

One student has a thriving cake-decorating business, in fact she employs fifteen students to deliver the cakes.

1. business. In fact, she
2. business; in fact, she
3. business, and, in fact, she
4. business that employs fifteen students to deliver the cakes.

All over the country, student businesses are selling everything from tennis balls to bagels, the student owners are making impressive profits.

1. bagels. The student owners
2. bagels; the student owners
3. bagels, and the student owners
4. bagels, which make impressive profits for the student owners.

Formal courses at the graduate as well as undergraduate level are attracting more business students than ever, several business schools (such as Baylor University, the University of Southern California, and Babson College) even offer degree programs in entrepreneurship.

continued on following page

14c Revising with Coordinating Conjunctions

See ◀
21g
If two closely related clauses are of equal importance, you can use an appropriate coordinating conjunction to indicate whether the clauses are linked by addition *(and)*, contrast *(but, yet)*, causality *(for, so)*, or a choice of alternatives *(or, nor)*.

> **COMMA SPLICE:** Elias Howe invented the sewing machine, Julia Ward Howe was a poet and social reformer.
>
> **FUSED SENTENCE:** Elias Howe invented the sewing machine Julia Ward Howe was a poet and social reformer.
>
> **REVISED:** Elias Howe invented the sewing machine, but Julia Ward Howe was a poet and social reformer. (Coordinating conjunction *but* shows emphasis is on contrast.)

14d Revising with Subordinating Conjunctions or Relative Pronouns

See ◀
9b
When the ideas in two clauses are not of equal importance, correct the comma splice or fused sentence by subordinating the less important idea to the more important one, placing the less important idea in a dependent clause. The subordinating conjunction or relative pronoun establishes the nature of the relationship between the clauses.

> **COMMA SPLICE:** Stravinsky's ballet *The Rite of Spring* shocked Parisians in 1913, its rhythms and the dancers' movements seemed erotic.
>
> **FUSED SENTENCE:** Stravinsky's ballet *The Rite of Spring* shocked Parisians in 1913 its rhythms and the dancers' movements seemed erotic.
>
> **REVISED:** Because its rhythms and the dancers' movements seemed erotic, Stravinsky's ballet *The Rite of Spring* shocked Parisians in 1913. (Subordinating conjunction *because* has been added to make the original sentence's second clause subordinate to its first; the result is one complex sentence.)
>
> **COMMA SPLICE:** Lady Mary Wortley Montagu had suffered from smallpox herself, she helped spread the practice of inoculation against the disease in eighteenth-century England.
>
> **REVISED:** Lady Mary Wortley Montagu, who had suffered from smallpox herself, helped spread the practice of inoculation against the disease in eighteenth-century England. (Relative pronoun *who* has been added to make the original sentence's first clause subordinate to its second; the result is one complex sentence.)

CLOSE-UP

COMMA SPLICES AND FUSED SENTENCES

In rare cases comma splices are considered acceptable. For instance, a comma is used in dialogue between a statement and a tag question, even though each is a separate independent clause.

This is Ron's house, isn't it?

I'm not late, am I?

In addition, commas may connect two short balanced independent clauses or two or more short parallel independent clauses, especially when one clause contradicts the other.

Commencement isn't the end, it's the beginning.

EXERCISE 1

Find the comma splices and fused sentences in the following paragraph. Correct each in *two* of the four possible ways listed on page 259. If a sentence is correct, leave it alone.

EXAMPLE: The fans rose in their seats, the game was almost over.

The fans rose in their seats; the game was almost over.

The fans rose in their seats, for the game was almost over.

Entrepreneurship is the study of small businesses, college students are embracing it enthusiastically. Many schools offer one or more courses in entrepreneurship these courses teach the theory and practice of starting a small business. Students are signing up for courses, moreover they are starting their own businesses. One student started with a car-waxing business, now he sells condominiums. Other students are setting up catering services they supply everything from waiters to bartenders. One student has a thriving cake-decorating business, in fact she employs fifteen students to deliver the cakes. All over the country, student businesses are selling everything from tennis balls to bagels, the student owners are making impressive profits. Formal courses at the graduate as well as undergraduate level are attracting more business students than ever, several business schools (such as Baylor University, the University of Southern California, and Babson College) even offer degree programs in entrepreneurship. Many business school students are no longer planning to be corporate executives instead they plan to become entrepreneurs.

continued from previous page

1. than ever. Several business schools
2. than ever; several business schools
3. than ever, so several business schools
4. Since formal courses at the graduate as well as undergraduate level are attracting more business students than ever, several business schools . . .

Many business school students are no longer planning to be corporate executives, instead they plan to become entrepreneurs.

1. executives. Instead, they
2. executives; instead, they
3. executives, but, instead, they
4. Many business school students, who are no longer planning to become corporate executives, plan, instead, to become entrepreneurs.

ANSWERS TO EXERCISE 2

Your students' answers may vary.

1. Several recent studies indicate that many American high school students have a poor sense of history; this is affecting our future as a democratic nation and as individuals.
2. Surveys show that nearly one-third of American 17-year-olds cannot identify the countries the United States fought against in World War II, and one-third think Columbus reached the New World after 1750.
3. Several reasons have been given for this decline in historical literacy, but the main reason is the way history is taught.

continued on following page

continued from previous page

4. Although this problem is bad news, the good news is that there is increasing agreement among educators about what is wrong with current methods of teaching history.

5. History can be exciting and engaging, but too often it is presented in a boring manner.

6. Students are typically expected to memorize dates, facts, and figures; history as adventure—as a "good story"—is frequently neglected.

7. One way to avoid this problem is to use good textbooks, which should be accurate, lively, and focused.

8. Another way to create student interest in historical events is to use primary sources instead of so-called "comprehensive" textbooks, for autobiographies, journals, and diaries can give students insight into larger issues.

9. Students can also be challenged to think about history by taking sides in a debate, and they can learn more about connections between historical events by writing essays rather than taking multiple-choice tests.

10. Finally, history teachers should be less concerned about specific historical details, and they should be more concerned about conveying the wonder of history.

STUDENT WRITER AT WORK

Comma Splices/Fused Sentences

Possible Revision

In June 1934 Franklin D. Roosevelt selected Frances Perkins to head the new committee on Economic Security. Its report was the basis of our

continued on following page

EXERCISE 2

Combine each of the following sentence pairs into one sentence without creating comma splices or fused sentences. In each case, either connect the clauses into a compound sentence with a semicolon or with a comma and a coordinating conjunction, or subordinate one clause to the other to create a complex sentence. You may have to add, delete, reorder, or change words or punctuation.

1. Several recent studies indicate that many American high school students have a poor sense of history. This is affecting our future as a democratic nation and as individuals.

2. Surveys show that nearly one-third of American seventeen-year-olds cannot identify the countries the United States fought against in World War II. One-third think Columbus reached the New World after 1750.

3. Several reasons have been given for this decline in historical literacy. The main reason is the way history is taught.

4. This problem is bad news. The good news is that there is increasing agreement among educators about what is wrong with current methods of teaching history.

5. History can be exciting and engaging. Too often it is presented in a boring manner.

6. Students are typically expected to memorize dates, facts, and figures. History as adventure—as a "good story"—is frequently neglected.

7. One way to avoid this problem is to use good textbooks. Texts should be accurate, lively, and focused.

8. Another way to create student interest in historical events is to use primary sources instead of so-called comprehensive textbooks. Autobiographies, journals, and diaries can give students insight into larger issues.

9. Students can also be challenged to think about history by taking sides in a debate. They can learn more about connections among historical events by writing essays rather than taking multiple-choice tests.

10. Finally, history teachers should be less concerned about specific historical details. They should be more concerned about conveying the wonder of history.

STUDENT WRITER AT WORK

COMMA SPLICES AND FUSED SENTENCES

Read the following answer to an economics examination question that asked students to discuss the provisions of the 1935 Social Security Act; then, correct all comma splices and fused sentences. After you have corrected the errors, go over the answer again and, if necessary, revise further to strengthen coherence, unity, and style.

In June of 1934 Franklin D. Roosevelt selected Frances Perkins to head the new Committee on Economic Security, its report was the basis of our current Social Security program. The committee formulated two policies, one dealt with the employable the other with the unemployable. Roosevelt insisted that these programs be self-financing, as a result both employer and employee social insurance were required. In 1935 the Social Security Act was passed it attempted to categorize the poor and provided for federal sharing of the cost, but under local control (The Social Security Act did not include a public works program, this feature of the New Deal was eliminated.)

Unemployment insurance was one major part of the Act. Funds were to be payable through public employment offices, also the money was to be paid into a trust fund. It was to be used solely for benefits an individual could not be denied funds even if work were available. The program provided for payroll taxes, in addition separate records were to be kept by each state. Old Age Survivor Insurance, another major provision of the Act, was for individuals over sixty-five it was amended in 1939 to cover dependents. One-quarter of the recipients were disabled. Public Assistance was the third major part of the Act this program was designed to help children left alone by the death or absence of the parents and children with mental or physical disabilities. General assistance covered everything not included under the Public Assistance Program this coverage varied from state to state.

The Social Security Act stressed public administration of federal emergency relief assistance thus, it forced reorganization of public assistance. These efforts differed from previous efforts earlier there were no clear guidelines defining which individuals should get aid and why. The Social Security Act attempted to eliminate gaps and overlaps in services.

continued from previous page

current social security program. The committee formulated two policies; one dealt with the employable and the other with the unemployable. Roosevelt insisted that these programs be self-financing; as a result, employer and employee social insurance were required. In 1935 the Social Security Act was passed. It attempted to categorize the poor and provided for federal sharing of the cost, but under local control. (The Social Security Act did not include a public works program; this feature of the New Deal was eliminated.)

Unemployment insurance was one major part of the Act. Funds were to be payable through public employment offices; also, the money was to be paid into a trust fund. It was to be used solely for benefits, but an individual could not be denied funds even if work were available. The program provided for payroll taxes; and, in addition, separate records were to be kept by each state. Old Age Survivor Insurance, another major provision of the Act, was for individuals over sixty-five; it was amended in 1939 to cover dependents. One quarter of the recipients were disabled. Public Assistance was the third major part of this Act. This program was designed to help children left alone by the death or absence of the parents and children with mental or physical disabilities. General assistance covered everything not included under the Public Assistance Program. This coverage varied from state to state.

The Social Security Act stressed public administration of federal emergency relief assistance; thus, it forced reorganization of public assistance. These efforts differed from previous efforts because earlier there were no clear guidelines defining which individuals should get aid and why. The Social Security Act attempted to eliminate gaps and overlaps in services.

CHAPTER 15

Faulty Modification

A **modifier** is a word, phrase, or clause that acts as an adjective or an adverb. Thus, a modifier describes, limits, or qualifies another word, phrase, or clause in a sentence. In general, a modifier is placed close to its **headword**, the word or phrase it modifies. **Faulty modification** is the awkward or confusing placement of modifiers or the modification of nonexistent words. Faulty modification takes two forms: *misplaced modifiers* and *dangling modifiers*.

15a   Revising Misplaced Modifiers

A **misplaced modifier** is a word or word group whose placement indicates that it modifies one word or phrase when it is intended to modify another.

Faster than a speeding bullet, the citizens of Metropolis saw Superman flying overhead.

The placement of the introductory phrase makes it appear to modify *citizens*, yet it should modify *Superman*. Here is a corrected version.

The citizens of Metropolis saw Superman flying overhead, faster than a speeding bullet.

When writing and revising, take care to put modifying words, phrases, and clauses in a position that clearly identifies the headword and that does not awkwardly interrupt a sentence.

266

(1) Revising misplaced words

Readers expect to find modifiers directly before or directly after their headwords.

Dark and threatening, Wendy watched the storm. (Incorrectly placed adjectives *dark* and *threatening* seem to describe Wendy instead of the storm.)

Wendy watched the storm, dark and threatening. (Correct placement of modifying words clarifies meaning.)

Certain modifiers—such as *almost, only, even, hardly, merely, nearly, exactly, scarcely, just,* and *simply*—should always immediately precede the words they modify. Different placements of these modifiers change the meaning of a sentence.

Nick *just* set up camp at the edge of the burned-out town. (He set up camp just now.)

Just Nick set up camp at the edge of the burned-out town. (He set up camp alone.)

Nick set up camp *just* at the edge of the burned-out town. (His camp was precisely at the edge.)

The imprecise placement of modifiers like these sometimes produces a **squinting modifier**, one that seems to modify either a word before it or one after it and to convey a different meaning in each case. To avoid ambiguity, place the modifier so it clearly modifies its headword.

SQUINTING:
The life that everyone thought would fulfill her totally bored her. (Was she supposed to be totally fulfilled, or is she totally bored?)

REVISED:
The life that everyone thought would totally fulfill her bored her. (Everyone expected her to be totally fulfilled.)

REVISED:
The life that everyone thought would fulfill her bored her totally. (She was totally bored.)

EXERCISE 1

In the following sentence pairs, the modifier in each sentence points to a different headword. Underline the modifier and draw an arrow to the word it limits. Then explain the meaning of each sentence.

EXAMPLE: She just came in wearing a hat. (She just now entered.)

She came in wearing just a hat. (She wore only a hat.)

ANSWERS TO EXERCISE 1

1. He wore his almost new jeans. [He wore his nearly new jeans.]

   He almost wore his new jeans. [He decided at the last minute not to wear his new jeans.]

2. He only had three dollars in his pocket. [Besides the three dollars, he had nothing else in his pocket.]

   Only he had three dollars in his pocket. [He alone had this amount of money in his pocket.]

3. I don't even like freshwater fish. [I really dislike them.]

   I don't like even freshwater fish. [I dislike all kinds of fish.]

4. I go only to the beach on Saturdays. [one place]

   I go to the beach only on Saturdays. [one day per week]

5. He simply hated living. [He absolutely hated living.]

   He hated simply living. [He hated merely living.]

1. He wore his almost new jeans.
   He almost wore his new jeans.
2. He only had three dollars in his pocket.
   Only he had three dollars in his pocket.
3. I don't even like freshwater fish.
   I don't like even freshwater fish.
4. I go only to the beach on Saturdays.
   I go to the beach only on Saturdays.
5. He simply hated living.
   He hated simply living.

(2) Revising misplaced phrases

See ◄
8e1,
8e2

Placing a modifying verbal or prepositional phrase incorrectly can change the meaning of a sentence or create an unclear or confusing sentence.

Misplaced Verbal Phrases Certain verbal phrases act as modifiers. As a rule, place them directly *before* or directly *after* the nouns or pronouns they modify.

Roller-skating along the shore, Jane watched the boats.

She watched the car rolling down the hill.

The incorrect placement of a verbal phrase can make a sentence convey an entirely different meaning or make no sense at all.

Jane watched the boats roller skating along the shore.

Rolling down the hill, she watched the car.

Misplaced Prepositional Phrases When a prepositional phrase is used as an adjective, it nearly always directly *follows* the word it modifies.

This is a Dresden figurine from Germany.

Created by a famous artist, Venus de Milo is a statue with no arms.

Incorrect placement of such modifiers can give rise to confusion or even unintended humor.

Venus de Milo is a statue created by a famous artist with no arms.

Revising Misplaced Modifiers

When a prepositional phrase is used as an adverb, it also usually *follows* its head word.

Cassandra looked into the future.

As long as the meaning of the sentence is clear, however, and as long as the headword is clearly identified, you can place an adverbial modifier in other positions.

He had been waiting anxiously at the bus stop for a long time.

CLOSE-UP

FAULTY MODIFICATION

Be careful to avoid ambiguous placement of prepositional phrases serving as adverbs.

MISPLACED: She saw the house she built in her mind.

REVISED: In her mind, she saw the house she built.

REVISED: She saw in her mind the house she built.

EXERCISE 2

Underline the modifying verbal or prepositional phrases in each sentence and draw arrows to their headwords.

EXAMPLE: Calvin is the democrat running for town council.

1. The bridge across the river swayed in the wind.
2. The spectators on the shore were involved in the action.
3. Mesmerized by the spectacle, they watched the drama unfold.
4. The spectators were afraid of a disaster.
5. Within the hour, the state police arrived to save the day.
6. They closed off the area with roadblocks.
7. Drivers approaching the bridge were asked to stop.
8. Meanwhile, on the bridge, the scene was chaos.
9. Motorists in their cars were paralyzed with fear.
10. Struggling against the weather, the police managed to rescue everyone.

ANSWERS TO EXERCISE 2

1. The bridge across the river swayed in the wind.

2. The spectators on the shore were involved in the action.

3. Mesmerized by the spectacle, they watched the drama unfold.

4. The spectators were afraid of a disaster.

5. Within the hour, the state police arrived to save the day.

6. They closed off the area with roadblocks.

7. Drivers approaching the bridge were asked to stop.

8. Meanwhile, on the bridge, the scene was chaos.

9. Motorists in their cars were paralyzed with fear.

10. Struggling against the weather, the police managed to rescue everyone.

ANSWERS TO EXERCISE 3

1. The lion, <u>watching Jack</u>, paced up and down in its cage, ignoring the crowd.

2. <u>Nervous yet curious</u>, Jack stared back at the lion.

3. <u>Anxious to see what would happen</u>, the crowd around them grew.

4. Suddenly Jack heard a <u>terrifying</u> growl from deep in the lion's throat.

5. <u>Scared to death</u>, Jack ran from the zoo, leaving the lion behind.

EXERCISE 3

Use the word or phrase that follows each sentence as a modifier in that sentence. Then draw an arrow to indicate its headword.

EXAMPLE: He approached the lion. (timid)

<u>Timid</u>, he approached the lion.

1. The lion paced up and down in his cage, ignoring the crowd. (watching Jack)
2. Jack stared back at the lion. (nervous yet curious)
3. The crowd around them grew. (anxious to see what would happen)
4. Suddenly Jack heard a growl from deep in the lion's throat. (terrifying)
5. Jack ran from the zoo, leaving the lion behind. (scared to death)

(3) Revising misplaced dependent clauses

Dependent clauses that serve as modifiers—adjective clauses and adverb clauses—must be clearly related to their headwords. Adjective clauses usually appear immediately *after* the words they modify.

During the Civil War, Lincoln was the president <u>who governed the United States</u>.

An adverb clause can appear in any of several positions, as long as the relationship to the clause it modifies is clear and its position conveys the intended emphasis.

<u>During the Civil War</u> Lincoln was president.

Lincoln was president <u>during the Civil War</u>.

To correct misplaced dependent clauses, make the relationship between modifier and headword clear.

MISPLACED ADJECTIVE CLAUSE: This diet program will limit the consumption of possible carcinogens, <u>which will benefit everyone</u>. (Will carcinogens benefit everyone?)

REVISED: This diet program, <u>which will benefit everyone</u>, will limit the consumption of possible carcinogens.

MISPLACED ADVERB CLAUSE: The parents checked to see that the children were sleeping <u>after they finished the wine</u>. (The children drank the wine and then went to sleep?)

REVISED: <u>After they finished the wine</u>, the parents checked to see that the children were sleeping.

EXERCISE 4

Relocate the misplaced verbal or prepositional phrases or dependent clauses so that they clearly point to the words or word groups they modify.

> **EXAMPLE:** *Silent Running* is a film about a scientist left alone in space with Bruce Dern.
>
> *Silent Running* is a film with Bruce Dern about a scientist left alone in space.

1. She realized that she had married the wrong man after the wedding.
2. *The Prince and the Pauper* is a novel about an exchange of identities by Mark Twain.
3. The energy was used up in the ten-kilometer race that he was saving for the marathon.
4. He loaded the bottles and cans into his new Porsche, which he planned to leave at the recycling center.
5. The manager explained the sales figures to the board members using a graph.

(4) Revising intrusive modifiers

An **intrusive modifier** interrupts a sentence, making it difficult for readers to see the connections between parts of verb phrases, between parts of infinitives, or between subjects and their verbs or verbs and their objects or complements.

Interrupted Verb Phrases Revise when modifiers come between an auxiliary and a main verb.

AWKWARD: She had, without giving it a second thought or considering the consequences, planned to reenlist.

REVISED: Without giving it a second thought or considering the consequences, she had planned to reenlist.

AWKWARD: He will, if he ever gets his act together, be ready to leave on Friday.

REVISED: If he ever gets his act together, he will be ready to leave on Friday.

If a modifier is brief, it can usually interrupt a verb phrase.

She had always planned to reenlist.

He will, therefore, be ready to leave on Friday.

Longer interruptions, however, may obscure your meaning.

Split Infinitives Revise when modifiers awkwardly interrupt an infinitive. As a rule, the parts of an infinitive should be pre-

1. She realized after the wedding that she had married the wrong man.
2. *The Prince and the Pauper* by Mark Twain is a novel about an exchange of identities.
3. The energy that he was saving for the marathon was used up in the 10-kilometer race.
4. He loaded the bottles and cans, which he planned to leave at the recycling center, into his new Porsche.
5. Using a graph, the manager explained the sales figures to the board members.

"Intervening words between subject and verb should not distract from the meaning of the sentence. Even with the presence of a phrase mucking up the middle ground between them, a subject and verb must agree.

An *entourage* of hangers-on *was* sprawling in the lounge.

A *coven* of baby witches and warlocks *was* spotted on the moor."

–Karen Elizabeth Gordon, *The Transitive Vampire: A Handbook of Grammar for the Innocent, the Eager, and the Doomed* sented together. When a modifier splits an infinitive—that is, comes between the word *to* and the base form of the verb—the sentence often becomes awkward.

> **AWKWARD:** He hoped <u>to</u> quickly and easily <u>defeat</u> his opponent.
> **REVISED:** He hoped <u>to defeat</u> his opponent quickly and easily.

CLOSE-UP

FAULTY MODIFICATION

Although the general rule in writing has been never to split an infinitive, it is sometimes necessary to do so. When the intervening modifier is short, and when the alternative is awkward or ambiguous, a split infinitive is acceptable. In the following sentence, for example, a reader would have no trouble connecting the parts of the infinitive.

She expected <u>to</u> not quite <u>beat</u> her previous record.

Moreover, any revision using the same words is awkward or incoherent.

She expected not quite to beat her previous record.

She expected to beat not quite her previous record.

The only way to avoid a split infinitive in this case is to re-word the original sentence.

She expected her score to be close to her previous record.

Interrupted Subjects and Verbs or Verbs and Objects or Complements Revise if you have any doubts about letting a modifier stand between subject and verb or verb and object or complement. It is standard practice to place even a complex or lengthy adjective phrase or clause between a subject and a verb or between a verb and its object or complement.

> **ACCEPTABLE:** Major <u>films</u> that were financially successful in the 1930s <u>include</u> *Gone with the Wind* and *The Wizard of Oz.* (Adjective clause between subject and verb does not obscure sentence's meaning.)

An adverb phrase or clause in this position, however, may not be natural sounding or clear.

CONFUSING: The election, because officials discovered that some people voted twice, was contested. (Adverb clause intrudes between subject and verb.)

REVISED: Because officials discovered that some people voted twice, the election was contested. (Subject and verb are no longer separated.)

CONFUSING: A. A. Milne wrote, when his son Christopher Robin was a child, *Winnie-the-Pooh*. (Adverb clause intrudes between verb and object.)

REVISED: When his son Christopher Robin was a child, A. A. Milne wrote *Winnie-the-Pooh*. (Verb and object are no longer separated.)

EXERCISE 5

Revise these sentences so the modifying phrases or clauses do not interrupt the parts of a verb phrase or infinitive or separate a subject from a verb or a verb from its object or complement.

> **EXAMPLE:** A play can sometimes be, despite the playwright's best efforts, mystifying to the audience.
>
> Despite the playwright's best efforts, a play can sometimes be mystifying to the audience.

1. The people in the audience, when they saw the play was about to begin and realized the orchestra had finished tuning up and had begun the overture, finally quieted down.
2. They settled into their seats, expecting to very much enjoy the first act.
3. However, most people were, even after watching and listening for twenty minutes and paying close attention to the drama, completely baffled.
4. In fact, the play, because it had nameless characters, no scenery, and a rambling plot that didn't seem to be heading anywhere, puzzled even the drama critics.
5. Finally one of the three major characters explained, speaking directly to the audience, what the play was really about.

1. The people in the audience finally quieted down when they saw the play was about to begin and realized the orchestra had finished tuning up and had begun the overture.
2. Expecting to enjoy the first act very much, they settled into their seats.
3. However, most people were completely baffled, even after watching and listening for twenty minutes and paying close attention to the drama.
4. In fact, because it had nameless characters, no scenery, and a rambling plot that didn't seem to be heading anywhere, the play puzzled even the drama critics.
5. Finally, speaking directly to the audience, one of the three major characters explained what the play was really about.

15b Revising Dangling Modifiers

A **dangling modifier** is a word or phrase that cannot logically describe, limit, or restrict any word or word group in the sentence. In fact, its true headword does not appear in the sentence. Consider the following example.

Many undesirable side effects are experienced using this drug.

Dangling Modifiers

A modifier "dangles" when what it is supposed to modify is left out of the sentence. Even the most experienced writers are guilty of danglers—and they are the curse of student writers, who are more likely than experienced writers to get lost in the complexities of a long sentence.

All too many dangling modifiers can be traced back to fuzzy thinking: If the writer is not sure precisely what he or she is talking about, it is sometimes simpler not to worry about matching subjects and their modifiers. Witness the following sentence: "Looking at recent statistics on gun ownership, more American women are taking up arms." *Who* is "looking at recent statistics"? Does the writer mean women are buying guns because so many OTHER people own them—or has the writer been reading gender-specific surveys?

As an in-class exercise, you might distribute copies of a paragraph from an essay or magazine article and ask students to underline all modifiers—and draw arrows from them to the words they modify. You may want to follow up by asking students to read one of their own recently completed essays, looking exclusively for dangling modifiers of the types discussed in section 15b.

Using this drug appears to modify *side effects,* but this interpretation makes no sense. Because its true headword does not appear in the sentence, the modifier dangles.

One way to correct the dangling modifier is to add a word or word group that it can logically modify. To do so, you usually must change the subject of the main clause.

Patients using this drug experience many undesirable side effects.

The original incorrect sentence, like many sentences that include dangling modifiers, is in the passive voice. The true headword is absent because the passive construction *many undesirable side effects are experienced* does not tell *who* experiences the side effects. Changing the passive construction to active voice corrects the dangling modifier by changing the subject of the sentence's main clause from *side effects* to *patients,* a word the dangling modifier can logically modify. Sometimes, however, passive voice is a desirable stylistic option. In such cases you may correct the dangling modifier by supplying the subject while retaining the passive voice.

See ◄
231

Many undesirable side effects are experienced by patients using this drug.

Another way to correct the dangling modifier is to change it into a dependent clause.

Many undesirable side effects are experienced when this drug is used.

REVISING DANGLING MODIFIERS

- Supply a word or word group that the dangling modifier can logically modify.
- Change the dangling modifier into a dependent clause.

The three most common kinds of dangling modifiers are *dangling verbal phrases*, *dangling prepositional phrases*, and *dangling elliptical clauses*.

(1) Revising dangling verbal phrases

Verbal phrases used as modifiers sometimes dangle in a sentence.

DANGLING PARTICIPIAL PHRASE: Using a pair of forceps, the skin of the rat's abdomen was lifted, and a small cut was made into the body with scissors. (Participial phrase cannot logically modify *skin.*)

REVISED: Using a pair of forceps, the technician lifted the skin of the rat's abdomen and made a small cut into the body with scissors. (Subject of main clause has been changed from *the skin* to *the technician,* a headword the participial phrase can logically modify.)

DANGLING PARTICIPIAL PHRASE:

Paid in three installments, Brad's financial situation seemed stable. (Participial phrase cannot logically modify *Brad's financial situation.*)

REVISED: Because the grant was paid in three installments, Brad's financial situation seemed stable. (Modifying phrase is now a dependent clause.)

DANGLING INFINITIVE PHRASE: To make his paper accurate, all references were checked twice. (Infinitive phrase cannot logically modify *references.*)

REVISED: To make his paper accurate, Don checked all references twice. (Subject of main clause has been changed from *references* to *Don,* a headword the infinitive phrase can logically modify.)

DANGLING INFINITIVE PHRASE: To implement a plus-minus grading system, all students were polled. (Infinitive phrase cannot logically modify *students.*)

REVISED: Before a plus-minus grading system was implemented, all students were polled. (Infinitive phrase is now a dependent clause.)

(2) Revising dangling prepositional phrases

Prepositional phrases can also dangle in a sentence.

DANGLING: With fifty pages left to read, *War and Peace* was absorbing. (Prepositional phrase cannot logically modify *War and Peace.*)

REVISED: With fifty pages left to read, Meg found *War and Peace* absorbing. (Subject of main clause has been changed from *War and Peace* to *Meg,* a word the prepositional phrase can logically modify.)

DANGLING: On the newsstands only an hour, its sales surprised everyone. (Prepositional phrase cannot logically modify *sales.*)

REVISED: Because the magazine had been on the newsstands only an hour, its sales surprised everyone. (Prepositional phrase is now a dependent clause.)

Student revisions may vary.

1. Most people agree that Buckminster Fuller's geodesic dome, although architecturally unusual, is well designed.
2. As an out-of-state student without a car, Joe had difficulty getting to off-campus cultural events.
3. To build a campfire, one needs kindling.
4. With every step we took upward, the trees became sparser.
5. Because I am an amatuer tennis player, my backhand is weaker than my forehand.
6. When you are exiting the train, the station will be on your right.
7. Driving through the Mojave, I found the landscape bleak and oppresive.
8. The air quality will get better if we require auto manufacturers to further improve emission control devices.
9. Using a piece of filter paper, the lab technician dried the ball of sodium as much as possible and then placed it in a dry test tube.
10. Because I missed work for seven days straight, my job was in jeopardy.

Faulty Modification

Possible Revision

   Designed to inject liquids into, or withdraw them from, any vessel or cavity, a syringe is often used to inject drugs into the body or withdraw blood from it. Syringes are also used to measure precisely amounts of drugs or electrolytes which must be added to intravenous solutions.

(3) Revising dangling elliptical clauses

See ◄
8c2

Elliptical clauses are dependent clauses from which part of the subject or predicate or the entire subject or predicate is missing. The absent words, therefore, must be inferred from the context. When such a clause cannot logically modify the subject of the sentence's main clause, it too dangles.

DANGLING: While still in the Buchner funnel, you should press the crystals with a clear stopper to eliminate any residual solvent. (Elliptical clause cannot logically modify subject of main clause.)

REVISED: While still in the Buchner funnel, the crystals should be pressed with a clear stopper to eliminate any residual solvent. (Subject of main clause has been changed from *you* to *crystals,* a word the elliptical clause can logically modify.)

DANGLING: Though a high-pressure field, I find great personal satisfaction in nursing. (Elliptical clause cannot logically modify subject of main clause.)

REVISED: Though it is a high-pressure field, I find great personal satisfaction in nursing. (Elliptical clause has been expanded into a complete dependent clause.)

EXERCISE 6

Eliminate the dangling modifier from each of the following sentences. Either supply a word or word group the dangling modifier can logically modify or change the dangling modifier into a dependent clause.

EXAMPLE: Skiing down the mountain, the restaurant seemed warm and inviting. (dangling modifier)

Skiing down the mountain, I thought the restaurant seemed warm and inviting. (logical headword added)

As I skied down the mountain, the restaurant seemed warm and inviting. (dependent clause)

1. Although architecturally unusual, most people agree that Buckminster Fuller's geodesic dome is well designed.
2. As an out-of-state student without a car, it was difficult to get to off-campus cultural events.
3. To build a campfire, kindling is necessary.
4. With every step upward, the trees became sparser.
5. Being an amateur tennis player, my backhand is weaker than my forehand.
6. When exiting the train, the station will be on your right.
7. Driving through the Mojave, the bleak landscape was oppressive.
8. By requiring auto manufacturers to further improve emission-control devices, the air quality will get better.

9. Using a piece of filter paper, the ball of sodium is dried as much as possible and placed in a test tube.
10. Having missed work for seven days straight, my job was in jeopardy.

STUDENT WRITER AT WORK

FAULTY MODIFICATION

Read this draft of a student's technical writing exercise, a description of a 10-cc syringe. Correct misplaced and dangling modifiers and revise again if necessary to strengthen coherence, unity, and style.

Designed to inject liquids into, or withdraw them from, any vessel or cavity, the function of a syringe is often to inject drugs into the body or withdraw blood from it. Syringes are also used to precisely measure amounts of drugs or electrolytes that must be added to intravenous solutions.

There are available on the market today many different types of syringes, but the one most commonly used in hospitals is the 10-cubic centimeter (cc) disposable syringe. Approximately 5 inches long, the primary composition of this particular syringe is transparent polyethylene plastic. The 10-cc syringe and the majority of other syringes all have a round plunger or piston within a barrel.

The barrel of a syringe is a round hollow cylinder about 4 1/2″ long with a diameter of 3/8″. The bottom end of the barrel has two extensions on its opposite sides, which are perpendicular to the cylinder. With a width equal to the diameter of the barrel, the length of these extensions is about 1/2″. The purpose of these extensions is to enable one to hold with the index finger and middle finger the barrel of the syringe while depressing the plunger with the thumb.

The barrel of the syringe is calibrated on the side in black ink subdivided into gradations of 2cc. At the top of the syringe the barrel abruptly narrows to a very small cylinder, 1/8″ in diameter and 1/4″ in length. This small cylinder is surrounded by another hollow cylinder with a slightly larger diameter. The inside wall of the outer cylinder is threaded like a corkscrew. The purpose of this thread is to keep the needle in place.

The other major part of the syringe is the plunger. The plunger is a solid round cylinder that fits snugly into the barrel made of plastic. At the bottom of the plunger is a plastic ring the size of a dime, which provides something to grasp while withdrawing the plunger. The body of the plunger connects the bottom rim with the tip of the plunger, which is made of black rubber.

Many different types of syringes are available on the market today, but the one most commonly used in hospitals is the 10-cubic centimeter (cc) syringe. Approximately 5 inches long, this particular syringe is composed primarily of transparent polyethylene plastic. The 10-cc syringe and the majority of other syringes all have a round plunger or piston within a barrel.

The barrel of a syringe is a round hollow cylinder 4 1/2″ long with a diameter of 3/8″. The bottom end of the barrel has two extensions perpendicular to the cylinder on its opposite sides. With a width equal to the diameter of the barrel, the extensions are about 1/2″ long. The purpose of these extensions is to enable one to hold the barrel of the syringe with the index and middle fingers while depressing the plunger with the thumb.

The barrel of the syringe, subdivided into gradations of 2cc, is calibrated on one side in black ink. At the top of the syringe the barrel abruptly narrows to a very small cylinder, 1/8″ in diameter and 1/4″ in length. This small cylinder is surrounded by another hollow cylinder, slightly larger in diameter. The inside wall of the outer cylinder is threaded like a corkscrew. The purpose of this thread is to keep the needle in place.

The other major part of the syringe is the plunger. The plunger is a solid, round plastic cylinder that fits snugly into the barrel. At the bottom of the plunger is a plastic ring the size of a dime, which provides something a person can grasp while withdrawing the plunger. The body of the plunger connects the bottom rim with the black rubber top of the plunger.

Commentary
The writer has eliminated misplaced and dangling modifiers. However, the essay is still wordy and repetitious, and many sentences need to be revised. In the next draft, the writer should focus on eliminating awkwardness and increasing clarity.

CHAPTER 16

Faulty Parallelism

Parallelism is the use of equivalent grammatical elements or matching sentence structures to express equivalent ideas. It ensures that elements sharing the same function also share the same grammatical form—for instance, that verbs match corresponding verbs in tense, mood, and number.

16a Using Parallelism

Effective parallelism adds force, unity, balance, and symmetry to your writing. It makes sentences clear and easy to follow and emphasizes relationships among equivalent ideas. It helps your readers keep track of ideas, and it makes sentences more emphatic, more concise, and more varied.

(1) With items in a series

Coordinate elements—words, phrases, or clauses—in a series should be presented in parallel form.

<u>Eat</u>, <u>drink</u>, and <u>be merry</u>.

<u>I came</u>; <u>I saw</u>; <u>I conquered</u>.

<u>Baby food consumption</u>, <u>toy production</u>, and <u>marijuana use</u> are likely to decline as the U.S. population grows older.

Three factors influenced him: <u>his desire to relocate</u>; <u>his need for greater responsibility</u>; and <u>his dissatisfaction with his current job</u>.

(2)   With paired items

Paired points or ideas (words, phrases, or clauses) should be presented in parallel terms. Parallelism conveys their equivalence and relates points to each other.

Her note was <u>short</u> but <u>sweet</u>.

<u>Roosevelt represented</u> the United States, and <u>Churchill represented</u> Great Britain.

The research focused on <u>muscle tissue</u> and <u>nerve cells</u>.

<u>Ask not what your country can do for you</u>; <u>ask what you can do for your country</u>. (John F. Kennedy, Inaugural Address)

Correlative conjunctions (such as *not only/but also, both/and, either/or, neither/nor,* and *whether/or*) are frequently used to link paired elements. These phrases convey equivalence, so the terms they introduce should be parallel.

The design team paid close attention not only <u>to color</u> but also <u>to texture</u>.

Either <u>repeat physics</u> or <u>take calculus</u>.

Both <u>cable television</u> and <u>videocassette recorders</u> continue to threaten the dominance of the major television networks.

Parallelism can also be used to highlight opposition between paired elements linked by the word *than*.

Richard Wright and James Baldwin chose <u>to live in Paris</u> rather than <u>to remain in the United States</u>.

(3)   In lists and outlines

Elements in a list should be expressed in parallel terms.

The Irish potato famine had four major causes:

1. The establishment of the landlord-tenant system
2. The failure of the potato crop
3. The reluctance of England to offer adequate financial assistance
4. The passage of the Corn Laws

Elements in a formal outline should also be parallel.

▶ See 41g

EXERCISE 1

Identify the parallel elements in these sentences by underlining parallel words and bracketing parallel phrases and clauses.

continued from previous page

King, Jr.'s "Letter from Birmingham Jail." Ask students to locate and discuss the parallel structures used in this essay.

TEACHING STRATEGY

Parallelism

If your students are intrigued by the flourishes of classical rhetoric, this might be a good place in the course to introduce some of the classical sentence schemes requiring parallel elements.

antithesis (opposites): "That's one small step for a man, one giant leap for mankind."

–Neil Armstrong

asyndeton (no conjunction): ". . . and that government of the people, by the people, for the people, shall not perish from the earth."

–Abraham Lincoln

polysyndeton (repeated conjunction): "Neither rain nor snow nor sleet nor gloom of night shall stay these couriers from the swift completion of their appointed rounds."

–Herodotus

anaphora (repeated phrases at the beginning of clauses): "But, in a larger sense, we can not dedicate—we can not consecrate—we can not hallow—this ground."

–Abraham Lincoln

John F. Kennedy enjoyed using a parallel device called *chiasmus*, in which elements of a first phrase or clause are inverted in a second "mirror-image" phrase or clause: "Ask not what your country can do for you; ask what you can do for your country . . . . Let us never negotiate out of fear; but never let us fear to negotiate."

ANSWERS TO EXERCISE 1

1. You have lived an American dream when you <u>begin the year</u> [<u>setting</u> pressure gauges for the Caterpillar Tractor Company] in Peoria and <u>end it</u> [<u>building</u> rocking chairs for your grandchildren].

2. The public image of the American woman in the magazine and television commercials is designed to sell <u>washing machines</u>, <u>cake mixes</u>, <u>deodorants</u>, <u>rejuvenating face creams</u>, <u>hair tints</u>.

3. [Surgery restores to function] <u>broken limbs</u> and <u>damaged hearts</u> with <u>amazing safety</u> and <u>little suffering</u>; [sanitation removes from our environment] many of the germs of disease; new drugs are constantly being developed [to relieve physical pain], [to help us sleep] [if we are restless,] [to keep us awake] [if we feel sleepy,] and [to make us oblivious] to worries.

4. <u>Theoretically</u>—and <u>secretly</u>, of course—I was [all for the Burmese] and [all against their oppressors], the British.

5. As nations, we can apply to affairs of state the realism of science: [holding to <u>what works</u>] and [discarding <u>what does not</u>].

ANSWERS TO EXERCISE 2

1. Originally there were five performing Marx Brothers: Groucho, Chico, Harpo, Gummo, and Zeppo.

2. Groucho, Chico, and Harpo were very well known, but first Gummo and later Zeppo dropped out of the act.

continued on following page

EXAMPLE: He is 81 now, [a tall <u>man</u> in a dark blue suit], [a smiling <u>man with</u> a nimbus of <u>snowy white hair</u>], and much of the world knows him [<u>as the proponent</u> of vitamin C to cure colds], [<u>as a quixotic</u>, vaguely ridiculed figure on the fringes of medicine]. (Maralyn Lois Polak, *Philadelphia Inquirer*)

1. You have lived an American dream when you begin the year setting pressure gauges for the Caterpillar Tractor Company in Peoria and end it building rocking chairs for your grandchildren. (*Newsweek*)

2. The public image [of the American woman] in the magazine and television commercials is designed to sell washing machines, cake mixes, deodorants, detergents, rejuvenating face creams, hair tints. (Betty Friedan, *The Feminine Mystique*)

3. Surgery restores to function broken limbs and damaged hearts with amazing safety and little suffering; sanitation removes from our environment many of the germs of disease; new drugs are constantly being developed to relieve physical pain, to help us sleep if we are restless, to keep us awake if we feel sleepy, and to make us oblivious to worries. (René Dubos, *Medical Utopias*)

4. Theoretically—and secretely, of course—I was all for the Burmese and all against their oppressors, the British. (George Orwell, "Shooting an Elephant")

5. As nations, we can apply to affairs of state the realism of science: holding to what works and discarding what does not. (Jacob Bronowski, *A Sense of the Future*)

EXERCISE 2

Combine each of the following sentence pairs or sentence groups into one sentence that uses parallel structure. Be sure all parallel words, phrases, and clauses are expressed in parallel terms.

1. Originally, there were five performing Marx Brothers. One was nicknamed Groucho. The others were called Chico, Harpo, Gummo, and Zeppo.

2. Groucho was very well known. So were Chico and Harpo. Gummo soon dropped out of the act. And later Zeppo did, too.

3. They began in vaudeville. That was before World War I. Their first show was called *I'll Say She Is*. It opened in New York in 1924.

4. The Marx Brothers' first movie was *The Coconuts*. The next was *Animal Crackers*. And this was followed by *Monkey Business*, *Horsefeathers*, and *Duck Soup*. Then came *A Night at the Opera*.

5. In each of these movies, the Marx Brothers make people laugh. They also establish a unique, zany comic style.

6. In their movies, each man has a set of familiar trademarks. Groucho has a mustache and a long coat. He wiggles his eyebrows and smokes a cigar. There is a funny hat that Chico always wears. And he affects a phony Italian accent. Harpo never speaks.

7. Groucho is always cast as a sly operator. He always tries to cheat people out of their money. He always tries to charm women.
8. In *The Coconuts* he plays Mr. Hammer, proprietor of the run-down Coconut Manor, a Florida hotel. In *Horsefeathers* his character is named Professor Quincy Adams Wagstaff. Wagstaff is president of Huxley College. Huxley also has financial problems.
9. In *Duck Soup* Groucho plays Rufus T. Firefly, president of the country of Fredonia. Fredonia was formerly ruled by the late husband of a Mrs. Teasdale. Fredonia is now at war with the country of Sylvania.
10. Margaret Dumont is often Groucho's leading lady. She plays Mrs. Teasdale in *Duck Soup*. In *A Night at the Opera* she plays Mrs. Claypool. Her character in *The Coconuts* is named Mrs. Potter.

16b Revising Faulty Parallelism

When elements that have the same function in a sentence are not presented in the same terms, the sentence is flawed by **faulty parallelism**. Consider the following sentence.

FAULTY PARALLELISM: Many people in developing countries suffer because the countries lack sufficient housing to accommodate them, sufficient food to feed them, and their health-care facilities are inadequate.

Because all three reasons are of equal importance and are presented in a series connected by the coordinating conjunction *and,* readers expect them to be expressed in parallel terms. The first two elements satisfy this expectation.

sufficient housing to accommodate them . . .

sufficient food to feed them . . .

The third item in the series, however, breaks this pattern.

their health-care facilities are inadequate.

To create a clear, emphatic sentence, all three elements should be presented in the same terms.

Many people in developing countries suffer because the countries lack sufficient housing to accommodate them, sufficient food to feed them, and sufficient health-care facilities to serve them.

(1) Repeating parallel elements

Faulty parallelism occurs when a writer does not use parallel elements in a series or in paired points. Nouns must be matched

continued from previous page

3. They began in vaudeville before World War I and opened their first show, "I'll Say She Is," in New York in 1924.
4. The Marx Brothers' first movie, *The Coconuts,* was followed by *Animal Crackers, Monkey Business, Horsefeathers, Duck Soup,* and *A Night at the Opera.*
5. In each of these movies, the Marx Brothers make people laugh and establish a unique, zany comic style.
6. In their movies, each man has a set of familiar trademarks: Groucho has a mustache and a long coat, wiggles his eyebrows, and smokes a cigar; Chico always wears a funny hat and affects a phony Italian accent; Harpo never speaks.
7. Groucho is always cast as a sly operator who tries to cheat people out of their money and to charm women.
8. In *The Coconuts* he plays Mr. Hammer, proprietor of the run-down Coconut Manor, a Florida hotel; in *Horsefeathers* he plays Professor Quincy Adams Wagstaff, president of Huxley College, which also has financial problems.
9. In *Duck Soup* Groucho plays Rufus T. Firefly, president of the country of Fredonia, which was formerly ruled by the late husband of a Mrs. Teasdale and is now at war with the country of Sylvania.
10. Margaret Dumont, often Groucho's "leading lady," plays Mrs. Teasdale in *Duck Soup,* Mrs. Claypool in *A Night at the Opera,* and Mrs. Potter in *The Coconuts.*

with nouns, verbs with verbs, phrases and clauses with similarly constructed phrases and clauses, and so on, in places where they are expected.

FAULTY PARALLELISM:	**REVISED:**
Popular exercises for men and women include aerobic dancing, weight lifters, and jogging. (*Dancing* and *jogging* are gerunds; *weight lifters* is a noun phrase.)	Popular exercises for men and women include aerobic dancing, weight lifting, and jogging. (three gerunds)
Some of the side effects are skin irritation and eye irritation, and mucous membrane irritation may also develop. (*Skin irritation* and *eye irritation* are noun phrases; *mucous membrane irritation may also develop* is an independent clause.)	Some of the side effects that may develop are skin, eye, and mucous membrane irritation. (three nouns used as modifiers of *irritation*)
I look forward to hearing from you and to have an opportunity to tell you more about myself. (*Hearing from you* is a gerund phrase; *to have an opportunity* is an infinitive phrase.)	I look forward to hearing from you and to having an opportunity to tell you more about myself. (two gerund phrases)

(2) Repeating signals of parallelism

Faulty parallelism also occurs when a writer does not repeat words that signal parallelism: prepositions, articles, the *to* that is part of the infinitive, or the word that introduces a phrase or clause. Although similar grammatical structures (verbs that match verbs, nouns that match nouns, and so on) may sometimes be enough to convey parallelism, sentences are clearer and more emphatic if other key words in parallel constructions are also parallel. Repeating these signals makes the boundaries of each parallel element clear. But be sure to include the same signals with *all* the elements in a series.

FAULTY PARALLELISM:	**REVISED:**
Computerization has helped industry by not allowing labor costs to skyrocket, increasing the speed of production, and improving efficiency. (Does *not*	Computerization has helped industry by not allowing labor costs to skyrocket, by increasing the speed of production, and by improving efficiency.

apply to all three phrases, or only the first?)

The United States suffered casualties in the Civil War, French and Indian War, Spanish-American War and Korean War. (Without the repeated definite article, it is hard for readers to distinguish the four different wars.)

It may be easier to try remodeling than sell a house. (Although *try* and *sell* correspond, the sentence does not highlight their parallel structure; in fact, *remodeling* and *sell* seem to be the paired elements.)

Koala bears are not as appealing as they look because they have fleas, they have bad breath, and a tendency to scratch. (Are *bad breath* and *a tendency to scratch* reasons why Koalas are unappealing, or are they incidental points?)

(Preposition *by* is repeated to clarify the boundaries of the three parallel phrases.)

The United States suffered casualties in the Civil War, the French and Indian War, the Spanish-American War, and the Korean War. (The article *the* is repeated for clarity and emphasis.)

It may be easier to try remodeling than to sell a house. (The *to* of the infinitive is repeated for clarity and emphasis.)

Koala bears are not as appealing as they look because they have fleas, they have bad breath, and they have a tendency to scratch. (The introductory words *they have* are repeated for clarity and emphasis.)

1. The world is divided between those who wear galoshes and those who discover continents.
2. Soviet leaders, members of Congress, and American Catholic bishops all pressed the president to limit the arms race. (C)
3. A national task force on education recommended improving public education by making the school day longer, by raising teachers' salaries, and by integrating more technology into the curriculum.
4. The fast-food industry is expanding to include many kinds of restaurants: those that serve pizza, those that serve fried chicken, those that serve Mexican-style food, and those that serve hamburgers.
5. Many Scotch drinkers in the United States are switching to wine or beer because of high prices, changing tastes, and increased health awareness.

(3) Repeating relative pronouns

Faulty parallelism occurs when a writer fails to use a clause beginning with the relative pronoun *who, whom,* or *which* before one beginning with *and who, and whom,* or *and which.*

Like correlative conjunctions, *who . . . and who* and similar expressions are always paired and always introduce parallel clauses. When you omit the first part of the expression, you throw readers off balance.

INCORRECT: *The Thing,* directed by Howard Hawks, and which was released in 1951, featured James Arness as the monster.

REVISED: *The Thing,* which was directed by Howard Hawks and which was released in 1951, featured James Arness as the monster.

In many cases, however, eliminating the relative pronouns produces a more concise sentence.

REVISED: *The Thing,* directed by Howard Hawks and released in 1951, featured James Arness as the monster.

Faulty Parallelism

Possible Revision

The average life span of Native Americans is lower than that of the general population. Not only is their infant mortality rate four times higher than that of the general population, but their suicide rate is twice as high as that of other races. Moreover, there are more homicides and more alcohol-related deaths among Native Americans than among people of other races. Medical care available for them does not meet their needs, and this presents a challenge for the health professionals who serve them.

The Indian Health Service (IHS), a branch of the U.S. Public Health Service, is responsible for providing medical care to Native Americans who live on reservations. The IHS has been criticized for its inability to deal with cultural differences between health professionals and Native Americans—cultural differences that interfere with adequate medical care. In order to diagnose disease states, to prescribe drug therapy, and to counsel patients, health professionals need to acquire an understanding of Native American culture. They must gain a working knowledge of Native Americans' ideas and feelings not only toward health but also toward God, relationships, and death. Only then can health professionals communicate their goals, provide quality medical care, and achieve patient compliance.

There are many obstacles to effective communication: hostility to white authority and whites' structured, organized society; language; Native

continued on following page

EXERCISE 3

Identify and correct faulty parallelism in these sentences. Then underline the parallel elements—words, phrases, and clauses—in your corrected sentences. If a sentence is already correct, mark it with a *C* and underline the parallel elements.

> **EXAMPLE:** Alfred Hitchcock's films include *North by Northwest, Vertigo, Psycho,* and he also directed *Notorious* and *Saboteur.*
>
> **REVISED:** Films directed by Alfred Hitchcock include <u>*North by Northwest*</u>, <u>*Vertigo*</u>, <u>*Psycho*</u>, <u>*Notorious*</u>, and <u>*Saboteur*</u>.

1. The world is divided between those with galoshes on and those who discover continents.
2. Soviet leaders, members of Congress, and the American Catholic bishops all pressed the president to limit the arms race.
3. A national task force on education recommended improving public education by making the school day longer, higher teachers' salaries, and integrating more technology into the curriculum.
4. The fast-food industry is expanding to include many kinds of restaurants: those that serve pizza, fried chicken chains, some offering Mexican-style menus, and hamburger franchises.
5. The consumption of Scotch in the United States is declining because of high prices, tastes are changing, and increased health awareness has led many whiskey drinkers to switch to wine or beer.

FAULTY PARALLELISM

In the following section of a draft of a paper written for a class in public health, a student discusses factors that must be taken into account by medical practitioners at the Indian Health Service. Correct the faulty parallelism and revise again if necessary to strengthen coherence, unity, and style.

The average life span of Native Americans is considerably lower than that of the general population. Not only is their infant mortality rate four times higher than that of the general population, but they also have a suicide rate that is twice as high as that of other races. Moreover, Native Americans both die in homicides more often than people of other races do and there are more alcohol-related deaths among Native Americans than among people of other races. Medical care available for them does not meet their needs and is presenting a challenge for the health professionals who serve them.

continued on the following page

continued from the previous page

The Indian Health Service (IHS), a branch of the U.S. Public Health Service, is responsible for providing medical care to Native Americans who live on reservations. The IHS has been criticized for its inability to deal with cultural differences between health professionals and Native Americans--cultural differences that interfere with adequate medical care. In order to diagnose disease states, for prescribing drug therapy, and to counsel patients, health professionals need to acquire an understanding of Native American culture. They must gain a working knowledge of Native Americans' ideas and feelings toward health and also God, relationships, and death. Only then can health professionals communicate their goals, provide quality medical care, and in addition they will be able to achieve patient compliance.

There are many obstacles to effective communication: hostility to white authority and whites' structured, organized society; language is another obvious barrier to communication; Native Americans' view of sickness, which may be different from Anglos'; and some Indian cultures' concept of time is also different from that of the Anglo health workers. Other problems are more basic: A physician cannot expect a patient to refrigerate medication if no refrigerators are available or dilute dosage forms at home or be changing wet dressings several times a day if clean water is not readily available nor quart/pint measuring devices to dilute stock solutions.

The defects in Native American health care cannot be completely solved by the improvement of communication channels or making these channels stronger. But the health professional's communication with the Native American patient can be effective enough so that medical staff can acquire an adequate medical history, monitor drug use, be alert for possible drug interactions, and to provide useful discharge counseling. If health professionals can communicate understanding and respect for Native American culture and concern for their welfare, they may be able to meet the needs of their Native American patients more effectively.

continued from previous page

Americans' view of sickness, which may be different from Anglos'; and some Indian cultures' concept of time, which is also different from that of the Anglo health workers. Other problems are more basic: A physician cannot expect a patient to refrigerate medication if there are no refrigerators available or to dilute dosage forms at home or to change wet dressings several times a day if neither clean water nor quart/pint measuring devices are readily available to dilute stock solutions.

The defects in Native American health care cannot be completely solved by improving communication channels or by making these channels stronger. But the health professional's communication with the Native American patient can be effective enough so that medical staff can acquire an adequate medical history, monitor drug use, be alert for possible drug interactions, and provide useful discharge counseling. If health professionals can communicate understanding of and respect for Native American culture and concern for their welfare, they may be able to meet the needs of their Native American patients more effectively.

Commentary on Possible Revision

Although the writer has corrected the faulty parallelism, she could revise further. Much of the information in this short essay, which is likely to be unfamiliar to readers, is presented as dense lists. Instead of crowding essential information into sentences that present a series of problems (as in paragraph 3), for instance, she could achieve greater clarity by introducing each problem in a separate sentence and explaining it more fully.

CHAPTER 17

Awkward or Confusing Sentences

Sentences that are carelessly or illogically constructed can be extremely hard to understand. For example, when a **shift**—an unwarranted change of tense, voice, mood, person, number, or type of discourse—occurs within or between sentences, it is likely to confuse readers. Other kinds of sentences are also awkward or confusing because they send mixed messages. For example, in **mixed constructions** and **faulty predication** two or more parts of a sentence do not fit logically together. Finally, **incomplete or illogical comparisons** are confusing because readers cannot tell what two items are being compared.

 17a **Shifts in Tense**

See ◀ 23f Verb tenses in a sentence or a related group of sentences should shift only with good reason—to indicate changes of time, for example.

> *The Wizard of Oz* is a film that has enchanted audiences since it <u>was made</u> in 1939. (acceptable shift from present to past)

Unwarranted shifts can mislead readers and obscure your meaning.

(1) Unwarranted tense shifts in a sentence

> **FAULTY:** The judge <u>told</u> the defendant he would not release him unless he <u>promises</u> to undergo therapy. (unwarranted shift from past to present)

REVISED: The judge <u>told</u> the defendant he would not release him unless he <u>promised</u> to undergo therapy. (both verbs in past tense)

FAULTY: *On the Road* <u>is</u> a novel about friends who <u>drove</u> across the United States in the 1950s. (unwarranted shift from present to past)

REVISED: *On the Road* <u>is</u> a novel about friends who <u>drive</u> across the United States in the 1950s. (work of literature discussed in both clauses; both verbs in present tense)

▶ See 47c

FAULTY: Medical researchers <u>know</u> that asbestos <u>caused</u> cancer. (unwarranted shift from present to past tense)

REVISED: Medical researchers <u>know</u> that asbestos <u>causes</u> cancer. (general truth stated in both clauses; both verbs in present tense)

▶ See 23c1

(2) Unwarranted tense shifts between sentences

FAULTY: Once I <u>was</u> driving late at night. Suddenly I <u>see</u> a dog right in the path of my car. I <u>slam</u> on my brakes and barely <u>avoid</u> hitting it. (unwarranted shift from past to present tense)

REVISED: Once I <u>was</u> driving late at night. Suddenly I <u>saw</u> a dog right in the path of my car. I <u>slammed</u> on my brakes and barely <u>avoided</u> hitting it. (all verbs in past tense)

FAULTY: Each day I <u>lie</u> on the beach and <u>let</u> the sun <u>drive</u> away my troubles. I <u>listened</u> to the sound of the surf and <u>watched</u> children playing by the water. (unwarranted shift from present to past)

REVISED: Each day I <u>lie</u> on the beach and <u>let</u> the sun <u>drive</u> away my troubles. I <u>listen</u> to the sound of the surf and <u>watch</u> children playing in the water. (discussion of an event that occurs regularly; all verbs in present tense)

▶ See 23c1

CLOSE-UP

SHIFTS IN VOICE

Shifts from active to passive voice may be necessary to give a sentence proper emphasis.

Although consumers protested, the tax on gasoline was increased.

▶ See 23k

Here the shift from active (*protested*) to passive (*was increased*) enables the writer to keep the focus on consumer groups and the issue they protested. To say *Congress increased the tax on gasoline* would change the emphasis of the sentence.

17b Shifts in Voice

Unwarranted shifts from active to passive voice can be confusing and misleading. In the following sentence, for instance, the shift from active (*wrote*) to passive (*was written*) creates ambiguity.

FAULTY: F. Scott Fitzgerald <u>wrote</u> *This Side of Paradise,* and later *The Great Gatsby* <u>was written.</u> (Who wrote *The Great Gatsby*?)

REVISED: F. Scott Fitzgerald wrote *This Side of Paradise* and later wrote *The Great Gatsby.* (consistent use of active voice)

17c Shifts in Mood

See ◄
23g–i

Unnecessary shifts in mood can also be confusing and annoying.

FAULTY: It is important that a student <u>buy</u> a dictionary and <u>uses</u> it. (shift from subjunctive to indicative mood)

REVISED: It is important that a student <u>buy</u> a dictionary and <u>use</u> it. (both verbs in subjunctive mood)

FAULTY: Next, <u>heat</u> the mixture in a test tube and <u>you should make sure</u> it does not <u>boil.</u> (shift from imperative to indicative mood)

REVISED: Next, <u>heat</u> the mixture in a test tube and <u>be sure</u> it does not boil (both verbs in imperative mood)

17d Shifts in Person and Number

Person indicates who is speaking (first person—*I, we*), who is spoken to (second person—*you*), and who is spoken about (third person—*he, she, it,* and *they*).

Faulty shifts between the second- and the third-person pronouns cause most errors.

FAULTY: When <u>one</u> looks for a loan, <u>you</u> compare the interest rates of several banks. (shift from third to second person)

REVISED: When <u>you</u> look for a loan, <u>you</u> compare the interest rates of several banks. (consistent use of second person)

REVISED: When <u>one</u> looks for a loan, <u>one</u> compares the interest rates of several banks. (consistent use of third person)

REVISED: When a <u>person</u> looks for a loan, <u>he or she</u> compares the interest rates of several banks. (consistent use of third person)

Number indicates one (singular—*novel, it*) or more than one (plural—*novels, they, them*). Singular pronouns should refer to singular antecedents and plural pronouns to plural antecedents.

▶ See 24b

> **FAULTY:** If a person does not study regularly, they will have a difficult time passing organic chemistry. (shift from singular to plural)
>
> **REVISED:** If a person does not study regularly, he or she will have a difficult time passing organic chemistry. (Singular pronouns refer to singular antecedents.)
>
> **REVISED:** If students do not study regularly, they will have a difficult time passing organic chemistry. (Plural pronoun refers to plural antecedent.)

CLOSE-UP

SHIFTS IN PERSON AND NUMBER

Although college writing follows standard conventions of pronoun-antecedent agreement, use of a plural pronoun referring to a singular antecedent is increasingly common in speech and in informal writing, when such use enables the speaker or writer to avoid sexist language.

Buddy Holly and Janis Joplin each made a significant contribution with their music.

▶ See 18f2

EXERCISE 1

Read the following sentences and eliminate any shifts in tense, voice, mood, person, or number. Some sentences are correct, and some will have more than one possible answer.

1. Gettysburg is a borough of southwestern Pennsylvania where some of the bloodiest fighting of the Civil War occurs in July 1863.
2. Giotto was born near Florence in 1267 and is given credit for the revival of painting in the Renaissance.
3. The early Babylonians divided the circle into 360 parts, and the volume of a pyramid could also be calculated by them.
4. During World War II General Motors expanded its production facilities, and guns, tanks, and ammunition were made.
5. Diamonds, the only gems that are valuable when colorless, were worn to cure disease and to ward off evil spirits.
6. First, clear the area of weeds, and then you should spread the mulch in a six-inch layer.

ANSWERS TO EXERCISE 1

1. Gettysburg is a borough of southwestern Pennsylvania where some of the bloodiest fighting of the Civil War occurred in July of 1863.
2. correct
3. The early Babylonians divided the circle into 360 parts and could calculate the volume of a pyramid.
4. During World War II General Motors expanded its production facilities and made guns, tanks, and ammunition.
5. correct
6. First clear the area of weeds, and then spread the mulch in a six-inch layer.
7. When one visits the Grand Canyon, one should be sure to notice the fractures and faults on the north side of the Kaibab Plateau.
8. For a wine grape, cool weather means it will have a higher acid content and a sour taste; hot weather means it will have a lower acid content and a sweet taste.
9. correct
10. When you look at the Angora goat, you will see it has an abundant undergrowth.

ANSWERS TO EXERCISE 2

1. In cases of possible sexual harassment, Ellen Goodman suggests a "reasonable woman standard" be applied. We should ask how a reasonable woman would interpret the situation and how a reasonable woman would behave.

2. Sally Thane Christensen, advocating the use of an endangered species of tree, the yew, as a treatment for cancer, asked whether a tree was worth a life.

3. Stephen Nathanson, considering the morality of the death penalty, asked what we would think if the death penalty actually saved lives.

4. Martin Luther King, Jr., said that he had a dream that one day this nation would rise up and live out the true meaning of its creed.

5. Mohandas K. Gandhi wrote that complete civil disobedience is a state of peaceful rebellion—a refusal to obey every single State-made law.

6. Benjamin Franklin once stated that the older he grew, the more apt he was to doubt his own judgment of others.

7. Speaking of the theater of the absurd in 1962, Edward Albee asked if it was, as it had been accused of being, obscure, sordid, destructive, anti-theater, perverse, and absurd (in the sense of foolish).

8. Thoreau said that the finest qualities of our nature, like the bloom on fruits, can be preserved only by the most delicate handling.

continued on following page

7. When one visits the Grand Canyon, you should be sure to notice the fractures and faults on the north side of the Kaibab plateau.
8. For a wine grape, cool weather means it will have a higher acid content and a sour taste; hot weather means they will have lower acid content and a sweet taste.
9. Mary Wollstonecraft wrote *Vindication of the Rights of Women,* and then she wrote *Vindication of the Rights of Men.*
10. When you look at the Angora goat, one will see it has an abundant undergrowth.

17e   Shifts from Direct to Indirect Discourse

Direct discourse reports the exact words of a speaker or writer. It is always enclosed in quotation marks and is often accompanied by an identifying tag (*he says, she said*). **Indirect discourse** summarizes the words of a speaker or writer. No quotation marks are used, and the reported statement is often introduced with the word *that* (or, for questions, *who, what, why, whether, how,* or *if*). Both pronouns and verb tenses in a reported statement are different from those in a directly quoted statement.

DIRECT DISCOURSE: My instructor said, "I want your paper by this Friday."

INDIRECT DISCOURSE: My instructor said that he wanted my paper by this Friday.

Statements and questions that shift between indirect and direct discourse are often confusing to readers.

FAULTY: During the trial, Captain Dreyfus repeatedly defended his actions and said that I am not guilty. (shift from indirect to direct discourse)

REVISED: During the trial, Captain Dreyfus repeatedly defended his actions and said, "I am not guilty." (direct discourse indicated with quotation marks and *that* eliminated)

REVISED: During the trial, Captain Dreyfus repeatedly defended his actions and said that he was not guilty. (indirect discourse used correctly)

FAULTY: My mother asked was I ever going to get a job. (sentence combines direct and indirect discourse)

REVISED: My mother asked whether I was ever going to get a job. (correct use of indirect discourse)

REVISED: My mother asked, "Are you ever going to get a job?" (correct use of direct discourse)

EXERCISE 2

Transform the direct discourse in the following sentences into indirect discourse. Keep in mind that general truths stay in the present.

> **EXAMPLE:** Anna Quindlen explained why she kept her maiden name when she married: "It was a political decision, a simple statement that I was somebody and not an adjunct of anybody, especially a husband."
>
> Anna Quindlen explained that she made a decision to keep her maiden name when she married because it expressed a simple political statement that she was somebody and not an adjunct to anybody, especially not to a husband.

1. In cases of possible sexual harassment, Ellen Goodman suggests a "reasonable woman standard" be applied: "How would a reasonable woman interpret this? How would a reasonable woman behave?"
2. Sally Thane Christensen, advocating the use of an endangered species of tree, the yew, as a treatment for cancer, asked, "Is a tree worth a life?"
3. Stephen Nathanson, considering the morality of the death penalty, asked, "What if the death penalty did save lives?"
4. Martin Luther King, Jr., said, "I have a dream that one day this nation will rise up and live out the true meaning of its creed."
5. Mohandas K. Gandhi wrote, "Complete civil disobedience is a state of peaceful rebellion—a refusal to obey every single State-made law."
6. Benjamin Franklin once stated, "The older I grow, the more apt I am to doubt my own judgment of others."
7. Speaking of the theater of the absurd in 1962, Edward Albee asked, "Is it, as it has been accused of being, obscure, sordid, destructive, anti-theater, perverse, and absurd (in the sense of foolish)?"
8. Thoreau said, "The finest qualities of our nature, like the bloom on fruits, can be preserved only by the most delicate handling."
9. In *Death Knocks*, Death asks Nat, "Who should I look like?"
10. In *Death of a Salesman*, Linda stands by her husband's grave after his funeral and asks, "Why didn't anybody come?"

continued from previous page

9. In *Death Knocks*, Death asks Nat who he should look like.
10. In *Death of a Salesman*, Linda stands by her husband's grave after his funeral and asks why no one came.

17f Mixed Constructions

A **mixed construction** occurs when a sentence begins with one grammatical strategy and then abruptly shifts to another. Typically, this occurs when a prepositional phrase or a dependent or independent clause is illogically used as the subject of a sentence.

MIXED: Because she studies every day explains why she gets good grades.

Mixed Constructions

Some of your students may find it easier to correct mixed construction "by ear"—reading their essays aloud to detect problems. You may need to work individually with other students. Instead of marking sentences on their papers with a nebulous "Awkward" or "Mixed," use a line to mark the point at which a sentence suddenly shifts to another construction, and identify the problem (dependent clause used as subject, etc.).

Remind students that a mixed construction is like a pair of mismatched shoes: The writer can change *either* the left or right shoe (the first or second part of the sentence), so long as the end result is a grammatical match.

ANSWERS TO EXERCISE 3

Your students' answers will vary.

1. Implementing the "motor voter" bill will make it easier for people to register to vote.
2. They won the game because she sank the basket.
3. Because of a defect in design, the roof of the Hartford Stadium collapsed.
4. Since he wants to make his paper clearer, he revises extensively.
5. Even though she works for a tobacco company, she should not necessarily be against prohibiting smoking in restaurants.

The writer begins with a dependent clause (*Because she studies every day*), which should logically be followed by an independent clause. Instead, the writer uses the dependent clause as the subject of *explains*. The two parts of the sentence are, therefore, at odds with each other. The writer can correct this sentence by deleting *explains why*, thus making the second clause independent.

REVISED: Because she studies every day, she gets good grades.

Each of the following groups of sentences illustrates a common kind of mixed construction and presents some options for revision.

MIXED: <u>By calling for information</u> is the way to learn more about the benefits of ROTC. (Prepositional phrase cannot logically serve as subject.)

REVISED: By calling for information you can learn more about the benefits of ROTC. (Subject *you* is provided.)

REVISED: Calling for information is the way to learn more about the benefits of ROTC. (Prepositional phrase has been changed to gerund phrase, which can serve as a sentence's subject.)

MIXED: <u>Even though he published a paper on the subject</u> does not mean he should get credit for the discovery. (A dependent clause cannot logically serve as subject.)

REVISED: Even though he published a paper on the subject, he should not get credit for the discovery. (Subject *he* is provided.)

REVISED: Publishing a paper on the subject does not necessarily mean he should get credit for the discovery. (Dependent clause has been changed to a gerund phrase, which can serve as a sentence's subject.)

MIXED: <u>He was late</u> was what made him miss Act I. (Independent clause cannot logically serve as subject.)

REVISED: Being late made him miss Act I. (Independent clause has been changed to a gerund phrase, which can serve as a sentence's subject.)

REVISED: Because he was later, he missed Act I. (Independent clause has been changed to a dependent clause, creating a complex sentence.)

EXERCISE 3

Revise the following mixed constructions so their parts fit together both grammatically and logically.

EXAMPLE: By investing in commodities made her rich.

Investing in commodities made her rich.

1. In implementing the "motor voter" bill will make it easier for people to register to vote.

2. She sank the basket was the reason they won the game.
3. Because of a defect in design was why the roof of the Hartford Stadium collapsed.
4. Since he wants to make his paper clearer is the reason he revises extensively.
5. Even though she works for a tobacco company does not mean that she should be against laws prohibiting smoking in restaurants.

 17g   Faulty Predication

Faulty predication occurs when a sentence's predicate does not logically complete its subject.

(1)   Incorrect use of *be*

Faulty predication is especially common in sentences that contain a *linking verb*—a form of the verb *be,* for example—and a subject complement.

FAULTY: Mounting costs and declining advertising revenue <u>were the demise</u> of *Look* magazine in 1971.

This sentence states incorrectly that mounting costs and declining advertising *were* the demise of *Look* magazine when, in fact, they were *the reasons* for its demise. You can correct this problem by revising the sentence so it includes a complement that can logically be equated with the subject.

REVISED: Mounting costs and declining advertising revenue <u>were the reasons for the demise</u> of *Look* magazine in 1971.

(2)   Intervening words

Faulty predication can also occur when intervening words obscure the connection between the subject and the verb. When you revise, be sure the connection between subject and verb is logical.

FAULTY: The <u>purpose</u> of Hitler's campaign <u>failed</u> because of the severity of the Russian winter. (Intervening words obscure the relationship between the subject, *purpose,* and the verb, *failed.*)

REVISED: <u>Hitler's campaign failed</u> because of the severity of the Russian winter. (*Hitler's campaign* is now the sentence's subject.)

REVISED: The <u>purpose</u> of Hitler's campaign <u>was thwarted</u> by the severity of the Russian winter. (The verb is now consistent with the subject.)

Faulty Predication
Not all subjects and verbs fit together. In your students' work, you are likely to find that illogical combinations of subject and verb occur most often.

1. when forms of the linking verb *be* are part of the sentence construction. (Review the idea of linking verbs mentioned in 24a8, noting that they cannot be used to connect things that are not equal.)
2. when students are writing sentences beginning with explanatory words—"A shivaree is when . . . " or "The reason is because. . . ." Suggest that students be sure that both the subject and the subject complement are nouns or groups of words that function as nouns. Better yet, encourage them to start cutting the "becauses" and "whens" from their writing.

Your students' answers will vary.

1. Inflation is a decline in the purchasing power of currency.
2. Hypertension is elevated blood pressure.
3. Machines containing computers, such as instant cash machines, have become part of our everyday lives.
4. Some people say that increasing violence in American cities results from guns being too easily available.
5. American cities are congested because too many people live too close together.

(3) *is when* **or** *is where*

Another kind of faulty predication occurs when a clause beginning with *where* or *when* follows *is* in a definition. When you revise a clause containing a definition, be sure to use a noun or a noun phrase as both the subject and the subject complement.

FAULTY: Taxidermy is where you construct a lifelike representation of an animal from its preserved skin. (In definitions, *be* must be preceded and followed by nouns or noun phrases.)

REVISED: Taxidermy is the construction of a lifelike representation of an animal from its preserved skin. (The subject complement is now a noun phrase.)

(4) *the reason . . . is because*

A similar type of faulty predication occurs when the phrase *the reason is* precedes *because*.

FAULTY: The reason we drive is because we are afraid to fly.

REVISED: The reason we drive is that we are afraid to fly.

REVISED: We drive because we are afraid to fly.

(5) Faulty appositive

A **faulty appositive** is a type of faulty predication in which an appositive is equated with a noun or pronoun it cannot logically modify.

FAULTY: The salaries are high in professional athletics, such as baseball players. (*Professional athletics* is not the same as *baseball players*.)

REVISED: The salaries are high for professional athletes, such as baseball players. (baseball players = professional athletes)

EXERCISE 4

Revise the following sentences to eliminate faulty predication. Keep in mind that each sentence may be revised in more than one way.

EXAMPLE: The reason there are traffic jams at 9 a. m. and 5 p.m. is because too many people work traditional rather than staggered hours.

REVISED: There are traffic jams at 9 a.m. and 5 p.m. because too many people work traditional rather than staggered hours.

1. Inflation is when the purchasing power of currency declines.
2. Hypertension is where the blood pressure is elevated.
3. Computers have become part of our everyday lives, such as instant cash machines.

4. Some people say the reason for the increasing violence in American cities is because guns are too easily available.
5. The reason there is congestion in American cities is because there are too many people living too close together.

 17h **Incomplete or Illogical Comparisons**

A comparison tells how two things are alike or unlike. When you make a comparison, be sure it is *complete* (that readers can tell what two items are being compared) and *logical* (that it equates two comparable items).

INCOMPLETE: Pharmacists worry about the side effects of medication as much as physicians. (Are the pharmacists worrying about physicians, or are they worrying about the side effects of medication?)

COMPLETE: Pharmacists worry about the side effects of medication as much as physicians <u>do</u>.

INCOMPLETE: My chemistry course is harder. (What two things are being compared?)

COMPLETE: My chemistry course is harder <u>than Craig's</u>. (Comparison is now complete.)

ILLOGICAL: The intelligence of a pig is greater than a dog. (illogically compares the intelligence of a pig to a dog)

LOGICAL: The intelligence of a pig is greater than <u>that of</u> a dog.

ILLOGICAL: The United States exports more movies than any country in the world. (The United States is a country in the world; therefore, the sentence illogically implies the United States exports more movies than itself.)

LOGICAL: The United States exports more movies than <u>any other</u> country in the world. (Use *any other* to compare items in the same class.)

ILLOGICAL: The United States exports more movies than any other country in Europe. (The United States is not, as the sentence illogically implies, a country in Europe.)

LOGICAL: The United States exports more movies than <u>any</u> country in Europe. (Use *any* to compare items in different classes.)

EXERCISE 5

Revise the following sentences to correct any incomplete or illogical comparisons.

INCOMPLETE: Technology-based industries are concerned about inflation as much as service industries.

COMPLETE: Technology-based industries are concerned about inflation as much as service industries are.

1. Opportunities in technical writing are more promising than those in business writing. (illogical comparison)
2. Technical writing is more challenging than business writing. (incomplete comparison)
3. In some ways, technical and business writing require more attention to correctness than other forms of writing and are, therefore, more difficult. (incomplete comparison)
4. Business writers are concerned about clarity as much as technical writers are. (illogical comparison)
5. Technology-based industries may one day create more writing opportunities than any other industry. (illogical comparison)

1. Opportunities in technical writing are more promising than business writing.
2. Technical writing is more challenging.
3. In some ways, technical and business writing require more attention to correctness and are, therefore, more difficult.
4. Business writers are concerned about clarity as much as technical writers.
5. Technology-based industries may one day create more writing opportunities than any industry.

STUDENT WRITER AT WORK

AWKWARD OR CONFUSING SENTENCES

Following is an excerpt from a draft of a student's research paper on immigrant factory workers in New York City in the early twentieth century. In this section of her paper, the student focuses on a fire that contributed to the creation of stricter fire safety codes. Read the paragraphs and correct any mixed constructions, faulty predication, unwarranted grammatical shifts, or incomplete or illogical comparisons. If necessary, revise further to strengthen coherence, unity, and style.

In one particularly compelling section of World of Our Fathers, Irving Howe describes the devastating 1911 fire at the Triangle Shirtwaist Company (304–6). In quoting an eyewitness account and contemporary reactions and by reproducing graphic photographs is how he added drama to his account of an already dramatic event. For instance, Howe quotes labor activist Rose Schneiderman, who said this is not the first time girls have been burned alive in this city, and "The life of men and women is so cheap and property is so sacred" (305).

In his book The Triangle Fire, Leon Stein suggests that the fire, in the ten-story Asch Building near Washington Square in New York City, probably began with a cigarette or spark in a rag bin. Some people in the building apparently tried dousing the flames, but because of rotted hoses and rusted water valves made their efforts useless (15).

The announcement of the many causes of the fire deaths was obvious. According to the investigating committee, the one fire escape visible from Green Street collapsed after fewer than twenty people escaped. In addition, although sprinkler systems had been invented in 1895 did not mean any were present in the Asch Building. They were considered more expensive. Records show

continued on the following page

STUDENT WRITER AT WORK

Awkward or Confusing Sentences

Possible Revision

In one particularly compelling section of *World of Our Fathers*, Irving Howe describes the devastating 1911 fire at the Triangle Shirtwaist Company (304–6). By quoting an eyewitness account and contemporary reactions and reproducing graphic photographs, he adds drama to his account of an already dramatic event. For instance, Howe quotes labor activist Rose Schneiderman, who said, "This is not the first time girls have been burned alive in New York City! The life of men and women is so cheap, and property is so sacred" (305).

In his book *The Triangle Fire*, Leon Stein suggests that the fire, in the ten-story Asch Building near Washington Square in New York City, probably began with a cigarette or spark in a rag bin. Some people in the building apparently tried dousing the flames, but rotted hoses and rusted water valves made their efforts useless (15).

The many causes of the fire deaths were obvious. According to the investigating committee, the one fire escape visible from Green Street collapsed after fewer than twenty people escaped. In addition, although sprinkler systems had been invented in 1895, none were present in the Asch Building. They were considered too expensive. Records show that, six months before the fire, the city cited the building as a firetrap. The owners' failure to make alterations was identified in the report as a cause of the fire. Failure to have regular fire drills and a lack of clearly marked exits also contributed to the high death toll (Stein 117–19).

continued on following page

Compar 17h

continued from the previous page

that six months before the fire, the building was cited as a fire-trap by the city. The owners failed to make alterations was what the report identified as a cause of the fire. Failure to have regular fire drills and a lack of clearly marked exits also contributed to the high death toll (Stein 117–19).

Because it was 4:30 P.M. on a Saturday when the fire broke out meant there were 650 workers in the building. The majority of these were young Jewish and Italian women, and there was no common language spoken by them. By not sharing a common language was one reason for the chaos among the workers (Stein 14–15). When those on the ninth floor found the fire exit doors locked, their panic peaked. With their exit blocked forced everyone to jump from the windows to avoid the intense fire that swept through the building. Although it took fire fighters only 18 minutes to bring the fire under control, 146 workers died.

continued from previous page

Because it was 4:30 PM on a Saturday when the fire broke out, there were 650 workers in the building. The majority of these were young Jewish and Italian women who spoke no common language. Their not sharing a common language was one reason for the chaos among the workers (Stein 14–15). When those on the ninth floor found the fire exit doors locked, their panic peaked. With their exit blocked, everyone was forced to jump from the windows to avoid the intense fire that swept through the building. Although it took fire fighters only 18 minutes to bring the fire under control, 146 workers died.

Commentary on Possible Revision

The writer corrects mixed constructions, faulty predication, and unwarranted grammatical shifts. However, some passive constructions still exist, and the passage lacks unity.

PART 5

Using Words Effectively

CHAPTER 18

Choosing Words

Choosing the right word is a complex process. Unfortunately, no clear rule exists for distinguishing the right word from the wrong one. The same word may be appropriate in one situation and inappropriate in another.

18a Choosing an Appropriate Level of Diction

Diction, which comes from the Latin word meaning *say*, denotes choice and use of words. Different audiences and situations call for different *levels* of diction. You would think it odd, for example, if your history textbook said that Julius Caesar was the *guy* who ruled Rome, even though *man* and *guy* essentially mean the same thing. When you know who your readers are and what they expect, you can determine the appropriate level of diction.

(1) Formal diction

When decorum is in order—in eulogies and some scholarly articles, for example—readers expect formal diction. **Formal diction** calls for words familiar to an educated audience: *impoverished* rather than *poor*, *wealthy* or *affluent* rather than *rich*, *intelligent* rather than *smart*, *automobile* rather than *car*. Contradictions, shortened word forms, and utility words such as *nice* generally do not appear in formal diction. Formal English is grammatically accurate and often maintains emotional distance from an audience by using the impersonal *one* or the collective *we* rather than the more personal *I* and *you*.

See ◄ 11a2

300

The following passage from John F. Kennedy's inaugural address illustrates these characteristics. Although formal, Kennedy's diction is not stiff or artificial; rather, it is eloquent, graceful, and clear, with parallelism heightening its impact.

> Let the word go forth from this time and place, to friend and foe alike, that the torch has been passed to a new generation of Americans—born in this century, tempered by war, disciplined by a hard and bitter peace, proud of our ancient heritage, and unwilling to witness or permit the slow undoing of those human rights to which this nation has always been committed, and to which we are committed today at home and around the world.

Academics frequently use formal diction aimed at a learned audience, as the psychologist B. F. Skinner does in the following paragraph. Note that he expects his readers to be familiar with the terminology of his field and that he uses the collective *we* and *our*.

> We learn to perceive in the sense that we learn to respond to things in particular ways because of the contingencies of which they are a part. We may perceive the sun, for example, simply because it is an extremely powerful stimulus, but it has been a permanent part of the environment of the species throughout its evolution, and more specific behavior with respect to it could have been selected by contingencies of survival (as it has been in many other species). (*Beyond Freedom and Dignity*)

(2) Informal diction

Whereas formal diction is primarily a language of writing and formal addresses, **informal diction** is the language people use daily in conversation. It includes *colloquialisms, slang, regionalisms,* and occasionally *nonstandard language.* Although you will encounter informal diction quite frequently in your reading, limit its use in your college writing to imitating speech or dialect or to giving an essay a conversational tone.

Colloquialisms **Colloquial diction** is the language of everyday speech. We use it when we are not concentrating on being grammatically correct, and it is perfectly acceptable in informal situations.

Contractions—*isn't, won't, I'm, he'd*—are typical colloquialisms, as are **clipped forms**—*phone* for *telephone, TV* for *television, dorm* for *dormitory.* Other colloquialisms include placeholders like *you know, sort of, kind of,* and *I mean* and utility words like *nice* for *good* or *acceptable, funny* for *odd,* and *great* for

▶ See
27a2

FURTHER READING

Linda Brodkey. "On the Subjects of Class and Gender in 'The Literacy Letters.'" *College English* 51.2 (February 1989): 125–41.
The author begins with a quotation by and discussion of Michel Foucault on various "subject positions." (Subject positions are determined by the particular language/discourse that we use. This language/discourse is not necessarily different because it is foreign [French, Spanish] but different because of positions of class and gender. A classroom teacher speaks in educational discourse, which denotes for the most part a position of social class; a student, especially in an Adult Basic Education Course [ABE], speaks in an untutored discourse, which denotes a position of class lower than that of the teachers. Although the teacher is in a position of power, the student, on occasion, has ways of resisting that power.)

The author reports on a series of letters that were written between teachers and students in an ABE class. The letters were written between six white, middle-class teachers (four women and two men) and six white, working-class students (all women). The letters demonstrate how the teachers' educational discourse speaks at cross purposes with the students' untutored discourse; in fact, the teachers' discourse displays "discursive hegemony," whereas the students' discourse displays "discursive resistance."

For example, in the letters between Don (teacher) and Dora (student), Don refers to himself (in his subject position) as being "lazy" and a "bum"

continued on following page

continued from previous page

because he likes to stay up late and watch rented movies. Later in a letter Dora says: "I wouldn't want to see you living in Kensington with the rest of us bums. Ha, ha." The word *lazy* means something very different for each letter-writer; in other words, the word conjures up, for Don and then Dora, a radically different subject position. For Don, *lazy* means "leisure time" that may be subject to the Puritan ethic, but leisure time nonetheless. (This is a subject position of the middle to upper class.) For Dora, being lazy and a bum has a literal referent. Working-class people who are out of work take on the subject position of being called "lazy" and actually being believed to be lazy, for the assumption by the privileged class is that working-class people out of work do not want to work but instead to be supported by society; and literally, not figuratively, the word *bum* not only is a subject position but also has a very real, material referent in the other term of *street people*, who populate Dora's immediate life. Again, in the exchange between Rita (teacher) and Esther (student), Rita writes: "What is it like where you live and what are the shopping areas like?" (These questions presuppose subject positions.) Esther, of course, being in the subject position that she is in, does not think in terms of "shopping areas"—be they malls or boutiques.

In conclusion, the author says that though the law of the land guarantees all citizen-students equal education, the particular discourse with its limited subject positions—that is, "educational discourse," or the discourse of the educated elite—undercuts the law and therefore, does not finally protect working-class or lower-class

continued on following page

almost anything. Colloquial English also includes verb forms like *get across* for *communicate*, *come up with* for *find*, and *check out* for *investigate*.

Slang   **Slang** is a vivid and forceful use of language that is often restricted to a single group—urban teenagers, rock musicians, or computer users, for example. Slang words are extremely informal. Whether they are inventions or existing words redefined, they emerge to meet a need. They add spice to spoken language and create group cohesion. Words like *high, spaced out, dove, hawk, hippie, uptight, groovy, heavy, be-in, happening*, and *rip off* emerged in the 1960s as part of the counterculture movement. During the 1970s, technology, music, politics, and feminism influenced slang, giving us expressions like *high tech, hacker, input, disco, stonewalling, Watergate, nuke, burnout, macho*, and *male chauvinism*. Other 1970s words that are still with us include *humongous* and *lifestyle*. The 1980s contributed *rap music, yuppie, fax, chocoholic*, and *crack*. Slang in the 1990s includes expressions such as *hip-hop, wonk, spin doctor*, and *P.C.*

CLOSE-UP

CHOOSING WORDS

Slang varies with time and place and can become dated quickly. In time, slang can either become part of the language or fade into disuse. (Consider *beatnik, daddy-o*, and *hepcat*.) In college writing, use slang only when imitating speech or dialect.

Regionalisms   **Regionalisms** are words, expressions, and idiomatic forms commonly used in certain geographical areas that may not be understood or sound natural to a general audience. Regionalisms include words like *overshoe, fried cake*, and *angledog* and expressions like *take sick* and *come down with a cold*. In eastern Tennessee, for example, a paper bag is a *poke*, and empty soda bottles are *dope bottles*. In the Ozarks *I suspicioned* is used to mean *I suspected*. In Lancaster, Pennsylvania, which has a large Amish population, it is not unusual to hear an elderly person saying *darest* for *dare not* or *daresome* for *adventurous*. Similarly, New Yorkers stand *on* line, while people in most other parts of the country stand *in* line.

CLOSE-UP

CHOOSING WORDS

Regionalisms can make your writing more vivid. However, you should use them only when writing informally for a local audience who will be familiar with a particular region's dialect or when you want to give a regional flavor to your writing.

Nonstandard Language   **Nonstandard** diction refers to words and expressions not generally considered a part of standard English, even though many intelligent individuals use them when speaking. Included are words like *ain't, nohow, anywheres, nowheres, hisself, theirselves,* and *wait on* (instead of *wait for*).

No absolute rules distinguish standard from nonstandard usage. In fact, this issue is currently the subject of much debate. Some linguists reject the idea of nonstandard usage altogether, arguing that this designation serves only to relegate both the language and those who use it to second-class status. Dictionaries and handbooks can at best only attempt to define the norms of language. In the end, you will have to supplement the guidelines these books provide with your own assessment of what is appropriate usage in a particular writing situation.

▶ See 19b9

(3)   Popular diction

Popular diction is the language of mass-audience magazines, newspapers, bestsellers, and editorials. Conversational in tone, it often includes colloquialisms, contractions, and the first person. Even so, popular diction generally uses correct grammar, avoiding slang and nonstandard language.

The following passage from *Esquire* magazine illustrates many of these characteristics.

> Decade after decade of mediocre Disney films triumphed and stoked our sensibilities because the art itself, animation, is such a delight. It has been right from the start. In 1928, when Mickey Mouse as Steamboat Willie joins prototype Minnie to crank up a goat's tail in order to see musical notes leave the beast's mouth and dance to the tune of "Turkey in the Straw," the magic is already in full bloom. When Dumbo's ears flap and the ungainly mass finally soars, nobody cares where he's going. We love this creature; indeed we fall for Disney the way we fall in love: what the eye sees is more important than what the eye judges. (Max Apple, "Uncle Walt")

continued from previous page

students from the dialect of this discourse. To be sure, teachers work exceptionally hard at providing equal education, but (again) in the *class*room the very discourse that the teachers use excludes many students. "Discursive hegemony of teachers over students is usually posed and justified in developmental terms—as cognitive deficits, emotional or intellectual immaturity, ignorance, and most recently, cultural literacy—any one of which would legitimate asymmetrical relationships between its knowing subjects, teachers, and its unknowing subjects, students."

TEACHING STRATEGY

Audience

Experts sometimes argue that "college writing" often takes on a flat, "dead" tone—perhaps because the words are not being chosen with a specific reader in mind. In some class assignments, you may find it interesting to specify a target readership and see what happens.

Bring to class some samples of good popular writing—for example, features from magazines such as *Esquire, New Yorker, Texas Monthly,* or *Vanity Fair.* As an exercise, ask students to rewrite the paragraph on this page as though they were writing it for submission to a "trendy" magazine. The rough draft need not be entirely fleshed out; students may want to jot down notes of the kind of material they might use (quotations, anecdotes, etc.). But the final results should reveal how a change in audience affects diction, tone, and word choice.

ANSWERS TO EXERCISE 1

Manny was supposed to be crazy. <u>That was his story.</u> To say <u>you</u> were bad <u>put some people off.</u> <u>But</u> to say <u>you</u> were crazy, <u>well,</u> <u>you</u> were officially not to be <u>messed with.</u> <u>So that was his story.</u> On the other hand, after <u>I</u> called him what <u>I</u> called <u>him</u> and said <u>a few choice things</u> about his mother, his face did <u>go through some piercing changes.</u> <u>And</u> I did <u>kind</u> of wonder if <u>maybe</u> he sure was nuts. I <u>didn't</u> wait to find out. <u>I got in the wind.</u> <u>And</u> then he waited for me on my stoop all day and all night, <u>not hardly</u> speaking to the people going in and out. <u>And</u> he was there all day Saturday, with his sister bringing him peanut-butter sandwiches and cream sodas. He <u>must've</u> gone to the bathroom right <u>there</u> <u>cause</u> every time I looked out the kitchen window, there he was. <u>And Sunday, too.</u> <u>I got to</u> <u>thinking</u> the boy was mad.

Rewritten Sample

Manny was supposed to be crazy. He realized that some people simply feel uncomfortable around someone with a bad reputation, but, without a doubt, they know not to challenge anyone who is crazy. Although I doubted that he was actually crazy, I began to reconsider after his reaction when I called him by an objectionable name and went on to make some rather unflattering remarks about his mother as well. Immediately, his facial expression changed dramatically, making me decide not to wait around to see how he would further react; instead, I ran home as quickly as possible. He soon appeared on my stoop, where he waited for me all day and all night. Obviously, his entire focus was on me because he barely spoke

continued on following page

The contraction *he's* and the colloquial expressions *in full bloom*, *crank up*, and *fall for* give this piece a relaxed, conversational tone. However, the author also uses relatively formal words like *decade, prototype,* and *indeed* and the collective *we* and *our* instead of the more informal *I* and *you*.

(4) College writing

The level of diction appropriate for college writing depends on your assignment and your audience. A personal experience essay calls for a natural, informal style, but a research paper, an examination, or a report calls for a more formal vocabulary and a detached tone. In general most college writing falls somewhere between formal and informal English, using a conversational tone but maintaining grammatical correctness and using a technical vocabulary when the situation requires it.

The following student passage, which occasionally uses formal diction and includes no nonstandard language, contractions, or shortened word forms, illustrates the level of diction typical of much college writing.

> Deaf students face many problems. Their needs are greater than those of other students. Often ignored by hearing students, deaf students may develop a severe inferiority complex. Their inability to mix in a large group is another cause of this problem. Because deaf students communicate by sign language or by lip reading, they usually interact on a one-to-one basis. It is not at all unusual for deaf students to be in a classroom and not even realize that someone in the class is speaking. Other problems occur when people in the class first find out that a person has a hearing problem. Often, in an effort to try to help, people begin exaggerating their lip movements. This exaggerated lip movement, called "mouthing," makes it impossible for many deaf students to read lips. For this reason, many deaf students find it better to conceal a hearing impairment than to disclose it.

EXERCISE 1

This paragraph, from Toni Cade Bambara's short story "The Hammer Man," is characterized by informal diction. Representing the speech of a young girl, it includes slang expressions and grammatical inaccuracies. Underline the words that identify the diction of this paragraph as informal. Then rewrite the paragraph, using popular diction.

> Manny was supposed to be crazy. That was his story. To say you were bad put some people off. But to say you were crazy, well, you were officially not to be messed with. So that was his story. On the

other hand, after I called him what I called him and said a few choice things about his mother, his face did go through some piercing changes. And I did kind of wonder if maybe he sure was nuts. I didn't wait to find out. I got in the wind. And then he waited for me on my stoop all day and all night, not hardly speaking to the people going in and out. And he was there all day Saturday, with his sister bringing him peanut-butter sandwiches and cream sodas. He must've gone to the bathroom right there cause every time I looked out the kitchen window, there he was. And Sunday, too. I got to thinking the boy was mad.

*EXERCISE 2

After reading the following paragraphs, underline the words and phrases that identify each as formal diction. Then choose one paragraph and rewrite it using a level of diction that you would use in your college writing. Use a dictionary if necessary.

1. In looking at many small points of difference between species, which, as far as our ignorance permits us to judge, seem quite unimportant, we must not forget that climate, food, etc., have no doubt produced some direct effect. It is also necessary to bear in mind that owing to the law of correlation, when one part varies and the variations are accumulated through natural selection, other modifications, often of the most unexpected nature, will ensue. (Charles Darwin, *The Origin of Species*)

2. I hope you are able to see the distinction I am trying to point out. In no sense do I advocate evading or defying the law, as would the rabid segregationist. That would lead to anarchy. One who breaks an unjust law must do so openly, lovingly, and with a willingness to accept penalty. I submit that an individual who breaks a law that conscience tells him is unjust, and who willingly accepts the penalty of imprisonment in order to arouse the conscience of the community over its injustice, is in reality expressing the highest respect for law. (Martin Luther King, Jr., "Letter from Birmingham Jail")

18b Choosing the Right Word

According to Mark Twain, the difference between the right word and almost the right word is the difference between the lightning and the lightning bug. If you use the wrong words—or even *almost* the right ones—you run the risk of understating or misrepresenting your feelings and ideas.

continued from previous page

to the other people going in and out of the building. On Saturday, he was still there, waiting. I realized that he must not have even left to go to the bathroom because every time I looked out the kitchen window, he was sitting in the same spot on the stoop. The only interruption in his vigil was the short time it took him to devour the peanut-butter sandwiches and cream sodas his sister brought him. By the time Sunday arrived and Manny was still there, I began to think the boy was, indeed, quite mad.

ANSWERS TO EXERCISE 2

There will be some variation in students' answers here. Discussing their ideas of formal diction in class will help you and your students agree on what formal language is. King's "Letter from Birmingham Jail" is formal, in spite of its use of the first person ("I submit that . . ."). A possible revision of his paragraph appears below.

This distinction is important: Evading or defying the law, as segregationists have done, would lead to anarchy. Anyone who breaks an unjust law must do so openly, accepting the consequences. A person who breaks an unjust law and goes to prison to call attention to an injustice expresses great respect for law.

QUOTATIONS

"Only presidents, editors and people with tapeworm have the right to use the editorial 'we'."

—Mark Twain

continued on following page

continued from previous page

"Say all you have to say in the fewest possible words, or your reader will be sure to skip them; and in the plainest possible words or he will certainly misunderstand them."

–John Ruskin

"The difference between the almost right word and the right word is really a large matter—'tis the difference between the lightning bug and the lightning."

–Mark Twain

IN THE CLASSROOM

As an example of the difference between a word's denotation and its connotation, offer the term "housewife." Give the dictionary meaning, and then ask students to discuss, for example, what the word might connote to a mother at home with young children (having temporarily sidelined a prospering career) who is asked the inevitable question: "And what do YOU do?" Ask students for other examples of words loaded with connotations.

(1) Denotation and connotation

A word's **denotation** is its explicit dictionary meaning, what it stands for without any emotional associations. A word's **connotations** are the emotional, social, and political associations it has in addition to its denotative meaning.

DENOTATION AND CONNOTATION

Word	Denotation	Connotation
Politician	Someone who holds a political office	Opportunist; wheeler-dealer

---

You would think that determining the denotative meaning of a word would present few problems, but this is not always so. Words can have different denotations for different people. The linguist S. I. Hayakawa points out that in England and in the United States the word *robin* denotes entirely different species of birds and that what the English call a *sparrow* is what North Americans call a *weaver finch*. Moreover, words can have similar but not identical meanings, and this too can cause confusion. For example, you make an error in denotation when you say *molecule* when you mean *atom* or *compound* when you mean *mixture*.

Selecting a word with the appropriate connotation is also not easy. Slang and colloquialisms sometimes give ideas unintended connotations. For instance, the sentence "In school we *mess around* with computers" gives the impression that your involvement with computers is not serious or important. If your involvement is serious, it would be far better to say that you *work* with computers. Words that contain built-in judgments can be a particular source of unintended connotations. For example, *mentally ill, insane, neurotic, crazy, psychopathic,* and *disturbed* have different social and political connotations that color the way people respond. If you use such terms without considering their connotations, you run the risk of undercutting your credibility, to say nothing of confusing and possibly angering your readers.

CONSIDERING CONNOTATIONS

Positive	Neutral	Negative
Assertive	Aggressive	Pushy
Bold; confident; aggressively self-assured	Enterprising	Disagreeably aggressive or forward

---

Most of the writing you do requires a combination of connotative and denotative language. Technical reports will sometimes use an analogy to explain an unfamiliar concept or situation, and literary analyses routinely employ precise denotative language to describe a scene or character. Used appropriately, both connotative and denotative language enable you to express your ideas and emotions clearly and effectively.

*EXERCISE 3

The following words have negative connotations. For each, list one word with a similar meaning whose connotation is neutral and another whose connotation is favorable.

EXAMPLE: *Negative* skinny
Neutral thin
Favorable slender

1. deceive
2. antiquated
3. egghead
4. pathetic
5. cheap
6. blunder
7. weird
8. politician
9. shack
10. stench

*EXERCISE 4

Think of a trip you took. First, write a one-paragraph description that would discourage anyone from taking the same trip. Next, rewrite this paragraph, describing your trip favorably. Finally, rewrite your paragraph again, using neutral words that convey no judgments. In all three versions of your paragraph, underline the words that helped you to convey your impressions to your readers.

(2) Euphemisms

A **euphemism** is a term used in place of a blunt term that describes a subject society considers disagreeable, frightening, or offensive.

In the Victorian era, direct reference to the body and its functions was disdained. Consequently, table supports were delicately referred to as *limbs,* and (to avoid saying *leg* and *breast*) people referred to the meat of a turkey as *dark* or *light.*

ANSWERS TO EXERCISE 3

This exercise could be used to initiate a discussion of connotation. Because students often misuse the thesaurus, which usually gives no indication of connotations, you might bring this reference work to class and demonstrate its proper use. Sample answers to the exercise appear below.

Negative
Neutral
Favorable

1. deceive
   mislead
   beguile
2. antiquated
   old
   antique
3. egghead
   intellectual
   genius
4. pathetic
   unfortunate
   touching
5. cheap
   inexpensive
   economical
6. blunder
   error
   mistake
7. weird
   unusual
   unique
8. politician
   public servant
   statesman
9. shack
   cabin
   cottage
10. stench
    smell
    scent

ANSWERS TO EXERCISE 4

When the students have written all three paragraphs, they may find it interesting to see what others have written. They can share their examples with one another in small groups or by reading them aloud. You may also duplicate copies for the class or use an overhead projector so that the class can discuss them.

"The habit of creating *euphemisms* dates back at least to the Norman Conquest of England in 1066. At that time the community began to make a distinction . . . between the Latinate words of the upper class and the lusty Anglo-Saxon of the lower. That is why a duchess *perspired* and *expectorated* and *menstruated*—while a kitchen maid *sweated* and *spat* and *bled*. The linguistic gulf between Norman-derived and native Anglo-Saxon words remains as wide as ever after nine hundred years. The farmer today still looks after his Anglo-Saxon *cows, calves, swine,* and *sheep*—but once they are served up appetizingly in a restaurant or super-market, they become French *beef, veal, pork,* and *mutton.* And whenever the speech community must discuss anything it deems unpleasant, the discussion is acceptable on the condition that it is carried on in the elegant vocabulary bestowed on English by the Normans."

–Peter Farb,
Word Play

Place students in groups of five or six for this oral exercise. Give each group several relatively general words (such as *computer, cat, cassette, car*). Going around in a circle, each student makes one change to the word, trying to make it more and more specific. (For example, *computer, MacIntosh, Mac Classic, Mac Classic with 40 megabytes,* etc.)

People still avoid discussing certain subjects, such as bodily functions, death, and some social problems. Therefore, toilets are *lounges, bathrooms,* or *powder rooms.* We say that the dead have *passed on, gone to their reward,* or *departed*, and we call graveyards *resting places* or *memorial parks.* We refer to divorce as *marital dissolution,* adultery as an *affair,* the poor as *deprived* or *disadvantaged,* and retarded children as *developmentally delayed.*

College writing is no place for euphemisms. Say what you mean—*pregnant,* not *expecting; died,* not *passed away; strike,* not *work stoppage; drunk,* not *inebriated;* and *used car,* not *preowned automobile.*

(3) Specific and general words

Specific words refer to particular persons, items, or events; **general** words signify an entire class or group. *Queen Elizabeth II,* for example, is more specific than *monarch; jeans* is more specific than *clothing;* and *Jeep* is more specific than *automobile.* General words are, of course, useful. Statements that use general words to describe entire classes of items or events are often necessary to convey a point. But such statements must also include specific words for support and clarity. The more specific your choice of words, the more vivid your writing will be.

Whether a word is general or specific is relative, determined by its relationship to other words. The following word chains illustrate increasing specificity, with the word farthest to the left denoting a general category or class and the one farthest to the right naming a specific, tangible member of that class.

GENERAL → SPECIFIC

history—American history—Civil War history—History 263
apparel—accessory—tie—my Three Stooges tie
human being—official—president—Bill Clinton
reading matter—book—novel—*The Bluest Eye*
machine—vehicle—train—*Orient Express*

---

(4) Abstract and concrete words

Abstract words—*beauty, truth, justice,* and so on—refer to ideas, qualities, or conditions that cannot be perceived by the senses. **Concrete** words, on the other hand, convey a vivid picture by naming things that readers can *see, hear, taste, smell,* or *touch*. As with general and specific words, whether a word is abstract or concrete is relative. The more concrete your words and phrases, the more vivid the image you evoke in the reader.

ABSTRACT AND CONCRETE WORDS

ABSTRACT: The night I stayed too late I was spellbound by the beautiful sights.

CONCRETE: The night I stayed too late I was hunched on the log staring spellbound at spreading, reflected stains of lilac on the water. A cloud in the sky suddenly lighted as if turned on by a switch; its reflection just as suddenly materialized on the water upstream, flat and floating, so that I couldn't see the creek bottom, or life in the water under the cloud. Downstream, away from the cloud on the water, water turtles smooth as beans were gliding down with the current in a series of easy, weightless pushoffs, as men bound on the moon. (Annie Dillard, *Pilgrim at Tinker Creek*)

Of course we need abstract words to discuss concepts. The works of many great writers examine abstractions such as *truth, faith,* and *beauty,* but these writers know they must go on to describe abstractions with concrete details. However, abstract terms can create problems for beginning writers, who may use them—without concrete supporting detail—to conceal fuzzy, inexact thinking.

CLOSE-UP

CHOOSING WORDS

Imprecise diction often occurs when you use abstract terms such as *nice, great,* and *terrific* that say nothing and could be used in almost any sentence. These **utility words** indicate only enthusiasm. Replace them with more specific words.

VAGUE: The movie was <u>good</u>.

BETTER: The movie was <u>an exciting and suspenseful mystery</u>.

▶ See 11a2

▶ See 11a2

TEACHING STRATEGY

Sensory Writing
Teaching students how to move from general to specific words and from abstract to concrete language often means persuading them to think in terms of *sensory writing*: writing that chooses words that make the reader see, hear, taste, smell, or touch.

"The more specific [a picture] gets, the harder it is for you not to think about it," says writing teacher Gary Provost in *Make Every Word Count.* Ask your students to NOT think about an old woman. Then ask them to NOT think about an old woman sitting on a park bench. Then ask them to NOT think about an old woman on a park bench in a huge flowered hat and a red dress. Get the idea?

IN THE CLASSROOM

Write several general statements on the board (use those below or invent your own) and ask students to write one or two paragraphs that move these statements in the direction of specific, concrete language.

1. I was bored.
2. We had a good time.
3. I found the movie interesting.
4. It is my favorite piece of clothing.
5. I dislike his attitude.

ANSWERS TO EXERCISE 5

Student answers may vary.
—— = relatively specific and concrete
== = relatively general and abstract

I grew up slowly beside the tides and marshes of Colleton; my arms were tawny and strong from working long days on the boat in the blazing South Carolina heat. Because I was a Wingo, I worked as soon as I could walk; I could pick a blue crab clean when I was five. I had killed my first deer by the age of seven, and at nine was regularly putting meat on my family's table. I was born and raised on a Carolina sea island and I carried the sunshine of the low-country, inked in dark gold, on my back and shoulders. As a boy I was happy above the channels, navigating a small boat between the sandbars with their quiet nation of oysters exposed on the brown flats at the low watermark. I knew every shrimper by name, and they knew me and sounded their horns when they passed me fishing in the river.

ANSWERS TO EXERCISE 6

This exercise might be assigned in connection with Chapter 48 ("Practical Writing").

Part-time jobs I have held include waiting tables, landscaping, and selling stereo equipment. Each of these jobs requires strong communications skills. In my most recent position, I sold automobile stereos. My supervisor was impressed with my ability to help customers select components that fit their budgets and preferences. With my proven success in sales—being named Salesperson of the

continued on following page

In the following paragraph the overuse of abstract words creates a general and not very vivid description.

TOO ABSTRACT: *The Balzac Monument* is Rodin's most daring creation. The figure is unusual. Balzac is wrapped in a cloak in an interesting way. He has a strange look in his eyes.

MORE CONCRETE: *The Balzac Monument* is Rodin's most daring piece of sculpture. The figure resembles a ghost or specter. Balzac seems to tower above us so that from a distance we see only his great size. Upon closer inspection we see that Balzac is wrapped in a cloak. From the indistinct lines of the cloak, Balzac's head emerges godlike, with eyes that stare off into the distance.

EXERCISE 5

Effective writing usually mixes specific and general words and abstract and concrete words. Read the following passage and underline words that are relatively specific and concrete. How do they make the paragraph more effective? Are any very general or abstract words used? How do they function? What impression does the writer want to convey?

I grew up slowly beside the tides and marshes of Colleton; my arms were tawny and strong from working long days on the shrimp boat in the blazing South Carolina heat. Because I was a Wingo, I worked as soon as I could walk; I could pick a blue crab clean when I was five. I had killed my first deer by the age of seven, and at nine was regularly putting meat on my family's table. I was born and raised on a Carolina sea island and I carried the sunshine of the low-country, inked in dark gold, on my back and shoulders. As a boy I was happy above the channels, navigating a small boat between the sandbars with their quiet nation of oysters exposed on the brown flats at the low watermark. I knew every shrimper by name, and they knew me and sounded their horns when they passed me fishing in the river. (Pat Conroy, *The Prince of Tides*)

*EXERCISE 6

Revise this paragraph from a job application letter by substituting specific, concrete language for general or abstract words and phrases.

I have had several part-time jobs lately. Some of them would qualify me for the position you advertised. In my most recent job, I sold products in a store. My supervisor said I was a good worker who possessed a number of valuable qualities. I am used to dealing with different types of people in different types of settings. I feel that my qualifications would make me a good candidate for your job.

18c   Avoiding Unoriginal Language

(1)   Jargon

Jargon refers to the specialized or technical vocabulary of a trade, profession, or academic discipline. Jargon is useful in the field for which it was developed, but outside that field it is often imprecise and confusing. Medical doctors tell patients that a procedure is *contraindicated* or that they are going to carry out a *differential diagnosis*. Business executives ask for *feedback* and want departments to *interface* effectively. On a recent television talk show, a sociologist spoke about the need for *perspectivistic thinking* to achieve organizational goals. Is it any wonder that the befuddled host asked his guest to explain this term to the audience?

Jargon is often accompanied by overly formal diction, the passive voice, and wordy constructions. The following sentences contrast a passage choked by jargon with a simplified version.

> **ORIGINAL:** Procedures were instituted to implement changes in the parameters used to evaluate all aspects of the process.

> **TRANSLATION:** We began using different criteria to judge the process.

Although many people deliberately use jargon to impress their audience, its effect is usually just the opposite. When you write, avoid jargon and concentrate on a vocabulary that is appropriate for your audience and purpose.

(2)   Neologisms

Neologisms are newly coined words that are not part of standard English. New situations call for new words, and frequently such words become a part of the language. The rapid development of the media and growth of scientific knowledge, for example, have introduced thousands of new words that have become part of standard English.

NEOLOGISMS

quark: A hypothetical subatomic particle

televangelist: A member of the clergy whose primary ministry is a television audience

E-Mail: A system that distributes messages to people along a computer network

---

continued from previous page

Month three times last year—I believe I am well-qualified for a position in sales management.

IN THE CLASSROOM

As an example of jargon, read the following from "Housing: An Overview," (from *Housing: Prospects and Problems* by George Sternlieb and James W. Hughes). Ask students to identify and define the jargon used in the excerpt. Ask pairs of students to rewrite the excerpt eliminating all jargon and compare the results.

> But as the morale of America's citizenry sags in the context of the broader sweep of events, the evolution of the nation's housing markets—while beset by rising cost thresholds frustrating to shelter seekers—also has become a unique source of solace for many. Despite somber perceptions, the realities of housing production and the changes in the overall supply parameters appear much more sanguine upon detailed scrutiny. While serious reservations are voiced on the future, a number of basic trendlines have been substantially positive in nature.

QUOTATION

"First and most important: TRANSLATE. I mean by this replace every current jargon phrase by the word for its root idea: learning experience = learning; teaching profession = teachers; weather conditions = weather . . . precipitation = rainfall; nutritional routine = eating habits; of a calcareous nature = chalky . . . Get rid of

continued on following page

continued from previous page

them and the effect will be to lay bare what you have actually said. It is then easier to notice gaps in thought, insufficient or repetitious details, apparent contradictions, needless variants, and all the other faults writing is liable to."

–Jacques Barzun,
A Word or Two Before You Go

Writer Theodore Bernstein enjoyed coining words for the amusement of colleagues.

"A *monologophobe* (you won't find it in the dictionary) is a writer who would rather walk naked in front of Saks Fifth Avenue than be caught using the same word more than once in three lines. What he suffers from is *synonymomania* (you won't find that one, either), which is a compulsion to call a spade successively a *garden implement* and an *earth-turning* tool."

–from *The Careful Writer*

"Clichés are *the last word* in bad writing, and it's *a crying shame* to see all you *bright young things* spoiling your *deathless prose* with phrases that are *as old as the hills.*"

–Jane Burroway

Other coined words, however—even some that are very popular in speech or in specific contexts—may never be accepted. Many of these questionable neologisms are created when the suffixes *-wise* and *-ize* are added to existing words. Businesspeople *prioritize* before they *finalize* things *investmentwise*. Popular journalists and news commentators seem to attach these suffixes to many words—creating new words like *weatherwise, sportswise, timewise,* and *productwise*.

If you are not sure whether to use a term, look it up in a current college dictionary. If it is not there, it is probably not standard English.

(3)   Pretentious diction

Pretentious diction is language that is inappropriately elevated and wordy. In an effort to impress readers, beginning writers sometimes use pretentious diction: They elevate their style, overusing adjectives and adverbs, polysyllabic words, complex sentences, and poetic devices. Good writing is clear writing, and pompous or flowery language is no substitute for clarity.

> **PRETENTIOUS DICTION:** The expectations of offspring may be appreciably different from their parents'.
>
> **REVISED:** Children's goals may be different from their parents'.
>
> **PRETENTIOUS DICTION:** As I fell into slumber, I cogitated about my day ambling through the splendor of the Appalachian Mountains.
>
> **REVISED:** As I fell asleep, I thought about my day hiking through the Appalachian Mountains.

Pretentious diction is not formal diction used in an inappropriate situation; it is always out of place. By calling attention to itself, it gets in the way of meaning and draws readers away from the point you are making.

(4)   Clichés

Clichés are expressions that through overuse have lost all meaning. Familiar sayings like "Happily ever after," "We're in the same boat," "That's the last straw," and "Water under the bridge," for example, have lost their concrete associations and are now virtually meaningless.

Similarly, in many pat phrases, words become bound to other words. The phrases that result become so familiar that they lose their original impact. Thus, a campfire is always a *roaring campfire*, and childhood is invariably full of *fun-filled days* and *carefree moments* in which children *play their hearts out* and run *as fast as their little legs can carry them, free as the birds*. Similarly, political, social, or economic situations are often described as *rapidly deteriorating*. *Root causes* need to be uncovered, so *options are explored, Herculean efforts* are made, and sometimes *mutually agreeable solutions* are found. If not, *viable alternatives* may allow the two sides to *peacefully coexist*.

The purpose of college writing is always to convey information clearly; clichés do just the opposite.

*EXERCISE 7

Rewrite the following passage, eliminating jargon, neologisms, pretentious diction, and clichés. Feel free to add words and phrases and to reorganize sentences to make their meaning clear. If you are not certain about the meaning or status of a word, consult a dictionary.

> At a given point in time there coexisted a hare and a tortoise. The aforementioned rabbit was overheard by the tortoise to be blowing his horn about the degree of speed he could attain. The latter quadruped thereupon put forth a challenge to the former by advancing the suggestion that they interact in a running competition. The hare acquiesced, laughing to himself. The animals concurred in the decision to acquire the services of a certain fox to act in the capacity of judicial referee. This particular fox was in agreement, and consequently implementation of the plan was facilitated. In a relatively small amount of time the hare had considerably outdistanced the tortoise and, after ascertaining that he himself was in a more optimized position distancewise than the tortoise, he arrived at the unilateral decision to avail himself of a respite. He made the implicit assumption in so doing that he would anticipate no difficulty in overtaking the tortoise when his suspension of activity ceased. An unfortunate development racewise occurred when the hare's somnolent state endured for a longer-than-anticipated time frame, facilitating the tortoise's victory in the contest and affirming the concept of unhurriedness and firmness triumphing in competitive situations. Thus, the hare was unable to snatch victory out of the jaws of defeat. Years later he was still ruminating about the exigencies of the situation.

EXERCISE 8

Go through a newspaper or magazine and list the jargon, neologisms, pretentious diction, or clichés you find. Then substitute more original words for the ones you identified. Be prepared to discuss your interpretation of each word and of the word you chose to put in its place.

"It's impossible to draw a line between those [clichés] you should avoid and those that are indispensable because there's no better way of saying what it meant. You'll have to use your own judgment. For example, I wouldn't know what to say instead of *conspicuously absent* if I wanted to express just that idea; on the other hand, I'd never write *in the arms of Morpheus* instead of *asleep*. But you may feel differently. If in doubt whether an expression is a cliché or not, assume that it is and avoid it."

–Rudolph Flesch,
The ABC of Style

ANSWERS TO EXERCISE 7

The original, more readable story of the tortoise and the hare is well known. Students should enjoy exercises such as these.

STUDENTS WORKING TOGETHER

In groups of four or five, students select a piece of writing from one member's portfolio. They then rewrite the piece, using as many clichés as possible. Each group exchanges their clichéd piece with another group. Their task is to find and eliminate all clichés. Groups compare to see if all clichés were discovered and eliminated.

Your students' answers will vary. You may wish to clarify terminology in this exercise by using examples from Exercise 7.

jargon: aforementioned, quadruped, etc.

neologisms: distancewise, racewise, etc.

pretentious diction: At a given point in time there coexisted . . . , etc.

clichés: blowing his horn, to snatch victory out of the jaws of defeat, etc.

STUDENTS WORKING TOGETHER

Students develop a feel for figurative language by experimenting with it. A freewheeling, collaborative approach can be just what you need to stimulate your students' imaginations and expand their willingness to "try out" figures of speech in their own writing.

Divide the class into small groups. You might 1) ask each group to compose its own "figurative" version of each of the five sentences below; 2) assign one kind of figure of speech to each of the groups, asking one group to write metaphors for each of the sentences, another to write similes, and so on; or 3) ask each group to compose two, three, or four versions of each sentence using different figures of speech. Then ask each group to read a few of their creations out loud, identifying the figure of speech used. You may want to write some of the sentences on the board and discuss the class reaction to them.

continued on following page

18d Using Figurative Language

Language that conforms to the standard meaning of words is called **literal language**. But writers must often go beyond literal meanings to **figurative language**—language that uses imaginative comparisons called **figures of speech**. Although you should not overuse figurative language, do not be afraid to use it when you think it will help you communicate with your readers.

(1) Similes

A **simile** is a comparison between two essentially unlike items on the basis of a shared quality. Similes are introduced by *like* or *as*.

Like travelers with exotic destinations on their minds, the graduates were remarkably forceful. (Maya Angelou, *I Know Why the Caged Bird Sings*)

A cloud in the sky suddenly lighted as if turned on by a switch. (Annie Dillard, *Pilgrim at Tinker Creek*)

CLOSE-UP

CHOOSING WORDS

A simile must compare two *dissimilar* things. The first sentence here is not a simile, but the second one is.

My dog is like your dog.

My dog is as sleek as an otter.

(2) Metaphors

A **metaphor** also compares two essentially dissimilar things, but instead of saying that one thing is *like* another, it *equates* them.

In its first days of operation, a new telescope orbiting the earth has returned infrared images showing previously unobserved features of distant galaxies and revealing cosmic "maternity wards" where clouds of interstellar gas appear at various stages of giving birth to stars. (John Noble Wilford, *New York Times*)

Perhaps it is easy for those who have never felt the stinging darts of segregation to say, "Wait." (Martin Luther King, Jr., "Letter from Birmingham Jail")

CLOSE-UP

CHOOSING WORDS

For a metaphor to work, it has to employ images with which readers are familiar. If the comparison is too remote, readers will miss the point entirely. If it is too common, it will become a cliché.

▶ See 18c4

(3) Analogies

An **analogy** explains an unfamiliar object or idea by comparing it to a more familiar one.

▶ See 4e6

An atom is like a miniature solar system.

According to Robert Frost, writing free verse is like playing tennis without a net.

The circulatory system runs through the body like a network of rivers and streams.

CLOSE-UP

CHOOSING WORDS

Analogies work only when the subjects you are comparing have something in common. For example, drawing an analogy between tables and ants would not be useful because the items compared are too dissimilar. But explaining ants by comparing them to people—other social animals—makes good sense.

(4) Personification

Personification gives an idea or inanimate object human attributes, feelings, or powers. We use personification every day in expressions such as *The engine died*.

continued from previous page

EXAMPLE: The sentence was long and confusing. → The sentence encircled and choked the meaning like a giant boa constrictor.

1. His opponent in the tennis match was fearsome.
2. The book was boring.
3. The plane ride was extremely rough.
4. Hate made him irrational.
5. The hurricane destroyed the town.

STUDENTS WORKING TOGETHER

Divide the class into small groups and ask each group to come up with some interesting and effective examples of analogy. Ask each group to share its best example with the class. Or, have the whole class produce a list of 10 nouns associated with some subject (e.g., libraries) and a list of 10 verbs associated with another subject (e.g., cars). Put lists on the board. Then, have small groups take a word from each list and combine them into imaginative sentences. ("Books drive you to knowledge.")

315

Personification can make an entity that is abstract or hard to describe more concrete and familiar. By doing so, it also makes your writing more precise and more vivid.

> Truth strikes us from behind, and in the dark, as well as from before in broad daylight. (Henry David Thoreau, *The Journals*)

> Institutions, no longer able to grasp firmly what is expected of them and what they are, grow slovenly and misshapen and wander away from their appointed tasks in the Constitutional scheme. (Jonathan Schell, *The Time of Illusion*)

(5) Allusion

An **allusion** is a reference—which readers are expected to recognize—to a well-known historical, biblical, or literary person or event. Allusion enriches your readers' understanding by suggesting a relationship between your writing and something outside it. For example, suppose you title an essay you have written about your personal goals "Miles to Go Before I Sleep." By reminding your readers of the concluding lines of Robert Frost's poem "Stopping by Woods on a Snowy Evening," you suggest your determination and self-discipline.

CLOSE-UP

CHOOSING WORDS

An allusion will work only when readers recognize what you are alluding to. Family jokes, expressions that your friends use, and esoteric references mean nothing to a general audience.

(6) Overstatement (hyperbole)

Overstatement, or **hyperbole**, is an intentional exaggeration for emphasis. Its effect can be serious or humorous. For example, Sylvia Plath uses hyperbole in her poem "Daddy" when she compares her father to a Nazi storm trooper, and Jonathan Swift does so in his essay "A Modest Proposal" when he suggests that eating Irish babies would help the English solve their food shortage.

(7) Understatement

Whereas overstatement exaggerates, **understatement** downplays a situation or sentiment by saying less than what is really meant. When Mao-Tse Tung, former chairman of the People's Republic of China, said that a revolution was not a tea party, he was using understatement. When, after a hurricane, a survivor interviewed by a reporter said, "Well, that was quite a breeze," she too was using understatement.

CLOSE-UP

CHOOSING WORDS

Experienced writers intentionally use overstatement and understatement for emphasis. Carelessly used, however, such language can make your ideas seem inappropriately exaggerated or trivial.

Moby-Dick is the greatest literary work ever written.

Things got a little difficult for George Washington at the Battle of Brandywine.

18e   Avoiding Ineffective Figures of Speech

Effective figures of speech enrich your diction; ineffective figures of speech weaken it.

(1) Dead metaphors and similes

Metaphors and similes stimulate thought by calling up vivid images in a reader's mind. A **dead metaphor** or **simile**, however, has been so overused it has become a pat, meaningless expression that evokes no particular visual image. Here are some examples.

beyond a shadow of a doubt	off the track
off the beaten path	the last straw
sit on the fence	a shot in the arm
green with envy	smooth sailing

IN THE CLASSROOM

Bring a collection of resource books to class and ask small groups of students to search for the origins of some of the more unusual dead metaphors and similes on the list. (If students have personal favorites, include them in the search.) Useful books might include the shorter *Oxford English Dictionary*, Eric Partridge's *Dictionary of Slang* (or other Partridge works), or more recent popular books such as *A Hog on Ice and other Curious Expressions* by Charles E. Funk. Discuss the findings and ask students to offer opinions on why metaphors and similes "pass away."

See ◄
18c4

QUOTATION

The mixed metaphor does, on occasion, trap the greatest of us.

"Or to take arms against a sea of troubles . . ."

–William Shakespeare,
Hamlet

ANSWERS TO EXERCISE 9

Some examples appear below. You and your students will be able to find many others.

simile: shone like silver

metaphor: a broad expanse of river was turned to blood

analogy: as many-tinted as an opal

personification: language of this water

Avoid dead metaphors and similes; instead, take the time to think of images that make your writing fresher and more vivid.

(2) Mixed metaphors

A **mixed metaphor** results when you combine two or more incompatible images in a single figure of speech. Mixed images leave readers wondering what you are trying to say—or leave them laughing. When you revise mixed metaphors, make your imagery consistent.

> **MIXED:** Management extended an olive branch in an attempt to break some of the ice between the company and the striking workers.

> **REVISED:** Management extended an olive branch with the hope that the striking workers would pick it up.

EXERCISE 9

Read the following paragraph from Mark Twain's *Life on the Mississippi* and identify as many figures of speech as you can.

Now when I had mastered the language of this water, and had come to know every trifling feature that bordered the great river as familiarly as I knew the letters of the alphabet, I had made a valuable acquisition. But I had lost something, too. I had lost something which could never be restored to me while I lived. All the grace, the beauty, the poetry, had gone out of the majestic river! I still keep in mind a certain wonderful sunset which I witnessed when steamboating was new to me. A broad expanse of the river was turned to blood; in the middle distance the red hue brightened into gold, through which a solitary log came floating black and conspicuous; in one place a long, slanting mark lay sparkling upon the water; in another the surface was broken by boiling, tumbling rings, that were as many-tinted as an opal; where the ruddy flush was faintest, was a smooth spot that was covered with graceful circles and radiating lines, ever so delicately traced; the shore on our left was densely wooded, and the somber shadow that fell from this forest was broken in one place by a long, ruffled trail that shone like silver; and high above the forest wall a clean-stemmed dead tree waved a single leafy bough that glowed like a flame in the unobstructed splendor that was flowing from the sun. There were graceful curves, reflected images, woody heights, soft distances; and over the whole scene, far and near, the dissolving lights drifted steadily, enriching it every passing moment with new marvels of coloring.

*EXERCISE 10

Rewrite the following sentences, adding one of the figures of speech just discussed to each sentence to make the ideas more vivid and exciting.

Identify each figure of speech you use.

> **EXAMPLE:** The room was cool and still.
>
> The room was cool and still like the inside of a cathedral. (simile)

1. The child was small and carelessly groomed.
2. I wanted to live life to its fullest.
3. The December morning was bright and cloudy.
4. As I walked I saw a cloud floating in the sky.
5. House cats stalk their prey.
6. The sunset turned the lake red.
7. The street was quiet except at the hour when the school at the corner let out.
8. A shopping mall is a place where teenagers like to gather.
9. The President faced an angry Senate.
10. Education is a long process that takes much hard work.

18f   Avoiding Offensive Language

The language we use not only expresses ideas but also shapes our thinking. Because language reflects the ideas and attitudes of the culture in which we live, it should come as no surprise that it tends to reflect a white, male perspective. These attitudes emerge whenever we use *white* to mean good ("white lie") and *black* to mean bad ("black sheep") or the pronoun *he* to refer to females as well as males. For this reason we all should be aware of the influence of language on our perceptions. Although we may not be able to change the language, we can avoid using words that insult or degrade others.

(1)   Racial, ethnic, and religious labels

When referring to any racial, ethnic, or religious group, use words with neutral connotations or words that the groups use in *formal* speech or writing to refer to themselves. This is not always an easy task, for the preferred names for specific groups change over time. For example, *African American* is now preferred by many Americans of African ancestry over *black*, which itself replaced *Negro* in the 1960s. People from East Asia—once called *Orientals*—now refer to themselves as Asian Americans or by country of origin. Many of America's native peoples prefer *Native Americans* although some call themselves *Indians* and others

---

Some sample sentences to get the students started appear below.

1. The child was as small and as carelessly groomed as an old rag doll. [simile]
2. I want to live life to its fullest, the way Pepsi drinkers do. [allusion]
3. The December morning was moody: first bright, then cloudy, then bright again. [personification]
4. As I walked, I saw clouds floating in the sky, their sails unfurled, traveling at several knots. [metaphor]
5. House cats, like tigers, stalk their prey slowly and silently, moving quickly only when it's too late for the catch to get away. [simile]

In groups of three to five, students select a paragraph from their own writing. The group then rewrites each paragraph, adding figures of speech to each one. Groups exchange sets of paragraphs with other groups. Their task is to rank paragraphs according to the "naturalness" of the writing.

identify themselves as members of a particular tribe—for example, Kiowa or Navajo. Native Americans in Canada and Alaska, who consider *Eskimo* demeaning, have adopted *Inuit*. The preferences of people of Spanish descent vary according to their national origin. *Hispanic*—a term invented by the U.S. Bureau of the Census—is often used to refer to anyone of Spanish descent, as are *Latino* and *Latina*. But many individuals prefer other designations—for example, *Chicano* and *Chicana* for people from Mexico and *Puerto Rican* for people from Puerto Rico. A large number of Americans of Spanish descent, however, prefer to use names that emphasize their dual heritages—*Cuban American*, *Mexican American*, *Dominican American*, and so on.

As you can see, choosing a word that will not offend the sensibilities of others can be difficult. If you have questions, consult a current dictionary, such as *The American Heritage College Dictionary*, third edition—or ask your instructor for advice.

In addition to racial, ethnic, and religious stereotypes, other types of labels for groups of people can also be insulting.

Age   Unwarranted assumptions based on age can offend readers. Do not assume, for example, that all people over a certain age are forgetful, sickly, or inactive. Also do not express surprise or shock at the ability of an older person to do something that would be perfectly natural for someone younger to do.

Class   Your readers may not share your background or your assumptions about people. Do not demean certain jobs because they are low paying or praise others because they have impressive titles. Similarly, do not use words—*hick, cracker, redneck, white trash*, or *WASP*, for example—that denigrate people based on their class.

Geographical Area   Avoid making unwarranted assumptions about people based on where they live. People who live in the East are not all loud and pushy, just as people who live in the Midwest are not all soft-spoken and polite. Not all Californians eat health food and follow the latest fads, and not all Texans wear cowboy boots and speak with a drawl. Also remember that *American* can refer to anyone living in the Western Hemisphere—not just to someone living in the United States.

Physical Ability   There is a real difference between saying someone has a physical disability and describing someone as a disabled person. In the first case, the disability is part of how a person is characterized, but in the second, the focus is totally on the disability. In general, you should not even mention a person's disability unless it is relevant to the discussion. If it is relevant, use the preferred term—*disability*, not *handicap*, for example.

Sexual Orientation You should not mention an individual's sexual orientation unless it is relevant to your subject. For example, the sexual orientation of a retired army colonel who made her lesbian lifestyle public would be relevant if you were writing an essay about gays in the military, but not if you were writing an article about an award she won for heroism.

(2) Sexist language

Insulting to both men and women, sexist language entails much more than the use of derogatory words such as *broad*, *hunk*, *chick*, and *bimbo*. Assuming that some professions are exclusive to one gender—that, for instance, *nurse* denotes only women and *doctor* denotes only men—is also sexist. So is the use of job titles such as *postman* for *letter carrier*, *fireman* for *firefighter*, and *policeman* for police officer. Habits of thought and language change slowly, but they do change.

Sexist language also occurs when a writer fails to use the same terminology when referring to men and women. For example, you should refer to two scientists with Ph.D's not as Dr. Sagan and Mrs. Yallow, but as Dr. Sagan and Dr. Yallow. You should refer to two writers as James and Wharton, not Henry James and Mrs. Wharton.

In your writing, always use *women*—not *girls*—when referring to adult females. Use *Ms.* as the form of address when a woman's marital status is unknown or irrelevant. If the woman you are addressing refers to herself as *Mrs.* or *Miss*, however, use the form of address she prefers. Finally, avoid using the generic *he* or *him* when your subject could be either male or female. Use the third-person plural or the phrase *he or she* (not *he/she*).

> **TRADITIONAL:** Before boarding, each passenger should make certain that he has his ticket.
>
> **REVISED:** Before boarding, passengers should make certain that they have their tickets.
>
> **REVISED:** Before boarding, each passenger should make certain that he or she has a ticket.

Remember, however, not to overuse *his or her* or *he or she* constructions, which can make your writing repetitive and wordy.

TEACHING STRATEGY

Sexist Language
Use exercises 11–16 to stimulate discussion about how different individuals respond to the same language. Some women, for example, may object to the term "Mother Nature" because it places females in the traditional, stereotypical role of nurturer; others may like the fact that the term suggests a powerful female force. Still others will see it as neutral or as an essentially meaningless cliché. Some other terms and alternatives to discuss might include the following.

craftsman: artisan, craftsperson

man-hours: work hours, time

spokesman: speaker, spokesperson, representative

freshman: first-year student

sportsmanship: fair play, fairness

fatherland: native land, homeland

ladylike: polite, mannerly

maiden voyage: first voyage

airman/aviatrix: flier, pilot

CLOSE-UP

CHOOSING WORDS

When trying to avoid sexist use of *he* and *him*, be careful not to create ungrammatical constructions such as the following.

UNGRAMMATICAL: Before the publication of Richard Wright's novel *Native Son*, any unknown African-American <u>writer</u> had trouble getting <u>their</u> work published.

Although many educated speakers do use *they* and *their* in cases such as the one above, you should avoid using a singular noun along with a plural pronoun in your college writing. Instead, use a plural noun.

GRAMMATICAL: Before the publication of Richard Wright's novel *Native Son*, unknown African-American <u>writers</u> had trouble getting <u>their</u> work published.

REVISING SEXIST LANGUAGE

Sexist Usage	Possible Revisions
1. A student should choose <u>his</u> courses carefully.	Students should choose <u>their</u> courses carefully. A student should choose <u>his</u> or <u>her</u> courses carefully.
2. Equality is a desirable goal for <u>all men</u>.	Equality is a desirable goal for <u>everyone</u>.
3. Mankind Man's accomplishments Man-made	People, human beings Human accomplishments Synthetic
4. The librarian . . . <u>she</u> The lawyer . . . <u>he</u>	The librarian . . . <u>he or she</u>; the librarians . . . <u>they</u> The lawyer . . . <u>he or she</u>; the lawyers . . . <u>they</u>
5. Female doctor (lawyer, painter, etc.), male nurse	Doctor (lawyer, painter, etc.), nurse

continued on the following page

Avoiding Offensive Language

continued from the previous page

6. Mailman — Letter carrier
   Fireman — Firefighter
   Policeman/woman — Police officer
   Salesman/woman/girl — Salesperson/representative
   Businessman/woman — Businessperson/executive
   Steward/Stewardess — Flight attendant

7. The <u>girl</u> at the desk answered the phone. — The person, the receptionist, the secretary. . . .

8. Women's libber — Feminist
   Women's lib — Women's movement

9. Phyllis Knable, <u>wife of Dr. Peter</u> Knable, is an outstanding surgeon. — Dr. Phyllis Knable is an outstanding surgeon.

10. Linda Richards, <u>mother of three</u>, has been elected to the Jaycees Hall of Fame. — Linda Richards has been elected to the Jaycees Hall of Fame.

11. A professor works hard to get tenure. As a result, <u>his wife</u> must be supportive. — A professor works hard to get tenure. As a result, <u>his or her spouse</u> must be supportive; <u>Professors</u> work hard to get tenure. As a result, <u>their spouses</u> must be supportive.

12. <u>Everyone</u> should complete <u>his</u> application by Tuesday. — Everyone should complete <u>his or her</u> application by Tuesday; All <u>students</u> should complete <u>their</u> applications by Tuesday.

ANSWERS TO EXERCISE 11

1. Implies that women as a group are poor drivers.
2. Suggests that all medical students are male.
3. United States citizens of East Asian descent usually prefer the term Asian Americans.
4. Denigrates people based on class.
5. A person with a disability might see a wheelchair as liberating, not confining.
6. Implies that hairdressing is an occupation associated with a particular sexual preference.
7. Suggests that people from large cities are unfriendly.
8. Adult females should be referred to as women, not as girls.

EXERCISE 11

Identify the stereotypes in the following sentences.

1. Uncle Max wasn't surprised to find the driver of the car that rear ended his new BMW was a woman.
2. A new medical school student knows that he faces several years of hard work.
3. California has a large Oriental population.
4. She couldn't wait to leave her hometown because it was populated with rednecks and hicks.
5. Because of a back injury, he was confined to a wheelchair.

Your students' answers and comments will vary. These are some examples.

congressman—member of Congress

forefathers—ancestors

men at work—work crews

You may wish to have your students answer the questions in these exercises in group or class discussions.

6. I was surprised to notice that my hairdresser wears a wedding ring.
7. My roommate is from New York, but she is friendly.
8. Next Tuesday is her day to have the girls over for bridge.

EXERCISE 12

Suggest possible alternative forms for any of the following constructions you consider sexist. In each case, comment on the advantages and disadvantages of the alternative you recommend. If you feel that a particular term is not sexist, explain why.

forefathers	longshoreman
man-eating shark	committeeman
manpower	(to) man the battle stations
workman's compensation	Girl Friday
men at work	Board of Selectmen
copy boy	stock boy
bus boy	cowboy
first baseman	man overboard
corpsman	fisherman
congressman	foreman

EXERCISE 13

These terms denote professions that have traditionally been associated with a particular gender. Now that the professions are open to both sexes, are any new terms needed? If so, suggest possible terms. If not, explain why not.

miner	farmer
mechanic	barber
soldier	rabbi
sailor	minister
rancher	bartender

EXERCISE 14

The following terms have emerged in the last few years as possible alternatives for older gender-specific words. Which usages do you believe are likely to become part of the English language? Which do you expect to disappear? Explain.

Coinage	Older Form
• waitperson	waiter or waitress
server	

- househusband   housewife
  homemaker
- weather forecaster   weatherman, weathergirl
- chair   chairman
  chairwoman
  chairperson
- spokesperson   spokesman

EXERCISE 15

Each of the following pairs of terms includes a feminine form that was at one time in wide use; all are still used to some extent. Which do you think are likely to remain in our language for some time, and which do you think will disappear? Explain your reasoning.

heir/heiress	author/authoress
benefactor/benefactress	poet/poetess
murderer/murderess	tailor/seamstress
actor/actress	comedian/comedienne
hero/heroine	villain/villainess
host/hostess	prince/princess
aviator/aviatrix	widow/widower
executor/executrix	

EXERCISE 16

In recent years the word *parenting* has been introduced as a gender-free equivalent of *mothering*. How do the connotations of *parenting* differ from those associated with *mothering*? With *fathering*? Given those differences, what is your prediction about the continued use of *parenting* in years to come?

STUDENT WRITER AT WORK

CHOOSING WORDS

The following excerpt is from a draft written for a communications class. Underline any words and phrases you think are not appropriate, accurate, or fresh. Then revise the essay, changing words and sentences as you see fit. If necessary, revise again to strengthen coherence, unity, and style.

Saying It with a Smile

Our first impressions of a local television news program derive from the newscasters, those smiling people who converse

continued on the following page

Choosing Words

Possible Revision

Saying It with a Smile

Our first impressions of a local television news program come from the newscasters, those smiling reporters who talk to us every evening through our television sets. Local reporters must be conscientious and have an educational background in journalism, some experience in broadcasting, and a keen interest in exposing the truth. However, above all, broadcasters must have a pleasing appearance.

Using familiar faces has become an effective marketing strategy for the news. Many station managers have seen their market shares increase after hiring a newscaster with whom the audience identified. Market research has shown that viewers like to see the newscaster as one of the family. Therefore, reporters chosen for the job must look wholesome and attractive. Promotional advertisements for the local news reinforce this healthy image. In one spot, for example, we see a reporter walking through an economically depressed urban neighborhood playing ball with children and shaking hands with the children's parents. In another spot, we see an anchor bringing her son and some of his classmates to the television station. These seventh grade children's eyes are wide with wonder, and they use words like "cool" and "awesome" as they talk about the evening news. Clearly, this marketing strategy is calculated to present the reporters to the viewers as average people.

continued on following page

continued from previous page

Unfortunately, just as tranquilizers dull the senses, these newscasters dull our reactions to the actual news of the day. We are so conditioned to identify with the reporters that we tend not to think about the actual content of the news. Instead, we are inclined to remember those stories that consume the most time in the half-hour broadcast. Except in the case of national emergencies, these "big" stories, which cover local events such as rock concerts, ice skating on the local pond, and pizza-eating contests, are reported with happy chatter and have almost no news value. The important stories, such as the announcement of budget cuts or the formation of a task force on education, are lost in the process. The "fluff" stories must be what we want to know about; after all, we do tune in. In fact, ratings for local news shows have never been higher. Tomorrow night will come, and we will ritualistically tune in to hear our favorite news anchor report on the big story: "Beer drinking banned at the baseball stadium."

Commentary on Possible Revision

Although the writer has revised the paper for appropriateness, accuracy, and freshness, the essay's content can still be strengthened with additional examples. The writer also needs to broaden the thesis to reflect the actual content of the essay. Finally, the writer should develop a separate conclusion, perhaps using some material from the essay's last paragraph.

continued from the previous page

with us every evening. Undoubtedly, there are many qualifications local reporters must have, including a superlative educational background and some time in broadcasting. They should also be sharp and have a keen interest in exposing the real truth. Above all, however, broadcasters must have a pleasing appearance.

Using familiar faces has become an effective marketing strategy for the news. A number of station managers have seen their market share increase and their negatives decrease when they slotted a newscaster with whom the audience identified. Market research has shown that viewers like to visualize the newscaster as if he were one of the family. Therefore, people chosen for the job must look nice--just like the boy or girl next door. Promos for the local news reinforce this squeaky clean image. In one spot we see a newsman walking through a deteriorating urban neighborhood playing ball with lower-class kids and shaking hands with their moms and dads. In another, we see a female anchor bringing her son and some schoolmates to the station. Their eyes are as big as saucers as they look around. These kids are in seventh grade and they use words like "cool" and "awesome" as they talk about the evening news. Clearly, this rather blatant tactic is calculated to present the reporters to the viewers as everyday people.

Just as tranquilizers calm someone's nerves, these newscasters lull us into not caring and then feed us empty calories. We are so conditioned to identify with the reporters that we never think about the actual content of the news. Beyond a shadow of a doubt we are inclined to remember those stories that consume the most time in the half-hour broadcast. Except for national emergencies, the "big" stories are happy talk about local issues. Many of these have almost no news value--a rock concert, ice skating at a pond, and a pizza-eating contest, for example. This is obviously what we want to know and why we tune in. It must be, for ratings have never been higher. There is no mention of the unsolved hit-and-run murder of a young boy in my neighborhood. This is outrageous! But what the heck, that's what we expect anyhow. Tomorrow night we will ritualistically tune in and hear our favorite newsman tell us the big story: "Beer drinking banned at the baseball stadium."

CHAPTER 19

Using the Dictionary and Building a Vocabulary

19a   Exploring the Contents of a Dictionary

Like tools, different dictionaries are designed for different tasks. The most widely used type of dictionary, the one-volume **abridged dictionary** (also called a **desk dictionary** or a **college dictionary**), is ideal for daily use. Other types of dictionaries—including multivolume **unabridged dictionaries** and the many **special-purpose dictionaries**—have their own uses.

A dictionary is usually divided into three parts: the *front matter,* the *alphabetical listing,* and the *back matter.*

The **front matter** differs in each dictionary but generally contains a preface explaining how the dictionary is set up and guides for pronunciation and abbreviations. Some dictionaries include special material in the front matter—articles focusing on the history of the language, for example.

The **alphabetical listing** is the largest section of every dictionary. Entries tell how a word is spelled, pronounced, and used, and some dictionaries also include information on stress, grammatical function, and *etymology* (history and origin of the word). On the basis of hundreds of examples of usage, lexicographers also describe words with terms such as *regional, slang,* and *preferred.* Dictionaries may disagree regarding these designations, however.

Meanings under each entry may also be arranged differently from one dictionary to another. Some dictionaries list meanings in historical order, with the oldest usage coming first; others list meanings in order of most frequent usage. Therefore, you cannot

---

continued from previous page

edition of *Webster's New International Dictionary*—as he throws it into the fire, page by page.

QUOTATIONS

"The greatest masterpiece in literature is just a dictionary out of order."

–Jean Cocteau

"**Lexicographer**, n. A pestilent fellow who, under the pretense of recording some particular stage in the development of a language, does what he can to arrest its growth, stiffen its flexibility, and mechanize its methods. . . . [The] bold and discerning writer who . . . makes new words and uses the old in an unfamiliar sense has no following and is tartly reminded that 'it isn't in the dictionary'—although down to the time of the first lexicographer (Heaven forgive him!) no author ever had used a word that *was* in the dictionary."

–Ambrose Bierce,
The Devil's Dictionary

assume that the first definition under an entry is the preferred usage. You must look in the front matter to find out which principle of arrangement the lexicographers used.

The **back matter** also differs in each dictionary. Some dictionaries contain a list of weights and measures; others, an essay on punctuation; and still others, a glossary of foreign words. To get the most out of your dictionary, acquaint yourself with its back matter so you can refer to it when the need arises.

CLOSE-UP

USING THE DICTIONARY

Because content differs from dictionary to dictionary, you should become familiar with your dictionary's special features. Knowing exactly what your dictionary has to offer can help you get the most out of it as you write and revise.

19b   Using a Dictionary

To fit a lot of information into a small space, dictionaries use a system of symbols, abbreviations, and typefaces. Each dictionary uses a slightly different system, so you should consult the front matter of your dictionary to determine how its system operates.

A labeled entry from *The American Heritage College Dictionary* is shown in Figure 1.

(1)   Guide words

To help you locate words, all dictionaries include a pair of **guide words** at the top of each page.

Couperin/court-martial

shriek · shut

The guide words indicate the first and last entries appearing on that page. The word on the left shows the first entry on the page, and the word on the right shows the last. All entries on the page fall alphabetically between these words.

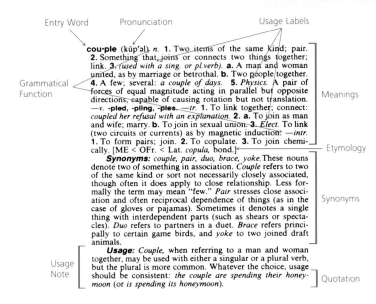

Figure 1 *Sample Dictionary Entry*

(2) Entry word

The **entry word**, which appears in boldface at the beginning of the entry, gives the spelling of a word and any variant forms.

col · or n. Also chiefly British **col · our**

(3) Guide to pronunciation

A guide to the pronunciation of a word appears in parentheses or between slashes [/ /] after the main entry. Dictionaries use symbols to represent sounds, and an explanation of these symbols usually appears at the bottom of each page or across the bottom of facing pages throughout the alphabetical listing. (A full guide to pronunciation appears as part of the front matter of the dictionary.) The stressed syllable of a word is indicated by an accent mark ('). A secondary stress is indicated with a similar but lighter mark (').

(4) Part-of-speech labels

Dictionaries use abbreviations to indicate parts of speech and grammatical forms.

If a verb is regular, the entry provides only the base form of the verb. If a verb is irregular, the entry lists the principal parts of the verb.

ANSWERS TO EXERCISE 1

(Using *The American Heritage Dictionary*, 3rd ed.)

1. drank, drunk, drinking, drinks
   deified, deified, deifying, deifies
   caroled, caroled, caroling, carols;
   also carolled, carolled, carolling,
   carols
   drew, drawn, drawing, draws
   rang, rung, ringing, rings
2. canter, command, lord, minister,
   mother
3. silos, sheep, seeds or seed, mice,
   scissors (a plural noun used with
   either singular or plural verb),
   genetics (a plural noun used with
   a singular verb), alchemies (no
   plural given, so it forms regularly)
4. fast, faster, fastest
   airy, airier, airiest
   good, better, best
   mere, no comparative, merest
   homey, homier, homiest
   unlucky, unluckier, unluckiest
5. bias: transitive, with no phrase
   halt: transitive or intransitive,
   with no phrase
   dissatisfy: transitive, with no phrase
   die: intransitive, *died of embarrassment* or *sunlight died in west;*
   transitive, no phrase
   turn: transitive, *turn a knob;* intransitive, *wheels turning at a
   rapid rate*

with · draw . . . v. -drew; -drawn; -drawing

In addition, the entry indicates whether a verb is transitive (*tr.*), intransitive (*intr.*), or both.

Part-of-speech labels also indicate the plural forms of irregular nouns.

to · ma · to . . . n. pl. -toes
moth · er-in-law . . . n. pl. moth · ers-in-law

When the plural form is regular, it is not shown.

Finally, dictionaries usually show the comparative and superlative forms of both regular and irregular adjectives and adverbs.

red . . . adj. redder; reddest

bad . . . adv. worse; worst

EXERCISE 1

Use your college dictionary to answer the following questions about grammatical form.

1. What are the principal parts of the following verbs: *drink, deify, carol, draw,* and *ring*?
2. Which of the following nouns can be used as verbs: *canter, minister, council, command, magistrate, mother,* and *lord*?
3. What are the plural forms of these nouns: *silo, sheep, seed, scissors, genetics,* and *alchemy*?
4. What are the comparative and superlative forms of the following adverbs and adjectives: *fast, airy, good, mere, homey,* and *unlucky*?
5. Are the following verbs transitive, intransitive, or both: *bias, halt, dissatisfy, die,* and *turn*? Copy the phrase or sentence from the dictionary that illustrates the use of each verb.

(5)   Etymology

The **etymology** of a word is its history, its evolution over the years. This information appears in brackets either before or after the list of meanings. The etymology traces a word back to its roots and shows its form when it entered English. For instance, *The American Heritage Dictionary* shows that *couple* came into Middle English (ME) from Old French (OF) and into Old French from Latin (L).

(6)   Meanings

Some dictionaries give the most common meaning first and then list less common ones. Others begin with the oldest meaning and move to the most current ones. Check the front matter of your dictionary to find out its principle of arrangement.

Remember that a dictionary records the meanings that appear most regularly in speech and in writing. If a word is in the process of acquiring new meanings, the dictionary may not yet include them. Also remember that a dictionary meaning is primarily a record of the **denotations**, or exact meanings, of a word. A word's emotional associations, or **connotations**, are not always among the meanings listed, although some dictionaries do attempt to suggest them.

▶ See
18b1

(7)   Synonyms and antonyms

A dictionary entry often lists synonyms and occasionally antonyms in addition to definitions. **Synonyms** are words that have similar meanings, such a *well* and *healthy*. **Antonyms** are words that have opposite meanings, such as *courage* and *cowardice*. Dictionaries present synonyms—and sometimes antonyms—because they clarify the meanings of a word and also because they are useful to writers who want to vary their choice of words. However, no two words are exactly equivalent, so you must use synonyms carefully, making certain the connotations of the synonym is as close as possible to that of the original word.

(8)   Idioms

Dictionary entries often show how certain words are used with other words in set expressions called **idioms**. Such phrases present problems for some native speakers and especially for non-native speakers, as you may know from your own study of other languages. For example, what are we to make of the expression *from the shoulder*? That it means "in a direct or outspoken manner of telling" is not at all apparent from its words.

Dictionaries also indicate the idiomatic use of prepositions. For example, we never say *acquiesce with* a decision. Most dictionaries will indicate that when it takes a preposition *acquiesce* is usually used with *in* and sometimes with *to*. Similarly, we do not say that we *abide with* a decision or that we *interfere on* a performance. We say *abide by* and *interfere with*. The idiomatic use of prepositions is a matter of custom.

Understanding Synonyms and Antonyms
Introduce your students to a thesaurus. Help them understand that the thesaurus is not a general usage dictionary and that the words listed under a heading (for example, *little* or *strange*) are rarely exact synonyms. It might be fun to create one sentence and then try to "plug in" each of the listings interchangeably. "Synonyms" sometimes carry very different connotative meanings; caution the students to use their dictionaries to check words found in the thesaurus before using those words in their writing.

(9) Usage labels

Dictionaries use **usage labels** to indicate restrictions on word meanings. Where such labels involve value judgments, dictionaries differ.

USAGE LABELS

Label	Definition	Example
1. Nonstandard	A word that is in wide use but not considered part of standard usage	*ain't*
2. Informal/ Colloquial	A word that is part of the language of conversation and acceptable in informal writing	*I've* for *I have* *sure* for *surely*
3. Slang	A word appropriate only in extremely informal situations but, unlike regionalisms, widely used	*rip off* *prof* for *professor*
4. Dialect/ Regional	A word or meaning of a word limited to a certain geographical region	*arroyo,* a word used in the Southwest to mean "deep gully" *potlatch,* a word used in the Northwest to mean "celebration"
5. Vulgar	A word that is offensive. This category includes words labeled **obscene**, which are extremely offensive, and words labeled **profane**,	*crap* *egal* meaning

continued on the following page

continued from the previous page

	which show disrespect for the deity	
6. Obsolete	A word that is no longer in use; this label applies only to words that have disappeared from the language	"equal"
7. Archaic/Rare	A word or meaning of a word that was once common but now is seldom used; **archaic** differs from **rare.** which means that a word was never in common usage	*affright* meaning "to arouse fear or terror" (archaic) *nocent* meaning "guilty" or "harmful" (rare)
8. Poetic	A word used commonly only in poetry	*eve* for *evening* *o'er* for *over*
9. Foreign Language Labels	These identify expressions or words from other languages that are commonly used in English but are not considered part of the language.	*Adios,* Spanish for "goodbye" *Sine qua non,* Latin for "something that is essential"
10. Field Labels	These indicate that a word or meaning of a word is limited to a certain field or discipline (mathematics, biology, military, etc.).	*couple* (physics and electrical engineering)

ANSWERS TO EXERCISE 2

1. [note spelling] nonstandard
2. term in architecture, especially church architecture, and in astronomy
3. informal
4. British: truck; more generally, horse-drawn wagon
5. Scottish: church; British: Presbyterian Church of Scotland
6. slang
7. British: a conjunction
8. mathematical
9. British regional term meaning unleavened bread
10. slang

ANSWERS TO EXERCISE 3

1. 1978
2. after George W. G. Ferris, 1859–96
3. 32.064
4. Jósef Teodor Konrad Nalecz Korzeniowski
5. a modern movement in art and literature in which an attempt is made to portray or interpret the workings of the unconscious mind as manifested in dreams: it is characterized by an irrational, noncontextual arrangement of material.

EXERCISE 2

Use your college dictionary to find the restrictions on the use of the following words.

1. irregardless
2. apse
3. flunk
4. lorry
5. kirk
6. gofer
7. whilst
8. sine
9. bannock
10. blowhard

(10) General information

In addition to containing information about words, your dictionary is an excellent source of general information. If you wanted to find out the year in which John Glenn orbited the earth, you might look up the entry *Glenn, John.* If you needed to find out after whom the Davis Cup is named, you would look up *Davis Cup.* When no other references are handy, your dictionary can be extremely useful.

EXERCISE 3

To test the research capability of your dictionary, use it to answer the following questions.

1. In what year did Anwar el-Sadat win the Nobel Peace Prize?
2. After whom was the Ferris wheel named?
3. What is the atomic weight of sulfur?
4. What was Joseph Conrad's original name?
5. What is surrealism?

19c Surveying Abridged Dictionaries

An **abridged dictionary** is one that is condensed from a more complete collection of words and meanings. A good hardback abridged dictionary will contain about 1,500 pages and about 150,000 entries. Paperback dictionaries usually contain fewer entries, treated in less detail. A paperback dictionary is adequate for checking spelling, but for reference, you should consult one of the following hardback abridged dictionaries.

The American Heritage College Dictionary, 3rd ed. Boston: Houghton Mifflin, 1993. This extensively illustrated dictionary contains more than 180,000 entries. Within the alphabetical listing the principal and most current meaning appears first, with other meanings following. Throughout the dictionary, the editors have provided *usage notes* and short essays on word history. Synonyms and sometimes antonyms are also listed.

The Concise Oxford Dictionary of Current English, 8th ed. New York: Oxford University Press, 1990. This no-nonsense dictionary contains no illustrations and gives little guidance on usage. It lists meanings according to the most common usage, includes illustrative quotations, and gives British as well as American spellings. Front and back matter is sparse.

The Random House College Dictionary, rev. ed. New York: Random House, 1991. This is an abridged version of the larger unabridged *Random House Dictionary of the English Language.* It lists meanings according to frequency of use and indicates informal and slang usage. It also gives synonyms and antonyms as well as geographical and biographical names. Back matter includes a manual of style.

Marrion-Webster's Collegiate Dictionary, 10th ed. Springfield, MA: Merriam, 1993. Like *The Random House College Dictionary,* this illustrated dictionary is an abridged version of a larger unabridged dictionary. Words are defined by their current usage rather than their original definitions. Many entries are followed by usage notes.

The front matter of this dictionary contains a detailed essay on the English language commissioned for this edition. Foreign words and phrases as well as biographical and g geographical names appear in separate sections in the back matter.

Webster's New World Dictionary, 3rd coll. ed. Englewood Cliffs, NJ: Prentice-Hall, 1988. This good basic dictionary presents meanings in historical order and indicates frequent usage. Foreign terms and geographical and biographical names are included in the alphabetical listing.

NOTE: The name *Webster,* referring to the great lexicographer Noah Webster, is in the public domain. Because it cannot be copyrighted, it appears in the titles of many dictionaries of varying quality.

19d Surveying Unabridged Dictionaries

In some situations you may need more information than your college dictionary offers. When you are looking for a detailed history of a word or when you want to look up an especially rare usage, you need to consult an unabridged dictionary in the reference section of your library. An **unabridged dictionary** attempts to present a comprehensive survey of all words in a language. Consequently, it gives a wider and more detailed treatment of entries than an abridged dictionary, with listings that may extend over several volumes.

The Random House Dictionary of the English Language, 2nd ed., unabridged. New York: Random House, 1987. This short unabridged dictionary contains about 315,000 entries. In the process of updating the first (1966) edition, the editors have added 50,000 new words and 75,000 new definitions of old words.

Webster's Third New International Dictionary of the English Language. Springfield, MA: Merriam, 1986. This unabridged dictionary contains over 450,000 entries along with illustrations, some in color. Meanings appear in chronological order and are extensively illustrated with quotations. This dictionary does not, however, give much guidance on usage.

The Oxford English Dictionary. New York: Oxford University Press, 1933, 1986. Consisting of twelve volumes plus four supplements, *The Oxford English Dictionary* offers over 500,000 definitions, historically arranged, and 2 million supporting quotations. The quotations begin with the earliest recorded use of a word and progress through each century either until the word becomes obsolete or until its latest meaning is listed. For this reason, many scholars consider *The Oxford English Dictionary* the best place to find the history of a word or to locate illustrations of its usage. However, *The Oxford English Dictionary* emphasizes British usage and does not treat American usage fully.

Figure 2, an excerpt from a long entry in *The Oxford English Dictionary,* illustrates the in-depth coverage offered by an unabridged dictionary.

courage ('kʌrɪdʒ), *sb.* Forms: 4-7 corage, curage, (4-6 corrage, 5 curag, coreage, 6 currage, courra(d)ge, 7 coregg), 5- courage. [ME. *corage,* a. OF. *corage, curage,* later *courage* = Pr. and Cat. *coratge,* Sp. *corage,* It. *coraggio,* a Common Romanic word, answering to a L. type **corāticum,* f. *cor* heart. Cf. the parallel *ætāticum* from *ætāt-em* (AGE); and see -AGE.]

†1. The heart as the seat of feeling, thought, etc.; spirit, mind, disposition, nature. *Obs.*

*c*1300 K. *Alis.* 3559 Archelaus, of proud corage. *c*1386 CHAUCER *Prol.* 11 Smale fowles maken melodie . . So priketh hem nature in here corages. *c*1430 *Pilgr. Lyf Munhode* I. xxxiii. (1869) 20 What thinkest in thi corage? *c*1430 *Stans Puer* 5 To all norture thi corage to enclyne. *c*1500 *Knt.*

Curtesy 407 in Ritson *Met. Rom.* III. 213 in his courage he was full sad. 1593 SHAKS. *3 Hen. VI,* ii. ii. 57 This soft courage makes your Followers faint. 1638 DRUMM. OF HAWTH. *Irene* Wks. (1711) 163 Men's courages were growing hot, their hatred kindled. 1659 B. HARRIS *Parival's Iron Age* 41 The Spaniards . . attacked it with all the force and maistry the greatest courages were able to invent.

†b. *transf.* Of a plant. *Obs.* (Cf. 'To bring a thing into *good heart.*')

*c*1420 *Palladius on Husb.* XI. 90 In this courage Hem forto graffe is goode.

†c. Applied to a person: *cf. spirit. Obs.*

1561 T. HOBY tr. *Castiglione's Courtyer* (1577) Vj b, The prowes of those diuine courages [viz. Marquesse of Mantua, etc.]. 1647 W. BROWNE *Polex.* II. 197 These two great courages being met, and followed by a small companie of the most resolute pirates.

Figure 2 *Excerpt from* Oxford English Dictionary *Entry*

19e   Analyzing Your Vocabulary

Looking up a word in the dictionary is the first step toward learning its meaning and making it part of your vocabulary. In a sense, your vocabulary is composed of all the words you know. But this definition oversimplifies the situation. Actually you have four overlapping vocabularies that you use in various situations.

YOUR FOUR VOCABULARIES

Speaking vocabulary: the words you use in general conversation. For most people, this vocabulary consists of a few hundred words.

Writing vocabulary: the words you use when writing. This vocabulary is considerably larger than your speaking vocabulary, consisting of about 10,000 to 45,000 words. Many of the words in your spoken vocabulary are also part of your written vocabulary. But certain words—such as *satire, analogy, positron,* and *logarithm*—belong almost exclusively to your writing vocabulary.

Reading vocabulary: the words whose meanings you know but that you do not necessarily use in writing or in conversation. Words such as *plutocrat, elucidate,* and *equivocate* might fall into this category.

Guess vocabulary: words whose meanings you do not know exactly but can infer because the words are similar to ones you already know or because their context gives clues to their meanings.

Beyond these four vocabularies are words you do not know and cannot figure out. These words you must look up in a dictionary.

19f   Building a Better Vocabulary

A good vocabulary not only strengthens your performance on written examinations, in papers, and in oral presentations, but also increases your ability to understand reading material and your instructors' statements in class. Without an extensive vocabulary, your ability to learn is limited. Broadly speaking, then, edu-

---

Building a Vocabulary

Writing theorists sometimes make a distinction between a person's *processing vocabulary* (words whose meaning the person understands or infers) and *producing vocabulary* (words actually used in writing or speaking). In a writing class, the instructor asks the student to move from a more familiar role (word processor) to a less familiar role (word producer)—in other words, from reader to writer. To help your students begin to build a vocabulary they can use—a *producing* vocabulary—you should do the following.

1. Impress upon the students that vocabulary building is not a one-time assignment, but the process of a lifetime. They will work throughout their academic careers—and thereafter—to acquire the vocabulary expected of an educated person or a specialist in their field of study.

2. To that end, suggest that students keep a vocabulary journal of new words as a long-term project. Ask them to set up categories that make sense for their own lives; a typical division might include categories for words encountered in each of their courses or fields of study, words found in literature and poetry, words encountered in "popular" reading or speech, and so on.

1. **mountebank:** from Italian *monta in banco,* meaning "he stands on the bench"; refers to the bench stood on by hawkers of medicines at medieval fairs.
2. **pyrrhic:** if referring to the meter, from Greek *purrhikhios,* a war dance; if referring to the victory, from King Pyrrhos, whose army defeated the Romans in 279 B.C. but was nearly destroyed itself.
3. **pittance:** from Middle English *pitaunce,* a donation to a monastery.
4. **protean:** from Proteus, the Greek sea god who could change shape at will.
5. **gargantuan:** from Gargantua, the gigantic king with enormous appetites in Rabelais' farce *Gargantua and Pantagruel.*
6. **cicerone:** after the famous Roman author Cicero.
7. **fathom:** from Middle English *fathme,* meaning "outstretched arms."
8. **gossamer:** from Middle English *gos* and *somer,* referring to the softness of goose feathers plucked in summertime.
9. **rigmarole:** from Middle English *rageman rolle,* a scroll used in a game of chance.
10. **maudlin:** corruption of Magdalen as in Mary Magdalen, who was often portrayed in Middle English literature as a weeping penitent.

cation is the learning of a new vocabulary, the words with which you express new ideas.

Learning new words takes work and, at first, a good deal of time. The following steps, however, can make your task easier.

(1) Become a reader

Reading in itself will not increase your vocabulary. But focusing on words as you read is one of the best ways of learning new words. Seeing words in context, remembering the sentences in which they appear and the ideas with which they are associated, helps you recall them later. Get into the routine of looking up new words as you encounter them and then writing them down along with their meanings. As your vocabulary grows, you will have to do this less often.

(2) Learn the histories of words

Many of the words you encounter have interesting histories, or **etymologies**. Knowing the etymology of a word can help you remember its definition. For example, *cliche,* meaning a worn-out expression, is a French word that refers to a plate used for printing. This meaning suggests the idea of being cast in metal from a mold. Hence a *cliche* is a fixed form of expression.

You can find the history of a word in any good college dictionary, or you can consult a specialized dictionary of etymology. As you build your vocabulary, consider whether the history of a word provides associations that help you remember it.

EXERCISE 4

Using your college dictionary, look up the histories of the following words. How does the history of each word help you remember its definition?

1. mountebank	6. cicerone
2. pyrrhic	7. fathom
3. pittance	8. gossamer
4. protean	9. rigmarole
5. gargantuan	10. maudlin

(3) Become familiar with roots, prefixes, and suffixes

The words you encounter in your college studies are sometimes long and complex. Usually, however, these words can be broken down into smaller units that give you clues to their meanings.

Roots A **root** is a word from which other words are formed. Both *hypodermic* and *dermatologist,* for example, come from the Greek root *derma,* meaning "skin." *Manual* means working by hand, *manuscript* refers to a handwritten draft of a book, and *manufacture* literally means making a product by hand. All these words derive from the Latin root *manus,* meaning "hand."

Many scientific words are based on Latin and Greek roots. *Biograph* contains the Greek root *graph,* meaning "to write," and *vacuum* contains the Latin root *vac,* meaning "empty." *Biology* contains the Greek root *bios,* meaning "life," as do *biosphere, biophysics, bionics,* and *biopsy.* When you come across words that contain *bio,* you already know half their meaning.

Here are some common Latin and Greek roots whose meanings can help you identify and remember new words.

COMMON LATIN ROOTS

Latin Roots	Meanings	Examples
1. æquus	equal	equivocal, equinox
2. amare, amatum	to love	amiable
3. annus	year	annual
4. audire	to hear	audible
5. capere, captum	to take	capture
6. caput	head	caption, capital
7. dicere, dictum	to say, to speak	edict, diction
8. duco, ductum	to lead	aqueduct
9. facere, factum	to make, to do	manufacture
10. loqui, locutum	to speak	eloquence
11. lucere	to light	elucidate, translucent
12. manus	hand	manual, manuscript
13. medius	middle	mediate
14. mittere, missum	to send	admit, permission

continued on the following page

Continuing the Program
Encourage students to feed their interest in the English language by reading and thinking about its history—and its future. For a fascinating look at the development of this hybrid language, suggest students read *The Story of English* by Robert McCrum, William Cran, and Robert McNeil (New York: Penguin, 1987). Based on the PBS television series, the book includes an epilogue on the future of the language, "New Year's Words."

Where is the language heading? What are "next year's words"? Suggest students read and discuss some of Tom Wolfe's collected essays, or the books of columnist William Safire—in particular, *You Could Look It Up.*

continued from the previous page

STUDENTS WORKING TOGETHER

Give your students practice in the "educated guesswork" of inferring word meaning. Write and distribute a series of sentences or paragraphs that include some difficult vocabulary words drawn from the text listings of roots and prefixes/suffixes. As a whole-class exercise, ask students to guess what the words mean and then have volunteers check the guesses in a dictionary.

15. omnis	all	omnipotent
16. plicare, plicatum	to fold	implicate
17. ponere, positum	to place	post, depose
18. portare, portatum	to carry	porter
19. quarere, quaesitum	to ask, to question	inquire
20. rogare, rogatum	to ask	interrogate
21. scribere, scriptum	to write	scribble
22. sentire, sensum	to feel	sense
23. specere, spectrum	to look at	inspect
24. spirare, spiratum	to breathe	inspire, conspire
25. tendere, tentum	to stretch	extend, attend
25. verbum	word	verb, verbiage

COMMON GREEK ROOTS

Greek Roots	Meanings	Examples
1. bios (bio-)	life	biology, biography
2. chronos (chrono-)	time	chronology
3. derma (derma-, -dermic)	skin	dermatologist, hypodermic
4. ethos (ethno-)	race, tribe	ethnic
5. gamos (-gamy, -gamous)	marriage, union	bigamy, bigamous
6. genos (gene-)	race, kind, sex	genetics, genealogy
7. geo-	earth	geology, geography

continued on the following page

continued from the previous page

8. graphein (-graph)	to write	paragraph
9. helios (helio-)	sun, light	heliotrope
10. krates (-crat)	member of a group	plutocrat, democrat
11. kryptos (crypto-)	hidden, secret	cryptic, cryptogram
12. metron (-meter, metro-)	to measure	barometer, metronome
13. morphe (morph)	form	morphology
14. osteon (0steo-)	bone	osteopath, osteomyelitis
15. pathos (patho-, -pathy)	suffering, feeling	sympathy
16. phagein (phag)	to feed, to consume	bacteriophage
17. philos (philo-, -phile)	loving	bibliophile, philosophy
18. phobos (-phobe, phobia)	fear	Anglophobe, claustrophobia
19. photos (photo-)	light	photograph
20. pneuma	wind, air	pneumatic
21. podos (-pod, -poda)	foot	tripod, podiatrist
22. pseudein (pseudo-)	to deceive	pseudonym
23. pyr (pyro-)	fire	pyrotechnical
24. soma	body	psychosomatic
25. tele-	distant	telephone
26. therme (thermo-, -therm)	heat	thermometer

Prefixes A **prefix** is a letter or group of letters put before a root or word that adds to or modifies it. The prefix *anti,* for example, means "against." When combined with other words, it forms new words—*antiaircraft, antibiotic,* and *anticoagulant,* for example.

CLOSE-UP

BUILDING A VOCABULARY

Be careful when making generalizations based on roots, prefixes, and suffixes. Some words appear to have the same root but do not—for instance, *homosexual* (from the Greek *homos*, meaning "same") and *homo sapiens* (from the Latin *homo*, meaning "man").

Here is a list of prefixes which, combined with different roots, form thousands of words.

PREFIXES INDICATING NUMBER

Prefix	Meaning	Example	Definition
uni-	one	unify	To make into a unit
bi-	two	bimonthly	Every two months
duo-	two	duotone	Printed in two tones of the same color
tri-	three	triad	A group of three
quadri-	four	quadruped	A four-footed animal
tetra-	four	tetrachloride	A chemical with four chlorine atoms
quint-	five	quintuplets	Five offspring born in a single birth
pent-	five	pentagon	A five-sided figure
multi-	many	multilateral	Having many sides
mono-	one	monogamy	Having one spouse
poly-	many	polygamy	Having many spouses
omni-	all	omnivore	Eating all kinds of food

PREFIXES INDICATING SMALLNESS

Prefix	Meaning	Example	Definition
micro-	small	microscope	An instrument for observing small things
mini-	small	minibus	A small bus

PREFIXES INDICATING SEQUENCE AND SPACE

Prefix	Meaning	Example	Definition
ante-	before	antebellum	Before the war
pre-	before	prehistory	Before history
intro-	within	introspective	Looking into oneself
post-	after	postscript	A message written after the body of the letter
re-	back, again	review	To look at again
sub-	under	submarine	An underwater ship
super-	above	supervise	To look over the performance of others
inter-	between	international	Between nations
intra-	within	intramural	Within the bounds of an institution
in-	in, into	incorporate	To form into a body
ex-	out, from	exhale	To breathe out
circum-	around	circumnavigate	To sail or fly around (the earth, an island, etc.)
con-	with, together	congregate	To come or bring together

PREFIXES INDICATING NEGATION

Prefix	Meaning	Example	Definition
non-	not	nonpartisan	Not affiliated
in-	not	inactive	Not active
un-	not, the opposite of, against	unequal	Not equal
anti-	against	antiseptic	Free from germs
counter-	opposing	countermand	To revoke an order with another order
contra-	against	contradict	To speak against
dis-	not, the opposite of	dislike	To not like
mis-	wrong, ill	mislead	To give bad advice
mal-	bad, wrong, ill	malformed	Incorrectly shaped
pseudo-	false	pseudonym	A false name

Suffixes **Suffixes** are syllables added to the end of a word or root that change its part of speech. For example, suffixes added to the verb *believe* form two nouns, an adjective, and an adverb.

believe	(verb)
believer	(noun)
believability	(noun)
believable	(adjective)
believably	(adverb)

VERB SUFFIXES

Suffix	Meaning	Example
-en	to cause to or become	cheapen, redden
-ate	to cause to be	activate, animate
-ify, fy	to make or cause to be	fortify, magnify
-ize	to make, to give, to practice	memorize, modernize

ADVERB SUFFIXES

The only regular suffix for adverbs is *-ly*, as in *slowly*, *wisely*, and *casually*.

ADJECTIVE SUFFIXES

Suffix	Meaning	Example
-al	capable of, suitable for	comical
-ial	pertaining to	managerial
-ic	pertaining to	democratic
-ly	resembling	sisterly
-ly	at specific intervals	hourly
-ful	abounding in	colorful
-ous, -ose	full of	porous, verbose
-ive	quality of	creative, adaptive
-less	lack of, free of	toothless
-ish	having the qualities of, preoccupied with	childish, bookish

NOUN SUFFIXES

Suffix	Meaning	Example
-ance, -ence	quality or state of	insurance, competence
-acy	quality or state of	piracy, privacy
-or	one who performs an action	actor
-arium, -orium	place for	aquarium, auditorium
-ary	place for, pertaining to	dictionary
-cide	kill	suicide, homicide
-icle, -cle	a diminutive ending	icicle, corpuscle
-hood	state or condition of	childhood

continued on the following page

345

continued from the previous page

-ism	quality or doctrine of	Marxism, conservatism
-ity	quality or state of	acidity
-itis	inflammation of	appendicitis
-ics	the science or art of	economics
-ment	act or condition of	resentment
-mony	resulting condition	testimony
-ology	the study of	biology, psychology

(4) Learn words according to a system

Learning related words according to a system is far more effective than memorizing words at random. Consider the following pair of words.

duct	a tubular passage
aqueduct	a conduit designed to transport water

Both words share the Latin root *ducere* (to lead). Because you are already familiar with the word *duct,* you have a clue to the meaning of *aqueduct.* If you know that *aqua* is a Latin word meaning "water," you can easily remember the definition of the word.

Other words also share the root *ducere.*

conduct	to direct the course of
induct	to install, to admit as a member
viaduct	a series of spans used to carry a railroad over a valley or other roads
abduct	to carry off
deduct	to take away
ductile	capable of being fashioned into a new form

Learning these related words together would clearly be easier than learning them at random.

See ◀
20c4
Other word groupings also facilitate learning. You can, for example, study together words that are confusing beause they sound so much alike.

imminent	about to occur
eminent	prominent

ascent	a rise
assent	agreement

You can also study together words that are confusing because they look somewhat alike.

marital	referring to marriage
martial	referring to war

descent	a downward movement
decent	characterized by good taste or morality

Finally, you can study together words that are confusing because their meanings are so closely associated with each other.

imply	to suggest
infer	stated outright

explicit	to conclude
implicit	implied, unsaid

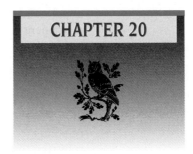

CHAPTER 20

Improving Spelling

Spelling errors distract readers and make it difficult for them to understand what you are trying to say. In some cases, incorrectly spelled words misrepresent your meanings, as when you use *equivalents* for *equivalence* or *benzene* for *benzoin.*

If spelling has always given you trouble, you will probably not become a good speller overnight. But the situation is not hopeless. Most people can spell even difficult words "almost" correctly; usually only a letter or two is wrong. For this reason, memorizing a few simple rules and their exceptions and learning the correct spelling of the most commonly misspelled words can make a big difference.

20a Understanding Spelling and Pronunciation

Sound alone does not necessarily indicate a word's spelling. For instance, *gh* is silent in *light* but pronounced *f* in *cough;* the $\overline{o}$ sound is spelled differently in *mow, toe, though, sew,* and *beau.* These and other inconsistencies create a number of problem areas to watch for.

(1) Vowels in unstressed positions

Many unstressed vowels sound exactly alike when we say them. For instance, it is hard to tell from pronunciation alone that the *i* in *terrible* is not an *a.* In general, the vowels *a, e,* and *i* See ◄ 20b7 are impossible to distinguish in the suffixes *-able* and *-ible, -ance* and *-ence,* and *-ant* and *-ent.*

comfort*able*	brilli*ance*	serv*ant*
compat*ible*	excell*ence*	independ*ent*

Memorize such words and use a dictionary or spell checker whenever you are uncertain about how to spell a word with an unstressed vowel.

(2) Silent letters

Some English words contain silent letters. The *b* in *climb* is silent, as is the *t* in *mortgage*. Silent letters at the beginning of a word are especially bothersome: You cannot look up the spelling of *gnu* (pronounced *new*) in a dictionary if you do not already know that it begins with a *g*. The spelling of words with silent letters follows no rules, so you have to memorize such words as you encounter them.

aisle	depot	silhouette
condemn	knight	sovereign
climb	pneumonia	

(3) Words that are often pronounced carelessly

Most people pronounce words rather carelessly in everyday speech. Consequently, when spelling, they leave out, add, or transpose letters. The following words are often misspelled because they are pronounced incorrectly.

literature	recognize	probably
February	nuclear	specific
candidate	environment	surprise
library	disastrous	supposed to
government	hundred	used to
quantity	lightning	

(4) Variant forms of the same word

Spelling problems can occur when the verb and noun forms of a word are spelled differently.

advise (v)	advice (n)
renounce (v)	renunciation (n)
announce (v)	annunciation (n)
describe (v)	description (n)
omit (v)	omission (n)

In addition, some words are spelled one way in the United States and another way in Great Britain and the Commonwealth

continued from previous page

Spelling for the Millions.) Possible avenues of exploration might include the following.

- The gap between sound and symbol in English. Over the centuries, we have made drastic changes in the way we *say* words—but their spellings have, in many cases, remained the same. The *k* in *knight* and the *gh* in *right*, *tight*, and *sight* were once pronounced.

- The insistence of 16th- and 17th-century printers on standardized spelling, which, in effect, "froze" the spelling of words at a time when their pronunciation was changing radically. In Chaucer's time, the word *moon* was pronounced *moan* and the word *house*, *hoose*; in Shakespeare's time, the word *reason* was pronounced *raisin*. (Many of these archaic pronunciations survive in English dialects in Great Britain, Scotland, Ireland, and the United States.)

- The effect of determined lexicographers on spelling standards. In this country, Noah Webster "Americanized" the spelling of many words by altering the English word endings *-our* (*honour*, *colour*), *-re* (*centre*, *theatre*), and *-ck* (*magick*).

nations. If you have trouble with these, consult a dictionary to determine the preferred American spelling.

American	*British*
color	colour
defense	defence
judgment	judgement
theater	theatre
traveled	travelled

(5) Homophones

See ◀
20c4

Homophones are words—such as *accept* and *except*—that are pronounced alike but spelled differently.

20b Learning Spelling Rules

The few reliable rules that govern English spelling can help you overcome the general inconsistency between pronunciation and spelling. These rules have exceptions, but they are still useful guides.

(1) The *ie/ei* combinations

The old rule still stands: Use *i* before *e* except after *c* or when pronounced *ay* as in *neighbor*.

i before *e*	*ei* after *c*	*ei* pronounced *ay*
belief	ceiling	neighbor
chief	conceit	weigh
field	deceit	freight
niece	receive	eight
piece	perceive	
friend		

There are a few exceptions to this rule: *either, neither, foreign, leisure, weird,* and *seize.* In addition, if the *ie* combination is not pronounced as a unit, the rule does not apply: *atheist, science.*

EXERCISE 1

Fill in the blanks with the proper *ie* or *ei* combination. After completing the exercise, use your dictionary to check your answers.

EXAMPLE: conc_ei_ve

1. rec____pt
2. var____ty
3. caff____ne
4. ach____ve
5. kal____doscope

6. misch____f
7. effic____nt
8. v____n
9. spec____s
10. suffic____nt

1. receipt
2. variety
3. caffeine
4. achieve
5. kaleidoscope
6. mischief
7. efficient
8. vein
9. species
10. sufficient

(2) Doubling final consonants

Some words double their final consonants before a suffix that begins with a vowel (*-ed, -ing*); others do not. Fortunately, there is a rule to distinguish them. The only words that double their consonants in this situation are those that meet the following criteria.

1. They have one syllable or are stressed on the last syllable.
2. They contain only one vowel in the last syllable.
3. They end in a single consonant.

The word *tap* satisfies all three conditions: It has only one syllable, it contains only one vowel (*a*), and it ends in a single consonant (*p*). Therefore, the final consonant doubles before a suffix beginning with a vowel (*tapped, tapping*). The word *relent* meets two of the above criteria (it is stressed on the last syllable and it has one vowel in the last syllable), but it does not end in a single consonant. Therefore, its final consonant is not doubled (*relented, relenting*).

(3) Prefixes

The addition of a prefix never affects the spelling of the root.

un + acceptable = unacceptable
dis + agree = disagree
mis + spell = misspell
dis + joint = disjoint

▶ See
19f 3

Some prefixes can cause spelling problems because they are easily confused or because they are pronounced alike although they are not spelled alike. Be especially careful of the prefixes *ante-/anti-, en-/in-, per-/pre-,* and *de-/di-*.

antebellum	antiaircraft
encircle	integrate
perceive	prescribe
deduct	direct

351

ANSWERS TO EXERCISE 2

1. surprising
2. surely
3. forcible
4. manageable
5. duly
6. outrageous
7. serviceable
8. awful
9. shaming
10. shameless

ANSWERS TO EXERCISE 3

1. journeying
2. studied
3. carrying
4. shyly
5. studying
6. sturdiness
7. merriment
8. likelihood
9. plentiful
10. supplier

TEACHING STRATEGY

Latin-Root Words

As the text notes, many English nouns form plurals in exceptional ways. Writers often have particular trouble with Latin-root words that follow singular-plural rules different from those of common English nouns. Today, English grammar seems to be in a "transitional mode," and many Latin-root words now have two acceptable plural endings. (For example, both the traditional *appendices* and the more recent *appendixes* are now acceptable plural forms for *appendix*.) Review the irregular plural endings of the Latin-root words below.

continued on following page

(4) Silent *e* before a suffix

When a suffix that starts with a consonant is added to a word ending in silent *e*, the *e* is generally kept: *hope/hopeful; lame/lamely; bore/boredom.*

Familiar exceptions include *argument, truly, ninth, judgment,* and *abridgment.*

When a suffix that starts with a vowel is added to a word ending in silent *e*, the *e* is generally dropped: *hope/hoping; trace/traced; grieve/grievance; love/lovable.*

Familiar exceptions include *changeable, noticeable,* and *courageous.* In these cases the *e* is kept so the *c* or *g* will be pronounced like the initial consonants in *cease* or *gem* and not like the initial consonants in *come* or *game.*

EXERCISE 2

Combine the following words with the suffixes in parentheses. Determine whether to keep or drop the silent *e;* be prepared to explain your choice.

> **EXAMPLE:** fate (al)
>
> fatal

1. surprise (ing)
2. sure (ly)
3. force (ible)
4. manage (able)
5. due (ly)
6. outrage (ous)
7. service (able)
8. awe (ful)
9. shame (ing)
10. shame (less)

(5) *y* before a suffix

When a word ends in a consonant plus *y*, the *y* generally changes to an *i* when a suffix is added (beauty + ful = beautiful). The *y* is retained, however, when the suffix *-ing* is added (tally + ing = tallying). It is also retained when the *y* ends a proper name (McCarthy + ite = McCarthyite) and in some one-syllable words (dry + ness = dryness). Exception: city + scape = cityscape.

When a word ends in a vowel plus *y*, the *y* is retained (joy + ful = joyful; employ + er = employer). Exception: day + ly = daily.

EXERCISE 3

Add the endings in parentheses to the following words. Change or keep the final *y* as you see fit; be prepared to explain your choice.

> **EXAMPLE:** party (ing)
>
> partying

Spell 20b

1. journey (ing)
2. study (ed)
3. carry (ing)
4. shy (ly)
5. study (ing)
6. sturdy (ness)
7. merry (ment)
8. likely (hood)
9. plenty (ful)
10. supply (er)

(6) *seed* endings

Endings with the sound *seed* are nearly always spelled *cede,* as in *precede, intercede, concede,* and so on. The only exceptions are *supersede, exceed, proceed,* and *succeed.*

(7) -*able,* -*ible*

These endings sound alike, and they often cause spelling problems. Fortunately, there is a rule that can help you distinguish them. If the stem of a word is itself an independent word, the suffix -*able* is most commonly used. If the stem of a word is not an independent word, the suffix -*ible* is most often used.

*comfort*able	*compat*ible
*agree*able	*incred*ible
*dry*able	*plaus*ible

(8) Plurals

Most nouns form plurals by adding -*s.*

savage/savages	tortilla/tortillas
girl/girls	gnu/gnus
boat/boats	taxi/taxis

There are, however, a number of exceptions.

***Words Ending in* f *or* fe** Some words ending in *f* or *fe* form plurals by changing the *f* to *v* and adding -*es* or -*s.*

knife/knives	wife/wives
life/lives	self/selves

Other words ending in *f* or *fe* just add -*s.*

belief/beliefs	proof/proofs

A few such words can form plurals either by adding -*s* or by substituting -*ves* for *f.*

scarf/scarfs/scarves	hoof/hoofs/hooves

Words ending in double *f* take -*s* to form plurals (*tariff/tariffs*).

---

continued from previous page

Singular	Plural
alga	algae
analysis	analyses
antenna	antennas (TV); antennae (bugs)
apparatus	apparatus
bacterium	bacteria
criterion	criteria
curriculum	curriculums *or* curricula
datum	data (see below)
focus	focuses *or* foci
genus	genera
index	indexes (pub.); indices (math)
matrix	matrixes *or* matrices
medium	media
radius	radii
vertebra	vertebrae

The fading controversy surrounding the use of *data* is a good way to illustrate the effect of "popular demand" on the rules of grammatical usage. William Safire: "I think 'the data is' is widely accepted and 'the data are' is passé. . . . 'The data shows' comes more readily to the tongue than 'the data show.'" The correct singular form of the word, *datum,* has fallen almost completely from common usage.

Ask students to come up with other examples of homophone pairs from their own experience or from a dictionary.

Sound-alike verbs can be a source of trouble for students who are not familiar with the written form of each word. As an additional exercise, write some or all of the following sentences on the board and ask students to tell you what's wrong.

The criminal <u>alluded</u> his pursuers.

The law will <u>persecute</u> the guilty.

I will <u>except</u> your wonderful offer.

My parents have <u>ensured</u> my car.

As a teacher, he <u>effects</u> many lives.

I want to <u>complement</u> you on your speech.

The Senate voted to <u>censor</u> their colleague for his actions.

Words Ending in y   Most words that end in a consonant followed by *y* form plurals by changing the *y* to *i* and adding *-es*.

baby/babies   blueberry/blueberries
seventy/seventies worry/worries

Proper nouns, however, are exceptions: the *Kennedys* (never the *Kennedies*).

Words that end in a vowel followed by a *y* form plurals by adding *-s*.

monkey/monkeys   key/keys
turkey/turkeys   day/days

Words Ending in o   Words that end in a vowel followed by *o* form the plural by adding *-s*.

radio/radios   stereo/stereos   zoo/zoos

Most words that end in a consonant followed by *o* add *-es* to form the plural.

tomato/tomatoes   hero/heroes   potato/potatoes

Some, however, add *-s*.

silo/silos   piano/pianos
memo/memos   soprano/sopranos

Still other words that end in *o* add either *-s* or *-es* to form plurals.

memento/mementos/mementoes
mosquito/mosquitos/mosquitoes

Words Ending in s, ss, sh, ch, x, *and* z   These words form plurals by adding *-es*.

Jones/Joneses   rash/rashes   box/boxes
mass/masses   lunch/lunches   buzz/buzzes

NOTE: Some one-syllable words that end in *s* or *z* double their final consonants when forming plurals (*quiz/quizzes*).

Compound Nouns   Compound nouns—nouns formed from two or more words—usually conform to the rules governing the last word in the compound construction.

welfare state/welfare states   snowball/snowballs

However, in compound nouns where the first element of the construction is more important than the others, the plural is formed with the first element (*sister-in-law/sisters-in-law; attorney general/attorneys general*).

Irregular Plurals Some words in English have irregular plural endings. No rules govern these plurals, so you have to memorize them.

child/children	ox/oxen
woman/women	louse/lice
goose/geese	mouse/mice

Foreign Plurals Some words, especially those borrowed from Latin or Greek, keep their foreign plurals. When you use these words, you must look up their plural forms in your college dictionary if you do not know them.

Singular	*Plural*
criterion	criteria
larva	larvae
memorandum	memoranda
stimulus	stimuli

Some foreign words have a regular English plural as well as the one from their language of origin.

Singular	*Plural*
hippopotamus	hippopotami, hippopotamuses
antenna	antennae, antennas

No Plural Forms A few words use the same form for both the singular and the plural.

Singular	*Plural*
apparatus	apparatus
deer	deer
fish	fish (*also* fishes)
sheep	sheep
species	species

20c Developing Spelling Skills

To develop good spelling skills, you must invest time and effort. In addition to studying the rules outlined above, you can do the following things to help yourself become a better speller.

Developing Spelling Skills
Use some class time to let students create *mnemonic* devices (Mnemosyne was the Greek goddess of memory) to help them remember words they find hard to spell. Some examples follow.

Right:	all right
Wrong:	alright
Mnemonic clue:	You'd say *all wrong,* so. . . .
Right:	cemetery
Wrong:	cemetary
Mnemonic clue:	"EEEK!" It's all e's!
Right:	embarrass
Wrong:	embarass
Mnemonic clue:	It makes you want to hide in a *baRRel.*

Suggest that students keep an ongoing spelling list in a journal.

Using Spell Checkers Effectively
Whether it's "good" or "bad" for students to use spell checkers, whether spell checkers cause spelling skills to atrophy—to a great extent, these are irrelevant questions. Spell checkers are here to stay. What you can do, however, is ensure that your students use them wisely and well.

Students need to understand that a spell checker doesn't find "mistakes"; it simply finds character strings that it doesn't recognize. Since a spell checker won't recognize *there* in "They forgot there books on the table" as incorrect, nor spot the typo in "Whatever he wanted to do, he dad," students need to understand that they still must proofread their papers carefully.

CLOSE-UP

IMPROVING SPELLING

If you write with a computer and use a spell checker, remember that spell checkers will not identify a word that might be spelled correctly but used incorrectly—*then* for *than* or *its* for *it's*, for example—or a typo that results in another word such as *form* for *from*. For this reason you should carefully proofread your papers even after you have run a spell check.

(1) Make your own spelling list

Compile a list of your own problem words. When you write a first draft, circle any words whose spelling you are unsure of. Then, look them up in your dictionary as you revise and add them all (even those you have spelled correctly) to your list. When your instructor returns a paper, record any words you have misspelled. In addition, record problem words that you encounter when reading, including those from class notes and textbooks. It is especially important that you master the spellings of words that are basic to a course or field of study.

(2) Uncover patterns of misspelling

Review your spelling list to identify any patterns of misspelling. Do you consistently have a problem with plurals or with *-ible/-able* endings? If so, review the spelling rules that apply to these particular problems. By using this strategy, you can eliminate the need to memorize single words.

(3) Fix each word in your mind

Once you isolate the words you consistently misspell, think of associations that will help fix the correct spellings in your mind. For example, you can arrive at the correct spelling of *definite* (often misspelled *definate*) by remembering that it contains the word *finite*, which suggests the concept of limit, as does *definite*. You can recall the *a* in *brilliance* (often misspelled *brillience*) by remembering that brilliant people often get *A*'s in their classes. You can master the spelling of *criticism* by remembering that it contains the word *critic*.

Another way of fixing words in your mind is to write them down. When you review your spelling list, do not just *read* the words on it; *write* them. This repeated copying will help you remember the correct spellings.

(4) Learn to distinguish commonly confused words

Following is a list of commonly confused **homophones** (words that sound exactly alike but have different spellings and meanings, such as *night* and *knight*) and words that create problems because they sound similar (for example, *accept* and *except*). It is a good idea to take a group of these pairs of words—say, ten each day—and learn them.

accept	to receive
except	other than
advice	recommendation
advise	to recommend
affect	to have an influence on (*verb*); an emotional response (*noun*)
effect	result (*noun*); to cause (*verb*)
all ready	prepared
already	by or before this or that time
allude	to refer to indirectly
elude	to avoid
allusion	indirect reference
illusion	false belief or perception
ascent	movement upward
assent	agreement
bare	uncovered
bear	to carry (*verb*); an animal (*noun*)
board	a wooden plank (*noun*); to get on an airplane, etc. (*verb*)
bored	uninterested
born	brought into life
borne	carried
brake	device for stopping
break	destroy, smash

357

buy	purchase
by	next to; near
capital	the seat of government; monetary assets
capitol	government building
cite	to quote, refer to
sight	the ability to see
site	a place
coarse	rough
course	path; class
complement	to complete or add to (*verb*); something that completes (*noun*)
compliment	praise
conscience	sense of right and wrong
conscious	mentally awake
council	governing body
counsel	advice (*noun*); to give advice (*verb*)
descent	downward movement
dissent	disagreement
desert	to abandon
dessert	sweet course at the end of a meal
device	an implement; a plan
devise	to invent
die	to lose life
dye	to change the color of something
discreet	reserved
discrete	individual, distinct
elicit	to draw out, evoke
illicit	unlawful; forbidden
eminent	prominent
immanent	inherent
imminent	about to happen
fair	equitable; light-complected
fare	a fee for transportation

forth	forward
fourth	referring to the number 4
gorilla	the animal
guerilla	a type of soldier or warfare
hear	to perceive by ear
here	in this place
heard	past tense of *hear*
herd	group of animals
hole	an opening
whole	complete
its	possessive of *it*
it's	contraction of *it is*
later	after a time
latter	the last in a series
lead	to guide or direct (*verb*); the metal (*noun*)
led	past tense of *lead*
lessen	to reduce
lesson	something learned
loose	not tight; unbound
lose	to misplace
maybe	perhaps
may be	might be
meat	flesh
meet	encounter
no	negative
know	to be certain
passed	past tense of *pass*
past	a previous time; a time gone by
patience	forbearance
patients	persons receiving medical care
peace	the absence of war; quiet
piece	a portion of something

persecute	to harass or worry
prosecute	to institute criminal proceedings against
personal	private; one's own
personnel	employees
plain	unadorned
plane	an aircraft; a carpenter's tool
precede	to come before
proceed	to continue
principal	most important (*adjective*); head of a school (*noun*)
principle	a basic truth; rule of conduct
quiet	silent
quite	very
rain	precipitation
reign	to rule
rein	a strap, e.g., for a horse's bridle
raise	to build up
raze	to tear down
right	correct
rite	a ritual
write	to put words on paper
road	street, highway
rode	past tense of *ride*
scene	place of action; section of a play
seen	viewed
sense	perception, understanding
since	from a time in the past up to the present
stationary	standing still
stationery	writing paper
straight	unbending
strait	a water passageway
than	as compared with
then	at that time; next

their	possessive of *they*
there	in that place
they're	contraction of *they are*
through	finished; into and out of
threw	past tense of *throw*
thorough	complete
to	toward
too	also; more than sufficient
two	the number
waist	the middle of the body
waste	discarded material (*noun*); to squander (*verb*)
weak	not strong
week	seven days
weather	atmospheric conditions
whether	in either case
which	one of a group
witch	female sorcerer
who's	contraction of *who is*
whose	possessive of *who*
your	possessive of *you*
you're	contraction of *you are*

CHAPTER 21

Identifying the Parts of Speech

The eight basic parts of speech—the building blocks for all English sentences—are *nouns, pronouns, verbs, adjectives, adverbs, prepositions, conjunctions,* and *interjections.* The part of speech to which a word belongs depends on its function in a sentence.

 21a Nouns

Nouns name people, places, things, ideas, actions, or qualities.

A **common noun** names any of a class of people, places, or things: *artist, judge, building, event, city.*

A **proper noun**, always capitalized, refers to a particular person, place, or thing: *Mary Cassatt, Learned Hand, World Trade Center, Crimean War, St. Louis.*

See ◄ 33b

A **mass noun** names a quantity that is not countable: *time, dust, work, gold.* Mass nouns are generally treated as singular.

A **collective noun** designates a group of people, places, or things thought of as a unit: *committee, class, navy, band, family.* Collective nouns are generally singular unless the members of the group are referred to as individuals.

See ◄ 24a5

An **abstract noun** refers to an intangible idea or quality: *love, hate, justice, anger, fear, prejudice.*

 21b Pronouns

Pronouns are words used in place of nouns. The noun for which a pronoun stands is called its **antecedent**. There are eight
364

types of pronouns. Different types of pronouns may have exactly the same forms, but they are distinguished by their functions in the sentence.

A **personal pronoun** stands for a person or thing: *I, me, we, us, my, mine, our, ours, you, your, yours, he, she, it, its, him, his, her, hers, they, them, their, theirs.*

They made her an offer she couldn't refuse.

An **indefinite pronoun** functions in a sentence as a noun but does not refer to any particular person or thing. For this reason, indefinite pronouns do not require antecedents. Indefinite pronouns include *another, any, each, few, many, some, nothing, one, anyone, everyone, everybody, everything, someone, something, either,* and *neither.*

Many are called, but few are chosen.

A **reflexive pronoun** ends with -*self* and refers to a recipient of the action that is the same as the actor: *myself, yourself, himself, herself, itself, oneself, themselves, ourselves, yourselves.*

They found themselves in downtown Pittsburgh.

Intensive pronouns have the same form as reflexive pronouns; an intensive pronoun emphasizes a preceding noun or pronoun.

Darrow himself was sure his client was innocent.

A **relative pronoun** introduces an adjective or noun clause in a sentence: *which, who, whom, that, what, whose, whatever, whoever, whomever, whichever.*

Gandhi was the charismatic man who helped lead India to independence. (introduces adjective clause)

Whatever happens will be a surprise. (introduces noun clause)

An **interrogative pronoun** introduces a question: *who, which, what, whom, whose, whoever, whatever, whichever.*

Who was that masked man?

A **demonstrative pronoun** points to a particular thing or group of things: *this, that, these, those.*

This is one of Shakespeare's early plays.

A **reciprocal pronoun** denotes a mutual relationship: *each other, one another.*

Each other generally indicates a relationship between two individuals; *one another* generally denotes a relationship among more than two.

TEACHING STRATEGY

Ask students to find a recent piece of their own writing and to rewrite one or two paragraphs without using any pronouns. Discuss the impact on the fluidity and "readability" of their work.

Romeo and Juliet declared their love for each other.
Concertgoers jostled one another in the ticket line.

21c   Verbs

(1)   Recognizing verbs

A verb may express either action or a state of being.

He ran for the train. (action)

Elizabeth II became queen after the death of her father, George VI. (state of being)

Verbs can be classified into two groups: *main verbs* and *auxiliary verbs*.

Main Verbs   A **main verb** carries most of the meaning in the sentence or clause in which it appears.

Bulfinch's *Mythology* contains a discussion of Greek mythology.

Emily Dickinson anticipated much of twentieth-century poetry.

A main verb is a **linking verb** when it is followed by a **subject complement**, a word or phrase that defines or describes the subject.

Carbon disulfide smells bad.

COMMONLY USED LINKING VERBS			
appear	believe	prove	smell
be	feel	remain	taste
become	grow	seem	turn

Auxiliary Verbs   Auxiliary verbs, such as *be* and *have*, combine with main verbs to form **verb phrases**. The auxiliary verbs, also called **helping verbs**, indicate tense, voice, or mood.

[auxiliary] [main verb]   [auxiliary] [main verb]

The train has started. We are leaving soon.

[verb phrase]   [verb phrase]

Certain auxiliary verbs, known as **modal auxiliaries**, indicate necessity, possibility, willingness, obligation, or ability.

In the near future farmers might cultivate seaweed as a food crop.

Coal mining would be safer if dust were controlled in the mines.

MODAL AUXILIARIES

can	might	ought [to]	will
could	must	shall	would
may	need [to]	should	

---

(2) Recognizing verbals

Verbals, such as *known* or *running* or *to go*, are verb forms that act as adjectives, adverbs, or nouns. A verbal can never serve as a sentence's verb. Verbals include *participles, infinitives,* and *gerunds.*

Participles   Virtually every verb has a **present participle**, which ends in *-ing (loving, learning, going, writing),* and a **past participle**, which usually ends in *-d* or *-ed (agreed, learned).* Some verbs have irregular past participles *(gone, begun, written).* Participles may function in a sentence as adjectives or as nouns.

▶ See 23a2

Twenty brands of running shoes were displayed at the exhibition. (Present participle *running* serves as adjective modifying noun *shoes.*)

The crowded bus went right by those waiting at the corner. (Past participle *crowded* serves as adjective modifying noun *bus.*)

The wounded were given emergency first aid. (Past participle *wounded* serves as sentence's subject.)

Infinitives   An **infinitive**—the base form of the verb preceded by *to*—may serve as an adjective, an adverb, or a noun.

Ann Arbor was clearly the place to be. (Infinitive *to be* serves as adjective modifying noun *place.*)

They say that breaking up is hard to do. (Infinitive *to do* serves as adverb modifying adjective *hard.*)

Carla went outside to think. (Infinitive *to think* serves as adverb modifying verb *went.*)

To win was everything. (Infinitive *to win* serves as sentence's subject.)

Gerunds   **Gerunds**, special forms of verbs ending in *-ing*, are always used as nouns.

Splitting Infinitives
When the subject of infinitives comes up, so do students' hands: "When is it OK to split an infinitive?" Note that while infinitives should, in most cases, remain intact, they may be split (preferably by just one word) whenever it will improve the clarity of meaning or "readability" of a sentence (see 15a4). Of the two sentences below, for example, the second (with a split infinitive) has a more natural, readable style.

Their objective is almost to double coffee consumption in Russia in this decade.

Their objective is to almost double coffee consumption in Russia in this decade.

"Avoid the split infinitive wherever possible," advises Eric Partridge in *Usage and Abusage*, "but if it is the clearest and the most natural construction, use it boldly. The angels are on our side."

Conjunctive Adverbs
Words can tell readers what to expect: *but* or *however* warns that you are changing direction; *therefore* spotlights a conclusion. The list of transitional words and phrases in 4d2 can help students use appropriate transitions.

<u>Seeing</u> is <u>believing</u>. (Gerund *seeing* serves as sentence's subject; gerund *believing* serves as subject complement.)

He worried about <u>interrupting</u>. (Gerund *interrupting* is object of preposition *about*.)

Andrew loves <u>skiing</u>. (Gerund *skiing* is direct object of verb *loves*.)

CLOSE-UP   **IDENTIFYING THE PARTS OF SPEECH**

When the *-ing* form of a verb is used as a noun, it is considered a *gerund*; when it is used as a modifier, it is a *present participle*.

21d   Adjectives

Adjectives are words that describe, limit, qualify, or in any other way modify nouns or pronouns.

Descriptive adjectives, the largest class of adjectives, name a quality of the noun or pronoun they modify.

After the game, they were <u>exhausted</u>.

They ordered a <u>chocolate</u> soda and a <u>butterscotch</u> sundae.

Some descriptive adjectives are formed from common nouns or from verbs *(friend/friendly, agree/agreeable)*. Others, called **proper adjectives**, are formed from proper nouns.

Eubie Blake was a talented <u>American</u> musician who died in 1983.

The <u>Shakespearean</u> or <u>English</u> sonnet consists of an octave and a sestet.

See ◄
35b1
Two or more words may be joined, with or without a hyphen, to form a **compound adjective** *(foreign born, well-read)*.

Articles, pronouns, and numbers are considered adjectives when they limit or qualify nouns.

Articles (*a, an, the*)

<u>The</u> boy found <u>a</u> four-leaf clover.

Possessive adjectives (the personal pronouns *my, your, his, her, its, our, their*)

<u>Their</u> lives depended on <u>my</u> skill.

Demonstrative adjectives (*this, these, that, those*)

<u>This</u> song reminds me of <u>that</u> song we heard yesterday.

Interrogative adjectives (*what, which, whose*)

<u>Whose</u> book is this?

Indefinite adjectives (*another, each, both, many, any, some,* and so on)

<u>Both</u> candidates agreed to return <u>another</u> day.

Relative adjectives (*what, whatever, which, whichever, whose, whosever*)

I forgot <u>whatever</u> reasons I had for leaving.

Numerical adjectives (*one, two, first, second,* and so on)

The <u>first</u> time I played I only got <u>one</u> hit.

21e   Adverbs

Adverbs describe the action of verbs or modify adjectives, other adverbs, or complete phrases, clauses, or sentences. They answer the questions "How?" "Why?" "Where?" "When?" and "To what extent?"

He walked <u>rather hesitantly</u> toward the front of the room.

It seems <u>so</u> long since we met <u>here yesterday</u>.

<u>Unfortunately</u>, the program didn't run.

Cody's brother and sister were <u>staggeringly</u> unobservant. (Anne Tyler, *Dinner at the Homesick Restaurant*)

Interrogative adverbs—the words *how, when, why,* and *where*—introduce questions.

<u>Why</u> did the compound darken?

Conjunctive adverbs join and relate independent clauses.
Conjunctive adverbs may appear in any of several positions in a sentence.

Jason forgot to register for chemistry. <u>However</u>, he managed to sign up during the drop/add period. (conjunctive adverb placed at beginning of sentence)

Jason forgot to register for chemistry; however, he managed to sign up during the drop/add period. (conjunctive adverb placed at beginning of clause)

Jason forgot to register for chemistry. He managed, however, to sign up during the drop/add period. (conjunctive adverb placed within sentence)

Jason forgot to register for chemistry. He managed to sign up during the drop/add period, however. (conjunctive adverb placed at end of sentence)

FREQUENTLY USED CONJUCTIVE ADVERBS

accordingly	furthermore	meanwhile	similarly
also	hence	moreover	still
anyway	however	nevertheless	then
besides	incidentally	next	thereafter
certainly	indeed	nonetheless	therefore
consequently	instead	now	thus
finally	likewise	otherwise	undoubtedly

FREQUENTLY USED PREPOSITIONS

about	beneath	inside	since
above	beside	into	through
across	between	like	throughout
after	beyond	near	to
against	by	of	toward
along	concerning	off	under
among	despite	on	underneath
around	down	onto	until
as	during	out	up
at	except	outside	upon
before	for	over	with
behind	from	past	within
below	in	regarding	without

21f Prepositions

A **preposition** introduces a word or word group consisting of one or more nouns or pronouns or of a phrase or clause functioning in the sentence as a noun. The word or word group the preposition introduces is called its **object**.

prep   obj   prep   obj
They received a postcard from Bobby telling about his trip

prep   obj

to Canada.

21g Conjunctions

Conjunctions are words used to connect words, phrases, clauses, or sentences.

Coordinating conjunctions *(and, or, but, nor, for, so, yet)* connect words, phrases, or clauses of equal weight.

He had to choose pheasant or venison. (*Or* links two nouns.)

. . . of the people, by the people, and for the people (*And* links three prepositional phrases.)

Thoreau wrote *Walden* in 1854, and he died in 1862. (*And* links two independent clauses.)

Correlative conjunctions, always used in pairs, also link items of equal weight.

▶ See 9a3

FREQUENTLY USED CORRELATIVE CONJUNCTIONS

both . . . and	neither . . . nor
either . . . or	not only . . . but also
just as . . . so	whether . . . or

Both Hancock and Jefferson signed the Declaration of Independence. (Correlative conjunctions link two nouns.)

Either I will renew my lease, or I will move. (Correlative conjunctions link two independent clauses.)

TEACHING STRATEGY

Prepositions

"The proper preposition," said former *New York Times* editor Theodore M. Bernstein, "is a matter of idiom." All of us run into occasional trouble trying to decide which preposition to use with a certain word. Below is a list of some combinations often cited as correct or incorrect, though some ("different *from*" versus "different *than*," for instance) are the focus of continuing disagreement among grammarians.

Incorrect	*Correct*
careful with (money, reputation)	careful of
center around	center on
different than	different from
embarrassed at	embarrassed by
end by	end with
equal with	equal to
involved by	involved in
protest at	protest against
tendency for	tendency to

Subordinating conjunctions (*since, because, although, if, after, when, while, before, unless,* and so on) introduce adverb clauses. Thus, a subordinating conjunction connects the sentence's main (independent) clause with a subordinate (dependent) See ◄
9b clause.

Although drug use is a serious concern for parents, many parents are afraid to discuss it with their children.

It is best to diagram your garden before you start to plant it.

Conjunctive adverbs, also known as adverbial conjunctions, are discussed in 21e.

21h Interjections

Interjections are words used as exclamations: *Oh! Ouch! Wow! Alas! Hey!* These words, which express emotion, are grammatically independent; that is, they do not have a grammatical function in a sentence. Interjections may be set off in a sentence by commas.

The message, alas, arrived too late.

Or, for greater emphasis, they can be punctuated as independent units, set off with an exclamation point.

Alas! The message arrived too late.

Other words besides interjections are sometimes used in isolation. These include words such as *yes, no, hello, good-bye, please,* and *thank you.* All such words, including interjections, may be collectively referred to as **isolates**.

CHAPTER 22

Nouns and Pronouns

22a Case

Case is the form a noun or pronoun takes to indicate how it functions in a sentence. English has three cases: *subjective*, *objective*, and *possessive*.

As the English language developed, nouns generally lost their case distinctions and now change form only in the possessive case: the *cat's* eyes, *Molly's* book. Pronouns, however, have retained many of their case distinctions.

PRONOUN CASE FORMS

Subjective

I	he, she	it	we	you	they	who whoever

Objective

me	him, her	it	us	you	them	whom whomever

Possessive

my mine	his, her hers	its	our ours	your yours	their theirs	whose

---

(1) Using the subjective case

SUBJECT OF A VERB: David and I bought the same kind of mountain bike. (*I* is the subject of a verb.)

SUBJECT COMPLEMENT: It was <u>he</u> the men were looking for. (*He* is the subject complement.)

SUBJECT OF A CLAUSE: The sergeant asked <u>whoever</u> wanted to volunteer to step forward. (*Whoever* is the subject of the noun clause *whoever wanted to step forward*.)

APPOSITIVE IDENTIFYING SUBJECT: Both scientists, <u>Oppenheimer and he</u>, worked on the atomic bomb. (*Oppenheimer and he* is an appositive identifying the subject *both scientists*.)

CLOSE-UP

NOUNS AND PRONOUNS

Using the proper case for subject complements sometimes creates forced-sounding constructions. Most people feel silly saying "It is I" or "It is he," and they use the more natural colloquial constructions "It's me" or "It's him" in speech or informal writing. In college writing, however, you should be careful to use proper pronoun case.

(2) Using the objective case

DIRECT OBJECT: Our sociology teacher likes Adam and <u>me</u>. (*Me* is a direct object of the verb *likes*.)

INDIRECT OBJECT: During the 1950s the Kinsey report gave <u>them</u> quite a shock. (*Them* is the indirect object of the verb *gave*.)

CLOSE-UP

NOUNS AND PRONOUNS

Discard the mistaken idea that *I* is always somehow more appropriate than *me*. In compound constructions like the following, *me* is correct.

He told Jason and <u>me</u> [not <u>I</u>] to put our dog on a leash. (*Me* is the object of the verb *told*.)

Between you and <u>me</u> [not <u>I</u>] we own ten shares of stock. (*Me* is the object of the preposition *between*.)

Let's you and <u>me</u> [not <u>I</u>] go to the art museum. (*Me* is an appositive identifying *us*, direct object of the verb *let*.)

TEACHING STRATEGY

Case

Case is not as important a part of the English language as it is of other languages. Latin has six cases; some modern languages (Hungarian, to name one) have as many as 25 or 30. Old English was a "case heavy" language—the technical term would be "highly inflected"—resembling modern German. (If your students have studied foreign languages, they may know that an *inflection* is a set of word forms showing function or meaning in a sentence. The set of inflections of a noun is called a *declension;* the set of inflections of a verb is called a *conjugation.*)

Old English nouns had (and still have) five cases. Over a period of about 1,000 years, these five were reduced to only two: a *common* or "all purpose" form of the noun, and a *possessive* form. You may want to compare the relative simplicity of English noun case forms to those of Latin or another language.

Object of a Preposition: In 1502 Leonardo da Vinci designed the fortifications of the city for <u>him</u>. (*Him* is the object of the preposition *for*.)

Object of an Infinitive: Hoover did not want to anger Alfred E. Smith or <u>him</u>. (*Him* is the object of the infinitive *to anger*.)

Object of a Gerund: Finding <u>her</u> was not easy for Marlow. (*Her* is the object of the gerund *finding*.).

Subject of an Infinitive: They told <u>him</u> to defend his flank against the French cavalry. (*Him* is the subject of the infinitive *to defend*.)

Appositive Identifying an Object: Rachel discussed both authors, <u>Hannah Arendt and her</u>. (*Hannah Arendt and her* is an appositive identifying the object *authors*.)

(3) Using the possessive case

A pronoun takes the **possessive case** when it indicates ownership (*our* car, *your* book). Be sure to distinguish gerunds, which always function as nouns, from present participles functioning as adjectives. Use the possessive, not the objective, case before a gerund.

▶ **See 21c2**

Napoleon approved of <u>their</u> [not *them*] ruling Naples. (*Ruling* is a gerund.)

CLOSE-UP

NOUNS AND PRONOUNS

Remember the distinction between the possessive pronoun *its* and the contraction *it's*. *Its* designates ownership (*its* leg), while *it's* is the contraction of *it is* ("*It's* a nice day") or *it has* ("*It's* faded in the sunlight").

EXERCISE 1

Choose the correct form of the pronoun within the parenthesis. Be prepared to explain why you chose each form.

> **EXAMPLE:** Toni Morrison, Alice Walker, and (<u>she</u>, her) are perhaps the most widely recognized African-American women writing today.

1. Both Walt Whitman and (he, him) wrote a great deal of poetry about nature.

QUOTATION

Columnist and political commentator William Safire includes the following letter in his book, *On Language*.

"Mr. Safire,
As regards your recent statement that you are 'down on whom,' even in instances where its use would be correct, I beg you to remember what happens when one asks 'For Who the Bell Tolls.'
It tolls for he."

2. Randall Jarrell, Wilfred Owen and (he, him) wrote about their experiences in war.
3. Both (she, her) and the Pulitzer Prize-winning poet Gwendolyn Brooks were born in Kansas but moved early in life to Chicago's South Side.
4. Our instructor gave Matthew and (me, I) an excellent idea for our project.
5. The philosophy class made (we, us) more aware of how we form our values.
6. The sales clerk objected to (me, my) returning the sweater.
7. (We, Us) students have the opportunity to attend the concert free of charge.
8. Ezra Pound, Sylvia Plath and (he, him) were plagued with bouts of mental illness.
9. I understand (you, your) being unavailable to work tonight.
10. The waitress asked Michael and (me, I) to move to another table.

22b Revising Common Errors: Case

(1) Implied comparisons with *than* or *as*

When a sentence containing an implied comparison ends with a pronoun, your meaning dictates your choice of pronoun.

Darcy likes John more than I. (*more than I like John*)

Darcy likes John more than me. (*more than she likes me*)

Alex helps Dr. Elliott as much as I. (*as much as I help Dr. Elliott*)

Alex helps Dr. Elliott as much as me. (*as much as he helps me*)

(2) *Who* and *whom*

The case of the pronouns *who* and *whom* depends on their function *within their own clause*. When a pronoun serves as the subject, use *who* or *whoever*; when it functions as an object, use *whom* or *whomever*.

The Salvation Army gives food and shelter to whoever is in need. (*Whoever* is the subject of the verb *is* in the dependent clause *whoever is in need*.)

Shortly after leaving Oklahoma the Joads were reminded who they were and where they came from. (*Who* is the subject of the dependent clause *who they were and where they came from*.)

I wonder whom jazz musician Miles Davis influenced. (*Whom* is the object of *influenced* in the dependent clause *whom jazz musician Miles Davis influenced*.)

Whomever Stieglitz photographed, he revealed. (*Whomever* is the object of *photographed* in the dependent clause *Whomever Stieglitz photographed*.)

CLOSE-UP

NOUNS AND PRONOUNS

To determine the case of *who* at the beginning of a question, answer the question using a personal pronoun.

Who wrote *The Age of Innocence*? She wrote it. (Subject)

Whom do you support for mayor? I support her. (Object)

To whom is the letter addressed? It is addressed to them. (Object of a preposition)

If intervening phrases such as *I think, we know,* or *she* or *he says* cause problems, read the sentence without the intervening phrase.

Albert Einstein is the man who [we know] revolutionized physics. (*Who* is the subject of the clause *who we know revolutionized physics*.)

CLOSE-UP

NOUNS AND PRONOUNS

Although formal writing requires that *whom* be used for all objects, strict adherence to this rule can result in stilted constructions. In all but the most formal situations, current usage accepts *who* at the beginning of questions.

Who do you support for mayor?

Who is the letter addressed to?

1. Edith Wharton was a celebrated American writer who wrote in an elegant, forceful prose style.
2. Jonathan Swift was a satirist who his cousin, John Dryden, believed would be a failure as a poet.
3. Charles Lutwidge Dodgson, a noted professor of mathematics who wrote *Alice's Adventures in Wonderland*, used the pen name Lewis Carroll.
4. William Carlos Williams was a prolific writer who influenced a whole generation of younger writers.
5. Sylvia Plath met the poet Ted Hughes, whom she later married, at Cambridge University in England.

EXERCISE 2

Using the word in parentheses, combine each pair of sentences into a single sentence. You may change word order and add or delete words.

EXAMPLE: Charles Dickens was a popular author of the nineteenth century. He wrote *David Copperfield*. (who)

Charles Dickens, who wrote *David Copperfield*, was a popular author of the nineteenth century.

1. Edith Wharton was a celebrated American writer. She wrote in an elegant, forceful prose style. (who)
2. Jonathan Swift was a satirist. His cousin, John Dryden, believed Swift would be a failure as a poet. (who)
3. Charles Lutwidge Dodgson, a noted professor of mathematics, used the pen name Lewis Carroll. He wrote *Alice's Adventures in Wonderland*. (who)
4. William Carlos Williams was a prolific writer. He influenced a whole generation of younger writers. (who)
5. Sylvia Plath met the poet Ted Hughes at Cambridge University in England. She later married him. (whom)

(3) Appositives

An **appositive** is a noun or noun phrase that renames the word it follows. The case of the pronoun in an appositive depends on the function of the word it describes.

See ◀ 8e4

Motown Records recorded two particularly great male singers, Smokey Robinson and <u>him</u>. (*Male singers* is the object of the verb *recorded*, so the pronoun in the appositive phrase *Smokey Robinson and him* takes the objective case.)

Two particularly great male singers, Smokey Robinson and <u>he</u>, recorded for Motown Records. (*Male singers* is the subject of the sentence, so the pronoun in the appositive phrase *Smokey Robinson and he* takes the subjective case.)

(4) *We* and *us* before a noun

When a first-person plural pronoun precedes a noun, the case of the pronoun depends on the way the noun functions in the sentence.

<u>We</u> women must stick together. (*Women* is the subject of the sentence; therefore, the pronoun we must be in the subjective case.)

Teachers make learning easy for <u>us</u> students. (*Students* is the object of the preposition *for*; therefore, the pronoun us must be in the objective case.)

378

22c Pronoun Reference

An **antecedent** is the word or word group to which a pronoun refers. Pronoun reference is clear when readers can easily identify the noun or pronoun to which it refers. In the following passage, notice how the underscored pronouns point clearly to the noun *warts.*

> Warts are wonderful structures. <u>They</u> can appear overnight on any part of the skin, like mushrooms on a damp lawn, full grown and splendid in the complexity of <u>their</u> architecture. Viewed in stained sections under a microscope, <u>they</u> are the most specialized of cellular arrangements, constructed as though for a purpose. <u>They</u> sit there like turreted mounds of dense impenetrable horn, impregnable, designed for defense against the world outside. (Lewis Thomas, *The Medusa and the Snail*)

22d Revising Common Errors: Pronoun Reference

(1) Ambiguous antecedents: *this, that, which,* and *it*

The meaning of a pronoun is ambiguous if the pronoun appears to refer to more than one antecedent. The pronouns *this, that, which,* and *it* are most likely to invite this kind of confusion. To ensure clarity make sure each pronoun points to a specific antecedent.

AMBIGUOUS: The accountant took out his calculator and completed the tax return. Then, he put <u>it</u> in his briefcase. (The pronoun *it* can refer either to *calculator* or to *tax return.*)

CLEAR: The accountant took out his calculator and completed the tax return. Then, he put <u>the calculator</u> in his briefcase.

AMBIGUOUS: Some one-celled organisms contain chlorophyll and are considered animals. <u>This</u> illustrates the difficulty of classifying single-celled organisms as either animals or plants. (*This* can refer either to the fact that some one-celled organisms are animals or to the fact that they contain chlorophyll.)

CLEAR: Some one-celled organisms contain chlorophyll and are considered animals. <u>This paradox</u> illustrates the difficulty of classifying single-celled organisms as either animals or plants.

Pronoun References

Students often have trouble with confused pronoun references because a) they put the pronoun too far away from its antecedent, or b) the antecedent and pronoun are separated by one, two or more intervening nouns (causing the reader to wonder if one of these "interlopers" might be the antecedent to which the pronoun refers).

Remind your students that even though pronouns are considered "economical" words, saving space at the expense of clear meaning is not a good idea. If readers are struggling to trace pronouns back to their antecedents, they are not paying attention to the "sense" of the words—and the writer should consider putting back some of the nouns he or she left out.

"The noun for which the pronoun is stepping in is called its antecedent.

> *The chap with the long face* is buying *his* insurance policy.
> *The guys at the gym* were snapping *their* towels.
> There was *a lull in the conversation. It* was embarrassing.
> What is the name of *that surly bloke*? I'm dying to meet *him.*

The pronoun stands for the other words accompanying the noun as well as the noun itself. Thus . . . *him* replaces not only the bloke but his surliness, too."

–Karen Elizabeth Gordon,
*The Transitive Vampire:
A Handbook of Grammar for the
Innocent, the Eager, and the Doomed*

CLOSE-UP

NOUNS AND PRONOUNS

If you are certain that no misunderstanding will occur, you can use *this, that, which,* or *it* to refer to a previous clause.

Visitors would constantly interrupt Edison. <u>This</u> made him angry.

Be careful, however. General references often invite confusion. For this reason, it is a good idea to avoid them in your college writing.

(2) Remote antecedents

The farther a pronoun is from its antecedent, the more difficult it is for readers to make a connection between them. As a result, readers lose track of meaning and must reread a passage to determine the connection.

AMBIGUOUS: Rumors of gold, letters from friends and relatives, and newspaper articles praising democracy persuaded many Czechs to come to America. By 1860 about 23,000 Czechs had left their country. By 1900, 13,000 Czech immigrants were coming to <u>its</u> shores each year.

The pronoun *its* in the last sentence is so far removed from its antecedent, *America,* that the reference cannot easily be understood. For clarity, the antecedent should be restated in the final sentence.

CLEAR: By 1900, 13,000 Czech immigrants were coming to <u>America's</u> shores each year.

(3) Nonexistent antecedents

An unclear pronoun reference also occurs when a pronoun refers to a nonexistent antecedent.

AMBIGUOUS: Our township has decided to build a computer lab in the elementary school. <u>They</u> feel that children should learn to use computers in fourth grade.

In the second sentence *they* seems to refer to *township* as a collective noun. Actually *they* refers to an antecedent that the writer has neglected to mention. Supplying the noun *teachers* eliminates the confusion.

> **CLEAR:** Our township has decides to build a computer lab in the elementary school. <u>Teachers</u> feel that children should learn to use computers in fourth grade.

CLOSE-UP

NOUNS AND PRONOUNS

References such as "*It* says in the paper" and "*They* said on the news" refer to unidentified antecedents and therefore are not acceptable in college writing. To correct this problem, substitute the appropriate noun for the unclear pronoun. "*The article* in the paper says. . . ." and "On the news, *Cokie Roberts* said. . . ."

(4) *Who, which* and *that*

In general, *who* refers to people or to animals that have names. *Which* and *that* usually refer to objects, events, or unnamed animals and sometimes to groups of people.

> David Henry Hwang, <u>who</u> wrote the Tony Award–winning play *M. Butterfly*, also wrote *Family Devotions* and *FOB*.

> Bucephalus, <u>who</u> was Alexander the Great's horse, was famous for his bravery in battle.

> The spotted owl, <u>which</u> lives in old-growth forests, is in danger of extinction.

> Houses <u>that</u> are built today are more energy efficient than those built twenty years ago.

CLOSE-UP

NOUNS AND PRONOUNS

When you revise, make certain that you use *which* in nonrestrictive clauses, which are always set off with commas. In most cases, use *that* in restrictive clauses. *Who* may be used in both restrictive and nonrestrictive clauses.

▶ See
28d1

Your students' answers will vary.

1. ambiguous antecedent
   The Louisiana Purchase increased the need for this information.
2. nonexistent antecedent
   Lewis and Clark trained the men in Illinois, the starting point.
3. ambiguous antecedent
   The campsite was beautiful and wild.
4. ambiguous antecedent
   Both Jefferson and Lewis had faith that Lewis would be successful in his transcontinental journey.
5. remote and ambiguous antecedent
   The fact that only one man died was extraordinary.

STUDENT WRITER AT WORK

Nouns and Pronouns

Possible Revision

   John Updike's "A & P," a short story that appears in his book *Pigeon Feathers*, takes place in a small town similar to Updike's hometown. The character who has the significant role in "A & P" is Sammy, a cashier at the supermarket. Sammy is a nineteen-year-old boy who is just out of high school. He analyzes everyone who comes to the A & P to shop. It is he who is the narrator of the story.

   The story takes place on a Thursday afternoon when three girls in bathing suits walk into the store. The girls are different from the other shoppers. Their manner and the way they walk make them different from them. Sammy notices that one of the girls, whom he calls Queenie, leads the other girls. This characteristic appeals to him. He identifies with her

continued on following page

EXERCISE 3

Analyze the pronoun errors in each of the following sentences. After doing so, revise each sentence by substituting an appropriate noun or noun phrase for the underlined pronoun.

EXAMPLE: Jefferson asked Lewis to head the expedition, and Lewis selected <u>him</u> as his associate.

ANALYSIS: *Him* refers to a nonexistent antecedent.

REVISION: Jefferson asked Lewis to head the expedition, and Lewis selected Clark as his associate.

1. The purpose of the expedition was to search out a land route to the Pacific and to gather information about the West. The Louisiana Purchase increased the need for <u>it</u>.
2. The expedition was going to be difficult. <u>They</u> trained the men in Illinois, the starting point.
3. Clark and most of the men that descended the Yellowstone River camped on the bank. <u>It</u> was beautiful and wild.
4. Both Jefferson and Lewis had faith that <u>he</u> would be successful in this transcontinental journey.
5. The expedition was efficient, and only one man was lost. <u>This</u> was extraordinary.

STUDENT WRITER AT WORK

NOUNS AND PRONOUNS

   Following is part of a draft of a student essay about John Updike. This section of the essay gives a plot summary of Updike's short story "A & P." Read the draft and revise it to correct errors in case and to eliminate inexact pronoun reference. After you have corrected the errors, go over the draft again and, if necessary, revise further to strengthen coherence, unity, and style.

   John Updike's "A & P," a short story that appears in the book Pigeon Feathers, takes place in a small town similar to Updike's hometown. The character which has the significant role in "A & P" is Sammy, a cashier at the supermarket. Sammy is a nineteen-year-old boy that is just out of high school. He analyzes everyone who comes to the A & P to shop. It is him who is the narrator of the story.

   The story takes place on a Thursday afternoon when three girls in bathing suits walk into the store. They are different from the other shoppers. Their manner and the way they walk make them different from them. Sammy notices that one of the girls,

continued on the following page

continued from the previous page

who he calls Queenie, leads the other girls. This appeals to him. He identifies with her because he feels that he too is a leader.

When the girls come to his check-out counter, he rings up their purchase. Suddenly the store manager, Lengel, begins scolding the girls for coming into the store in bathing suits. Sammy feels sorry for them, and in a gesture of defiance he quits. Sammy feels that him quitting is a rejection of him and all that he stands for. To Sammy, Lengel is a person that represents the narrow morality of the town.

Sammy's quitting is the climax of the story. Sammy chooses to follow his conscience and in doing so pays the price. He feels that not following his ideals would be bad. Because he is young, however, he does not realize the significance of the act which he commits. For a moment Lengel and Sammy face each other, but he does not change his mind. Sammy feels that he has won his freedom. His confidence is short-lived, though. When he walks out into the parking lot, the girls are gone, and he is alone. It is then he realizes that the world is going to be hard for him from this point on.

continued from previous page

because he feels that he too is a leader.

When the girls come to his check-out counter, he rings up their purchase. Suddenly the store manager, Lengel, begins scolding the girls for coming into the store in bathing suits. Sammy feels sorry for the girls, and in a gesture of defiance, he quits. Sammy feels that his quitting is a rejection of Lengel and all that he stands for. To Sammy, Lengel is a person who represents the narrow morality of the town.

Sammy's quitting is the climax of the story. Sammy chooses to follow his conscience, and in doing so pays the price. He feels that not following his ideals would be bad. Because he is young, however, he does not realize the significance of the act that he commits. For a moment Lengel and Sammy face each other, but Sammy does not change his mind. Sammy feels that he has won his freedom. His confidence is short-lived, though. When he walks out into the parking lot, the girls are gone, and he is alone. It is then he realizes that the world is going to be hard for him from that point on.

Commentary on Possible Revision

The student has edited to eliminate case errors and inexact or unclear pronoun references, but he could edit further to make his sentences smoother and more concise.

CHAPTER 23

Verbs

VERBS: KEY TERMS

Form The spelling of a verb that conveys tense, person, number, and so on.

Tense The form a verb takes to indicate when an action occurs or when a condition exists—*present, past, future,* and so on.

Person The form a verb takes to indicate whether a person is speaking (*first person*), is spoken to (*second person*), or is spoken about (*third person*).

Number The form a verb takes to indicate whether the verb is singular (The child *reads*) or plural (The children *read*).

Mood The form a verb takes to indicate a writer's attitude—for example, whether he or she is making a statement, giving a command, or making a recommendation (I *read* the book; *Read* the book! I suggest you *read* the book).

Voice The form a verb takes to indicate whether the subject acts or is acted upon (He *wrote* the book; The book *was written* by him).

23a Verb Forms

All verbs have four **principal parts** from which their tenses are derived: a **base form** (the form of the verb used with *I, we, you,* and *they* in the present tense),* a **present participle**, a **past tense form**, and a **past participle**. Most verbs in English are **regular** and form their principal parts with *-ing* and *-ed* or *-d* added to the base form. Consult a dictionary whenever you are uncertain about the form of a verb.

(1) Principal parts of regular verbs

If the dictionary lists only the base form, the verb is regular and forms both past tense and past participle by adding *-d* or *-ed*.

PRINCIPAL PARTS OF REGULAR VERBS			
Base Form	*Present Participle*	*Past*	*Past Participle*
smile	smiling	smiled	smiled
talk	talking	talked	talked
jump	jumping	jumped	jumped

(2) Principal parts of irregular verbs

Irregular verbs are those that do not follow the pattern outlined above. If a verb is irregular, the dictionary lists its principal parts, three if the past tense and past participle are different and two if the past tense and past participle are the same.

hide, v (past hid, pp hidden)
fall, v (past fell, pp fallen)
make, v (made)
spin, v (spun)

Many irregular verbs change an internal vowel in the past tense and past participle.

Base Form	*Present Participle*	*Past*	*Past Participle*
begin	beginning	began	begun
come	coming	came	come

*NOTE: The verb *be* is so irregular that it is the one exception to this definition; its base form is *be*.

TEACHING STRATEGY

Verb Forms

Modern English has only about 200 irregular verbs—but many of them are the verbs we use most frequently: *do, go, see, run, give, eat,* and so on. Remind students that most irregular verbs change an internal vowel to form the past tense or the past participle—a holdover from Anglo-Saxon verb forms.

Some students have problems with verb endings because they do not pronounce those endings and have not heard them pronounced frequently. Devise appropriate listening and speaking drills to help these students increase their awareness of verb endings and the distinctions among them.

If your students have trouble with the forms of the verb *be,* or tend to drop *-s, -d,* or *-ed* endings (or add these endings where they are not grammatically appropriate), you might need an informal discussion of "standard" and "nonstandard" verb forms ("I talk" vs. "I talks," for example). If *be* is a problem, you may want to note that because the verb has eight forms (three more than any other English verb), memorizing them may be the simplest way for students to be sure they know when to use a given form.

When you refer students to this chapter, be sure to refer them as well to other chapters with pertinent sections on voice and tense.

"Verb may mean 'the word' in Latin, but it stands for *power*. A verb is the engine of a sentence; without it, your carefully organized words stand still. . . . Using the correct verb form is a matter of grammar; choosing the right verb is a matter of conciseness and clarity. Our language has thousands of strong, concrete, precise verbs. We should learn to use them rather than saddling an imprecise verb with a load of modifiers. If you must tack on an adverb or a series of adverbs to clarify the meaning of the verb you have chosen, perhaps you should reconsider your original choice."

–Lauren Kessler and
Duncan McDonald,
When Words Collide

Other irregular verbs not only change an internal vowel in the past tense but also add *-n* or *-en* in the past participle.

Base Form	Present Participle	Past	Past Participle
fall	falling	fell	fallen
rise	rising	rose	risen

Still other irregular verbs take the same form in both the past and the past participle forms.

Base Form	Present Participle	Past	Past Participle
bet	betting	bet (betted)	bet
have	having	had	had

COMMON IRREGULAR VERBS

Base Form	Present Participle	Past	Past Participle
arise	arising	arose	arisen
awake	awaking	awoke, awaked	awoke, awaked
be	being	was/were	been
bear (carry)	bearing	bore	borne
beat	beating	beat	beaten
begin	beginning	began	begun
bend	bending	bent	bent
bet	betting	bet, betted	bet
bid	bidding	bid	bid
bind	binding	bound	bound
bite	biting	bit	bitten
bleed	bleeding	bled	bled
blow	blowing	blew	blown
break	breaking	broke	broken
bring	bringing	brought	brought
build	building	built	built
burst	bursting	burst	burst
buy	buying	bought	bought
catch	catching	caught	caught
choose	choosing	chose	chosen
cling	clinging	clung	clung
come	coming	came	come
creep	creeping	crept	crept

continued on the following page

Verb Forms

continued from the previous page

Base Form	Present Participle	Past	Past Participle
cut	cutting	cut	cut
deal	dealing	dealt	dealt
dig	digging	dug	dug
dive	diving	dived, dove	dived
do	doing	did	done
drag	dragging	dragged	dragged
draw	drawing	drew	drawn
drink	drinking	drank	drunk
drive	driving	drove	driven
eat	eating	ate	eaten
fall	falling	fell	fallen
feed	feeding	fed	fed
feel	feeling	felt	felt
fight	fighting	fought	fought
find	finding	found	found
fling	flinging	flung	flung
fly	flying	flew	flown
forbid	forbidding	forbade, forbad	forbidden, forbid
forget	forgetting	forgot	forgotten, forgot
forsake	forsaking	forsook	forsaken
freeze	freezing	froze	frozen
get	getting	got	gotten
give	giving	gave	given
go	going	went	gone
grind	grinding	ground	ground
grow	growing	grew	grown
hang (suspend)	hanging	hung	hung
hang (execute)	hanging	hanged	hanged
have	having	had	had
hear	hearing	heard	heard
hit	hitting	hit	hit
keep	keeping	kept	kept
know	knowing	knew	known
lay	laying	laid	laid
lead	leading	led	led
leap	leaping	leaped, leapt	leaped, leapt
lend	lending	lent	lent

continued on the following page

continued from the previous page

Base Form	Present Participle	Past	Past Participle
let	letting	let	let
lie (recline)	lying	lay	lain
lie (tell an untruth)	lying	lied	lied
light	lighting	lighted, lit	lighted, lit
mow	mowing	mowed	mowed, mown
plead	pleading	pleaded, pled	pleaded, pled
prove	proving	proved	proved, proven
put	putting	put	put
read	reading	read	read
rid	ridding	rid, ridded	rid, ridded
ride	riding	rode	ridden
ring	ringing	rang	rung
rise	rising	rose	risen
run	running	ran	run
see	seeing	saw	seen
seek	seeking	sought	sought
set	setting	set	set
shake	shaking	shook	shaken
shed	shedding	shed	shed
shine	shining	shone	shone
shoe	shoeing	shod, shoed	shod, shoed
shrink	shrinking	shrank, shrunk	shrunk, shrunken
sing	singing	sang	sung
sink	sinking	sank	sunk
sit	sitting	sat	sat
slay	slaying	slew	slain
sneak	sneaking	sneaked	sneaked
sow	sowing	sowed	sowed, sown
speak	speaking	spoke	spoken
speed	speeding	sped, speeded	sped, speeded
spin	spinning	spun	spun
spring	springing	sprang	sprung
stand	standing	stood	stood
steal	stealing	stole	stolen

continued on the following page

continued from the previous page

Base Form	Present Participle	Past	Past Participle
stick	sticking	stuck	stuck
strike	striking	struck	struck, stricken
swear	swearing	swore	sworn
swim	swimming	swam	swum
swing	swinging	swung	swung
take	taking	took	taken
teach	teaching	taught	taught
tear	tearing	tore	torn
think	thinking	thought	thought
throw	throwing	threw	thrown
tread	treading	trod	trodden, trod
wake	waking	woke, waked	waked, woken
wear	wearing	wore	worn
weave	weaving	wove	woven
wed	wedding	wed, wedded	wed, wedded
weep	weeping	wept	wept
win	winning	won	won
wind	winding	wound	wound
wring	wringing	wrung	wrung
write	writing	wrote	written

EXERCISE 1

Complete the sentences in the following paragraph with an appropriate form of the verbs in parentheses.

> **EXAMPLE:** Many writers have —————— stories of intrigue and suspense. (weave)
>
> Many writers have <u>woven</u> stories of intrigue and suspense.

Eric Blair, a well-known British novelist and essayist, —————— (choose) to use the pseudonym George Orwell. His essay "A Hanging," —————— (take) from his book *Shooting an Elephant and Other Essays,* is —————— (write) in narrative form. It describes a morning in Burma when Orwell observed the execution of a prisoner who was to be —————— (hang) for his crime. One of the most startling scenes in the essay comes at the end when the onlookers, —————— (forsake) the seriousness of the moment, leave the prison yard after the event, laughing on their way to have a drink.

1. set
2. laying
3. sits
4. lying
5. sit

CLOSE-UP

IRREGULAR VERB FORMS

See ◄
8b

The irregular verbs *lie* and *lay* and *sit* and *set* frequently give writers trouble because each pair of verbs has similar sounding forms with similar meanings. You can keep them straight only by memorizing the forms and remembering that one takes an object and the other does not.

Lie means "to recline" and does not take an object (He likes to *lie* on the floor); *lay* means "to place" or "to put" and does take an object (He wants to *lay* a rug on the floor).

Base Form	Present Participle	Past	Past Participle
lie	lying	lay	lain
lay	laying	laid	laid

Sit means "to assume a seated position" and does not take an object (She wants to *sit* on the table); *set* means "to place" or "to put" and usually takes an object (She wants to *set* a vase on the table).

Base Form	Present Participle	Past	Past Participle
sit	sitting	sat	sat
set	setting	set	set

EXERCISE 2

Complete the following sentences with an appropriate form of the verbs in parentheses.

> **EXAMPLE:** Mary Cassatt ———— down her paintbrush. (lie, lay)
>
> Mary Cassatt <u>lay</u> down her paintbrush.

1. Impressionist artists of the nineteenth century preferred everyday subjects and used to ———— fruit on a table to paint. (sit, set)
2. They were known for their technique of ———— dabs of paint quickly on canvas, giving an "impression" of a scene, not extensive detail. (lying, laying)
3. Claude Monet's "Women in the Garden" featured one woman in the foreground who ———— on the grass in a garden. (sat, set)
4. In Pierre Auguste Renoir's "Nymphs" two nude figures talk while ———— on flowers in a garden. (lying, laying)

5. Paul Cézanne liked to —————— in front of his subject as he painted and often completed paintings out of doors rather than in a studio. (sit, set)

23b Tense

Tense is the form a verb takes to indicate when an action occurred or when a condition existed. Tense, however, is not the same as time. The present tense, for example, indicates present time, but it can also indicate future time or a generally held belief.

ENGLISH VERB TENSES

Simple Tenses
Present (I finish; she or he finishes)
Past (I finished)
Future (I will finish)

Perfect Tenses
Present perfect (I have finished; she or he has finished)
Past perfect (I had finished)
Future perfect (I will have finished)

Progressive Tenses
Present progressive (I am finishing; she or he is finishing)
Past progressive (I was finishing)
Future progressive (I will be finishing)
Present perfect progressive (I have been finishing)
Past perfect progressive (I had been finishing)
Future perfect progressive (I will have been finishing)

23c Tense: Using the Simple Tenses

The **simple tenses** include *present*, *past*, and *future*.

(1) Present tense

The **present tense** usually indicates an action taking place at the time it is expressed in speech or writing. With subjects other than singular nouns or third-person singular pronouns, the

Tense
As you proceed through sections 23c–23e, ask individual students to create other examples of sentences using each of the six tenses given. If needed, spend extra time reinforcing the concept of *perfect tenses*, which indicate that the action named is completed or finished before a given point in time.

Students sometimes find the *present perfect tense* difficult to grasp. This quotation from the late *New York Times* editor Theodore M. Bernstein may be helpful: "The *present perfect tense* in a sense straddles the past and the present; it speaks of actions that are thought of as not wholly over and done with but closely associated with the present . . .: 'The United States repeatedly *has declared* that the presence of Allied garrisons in West Berlin is not a negotiable issue.'"

To reinforce student awareness of tense, select a paragraph (from a book, magazine, or newspaper) that begins in perfect tense. Ask students to rewrite the paragraph three times, using three different tenses to begin the paragraph and then revising all other tenses as appropriate. You may want to ask students to label each tense once they have completed the rewrites.

In groups of four or five, students change an essay written in one tense to another tense (e.g., from past to present). Groups exchange rewrites and read each other's versions, looking for inappropriate shifts in verb tense.

present tense uses only the base form of the verb. With singular nouns or third-person singular pronouns, *-s* or *-es* is added to the base form.

I <u>smile</u>.	She <u>smiles</u>.
They <u>hear</u>.	He <u>hears</u>.

In addition to expressing action in the present, the present tense has some special uses.

SPECIAL USES OF THE PRESENT TENSE

The rector <u>opens</u> the chapel every morning at six o'clock. (indicates that something occurs regularly)

The grades <u>arrive</u> next Thursday. (indicates future time)

Studying <u>pays</u> off. (states a generally held belief)

An object at rest <u>tends</u> to stay at rest. (states a scientific truth)

In *Family Installments* Edward Rivera <u>tells</u> the story of a family's journey from Puerto Rico to New York City. (discusses a literary work)

Notice that in some cases words like *every* and *next* help to indicate time.

---

(2) Past tense

The **past tense** indicates that an action has already taken place. It is formed by adding *-d* or *-ed* to the base form or, for irregular verbs, by changing the form of the verb. The past tense has two uses.

John Glenn <u>orbited</u> the earth three times on February 20, 1962. (indicates an action completed in the past)

When he <u>was</u> young, Mark Twain <u>traveled</u> through the mining towns of the Southwest. (indicates actions that recurred in the past but did not extend into the present)

(3) Future tense

The **future tense** indicates that an action will take place. A number of constructions can be used to indicate future action, including the present tense, but here we discuss future tense verb forms. These verb forms consist of the auxiliaries *will* or *shall* plus the present tense. The future tense has several uses.

Halley's Comet <u>will reappear</u> in 2061. (indicates a future action that will definitely occur)

The dean has announced that the college <u>will require</u> all freshmen to buy computers. (indicates intention)

The land boom in Nevada <u>will</u> most likely <u>continue</u>. (indicates probability)

CLOSE-UP

VERB TENSE

At one time *will* was used exclusively for the second- and third-person future tense of a verb, and *shall* was used for the first person. Except in formal usage, however, *shall* is now rare.

23d Tense: Using the Perfect Tenses

The **perfect tenses** designate actions that were or will be completed before other actions or conditions. The perfect tenses are formed with the appropriate tense form of the auxiliary verb *have* plus the past participle.

(1) Present perfect tense

The **present perfect** tense can indicate either of two types of continuing action beginning in the past.

Dr. Kim <u>has finished</u> studying the effects of BHA on rats. (indicates an action that began in the past and is finished at the present time)

My mother <u>has invested</u> her money wisely. (indicates an action that began in the past and extends into the present)

(2) Past perfect tense

The **past perfect** tense has three uses.

By 1946 engineers <u>had built</u> the first electronic digital computer. (indicates an action occurring before a certain time in the past)

By the time Alfred Wallace wrote his paper on evolution, Darwin <u>had</u> already <u>published</u> *The Origin of Species*. (indicates that one action was finished before another one started)

We <u>had hoped</u> to visit Disney World on our trip to Florida. (indicates an unfulfilled desire in the past)

(3) Future perfect tense

The **future perfect** tense has two uses.

By Tuesday the transit authority will have run out of money. (indicates that an action will be finished by a certain future time)

By the time a commercial fusion reactor is developed, the government will have spent billions of dollars on research. (indicates that one action will be finished before another occurs in the future)

23e Tense: Using the Progressive Tenses

The **progressive tenses** express continuing action. They consist of the appropriate tense of the verb *be* plus the present participle.

(1) Present progressive tense

The **present progressive** tense has two uses.

The volcano is erupting and lava is flowing toward the town. (indicates that something is happening at the time it is expressed in speech or writing)

Law is becoming an overcrowded profession. (indicates that an action is happening even though it may not be taking place at the time it is expressed in speech or writing)

(2) Past progressive tense

The **past progressive** tense has two uses.

Roderick Usher's actions were becoming increasingly bizarre. (indicates a continuing action in the past)

The French revolutionary Jean-Paul Marat was stabbed to death while he was bathing. (indicates an action occurring at the same time in the past as another action)

(3) Future progressive tense

The **future progressive** tense has two uses.

The secretary of the treasury will be carefully monitoring the money supply. (indicates a continuing action in the future)

Next month NATO forces will be holding military exercises. (indicates a continuing action at a specific future time)

Verbs 23e

(4) Present perfect progressive tense

The **present perfect progressive** tense has only one use.

The number of women getting lung cancer <u>has been increasing</u> dramatically. (indicates action continuing from the past into the present and possibly into the future)

(5) Past perfect progressive tense

The **past perfect progressive** tense has only one use.

Before President Kennedy was assassinated, he <u>had been working</u> on civil rights legislation. (indicates that one past action went on until a second occurred)

(6) Future perfect progressive tense

The **future perfect progressive** tense has only one use.

By eleven o'clock we <u>will have been driving</u> for seven hours. (indicates that an action will continue until a certain future time)

*EXERCISE 3

A verb is missing from each of the following sentences. Fill in the form of the verb indicated in parentheses after each sentence.

EXAMPLE: The Outer Banks —————— (stretch: present) along the North Carolina coast for more than 175 miles.

The Outer Banks <u>stretch</u> along the North Carolina coast for more than 175 miles.

1. Many portions of the Outer Banks of North Carolina —————— (give: present) the visitor a sense of history and timelessness.
2. Many students of history ————— (read: present perfect) about the Outer Banks and its mysteries.
3. It was on Roanoke Island in the 1580s that English colonists ————— (establish: past) the first settlement in the New World.
4. That colony vanished soon after it was settled, ————— (become: present participle) known as the famous "lost colony."
5. By 1718, the pirate Blackbeard ————— (made: past perfect) the Outer Banks a hiding place for his treasures.
6. It was at Ocracoke, in fact, that Blackbeard ————— (meet: past) his death.
7. Even today, fortune hunters ————— (search: present progressive) the Outer Banks for Blackbeard's hidden treasures.
8. The Outer Banks are also famous for Kitty Hawk and Kill Devil Hills; even as technology has advanced into the space age, the number of tourists flocking to the site of the Wright brothers' epic flight —————. (grow: present perfect progressive)

ANSWERS TO EXERCISE 3

1. give
2. have read
3. established
4. becoming
5. had made
6. met
7. are searching
8. has been growing
9. had claimed
10. will be enjoying

395

9. Long before that famous flight, however, the Outer Banks _____ (claim: past perfect) countless ships along its ever-shifting shores, resulting in its nickname—the "Graveyard of the Atlantic."

10. If the Outer Banks continue to be protected from the ravages of overdevelopment and commercialization, visitors _____ (enjoy: future progressive) the mysteries of this tiny finger of land for years to come.

23f Using the Correct Sequence of Tenses

The relationship among the verb tenses in a sentence is called the **sequence of tenses**. If the actions of all the verbs in a sentence occur at approximately the same time, the tenses should be the same.

> When Katharine Hepburn <u>walked</u> on stage, the audience <u>rose</u> and <u>applauded</u>.

Often, however, a single sentence contains several verbs describing actions that occur at different times. The tenses of the verbs must therefore shift. Which tense to use depends both on the sentence's meaning and on the nature of the clauses in which the verbs occur.

(1) Verbs in independent clauses

The tense of verbs that appear in adjacent independent clauses can shift as long as their relationships to their subjects and to each other are clear.

> The debate <u>was</u> not impressive, but the election <u>will determine</u> the winner.

(2) Verbs in dependent clauses

When a verb appears in a dependent clause, its tense depends on the tense of the main verb in the independent clause. When the main verb in the independent clause is in any tense except the past or past perfect, the verb in the dependent clause may be in any tense needed to convey meaning.

MAIN VERB	VERB IN DEPENDENT CLAUSE
Ryan <u>knows</u>	that the Beatles <u>acted</u> in three movies.
Senator Mikulski <u>will explain</u>	why she <u>changed</u> her position.

When the main verb is in the past tense, the verb in the dependent clause is usually in the past or past perfect tense. When the main verb is in the past perfect, the verb in the dependent clause is usually in the past tense.

MAIN VERB	VERB IN DEPENDENT CLAUSE
George Hepplewhite <u>was</u> an English cabinetmaker	who <u>designed</u> distinctive chair backs.
The battle <u>had ended</u>	by the time reinforcements <u>arrived</u>.

(3)   Infinitives in verbal phrases

When an **infinitive** appears in a verbal phrase, the tense it expresses depends on the tense of the main verb in the independent clause. The *present infinitive* (*to* plus the base form of the verb) indicates an action happening at the same time as or later than the main verb. The *perfect infinitive* (*to have* plus the past participle) indicates action happening earlier than the main verb.

MAIN VERB	INFINITIVE
I <u>went</u>	<u>to see</u> the Eagles play last week. (The going and the seeing occurred at the same time.)
I <u>want</u>	<u>to see</u> the Eagles play tomorrow. (Wanting is in the present, and seeing is in the future.)
I would <u>like</u>	<u>to have seen</u> the Eagles play. (Liking occurs in the present, and seeing would have occurred in the past.)

(4)   Participles in verbal phrases

When a **participle** appears in a verbal phrase, its tense depends on the tense of the main verb in the independent clause. The *present participle* indicates action happening at the same time as the action of the main verb. The *past participle* or the *present perfect participle* indicates action occurring before the action of the main verb.

PARTICIPLE	MAIN VERB
<u>Addressing</u> the 1896 Democratic Convention,	William Jennings Bryan <u>delivered</u> his Cross of Gold speech. (The addressing and the delivery occurred at the same time.)

397

1. declares
2. has learned
3. was
4. have given
5. became (happened before the action of the play)
6. married (happened before the action of the play)
7. accuses
8. learns

Having <u>written</u> her term paper, Camille <u>studied</u> for her history final. (The writing occurred before the studying.)

EXERCISE 4

From inside each pair of parentheses, choose the correct verb form. Make certain you use the correct sequence of verb tenses and are able to explain your choices.

EXAMPLE: Sophocles ———— (won, had won) his first victory in the Athenian spring drama competition in 468 B.C.

Sophocles <u>won</u> his first victory in the Athenian spring drama competition in 468 B.C.

1. In Sophocles' famous tragedy *Oedipus Rex*, Oedipus ———— (declares, declared) that the murderer of King Laios, his predecessor to the throne, will be found and removed from the city of Thebes.
2. His declaration comes after he ———— (learned, has learned) that the presence of the murderer has caused the plague on the city.
3. Sophocles ———— (was, is) one of the three great ancient Greek writers of tragedy; in keeping with the characteristics of tragedy, he portrayed Oedipus as a character with a tragic flaw.
4. By the time Oedipus learns of the presence of the murderer in the city, the citizens ———— (gave, have given) up hope of restoring the city to its former glory.
5. Oedipus came to the city just after King Laios' death, and when he solved the riddle of the Sphinx, he ———— (becomes, became) the new king.
6. Having been widowed as a result of the king's death, Queen Iocaste ———— (had married, married) Oedipus.
7. When the blind prophet Tiresias says that Oedipus is the murderer being sought, Oedipus ———— (accused, accuses) Tiresias of being involved in a plot against him.
8. In spite of his protestations, Oedipus ———— (learned, learns) that he is indeed the murderer and, worse, the son of his wife.

23g Mood

Mood is the form a verb takes to indicate whether a writer is making a statement, asking a question, giving a command, or expressing a wish or a contrary-to-fact statement.

INDICATIVE, IMPERATIVE, AND SUBJUNCTIVE MOOD

Indicative mood

Use

To express an opinion, state a fact, or ask a question

Example

Margaret Sanger <u>thought</u> that family planning <u>was necessary</u> for social progress.
<u>Did</u> the Phonecians <u>develop</u> a phonetic alphabet?

Imperative mood

Use

To give commands and make direct requests

Example

[You] <u>Read</u> the next chapter.
[You] Please <u>finish</u> today.
Let us <u>discuss</u> the images of minorities created by television.
<u>Let's check</u> our lab report.

Subjunctive mood

Use

With *that* clauses, contrary-to-fact statements, and certain idiomatic expressions

Example

The committee recommended that Houston <u>be</u> the site of next year's meeting.
Hamlet acted as if he <u>were</u> mad.

Mood

Students generally have little trouble with the *indicative* and *imperative* moods, but may struggle with the *subjunctive*, which today is used primarily in formal written English and less in "everyday" speech. For example, although we may write the sentence using subjunctive mood—"If that tree *were* taller, it would shade the house"—we are more likely to *say* it as follows: "If that tree *was* taller, it would shade the house." The subjunctive survives in many longstanding English catch phrases and idioms: *Be that as it may . . . Long live the Queen . . . Far be it for me . . . Lest it be thought . . . ,* and in "poetical" language *"Would that I had wings!"*

Make it clear that students should NOT assume that every *if* clause must be followed by a subjunctive. *If* clauses sometimes introduce a supposition, concede a point, or present a hypothetical situation; in most cases the verb is in the indicative mood. For example, the *if* in the following sentence introduces a *supposition:* "If my efforts *are* (NOT *be*) successful, I will win the fellowship next year."

23h   Using the Indicative Mood

The **indicative** mood expresses an opinion, states a fact, or asks a question. It may be used along with a form of *do* for emphasis.

Jackie Robinson <u>had</u> an impact on American professional baseball.

<u>Did</u> Margaret Mead <u>say</u> that behavioral differences are rooted in culture?

Shifts in Tense and Mood
Student essays and research papers often suffer from awkward or illogical shifts in tense and mood. Ask students to collect recent samples of their writing and to review them for unintentional shifts of this nature. (A review of pertinent sections of Chapter 17 may be useful.) You may want to highlight the following two errors, which are found all too often in student writing.

- *unnecessary shifts from indicative to imperative mood, particularly when offering "examples"*: "The Richter scale is a measure of the energy released by an earthquake. Take, for example, the Bay area 'temblor' of October, 1988." Ask students to look for and correct similar shifts in their own papers—or to defend mood shifts they view as being necessary and logical.

- *shifts in tense when describing or discussing narrative fiction*: "Hamlet *finds* the king praying, but he *was* unable to act, and *lets* his best opportunity for revenge slip away." Because novels and short stories are told in past tense, it's easy to start in the present tense, and then slip into the past. Ask students to review their own writing and to correct faulty shifts in tense.

23i Using the Imperative Mood

The **imperative** mood is used in commands and direct requests. Usually the imperative includes only the base form of the verb without a subject. Writers including themselves in a command use *let's* or *let us* before the base form of the verb.

[You] <u>Use</u> a dictionary.
[You] Please <u>vote</u> today.
<u>Let us examine</u> Machiavelli's view of human nature.

23j Using the Subjunctive Mood

The **subjunctive** mood is used in certain *that* clauses, contrary-to-fact statements, and certain idiomatic expressions. The *present subjunctive* uses the base form of the verb, regardless of the subject. The *past subjunctive* has the same form as the past tense of the verb. (The auxiliary verb *be*, however, takes the form *were* regardless of the number or person of the subject.) The *past perfect subjunctive* has the same form as the past perfect.

Dr. Gorman suggested that I <u>study</u> the Cambrian period. (present subjunctive)
I wish I <u>were</u> going to Europe. (past subjunctive)
I wish I <u>had gone</u> to the review session. (past perfect subjunctive)

(1) *That* clauses

Use the subjunctive in *that* clauses after words such as *ask, suggest, require, recommend,* and *demand*.

The report <u>recommended that</u> juveniles <u>be</u> given mandatory counseling.
Captain Ahab <u>insisted that</u> his crew <u>hunt</u> the white whale.

CLOSE-UP

USING THE SUBJUNCTIVE MOOD

The subjunctive mood is not used as much as it once was. Because the subjunctive mood can seem overly formal, many people use the indicative mood in informal writing or speech. In your college writing, however, you should use the subjunctive mood when it is called for.

INFORMAL: I wish I was a better driver.

REVISED: I wish I were a better driver.

(2) Contrary-to-fact statements

Use the subjunctive in conditional statements that are contrary to fact, including statements that express a wish. A **conditional statement** begins with a dependent *if* clause that presents a condition and concludes with an independent clause that presents the effect of that condition. If the effect is even slightly possible, use the indicative mood for the verb in the *if* clause.

If a peace treaty is signed, the world will be safer. (A peace treaty is possible.)

If the condition is impossible or **contrary to fact**, use the subjunctive mood for the verb in the *if* clause.

If John were there, he would have seen Marsha. (John was not there.)

NOTE: A conditional clause beginning with *as if* is contrary to fact and should be in the subjunctive mood.

The father acted as if he were having the baby. (The father couldn't be having the baby.)

A **wish** is a condition that does not exist; therefore, it should be expressed in the subjunctive mood.

I wish I were more organized.

(3) Idiomatic expressions

The subjunctive is used in some special expressions.

If need be, we will stay up all night to finish the report.

Come what may, they will increase their steel production.

401

Far <u>be</u> it for me to correct an expert.

Special interest groups have, as it <u>were</u>, shifted the balance of power.

EXERCISE 5

Complete the sentences in the following paragraph by inserting the appropriate form (indicative, imperative, or subjunctive) of the verb in parentheses. Be prepared to explain your choices.

Harry Houdini was a famous escape artist. He _____ (perform) escapes from every type of bond imaginable: handcuffs, locks, straitjackets, ropes, sacks, and sealed chests underwater. In Germany workers _____ (challenge) Houdini to escape from a packing box. If he _____ (be) to escape, they would admit that he _____ (be) the best escape artist in the world. Houdini accepted. Before getting into the box he asked that the observers _____ (give) it a thorough examination. He then asked that a worker _____ (nail) him in the box. "_____ (place) a screen around the box," he ordered after he had been sealed inside. In a few minutes Houdini _____ (step) from behind the screen. When the workers demanded that they _____ (see) the box, Houdini pulled down the screen. To their surprise they saw the box with the lid still nailed tightly in place.

23k Voice

Voice is the form a verb takes to indicate whether its subject acts or is acted upon. When the subject of a verb does something—that is, acts—the verb is in the **active voice**. When the subject of a verb receives the action—that is, is acted upon—the verb is in the **passive voice**.

ACTIVE VOICE: <u>Buffalo Bill killed</u> over four thousand buffalo in one year.

PASSIVE VOICE: <u>Over four thousand buffalo were killed</u> in one year by Buffalo Bill.

23l Using the Passive Voice

(1) When the doer is unknown or unimportant

The passive voice enables you to emphasize what happened when the person or thing acting is unknown or unimportant.

DDT <u>was found</u> in local soil samples. (Passive voice emphasizes finding of DDT, not who found it.)

We <u>were required</u> to embroider and I had trunkfuls of colorful dish towels, pillowcases, runners, and handkerchiefs to my credit. (Maya Angelou, *I Know Why the Caged Bird Sings*) (Passive voice emphasizes what they had to do, not who required them to do it.)

CLOSE-UP

VERBS

Because the active voice emphasizes the doer of an action, it is usually briefer, clearer, and more emphatic than the passive voice. For this reason, you should use active constructions in your college writing. Some situations, however, require use of the passive voice. Keep in mind that you should use passive constructions only when you have good reason to do so.

▶ **See 10e**

Passive constructions place emphasis on the subjects of sentences 2, 4, and 5—*colors*, *combinations*, and *colors*, respectively. Passive construction (*was to be added*) in sentence 5 functions as an adjective modifying *color*. Sentences 2 and 4 contain passive verbs (*were limited*, *were discovered*).

(2) When the recipient of the action should logically receive the emphasis

The passive voice also enables you to emphasize the logical object of the action, which becomes the subject of the sentence.

Malted Milk <u>was patented</u> in 1881 by James and William Horlick as a food supplement for infants. (Passive voice emphasizes Malted Milk, not those who patented it.)

King Tut's tomb, which was filled with gold and jewels, <u>was discovered</u> in 1922 by E. S. M. Herbert. (Edith Hamilton, *The Greek Way*) (Passive voice emphasizes King Tut's tomb, not who discovered it.)

Grits, a dish made of ground corn, <u>is eaten</u> by people throughout the southern United States. (Passive voice emphasizes grits, not those who eat it.)

*EXERCISE 6

Read the following paragraph and determine which verbs are active and which are passive. Comment if you can on why the author used the passive voice in each case.

By the beginning of the seventeenth century, European fireworks technicians could create elaborate flares that exploded into historic scenes and figures of famous people, a costly and lavish entertainment that was popular at the French royal place at Versailles. For eight cen-

The Chinese invented rockets about 1000 A.D. They packed gunpowder into bamboo tubes and ignited it by means of a fuse. Soldiers fired these rockets at enemy armies and usually caused panic. In thirteenth-century England Roger Bacon introduced an improved form of gunpowder. As a result, soldiers used rockets as a common—although unreliable—weapon in battle. In the early eighteenth century William Congreve, an English artillery expert, constructed a twenty-pound rocket that traveled almost two miles. By the late nineteenth century the physicist Ernst Mach gave thought to supersonic speeds. He predicted the sonic boom. In America, Robert Goddard launched the first liquid fuel rocket in 1926. He wrote a pamphlet that anticipated almost all future rocket developments. As a result of his pioneering work, he is called the father of modern rocketry.

turies, though, the colors of firework explosions were limited mainly to yellows and reddish amber. It was not until 1830 that chemists produced metallic zinc powders that yield a greenish-blue flare. Within the next decade, combinations of chemicals were discovered that gave starlike explosions in, first pure white, then bright red, and later a pale whitish blue. The last and most challenging basic color to be added to the fireworks palette, in 1845, was a brilliant pure blue. By midcentury, all the colors we enjoy today had arrived.

23m Changing from Passive to Active Voice

You can change a verb from passive to active voice by making the subject of the passive verb the object of the active verb. The person or thing performing the action then becomes the subject of the new sentence.

PASSIVE: The novel *Frankenstein* <u>was written</u> by Mary Shelley.

ACTIVE: Mary Shelley <u>wrote</u> the novel *Frankenstein*.

You can easily change a verb from passive to active if the sentence contains an *agent* that performs the action. Often the word *by* follows the passive verb *(written by Mary Shelley)*, indicating the agent that can become the subject of an active verb.

If a passive verb has no agent, supply a subject for the active verb; if you cannot, keep the passive construction.

PASSIVE: Baby elephants are taught to avoid humans. (By whom are baby elephants taught?)

ACTIVE: <u>Adult elephants</u> teach baby elephants to avoid humans.

EXERCISE 7

Determine which verbs in the following paragraph should be changed from the passive to the active voice. Rewrite the sentences containing these verbs and be prepared to explain your changes.

Rockets were invented by the Chinese about A.D. 1000. Gunpowder was packed into bamboo tubes and ignited by means of a fuse. These rockets were fired by soldiers at enemy armies and usually caused panic. In thirteenth-century England an improved form of gunpowder was introduced by Roger Bacon. As a result, rockets were used in battles and were a common—although unreliable—weapon. In the early eighteenth century a twenty-pound rocket that traveled almost two miles was constructed by William Congreve, an English artillery expert. By the late nineteenth century thought was given to supersonic speeds by the physicist Ernst Mach. The sonic boom was predicted by him. The first liquid fuel rocket was launched by the American Robert Goddard in 1926. A pamphlet written by him anticipated almost all future rocket developments. As a result of his pioneering work, he is called the father of modern rocketry.

23n Changing from Active to Passive Voice

You can change verbs from active to passive voice by making the object of the active verb the subject of the passive verb. The subject of the active verb then becomes the object of the passive verb.

ACTIVE: Sir James Murray <u>compiled</u> the *Oxford English Dictionary.*

PASSIVE: The *Oxford English Dictionary* <u>was compiled</u> by Sir James Murray.

Remember that an active verb must have an object or it cannot be put into the passive voice. If an active verb has no object, supply one. This will become the subject of the passive sentence.

ACTIVE: Jacques Cousteau invented.

Cousteau invented _____?_____

PASSIVE: _____?_____ was invented by Jacques Cousteau.

The self-contained underwater breathing apparatus (scuba) was invented by Jacques Cousteau.

EXERCISE 8

Determine which sentences in the following paragraph would be more effective in the passive voice. Rewrite those sentences, making sure that you can explain the reasons for your choices.

The Regent Diamond is one of the world's most famous and coveted jewels. A slave discovered the 410-carat diamond in 1701 in an Indian mine. Over the years, people stole and sold the diamond several times. In 1717, the Regent of France bought the diamond for an enormous sum, but during the French Revolution, it disappeared again. Someone later found it in a ditch in Paris. Eventually, Napoleon had the diamond set into his ceremonial sword. At last, when the French monarch fell, the government placed the Regent Diamond in the Louvre, where it still remains to be enjoyed by all.

Your students answers will vary.

The Regent Diamond is one of the world's most famous and coveted jewels. The 410-carat diamond was discovered by a slave in 1701 in an Indian mine. (Emphasis is on the diamond rather than on who discovered it.) Over the years, it was stolen and sold several times. (Emphasis is on what happened rather than on *people*.) In 1717, the Regent of France bought the diamond for an enormous sum, but during the French Revolution, it disappeared again. It was later discovered in a ditch in Paris. (Emphasis is on its eventual discovery rather than on who discovered it.) Eventually, Napoleon had the diamond set into his ceremonial sword. At last, when the French monarch fell, the Regent Diamond was placed in the Louvre where it still remains to be enjoyed by all. (Emphasis is on where it was placed rather than on who placed it there.)

Verbs

Possible Revision

How Fast Did That Piece of Paper Fall?

Most people who have never studied physics assume that heavier objects fall faster than light objects. Aristotle, the brilliant philosopher of Ancient Greece, made the same assumption. He believed that heavy objects naturally tend to be closer to the ground than light objects. For this reason, they must fall faster than light objects. Aristotle assumed this explanation conformed to common sense. Is it true that heavier bodies fall faster than light ones do?

In the seventeenth century, Galileo Galilei, an Italian scientist, laid the groundwork for modern physics. He recognized that heavier objects do not always fall faster than lighter objects. As a result of Galileo's analysis of falling bodies, students now know that in the absence of air resistance, all objects fall at the same rate.

By repeating Galileo's experiments in the classroom, one can test his principle and prove the validity of his ideas while disproving Aristotle's assumptions. When two objects—a stone and a flat piece of paper—that are lying on a table are raised to equal distances above the ground and let go, the heavier object will reach the ground first. This experiment seems to confirm Aristotle's belief that heavier objects fall faster than light objects. However, when the experiment is repeated, this time crumpling the flat piece of paper into a wad, both objects will reach the ground at the same time. How can this difference be explained?

continued on following page

VERBS

This is a draft of a paper written for a technical writing class. The student was told to write an essay in which she explained a basic scientific principle to readers who had little or no understanding of science. As you read, look for inaccuracies or inconsistencies in verb form, tense, and mood, and finally, make sure the writer has made effective use of both passive and active voice. You may add and delete words and phrases as well as rearrange sentences. After you have corrected this draft, go over it again and, if necessary, revise further for coherence, unity, and style.

How Fast Did That Piece of Paper Fall?

Most people who never studied physics assume that heavier objects fall faster than light objects will. This assumption was also made by Aristotle, the brilliant philosopher of ancient Greece. He believed that heavy objects naturally tended to be closer to the ground than light objects. For this reason, they must have fell faster than light objects. This explanation was assumed by Aristotle to conform to common sense. Was it true that heavier bodies fell faster than lighter ones did?

In the seventeenth century, Galileo Galilei, an Italian scientist, laid the groundwork for modern physics. It was recognized by him that heavier objects did not always fall faster than lighter objects. As a result of Galileo's analysis of falling bodies, it is now known by students that in the absence of air resistance, all objects fall at the same rate.

By repeating his experiments in the classroom, you can test Galileo's principle. At the same time you prove the validity of Galileo's ideas, the assumptions of Aristotle can be disproved. When you raise two objects--a stone and a flat piece of paper--that were laying on a table to equal distances above the ground and let go, you should see the heavier object reach the ground first. This experiment would seem to confirm Aristotle's belief that heavier objects fall faster than light ones. Repeat the experiment, this time crumpling the flat piece of paper into a wad. Now the two objects should reach the ground at the same time. How is this difference explained?

The key to explaining the difference lies in Galileo's principle. The paper, whose weight was the same whether it was flat or crumpled up, falls faster when it is in a wad because it has a smaller cross-sectional area. It is the cross-sectional area of an object that determines the rate at which it will drop: The larger

continued on the following page

continued from the previous page

the area, the more air resistance the object encountered and the slower it will have dropped. (Keep in mind, however, that the air resistance factor is significant only when an object has a large cross-section compared to its weight.) If the flat piece of paper was dropped in a vacuum--where there is no air resistance--it would have dropped as fast as the heavy object. This, then, is why the phrase "in the absence of air resistance" is added to Galileo's principle.

continued from previous page

The key to explaining the difference lies in Galileo's principle. The paper, whose weight is the same whether it is flat or crumpled up, falls faster when it is in a wad because it has a smaller cross-sectional area. It is the cross-sectional area of an object that determines the rate at which it will drop: The larger the area, the more air resistance the object encounters and the more slowly it drops. (Keep in mind, however, that the air resistance factor is significant only when an object has a large cross-sectional area compared to its weight.) If the flat piece of paper is dropped in a vacuum—where there is no air resistance—it will drop as quickly as the heavy object. This, then, is why the phrase "in the absence of air resistance" is added to Galileo's principle.

Commentary on Possible Revision
Although the writer has revised the paper for inconsistencies in verb tense, form, and mood and has made effective use of active and passive voice, the essay's style and clarity can still be improved.

CHAPTER 24

Agreement

Agreement is the correspondence between words in number, gender, or person. Subjects and verbs agree in **number** (singular or plural) and **person** (first, second, or third); pronouns and their antecedents agree in number, person and **gender** (masculine, feminine, or neuter).

 Subject-Verb Agreement

Verbs should agree in number and person with their subjects: Singular subjects have singular verbs, and plural subjects have plural verbs.

> **SINGULAR:** Hydrogen peroxide is an unstable compound.

> **PLURAL:** The characters are not well developed in most of O. Henry's short stories.

Present tense verbs, except *be* and *have*, add *-s* or *-es* when the subject is third-person singular. Third-person singular subjects include nouns; the personal pronouns *he, she, it,* and *one;* and many indefinite pronouns.

> The president has the power to veto congressional legislation.

> She frequently cites statistics to support her assertions.

> In every group somebody emerges as a natural leader.

Present tense verbs do not add *-s* or *-es* when the subject is first-person singular *(I)*, first-person plural *(we)*, second-person singular or plural *(you)*, or third-person plural *(they)*.

I <u>recommend</u> that dieters avoid processed meat because of its high salt content.

In our Bill of Rights, <u>we</u> <u>guarantee</u> all defendants the right to a speedy trial.

At this stratum, <u>you</u> <u>see</u> rocks dating back fifteen million years.

<u>They</u> <u>say</u> that some wealthy people default on their student loans.

Subject-verb agreement is generally straightforward, but some situations can be troublesome.

(1) Intervening phrases

If a modifying phrase comes between subject and verb, the verb should agree with the subject, not with a word in the intervening phrase.

The <u>sound</u> of the drumbeats <u>builds</u> in intensity in *The Emperor Jones*.

The <u>games</u> won by the intramural team <u>are</u> usually few and far between.

CLOSE-UP

SUBJECT-VERB AGREEMENT

When phrases introduced by *along with, as well as, in addition to, including,* and *together with* come between subject and verb, the intervening phrases do not change the subject's number.

Heavy <u>rain</u>, together with high winds, <u>causes</u> hazardous driving conditions along the Santa Monica Freeway.

(2) Compound subjects joined by *and*

Compound subjects joined by *and* usually take plural verbs.

<u>Air bags and antilock brakes</u> <u>are</u> available on all models.

There are, however, two exceptions to this rule.

- Some compound subjects joined by and stand for a single idea or person. These should be treated as a unit and given singular verbs.

<u>Rhythm and blues</u> <u>is</u> a forerunner of rock and roll.

QUOTATION

"Here the writer is in the position of the golfer about to tee off whose eye is caught by a waving dandelion or by an ant crawling up a blade of grass; he takes his eye off the ball and flubs his stroke. Watch a couple of duffers in action: 'Scientists say it is difficult to convert horsepower into pounds of thrust since the velocity [*the ball—keep your eye on it*] of the rockets [*the ant—forget it*] are [*swoosh*] not known. . . .'"

–Theodore M. Bernstein,
The Careful Writer

IN THE CLASSROOM

Ask students to compose several sentence openings with indefinite pronouns as subjects. Write each partial sentence on the board, and ask another student to complete it. For example:

Write:	Complete:
Both of the teachers . . .	*are* challenging.
Anyone 18 or older . . .	*is* eligible.
Something about it . . .	*seems* odd.

As the students find it easier to choose the correct singular or plural verb form, ask for more challenging examples (for instance, with modifying phrases intervening between subject and verb).

• When *each* or *every* precedes a compound subject joined by *and*, the subject also takes a singular verb.

Every nook and cranny was searched before the purloined letter was found in plain sight on the mantel.

(3)   Compound subjects joined by *or*

Compound subjects joined by *or* or by *either . . . or* or *neither . . . nor* may take singular or plural verbs.

If both subjects are singular, use a singular verb; if both subjects are plural, use a plural verb.

Either radiation or chemotherapy is combined with surgery for the most effective results. (Both parts of the compound subject, *radiation* and *chemotherapy*, are singular, so the verb is singular.)

Either radiation treatments or chemotherapy sessions are combined with surgery for the most effective results. (Both parts of the compound subject, *treatments* and *sessions*, are plural, so the verb is plural.)

When a singular and a plural subject are linked by *or,* or by *either . . . or, neither . . . nor,* or *not only . . . but also,* the verb should agree with the subject that is nearer to it.

Either radiation treatments or chemotherapy is combined with surgery for the most effective results. (Singular verb agrees with *chemotherapy,* the part of the compound subject closer to it.)

Either chemotherapy or radiation treatments are combined with surgery for the most effective results. (Plural verb agrees with *treatments,* the part of the compound subject closer to it.)

CLOSE-UP

SUBJECT-VERB AGREEMENT

When a compound subject is made up of nouns and pronouns that differ in person, the verb should agree in person as well as in number with the nearest element of the compound subject.

Neither my running mate nor I wish to contest the election.

Neither I nor my running mate wishes to contest the election.

(4) Indefinite pronouns

In most cases, use a singular verb with an indefinite pronoun as a subject. Although some **indefinite pronouns**—*both, many, few, several, others*—are always plural, most—*another, anyone, everyone, one, each, either, neither, anything, everything, something, nothing, nobody,* and *somebody*—are singular.

> Anyone is welcome to apply for a grant, providing certain financial qualifications are met.

> Each of the chapters includes a review exercise.

CLOSE-UP

SUBJECT-VERB AGREEMENT

Some indefinite pronouns—*some, all, any, more, most,* and *none*—can be singular or plural. In these cases the noun to which the pronoun refers determines whether the verb form should be singular or plural.

> Of course, some of this trouble is to be expected. (*Some* refers to *trouble;* therefore, the verb is singular.)

> Some of the spectators are getting restless. (*Some* refers to *spectators;* therefore, the verb is plural.)

(5) Collective nouns

A **collective noun** names a group of persons or things—for instance, *navy, union, association, band.* A collective noun is always singular in form. When it refers to a group as a unit, it takes a singular verb; when it refers to the individuals or items that make up the group, it takes a plural verb.

> To many people the royal family symbolizes Great Britain. (The family, as a unit, is the symbol.)

> The family all eat at different times. (Each member eats separately.)

Sometimes, however, even when usage is correct, a plural verb sounds awkward with a collective noun. If this is the case, rewrite the sentence to eliminate the awkwardness.

> The family members all eat at different times.

Phrases that name a fixed amount—*three-quarters, twenty dollars, the majority*—are treated like collective nouns. When the amount is considered as a unit, it takes a singular verb; when it denotes part of the whole, it takes a plural verb.

Three-quarters of his usual salary is not enough. (Three-quarters denotes a unit.)

Three-quarters of workshop participants improve dramatically. (Three-quarters denotes part of the group.)

CLOSE-UP

SUBJECT-VERB AGREEMENT

The number is always singular and *a number* is always plural.

The number of voters has declined.

A number of students have missed the opportunity to pre-register.

(6)   Singular subjects with plural forms

Be sure to use a singular verb with a singular subject, even if the form of the subject is plural.

Politics makes strange bedfellows.

Statistics deals with the collection, classification, analysis, and interpretation of data.

In certain contexts, however, some of these words may actually have plural meanings. In these cases, a plural verb should be used.

Her politics are too radical for her parents. (*Politics* refers not to the science of political government but to political principles or opinions.)

The statistics prove him wrong. (*Statistics* denotes not a body of knowledge but the numerical facts or data themselves.)

Be sure that titles of individual works take singular verbs, even if their form is plural.

The Grapes of Wrath describes the journey of migrant workers and their families from the Dust Bowl to California.

This convention also applies to words referred to as words, even if they are plural.

Good *vibes* is a 1960s slang term meaning "positive feelings."

CLOSE-UP

SUBJECT-VERB AGREEMENT

Remember that plurals of foreign words do not always look like English plural forms. Be particularly careful to use the correct verb with those nouns.

criterion	is
criteria	are
media	are

▶ **See 20b8**

(7) Inverted subject-verb order

Be sure a verb agrees with its subject, even when the verb precedes the subject, as in questions and sentences beginning with *there is* or *there are*.

Is either answer correct?

There is a monument to Emiliano Zapata in Mexico City.

There are currently twelve circuit courts of appeals in the federal court system.

(8) Linking verbs

Be sure **linking verbs** agree with their subjects, not with the subject complement.

The problem was termites.

Here the verb *was* agrees with the subject *problem,* not with the subject complement *termites.* If *termites* were the subject, the verb would be plural.

Termites were the problem.

1. Subject and verb agree in spite of intervening words.
2. Compound subject preceded by *every* takes singular verb.
3. Singular subject takes singular verb despite plural form.
4. Collective noun referring to the group as a unit takes a singular verb.
5. Verb agrees with the singular antecedent of the relative pronoun.
6. Singular indefinite pronoun takes singular verb.
7. Linking verb agrees with subject, not with subject complement.
8. Linking verb agrees with subject, not with subject complement.
9. Singular subject in plural form takes singular verb.
10. Subject and verb agree in inverted word order.

(9) Relative pronouns

Because **relative pronouns** have the same form for singular and plural, they provide no clues to subject-verb agreement. When you use a relative pronoun (*who, which, that*) to introduce a dependent clause, the verb in that clause should agree in number with the pronoun's antecedent.

The farmer is among the ones who suffer during a grain embargo.

Here the verb *suffer* agrees with the antecedent *ones* of the relative pronoun *who*. Compare the following sentence.

The farmer is the only one who suffers during the grain embargo.

Now the verb agrees with the antecedent *one,* which is singular.

EXERCISE 1

Each of these ten correct sentences illustrates one of the conventions just explained. Read the sentences carefully and explain why each verb form is used in each case.

> **EXAMPLE:** *Harold and Maude* is a popular cult film. (The verb is singular because the subject *Harold and Maude* is the title of an individual work, even though it is plural in form.)

1. Jack Kerouac, along with Allen Ginsberg and William S. Burroughs, was a major figure in the "beat" movement.
2. Every American boy and girl needs to learn basic computational skills.
3. Aesthetics is not an exact science.
4. The audience was restless.
5. The Beatles' *Sergeant Pepper* album is one of those albums that remain popular long after the time they are issued.
6. All is quiet.
7. The subject was roses.
8. When he was young, Benjamin Franklin's primary concern was books.
9. Fifty dollars is too much to spend on one concert ticket.
10. "There are more things in heaven and earth, Horatio, than are dreamt of in your philosophy."

EXERCISE 2

Some of these sentences are correct, but others illustrate common errors in subject-verb agreement. If a sentence is correct, mark it with a *C*: if it has an error, correct it.

1. *I Love Lucy* is one of those television shows that almost all Americans have seen at least once.
2. The committee presented its findings to the president.
3. Neither Western novels nor science fiction appeal to me.
4. Stage presence and musical ability makes a rock performer successful today.
5. *It's a Wonderful Life,* like many old Christmas movies, seems to be shown on television almost daily from Thanksgiving to New Year's Day.
6. Hearts are my grandmother's favorite card game.
7. The best part of B. B. King's songs are the guitar solos.
8. Time and tide waits for no man.
9. Sports are my main pastime.
10. *Vincent and Theo* is Robert Altman's movie about the French impressionist painter Van Gogh and his brother.

24b Pronoun-Antecedent Agreement

A pronoun must agree with its **antecedent**—the word or word group to which the pronoun refers. Singular pronouns—such as *he, him, she, her, it, me, myself,* and *oneself*—should refer to singular antecedents. Plural pronouns—such as *we, us, they, them,* and *their*—should refer to plural antecedents.

▶ See
21b

(1) More than one antecedent

In most cases use a plural pronoun to refer to two or more antecedents connected by *and,* even if one or more of the antecedents is singular.

<u>Mormonism</u> and <u>Christian Science</u> were influenced in <u>their</u> beginnings by Shaker doctrines.

However, if the compound antecedent denotes a single unit—one person, thing or idea—use a singular pronoun to refer to the compound antecedent.

In 1904 <u>the husband and father</u> brought <u>his</u> family from Poland to America.

When the compound antecedent is preceded by *each* or *every,* use a singular pronoun.

<u>Every programming language and software package</u> has <u>its</u> limitations.

Students use the construction "everybody . . . they" in everyday speech and may resist the idea that the construction (though accepted in conversation) is usually inappropriate in written English. Write sentences on the board to show students that the ending *-one* or *-body* requires a singular pronoun.

In some cases, the overuse of "they" is an effort to avoid sexist language. You might want to briefly review section 18f2 for ways to avoid sexism in the use of pronouns.

Use a singular pronoun to refer to two or more singular antecedents linked by *or* or *nor*.

Neither Thoreau nor Whitman lived to see his work read widely.

When one antecedent is singular and one is plural, however, the pronoun agrees in person and number with the closer antecedent.

Neither Great Britain nor the Benelux nations have experienced changes in their borders in recent years.

(2)   Collective noun antecedents

Occasionally collective noun antecedents may require plural pronouns. If the meaning of the collective noun antecedent is singular, use a singular pronoun. If its meaning is plural, use a plural pronoun.

The teachers' union was ready to strike for the new contract its members had been promised. (All the members act as one.)

When the whistle blew, the team left their seats and moved toward the court. (Each member acts individually.)

CLOSE-UP

PRONOUN-ANTECEDENT AGREEMENT

Within any one sentence a collective noun should be treated consistently as either singular or plural. When a single collective noun serves as both the subject of a verb and the antecedent of a pronoun, both verb and pronoun must agree with the noun.

INCORRECT: The teachers' union, ready to strike for the new contract its members had been promised, were still willing to negotiate.

Here the collective noun *union* is singular; the verb *were* is plural and, therefore, incorrect.

(3)   Indefinite pronoun antecedents

See ◄
24a4

Most indefinite pronouns—*each, either, neither, one, anyone,* and the like—are singular in meaning and should take singular pronouns. (Others may be plural and require plural pronouns.)

Neither of these men had his proposal ready by the application deadline.

Each of these neighborhoods is like a separate nation, with its own traditions and values.

Everyone will get basic instructions in the modern foreign language of his choice.

CLOSE-UP

PRONOUN-ANTECEDENT AGREEMENT

Everyone presents some special problems for writers. Because *everyone* is singular in meaning and does not specify gender, convention says that it should be referred to by the singular pronoun *his*. But indefinite pronouns really denote members of both sexes, so many writers feel that using *his* is inaccurate. In speech and in informal writing, it is common to use the plural pronouns *they* or *their* to refer to *everyone*. In college writing, however, this is not acceptable.

Though it can be somewhat cumbersome if overused, one solution to this problem is to use both the masculine and feminine pronouns.

▶ See 18f2

Everyone will get basic instruction in the modern foreign language of his or her choice.

Another solution is to change the subject and use a plural pronoun.

All students will get basic instruction in the modern foreign language of their choice.

EXERCISE 3

Find and correct any errors in subject-verb and/or pronoun-antecedent agreement.

1. The core of a computer is a collection of electronic circuits that are called the central processing unit.
2. Computers, because of advanced technology which allows the central processing unit to be placed on a "chip," a thin square of semiconducting material about 1/4 inch on each side, has been greatly reduced in size.

1. The core of a computer is a collection of electronic circuits that *is* called the central processing unit.
2. Computers, because of advanced technology which allows the central processing unit to be placed on a "chip," a thin square of semiconducting material about 1/4 inch on each side, *have* been greatly reduced in size.
3. No error
4. Pressing of keys . . . *generates* electronic signals that are "input" for the computer.
5. Computers have built-in memory storage, and equipment such as disks or tapes *provides* external memory.
6. RAM (random-access memory), the erasable and reusable computer memory, *holds* the computer program, the computations executed by the program, and the results.
7. No error
8. ROM (read-only memory), the permanent memory which is "read" by the computer but cannot be changed, *is* used to store programs that are needed frequently.
9. A number of arcade-style video games with sound and color *are* available for home computers.
10. Although some computer users write their own programs, most buy ready-made software programs such as the ones that *allow* a computer to be used as a word processor.

ANSWERS TO EXERCISE 4

You might wish to refer students to the section on sexist language (18f2) in conjunction with this exercise.

1. The governess and the cook are seemingly pursued by evil as they try to protect Miles and Flora from those seeking to possess the children's souls.
2. The insulin-dependent diabetic is now able to take advantage of new technology that can help alleviate his or her symptoms.
3. Every homeowner in shore regions worries about the possible effects of a hurricane on his or her property.
4. Federally funded job-training programs offer the unskilled worker an opportunity to acquire skills he or she can use to secure employment.
5. The foreign import poses a major challenge to the American automobile market.
6. *Brideshead Revisited* tells how one family affects Charles Ryder.
7. *Writer's Digest* and *The Writer* are designed to aid the writer as he or she seeks markets for his or her work.
8. Almost every American family has access to television; in fact, more have televisions than have indoor plumbing.
9. In Montana, it seems as though all the towns' elevations are higher than their populations.
10. Women without men are like fish without bicycles.

3. Computers can "talk" to each other over phone lines through a modem, an acronym for modulator-demodulator.
4. Pressing of keys on keyboards resembling typewriter keyboards generate electronic signals that are "input" for the computer.
5. Computers have built-in memory storage, and equipment such as disks or tapes provide external memory.
6. RAM (random-access memory), the erasable and reusable computer memory, hold the computer program, the computations executed by the program, and the results.
7. After computer programs are "read" from a disk or tape, the computer uses the instructions as needed to execute the program.
8. ROM (read-only memory), the permanent memory which is "read" by the computer but cannot be changed, are used to store programs that are needed frequently.
9. A number of arcade-style video games with sound and color is available for home computers.
10. Although some computer users write their own programs, most buy ready-made software programs such as the ones that allows a computer to be used as a word processor.

EXERCISE 4

The following ten sentences illustrate correct subject-verb and pronoun-antecedent agreement. After following the instructions in parentheses after each sentence, revise each so its verbs and pronouns agree with the newly created subject.

> **EXAMPLE:** One child in ten suffers from a learning disability. (Change *one child in ten* to *ten percent of all children.*)
>
> Ten percent of all children suffer from a learning disability.

1. The governess is seemingly pursued by evil as she tries to protect Miles and Flora from those she feels seek to possess the children's souls. (Change *The governess* to *The governess and the cook.*)
2. Insulin-dependent diabetics are now able to take advantage of new technology that can help alleviate their symptoms. (Change *diabetics* to *the diabetic.*)
3. All homeowners in shore regions worry about the possible effects of a hurricane on their property. (Change *All homeowners* to *Every homeowner.*)
4. Federally funded job-training programs offer unskilled workers an opportunity to acquire skills they can use to secure employment. (Change *workers* to *the worker.*)
5. Foreign imports pose a major challenge to the American automobile market. (Change *Foreign imports* to *The foreign import.*)
6. *Brideshead Revisited* tells how one family and its devotion to its Catholic faith affect Charles Ryder. (Delete *and its devotion to its Catholic faith.*)

7. *Writer's Digest* and *The Writer* are designed to aid writers as they seek markets for their work. (Change *writers* to *the writer*.)
8. Most American families have access to television; in fact, more have televisions than have indoor plumbing. (Change *Most American families* to *Almost every American family*.)
9. In Montana it seems as though every town's elevation is higher than its population. (Change *Every town's elevation* to *All the towns' elevations*.)
10. A woman without a man is like a fish without a bicycle. (Change *A woman/a man* to *Women/men*.)

STUDENT WRITER AT WORK

AGREEMENT

Read the following draft of an English composition essay carefully, correcting all errors in subject-verb and pronoun-antecedent agreement. After you have corrected the errors, go over the draft again and, if necessary, revise further to strengthen coherence, unity, and style.

Marriage in the Ashanti Tribe

The Ashanti tribe is the largest in the small West African country of Ghana. The language of the Ashantis, Akan, is the most widely spoken in the country. The unity in the Ashanti tribe is derived from a golden stool which the Ashantis believe descended from the skies at the command of their chief priest. This unity has encouraged the Ashantis to create a system in which the family is so strong that the tribe has little need for formal support services. For instance, the tribe need few institutions to care for their orphans or homeless people. Among the Ashanti people, home and family means plenty of relatives, living and working and playing and worrying as well-knit units who live in single or neighboring households. Marriage among members of the Ashanti tribe is therefore a union of two families as well as of two individuals.

In Ashanti, marriage is less an agreement entered into by two individuals before God or the justice of the peace than it is a social contract between two families, each of which are represented by a partner to the marriage. Because a marriage binds two families together, it is not to be entered into hurriedly. In fact, everyone in the tribe fear the social consequences of an ill-conceived union. The families of both of the young people are active counselors during the courtship, and its wholehearted approval and

continued on the following page

STUDENT WRITER AT WORK

Agreement

Possible Revision

Marriage in the Ashanti Tribe

The Ashanti tribe is the largest in the small West African country of Ghana. The language of the Ashantis, Akan, is the most widely spoken in the country. The unity in the Ashanti tribe is derived from a golden stool which the Ashantis believe descended from the skies at the command of the chief priest. This unity has encouraged the Ashantis to create a system in which the family is so strong that the tribe has little need for formal support services. For instance, the tribe needs few institutions to care for its orphans or homeless people. Among the Ashanti people, home and family mean plenty of relatives, living and working and playing and worrying as well-knit units who live in single or neighboring households. Marriage among members of the Ashanti tribe is therefore a union of two families as well as of two individuals.

In Ashanti, marriage is less an agreement entered into by two individuals before God or the justice of the peace than it is a social contract between two families, each of which is represented by a partner to the marriage. Because a marriage binds two families together, it is not to be entered into hurriedly. In fact, everyone in the tribe fears the social consequences of an ill-conceived union. The families of both of the young people are active counselors during the courtship, and their wholehearted approval and endorsement are essential to the success of the marriage. Each family seeks the answers to many questions. For in-

continued on following page

continued from previous page

stance, are the bride and bridegroom of similar age? Has either been married before? If so, why did the previous marriage fail? What is the history of the family? Is the family in debt? Most important, of what clan is the family?

When all the questions have been answered satisfactorily, the man and woman are married. The respective troths—for bride and groom, for bride's family and groom's family—are plighted in a quiet ceremony without benefit of either clergy or justice of the peace. The crucial part of the ceremony is the giving of a small sum of money and various gifts and drinks by the family of the groom to that of the bride. The actual value of such payments is often small, amounting to about fifty dollars. (A royal family gives more and receives more.) This money is sometimes referred to as "bridewealth." It constitutes only a token of the agreement reached between bride and groom and between their families. In the giving and receiving of the gifts the young people and their families mutually pledge their faithfulness and support. When this transaction has been witnessed by both families, the man and woman are joined together as husband and wife. For better or for worse, they are married.

Commentary on Possible Revision

The writer has done a good job of editing the draft for errors of agreement. However, the draft still needs editing to make it smoother and more coherent. (Sentence 6 in paragraph 1, for example, is quite hard to follow.)

continued from the previous page

endorsement is essential to the success of the marriage. The family seek the answers to many questions. For instance, are the bride and bridegroom of similar age? Have either been married before? If so, why did the previous marriage fail? What is the history of the family? Is the family in debt? Most important, of what clan is the family?

When all the questions have been answered satisfactorily, the man and the woman are married. The respective troths—for bride and groom, for bride's family and groom's family—are plighted in a quiet ceremony without benefit of either clergy or justice of the peace. The crucial part of the ceremony is the giving of a small sum of money and various gifts and drinks by the family of the groom to that of the bride. The actual value of such payments are often small, amounting to about fifty dollars. (A royal family gives more and receives more.) This money is sometimes referred to as "bridewealth." It constitutes only a token of the agreement reached between bride and groom and between their families. In the giving and receiving of the gifts the young people and their families mutually pledge their faithfulness and support. When this transaction has been witnessed by both families, the man and woman are joined together as husband and wife. For better or for worse, they are married.

CHAPTER 25

Adjectives and Adverbs

25a Understanding Adjectives and Adverbs

Adjectives modify nouns and pronouns. **Adverbs** modify verbs; adjectives; other adverbs; or entire phrases, clauses, or sentences. Both adjectives and adverbs describe, limit, or qualify other words, phrases, or clauses.

The *function* of a word, not its *form,* determines whether it is classified as an adjective or an adverb. Although many adverbs (like *immediately* and *hopelessly*) end in *-ly,* others (like *almost* and *very*) do not. Moreover, some adjectives (like *lively*) end in *-ly.* Only by locating the modified word and determining its part of speech can you identify a modifier as an adjective or adverb.

25b Using Adjectives

Be sure to use an adjective—not an adverb—as a subject complement or an object complement. A **subject complement** is a word that follows a **linking verb** and modifies the sentence's subject, not its verb.

▶ See 21d

Because a subject complement modifies the subject—a noun or pronoun—it must be an adjective. Compare these two sentences.

Michelle seemed <u>brave</u>. (*Seemed* shows no action and is therefore a linking verb. Because *brave* is a subject complement that modifies the noun *Michelle,* the adjective form is used.)

Michelle smiled <u>bravely</u>. (*Smiled* shows action, so it is not a linking verb. *Bravely* modifies *smiled,* so it takes the adverb form.)

"Too many qualifiers weaken writing. E. B. White dramatically described such words as *rather, very, little,* and *pretty* as 'leeches that infest the pond of prose, sucking the blood of words.' No hedging there!"

–Jan Venolia,
Rewrite Right!

TEACHING STRATEGY

Using Adjectives and Adverbs
Compose a bare-bones paragraph, distribute copies, and ask students to add appropriate adjectives and adverbs. This could be a good collaborative exercise. Read some of the paragraphs to the class and discuss them. Which "new and improved" versions do the students like best? Are there any particularly striking or memorable sentences? Are any of the rewritten paragraphs guilty of (perhaps intentional) overblown imagery?

Discuss the danger of relying too heavily on adjectives and adverbs. Adjectives, says author William Zinsser in his book *On Writing Well,* should never "clutter" a sentence or be used to do the work of a flabby, poorly chosen noun or verb. Why write "attractive, virile young man" when you can write "hunk"? Why write "walked slowly and listlessly" when you can write "shuffled"? The rule: *look first for strong nouns and verbs*—and then for adjectives and adverbs that add something to them.

LINKING VERBS
Linking verbs show no action; their function is to connect a sentence's subject and complement. Words that are or can be used as linking verbs include *seem, appear, believe, become, grow, turn, remain, prove, look, sound, smell, taste, feel,* and forms of the verb *be.*

PLACEMENT OF ADJECTIVES
Adjectives are commonly placed close to the nouns or pronouns they modify. They most often appear immediately *before* nouns and directly *after* linking verbs, direct objects, and indefinite pronouns.

They bought two shrubs for the yard. (before noun)

The name seemed familiar. (after linking verb)

The coach ran them ragged. (after direct object)

Anything sad makes me cry. (after indefinite pronoun)

Two or more adjectives can be placed *after* the noun or pronoun they modify.

The expedition, long and arduous, ended in triumph.

PLACEMENT OF ADVERBS
Adverbs are also usually located close to the words they modify. However, they may occur in a greater variety of positions.

He walked slowly across the room.

Slowly he walked across the room.

He slowly walked across the room.

He walked across the room slowly.

Sometimes the same verb can either serve as a linking verb or convey action. Compare these two sentences.

He remained <u>stubborn</u>. (He was still stubborn.)

He remained <u>stubbornly</u>. (He remained, in a stubborn manner.)

When a word following a sentence's direct object modifies that object and not the verb, it is an **object complement**. Objects are nouns or pronouns, so their modifiers must be adjectives.

Most people called him <u>timid</u>. (*Timid* modifies *him*, the sentence's direct object, so the adjective form is correct.)

Most people called him <u>timidly</u>. (*Timidly* modifies the verb *called*—not the object—so the adverb form is correct.)

25c Using Adverbs

Be sure to use an adverb—not an adjective—to modify verbs; adjectives; other adverbs; or entire phrases, clauses, or sentences.

▶ See 21e

FAULTY: The majority of the class did <u>great</u> on the midterm. (adjective form used to modify verb)

My parents dress a lot more <u>conservative</u> than my friends do. (adjective form used to modify verb)

REVISED: The majority of the class did <u>well</u> (or <u>very well</u>) on the midterm.

My parents dress a lot more <u>conservatively</u> than my friends do.

CLOSE-UP

ADJECTIVES AND ADVERBS

In informal speech adjective forms such as *good, bad, sure, real, slow, quick,* and *loud* are often used to modify verbs, adjectives, and adverbs. In college writing, however, be sure to avoid these informal modifiers and to use adverbs to modify verbs, adjectives, and other adverbs.

FAULTY: The program ran <u>good</u> the first time we tried it, but the new system performed <u>bad</u>.

REVISED: The program ran <u>well</u> the first time we tried it, but the new system performed <u>badly</u>.

TEACHING STRATEGY

Adverbs

In everyday conversation, the same word is often used as both an adjective and an adverb: "She has a *loud* voice . . ." "Talk *loud* or I won't be able to hear you . . ." "You have a *quick* mind . . ." "Come *quick* and help me!" Some grammarians argue that these informal usages date from early English tradition, in which many adverbs were formed by adding an *e* to the adjective (*loud/loude, quick/quicke*); in time, the final *e* was dropped, but the words remained in use as adverbs.

Nonetheless, modern English usage decrees that these spoken conventions not be used in writing. If your students are accustomed to hearing a colloquial use of adjectives in place of adverbs, teaching correct usage may involve creating new habits.

ANSWERS TO EXERCISE 1

The most popular self-help trend in the United States today is subliminal tapes. These tapes, with titles like "How to Attract Love," "Freedom from Acne," and "I Am a Genius," are intended to solve every problem known to modern society—quickly and easily. The tapes are said to work because their "hidden messages" bypass conscious defense mechanisms. The listener hears only music or relaxing sounds, like waves rolling slowly and steadily. At decibel levels perceived only subconsciously, positive words and phrases are embedded, usually by someone who speaks deeply and rhythmically. The top-selling cassettes are those to help you lose weight or quit smoking. The popularity of such tapes is not difficult to understand. They promise easy solutions to complex problems. But the main benefit of these tapes appears to be for the sellers, who are accumulating profits really quickly.

ANSWERS TO EXERCISE 2

1. David seemed tired.
   Jerry was anxious.
   Lienne appeared happy.
   Maggie is depressed.
   Chris remained confident.
2. Julie finds her job interesting.
   I thought his speech dull.
   The reviewers deemed the play fascinating.
   The students found the course helpful.
   Charles proved the problem unsolvable.

EXERCISE 1

Revise each of the incorrect sentences in this paragraph so that only adjectives modify nouns and pronouns and only adverbs modify verbs, adjectives, or other adverbs. Be sure to eliminate informal forms.

The most popular self-help trend in the United States today is subliminal tapes. These tapes, with titles like "How to Attract Love," "Freedom from Acne," and "I Am a Genius," are intended to solve every problem known to modern society—quick and easy. The tapes are said to work because their "hidden messages" bypass conscious defense mechanisms. The listener hears only music or relaxing sounds, like waves rolling slow and steady. At decibel levels perceived only subconsciously, positive words and phrases are embedded, usually by someone who speaks deep and rhythmic. The top-selling cassettes are those to help you lose weight or quit smoking. The popularity of such tapes is not hard to understand. They promise easy solutions to complex problems. But the main benefit of these tapes appears to be for the sellers, who are accumulating profits real fast.

EXERCISE 2

Being careful to use adjectives—not adverbs—as subject complements and object complements, write five sentences in imitation of each of the following. Be sure to use five different linking verbs in your imitations of each sentence.

1. Julie looked worried.
2. Dan considers his collection valuable.

25d	**Distinguishing between Comparative and Superlative Forms**

COMPARATIVE AND SUPERLATIVE FORMS

Form	Function	Example
Positive	Describes a quality; indicates no comparisons	big
Comparative	Indicates comparisons between *two* qualities (greater or lesser)	bigger
Superlative	Indicates comparisons among *more than two* qualities (greatest or least)	biggest

CLOSE-UP

ADJECTIVES AND ADVERBS

Some adverbs, particularly those indicating time, place, and degree (*almost, very, here, yesterday,* and *immediately*), do not have comparative or superlative forms.

(1) Comparative forms

Comparative forms indicate a greater or lesser degree.

Adjectives To indicate a *greater* degree, all one-syllable adjectives and many two-syllable adjectives (particularly those that end in *-y, -ly, -le, -er,* and *-ow*) add *-er.*

slow   slower
funny   funnier
lovely   lovelier

(Note that a final *y* becomes *i* before *-er* is added.)

Other two-syllable adjectives and all long adjectives indicate a greater degree with *more.*

famous   more famous
incredible   more incredible

NOTE: Many two-syllable adjectives can indicate a greater degree with either *more* or *-er*—for example, *more lovely* or *lovelier.*

All adjectives indicate a lesser degree with *less.*

lovely   less lovely
famous   less famous

Adverbs Adverbs ending in *-ly* indicate a greater degree with *more.*

slowly   more slowly

Other adverbs use the *-er* ending to indicate a greater degree.

soon   sooner

All adverbs indicate a lesser degree with *less.*

less slowly   less soon

IN THE CLASSROOM

Give students 15 to 20 minutes to write a comparison/contrast paragraph in which they first write about the likenesses, differences, and qualities of two things (automobiles, vacation spots, restaurants, etc.) and then about how this "thing" compares to all others of its class. Caution students to use the comparative and superlative forms of modifiers carefully.

CLOSE-UP

ADJECTIVES AND ADVERBS

Never use both *more* and *-er* to form the comparative degree.

FAULTY: Nothing could have been <u>more easier</u>.

REVISED: Nothing could have been <u>easier</u>.

FAULTY: The package arrived <u>more later</u> than the letter.

REVISED: The package arrived <u>later</u> than the letter.

(2) Superlative forms

Superlative forms indicate the greatest or least degree.

Adjectives   Adjectives that form the comparative (a greater degree) with *-er* add *-est* to form the superlative (the *greatest* degree).

<u>nicer</u>	<u>nicest</u>
<u>funnier</u>	<u>funniest</u>

Adjectives that indicate the comparative with *more* use *most* to indicate the superlative.

<u>more</u> famous	<u>most</u> famous
<u>more</u> challenging	<u>most</u> challenging

All adjectives indicate the least degree with *least*.

<u>least</u> interesting   <u>least</u> enjoyable

Adverbs   The majority of adverbs are preceded by *most* to indicate the greatest degree.

<u>most</u> quickly   <u>most</u> helpfully   <u>most</u> efficiently

Others use the *-est* ending to indicate the greatest degree.

<u>soonest</u>

All adverbs use *least* to indicate the least degree.

<u>least</u> willingly   <u>least</u> fashionably

426

CLOSE-UP

ADJECTIVES AND ADVERBS

Never use both *most* and *-est* to form the superlative degree.

FAULTY: Jack is the <u>most meanest</u> person in town.

REVISED: Jack is the <u>meanest</u> person in town.

(3) Irregular comparatives and superlatives

Some adjectives and adverbs do not conform to the rules presented above. Instead of adding a word or an ending to the positive form, they use different words to indicate each degree.

	IRREGULAR COMPARATIVES AND SUPERLATIVES		
	Positive	**Comparative**	**Superlative**
ADJECTIVES	good	better	best
	bad	worse	worst
	a little	less	least
	many, some, much	more	most
ADVERBS	well	better	best
	badly	worse	worst

(4) Illogical comparisons

Many adjectives and adverbs have absolute meanings—that is, they can logically exist only in the positive degree. Words like *perfect, unique, excellent, impossible,* and *dead* can never be used in the comparative or superlative degree.

FAULTY: The vase is the <u>most unique</u> piece in her collection.

REVISED: The vase in her collection is <u>unique</u>.

Absolutes can, however, be modified by words that suggest approaching the absolute state—*nearly* or *almost,* for example.

He revised until his draft was <u>almost perfect</u>.

ANSWERS TO EXERCISE 3

1. many/more/most
2. eccentric/more eccentric/most eccentric
3. confusing/more confusing/most confusing
4. bad/worse/worst
5. mysterious/more mysterious/most mysterious
6. softly/more softly/most softly
7. embarrassing/more embarrassing/most embarrassing
8. well/better/best
9. often/more often/most often
10. tiny/tinier/tiniest

STUDENTS WORKING TOGETHER

Instruct each student in a group of four to compose a paragraph on a given subject (school, homework, etc.). Tell students to use noun modifiers effectively. Students should then read these paragraphs within the group and rank them from most to least effective in use of noun modifiers. Collect the paragraph with the most effective use of noun modifiers from each group and make copies for other groups. Each group should then rank the paragraphs according to 1) overall quality and 2) use of noun modifiers.

ANSWERS TO EXERCISE 4

Part I: Nouns used as adjectives are underlined.

The <u>student government business management trainee</u> program is extremely popular on campus. The <u>student</u> government donated some of the <u>seed</u> money to begin this <u>management trainee</u> program, which is one of the most successful the <u>university business</u> school has ever of-

continued on following page

CLOSE-UP

ADJECTIVES AND ADVERBS

Although comparative and superlative forms of absolutes should be avoided in college writing, they are sometimes used in informal conversation or writing.

He revised eight times, always looking for the <u>most perfect</u> draft.

It was the <u>most impossible</u> course I ever took.

EXERCISE 3

Supply the correct comparative and superlative forms for each of the following adjectives or adverbs. Then use each form in a sentence.

> **EXAMPLE:** strange   stranger   strangest
>
> The story had a *strange* ending. The explanation sounded *stranger* each time I heard it. This is the *strangest* gadget I have ever seen.

1. many
2. eccentric
3. confusing
4. bad
5. mysterious
6. softly
7. embarrassing
8. well
9. often
10. tiny

25e   Using Nouns as Adjectives

Nouns can function as adjectives in a sentence.

He made a sandwich of <u>turkey</u> bologna, <u>egg</u> salad, and <u>tomato</u> slices on <u>wheat</u> bread.

Many familiar phrases, such as *space station, art history,* and *amusement park,* consist of one noun modifying another. In such cases using a noun as a modifier saves words. Overusing nouns as modifiers, however, can create clumsy—even incoherent—sentences.

CONFUSING: The Chestnut Hill Fathers' Club Pony League beginners spring baseball clinic will be held Saturday.

To revise such cluttered sentences, restructure to break up a long series of nouns. You can also substitute equivalent adjective forms, where such forms exist, or possessives for some of the nouns used as modifiers.

IMPROVED: The Chestnut Hill Fathers' Club's spring baseball clinic for beginning Pony League players will be held Saturday.

Of course, eliminating the passive voice would make this sentence even clearer.

IMPROVED: On Saturday, the Chestnut Hill Fathers' Club will hold its spring baseball clinic for beginning Pony League players.

*EXERCISE 4

Identify every noun used as a modifier in the following passage. Then revise where necessary to eliminate clumsy or unclear phrasing created by overuse of nouns as modifiers. Try substituting adjective or possessive forms and rearrange word order where you feel it is indicated.

The student government business management trainee program is extremely popular on campus. The student government donated some of the seed money to begin this management trainee program, which is one of the most successful the university business school has ever offered to undergraduate students. Three core courses must be taken before the student intern can actually begin work. First, a management theory course is given every spring semester in conjunction with the business school. Then, the following fall semester, students in the trainee program are required to take a course in personnel practices, including employee benefits. Finally, they take a business elective.

During the summer, the student interns are placed in junior management positions in large electronics, manufacturing, or public utility companies. This job experience is considered the most valuable part of the program because it gives students a taste of the work world.

continued from previous page

fered to <u>undergraduate</u> students. Three <u>core</u> courses must be taken before the <u>student</u> intern can actually begin work. First, a <u>management the-ory</u> course is given every <u>spring</u> semester in conjunction with the <u>business</u> school. Then, the following <u>fall</u> semester, students in the <u>trainee</u> program are required to take a course in <u>personnel</u> practices, including <u>employee</u> benefits. Finally, they take a <u>business</u> elective.

During the summer, the <u>student</u> interns are placed in <u>junior manage-ment</u> positions in large <u>electronics</u>, <u>manufacturing</u>, or <u>public utility</u> companies. This <u>job</u> experience is considered the most valuable part of the program because it gives students a taste of the <u>work</u> world.

Part II (revised paragraph): Your students' answers will vary.

STUDENT WRITER AT WORK

Adjectives and Adverbs

Possible Revision

Working

My attitude toward work was shaped by my grandfather when I was very young, around the age of four. He did everything he could to encourage me to choose a job where I would use my brain, not my hands or my back. He tried hard to make his feelings really clear to me. Still, it took a long time before I realized what he was telling me. Eventually, I learned that I had two choices: I could take it easy and drift into a job, or I could work hard and train for a career. I chose the more difficult of the two alternatives.

Every morning, my mother would drop me off at my grandparents' house really early, on her way to work. I would eat breakfast, and my

---

STUDENT WRITER AT WORK

ADJECTIVES AND ADVERBS

Read this draft of an essay and correct errors in the use of adjectives and adverbs. Check to be sure adjectives modify nouns or pronouns and adverbs modify verbs, adjectives, or other adverbs; make sure the correct comparative and superlative forms are used; and eliminate any overuse of nouns as modifiers. After you have corrected the errors, go over the draft again and, if necessary, revise further to strengthen coherence, unity, and style.

continued on following page

continued from previous page

grandfather would tell me stories about his life in the mines. He would tell me about his three friends who were crushed by a cave-in, and about a terrifying gas explosion that nearly took his life. His most common stories were about the long, hard hours he had spent in the mines working for minimum wage, which was just a couple of cents an hour at that time. He didn't tell me these stories to scare me, but to make me think hard about the kind of job I might get when I grew up.

Years later, around the time of my sixteenth birthday, I needed money quickly. I needed spending money when I went out with my friends, and I had to start saving regularly for college. I decided to get a job, and I soon found one at Insalaco's supermarket. There I hauled heavy boxes of canned goods and unpacked them, stocked shelves, and labeled cans and boxes. This work was monotonous, and as time went on it grew more and more tedious. At the end of each day, I really felt very bad. In fact, every bone in my body ached. This was without a doubt the worst job I could imagine.

Everyone in my family had always considered me intelligent, and they thought I should go to college. To me, the thought of studying and doing homework for four more years after high school had never been very appealing. After working at Insalaco's, however, I knew my family was right, and I understood what my grandfather had been trying to tell me.

Commentary on Possible Revision

The writer has corrected the errors in adjective and adverb usage. However, further revision is needed to eliminate wordiness and awkwardness and to add sentence variety. He could also add dialogue to paragraph 2 and, in paragraph 3, he could relate specific incidents that occurred at his job. Finally, he could revise to eliminate clichés like "every bone in my body" (paragraph 3).

Working

My attitude toward work was shaped by my grandfather when I was real young, around the age of four. He did everything he could to encourage me to choose a job where I would use my brain, not my hands or my back. He tried hard to make his feelings real clearly to me. Still, it took a long time before I realized what he was telling me. Eventually, I learned that I had two choices: I could take it easy and drift into a job, or I could work hard and train for a career. I chose the most difficult of the two alternatives.

Every morning, my mother would drop me off at my grandparents' house real early, on her way to work. I would eat breakfast, and my grandfather would tell me stories about his life in the mines. He would tell me about his three friends who were crushed by a cave-in, and about a terrifying gas explosion incident that nearly took his life. His most commonest stories were about the long, hard hours he had spent in the mines working for minimum wage, which was just a couple of cents an hour at that time. He didn't tell me these stories to scare me, but to make me think hard about the kind of job I might get when I grew up.

Years later, around the time of my sixteenth birthday, I needed money quick. I needed spending money when I went out with my friends, and I had to start saving regular for college. I decided to get a job, and I soon found one at Insalaco's supermarket. There I hauled heavy boxes of canned goods and unpacked them, stocked shelves, and labeled cans and boxes. This work was monotonous, and as time went on it grew more and more tediously. At the end of each day, I really felt very badly. In fact, every bone in my body ached. This was without a doubt the worse job I could imagine.

Everyone in my family had always considered me intelligently, and they thought I should go to college. To me, the thought of studying and doing homework for four more years after high school had never been very appealing. After working at Insalaco's, however, I knew my family was right, and I understood what my grandfather had been trying to tell me.

CHAPTER 26

Language Issues for International Students (ESL)

26a Writing for Native English Speakers

Writing for a native English-speaking audience involves much more than writing correct English sentences. Great differences can exist between the writing of native speakers of English and non-native-speaking writers in logic, organization, support, and even purpose and audience. In fact, the field of **contrastive rhetoric** is devoted to the study of the *cultural* differences, not just the *grammatical* differences, between various languages. The paragraphs that follow summarize what scholars in the field of contrastive rhetoric have to say about the primary differences between writing in English and writing in other languages.

English is linear. English usually follows a straight line of logic, from clear statement to supporting examples. English speakers tend to say what they mean immediately, without pleasantries or elaborate introductions. They may be impatient with ambiguity, digressions, and indirect statements of opinion. Compared to speakers of other languages, native speakers of English can seem curt and businesslike, perhaps even blunt and rude. Whereas business letters in many Asian languages may begin with observations about the weather and even inquiries about the health of one's family, business letters in English usually begin by stating the main point of the letter. Many English idioms and sayings express the direct quality of the language: "Get to the point," "Say what you mean," "Stop beating around the bush."

FURTHER READING

Kaplan, Robert B. "Cultural Thought Patterns in Inter-Cultural Education." *Language Learning* 17 (1966) 1–20
The author begins with the assertion that different cultures use different rhetorical patterns to organize and communicate ideas. "Logic" is not a universal quality of the human mind but a cultural phenomenon that varies from culture to culture and is embodied in the rhetorical patterns common to those cultures. The work of Sapir and Whorf points to interrelationships among language, culture, and rhetoric. Therefore, teachers of students whose native language is not English should understand that the problems these students have in organizing their English writing may be due to interference from rhetorical patterns common to their native languages.

Until recently, teachers of non-native speakers have assumed that proficiency in writing essays in the native language leads to proficiency in written English. However, they consistently see "errors" in the writing of international students such as loose organization, poor cohesion, and insufficient support for the thesis. In a study of 600 essays by students representing a variety of linguistic groups, the author concludes these qualities are not errors but evidence of alternate rhetorical patterns.

Essays in English are typically linear in logic, proceeding from main idea to subordinate ideas to supporting examples. Arabic, by contrast, tends to the proceed by means of a variety of parallel constructions; the King James translation of the Old Testament provides a

continued on following page

continued from previous page

good example of this logic. Oriental (Chinese and Korean) writing is characterized by indirection, a pattern in which the writer circles a topic, viewing it tangentially but rarely directly. Writing in Romance languages such as Spanish and French tends to be constructed around many more digressions than English usually allows. This tendency is even more pronounced in Russian, where the digressions may have little or no direct relation to the stated topic. The author represents these cultural thought patterns graphically.

English   Semitic   Oriental   Romance   Russian

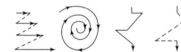

The author suggests several ways of making the rhetorical patterns of English clear to speakers of other languages. Students might be given a series of sentences and asked to put them in "logical" (English) order; they can check their work against the sentences as they appear in the original paragraph. Students can also be given ideas in outline form and asked to expand the outline with additional ideas and concrete examples.

The author acknowledges that teaching English rhetorical patterns in this way may be a kind of cultural imperialism that does a disservice even to the culture being taught; students may learn mindless imitation, not genuine rhetorical thinking. In the end, however, the author recommends teaching international students to imitate English rhetorical patterns in the interest of efficient communication with native speakers.

English-speaking audiences prize originality. In many cultures, the point of writing a public statement is to sound very much like other writers, both ancient and modern—to imitate their style, to echo their arguments. Native speakers of English, however, generally expect a writer to be highly individual in statements of opinion, choice of supporting details, and even style. Readers of English want to see novel variations on old themes, new twists on common ideas, surprising uses of language. They may even criticize overly familiar language as "trite" and object to the use of clichés.

Native-speaking English writers invite audience participation. In many cultures, audiences are expected to agree with the writer. By contrast, writers in English may challenge their readers to become actively involved. They may, for example, invite questions, criticism, or even disagreement. In some cultures, such responses by the audience might be considered inappropriate and rude, but such responses show English writers that the audience is reading closely and taking their ideas seriously.

Writers in English often question and argue with accepted ideas. Native speakers of English can assume they are operating within a democratic ideal of open debate. Every belief the audience holds, no matter how ancient or cherished, may be questioned by the writer. The English speaker's attitude of skepticism may disturb some members of some cultural groups. Indeed, to them, English speakers may appear to have no firm beliefs of their own at all. However, English speakers often use skepticism not to destroy belief, but to refine and reinforce it.

EXERCISE 1

Newspaper editorials or letters to the editor are often good examples of the qualities of English discussed in this section. Find a letter or editorial in a local newspaper and see how many of the qualities it displays. (For example, does it come to the point immediately? Does it challenge the truth of any accepted ideas?) Does it display qualities that are quite different from those a writer might write about in discussing the same subject in your native language?

26b   **Choosing a Topic**

See ◄
1C1

What kinds of **topics** do native speakers of English choose to write about? In general, they try either to *inform* their readers, giving new information that readers may not already know, or to *persuade* their readers, offering opinions that may not be popular and then convincing their readers to accept these opinions as valid. Therefore, they will choose a topic about which they know something or about which they can find information, or they will choose a controversial topic about which there is public disagreement.

Such topics may be quite different from those chosen by nonnative speakers of English. In some cultures, it is rude for a writer to claim to know more than an audience about a topic; a writer must be humble and perhaps even apologize to the audience for not knowing enough. In other cultures, controversial topics are intentionally avoided, sometimes for complex social and political reasons. However, too much humility may cause English audiences to lose their trust in the writer; avoiding controversial topics may make the writer seem indecisive and "wishy-washy."

At the same time, English audiences tend not to trust writers who are too dogmatic, who insist on being believed without offering examples to support their beliefs. Writers who simply use their social rank or reputation as support for their views are not convincing. Even the highest ranking members of English-speaking societies must still offer concrete facts to support their beliefs.

▶ See 2b2

26c Stating and Supporting a Thesis

CLOSE-UP

STATING A THESIS

An effective thesis statement will usually have the following characteristics.

- It is a statement of opinion or conjecture, not a self-evident statement of incontrovertible fact.
- It is not a familiar platitude such as "Honesty is the best policy" or "Look before you leap," but a more original, even unique, idea.
- It is not so general that there is no way to support it within the scope of the essay.
- It can be supported with concrete examples.

Bring in copies of a news magazine such as *Time* or *Newsweek*. Show the students the difference between "hard news" stories that make up most of the magazine and the opinion columns that also frequently appear. The opinion columns will have clear thesis statements—that the United States needs stricter gun control, that the United States needs fewer taxes—and even the hard news stories may put forth a thesis—that a war will start somewhere, that a politician will win election. Point out some of these thesis statements and ask students to examine the ways in which the writers support the statements.

(1) Stating a thesis

When writing in English, state your **thesis**—your main idea or point of view—as quickly as possible. You need not worry about offending or insulting your readers by doing this; in fact, this is just what they expect you to do.

You should also be quite specific in stating your thesis. You may begin with general statements, but keep in mind that an English-speaking audience will look for a specific thesis statement that states what you have to say about the topic.

(2) Supporting a thesis

See ◄ 2b1

Native speakers of English usually demand a very tight connection between the thesis and the support: The thesis states the writer's main point, and the support answers the question "Why?" or "How?" by giving reasons or examples. English speakers expect that all the support will be directly related to the thesis.

CLOSE-UP

SUPPORTING A THESIS

- Examples must be real and not hypothetical. It is better to point to the experiences of real people than to make up examples. It is better to say, "86 percent of Americans on welfare say they would prefer to have a job" than to say, "Few people would choose welfare over working if they had a choice."

- Examples must be relevant to the thesis. Most speakers of English want to see a direct connection between factual evidence and the position the writer is taking. If a writer asserts, for example, that cats make better pets than dogs, it makes no sense to say that pigs are also becoming popular pets in America. Instead, the writer can point out that cats are quieter and easier to care for than dogs.

- Examples must be sufficient to convince readers that a statement is true. Beware of false generalizations based on too few details. For example, the fact that your friend got a good job without going to college does not mean that no one needs to go to college to get a good job. Do enough people who do not go to college get good jobs to justify the assertion that college is not necessary?

Native speakers of English may have little patience for what they see as digressions or indirect discussions of the subject. In the classic essay patterns of many languages—the *ba-gu-wen* pattern of Chinese rhetoric, for example—digressions are just what the writer is supposed to write. In the eyes of Chinese essay writers, English essays may seem less imaginative, less creative, than the Chinese ideal. Even so, it is important that you meet the native-speaking audience's expectations of a direct, businesslike discussion.

Most native speakers of English support their points by citing events from the real world, facts and figures, and perhaps the opinions of authorities on the topic. What makes a person an authority for English speakers may be different from some of the criteria used in other cultures. It is not necessarily a person's social or political rank, nor his or her position as a political or religious leader, but rather the person's expertise in the subject.

EXERCISE 2

After each of the following thesis statements are statements of fact meant to support it. In each group, which factual statements best support the writer's position? Why?

1. Eating too much beef is bad for your health.
   a. People who eat a lot of beef have more heart attacks than people who do not.
   b. American beef in particular is full of unhealthy chemicals.
   c. Raising animals just to eat them is cruel.
   d. People who eat a lot of beef tend not to eat other foods that they need, such as grains and vegetables.
2. Standardized tests are a poor means of determining a student's academic abilities.
   a. Many students who perform well at other academic tasks do poorly on standardized tests.
   b. Standardized tests are too difficult.
   c. Other kinds of academic work are much more important than standardized tests.
   d. The conditions under which students takes standardized tests may damage their performance.
   e. Standardized tests may have content that is unfamiliar to some groups of otherwise able students.
3. Engineering is a very good field to enter today.
   a. There are too many doctors.
   b. There are many jobs open to engineers.
   c. Engineering is fun.
   d. Engineers are well paid.
   e. Engineers build things.

435

Students' answers will vary. The supporting statements they offer should be fact (rather than opinion), relevant to the thesis statement, and specific (rather than general).

Most international students studying in the United States have extensive educational experience in their home countries, and many have had writing instruction in their native languages. Organizational principles in particular are commonly taught in writing classrooms around the world. First, have students work together in small groups of people with similar national origins. They can pool their experiences and knowledge of writing instruction with particular attention to the organizational principles they were taught. Then, have students work in groups of mixed national origins, presenting the patterns they learned to other students and comparing the various patterns presented in their groups.

Students' answers will vary. Students might work together to chart similarities and differences among different rhetorical patterns on the board.

EXERCISE 3

For each of the following thesis statements, write at least three statements of fact that support it. If you strongly disagree with any of the statements, rewrite it so that you can better support it.

1. American women are the freest women in the world.
2. Knowing English well is a very important business skill.
3. Americans do not know enough about other countries.
4. Recycling is not enough to cure our pollution problems.

26d Organizing Ideas

(1) Introduction, body, conclusion

Native speakers of English are likely to expect to find a **thesis-and-support** organization of ideas. In addition, most Americans learn to use a very specific organizational pattern in their essays. The three-part beginning, middle, and end of a conventional English essay are the *introduction, body,* and *conclusion.* The **introduction** introduces the topic and states the thesis, the **body** presents the support for the thesis, and the **conclusion** sums up the writer's position.

See ◄ 2b5

Introduction	Thesis: Languages contain a lot of cultural information.
Body	Support: For example, a language can tell a lot about social relations among the people who speak it.
	Support: A language can also reveal the social values of particular groups.
	Support: Learning a language can even tell us about the material lives of its speakers.
	Support: On the deepest level, a language can contain ingrained cultural attitudes about time and space.
Conclusion	Summary: Therefore, when we study a language we are also studying the cultural attitudes of the people who speak it.

EXERCISE 4

Is there a specific strategy for organizing essays in your native language? If so, what is it? Is it similar to the English method for organizing ideas, or is it quite different from it? Make a list of similarities and differences.

(2) Writing in paragraphs

The **paragraph** is a basic unit of communication in English. Many other languages use paragraphs, of course, but some do not. Some languages are not even written, as English is, horizontally from left to right. In some other languages, even those that are written as English is, there is no convention of bringing together groups of sentences into a paragraph.

CLOSE-UP

WRITING IN PARAGRAPHS

An English paragraph is immediately recognizable.

▶ **See Ch 4**

- The first line is indented, giving a paragraph its characteristic shape on a page.
- Paragraphs vary in length, from a single sentence to many sentences, from a few lines to a page or more. Regardless of the paragraph's length, however, all its sentences focus on a single idea.
- A paragraph's main idea is usually stated explicitly as a topic sentence, most often (though not always) at the beginning.
- All the other sentences in the paragraph provide further explanation of the topic sentence or concrete examples to illustrate the topic sentence.
- The sentences in a paragraph are often linked with connecting devices such as the repetition of key words and the use of transitional words or phrases.

EXERCISE 5

Look for the elements listed above in each of the following paragraphs: a clearly stated main idea, examples that illustrate the topic sentence, connecting devices such as transitional phrases. If the main idea is not clearly stated, try stating it in your own words. Is every sentence in the paragraph clearly focused on that main idea?

1. Asian Americans are not one people but several—Chinese Americans, Japanese Americans, and Filipino Americans. Chinese and Japanese Americans have been separated by geography, culture, and history from China and Japan for seven and four generations respectively. They have evolved cultures and sensibilities distinctly not Chinese or Japanese and distinctly not white American. Even the Asian languages

Ask students to examine paragraphs in texts from other classes they may be enrolled in—science textbooks, history books, even novels. Ask them to determine if the paragraphs exhibit the qualities listed: indentation, focus on a single topic, a topic sentence (possibly near the beginning of the paragraph), concrete examples in support of the stated topic, transitional language between sentences. They might present particularly good examples to the rest of the class.

ANSWERS TO EXERCISE 5

1. The topic is stated in the first part of the first sentence ("Asian Americans are not one people but several . . ."). The support explains the history of Asians in America, with their different "cultures," "sensibilities," and "languages." The principal transitional device is the use of the pronouns *they* and *their* throughout the paragraph.
2. The first sentence states the topic ("sex differences in personality formation" can be seen "in studies of children's games"). The support offers the findings of researchers Mead and Piaget. The principal transitional device is the repetition of the word *games*: "Children's games," "In games," "In games."

See ◄
21a

TEACHING STRATEGY

Error Analysis

Most international students who are speakers of languages other than English are highly sensitive to errors in their spoken and written English. English is widely taught around the world as an exercise in error detection and avoidance, with sentence-level translation drills as the most common teaching technique. Some students are so anxious about committing errors that they pay little attention to the communicative function of language. Several strategies may help students see error in a more constructive way.

1. **Be systematic**. Identify only those types of errors that are pervasive in a student's writing. Categorize the errors, explaining grammar rules and pointing to specific examples in the student's writing. Show the student how to correct these errors with examples. Encourage the student to identify, categorize, and correct his or her own errors.

2. **Be positive**. "Error" in non-native English may in fact not be error as we understand it, but adherence to rules that apply to languages other than English. This is revealed in students' tendency to make the same error in the same pattern over and over. Ask students if certain pervasive errors relate to principles from their native languages. Explain the differences between the first and second languages.

3. **Be constructive**. Treat English grammar not as a list of "rights" and "wrongs," but as a repertoire of possibilities. When offering correct examples, offer more than one way to express a given idea. Rather than correctness, stress the differences in *meaning* between various grammatical possibilities.

as they exist today in America have been adjusted and developed to express a sensitivity created by a new experience. In America, Chinese and Japanese American culture and history have been inextricably linked by confusion, the popularization of their hatred for each other, and World War II. (Frank Chin, Jeffery Paul Chan, Lawson Fusao Inada, Shawn Wong, "Preface" to *Aiiieeeee*)

2. The sex differences in personality formation that Chodorow describes in early childhood appear during the middle childhood years in studies of children's games. Children's games are considered by George Herbert Mead and Jean Piaget as the crucible of social development during the school years. In games, children learn to take the role of the other and come to see themselves through another's eyes. In games, they learn respect for rules and come to understand the ways rules can be made and changed. (Carol Gilligan, *In a Different Voice*)

26e Writing Correct English

There is nothing shocking about grammar errors; people speaking or writing in a language other than their native language inevitably make mistakes. Grammar errors become a serious problem only when they get in the way of efficient communication. Therefore, although you may not be able to eliminate all grammar errors from your English, you should try to make it as correct as possible.

The study of grammatical differences between various languages is sometimes called **contrastive linguistics.** Some contrastive linguists make the following points about the differences between English and other languages.

In English, words may change their form according to their function. In some languages, words never change form: There are no plural forms of nouns, no past tense forms of verbs, and so on. Such information is communicated by means other than by changing the form of the word itself. In English, however, words may change their forms in many different ways. For example, **nouns**—words that name things—may change their form according to whether they name only one thing or more than one thing: one pickle or two pickles, one woman or two women. Nouns can also be transformed into other parts of speech—verbs, for example, or adjectives—if they change form to resemble them, and verbs may become other parts of speech—nouns or adjectives, among others—if they change form to resemble them.

Context is extremely important in understanding function. Sometimes it is impossible to name the function of a word without noting its context. In the following sentences, for instance, the very same words can perform different functions according to their relation to other words.

Juan and I are taking a <u>walk</u>. (*Walk* is a noun, a direct object of the verb *taking*, with an article, *a*, attached to it.)

If you <u>walk</u> whenever you can instead of driving, you will help conserve the earth's resources. (*Walk* is a verb, the predicate of the subject *you*.)

Jie was <u>walking</u> across campus when she met her chemistry professor. (*Walking* is part of the verb, predicate of the subject *Jie*.)

<u>Walking</u> a few miles a day will make you healthier. (*Walking* is a noun, the subject of the verb *will make*.)

Next summer we'll take a <u>walking</u> tour of southern Italy. (*Walking* is an adjective describing *tour*.)

Spelling in English is sometimes illogical. In many languages that use a phonetic alphabet or syllabary, such as Japanese, Korean, or Persian script, words are spelled exactly as they are pronounced. Spelling in English, however, is often a matter of memorization, not sounding out the words phonetically. For example, the "ough" sound in the words *tough*, *though*, and *thought* is pronounced quite differently in each case. In fact, spelling in English is related more to the history of the word and its origins in other languages than to the way the word is pronounced.

▶ See Ch. 20

Word order is extremely important in English sentences.

▶ See 26k

EXERCISE 6

As an international student of English, you have more conscious knowledge of English grammar than a native speaker who has never studied it as a foreign language. Moreover, as a speaker of at least two languages, you probably have a good sense of grammatical differences between languages.

Make a list of a few major differences between the grammar of English and the grammar of your native language and of any other languages you know. Be prepared to explain these differences to your classmates.

26f Nouns

The word *noun* has the same origin as the word *name*. (They both come from the Latin word *nomen*.) A **noun** *names* things: people, objects, places, feelings, ideas.

▶ See 21a

Nouns can be quite different in different languages. In some languages nouns have gender; that is, they may be *masculine* or *feminine*. In Spanish, the word for *moon* (*la luna*) is feminine, while the word for *sun* (*el sol*) is masculine. In other languages, there is no difference between the singular and plural forms of nouns. In Japanese, one person is *hito*, while many people are still *hito*. In some languages (including English), nouns may function as verbs: "A gray fog *blanketed* the campus." In other languages (again including English), nouns may be transformed into adjectives with special suffixes: "Henry is so *bookish;* he's always reading."

(1) Singular, plural, and non-count

In English nouns may have number; that is, they may change in form according to whether they name one thing or more than one thing. If a noun names only one thing, it is a *singular* noun; if a noun names more than one thing, it is a *plural* noun. Plural nouns are most often formed by adding *-s* to singular nouns: one cat, two cats; one pencil, a box of pencils. Some nouns require the addition of *-es* for phonetic reasons: lunch/lunches, flash/flashes. For most nouns ending in *y*, the final *y* is changed to *ie* before *-s* is added: story/stories, memory/memories. Some nouns are even more irregular, with plural forms that do not use *s*: man/men, mouse/mice.

Some English nouns do not have a plural form. These are called *non-count* nouns because the things they name usually are not counted. Words such as *love, justice,* and *money,* for example, represent qualities that cannot be counted. Understanding the distinction between count and non-count nouns is important in determining the correct use of articles with nouns.

EXERCISE 7

Underline all the nouns in the following passage. Then list the nouns in three columns (singular, plural, non-count).

The highway took me through Danville, where I saw a pillared antebellum mansion with a trailer court on the front lawn. Route 127 ran down a long valley of pastures and fields edged by low, rocky bluffs and split by a stream the color of muskmelon. In the distance rose the foothills of the Appalachians, old mountains that once separated the Atlantic from the sallow inland sea now the middle of America. The licks came out of the hills, the fields got smaller, and there

---

ANSWERS TO EXERCISE 7

singular	plural	non-count
highway	pastures	distance
Danville	fields	bluegrass
mansion	bluffs	fescue
court	foothills	dirt
lawn	Appalachians	glass
Route 127	mountains	
valley	mountains	
stream	licks	
color	hills	
color	fields	
muskmelon	sawmills	
Atlantic	hardwoods	
sea	pallets	
middle	crates	
paint	fenceposts	
America	houses	
	colors	
	pastels	
	iridescents	
	lawns	
	fragments	
	ornaments	
	birdbaths	
	flamingos	
	donkeys	
	automobiles	
	hoods	
	tombstones	
	porches	
	washers	
	sofas	
	doors	
	washtubs	
	coats of arms	

were little sawmills cutting hardwoods into pallets, crates, and fence-posts. The houses shrank, and their colors changed from white to pastels, to iridescents, to no paint at all. The lawns went from Vertagreen bluegrass to thin fescue to hard-packed dirt glinting with fragments of glass, and the lawn ornaments changed from birdbaths to plastic flamingos and donkeys to broken-down automobiles with raised hoods like tombstones. On the porches stood long-legged wringer washers and ruined sofas, and, by the front doors, washtubs hung like coats of arms. (William Least Heat Moon, *Blue Highways*)

EXERCISE 8

Each of the following sentences has one number error in a noun. Underline each noun in the sentence. Then locate the noun that has the incorrect number and correct it. Be prepared to explain the error and how you corrected it.

1. Donald arrived in New York with three suitcase and his aunt's telephone number.
2. Where is the magazines I lent you last month?
3. The United States has fifty state, one special district, and territories such as the Virgin Islands, Guam, and American Samoa.
4. There are more woman in the American military today than at any time in the past.
5. When Françoise came back from vacation, she was filled with happinesses.
6. The journey of a thousand mile begins with just one step.
7. John F. Kennedy was President of the United States for only three year.
8. Why do so many man say they are superior to women?
9. Most rock and roll bands have three guitar and one set of drums.
10. Except for the Native Americans, every American citizens is the descendant of immigrants.

(2)   Using articles with nouns

English has two **articles:** *a* and *the*. In some cases *a* is replaced by *an* purely for reasons of sound: If the word that follows begins with a vowel (*a, e, i, o,* or *u*) or a vowel sound, then the *a* is changed to *an*: a book, an apple; a magazine, an open door. Note that a word may begin with a vowel and still not begin with a vowel sound (a unique event), while another word may not begin with a vowel but still begin with a vowel sound (an honor).

The primary function of articles is to signal to the audience whether the noun being referred to is new to the conversation or has already been mentioned.

1. *Suitcases* should be plural to agree with the modifier *three*.
2. *Magazine* should be plural to agree with the verb *is*.
3. *States* should be plural to agree with the modifier *fifty*.
4. *Women* should be plural to agree with the verb *are*.
5. *Happiness* should be singular because it is a non-count noun.
6. *Miles* should be plural to agree with the modifier *thousand*.
7. *Years* should be plural to agree with the modifier *three*.
8. *Men* should be plural to agree with the verb *say*.
9. *Guitars* should be plural to agree with the modifier *three*.
10. *Citizen* should be singular to agree with the modifier *every*.

Exploring the Possibilities of Grammar

It is important to remember how odd article use in English may seem both to students whose native languages lack articles and to students whose native languages have articles that are deceptively similar to those in English. Misuse of articles is one of the most persistent errors made by even long-term residents of the United States if their native language is not English. Some knowledge of the rules governing article use may help, but use of articles is largely a matter of instinct based not on rules but on intended *meaning*.

To help non-native speakers develop a sense of the different meanings implied by various uses of articles, try a

continued on following page

continued from previous page

"grammatical exploration." Choose sample sentences from any text, or have students provide sentences, and write them on the board. Then, change the articles in the sentences one by one: Replace articles with other articles (*a* for *the*, *the* for *a*), create non-count nouns that do not require an article by making nouns plural, and so on. Explain the different *meanings* created by changing the articles. After explaining a few examples yourself, ask students to make their own guesses.

Students' answers should reflect their understanding of the rules governing article use. For example, in the opening sentence the words *plaza* and *cathedral* require the definite article because they represent information that has already been introduced to the reader (perhaps in the previous paragraph). Later in the paragraph, the words *zaguán* and *driveway* require the indefinite article because they are being introduced for the first time. In the sentence that follows, however, when the word *zaguán* is repeated (in the plural, *zaguánes*), since it has already been introduced, it requires the definite article.

Note that some article use is a matter of style and subtle shades of meaning. The definite article before *high front walls* and *street windows* is optional; the definite article before *cancel* could also be an indefinite article. The use of the definite article in these places communicates the idea that every rich home, to be a proper rich home, was obligated to have these items.

The indefinite article: *a* Use *a* with a noun when the reader has no reason to be familiar with the noun you are naming—when you are introducing the noun for the first time, for example. To say "There was a sidewalk in front of a building" signals to the audience that you are introducing the idea of the building into your speech or writing for the first time. The building is *indefinite*, or not specific, until it has first been identified.

The definite article: *the* Use *the* with a noun when the reader has already been introduced to the noun you are naming. Use of the *definite* article indicates that the noun introduced may already be familiar to the reader. To say, "There was a sidewalk in front of the building," signals to the reader that you are still referring to the same building you mentioned earlier. The building has now become specific and may be referred to by the definite article.

CLOSE-UP   **USING ARTICLES WITH NOUNS**

There are two main exceptions to the rules governing the use of articles with nouns.

1) Plural nouns do not require an indefinite article: "I love horses," not "I love a horses." (Plural nouns do, however, require definite articles: "I love the horses in the national park near my house.")
2) Non-count nouns may not require articles: "Love conquers all," not "A love conquers all" or "The love conquers all."

The use of articles in English is often more a matter of feeling and style than adherence to specific rules. The best method of learning the correct use of articles is to listen carefully to native speakers' spoken English and to read native speakers' written English with attention to the many different uses of articles.

EXERCISE 9

In the following passage, underline every noun and circle the article that accompanies it. Do your best to explain why the noun requires that article, or, if there is no article, why it does not require one.

Close about the plaza and the cathedral were the townhouses that intrigued me greatly. These were the homes of the rich, *los ricos*. The high front walls were neatly painted brown, grey, pink, or light cream.

The street windows were even with the sidewalk with long iron bars that reached also to the roof. Lace curtains, drapes, and wooden screens behind the bars kept people from looking in. Every town-house had a *zaguán* and a driveway cutting across the sidewalk, ramped and grooved so the carriages could roll in and out. On hot days the *zaguanes* were left wide open, showing a part of the patios with their fountains, rose gardens, and trees. The walls and the floors of the corridors were decorated with colored tile in solid colors and complicated designs. Between the open *zaguán* and the patio there was the *cancel*, a grill of wrought iron that was always kept closed and locked. (Ernesto Galarza, *Barrio Boy*)

26g Pronouns

Any English noun may be replaced a **pronoun**. Pronouns enable you to avoid repeating the same noun over and over. For example, *doctor* may be replaced by *he* or *she*, *books* by *them*, and *computer* by *it*.

▶ See 22a,b

Pronouns must be in proper grammatical **case** (subjective, objective, or possessive). (Possessive case in nouns is often indicated by an apostrophe followed by an *s*—Philip*'s* beard, England*'s* weather, a computer*'s* keyboard. These nouns too can be replaced by pronouns called *possessive pronouns*: *my, his/her/its, your, their*.)

Pronouns must agree with the nouns they replace; that is, they must be the same *number* (singular or plural) and *gender* (male, female, or neuter).

▶ See 24b

Finally, pronouns must be clearly connected to the nouns they replace.

▶ See 22c

EXERCISE 10

Underline all the pronouns in the following passage. Identify the noun each pronoun replaces.

In the olden days, both Land and Heaven were tight friends as they were once human-beings. So one day, Heaven came down from heaven to Land his friend and he told him to let them go to the bush and hunt for the bush animals; Land agreed to what Heaven told him. After that they went into a bush with their bows and arrows, but after they had reached the bush, they were hunting for animals from morning till 12 o'clock A.M., but nothing was killed in that bush, then they left that bush and went to a big field and were hunting till 5 o'clock in the evening and nothing was killed there as well. After that, they left there again to go to a forest and it was 7 o'clock before they could find

First sentence: *they* refers to the friends, Land and Heaven. Second sentence: *his* refers to Heaven, *he* refers to Heaven, *him* refers to Land, *them* refers to Land and Heaven, *him* refers to Land. Third sentence: *they* refers to Land and Heaven, *their* refers to Land and Heaven, *they* refers to Land and Heaven, *they* refers to Land and Heaven, *they* refers to Land and Heaven. Fourth sentence: *they* refers to Land and Heaven, *they* refers to Land and Heaven, *they* refers to Land and Heaven, *them* refers to Land and Heaven, *they* refers to Land and Heaven, *they* refers to Land and Heaven. Fifth sentence: *they* refers to Land and Heaven, *they* refers to Land and Heaven, *them* refers to Land and Heaven, *it* refers to mouse, *he* refers to Land, *it* refers to mouse, *he* refers to Heaven, *it* refers to mouse.

ANSWERS TO EXERCISE 11

There are many possible ways to rewrite the passage. One version follows.

The young couple seated across from Daniel at dinner the night before were newlyweds from Tokyo. *They* ate together with other guests of the inn at long, low tables in a large dining room with straw mat flooring. The man introduced himself immediately in English, shook Daniel's hand firmly, and, after learning that *he* was not a tourist but a resident working in Osaka, gave *him* a business card. The man had just finished college and was working at *his* first real job, clerking in a bank. Even in a sweatsuit, *he* looked ready for the office: chin closely shaven, bristly hair neatly clipped, nails clean and buffed. After a while *he* and Daniel exhausted *his* store of English and drifted into Japanese.

The man's wife, shy up until then, took over as *he* fell silent. *She* and Daniel talked about the new popularity of hot springs spas in the countryside around the inn, the difficulty of finding good schools for the children *she* hoped to have soon, the differences between food in Tokyo and Osaka. *Her* husband ate busily with an air of tolerating *her* prattling. From time to time *she* refilled *his* beer glass or served *him* radish pickles from a china bowl in the middle of the table, and then returned to the conversation.

a mouse and started to hunt for another, so that they might share them one by one, because the one they had killed already was too small to share, but they did not kill any more. After that they came back to a certain place with the one they had killed and both of them were thinking how to share it. But as this mouse was too small to divide into two and these friends were also greedy, Land said that he would take it away and Heaven said that he would take it away. (Amos Tutuola, *The Palm-Wine Drunkard*)

EXERCISE 11

There are no pronouns in the following passage. The repetition of the nouns again and again would seem strange to a native speaker of English. Rewrite the passage, replacing as many of the nouns as possible with appropriate pronouns. Be sure that the connection between the pronouns and the nouns they replace is clear.

The young couple seated across from Daniel at dinner the night before were newlyweds from Tokyo. The young couple and Daniel ate together with other guests of the inn at long, low tables in a large dining room with straw mat flooring. The man introduced himself immediately in English, shook Daniel's hand firmly, and, after learning that Daniel was not a tourist but a resident working in Osaka, gave Daniel a business card. The man had just finished college and was working at the man's first real job, clerking in a bank. Even in a sweatsuit, the man looked ready for the office: chin closely shaven, bristly hair neatly clipped, nails clean and buffed. After a while the man and Daniel exhausted the man's store of English and drifted into Japanese.

The man's wife, shy up until then, took over as the man fell silent. The woman and Daniel talked about the new popularity of hot springs spas in the countryside around the inn, the difficulty of finding good schools for the children the woman hoped to have soon, the differences between food in Tokyo and Osaka. The woman's husband ate busily with an air of tolerating the woman's prattling. From time to time the woman refilled the man's beer glass or served the man radish pickles from a china bowl in the middle of the table, and then returned to the conversation.

26h   Verbs

Verbs are words that describe *states of being* or *actions*. Some languages use two different verbs to describe permanent and impermanent **states of being**, such as *ser* and *estar* in Spanish, or two different verbs to describe animate and inanimate objects, such as *aru* and *iru* in Japanese. With its single verb *be,* English is in this instance simpler than many languages.

In Arabic, verbs can show whether the **action** they describe is complete or not. In Japanese, verbs can be conjugated to communicate the speaker's feeling about the action of the verb—for example, whether the action was overdone or "too much." Again, although English may communicate such concepts in other words, the forms of its verbs are simpler in some ways than verbs in other languages.

(1)   Tense and person

English verbs change their form according to *tense* and *person.* ▸ **See 23b-f**
Tense refers to when the action described by the verb takes place.
Person refers to who is performing the action described by the ▸ **See 17d**
verb (*I, you, she*).

In other languages, verbs may change their appearance according to different rules. In Japanese, for example, verbs are not conjugated according to the person performing the action, but according to the social relationship between the speaker and the listener. In Chinese, verbs themselves do not change form to express tense; the time when the action is performed is communicated through other words.

Many **irregular** English verbs do not change their form ac- ▸ **See 23a2**
cording to the usual rules governing tense and person, but change
in idiosyncratic ways. Unless you use the correct forms of the verbs in your sentences, you will confuse your English-speaking reader by communicating meanings you do not intend.

(2)   Subject–verb agreement

The **subject** of a verb is the person or thing that performs the ▸ **See 24a**
action expressed by the verb. Verbs must match their subjects, or
agree, in *person* (I, you, he) and *number* (he, they) so that in English we say *I read* but *she reads.* Be especially careful with irregular verbs: "I *am,* it *is,* they *are.*"

(3)   Auxiliary verbs

Many English verbs use other **auxiliary** verbs (also known as *helping verbs*) to communicate their meaning, often forms of the verb *to be* and *to have,* and also verbs such as *would, should, can,* ▸ **See 21c1**
and *want.* Such helping words are necessary because English

ANSWERS TO EXERCISE 12

owned (past part.) is used as an adjective; *owning* (present part.) is used as a noun; *to own* (infinitive) is used as a noun; *rusting* (present part.) is used as an adjective; *rotting* (present part.) is used as an adjective; *staying* (gerund) is used as a noun; *moving* (gerund) is used as a noun; *ending* (present part.) is used as an adjective; *rolling* (present part.) is used as an adjective; *personalized* (past part.) is used as an adjective; *contained* (past part.) is used as an adjective.

verbs do not change their form to express ideas such as the probability, desirability, or necessity of the action described by the verb.

(4) Using verbs as nouns and adjectives

In some cases English verbs may be used as nouns or adjectives, and this can confuse speakers of other languages, particularly if in their native language words do not change their function according to the way they are used in a sentence. Two particular forms of the verb may be used as nouns: **infinitives** (which always begin with *to*) and **gerunds** (which always end in *-ing*). **Present participles** (which also end in *-ing*) and **past participles** (which often end in *-ed, -t,* or *-en*) are frequently used as adjectives.

> To bite into this steak takes better teeth than mine. (infinitive used as noun)

> Cooking is one of my favorite hobbies. (gerund used as noun)

> Some people think raw fish is healthier than cooked fish. (past participle used as adjective)

> According to the Bible, God spoke to Moses from a burning bush. (present participle used as adjective)

EXERCISE 12

Identify all the infinitives (*to* _____), gerunds (_____-*ing*), present participles (_____-*ing*), and past participles (_____-*ed,* _____-*t,* _____-*en*) in the following passage. Define their function as either noun or adjective.

> The car is the quintessential American possession. The car, not the home, is the center of American life. Despite its central place in the mystique of the American Dream, the individually owned home is actually anti-American in many ways. Owning a car means freedom, progress, and individual initiative, the most basic American values. To own a house means rusting pipes and rotting roofbeams. Staying in one place implies stagnation and decay, while moving (always "forward") connotes energy, creativity, never-ending youth: "Moss doesn't grow on a rolling stone." And the way Americans move is in their cars, their personalized, self-contained, mobile units."

26i Adjectives and Adverbs

See ◀
25a

Adjectives and *adverbs* are words that modify other words, such as nouns and verbs.

Adjectives describe the qualities of nouns or modify other adjectives. A book might be *large* or *small, blue* or *red, difficult* or *easy, expensive* or *cheap.* Unlike adjectives in other languages, English adjectives change their form only to indicate degree (*fast, faster, fastest*). In English, adjectives do not have to agree in gender with the noun they describe, as adjectives must in French and German, for example. In Japanese, some adjectives are conjugated in past and present tenses, something English adjectives do not require.

▶ See 25b

▶ See 25d

Adverbs describe the qualities of verbs and sometimes help describe adjectives, other adverbs, and even whole clauses. Most adverbs in English end in *-ly,* making them easily identifiable. A person may walk *slowly* or *quickly, shyly* or *assuredly, elegantly* or *clumsily.*

▶ See 25c

(1) Position of Adjectives and Adverbs

In Arabic and in Romance languages such as Spanish, French, and Italian, adjectives typically follow the nouns they describe. In other languages, such as Japanese, Chinese, and English, adjectives usually appear before the nouns they describe. A native speaker of English would not say "Cars red and black are involved in more accidents than cars blue, green, or white," but would say instead, "Red and black cars are involved in more accidents than blue, green, or white cars."

Adverbs may appear before or after the verbs they describe, but they should be placed as close to the verb as possible: not "I *told* John that I couldn't meet him for lunch *politely,*" but "I *politely told* John that I couldn't meet him for lunch" or "I *told* John *politely* that I couldn't meet him for lunch." When an adverb describes an adjective, it usually comes before the adjective: "The essay has *basically sound* logic."

EXERCISE 13

Each sentence below is followed by a list of adjectives and adverbs that can be used in the sentence. Place the adjectives and adverbs where they belong in each sentence. (Adjectives and adverbs are listed in the order in which they should appear in the sentences.)

1. Researchers believe that tests are not reliable. (most, now, IQ, entirely)
2. Just as culture is derived from Greece, culture is derived from China. (European, ancient, Asian, ancient)

ANSWERS TO EXERCISE 13

1. *Most* researchers *now* believe that *IQ* tests are not *entirely* reliable.
2. Just as *European* culture is derived from *ancient* Greece, *Asian* culture is derived from *ancient* China.
3. According to the *old* Japanese proverb, for a *chipped* lid there's a *cracked* pot.
4. The *Cajun* people of *southern* Louisiana play a *lively* music called zydeco.
5. When you begin to exercise, start *slowly* and build *steadily* up to *strenuous* levels.
6. Because I transferred *many* credits from my *previous* school, I was able to graduate *early*.
7. *Many* signers of the *American* Declaration of Independence, which proclaimed the "*unalienable*" rights of *all* people, owned *African* slaves.
8. *Too* often, *American* feminists feel they are describing the condition of *all* women in the world, when they *really* are describing *only* women in America.
9. I *usually* work *many* hours on essays, but I remember to take *periodic* breaks.
10. The *same* word can have *different* meanings in *different* contexts.

3. According to the Japanese proverb, for a lid there's a pot. (old, cracked, chipped)

4. The people of Louisiana play a music called zydeco. (Cajun, southern, lively)

5. When you begin to exercise, start and build up to levels. (slowly, steadily, strenuous)

6. Because I transfered credits from my school, I was able to graduate. (many, previous, early)

7. Signers of the Declaration of Independence, which proclaimed the rights of people, owned slaves. (Many, American, "unalienable," all, African)

8. Often, feminists feel they are describing the condition of women in the world, when they are describing women in America. (too, American, all, really, only)

9. I work hours on essays, but I remember to take breaks. (usually, many, periodic)

10. The word can have meanings in contexts. (same, different, different)

(2) Order of adjectives

A single noun may be described by more than one adjective, perhaps even by a whole list of adjectives in a row. Determining the order in which these adjectives should be listed before the noun can be troublesome for speakers of languages other than English. Given a list of three or four adjectives, most native speak-

CLOSE-UP

ORDER OF ADJECTIVES

The following list indicates the order in which adjectives should be listed before a noun.

1. articles (a, an, the), demonstratives (this, those), and possessives (his, our, Maria's)
2. order (first, next, last), amounts (one, five, many, few)
3. personal opinions (nice, ugly, crowded, pitiful)
4. sizes and shapes (small, tall, straight, crooked)
5. age (young, old, modern, ancient)
6. colors (black, white, red, blue)
7. nouns functioning as adjectives to form a unit with the noun (soccer ball, cardboard box, history class)

See ◀
28b2

ers would arrange them in a sentence in the same order. If shoes are to be described as *green* and *big*, numbering *two*, and the type worn for playing *tennis*, a native speaker would say "two big green tennis shoes." Why?

Generally, the adjectives that are most important in completing the meaning of the noun are placed closest to that noun. For example, the most important fact in characterizing the shoes described above is that they are tennis shoes; their size and color are less important. Details about size generally precede details about color and texture. Details about number nearly always precede all others.

Another way of determining the correct order of adjectives is to make sure that each adjective describes all the words that follow. For example, in the phrase *fat black cat*, *black* describes the cat, while *fat* describes both the cat and its color.

EXERCISE 14

Write five original sentences in which two or three adjectives describe one noun. Be sure that the adjectives are in the right order.

26j Prepositions

In English, **prepositions** give meaning to nouns by linking them with other parts of the sentence. A preposition and its noun may function together as an adjective or an adverb. In the sentence "Frances gave me a book of poetry for my birthday," the phrase *of poetry*, consisting of the preposition *of* and the noun *poetry*, acts as an adjective to describe the noun *book*. In the sentence "I put my birthday present on the shelf," the phrase *on the shelf*, consisting of the preposition *of* and the noun *shelf* (with its article *the*), acts as an adverb to describe the verb *put*.

Learning to use prepositions correctly may cause problems for speakers of languages other than English. Prepositions may be used in quite different ways in other languages, or may exist in forms quite different from English, or may not exist at all.

For speakers of languages with prepositions very similar to those in English, there are different problems. Since speakers of Romance languages such as Italian and Romanian use prepositions in ways very similar to English, they may be tempted to

449

translate prepositional phrases from their own languages directly into English, though idiomatic uses of prepositions can vary widely.

CLOSE-UP

PREPOSITIONS

Non-native speakers of English may have trouble with prepositions in the following expressions.

Non-native speaker usage	*Native speaker usage*
according with	according to
apologize at	apologize to
appeal at	appeal to
apply with	apply to
believe at	believe in
different to	different from
for least, at most	at least, at most
refer at	refer to
relevant with	relevant to
similar with	similar to
subscribe with	subscribe to

Remember also that more than one preposition may be used with certain words, but they create different expressions with different meanings.

agree with, agree to
care for, take care of
from the beginning, at the beginning
look for, look at, look after
talk to, talk with
to the end, at the end
unfamiliar with, unfamiliar to

There is one construction incorporating a preposition that is peculiar to English. Infinitive forms of verbs are formed in English by adding *to* to the root form of the verb: *to write, to read, to sleep, to eat*. When these verbs are combined with other verbs, there can be confusion. In the sentence, "Ali is learning to read Braille," *to* is part of the verb *read,* not part of the verb *learning.* Thus, a native speaker would know to say, "Ali is learning Braille," not "Ali is learning to Braille," but a non-native speaker might not.

EXERCISE 15

Identify as many prepositions as you can in the following passage. Note their position and their use. How might similar information be communicated in your native language?

> In retrospect, the distinguishing feature of the post–World War II era was its remarkable affluence. From 1950 through 1970, by fits and starts, the American Gross National Product grew at an average annual rate of 3.9 percent, perhaps the best performance in the nation's history. Autos, chemicals, and electrically-powered consumer durables were the leading sectors driving the economy forward in the 1950s; housing, aerospace, and the computer industry, in the 1960s. In consequence, the average American commanded 50 percent more real income at the end of the period than at the beginning. Exuberant growth and dramatic changes in the standard of living were hardly novel in the American experience, and it was possible to view postwar economic developments as a mere extension of historic trends. But in one crucial respect the era was indeed different. Past increases in real income had mainly purchased improvements in the necessities of life—more and better food, clothing, shelter. After 1950 rising income meant that the mass of Americans, including many blue-collar workers, could, for the first time, enjoy substantial amounts of discretionary income—i.e., income spent not for essentials but for amenities. (Allen J. Matusow, *The Unraveling of America*)

EXERCISE 16

In the following passage, provide appropriate prepositions in the positions indicated.

> Everyone knows what is supposed to happen when two Englishmen who have never met before come face _____ face _____ a railway compartment—they start talking _____ the weather. _____ some cases this may simply be because they happen to find the subject interesting. Most people, though, are not particularly interested _____ analyses _____ climatic conditions, so there must be other reasons _____ conversations _____ this kind. One explanation is that it can often be quite embarrassing to be alone _____ the company _____ someone you are not acquainted _____ and *not* speak to them. If no conversation takes place the atmosphere can become rather strained. However, by talking _____ the other person _____ some neutral topic like the weather, it is possible to strike up a relationship _____ him without actually having to say very much. Railway-compartment conversations _____ this kind—and they do happen, although not of course as often as the popular myth supposes—are good examples of the sort _____ important social function that is often fulfilled _____ language. Language is not simply a means of communicating information _____ the weather or any other subject. It is also a very

ANSWERS TO EXERCISE 15

In retrospect, *of* the post-World War II era, *From* 1950 *through* 1970, *by* fits and starts, *at* an average annual rate *of* 3.9 percent, *in* the 1950s, *in* the 1960s, *in* consequence, *at* the end of the period, *at* the beginning, *in* the standard of living, *to* view, *as* a mere extension *of* historic trends, *in* one crucial respect, *in* real income; *in* the necessities *of* life; *of* Americans, *for* the first time, *of* discretionary income, *for* essentials, *for* amenities.

As for how similar information might be communicated in various languages, students' answers will vary. Students whose languages use prepositions that are deceptively similar to those in English may have the most trouble. A French speaker, for example, may want to say "*by* consequence" or "*for* consequence," since these resemble the French idiom; a Spanish speaker may want to say "*at* the first time" for the same reason.

ANSWERS TO EXERCISE 16

to, in, about, In, in, of, for, of, if, of, with, to, about, with, of, of, by, about, of, with

IN THE CLASSROOM

Encourage students to write a basic subject-verb-indirect object-direct object sentence and then change the order of the parts of speech in as many ways as they can and still have a sentence that makes sense. The most successful experiments can go on the board.

important means _____ establishing and maintaining relationships _____ other people. (Peter Trudgill, *Sociolinguistics: An Introduction to Language and Society*)

26k Word Order

The importance of word order varies from language to language. In English, word order is extremely important, contributing a good deal to the meaning of a sentence. Like Chinese, English is an "SVO" language, or one in which the most typical sentence pattern is "subject-verb-object." (Arabic, by contrast, is a "VSO" language.) A native speaker of English will understand that in the sentence "Dog bites man," the dog is doing the biting and the man is getting bitten, while in the sentence "Man bites dog," the opposite is true.

The importance of word order in English may confuse speakers of languages such as Japanese, in which word order does not identify words as subjects or objects of a sentence; instead, that function is served by preposition-like suffixes called **particles**. Since the Japanese particle *ga* would identify *dog* as the subject and the particle *o* would identify *man* as the object, the sentence would mean the same thing no matter what the word order was: "Man-ga bites dog-o," or "Dog-o bites man-ga."

Of course, there is some flexibility in the word order of English sentences. Modifying words, such as adjectives and adverbs (or phrases that function as adjectives or adverbs), may be placed in various positions in a sentence, depending upon the sentence's intended emphasis or meaning.

Word order in questions can be particularly troublesome for speakers of languages other than English, partly because there are so many different ways to form questions in English. For sentences using the verb *to be*, simply inverting the order of the subject and verb can be enough to form a question that elicits a "yes or no" answer:

Rasheem *is researching* the depletion of the ozone layer.
Is Rasheem researching the depletion of the ozone layer?

Inverting the subject and verb and then using a negative form of the verb can indicate that the person asking the question thinks that the answer to the question is probably *yes*.

Isn't Rasheem researching the depletion of the ozone layer?

To ask for more information, the question can use an interrogative word as well as inverting the subject and verb.

What is Rasheem researching?
Why is Rasheem researching the depletion of the ozone layer?

Questions can also be formed by adding *tag questions* to the ends of ordinary sentences.

Rasheem is researching the depletion of the ozone layer, *isn't he?*
Rasheem will write his dissertation about the depletion of the ozone layer, *won't he?*

EXERCISE 17

Write five ordinary declarative sentences. Then write as many questions as you can for each one. Pay particular attention to the word order of the questions.

 26l   Common Sentence Errors

Even native speakers of English make errors in their writing—in particular, common errors such as sentence fragments, run-on sentences, and excessive or unnecessary passive constructions. There are many reasons why native speakers make these errors, reasons that may also apply to speakers of other languages. For example, fragments and run-ons may be a result of the interference of spoken English (which tends to be less formal and less grammatically correct) with written English. Overuse of passive constructions may result from the opposite condition: writers trying too hard to be formal in their writing.

Speakers of languages other than English may make these errors in their writing for other reasons as well. In English a correct "complete" sentence must have at least one *independent clause* consisting of a subject and a verb that agree in number; if a sentence has more than one independent clause, the clauses must be joined in an appropriate way. Other languages, however, may define a sentence in other ways. Also, other languages make use of certain grammatical constructions much more often than English, so speakers of other languages may overuse such elements as the subjunctive mode of the verb ("If I were you") and passive constructions.

453

(1)   Fragments, comma splices, and fused sentences

See ◀
Ch 13 **Sentence fragments** occur in written English when some necessary element of a sentence is missing, or when the sentence uses a form of a verb that cannot stand alone. In languages such as Spanish, however, the conjugated verb form indicates the person performing the action. Thus, the subject does not have to be included as a separate word. "Estoy trabajando aqui," for example, means "I am working here," even when written without the subject "yo". English, however, does not allow ommission of the subject, as verb forms without corresponding subjects create fragments in English. "Am working here," for example, is not allowed. (The pronoun "I" is required to indicate who is working.)

See ◀
Ch 14 *Run-on sentences* occur when two or more independent clauses are "run together" without proper punctuation. (Independent clauses linked only by commas are called **comma splices**; those linked without any punctuation are called **fused sentences.** Both are incorrect.) Other languages, however, have no such punctuation rules. In Asian languages, which do not use the Roman alphabet, the use of punctuation elements (such as commas in Chinese and small circles to indicate periods in Japanese) is a result of contact between Asia and Europe. Even in languages that use the Roman alphabet, the use of periods to mark off sentences is a relatively recent development.

(2)   Excessive passive constructions

In many languages, a passive construction is a form of the verb itself, and in most languages other than English passive constructions are very common. They communicate information in a neutral way, without the added meaning that passives may have in English.

English forms its passives not by changing the verb but by adding auxiliary verbs (especially forms of the verb *be*), changing word order, and sometimes dropping the performer of the action altogether.

> **ACTIVE:** Douglas hurt Stephen's feelings.
>
> **PASSIVE:** Stephen's feelings were hurt by Douglas.
>
> **PASSIVE:** Stephen's feelings were hurt.

Passive constructions are less common in English than they are in other languages. They are used for very particular reasons, to communicate certain subtleties about the information in the sentence. Native speakers of English focus their attention on the subject of a sentence, assuming that it and the action it performs are the most important pieces of information in the sentence. In the sentences above, for example, changing the subject of the sentence shifts the focus from Douglas (focus on who hurt Stephen's feelings) to Stephen (focus on whose feelings were hurt).

In general, then, most writing in English should contain relatively few passive constructions, and those should be used for a specific purpose.

▶ See 231

EXERCISE 18

Examine the following sentences, whose authors chose very deliberately to use the passive voice. What meaning is communicated in these sentences that would not have been communicated if the authors had used the active voice?

1. "The Irish people will not be deceived." (Arthur Griffith, "The Irish Free State")
2. "We have been spurned, with contempt, from the foot of the throne." (Patrick Henry, "Give me liberty, or give me death!")
3. "Victor, vanquished and neutrals alike are affected physically, economically and morally." (Bernard Mannes Baruch, "Control of Atomic Weapons")
4. "Revolutions are never fought by turning the other cheek." (Malcolm X, "The Black Revolution")
5. "I felt smothered by their traditional values of how a Chinese girl should behave." (Kit Yuen Quan, "The Girl Who Wouldn't Sing")

ANSWERS TO EXERCISE 18

In general, the passive construction works to emphasize the noun that appears in the subject position, an emphasis that would be lost if that noun appeared in the object position. Thus, while "The English government will not deceive the Irish people" places the English government in the active role of deceiving, the passive construction "The Irish people will not be deceived" places the Irish people in the active role of refusing to be deceived, a subtle difference that communicates the writer's intention to make the Irish people active participants and not passive victims.

The other passive constructions in these sentences work the same way, focusing the reader's attention not on the King doing the spurning but on the American people being spurned; not on nuclear weapons but on the people who use them and have them used against them, and so on.

Understanding Punctuation and Mechanics

Overview of Sentence Punctuation: Commas, Semicolons, Colons, Dashes, Parentheses

(Further explanations and examples are located in the sections listed in parentheses after each example.)

Separating Independent Clauses

▶ ***With a Comma and a Coordinating Conjunction***

The year was 2081, and everybody was finally equal. (Kurt Vonnegut, Jr., "Harrison Bergeron") **(28a)**

▶ ***With a Semicolon***

Paul Revere's *The Boston Massacre* is an early example of traditional American Protest art; Edward Hicks's later "primitive" paintings are socially conscious art with a religious strain. **(29a)**

▶ ***With a Semicolon and a Coordinating Conjunction***

If such a world government is not established by a process of agreement among nations, I believe it will come anyway, and in a much more dangerous form; for war or wars can only result in one power being supreme and dominating the rest of the world by its overwhelming military supremacy. (Albert Einstein, *Einstein on Peace*) **(29b)**

▶ ***With a Semicolon and a Conjunctive Adverb***

Thomas Jefferson brought 200 vanilla beans and a recipe for vanilla ice cream back from France; thus, he gave America its all-time favorite ice-cream flavor. **(29c)**

Separating Items in a Series

▶ ***With Commas***

Chipmunk, *raccoon*, and *Mugwump* are Native American words. **(28b)**

▶ ***With Semicolons***

As ballooning became established, a series of firsts ensued: The first balloonist in the United States was 13-year-old Edward Warren, 1784; the first woman aeronaut was a Madame Thible who, depending on your source, either recited poetry or sang as she lifted off; the first airmail letter, written by Ben

Franklin's grandson, was carried by balloon; and the first bird's-eye photograph of Paris was taken by a balloon. (Elaine B. Steiner, *Games*) **(29d)**

Setting Off Illustrative Material

▶ *With a Colon*

Each camper should bring the following : a sleeping bag, a mess kit, a flashlight, and plenty of insect repellent. **(32a1)**

▶ *With a Dash*

Walking to school by myself, spending the night at a friend's house, getting my ears pierced, and starting to wear makeup— these were some of the milestones of my childhood and adolescence. **(32b2)**

Setting Off Nonessential Elements

▶ *With a Single Comma*

His fear increasing, he waited to enter the haunted house. **(28d4)** What do you think, Margie? **(28d5)**

▶ *With a Pair of Commas*

It was Roger Maris, not Mickey Mantle, who broke Babe Ruth's home run record. **(28d3)**

▶ *With a Pair of Dashes*

Although we are by all odds the most social of all social animals—more independent, more attached to each other, more inseparable in our behavior than bees—we do not often feel our conjoined intelligence. (Lewis Thomas, *Lives of a Cell*) **(32b1)**

▶ *With a Single Dash*

They could not afford to jump to conclusions—any conclusions. (Michael Crichton, *The Andromeda Strain*) **(32b1)**

▶ *With Parentheses*

It took Gilbert Fairchild two years at Harvard College (two academic years, from September, 1955, to June, 1957) to learn everything he needed to know. (Judith Martin, *Gilbert: A Comedy of Manners*) **(32c1)**

CHAPTER 27

End Punctuation

USING END PUNCTUATION

Use a period . . .

- To end a sentence **(27a1)**
- To mark an abbreviation **(27a2)**
- To mark divisions in dramatic, poetic, and Biblical references **(27a3)**

Use a question mark . . .

- To mark the end of a direct question **(27b1)**
- To mark questionable dates and numbers **(27b2)**

Use an exclamation point . . .

- To mark emphasis **(27c1)**

 27a Using Periods

Use **periods** to end declarative sentences (statements), mild commands, polite requests, and indirect questions. Also use periods in most familiar abbreviations and in dramatic and poetic references.

(1) Ending a sentence

Periods signal the end of a statement, a mild command or polite request, or an indirect question.

Something is rotten in the state of Denmark. (statement)

Be sure to have the oil checked before you start out. (mild command)

When the bell rings, please exit in an orderly fashion. (polite request)

They wondered whether it was safe to go back in the water. (indirect question)

(2) Marking an abbreviation

Periods appear in most abbreviations.

Mrs. Robinson	Captain Newman, M.D.	25 B.C.
Mr. Spock	George McGovern, Ph.D.	U.S.A.
Ms. J. R. Jones	Sue Barton, R.N.	9 P.M.
Dr. Kildare	221b Baker St.	etc.

CLOSE-UP

ABBREVIATIONS WITHOUT PERIODS

The following abbreviations do not require periods.

Acronyms (new words formed from the initial letters or first few letters of a series of words)

NATO	radar	OSHA	scuba
NOW	AIDS	SALT	CAT scan

Fannie Mae (Federal National Mortgage Association)
Gestapo (Geheime Staats Polizei)
Modem (modulator demodulator)
Soweto (Southwest townships)

Frequently Used Capital-Letter Abbreviations (familiar abbreviations of names of corporations or government agencies and scientific and technical terms)

CIA	NYU	IBM	FBI	WCAU-FM
EPA	CCC	DNA	IRA	AFT
MGM	RCA	HBO	UCLA	UFO

Clipped Forms (commonly accepted shortened forms of words)

gym   dorm   math

If you are unsure whether or not to use a period with a particular abbreviation, consult a dictionary.

ANSWERS TO EXERCISE 1

1. C
2. Dr. McLaughlin worked hard to earn his Ph.D.
3. Carmen was supposed to be at AFL-CIO headquarters by 2 P.M.; however, she didn't get there until 10 P.M.
4. After she studied the fall lineup proposed by NBC, she decided to work for CBS.
5. Representatives from the UMW began collective bargaining after an unsuccessful meeting with Mr. L. Pritchard, the coal company's representative.

PUNCTUATING WITH ABBREVIATIONS THAT TAKE PERIODS

At the End of a Sentence

If the abbreviation ends the sentence, do not add another period.

   FAULTY: He promised to be there at 6 A.M..

   REVISED: He promised to be there at 6 A.M.

However, do add a question mark after the abbreviations's final period if the sentence is a question.

   Did he arrive at 6 P.M.**?**

Within a Sentence

If the abbreviation falls within a sentence, use normal punctuation after the abbreviation's final period.

   FAULTY: He promised to be there at 6 P.M. but he forgot.

   REVISED: He promised to be there at 6 P.M.**,** but he forgot.

(3)   Marking divisions in dramatic, poetic, and Biblical references

Periods separate act, scene, and line numbers in plays; book and line numbers in long poems; and chapter and verse numbers in Biblical references. (Do not space between the periods and the elements they separate.)

   DRAMATIC REFERENCE: *Hamlet* II.ii.1–5 (or 2.2.1–5.)
   POETIC REFERENCE: *Paradise Lost* VII.163–67.
   BIBLICAL REFERENCE: Judges 4.14

EXERCISE 1

Correct these sentences by adding missing periods and deleting superfluous ones. If a sentence is correct, mark it with a *C*.

   EXAMPLE: Their mission changed the war

             Their mission changed the war**.**

1. Julius Caesar was killed in 44 B.C.
2. Dr. McLaughlin worked hard to earn his Ph.D..
3. Carmen was supposed to be at A.F.L.-C.I.O. headquarters by 2 P.M.; however, she didn't get there until 10 P.M.

4. After she studied the fall lineup proposed by N.B.C., she decided to work for C.B.S.
5. Representatives from the U.M.W. began collective bargaining after an unsuccessful meeting with Mr. L Pritchard, the coal company's representative.

27b   Using Question Marks

Use **question marks** at the end of direct questions or to indicate questionable dates or numbers.

(1)   Marking the end of a direct question

Use a question mark to signal the end of a direct question.

Who was that masked man **?** (direct question)

"Is this a silver bullet **?**" they asked. (declarative sentence opening with a direct question)

They asked, "Could he have been the Lone Ranger **?**" (declarative sentence closing with a direct question)

Who was it who asked, "Who was that masked man **?**" (question within a question)

Did he say where he came from, who his companion was, or where they were headed **?** (series of direct questions)

Did he say where he came from **?** Who his companion was **?** Where they were headed **?** (series of direct questions with each question asked separately)

Did he say where he came from **?** who his companion was **?** where they were headed **?** (series of direct questions; informal usage does not require capitalization of first word of each question)

A pair of dashes or a pair of parentheses is used around a direct question within a declarative sentence.

Someone—a disgruntled office seeker **?**—is sabotaging the campaign.

Part of the shipment (three dozen cases **?**) was delayed.

(2)   Marking questionable dates or numbers

Use a question mark in parentheses to indicate that a date or number is uncertain.

Aristophanes, the Greek playwright, was born in 448 (**?**) B.C. and died in 380 (**?**) B.C.

The clock struck five (**?**) and stopped.

1. He wondered whether he should take a nine o'clock class.
2. The instructor asked, "Was the Spanish-American War a victory for America?"
3. Are they really going to China?
4. He took a somewhat less than modest portion of dessert—half a pie.
5. "Is *data* the plural of *datum*?" he inquired.

(3) Editing misused or overused question marks

See ◀ 17e ***After an Indirect Question*** Use a period, not a question mark, with an indirect question.

> **FAULTY:** The personnel officer asked whether he knew how to type?
>
> **REVISED:** The personnel officer asked whether he knew how to type.

With Other Punctuation Do not use other punctuation marks along with question marks.

> **FAULTY:** "Can it be true?," he asked.
>
> **REVISED:** "Can it be true **?**" he asked.
>
> **FAULTY:** Can you believe this run of good luck?!
>
> **REVISED:** Can you believe this run of good luck **?**

With Another Question Mark Do not end a sentence with more than one question mark.

> **FAULTY:** You did what?? Are you crazy??
>
> **REVISED:** You did what **?** Are you crazy **?**

As an Indication of Attitude Do not use question marks to convey sarcasm. Instead, suggest your attitude through word choice.

> **FAULTY:** I refused his generous (?) offer.
>
> **REVISED:** I refused his not-very-generous offer **.**

In an Exclamation Do not use a question mark after an exclamation phrased as a question.

> **FAULTY:** Will you please stop that at once?
>
> **REVISED:** Will you please stop that at once **!**

EXERCISE 2

Correct the use of question marks and other punctuation in the following sentences.

> **EXAMPLE:** She asked whether Freud's theories were accepted during his lifetime?
>
> She asked whether Freud's theories were accepted during his lifetime **.**

1. He wondered whether he should take a nine o'clock class?
2. The instructor asked, "Was the Spanish-American War a victory for America?"?
3. Are they really going to China??!!
4. He took a modest (?) portion of dessert—half a pie.
5. "Is *data* the plural of *datum*?," he inquired.

"When the exclamation mark is used after mere statements it deserves the name, by which it is sometimes called, the mark of admiration; we feel that the writer is indeed lost in admiration of his own wit and impressiveness."

–H. W. and F. G. Fowler,
The King's English

27c   Using Exclamation Points

Use an **exclamation point** to convey strong feeling—astonishment, drama, shock, and the like—at the end of an emphatic statement, interjection, or command.

(1)   Marking emphasis

Use an exclamation point to signal the end of an emotional or emphatic statement, an emphatic interjection, or a forceful command.

Remember the *Maine* !

No! Don't leave !

Finish this job at once !

NOTE: An exclamation point can follow a complete sentence ("What big teeth you have !") or a phrase ("What big teeth !")

(2)   Editing misused or overused exclamation points

Exclamation points should not be used in the following situations.

With Mild Statements   Exclamation points are not used with mildly emphatic statements or with mild interjections or commands.

Please close the door behind you.

Stand by your man.

With Another Exclamation Point   It is incorrect to end a sentence with more than one exclamation point.

FAULTY: I could hardly believe my eyes!!!

REVISED: I could hardly believe my eyes !

ANSWERS TO EXERCISE 3

1. Are you kidding? I never said that!
2. When the cell divided, each of the daughter cells had an extra chromosome.
3. This is fantastic! I can't believe you bought this for me!
4. Wow! Just what I always wanted! A pink Cadillac!
5. "Eureka!" cried Archimedes as he sprang from his bathtub.

With Other Punctuation Do not use other punctuation marks along with exclamation points.

FAULTY: "Fire!," he shouted.

REVISED: "Fire!" he shouted.

FAULTY: You can't be serious?!

REVISED: You can't be serious!

As an Indication of Attitude Do not use an exclamation point to suggest sarcasm or humor. Use word choice and sentence structure instead.

FAULTY: The team's record was a near-perfect (!) 0 and 12.

REVISED: The team's record was a far-from-perfect 0 and 12.

> **CLOSE-UP**
>
> ### USING EXCLAMATION POINTS
>
> Except for recording dialogue, exclamation points are almost never appropriate in college writing. Even in informal writing, use exclamation points sparingly; too many exclamation points give readers the impression that you are overwrought, even hysterical.

EXERCISE 3

Correct the use of exclamation points and other punctuation in these sentences.

EXAMPLE: "My God," she cried. "I've been shot!!!"

"My God," she cried. "I've been shot!"

1. Are you kidding?! I never said that.
2. When the cell divided, each of the daughter cells had an extra chromosome!
3. This is fantastic. I can't believe you bought this for me.
4. Wow!! Just what I always wanted!! A pink Cadillac!!
5. "Eureka!," cried Archimedes as he sprang from his bathtub.

EXERCISE 4

Add appropriate punctuation to this passage.

Dr Craig and his group of divers paused at the shore, staring respectfully at the enormous lake Who could imagine what terrors lay beneath its surface Which of them might not emerge alive from this adventure Would it be Col Cathcart Capt Wilks, the MD from the naval base Her husband, P L Fox Or would they all survive the task ahead Dr Craig decided some encouraging remarks were in order

"Attention divers," he said in a loud, forceful voice "May I please have your attention The project which we are about to undertake—"

"Oh, no" screamed Mr Fox suddenly "Look out It's the Loch Ness Monster"

"Quick" shouted Dr Craig "Move away from the shore" But his warning came too late

ANSWERS TO EXERCISE 4

Dr. Craig and his group of divers paused at the shore, staring respectfully at the enormous lake. Who could imagine what terrors lay beneath its surface? Which of them might not emerge alive from this adventure? Would it be Col. Cathcart? Ms. Wilks, the M.D. from the naval base? Her husband, P. L. Fox? Or would they all survive the task ahead? Dr. Craig decided some encouraging remarks were in order.

"Attention, divers!" he said in a loud, forceful voice. "May I please have your attention? The project which we are about to undertake—"

"Oh, no!" screamed Mr. Fox suddenly. "Look out! It's the Loch Ness Monster!"

"Quick!" shouted Dr. Craig. "Move away from the shore!" But his warning came too late.

TEACHING STRATEGY

The Comma

Even more advanced writers may occasionally misplace commas; for this reason, work with this chapter may be ongoing throughout a composition course. When referring students to this chapter as part of your editorial comments on an essay or other writing assignment, be sure to specify the chapter, section, and number—for example "28c1" for a problem with commas and introductory adverb clauses.

CHAPTER 28

The Comma

USE COMMAS . . .

- To set off independent clauses **(28a)**
- To set off items in a series **(28b)**
- To set off introductory elements **(28c)**
- To set off nonessential elements **(28d)**
- In other conventional contexts **(28e)**
- To prevent misreading **(28f)**

28a Setting Off Independent Clauses

Use a comma when you form a compound sentence by linking two independent clauses with a **coordinating conjunction** (*and, but, or, nor, for, yet, so*).

> The year was 2081 **,** and everybody was finally equal. (Kurt Vonnegut, Jr., "Harrison Bergeron")

> The bride was not young **,** nor was she very pretty. (Stephen Crane, "The Bride Comes to Yellow Sky")

Use a comma before each coordinating conjunction, no matter how many independent clauses a compound sentence has.

> She had bonny children **,** yet she felt they had been thrust upon her **,** and she could not love them. (D. H. Lawrence, "The Rocking-Horse Winner")

468

NOTE: You may omit the comma if two clauses connected by a co-ordinating conjunction are very short.

Seek and ye shall find.

Love it or leave it.

Use a comma after the first clause when pairs of correlative conjunctions link two independent clauses.

▶ See 21g

<u>Just as it's fascinating to find out where your barber gets his hair cut or what the top chef eats</u>, so it can be worthwhile to find out how top brokers invest their own money. (*Money*)

QUOTATIONS

"All morning I worked on the proof of one of my poems, and I took out a comma; in the afternoon I put it back."

–Oscar Wilde

"Anyone who can improve a sentence of mine by the omission or placing of a comma is looked upon as my dearest friend."

–George Moore

CLOSE-UP

USING COMMAS

You may use a semicolon—not a comma—to separate two independent clauses linked by a coordinating conjunction when at least one clause already contains a comma.

▶ See 29b

The tour visited Melbourne , the capital of Australia ; and it continued on to Wellington , New Zealand.

You may also use a semicolon when one or more independent clauses are especially complex or when the second clause stands in sharp contrast to the first.

Helena wanted to marry her sister's brother-in-law in an elaborate outdoor wedding on June 24 ; but her parents thought she should wait until she turned 21 in September. (complex clauses)

The advent of TV has increased the false values ascribed to reading, since TV provides a vulgar alternative. But this piety is silly ; and most reading is no more cultural nor intellectual nor imaginative than shooting pool or watching *What's My Line*? (Donald Hall, "Four Kinds of Reading") (contrasting clauses)

In the sample sentences, you could delete the coordinating conjunctions that follow the semicolons; however, using both a semicolon and a coordinating conjunction makes the separation between the clauses more emphatic.

1. The Pope did not hesitate to visit his native Poland, nor did he hesitate to meet with Solidarity leader Lech Walesa.
2. Agents place brand-name products in prominent positions in films, so the products will be seen and recognized by large audiences.
3. Unisex insurance rates may have some drawbacks for women, or these rates may be very beneficial.
4. Cigarette advertising no longer appears on television, but it does appear in print media.
5. Dorothy Day founded the Catholic Worker movement more than fifty years ago, and today her followers still dispense free food, medical care, and legal advice to the needy.

EXERCISE 1

Combine each of the following sentence pairs into one compound sentence, adding commas where necessary.

EXAMPLE: Emergency medicine became an approved medical specialty in 1979. Now pediatric emergency medicine is become increasingly important. (and)

Emergency medicine became an approved medical specialty in 1979 **,** and now pediatric emergency medicine is becoming increasingly important.

1. The Pope did not hesitate to visit his native Poland. He did not hesitate to meet with Solidarity leader Lech Walesa. (nor)
2. Agents place brand-name products in prominent positions in films. The products will be seen and recognized by large audiences. (so)
3. Unisex insurance rates may have some drawbacks for women. These rates may be very beneficial. (or)
4. Cigarette advertising no longer appears on television. It does appear in print media. (but)
5. Dorothy Day founded the Catholic Worker movement more than fifty years ago. Today her followers still dispense free food, medical care, and legal advice to the needy. (and)

28b   Setting Off Items in a Series

(1)   Coordinate elements

Use commas with three or more coordinate elements (words, phrases, or clauses) in a series.

Chipmunk **,** *raccoon* **,** and *Mugwump* are Native American words. (series of words)

She is a child of her age **,** of depression **,** of war **,** of fear. (Tillie Olsen, "I Stand Here Ironing") (series of phrases)

Brazilians speak Portuguese, Colombians speak Spanish **,** and Haitians speak French and Creole. (series of clauses)

Do not use a comma to introduce or to close a series.

FAULTY: Three important criteria are, fat content, salt content, and taste.

REVISED: Three important criteria are fat content **,** salt content **,** and taste.

FAULTY: Quebec, Ontario, and Alberta, are Canadian provinces.

REVISED: Quebec **,** Ontario **,** and Alberta are Canadian provinces.

If phrases or clauses in a series already contain commas, separate the items with semicolons.

▶ See 29d

CLOSE-UP

USING COMMAS

In journalistic writing, the comma separating the last two items in a series is usually omitted. To avoid ambiguity, however, you should always use a comma between the last two items in a series—before the coordinating conjunction if the series includes one.

AMBIGUOUS: The party was made special by the company, the light from the hundreds of twinkling candles and the excellent hors d'oeuvres.

REVISED: The party was made special by the company **,** the light from the hundreds of twinkling candles **,** and the excellent hors d'oeuvres.

(2) Coordinate adjectives

Use a comma between two or more **coordinate adjectives**—adjectives that modify the same word or word group—unless they are joined by a conjunction.

She brushed her long **,** shining hair.

The fruit was crisp **,** tart **,** mellow—in short, good enough to eat.

The baby was tired and cranky and wet. (adjectives joined by conjunctions; no commas required)

Sometimes several adjectives will all seem to modify the same noun, pronoun, or noun phrase when in fact one or more of them modifies another word or word group. In the sentence *Ten red balloons fell from the ceiling,* for instance, the adjective *red* modifies the noun *balloons,* but the adjective *ten* modifies the word group *red balloons.* In this case, only one adjective modifies the noun; the adjectives are not coordinate, so no comma is used.

ANSWERS TO EXERCISE 2

1. Seals, whales, dogs, lions, and horses are all mammals.
2. C
3. Seals are mammals, but lizards and snakes and iguanas are reptiles, and newts and salamanders are amphibians.
4. Amphibians also include frogs and toads.
5. Eagles, geese, ostriches, turkeys, chickens, and ducks are classified as birds.

TESTS FOR COMMA USE WITH SERIES OF ADJECTIVES

1. If you can reverse the order of the adjectives, you need a comma.

She brushed her shining, long hair.
She brushed her long, shining hair.

If you cannot, the adjectives are not coordinate, and you should not use a comma.

Red ten balloons fell from the ceiling.
Ten red balloons fell from the ceiling.

2. If you can insert *and* between the adjectives without changing the meaning of the sentence, use a comma.

She brushed her long [and] shining hair.
She brushed her long, shining hair.

If you cannot, the adjectives are not coordinate, and you should not use a comma.

Ten [and] red balloons fell from the ceiling.
Ten red balloons fell from the ceiling.

(Numbers—such as *ten* in the preceding example—are not coordinate with other adjectives.)

EXERCISE 2

Correct the use of commas in the following sentences, adding or deleting commas where necessary. If a sentence is punctuated correctly, mark it with a *C*.

EXAMPLE: Neither dogs snakes bees nor dragons frighten her.

Neither dogs, snakes, bees, nor dragons frighten her.

1. Seals, whales, dogs, lions, and horses, are all mammals.
2. Mammals are warm-blooded vertebrates that bear live young, nurse them, and usually have fur.
3. Seals are mammals but lizards, and snakes, and iguanas are reptiles, and newts and salamanders are amphibians.
4. Amphibians also include frogs, and toads.
5. Eagles geese ostriches turkeys chickens and ducks are classified as birds.

EXERCISE 3

Add two coordinate adjectives to modify each of the following combinations, inserting commas where required.

EXAMPLE: classical music

strong, beautiful classical music

1. distant thunder
2. silver spoon
3. New York Yankees
4. miniature golf
5. Rolling Stones
6. loving couple
7. computer science
8. wheat bread
9. art museum
10. new math

28c   Setting Off Introductory Elements

Use a comma to separate introductory elements from the rest of the sentence.

(1)   Introductory adverb clauses

Introductory adverb clauses, including elliptical adverb clauses, are generally set off from the rest of the sentence by commas.

▶ **See 8c2**

Although the CIA used to call undercover agents *penetration agents*, they now routinely refer to them as *moles*.

When war came to Beirut and Londonderry and Saigon, the victims were the children.

While working in the mines, Paul longed for a better life.

If the adverb clause is short, you may omit the comma—*provided the sentence will be clear without it.*

When I exercise I drink plenty of water.

NOTE: When an adverb clause falls at the *end* of a sentence, a comma does not usually separate it from the independent clause.

▶ **See 28g7**

(2)   Introductory phrases

Introductory phrases are usually set off from the rest of the sentence by commas.

Thinking that this might be his last chance, Scott struggled toward the Pole. (introductory participial phrase)

To succeed in a male-dominated field, women engineers must work extremely hard. (introductory infinitive phrase)

When the most sagacious of Victorian culinarians, Mrs. Beeton, spoke rather cryptically of the "alliaceous tribe," she was referring to none other than the ancient and noble members of the lily family known in kitchens round the world as the onion, scallion, leek, shallot, clove, and garlic. I don't suppose it really matters that many cooks today are hardly aware of the close affinity the common bulb onion we take so much for granted has with these other vegetables of the *Allium* genus, but it does bother me how Americans underestimate the versatility of the onion and how so few give a second thought to exploiting the potential of its aromatic relatives. More often than not, the onion itself is considered no more than a flavoring agent to soups, stews, stocks, sauces, salads, and sandwiches. Though I'd be the last to deny that nothing awakens the gustatory senses or inspires the soul like the aroma of onions simmering in a lusty stew or the crunch of a few sweet, odoriferous slices on a juicy hamburger, it would be nice to see the onion highlighted in ways other than the all-too-familiar fried rings and creamed preparations.

During the worst days of the Depression, movie attendance rose dramatically. (introductory prepositional phrase)

CLOSE-UP

USING COMMAS

Gerund phrases and infinitive phrases that serve as subjects are not set off by commas.

FAULTY: Laughing out loud, can release tension.
REVISED: Laughing out loud can release tension.
FAULTY: To know him, is to love him.
REVISED: To know him is to love him.

If the introductory phrase is short and no ambiguity is possible, you may omit the comma.

For the first time Clint felt truly happy.

After the exam I took a four-hour nap.

(3) Introductory transitional expressions

When they begin a sentence, conjunctive adverbs or other transitional expressions are usually set off from the rest of the sentence with commas.

Certainly, any plan that is enacted must be fair.

In other words, we cannot act hastily.

EXERCISE 4

Add commas in this paragraph where they are needed to set off an introductory element from the rest of the sentence.

When the most sagacious of Victorian culinarians, Mrs. Beeton, spoke rather cryptically of the "alliaceous tribe" she was referring to none other than the ancient and noble members of the lily family known in kitchens round the world as the onion, scallion, leek, shallot, clove, and garlic. I don't suppose it really matters that many cooks today are hardly aware of the close affinity the common bulb onion we take so much for granted has with these other vegetables of the *Allium* genus, but it does bother me how Americans underestimate the versatility of the onion and how so few give a second thought to exploiting the potential of its aromatic relatives. More often than not the onion itself is considered no more than a flavoring agent to soups, stews, stocks, sauces, salads, and sandwiches. Though I'd be the last to deny that nothing awakens the gustatory senses or inspires the soul like the aroma of onions simmering in a lusty stew or the crunch of a few sweet, odoriferous slices on a juicy hamburger it would be nice to see the onion highlighted in ways other than the all-too-familiar fried rings and creamed preparations. (James Villas, "E Pluribus Onion")

28d Setting Off Nonessential Elements from the Rest of the Sentence

Certain elements at the beginning, middle, or end of a sentence are considered nonessential; in fact, they could be placed within parentheses. Although these words, phrases, or clauses do *contribute* to the meaning of the sentence, they are not *essential* to its meaning.

> Designer jeans are a contradiction in terms **,** like educational television. (Fran Lebowitz)
>
> There is no place I know of **,** other than the bathtub **,** where people should not have to worry about manners. (Judith Martin)

Commas should set off nonessential elements to keep them from blending into the rest of the sentence. (If the nonessential element falls at the beginning or end of a sentence, only one comma sets it off.)

(1) Nonrestrictive modifiers

Modifying phrases or clauses may be restrictive or nonrestrictive. **Restrictive modifiers** supply information essential to the meaning of the word or word group they modify and are *not* separated from it by commas. **Nonrestrictive modifiers**, which supply information essential to the meaning of the word or word group they modify, *are* set off by commas.

Compare these two sentences.

> Actors who have big egos are often insecure.
>
> Actors **,** who have big egos **,** are often insecure.

At first glance, the two sentences above seem to mean exactly the same thing. Upon closer examination, however, you can see that they convey different meanings. Look at the first sentence again.

> Actors who have big egos are often insecure.

In this sentence, *who have big egos* is **restrictive**; the writer's intention is to limit the noun *actors* only to those who have big egos. The sentence suggests that only those actors with big egos—not all actors—are insecure. The modifying phrase is essential to the

That is a restrictive modifier; *which,* a nonrestrictive modifier:

"It is a fact that the proportion of *thats* to *whichs* is far higher in speech than in writing; but the reason is not that the spoken *thats* are more properly converted into written *whichs,* but that the kind of clause properly begun with *which* is rare in speech with its short detached sentences, but very common in the more complex and continuing structure of writing, while the kind properly begun with *that* is necessary in both."

–H. W. Fowler,
Modern English Usage

"What kinds of clauses are 'properly' begun with *that* and what kinds with *which*? *That* is better used to introduce a limiting or defining clause, *which* to introduce a nondefining or parenthetical clause. Getting away from the grammatical jargon, we might take this as a guide: If the clause could be omitted without leaving the noun it modifies incomplete, or without materially altering the sense of what is said—or if it could reasonably be enclosed in parentheses—it would be better introduced by *which*; otherwise, by *that*. For example, 'The Hudson River, which flows west of Manhattan, is muddy.' (A nondefining clause: It could be omitted or parenthesized.) But: 'The river that flows west of Manhattan is the Hudson.' (The clause defines 'river' and could not be omitted.)"

–Theodore M. Bernstein,
The Careful Writer

475

meaning of *actors*, the noun it modifies, and therefore it is not set off by commas.

Now look at the second sentence again.

Actors , who have big egos , are often insecure.

In this sentence, the modifying phrase *who have big egos* is nonrestrictive because it suggests that *all* actors have big egos. Because the modifying phrase is not essential to the meaning of the noun *actors*, it is set off by commas.

As the following examples illustrate, commas set off only nonrestrictive modifiers—those that supply nonessential information—never restrictive modifiers, which supply essential information.

See ◄
8c2

Adjective Clauses

The artist who created Zap Comix during the 1960s has also had artwork displayed at the Whitney Museum of American Art. (restrictive; no commas used)

Robert Crumb , who created Zap Comix during the 1960s , is credited with popularizing the slogan "Keep on Truckin'." (nonrestrictive; commas used)

Speaking in public is something that most people fear. (restrictive; no comma used)

He ran for the bus , which was late as usual. (nonrestrictive; comma used)

See ◄
8e1

Prepositional phrases

The man with the gun demanded their money. (restrictive; no commas used)

The surgeon , with a smile on her face , pronounced the operation a success. (nonrestrictive; commas used)

See ◄
8e2

Verbal phrases

The candidates running for mayor have agreed to a debate. (restrictive; no commas used)

The marathoner , running as fast as he could , beat his previous time by several seconds. (nonrestrictive; commas used)

See ◄
8e4

Appositives

Orson Welles's film *Citizen Kane* has received great critical acclaim. (restrictive; no commas used)

Maya Angelou , a native of Arkansas , read an original poem at President Clinton's inauguration. (nonrestrictive; commas used)

TESTS FOR DETERMINING WHETHER A MODIFIER IS RESTRICTIVE OR NONRESTRICTIVE

1. Could the modifier be placed within parentheses (De-signer jeans are a contradiction in terms, *like educational television*.)? If so, it is nonrestrictive and needs commas.

2. Is the modifier essential to the meaning of the noun it modifies (*The artist who created Zap Comix*, not just any artist; *the man with the gun*, not just any man)? If so, it is restrictive and does not take commas. If not, commas should set off the modifier.

3. Is the modifier introduced by *that* (*something that most people fear*)? If so, it is restrictive. *That* cannot introduce a nonrestrictive clause.

4. Can you delete the relative pronoun without causing am-biguity or confusion (*something [that] most people fear*)? If so, the clause is restrictive and requires no commas.

5. Is the appositive more specific than the noun that pre-cedes it (*Orson Welles's film* Citizen Kane)? If so, it is re-strictive and should not be set off by commas.

CLOSE-UP

USING COMMAS

That is used to introduce only restrictive clauses.

I bought a used car <u>that</u> cost $2000.

Which can be used to introduce both restrictive and nonre-strictive clauses.

RESTRICTIVE: I bought a used car <u>which</u> cost $2000.

NONRESTRICTIVE: The used car I bought , <u>which</u> cost $2000 , broke down after a week.

Many writers, however, prefer to use *which* only to intro-duce nonrestrictive clauses. Remember, though, that when *which* introduces a nonrestrictive clause, it is preceded by a comma.

ANSWERS TO EXERCISE 5

The Statue of Liberty, which was dedicated in 1886, had undergone extensive renovation. Its supporting structure, whose designer was the French engineer Alexandre Gustave Eiffel, is made of iron. The Statue of Liberty, created over a period of nine years by sculptor Frédéric-August Bartholdi, stands 151 feet tall. The people of France, who were grateful for American help in the French Revolution, raised the money to pay the sculptor who created the statue. The people of the United States, contributing over $100,000, raised the money for the pedestal on which the statue stands.

EXERCISE 5

Insert commas where necessary to set off nonrestrictive phrases and clauses.

The Statue of Liberty which was dedicated in 1886 has undergone extensive renovation. Its supporting structure whose designer was the French engineer Alexandre Gustave Eiffel is made of iron. The Statue of Liberty created over a period of nine years by sculptor Frédéric-Auguste Bartholdi stands 151 feet tall. The people of France who were grateful for American help in the French Revolution raised the money to pay the sculptor who created the statue. The people of the United States contributing over $100,000 raised the money for the pedestal on which the statue stands.

(2) Conjunctive adverbs and other transitional expressions

See ◀
4d2

Conjunctive adverbs—words like *however, therefore, thus,* and *nevertheless*—and **transitional expressions** like *for example* and *on the other hand* qualify, clarify, and make connections explicit, but they are not essential to meaning.

When a conjunctive adverb or other transitional expression interrupts a clause, it is set off by commas.

> The House Ethics Committee recommended reprimanding two members of Congress. The House, however, overruled the recommendation and voted to censure them.

> The Outward Bound program, for example, is extremely safe.

A conjunctive adverb or similar expression at the *end* of a clause is still parenthetical, and it is separated from the rest of the sentence by a single comma.

> Some things were easier after school started. Other things were a lot harder, however.

Transitional expressions are also usually set off by commas when they fall at the *beginning* of a clause.

(3) Contradictory phrases

See ◀
28c3

A phrase that expresses contrast is usually set off by commas.

> This medication should be taken after a meal, never on an empty stomach.

> It was Roger Maris, not Mickey Mantle, who broke Babe Ruth's home run record.

CLOSE-UP

USING COMMAS

When a conjunctive adverb or other transitional expression separates two independent clauses, it is preceded by a semicolon or a period and followed by a comma.

▶ See 29c

Laughter is the best medicine; of course, penicillin also comes in handy sometimes.

(4) Absolute phrases

An **absolute phrase**, which usually consists of a noun plus a participle, is always set off by commas from the sentence it modifies.

▶ See 8e3

His fear increasing, he waited to enter the haunted house.

The Roanoke colonists vanished in 1591, their bodies never recovered.

(5) Miscellaneous nonessential elements

Wherever they appear, certain nonessential elements are usually separated from the rest of the sentence by commas.

Tag Questions (Auxiliary Verb + Pronoun Added to a Statement)

This is your first day on the job, isn't it?

It seems possible, does it not, that carrots may provide some protection against cancer?

Names in Direct Address

I wonder, Mr. Honeywell, whether Mr. Albright deserves a raise.

Freddie, what's your opinion?

What do you think, Margie?

Mild Interjections

Well, it's about time.

NOTE: Stronger interjections may be set off by dashes or exclamation points.

▶ See 32b

Yes and No

▶ See 27c1

Yes, we have no bananas.

No, we're all out of lemons.

479

ANSWERS TO EXERCISE 6

1. Kermit the Frog is a muppet, a cross between a marionette and a puppet.
2. The common cold, a virus, is frequently spread by hand contact, not by mouth.
3. C
4. C
5. The submarine *Nautilus* was the first to cross under the North Pole, wasn't it?
6. C
7. Superman was called Kal-El on the planet Krypton; on earth, however, he was known as Clark Kent, not Kal-El.
8. Its sales topping any of his previous singles, "Heartbreak Hotel" was Elvis Presley's first million seller.
9. Two companies, Nash and Hudson, joined in 1954 to form American Motors.
10. A firefly is a beetle, not a fly, and a prairie dog is a rodent, not a dog.

EXERCISE 6

Set off the nonessential elements in these sentences with commas. If a sentence is correct, mark it with a *C*.

> **EXAMPLE:** Piranhas like sharks will attack and eat almost anything if the opportunity arises.
>
> Piranhas , like sharks , will attack and eat almost anything if the opportunity arises.

1. Kermit the frog is a muppet a cross between a marionette and a puppet.
2. The common cold a virus is frequently spread by hand contact not by mouth.
3. The account in the Bible of Noah's Ark and the forty-day flood may be based on an actual deluge.
4. More than two-thirds of U.S. welfare recipients, such as children, the aged, the severely disabled, and mothers of children under six, are people with legitimate reasons for not working.
5. The submarine *Nautilus* was the first to cross under the North Pole wasn't it?
6. The 1958 Ford Edsel was advertised with the slogan "Once you've seen it, you'll never forget it."
7. Superman was called Kal-El on the planet Krypton; on earth however he was known as Clark Kent not Kal-El.
8. Its sales topping any of his previous singles "Heartbreak Hotel" was Elvis Presley's first million seller.
9. Two companies Nash and Hudson joined in 1954 to form American Motors.
10. A firefly is a beetle not a fly and a prairie dog is a rodent not a dog.

28e Using Commas in Other Conventional Contexts

(1) Around direct quotations

In most cases, use commas to set off a direct quotation from the **identifying tag**—the phrase that identifies the speaker (*he said, she answered*).

Emerson said to Thoreau , "I greet you at the beginning of a great career."

"I greet you at the beginning of a great career ," Emerson said to Thoreau.

"I greet you ," Emerson said to Thoreau , "at the beginning of a great career."

When the identifying tag comes between two complete sentences, however, the tag is introduced by a comma but followed by a period.

"Winning isn't everything ," Vince Lombardi said . "It's the only thing."

If the first sentence of an interrupted quotation ends with a question mark or exclamation point, do not use commas.

"Should we hold the front page?" she asked. "After all, it's a slow news day."

"Hold the front page!" he cried. "This is the biggest story of the decade."

▶ See 31e

(2) Between names and titles or degrees

Use a comma to set off a degree or title that follows a name.

Joycelyn Elders , M.D. , was named Surgeon General.

No movie has been made about Martin Luther King , Jr.

Hamlet , Prince of Denmark , is Shakespeare's most famous character.

If the title or degree precedes the name, however, no comma is required: Dr. Elders, Prince Hamlet.

No comma is used between a name and II, III, and so on.

Queen Elizabeth II

(3) In dates and addresses

Use commas to separate items in dates and addresses.

The space shuttle *Challenger* was launched on August 30 , 1983 , from Cape Canaveral. (30 August 1983—without commas—is also acceptable.)

Her address is 600 West End Avenue , New York , NY 10024.

NOTE: Commas are not used to separate the day from the month; when only the month and year are given, no commas are used (May 1968). No comma separates the street number from the street or the state name from the zip code. Within a sentence, a comma follows the last element of a date or address.

ANSWERS TO EXERCISE 7

1. India became independent on August 15, 1947.
2. The UAW has more than 1,500,000 dues-paying members.
3. Nikita Khrushchev, former Soviet premier, said, "We will bury you!"
4. Mount St. Helens, northeast of Portland, Oregon, began erupting on March 27, 1980, and eventually killed at least thirty people.
5. Located at 1600 Pennsylvania Avenue, Washington, D.C., the White House is a major tourist attraction.
6. In 1956, playing before a crowd of 64,519 fans in Yankee Stadium in New York, New York, Don Larsen pitched the first perfect game in World Series history.
7. Lewis Thomas, M.D., was born in Flushing, New York, and attended Harvard Medical School in Cambridge, Massachusetts.
8. In 1967, 2,000,000 people died of smallpox, but in 1977, only about twenty died.
9. "The reports of my death," Mark Twain remarked, "have been greatly exaggerated."
10. The French explorer Jean Nicolet landed at Green Bay, Wisconsin, in 1634, and in 1848, Wisconsin became the thirtieth state; it has 10,355 lakes and a population of more than 4,700,000.

(4)   In salutations and closings

Use commas in informal correspondence following salutations and closings and following the complimentary close in personal or business correspondence.

Dear John,	Love,
Dear Aunt Sophie,	Sincerely,

See ◄
48a2

In business correspondence use a colon, not a comma, after the salutation.

(5)   In long numbers

For a number of four digits or more, place a comma before every third digit, counting from the right. (The comma is optional with only four digits.)

1,200   (or 1200)
12,000
120,000
1,200,000

NOTE: Commas are not required in long numbers used in addresses, telephone numbers, zip codes, or years.

EXERCISE 7

Add commas where necessary to set off quotations, names, dates, addresses, and numbers.

1. India became independent on August 15 1947.
2. The UAW has more than 1500000 dues-paying members.
3. Nikita Krushchev, former Soviet premier, said "We will bury you!"
4. Mount St. Helens, northeast of Portland Oregon, began erupting on March 27 1980 and eventually killed at least thirty people.
5. Located at 1600 Pennsylvania Avenue Washington D.C., the White House is a popular tourist attraction.
6. In 1956, playing before a crowd of 64519 fans in Yankee Stadium in New York New York, Don Larsen pitched the first perfect game in World Series history.
7. Lewis Thomas M.D. was born in Flushing New York and attended Harvard Medical School in Cambridge Massachusetts.
8. In 1967 2000000 people worldwide died of smallpox, but in 1977 only about twenty died.
9. "The reports of my death" Mark Twain remarked "have been greatly exaggerated."

10. The French explorer Jean Nicolet landed at Green Bay Wisconsin in 1634, and in 1848 Wisconsin became the thirtieth state; it has 10355 lakes and a population of more than 4700000.

28f   Using Commas to Prevent Misreading

(1)   To clarify meaning

Use a comma to avoid ambiguity.

Those who can **,** sprint the final lap.

Without the comma, *can* appears to be an auxiliary verb ("Those who can sprint. . . .") and the sentence seems incomplete. The comma clarifies the sentence's meaning.

(2)   To indicate an omission

Use a comma to acknowledge the omission of a repeated word, usually a verb.

Pam carried the box; Tim **,** the suitcase.

Edwina went first; Marco **,** second.

(3)   To separate repeated words

Use a comma to separate words repeated consecutively within a sentence.

Everything bad that could have happened **,** happened.

EXERCISE 8

Add commas where necessary to prevent misreading.

> **EXAMPLE:** Whatever will be will be.
>
> Whatever will be **,** will be.

1. According to Bob Frank's computer has three disk drives.
2. Da Gama explored Florida; Pizarro Peru.
3. By Monday evening students must begin preregistration for fall classes.
4. Whatever they built they built with care.
5. When batting practice carefully.
6. Brunch includes warm muffins topped with whipped butter and freshly brewed coffee.

483

ANSWERS TO EXERCISE 9

1. Sometimes they did go shopping or to a movie, but sometimes they went across the highway, ducking fast across the busy road, to a drive-in restaurant where older kids hung out. The restaurant was shaped like a big bottle, though squatter than a real bottle, and on its cap was a revolving figure of a grinning boy who held a hamburger aloft.

2. His head whirled as he stepped into the thronged corridor, and he sank back into one of the chairs against the wall to get his breath. The lights, the chatter, the perfumes, the bewildering medley of color—he had, for a moment, the feeling of not being able to stand it.

3. It has been longed for, campaigned for, kissed and caressed on the top of its shiny, 24-karat-gold-plated bald head. It has inspired giddiness, wordiness, speechlessness, glee, and—in the case of the stars it has eluded—reactions ranging from wonder to rage.

4. Yes, society often did treat the elderly abysmally . . . they were sometimes ignored, sometimes victimized, sometimes poor and frightened, but so many of them were survivors.

5. The word success comes from the Latin verb *succedere,* meaning "to follow after."

continued on following page

EXERCISE 9

Commas have been intentionally deleted from some of the following sentences. Add commas where needed, and be prepared to explain why each is necessary. If a sentence is correct, mark it with a *C.*

1. Sometimes they did go shopping or to a movie but sometimes they went across the highway ducking fast across the busy road to a drive-in restaurant where older kids hung out. The restaurant was shaped like a big bottle though squatter than a real bottle and on its cap was a revolving figure of a grinning boy who held a hamburger aloft. (Joyce Carol Oates, "Where Are You Going, Where Have You Been?")

2. His head whirled as he stepped into the thronged corridor and he sank back into one of the chairs against the wall to get his breath. The lights the chatter the perfumes the bewildering medley of color—he had for a moment the feeling of not being able to stand it. (Willa Cather, "Paul's Case")

3. It has been longed for campaigned for kissed and caressed on the top of its shiny 24-karat-gold-plated bald head. It has inspired giddiness wordiness speechlessness glee, and—in the case of the stars it has eluded—reactions ranging from wonder to rage. (*Life,* in an article about the Oscar)

4. Yes society often did treat the elderly abysmally . . . they were sometimes ignored sometimes victimized sometimes poor and frightened but so many of them were survivors. (Katherine Barrett, "Old Before Her Time")

5. The word success comes form the Latin verb *succedere* meaning "to follow after." (Susan Ochshorn, "Economic Adventurers")

6. It was a big squarish frame house that had once been white decorated with cupolas and spires and scrolled balconies in the heavily lightsome style of the seventies set on what had once been our most select street. (William Faulkner, "A Rose for Emily")

7. The world as always is debating the issues of war and peace. (Sam Keen, "Faces of the Enemy")

8. About fifteen miles below Monterey on the wild coast the Torres family had their farm a few sloping acres above a cliff that dropped to the brown reefs and to the hissing white waters of the ocean. (John Steinbeck, "Flight")

9. I was looking for myself and asking everyone except myself questions which I and only I could answer. (Ralph Ellison, *Invisible Man*)

10. According to the Pet Food Institute a Washington-based trade association there were about 18 million more dogs than cats in the United States as recently as a decade ago but today there are 56 million cats and only 52 million dogs. (Cullen Murphy, "Going to the Cats")

28g Editing Misused or Overused Commas

continued from previous page

> **DO NOT USE COMMAS . . .**
> * Between two independent clauses **(28g1)**
> * Around restrictive elements **(28g2)**
> * Between inseparable grammatical constructions **(28g3)**
> * Between a verb and an indirect quotation or indirect question **(28g4)**
> * Between coordinate phrases that contain correlative conjunctions **(28g5)**
> * Between certain paired elements **(28g6)**
> * Before an adverb clause that falls at the end of a sentence **(28g7)**

(1) Between two independent clauses

Using just a comma to connect two independent clauses creates a comma splice.

▶ **See Ch. 14**

FAULTY: The season was unusually cool, nevertheless the orange crop was not seriously harmed.

REVISED: The season was unusually cool; nevertheless, the orange crop was not seriously harmed.

REVISED: The season was unusually cool. Nevertheless, the orange crop was not seriously harmed.

REVISED: The season was unusually cool, but the orange crop was not seriously harmed.

REVISED: Although the season was unusually cool, the orange crop was not seriously harmed.

(2) Around restrictive elements

Commas are not used to set off restrictive elements.

▶ **See 28d1**

FAULTY: Women, who seek to be equal to men, lack ambition.

REVISED: Women who seek to be equal to men lack ambition.

FAULTY: The film, *Malcolm X*, was directed by Spike Lee.

REVISED: The film *Malcolm X* was directed by Spike Lee.

6. It was a big, squarish frame house that had once been white, decorated with cupolas and spires and scrolled balconies in the heavily lightsome style of the seventies, set on what had once been our most select street.

7. The world, as always, is debating the issues of war and peace.

8. About fifteen miles below Monterey, on the wild coast, the Torres family had their farm, a few sloping acres above a cliff that dropped to the brown reefs and to the hissing white waters of the ocean.

9. I was looking for myself and asking everyone except myself questions which I, and only I, could answer.

10. According to the Pet Food Institute, a Washington-based trade association, there were about 18 million more dogs than cats in the United States as recently as a decade ago, but today there are 56 million cats and only 52 million dogs.

FAULTY: They planned a picnic, in the park.

REVISED: They planned a picnic in the park.

FAULTY: The word, *snafu,* is an acronym for "situation normal—all fouled up."

REVISED: The word *snafu* is an acronym for "situation normal—all fouled up."

(3) Between inseparable grammatical constructions

A comma should not be placed between a subject and its predicate; a verb and its complement or direct object; a preposition and its object; or an adjective and the noun, pronoun, or noun phrase it modifies. Placing a comma between such constructions interrupts the logical flow of a sentence.

FAULTY: We think that anyone who can walk a straight line out of the office before lunch, ought to be able to travel the same route on the way back. (Advertisement, Spirits Council of the United States) (comma between subject and predicate)

REVISED: We think that anyone who can walk a straight line out of the office before lunch ought to be able to travel the same route on the way back.

FAULTY: Louis Braille developed, an alphabet of raised dots for the blind. (comma between verb and object)

REVISED: Louis Braille developed an alphabet of raised dots for the blind.

FAULTY: They relaxed somewhat during, the last part of the obstacle course. (comma between preposition and object)

REVISED: They relaxed somewhat during the last part of the obstacle course.

FAULTY: Wind-dispersed weeds include the well-known and plentiful, dandelions, milkweed, and thistle. (comma between adjective and words it modifies)

REVISED: Wind-dispersed weeds include the well-known and plentiful dandelions, milkweed, and thistle.

(4) Between a verb and an indirect quotation or indirect question

Commas are not used between verb and dependent clause to set off indirect quotations or indirect questions.

FAULTY: Humorist Art Buchwald once said, that the problem with television news is that it has no second page.

REVISED: Humorist Art Buchwald once said that the problem with television news is that it has no second page.

FAULTY: The landlord asked, whether we would be willing to sign a two-year lease.

REVISED: The landlord asked whether we would be willing to sign a two-year lease.

(5) Between coordinate phrases that contain correlative conjunctions

Commas are not used between coordinate phrases that contain correlative conjunctions.

FAULTY: Thirty years ago, most college students had access to neither photocopiers, nor pocket calculators.

REVISED: Thirty years ago, most college students had access to neither photocopiers nor pocket calculators.

FAULTY: Both typewriters, and tape recorders were generally available, however.

REVISED: Both typewriters and tape recorders were generally available, however.

(6) Between certain paired elements

Commas are not used between two elements of a compound subject, predicate, object, complement, or auxiliary verb.

FAULTY: During the 1400s plagues, and pestilence were common. (Comma interrupts compound subject.)

REVISED: During the 1400s plagues and pestilence were common.

FAULTY: Many women thirty-five and older are returning to college, and tend to be good students. (Comma interrupts compound predicate.)

REVISED: Many women thirty-five and older are returning to college and tend to be good students.

FAULTY: Mattel has marketed a doctor's uniform, and an astronaut suit for its Barbie doll. (Comma interrupts compound object.)

REVISED: Mattel has marketed a doctor's uniform and an astronaut suit for its Barbie doll.

FAULTY: People buy bottled water because it is pure, and fashionable. (Comma interrupts compound complement.)

REVISED: People buy bottled water because it is pure and fashionable.

ANSWERS TO EXERCISE 10

1. A book is like a garden carried in the pocket.
2. Like the iodine content of kelp, air freight is something most Americans have never pondered.
3. Charles Rolls and Frederick Royce manufactured the first Rolls-Royce Silver Ghost in 1907.
4. The hills ahead of him were rounded domes of grey granite, smooth as a bald man's pate and completely free of vegetation.
5. C

ANSWERS TO EXERCISE 11

As the fruit starts to move along a concentrate plant's assembly line, it is first culled. In what some citrus people remember as "the old fresh-fruit days," before the Second World War, about forty percent of all oranges grown in Florida were eliminated at packinghouses and dumped in fields. Florida milk tasted like orangeade. Now, with the exception of the split and rotten fruit, all of Florida's orange crop is used. Moving up a conveyor belt, oranges are scrubbed with detergent before they roll on into juicing machines. There are several kinds of juicing machines, and they are something to see. One is called the Brown Seven Hundred. Seven hundred oranges a minute go into it and are split and reamed on the same kind of rosettes that are in the centers of ordinary kitchen reamers. The rinds that come pelting out the bottom are integral halves, just like the rinds of oranges squeezed in a kitchen.

FAULTY: She can, and will be ready to run in the primary. (Comma interrupts compound auxiliary verb.)

REVISED: She can and will be ready to run in the primary.

REVISED: She can , and will , be ready to run in the primary.

(7) Before an adverb clause that falls at the end of a sentence

Commas are generally not used before an adverb clause that falls at the end of a sentence.

FAULTY: Jane Addams founded Hull House, because she wanted to help Chicago's poor.

REVISED: Jane Addams founded Hull House because she wanted to help Chicago's poor.

EXERCISE 10

Unneeded commas have been intentionally added to some of the sentences that follow. Delete any unnecessary commas. If a sentence is correct, mark it with a *C*.

EXAMPLE: Spring fever, is a common ailment.

Spring fever is a common ailment.

1. A book is like a garden, carried in the pocket. (Arab proverb)
2. Like the iodine content of kelp, air freight, is something most Americans have never pondered. (*Time*)
3. Charles Rolls, and Frederick Royce manufactured the first Rolls Royce Silver Ghost, in 1907.
4. The hills ahead of him were rounded domes of grey granite, smooth as a bald man's pate, and completely free of vegetation. (Wilbur Smith, *Flight of the Falcon*)
5. Food here is scarce, and cafeteria food is vile, but the great advantage to Russian raw materials, when one can get hold of them, is that they are always fresh and untampered with. (Andrea Lee, *Russian Journal*)

EXERCISE 11

In the following passage, punctuation errors have been deliberately made. Delete excess commas and add any necessary ones. Be prepared to justify your revisions.

As the fruit starts to move along a concentrate plant's assembly line it is first culled. In what some citrus people remember, as "the old, fresh-fruit days" before the Second World War about forty percent of all oranges, grown in Florida, were eliminated at packinghouses, and dumped in fields. Florida milk tasted like orangeade. Now with the exception of the split, and rotten fruit, all of Florida's orange crop is used. Moving up a conveyor belt oranges are scrubbed with detergent before they roll, on into juicing machines. There are several kinds of juicing machines and they are something to see. One is called the Brown Seven Hundred. Seven hundred oranges a minute go into it, and are split, and reamed on the same kind of rosettes, that are in the centers of ordinary, kitchen reamers. The rinds, that come pelting out the bottom are integral halves just like the rinds of oranges squeezed in a kitchen. (Adapted from John McPhee's "Oranges")

CHAPTER 29

The Semicolon

The **semicolon** is weaker than the period and stronger than the comma. It signals a shorter pause than the period but a longer pause than the comma. The semicolon is used only between items of equal grammatical rank: two independent clauses, two phrases, and so on.

> **USE SEMICOLONS . . .**
> * To separate independent clauses **(29a)**
> * To separate complex, internally punctuated clauses **(29b)**
> * To separate clauses containing conjunctive adverbs **(29c)**
> * To separate items in a series **(29d)**

29a Separating Independent Clauses

Use a semicolon rather than a coordinating conjunction or a period between independent clauses that are closely related in meaning.

Paul Revere's *The Boston Massacre* is an early example of traditional American protest art Edward Hicks's later "primitive" paintings are socially conscious art with a religious strain. (clauses related by contrast)

In the preceding example, separate sentences marked by periods would be correct but would fail to convey the close relationships between the clauses.

490

; 29b

▶ See
Ch 14

CLOSE-UP

USING SEMICOLONS

Using only a comma or no punctuation at all between independent clauses will produce a comma splice or a fused sentence.

continued from previous page

And better still when it pivots a contrast:

Work when you work; play when you play. The semicolon is a stop sign; the colon, a green light."

–Sheridan Baker,
The Practical Stylist

29b Separating Complex, Internally Punctuated Clauses

Use a semicolon instead of a comma between two clauses joined by a coordinating conjunction if one clause contains internal punctuation or is long or complex.

> If such a world government is not established by a process of agreement among nations, I believe it will come anyway, and in a much more dangerous form **;** for war or wars can only result in one power being supreme and dominating the rest of the world by its overwhelming military supremacy. (Albert Einstein, *Einstein on Peace*)

The semicolon in the preceding sentence not only distinguishes the two clauses but also emphasizes the warning at the end by isolating it.

EXERCISE 1

Add semicolons, periods, or commas plus coordinating conjunctions where necessary to separate independent clauses. Reread the paragraph when you have finished to make certain no comma splices or fused sentences remain.

> **EXAMPLE:** *Birth of a Nation* was one of the earliest epic movies it was based on the book *The Klansman.*
>
> *Birth of a Nation* was one of the earliest epic movies **;** it was based on the book *The Klansman.*

During the 1950s movie attendance declined because of the increasing popularity of television. As a result, numerous gimmicks were introduced to draw audiences into theaters. One of the first of these was Cinerama, in this technique three pictures were shot side by side and projected on a curved screen. Next came 3-D, complete with special glasses, *Bwana Devil* and *The Creature from the Black Lagoon* were two early 3-D ventures. *The*

ANSWERS TO EXERCISE 1

During the 1950s, movie attendance declined because of the increasing popularity of television. As a result, numerous gimmicks were introduced to draw audiences into theaters. One of the first of these was Cinerama. In this technique three pictures were shot side by side and projected on a curved screen. Next came 3-D, complete with special glasses; *Bwana Devil* and *The Creature from the Black Lagoon* were two early 3-D ventures. *The Robe* was the first picture filmed in Cinemascope; in this technique a shrunken image was projected on a screen twice as wide as it was tall. Smell-O-Vision (or Aroma-rama) was a short-lived gimmick that enabled audiences to smell what they were viewing, but problems developed when it became impossible to get one odor out of the theater in time for the next smell to be introduced. William Castle's *Thirteen Ghosts* introduced special glasses for cowardly viewers who wanted to be able to control what they saw; the red part of the glasses was the "ghost viewer," and the green part was the "ghost remover." Perhaps the ultimate in movie gimmicks accompanied the film *The Tingler*. When this film was shown, seats in the

continued on following page

continued from previous page

theater were wired to generate mild electric shocks. Unfortunately, the shocks set off a chain reaction that led to hysteria in the theater. During the 1960s, such gimmicks all but disappeared, and viewers were able once again to simply sit back and enjoy a movie.

ANSWERS TO EXERCISE 2

1. Amusement parks offer exciting rides; they are thrill packed and flirt with danger.
2. Free Fall is located in Atlanta's Six Flags over Georgia; in this ride, riders travel up a 128-foot-tall tower and then plunge down at 55 miles per hour.
3. In the Sky Whirl ride at Great America, riders go 115 feet up in the air and circle about 75 times; Great America parks are in Gurnee, Illinois, and Santa Clara, California.
4. The Kamikaze Slide can be found at the Wet 'n Wild parks in Arlington, Texas, and Orlando, Florida; this ride is a slide 300 feet long that extends 60 feet in the air.
5. Parachuter's Perch, found at Great Adventure in Jackson, New Jersey, is another exciting ride; its chutes fall at 25 feet per second.
6. Astroworld in Houston, Texas, boasts Greezed Lightnin'; this ride is an 80-foot-high loop that goes from 0 to 60 miles per hour in four seconds and moves forward and backward.
7. The Beast is a wooden roller coaster at Kings Island near Cincinnati, Ohio; it has a 7400-foot track and goes 70 miles per hour.

Robe was the first picture filmed in Cinemascope in this technique a shrunken image was projected on a screen twice as wide as it was tall. Smell-O-Vision (or Aroma-rama) was a short-lived gimmick that enabled audiences to smell what they were viewing problems developed when it became impossible to get one odor out of the theater in time for the next smell to be introduced. William Castle's *Thirteen Ghosts* introduced special glasses for cowardly viewers who wanted to be able to control what they saw, the red part of the glasses was the "ghost viewer" and the green part was the "ghost remover." Perhaps the ultimate in movie gimmicks accompanied the film *The Tingler* when this film was shown seats in the theater were wired to generate mild electric shocks. Unfortunately, the shocks set off a chain reaction that led to hysteria in the theater. During the 1960s such gimmicks all but disappeared, viewers were able once again to simply sit back and enjoy a movie.

*EXERCISE 2

Combine each of the following sentence groups into one sentence that contains only two independent clauses. Use a semicolon to join the two clauses. You will need to add, delete, relocate, or change some words; keep experimenting until you find the arrangement that best conveys the sentence's meaning.

EXAMPLE: The Congo River Rapids is a ride at the Dark Continent in Tampa, Florida. Riders raft down the river. They glide alongside jungle plants and animals.

The Congo River Rapids is a ride at the Dark Continent in Tampa, Florida ; riders raft down the river, gliding alongside jungle plants and animals.

1. Amusement parks offer exciting rides. They are thrill packed. They flirt with danger.
2. Free Fall is located in Atlanta's Six Flags over Georgia. In this ride, riders travel up a 128-foot-tall tower. They plunge down at 55 miles per hour.
3. In the Sky Whirl riders go 115 feet up in the air and circle about 75 times. This ride is located in Great America. Great America parks are in Gurnee, Illinois, and Santa Clara, California.
4. The Kamikaze Slide can be found at the Wet 'n Wild parks in Arlington, Texas, and Orlando, Florida. This ride is a slide 300 feet long. It extends 60 feet in the air.
5. Parachuter's Perch is another exciting ride. It is found at Great Adventure in Jackson, New Jersey. Its chutes fall at 25 feet per second.
6. Astroworld in Houston, Texas, boasts Greezed Lightnin'. This ride is an 80-foot-high loop. The ride goes from 0 to 60 miles per hour in four seconds and moves forward and backward.
7. The Beast is at Kings Island near Cincinnati, Ohio. The Beast is a wooden roller coaster. It has a 7400-foot track and goes 70 miles per hour. (Adapted from *Seventeen*)

29c   Separating Independent Clauses Containing Conjunctive Adverbs

Use a semicolon between two closely related independent clauses when the second clause is introduced by a conjunctive adverb or transitional expression. (For a complete list of such expressions, **see 4d2.**)

> Thomas Jefferson brought 200 vanilla beans and a recipe for vanilla ice cream back from France ; <u>thus</u>, he gave America its all-time favorite ice-cream flavor.

NOTE: Within a clause, the position of a conjunctive adverb or transitional expression may vary, and so will the punctuation.

▶ **See 21e**

*EXERCISE 3

Combine each of the following sentence groups into one sentence that contains only two independent clauses. Use a semicolon and the conjunctive adverb or transitional phrase in parentheses to join the two clauses, adding commas within clauses where necessary. You will need to add, delete, relocate, or change some words. There is no one correct version; keep experimenting until you find the arrangement you feel is most effective.

> **EXAMPLE:** The Aleutian Islands are located off the west coast of Alaska. They are an extremely remote chain of islands. They are sometimes called America's Siberia. (in fact)
>
> The Aleutian Islands, located off the west coast of Alaska, are an extremely remote chain of islands ; in fact, they are sometimes called America's Siberia.

1. The Aleutians lie between the North Pacific Ocean and the Bering Sea. The weather there is harsh. Dense fog, 100-mile-per-hour winds, and even tidal waves and earthquakes are not uncommon. (for example)
2. These islands constitute North America's largest network of active volcanoes. The Aleutians boast some beautiful scenery. The islands are relatively unexplored. (still)
3. The Aleutians are home to a wide variety of birds. Numerous animals, such as fur seals and whales, are found there. These islands may house the largest concentration of marine animals in the world. (in fact)
4. During World War II, thousands of American soldiers were stationed on Attu Island. They were stationed on Adak Island. The Japanese eventually occupied both Attu and Adak Islands. (however)
5. The islands' original population of native Aleuts was drastically reduced in the eighteenth century by Russian fur traders. Today the total population is only about 8500. U.S. military employees comprise more than half of this. (consequently) (Adapted from *National Geographic*)

STUDENTS WORKING TOGETHER

Take a model essay that has good use of transitions and remove all transitions, leaving spaces where they were. Small groups should then try to figure out what the transitions should be. A few versions of the essay, with transitions inserted, can be read before discussion.

TEACHING STRATEGY

Semicolons
Emphasize the use of semicolons as an organizational tool for the "series" sentence. Initially, students may be confused about when to use semicolons to separate series items. Assure them that common-sense rules apply here: If the series is "simple"— that is, if each of its elements consists of only one word or a similarly constructed group of words, punctuating with comas will suffice.

> The pitcher, batter, and coach came out of the dugout together.

> We like to eat bread and butter, peaches and cream, or bacon and eggs.

But if the items in a series contain any "extras" (commas, modifying phrases, names or dates, explanations, etc.), semicolons will help readers separate series items into the proper subgroups.

ANSWERS TO EXERCISE 4

1. A quarter of a million Americans marched on Washington in 1983 to commemorate the twentieth anniversary of Dr. King's "I Have a Dream" speech; to remind the government that many Americans, even

continued on following page

29d Separating Items in a Series

Use semicolons between items in a series when one or more of those items include commas. Without semicolons it is difficult to distinguish the individual elements in the series.

> Three papers are posted on the bulletin board outside the building: a description of the exams ; a list of appeal procedures for students who fail ; and an employment ad from an automobile factory, addressed specifically to candidates whose appeals are turned down. (Andrea Lee, *Russian Journal*)

> As ballooning became established, a series of firsts ensued: The first balloonist in the United States was 13-year-old Edward Warren, 1784 ; the first woman aeronaut was a Madame Thible who, depending on your source, either recited poetry or sang as she lifted off ; the first airmail letter, written by Ben Franklin's grandson, was carried by balloon ; and the first bird's-eye photograph of Paris was taken by a balloon. (Elaine B. Steiner, *Games*)

CLOSE-UP

USING SEMICOLONS

Even when items in a series are brief, semicolons are required if any element in the series contains commas or other internal punctuation.

> Laramie, Wyoming ; Wyoming, Delaware ; and Delaware, Ohio, were the first three places they visited.

EXERCISE 4

Replace commas with semicolons where necessary to separate internally punctuated items in a series.

EXAMPLE: Luxury automobiles have some strong selling points: they are status symbols, some, such as the Corvette, appreciate in value, and they are usually comfortable and well appointed.

Luxury automobiles have some strong selling points: they are status symbols ; some, such as the Corvette, appreciate in value ; and they are usually comfortable and well appointed.

1. A quarter of a million Americans marched on Washington in 1983 to commemorate the twentieth anniversary of Dr. King's "I Have a Dream" speech, to remind the government that many Americans, even those who attended the 1963 march, are still without jobs and full equality, and to demonstrate for peace.
2. Steroids, used by some athletes, can be dangerous because they can affect the pituitary gland, causing a lowered sperm count, because they have been linked to the development of liver tumors, and because they can stimulate the growth of some cancers.
3. Tennessee Williams wrote *The Glass Menagerie*, which is about a handicapped young woman and her family, *A Streetcar Named Desire*, which starred Marlon Brando, and *Cat on a Hot Tin Roof*, which won a Pulitzer Prize.
4. New expressions like *outro*, used to designate the segment of a broadcast where the announcer signs off, *heavy breather*, used to describe a popular romance novel, *commuter marriage*, a union in which the partners live and work in different places, and *dentophobe*, a person who is afraid of dentists, are terms which have not yet appeared in most dictionaries.
5. Carl Yastrzemski is the former Red Sox star who replaced Ted Williams in 1961, joining Boston to play left field, who was chosen an all-star eighteen times, and who hit over 450 home runs in his career and was inducted into the Hall of Fame.

*EXERCISE 5

Combine each of the following sentence groups into one sentence that includes a series of items separated by semicolons. You will need to add, delete, relocate, or change words. Try several versions of each sentence until you find the most effective arrangement.

EXAMPLE: Collecting baseball cards is a worthwhile hobby. It helps children learn how to bargain and trade. It also encourages them to assimilate, evaluate, and compare data about major league ball players. Perhaps most important, it encourages them to find role models in the athletes whose cards they collect.

Collecting baseball cards is a worthwhile hobby because it helps children learn how to bargain and trade **;** encourages them to assimilate, evaluate, and compare data about major league ball players **;** and, perhaps most important, encourages them to find role models in the athletes whose cards they collect.

1. A good dictionary offers definitions of words, including some obsolete and nonstandard words. It provides information about synonyms, usage, and word origins. It also offers information on pronunciation and syllabication.
2. The flags of the Scandinavian countries all depict a cross on a solid background. Denmark's flag is red with a white cross. Norway's flag is also red, but its cross is blue, outlined in white. Sweden's flag is blue with a yellow cross.

495

continued from previous page

those who attended the 1963 march, are still without jobs and full equality; and to demonstrate for peace.
2. Steroids, used by some athletes, can be dangerous because they affect the pituitary gland, causing a lowered sperm count; because they have been linked to the development of liver tumors; and because they can stimulate the growth of some cancers.
3. Tennessee Williams wrote *The Glass Menagerie*, which is about a handicapped young woman and her family; *A Streetcar Named Desire*, which starred Marlon Brando; and *Cat on a Hot Tin Roof*, which won a Pulitzer Prize.
4. New expressions like *outro*, used to designate the segment of a broadcast where the announcer signs off; *heavy breather*, used to describe a popular romance novel; *commuter marriage*, a union in which the partners live and work in different places; and *dentophobe*, a person who is afraid of dentists, are terms which have not yet appeared in most dictionaries.
5. Carl Yastrzemski is the former Red Sox star who replaced Ted Williams in 1961, joining Boston to play left field; who was chosen an all-star eighteen times; and who hit over 450 home runs in his career and was inducted into the Hall of Fame.

ANSWERS TO EXERCISE 5

1. A good dictionary offers definitions of words, including some obsolete and nonstandard words; provides information about synonyms, usage, and word origins; and also offers information on pronunciation and syllabication.
2. The flags of the Scandinavian countries all depict a cross on a

continued on following page

continued from previous page

solid background: Denmark's flag is red with a white cross; Norway's flag is also red, but its cross is blue, outlined in white; and Sweden's flag is blue with a yellow cross.

3. Over one hundred international collectors' clubs are thriving today, including the Cola Clan, whose members buy, sell, and trade Coca-Cola memorabilia; the Cirtus Label Society; and a Cookie Cutter Collectors' Club.

4. Listening to the radio special, we heard "Shuffle Off to Buffalo" and "Moon over Miami," both of which are about eastern cities; "By the Time I Get to Phoenix" and "I Left My Heart in San Francisco," which mention western cities; and finally "The Star Spangled Banner," which seemed to be an appropriate finale.

5. There are three principal types of contact lenses: hard contact lenses, also called conventional lenses, which are easy to clean and handle and are quite sturdy; less durable soft lenses, which are easily contaminated and must be cleaned and disinfected daily; and gas-permeable lenses, which look and feel like hard lenses but are more easily contaminated and less durable.

3. Over one hundred international collectors' clubs are thriving today. One of these associations is the Cola Clan, whose members buy, sell, and trade Coca-Cola memorabilia. Another is the Citrus Label Society. There is also a Cookie Cutter Collectors' Club.

4. Listening to the radio special, we heard "Shuffle Off to Buffalo" and "Moon over Miami," both of which are about eastern cities. We heard "By the Time I Get to Phoenix" and "I Left My Heart in San Francisco," which mention western cities. Finally, we heard "The Star-Spangled Banner," which seemed to be an appropriate finale.

5. There are three principal types of contact lenses. Hard contact lenses, also called conventional lenses, are easy to clean and handle and quite sturdy. Soft lenses, which are easily contaminated and must be cleaned and disinfected daily, are less durable. Gas-permeable lenses, sometimes advertised as semihard or semisoft lenses, look and feel like hard lenses but are more easily contaminated and less durable.

29e Editing Misused or Overused Semicolons

Some inexperienced writers use semicolons because they think they are characteristic of a mature style. Others use them for variety. But semicolons are only called for in the contexts outlined in this chapter. Do not use semicolons in the following situations.

(1) Between items of unequal grammatical rank

Do not use a semicolon between a dependent and an independent clause or between a phrase and a clause.

> **FAULTY:** Because new drugs can now suppress the body's immune reaction; fewer organ transplants are rejected by the body. (Semicolon incorrectly used between dependent and independent clause.)

> **REVISED:** Because new drugs can now suppress the body's immune reaction, fewer organ transplants are rejected by the body.

> **FAULTY:** Increasing rapidly; computer crime poses a challenge for government, financial, and military agencies. (Semicolon incorrectly used between phrase and clause.)

> **REVISED:** Increasing rapidly, computer crime poses a challenge for government, financial, and military agencies.

(2) To introduce a list

See ◄ 32a1

Use a colon, not a semicolon, to introduce a list.

> **FAULTY:** The evening news is a battleground for the three major television networks; CBS, NBC, and ABC.

REVISED: The evening news is a battleground for the three major television networks : CBS, NBC, and ABC.

(3) To introduce a direct quotation

Do not use a semicolon to introduce a direct quotation.

▶ **See 31a**

FAULTY: Marie Antoinette may not have said; "Let them eat cake."

REVISED: Marie Antoinette may not have said , "Let them eat cake."

(4) To connect a string of clauses

Although semicolon use in the following paragraph is grammatically correct, the string of clauses connected by semicolons creates a monotonous passage.

> Art deco was a reflection of the jazz age of the roaring 1920s and more sober 1930s; at that time it was the dominant style, particularly in the United States. This was an art glorifying the machine; it was inspired by the speed of the automobile and airplane. It found expression in soaring skyscrapers and luxury ocean liners; streamlined statuettes, overstuffed furniture, and jukebox designs; and radio cabinets, toasters, and other kitchen gadgetry. Art deco was motivated by the vibrant energy released at the end of World War I; it was motivated by a faith in mechanized modernity; and it was also inspired by a joy in such new materials as glass, aluminum, polished steel, and shiny chrome.

Compare this paragraph, which uses semicolons selectively.

> Art deco was a reflection of the jazz age of the roaring 1920s and more sober 1930s when it was the dominant style, particularly in the United States. This was an art glorifying the machine and inspired by the speed of the automobile and airplane. It found expression in everything from soaring skyscrapers and luxury ocean liners to streamlined statuettes, overstuffed furniture, jukebox designs, radio cabinets, toasters, and other kitchen gadgetry. Art deco was motivated by the vibrant energy released at the end of World War I; a faith in mechanized modernity; and a joy in such new materials as glass, aluminum, polished steel, and shiny chrome. (William Fleming, *Arts & Ideas*)

EXERCISE 6

Read this paragraph carefully. Then add semicolons where necessary and delete excess or incorrectly used ones, substituting other punctuation where necessary.

> Barnstormers were aviators; who toured the country after World War I, giving people short airplane rides and exhibitions of stunt flying, in fact

---

"What the semicolon is not used for is to cover up the faults in a sentence out of control, as in this instance: 'The concern owns vast cacao, coffee, and palm plantations; a construction company that has built half the roads and new buildings in the territory; it imports machinery and equipment from all over the world and is constantly seeking to diversify and expand.'"

–Theodore M. Bernstein,
The Careful Writer

ANSWERS TO EXERCISE 6

Barnstormers were aviators who toured the country after World War I, giving people short airplane rides and exhibitions of stunt flying; in fact, the name *barnstormer* was derived from the use of barns as airplane hangars. Americans' interest in airplanes had all but disappeared after the war; planes had served their function in battle, but when the war ended, most people saw no future in aviation. The barnstormers helped popularize flying, especially in rural areas. Some of them were pilots who had flown in the war; others were just young men with a thirst for adventure. They gave people rides in airplanes, sometimes charging a dollar a minute. For most passengers, this was their first ride in an airplane; in fact, sometimes it was their first sight of one. In the early 1920s people grew bored with what the barnstormers had to offer, so groups of pilots began to stage spectacular—but often dangerous—stunt shows. Then, after Lindbergh's 1927 flight across the Atlantic, Americans

continued on following page

continued from previous page

suddenly needed no encouragement to embrace aviation. The barnstormers had outlived their usefulness, and an era ended.

the name *barnstormer* was derived from the use of barns as airplane hangars. Americans' interest in airplanes had all but disappeared after the war; planes had served their function in battle, but when the war ended, most people saw no future in aviation. The barnstormers helped popularize flying; especially in rural areas. Some of them were pilots who had flown in the war; others were just young men with a thirst for adventure. They gave people rides in airplanes; sometimes charging a dollar a minute. For most passengers, this was their first ride in an airplane, in fact, sometimes it was their first sight of one. In the early 1920s, people grew bored with what the barnstormers had to offer; so groups of pilots began to stage spectacular—but often dangerous—stunt shows. Then, after Lindbergh's 1927 flight across the Atlantic; Americans suddenly needed no encouragement to embrace aviation. The barnstormers had outlived their usefulness; and an era ended. (Adapted from William Goldman, *Adventures in the Screen Trade*)

CHAPTER 30

The Apostrophe

TEACHING STRATEGY

The Apostrophe
This chapter might be taught effectively with Chapter 22's discussion of case and personal pronouns. You may need to devote some class time to discussing the special problems posed by personal pronoun possessives and contractions.

Students frequently use apostrophes to form noun plurals. Discussing plural and possessive forms as homophones (*dogs/dog's*) may help clarify distinctions.

Use an Apostrophe
- To form the possessive case **(30a)**
- To indicate omissions in contractions **(30b)**
- To form plurals **(30c)**

30a Forming the Possessive Case

The possessive case indicates ownership. In English the possessive case of nouns and indefinite pronouns is indicated in two ways: either with a phrase that includes the word *of* (the hands *of* the clock) or with an apostrophe and, in most cases, an -*s* (the clock's hands). Personal pronouns do not use apostrophes to indicate the possessive case; instead, special forms are used to indicate the possessive.

(1) Singular nouns and indefinite pronouns

To form the possessive case of singular nouns and indefinite pronouns, add -*'s*.

"The Monk's Tale" is one of Chaucer's *Canterbury Tales.*

When we would arrive was anyone's guess.

(2) Singular nouns ending in -s

To form the possessive case of singular nouns that end in -*s*, add -*'s* in most cases.

499

History of the Apostrophe
Students may be interested in the history of the apostrophe. In early English, the endings of nouns were *inflected*—that is, they changed according to their function in the sentence. A noun used as a subject ended differently from the same noun used as an object. By the time of *The Second Shepherd's Play*, a Christmas play of the fourteenth century, Middle English had dropped almost all noun inflections—except for plural and possessive endings. Possessive was indicated by the endings *-es, -is,* or *ys,* as in this quotation from the play: "Of Godys [God's] son of hevyn he spak upward [spoke on high]. . . ." Sixteenth-century scholars mistakenly believed that these endings were contractions of *his* (that "Godys son" was actually, "God his son"). The "apostrophe *s*" form came into use, and remained even after later scholars scoffed at the theory behind it.

"While on this subject [of apostrophes], let me answer Karen Drew of Paducah, Ky., who has a son named Charles and wonders if she is 'Charles' Mother' or 'Charles's mother.' Dear Charley's Mother: You're 'Charles's mother,' because almost all names and words ending with *s* take an apostrophe as well as another *s* to form the possessive. When this rule causes you to trip over your tongue, for goodness' sake, break the rule."

–William Safire,
On Language

Reading Henry James's *The Ambassadors* was not Maris's idea of fun.
The class's time was changed to 8 A.M.

A few singular nouns ending in an *s* or *z* sound take only an apostrophe in the possessive case. If pronouncing the possessive ending as a separate syllable sounds awkward, use just an apostrophe.

For goodness' sake—are we really required to read both Aristophanes' *Lysistrata* and Thucydides' *History of the Peloponnesian War*?

An apostrophe is not used to form the possessive case of a title that already contains an *-'s* ending; use a phrase instead.

FAULTY: *A Midsummer Night's Dream's* staging
REVISED: the staging of *A Midsummer Night's Dream*

(3) Plural nouns ending in -s

To form the possessive case of regular plural nouns (those that end in *-s* or *-es*), add only an apostrophe.

The Readers' Guide to Periodical Literature is located in the reference section.
Two weeks' severance pay and three months' medical benefits were available to some workers.
The Lopezes' three children are identical triplets.

(4) Irregular plural nouns

To form the possessive case of nouns that have irregular plurals, add *-'s*.

Long after they were gone, the geese's honking could still be heard.
The Children's Hour is a play by Lillian Hellman; *The Women's Room* is a novel by Marilyn French.
The two oxen's yokes were securely attached to the cart.

NOTE: Plural nouns that are not possessive do not use apostrophes. (The *candidates* are speaking.)

(5) Compound nouns or groups of words

To form the possessive case of compound words or of word groups, add *-'s* to the last word.

The editor-in-chief's position is open.
He accepted the secretary of state's resignation under protest.
This is someone else's responsibility.

(6) Two or more items

To indicate individual ownership of two or more items, add -'s to each item. To indicate joint ownership, add -'s only to the last item.

INDIVIDUAL OWNERSHIP: Ernest Hemingway's and Gertrude Stein's writing styles have some similarities. (Hemingway and Stein have two separate writing styles.)

JOINT OWNERSHIP: Gilbert and Sullivan's operettas include *The Pirates of Penzance* and *H.M.S. Pinafore.* (Gilbert and Sullivan collaborated on both operettas.)

EXERCISE 1

In these examples change the modifying phrases that follow the nouns to possessive forms that precede the nouns.

EXAMPLE: the pen belonging to my aunt
my aunt's pen

1. the songs recorded by Ray Charles
2. the red glare of the rockets
3. the idea Warren had
4. the housekeeper Leslie and Rick hired
5. the first choice of everyone
6. the dinner given by Harris
7. furniture designed by William Morris
8. the climate of the Virgin Islands
9. the sport the Russells play
10. the role created by the French actress

EXERCISE 2

Wherever possible change each word or phrase in parentheses to its possessive form. In some cases you may have to use a phrase to indicate possession.

EXAMPLE: The (children) toys were scattered all over their (parents) bedroom.

The children's toys were scattered all over their parents' bedroom.

1. Jane (Addams) settlement house was called Hull House.
2. *(A Room of One's Own)* popularity increased with the rise of feminism.
3. The (chief petty officer) responsibilities are varied.
4. Vietnamese (restaurants) numbers have grown dramatically in ten (years) time.
5. (Charles Dickens) and (Mark Twain) works have sold millions of copies.

501

Joint Ownership

Students are sometimes confused when they deal with nouns that express joint, or shared, possession. You might want to spend a few minutes working with the examples below, and perhaps ask students to come up with additional sentences.

My brother's apartment (one brother, one apartment)

My brothers' apartment (two or more brothers, but still one apartment)

My brothers' apartments (two or more brothers, and two or more apartments)

The student's test score (one student, one score)

The students' test score (only if it was a group project!)

The students' test scores (two or more students, and two or more scores)

ANSWERS TO EXERCISE 1

1. Ray Charles's songs
2. The rockets' red glare
3. Warren's idea
4. Leslie and Rick's housekeeper
5. everyone's first choice
6. Harris's dinner
7. William Morris's furniture
8. the Virgin Islands' climate
9. the Russells' sport
10. the French actress's role

ANSWERS TO EXERCISE 2

1. Addams's
2. The popularity of *A Room of One's Own* . . .
3. chief petty officer's
4. restaurants'; years'
5. Dickens's; Twain's

1. Schaefers'; ours
2. colleges; outsiders
3. C
4. yours
5. favorites
6. hers
7. revolutions; changes
8. desks; dealers
9. ours
10. C

CLOSE-UP 🔍

USING APOSTROPHES

Do not use apostrophes with plural nouns that are not possessive or in the possessive case of personal pronouns.

FAULTY:	REVISED:
The Thompson's aren't at home.	The Thompsons aren't at home.
The down vest's are very warm.	The down vests are very warm.
The Philadelphia Seventy Sixer's sometimes play well.	The Philadelphia Seventy Sixers sometimes play well.
This ticket must be your's or her's.	This ticket must be yours or hers.
The next turn is their's.	The next turn is theirs.
The doll had lost it's right eye.	The doll had lost its right eye.
The next great moment in history is our's.	The next great moment in history is ours.

See ◄
30b1

Be especially careful not to confuse the possessive form of the personal pronoun *it* (its) with the contraction *it's* (it is).

EXERCISE 3

In the following sentences, correct any errors in the use of apostrophes to form noun plurals or the possessive case of personal pronouns. If a sentence is correct, mark it with a *C*.

EXAMPLE: Dr. Sampson's lecture's were more interesting than her's.

Dr. Sampson's lectures were more interesting than hers.

1. The Schaefer's seats are right next to our's.
2. Most of the college's in the area offer computer courses open to outsider's as well as to their own students.
3. The network completely revamped its daytime programming.
4. Is the responsibility for the hot dog concession Cynthia's or your's?
5. Romantic poets are his favorite's.
6. Debbie returned the books to the library, forgetting they were her's.
7. Cultural revolution's do not occur very often, but when they do they bring sweeping change's.
8. Roll-top desk's are eagerly sought by antique dealer's.
9. A flexible schedule is one of their priorities, but it isn't one of our's.
10. Is yours the red house or the brown one?

30b **Indicating Omissions in Contractions**

Apostrophes are used to mark the omission of letters or numbers.

(1) Omitted letters

Apostrophes replace omitted letters in frequently used contractions that combine a pronoun and a verb (*he + will = he'll*) or the elements of a verb phrase (*do + not = don't*).

COMMONLY USED CONTRACTIONS

it's (it is)	let's (let us)
he's (he is)	we've (we have)
who's (who is)	they're (they are)
isn't (is not)	we'll (we will)
wouldn't (would not)	I'm (I am)
couldn't (could not)	we're (we are)
don't (do not)	you'd (you would)
won't (will not)	we'd (we would)

Although common in speech and informal writing, contractions are not generally used in college writing except to reproduce dialogue or to deliberately create an informal tone.

CLOSE-UP

USING APOSTROPHES

Be careful not to confuse contractions with the possessive forms of personal pronouns.

▶ See 22a3

CONTRACTIONS	**POSSESSIVE FORMS**
Who's on first?	Whose book is this?
They're playing our song.	Their team is winning.
It's raining.	Its paws were muddy.
You're a real pal.	Your résumé is very impressive.

1. It's; you're
2. Who's
3. They're; their
4. Who's
5. its
6. You're; it's
7. your; you're
8. Their
9. C
10. Whose

(2) Omitted numbers

In informal writing an apostrophe may be used to represent the century in a year.

Crash of '29   Class of '97   '57 Chevy

In college writing, however, write out the year in full: *the Crash of 1929, the class of 1997, a 1957 Chevrolet.*

EXERCISE 4

In the following sentences correct any errors in the use of standard contractions or personal pronouns. If a sentence is correct, mark it with a *C*.

> **EXAMPLE:** Who's troops were sent to Korea?
>
> Whose troops were sent to Korea?

1. Its never easy to choose a major; whatever you decide, your bound to have second thoughts.
2. Olive Oyl asked, "Whose that knocking at my door?"
3. Their watching too much television; in fact, they're eyes are glazed.
4. Whose coming along on the backpacking trip?
5. The horse had been badly treated; it's spirit was broken.
6. Your correct in assuming its a challenging course.
7. Sometimes even you're best friends won't tell you your boring.
8. They're training had not prepared them for the hardships they faced.
9. It's too early to make a positive diagnosis.
10. Robert Frost wrote the poem that begins, "Who's woods these are I think I know."

30c Forming Plurals

The apostrophe plus -*s* is used to form plurals in some special cases. In these cases adding an -*'s* avoids a confusing combination that might obscure the fact that the word is plural.

FORMING PLURALS WITH APOSTROPHES

Plurals of Letters

The Italian language has no *j*'s or *k*'s.
Sesame Street helps children learn their ABC's.

Plurals of Numbers

Dick Button could make outstanding figure 8's.
Styles of the 1970's are back in fashion.

---

continued on the following page

continued from the previous page

Many writers prefer to omit the apostrophe with plurals of numbers; either style is acceptable.

Plurals of Symbols

The +'s and –'s indicate positive and negative numbers, respectively.

The entire paper used &'s instead of spelling out *and*.

Plurals of Abbreviations Followed by Periods

Only two R.N.'s worked the 7-to-11 shift.

Four Ph.D's and seven M.D.'s attended the high school reunion.

Plurals of Words Referred to as Words

The supervisor would accept no *if*'s, *and*'s, or *but*'s.

His first sentence contained three *therefore*'s.

CLOSE-UP

USING APOSTROPHES

Letters, numerals, and words spoken of as themselves are always set in italic type; the plural ending is further distinguished by being set in roman type. When you write or type, indicate italics by underlining. If you are typing on a computer, you can either use italics or underline. If no confusion is possible, you may omit the apostrophe, but you should still use italics where they are required.

▶ See 34c

The editor deleted all the *4*s in the report.

EXERCISE 5

In the following sentences, form correct plurals for the letters, numbers, and words in parentheses. Underline to indicate italics where necessary.

EXAMPLE: The word *bubbles* contains three (b).

The word *bubbles* contains three *b*'s.

1. She closed her letter with a row of (x) and (o) to indicate kisses and hugs.
2. The three (R) are reading, writing, and 'rithmetic.
3. The report included far too many (maybe) and too few (definitely).
4. His (4) and (9) were identical, and his (5) looked exactly like (s).
5. Her two (M.A.) were in sociology and history.

CHAPTER 31

Quotation Marks

USE QUOTATION MARKS . . .
- To set off direct quotations **(31a)**
- To set off titles **(31b)**
- To set off words used in a special sense **(31c)**
- To set off dialogue, prose passages, and poetry **(31d)**

31a Setting Off Direct Quotations

Use quotation marks with **direct quotations**—that is, when you reproduce exactly a word, phrase, or brief passage from someone else's speech or writing. Enclose the borrowed material in a pair of quotation marks.

> Gloria Steinem observed, "We are becoming the men we once hoped to marry."

See ◄ 17e Do not use quotation marks with **indirect quotations**—that is, when you report someone else's written or spoken words without quoting them exactly.

> Gloria Steinem observed that many women were becoming the men they had once hoped to marry.

Special punctuation problems occur when quoted material must be set off from phrases (such as *he said*) that identify its source. The following guidelines cover the most common situations.

506

CLOSE-UP

USING QUOTATION MARKS

Use single quotation marks to enclose a **quotation within a quotation**.

Claire noted, "It was Liberace who first said, ' I cried all the way to the bank.' "

(1) An identifying tag in the middle of a passage

Use a pair of commas to set off the identifying tag that interrupts a passage.

"In the future ," pop artist Andy Warhol once said , "everyone will be world-famous for 15 minutes."

If the identifying tag follows a completed sentence but the quoted passages continues, use a period after the tag, and begin the new sentence with a capital letter.

"Be careful ," Erin warned . "Reptiles can be tricky ."

(2) An identifying tag at the beginning

Use a comma after an identifying tag that introduces quoted speech or writing.

The Raven repeated , "Nevermore."

When you quote just a word or phrase, however, you may introduce it without any punctuation.

The meteorologist said he expected March to be "unpredictable."

Use a colon instead of a comma to formally introduce a quotation.

The ambassador announced the following : "All citizens should leave the area immediately. The instability of the military government and the acts of hostility toward Americans make this action necessary."

Use a colon before any quotation, even a brief one, if the introductory tag is an independent clause.

She gave her final answer : "No."

▶ See
32a3

"The use of quotation marks to say 'their word, not mine' is growing. . . . Editor Sol Steinmetz thinks that 'disbelieving quotation marks' first became popular during the Nazi era, and then were given a boost in the Vietnam years, especially around the word 'advisers.' . . .

Disdain now has its own punctuation. . . . [and] trendy critics can even put quotation signs around a spoken word by wiggling two fingers of each hand. I've been overworking the quotation-mark disdainer myself (and the hyphenated compound adjective, too) by knocking those self-appointed 'guardians' of English 'skills' or reviling the language slobs who 'enrich' our discourse with quotation 'marks.' Consider, instead, the style of Charles de Gaulle. He needed no attention-getting ironic quotes to put down the United Nations. He called that organization *'les nations dites* [said to be] *unies.'*"

–William Safire,
On Language

TEACHING STRATEGY

History of Quotation Marks
More background: British grammarians call quotation marks "inverted commas." Before the seventeenth century, direct quotations were introduced by phrases such as *she said*; gradually, writers and printers began placing a pair of commas before quotations (but not after). Over the years, another pair of commas was added behind each quotation (as in: ,, Alas, poor Yorick. . . ,,); eventually, printers began to raise the marks to their present position.

ANSWERS TO EXERCISE 1

1. Mr. Fox noted, "Few people can explain what Descartes' words 'I think, therefore I am' actually mean."
2. Gertrude Stein said, "You are all a lost generation."
3. "Freedom of speech does not guarantee anyone the right to yell 'Fire!' in a crowded theater," she explained.
4. Correct
5. "If everyone will sit down," the teacher announced, "the exam will begin."

(3) An identifying tag at the end

Use a comma to set off a quotation from an identifying tag that follows it.

"Be careful out there," the sergeant warned.

If the quotation ends with a question mark or an exclamation point, use that punctuation mark instead of the comma. The tag begins with a lowercase letter even though it follows end punctuation.

"Is Ankara the capital of Turkey?" she asked.

"Oh boy!" he cried.

If the quotation comes at the end of a sentence, use a period or other appropriate end punctuation after the quotation.

The principal said, "Education is serious business."

CLOSE-UP

USING QUOTATION MARKS

Note that commas and periods are always placed *before* quotation marks. For information on placement of other punctuation marks with quotation marks, **see 31e.**

EXERCISE 1

Add single and double quotation marks to these sentences where necessary to set off direct quotations from identifying tags. If a sentence is correct, mark it with a *C*.

> **EXAMPLE:** Wordsworth's phrase splendour in the grass was used as the title of a movie about young lovers.
>
> Wordsworth's phrase **"splendour in the grass"** was used as the title of a movie about young lovers.

1. Mr. Fox noted, Few people can explain what Descartes' words I think, therefore I am actually mean.
2. Gertrude Stein said, You are all a lost generation.
3. Freedom of speech does not guarantee anyone the right to yell fire in a crowded theater, she explained.
4. Dorothy kept insisting that there was no place quite like home.
5. If everyone will sit down the teacher announced the exam will begin.

31b **Setting Off Titles**

Titles of short works and titles of parts of long works are enclosed in quotation marks (other titles are set in italics; see **34a**).

TITLES REQUIRING QUOTATION MARKS

Articles in magazines, newspapers, and professional journals

"Why Johnny Can't Write" (*Newsweek*)

Essays

"Fenimore Cooper's Literary Offenses"

Short stories

"Flying Home"

Short poems (those not divided into numbered sections)

"The New Colossus"

Songs

"The Star-Spangled Banner"

Chapters or sections of books

"Miss Sharp Begins to Make Friends" (Chapter 10 of *Vanity Fair*)

Speeches

"How to Tell a Story"

Descriptive titles of speeches, such as Kennedy's Inaugural Address, are not enclosed in quotation marks.

Episodes of radio or television series

"Lucy Goes to the Hospital" (*I Love Lucy*)

CLOSE-UP

USING QUOTATION MARKS

When a title that is normally enclosed in quotation marks appears within a quotation, set the title off in *single* quotation marks.

I think what she said was, "Play it, Sam. Play 'As Time Goes By.'"

ANSWERS TO EXERCISE 2

1. "First Fig" and "Second Fig" are two of the poems in Edna St. Vincent Millay's *Collected Poems*.
2. In the article "Feminism Takes a New Turn," Betty Friedan reconsiders some of the issues first raised in her 1964 book *The Feminine Mystique.*
3. Edwin Arlington Robinson's poem "Richard Cory" was the basis for the song "Richard Cory" recorded by Paul Simon.
4. *Beside* means "next to," but *besides* means "except."
5. In an essay on the novel *An American Tragedy* published in *The Yale Review,* Robert Penn Warren noted, "Theodore Dreiser once said that his philosophy of love might be called 'Varietism.'"

31c Setting Off Words Used in a Special Sense

Words used in a special sense are enclosed in quotation marks.

It was clear that adults approved of children who were **"**readers,**"** but it was not at all clear why this was so. (Annie Dillard, *New York Times Magazine*)

It is often remarked that words are tricky—and that we were all prone to be deceived by **"**fast talkers,**"** such as high-pressure salesmen, skillful propagandists, politicians or lawyers. (S. I. Hayakawa, "How Words Change Our Lives")

Coinages—invented words—also take quotation marks.

After the twins were born, the station wagon became a **"**babymobile.**"**

See ◀ 34c When a word is referred to as a word, however, it is italicized.

How do you pronounce *trough*?

Compose and *comprise* cannot be used interchangeably.

CLOSE-UP

USING QUOTATION MARKS

When you quote a dictionary definition, put the word you are defining in italics and the definition in quotation marks.

To *infer* means "to draw a conclusion"; to *imply* means "to sug-gest.**"**

EXERCISE 2

Add quotation marks to the following sentences where necessary to set off titles and words. If italics are incorrectly used, substitute quotation marks.

EXAMPLE: To Tim, social security means a date for Saturday night.

To Tim, **"**social security**"** means a date for Saturday night.

1. *First Fig* and *Second Fig* are two of the poems in Edna St.Vincent Millay's *Collected Poems.*
2. In the article Feminism Takes a New Turn, Betty Friedan reconsiders some of the issues first raised in her 1964 book *The Feminine Mystique.*
3. Edwin Arlington Robinson's poem Richard Cory was the basis for the song Richard Cory written by Paul Simon.

4. *Beside* means next to, but *besides* means except.
5. In an essay on the novel *An American Tragedy* published in *The Yale Review*, Robert Penn Warren noted, Theodore Dreiser once said that his philosophy of love might be called Varietism.

 31d **Setting Off Dialogue, Prose Passages, and Poetry**

(1) Dialogue

When you record dialogue, begin a new paragraph for each new speaker. Be sure to enclose the quoted words in quotation marks.

"Sharp on time as usual," Davis said with his habitual guilty grin.

"My watch is always a little fast," Castle said, apologizing for the criticism which he had not expressed. "An anxiety complex, I suppose." (Graham Greene, *The Human Factor*)

Notice that the phrases identifying the speaker *(Davis said, Castle said)* appear in the same paragraph as the speaker's words.

CLOSE-UP

USING QUOTATION MARKS

When you are quoting several paragraphs of dialogue by one speaker, begin each new paragraph with quotation marks. However, use closing quotation marks only at the end of the *entire quoted passage,* not at the end of the paragraph.

EXERCISE 3

Add appropriate quotation marks to the dialogue in this passage, beginning a new paragraph whenever a new speaker is introduced.

The next time, the priest steered me into the confession box himself and left the shutter back [so] I could see him get in and sit down at the further side of the grille from me. Well, now, he said, what do they call you? Jackie, father, said I. And what's a-trouble to you, Jackie? Father, I said, feeling I might as well get it over while I had him in good humor, I had it all arranged to kill my grandmother.

The next time, the priest steered me into the confession box himself and left the shutter back [so] I could see him get in and sit down at the further side of the grille from me.

"Well, now," he said, "what do they call you?"

"Jackie, father," said I.

"And what's a-trouble to you, Jackie?"

"Father," said I, feeling I might as well get it over while I had him in good humor, "I had it all arranged to kill my grandmother."

He seemed a bit shaken by that, all right, because he said nothing for quite a while.

"My goodness," he said at last, "that'd be a shocking thing to do. What put that into your head?"

"Father," I said, feeling very sorry for myself, "she's an awful woman."

"Is she?" he asked. "What way is she awful?"

"She takes porter, father," I said, knowing well from the way Mother talked of it that this was a mortal sin, and hoping it would make the priest take a more favorable view of my case.

"Oh, my!" he said, and I could see that he was impressed.

"And snuff, father," said I.

"That's a bad case, sure enough, Jackie," he said.

Assign students to write an updated version of a favorite fairy tale or fable, using as much dialogue as they can. Then, ask pairs of students to exchange their stories. Peer criticism should focus on the use of quotation marks but may also include comments on style, grammar, and other forms of punctuation.

He seemed a bit shaken by that, all right, because he said nothing for quite a while. My goodness, he said at last, that'd be a shocking thing to do. What put that into your heard? Father, I said, feeling very sorry for myself, she's an awful woman. Is she? he asked. What way is she awful? She takes porter, father, I said, knowing well from the way Mother talked of it that this was a mortal sin, and hoping it would make the priest take a more favorable view of my case. Oh, my! he said, and I could see that he was impressed. And snuff, father, said I. That's a bad case, sure enough, Jackie, he said. (Frank O'Connor, "First Confession")

(2) Prose passages

When you quote a short prose passage, set if off in quotation marks.

Galsworthy describes Aunt Juley as **"**prostrated by the blow**"** (329).

However, do not place a prose passage of more than four lines in quotation marks. Instead, set it off by indenting it ten spaces from the left-hand margin. Double-space above and below the quotation, and double-space between lines within it. Introduce the passage with a colon.

The following portrait of Aunt Juley illustrates several of the devices Galsworthy uses throughout The Forsyte Saga, such as a journalistic detachment that is almost cruel in its scrutiny, a subtle sense of the grotesque, and an ironic stance:

> Aunt Juley stayed in her room, prostrated by the blow. Her face, discoloured by tears, was divided into compartments by the little ridges of pouting flesh which had swollen with emotion. . . . At fixed intervals she went to her drawer, and took from beneath the lavender bags a fresh pocket-handkerchief. Her warm heart could not bear the thought that Ann was lying there so cold. (329)

Similar characterizations appear throughout the book. . . .

CLOSE-UP

USING QUOTATION MARKS

When a prose passage longer than four lines is a single paragraph or less, do not indent the first line. When quoting two or more paragraphs, indent the first line of each paragraph three spaces.

If the passage you are quoting already includes material quoted by the author, enclose those words in double quotation marks.

(3) Poetry

One line of poetry is treated like a short prose passage—enclosed in quotation marks and run into the text.

> One of John Donne's best-known poems begins with the line, **"**Go and catch a falling star.**"**

Two or three lines of poetry are run into the text and separated by slashes (/).

▶ See 32e2

> Alexander Pope writes, **"**True Ease in Writing comes from Art, not Chance, **/** As those move easiest who have learned to dance.**"**

Four or more lines of poetry should be set off like a long prose passage. For special emphasis, fewer lines may also be set off in this manner. Punctuation, spelling, capitalization, and indentation are reproduced *exactly*.

▶ See 31d2

> Wilfred Owen, a poet who was killed in action in World War I, expressed the horrors of war with vivid imagery:
>
>> Bent double, like old beggars under sacks.
>> Knock-kneed, coughing like hags, we cursed through
>>   sludge.
>> Till on the haunting flares we turned our backs
>> And towards our distant rest began to trudge. (1–4)

31e Using Quotation Marks with Other Punctuation

Quotation marks frequently occur along with other punctuation marks. Sometimes the quotation marks are placed inside other punctuation, and sometimes they are placed outside.

(1) With final commas or periods

Quotation marks come *after* the comma or period at the end of a quotation.

> Many, like Frost, think about "the road not taken**,"** but not many have taken "the one less traveled by."
>
> Janis Joplin sang, "Freedom's just another word for nothing left to lose**."**

(2) With final semicolons or colons

Quotation marks come *before* a semicolon or colon at the end of a quotation.

Students who do not pass the functional-literacy test receive "certificates of completion "; those who pass are awarded diplomas.

Taxpayers were pleased with the first of the candidate's promised "sweeping new reforms ": a balanced budget.

(3) With question marks, exclamation points, and dashes

Quotation marks may be placed inside or outside a question mark, exclamation point, or dash at the end of a quotation, depending on the sentence's meaning.

If a question mark, exclamation point, or dash is part of the quotation, place the quotation marks *after* the punctuation.

"Who's there ?" she demanded.

"Stop !" he cried.

"Should we leave now, or—" Vicki paused, unable to continue.

If a question mark, exclamation point, or dash is *not* part of the quotation, place the quotation marks *before* the punctuation.

Did you finish reading "The Black Cat "?

Whatever you do, don't yell "Uncle "!

The first essay—George Orwell's "Politics and the English Language"— made quite an impression on the class.

If both the quotation and the sentence are questions or exclamations, place the quotation marks *before* the punctuation.

Who asked, "Is Paris Burning "?

31f Editing Misused or Overused Quotation Marks

Although some writers use quotation marks to indicate special attention or emphasis, quotation marks should not be used in the following situations.

(1) To convey emphasis

FAULTY: William Randolph Hearst's "fabulous" home is a castle called San Simeon.

REVISED: William Randolph Hearst's fabulous home is a castle called San Simeon.

(2) To set off nicknames or slang

FAULTY: The former "Lady Di" became the Princess of Wales when she married Prince Charles.

REVISED: The former Lady Diana Spencer became the Princess of Wales when she married Prince Charles.

FAULTY: Dawn is "into" running.

REVISED: Dawn is very involved in running.

CLOSE-UP

USING QUOTATION MARKS

Do not make the mistake of thinking that quotation marks will make nonstandard or slang terms acceptable in college writing. Avoid substituting nicknames for full names or slang for standard diction.

(3) To enclose titles of long works

FAULTY: "War and Peace" is even longer than "Paradise Lost."

REVISED: *War and Peace* is even longer than *Paradise Lost.*

NOTE: Titles of long works are set in italics.

▶ See 34a

(4) To set off terms being defined

FAULTY: The word "tintinnabulation," meaning the ringing sound of bells, was used by Poe in his poem "The Bells."

REVISED: The word *tintinnabulation,* meaning the ringing sound of bells, was used by Poe in his poem "The Bells."

NOTE: Words that are defined should be italicized.

▶ See 34c

(5) To set off technical terms

FAULTY: "Biofeedback" is sometimes used to treat migraine headaches.

REVISED: Biofeedback is sometimes used to treat migraine headaches.

(6) To set off the title of a student essay

Do not use quotation marks around the title of your paper, whether it appears on a title page or at the top of the first page.

ANSWERS TO EXERCISE 4

In her essay "The Obligation to Endure" from the book *Silent Spring*, Rachel Carson writes: "As Albert Schweitzer has said, 'Man can hardly even recognize the devils of his own creation.'" Carson goes on to point out that many chemicals have been used to kill insects and other organisms which, she writes, are "described in the modern vernacular as 'pests.'" Carson believes such supposedly advanced chemicals, by contaminating our environment, do more harm than good. In addition to *Silent Spring*, Carson is also the author of the book *The Sea Around Us*. This work, divided into three sections ("Mother Sea," "The Restless Sea," and "Man and the Sea about Him"), was first published in 1951.

FAULTY: "Images of Light and Darkness in George Eliot's *Adam Bede*"

REVISED: Images of Light and Darkness in George Eliot's *Adam Bede*

See ◄ **(7) To set off indirect quotations**
31a

FAULTY: Freud wondered "what a woman wanted."

REVISED: Freud wondered what a woman wanted.

REVISED: Freud wondered, "What does a woman want?"

EXERCISE 4

In the following paragraph, correct the use of single and double quotation marks to set off direct quotations, titles, and words used in a special sense. Supply the appropriate quotation marks where required and delete those not required. Be careful not to use quotation marks where they are not necessary.

> In her essay 'The Obligation to Endure' from the book "Silent Spring," Rachel Carson writes: As Albert Schweitzer has said, 'Man can hardly even recognize the devils of his own creation.' Carson goes on to point out that many chemicals have been used to kill insects and other organisms which, she writes, are "described in the modern vernacular as pests." Carson believes such "advanced" chemicals, by contaminating our environment, do more harm than good. In addition to "Silent Spring," Carson is also the author of the book "The Sea Around Us." This work, divided into three sections (Mother Sea, The Restless Sea, and Man and the Sea About Him) was published in 1951.

EXERCISE 5

Correct the use of quotation marks in the following sentences, making sure that the use and placement of any accompanying punctuation marks are consistent with accepted conventions. If a sentence is correct, mark it with a *C*.

> **EXAMPLE:** The "Watergate" incident brought many new terms into the English language.
>
> The Watergate incident brought many new terms into the English language.

1. Kilroy was here and Women and children first are two expressions *Bartlett's Familiar Quotations* attributes to "Anon."
2. Neil Armstrong said he was making a small step for man but a giant leap for mankind.
3. "The answer, my friend", Bob Dylan sang, "is blowin' in the wind".
4. The novel was a real "thriller," complete with spies and counterspies, mysterious women, and exotic international chases.

5. The sign said, Road liable to subsidence; it meant that we should look out for potholes.
6. One of William Blake's best-known lines—To see a world in a grain of sand—opens his poem Auguries of Innocence.
7. In James Thurber's short story The Catbird Seat, Mrs. Barrows annoys Mr. Martin by asking him silly questions like Are you tearing up the pea patch? Are you scraping around the bottom of the pickle barrel? and Are you lifting the oxcart out of the ditch?
8. I'll make him an offer he can't refuse, promised "the godfather" in Mario Puzo's novel.
9. What did Timothy Leary mean by "Turn on, tune in, drop out?"
10. George, the protagonist of Bernard Malamud's short story, A Summer's Reading, is something of an "underachiever."

ANSWERS TO EXERCISE 5

1. "Kilroy was here" and "Women and children first" are two expressions *Bartlett's Familiar Quotations* attributes to Anon.
2. Correct; indirect quotation
3. "The answer, my friend," Bob Dylan sang, "is blowin' in the wind."
4. The novel was a real thriller, complete with spies and counterspies, mysterious women, and exotic international chases.
5. The sign said, "Road liable to subsidence"; it meant that we should look out for potholes.
6. One of William Blake's best-known lines—"To see a world in a grain of sand"—opens his poem "Auguries of Innocence."
7. In James Thurber's short story "The Catbird Seat," Mrs. Barrows annoys Mr. Martin by asking silly questions like "Are you tearing up the pea patch?" "Are you scraping the bottom of the pickle barrel?" and "Are you lifting the oxcart out of the ditch?"
8. "I'll make him an offer he can't refuse," promised the godfather in Mario Puzo's novel.
9. What did Timothy Leary mean by "Turn on, tune in, drop out"?
10. George, the protagonist of Bernard Malamud's short story "A Summer's Reading," is something of an underachiever.

Other Punctuation Marks

Because the punctuation marks in this chapter have special, unrelated uses, you may find it most effective to discuss sections of this chapter as student assignments warrant—for example, discussing colons and ellipses in connection with the use of quotations in research papers.

Approach this grab-bag of punctuation from a stylistic viewpoint. Read and discuss the Tom Wolfe passage on the following page. Ask students to consider how this passage would be affected if punctuation were restricted to periods, commas, and an occasional exclamation point. The variety of English punctuation gives writers the chance to create prose of extraordinary flexibility of tone and pace.

Other Punctuation Marks

USING OTHER PUNCTUATION MARKS

Use a colon . . .
- To introduce material **(32a1–3)**
- Where convention requires it **(32a3)**

Use dashes . . .
- To set off nonessential material **(32b1)**
- To introduce a summary **(32b2)**
- To indicate an interruption **(32b3)**

Use parentheses . . .
- To set off nonessential material **(32c1)**
- In other conventional situations **(32c2)**

Use brackets . . .
- To set off comments within quotations **(32d1)**
- In place of parentheses within parentheses **(32d2)**

Use a slash . . .
- To separate one option from another **(32e1)**
- To separate lines of poetry run into the text **(32e2)**
- To separate the numerator from the denominator in fractions **(32e3)**

Use ellipses . . .
- To indicate an omission in a quotation **(32f1)**
- To indicate an omission within verse **(32f2)**
- To indicate unfinished statements **(32f3)**

32a Using Colons

The **colon** is a strong punctuation mark that points ahead to the rest of the sentence, linking the words that follow it to the words that precede it.

(1) To introduce lists or series

Colons set off lists or series, including those introduced by phrases like *the following* or *as follows.*

> He looked like whatever his beholder imagined him to be **:** a bank clerk, a high school teacher, a public employee, a librarian, a fussy custodian, an office manager. (Norman Katkov, *Blood and Orchids*)

> Like croup and whooping cough, it was treated with remedies Ida Rebecca compounded from ancient folk-medicine recipes **:** reeking mustard plasters, herbal broths, dosings of onion syrup mixed with sugar. (Russell Baker, *Growing Up*)

CLOSE-UP

USING COLONS

A colon introduces a list or series *only* when an independent clause precedes the colon.

FAULTY: Each camper should be equipped with: a sleeping bag, a mess kit, a flashlight, and plenty of insect repellent. (*Each camper should be equipped with* is not a grammatically complete independent clause.)

REVISED: Each camper should bring the following **:** a sleeping bag, a mess kit, a flashlight, and plenty of insect repellent.

---

(2) To introduce explanatory material

Colons often introduce material that explains, details, exemplifies, clarifies, or summarizes. Frequently this material is presented in the form of an **appositive**, a construction that identifies or describes a noun, pronoun, or noun phrase.

▶ **See 8e4**

> Diego Rivera painted a controversial mural **:** the one commissioned for Rockefeller Center in the 1930s.

The Colon
The colon is the Paul Revere of punctuation. *Heads up! Something's coming!* it cries. The colon points our attention to a list, series, or formal quotation that follows immediately. By using a colon, the writer can alert readers to get ready for an explanatory illustration of what has just been said. The clear connection between the two parts of the sentence below is destroyed when the colon is replaced with a period.

> She offered a flimsy excuse: Her car would not start.

> She offered a flimsy excuse. Her car would not start.

Ask students to compose similar sentences in which meaning or clarity is diminished when the colon is replaced with a period.

Her hair was the longest and strangest Mrs. Miller had ever seen: absolutely silver-white, like an albino's. (Truman Capote, "Miriam")

Sometimes a colon separates two independent clauses: one general clause and a subsequent, more specific one that illustrates or clarifies the first.

A *U.S. News and World Report* survey has revealed a surprising fact: Americans spend more time at shopping malls than anywhere else except at home and at work.

CLOSE-UP

USING COLONS

When a complete sentence follows a colon, that sentence may begin with either a capital or a lowercase letter. However, if the sentence introduced by a colon is a quotation, its first word is always capitalized, unless it was not capitalized in the source.

(3) To introduce quotations

A colon is used to set off a quotation introduced by a complete independent clause.

The French Declaration of the Rights of Man includes these words: "Liberty consists in being able to do anything that does not harm another person."

With great dignity, Bartleby repeated the words once again: "I prefer not to."

See ◀ 31d2 A quotation of more than four lines is always introduced by a colon.

OTHER CONVENTIONAL USES OF COLONS
To separate titles from subtitles
Family Installments: Memories of Growing Up Hispanic
To separate minutes from hours in numerical expressions of time
6:15 A.M.   3:30 P.M.

> **To separate place of publication from name of publisher in a works cited list (see 40b2)**
> Fort Worth: Harcourt, 1995.
> **After salutations in business letters (see 48a2)**
> Dear Dr. Evans:

(4)   Editing misused or overused colons

After* Such As, For Example, *and Similar Expressions Colons are not used after expressions like *such as, namely, for example,* or *that is.* Remember that a colon introduces a list or series only after an independent clause.

FAULTY: The Eye Institute treats patients with a wide variety of conditions, such as: myopia, glaucoma, and cataracts.

REVISED: The Eye Institute treats patients with a wide variety of conditions, such as myopia, glaucoma, and cataracts.

In Verb and Prepositional Constructions Colons should not be placed between verbs and their objects or complements or between prepositions and their objects.

FAULTY: James A. Michener wrote: *Hawaii, Centennial, Space,* and *Poland.*

REVISED: James A. Michener wrote *Hawaii, Centennial, Space,* and *Poland.*

FAULTY: Hitler's armies marched through: the Netherlands, Belgium, and France.

REVISED: Hitler's armies marched through the Netherlands, Belgium, and France.

EXERCISE 1

Add colons where required in the following sentences. If necessary, delete excess colons.

EXAMPLE: There was one thing he really hated getting up at 700 every morning.

There was one thing he really hated: getting up at 7:00 every morning.

TEACHING STRATEGY

Using Colons
Do your students want to slip in a colon in front of every list—or behind every *such as*? Emphasize the concept that a colon introduces a list or series only after an independent clause. As a reinforcement, write three sentences that require colons before a list, and three sentences that do NOT require colons. Ask a student to read the first three sentences out loud. Now ask another student to read the three sentences without colons. Do the sentences come to an almost full stop in front of the list or series?

1. Books about the late John F. Kennedy include the following: *A Hero for Our Time; Johnny, We Hardly Knew Ye; One Brief Shining Moment;* and *JFK: Reckless Youth.*
2. Only one task remained: to tell his boss he was quitting.
3. The story closed with a familiar phrase: "And they all lived happily ever after."
4. The sergeant requested [delete colon] reinforcements, medical supplies, and more ammunition.
5. She kept only four souvenirs: a photograph, a matchbook, a theater program, and a daisy pressed between the pages of *William Shakespeare: The Complete Works.*

1. Books about the late John F. Kennedy include the following *A Hero For Our Time; Johnny, We Hardly Knew Ye; One Brief Shining Moment;* and *JFK: Reckless Youth.*
2. Only one task remained to tell his boss he was quitting.
3. The story closed with a familiar phrase "And they all lived happily ever after."
4. The sergeant requested: reinforcements, medical supplies, and more ammunition.
5. She kept only four souvenirs a photograph, a matchbook, a theater program, and a daisy pressed between the pages of *William Shakespeare The Complete Works.*

32b Using Dashes

See ◀
28d2

Commas are the mark of punctuation most often used to set off nonessential elements, but dashes and parentheses also serve this function. While parentheses deemphasize the enclosed words, **dashes** tend to call attention to the material they set off.

CLOSE-UP

USING DASHES

When typing, indicate a dash with two unspaced hyphens; when writing, form a dash with an unbroken line about as long as two hyphens.

(1) To set off nonessential material

Explanations, qualifications, examples, definitions, and appositives may be set off by dashes for emphasis or clarity.

Use a pair of dashes to set off nonessential material within a sentence.

Although we are by all odds the most social of all social animals—more interdependent, more attached to each other, more inseparable in our behavior than bees—we do not often feel our conjoined intelligence. (Lewis Thomas, *Lives of a Cell*)

Use a single dash to set off material at the end of a sentence.

Most of the best-sellers are cookbooks and diet books—how not to eat it after you've cooked it. (Andy Rooney, *New Yorker*)

(2) To introduce a summary

A dash is used to introduce a statement that summarizes a list or series before it.

> Walking to school by myself, spending the night at a friend's house, getting my ears pierced, and starting to wear makeup—these were some of the milestones of my childhood and adolescence.

> "Study hard," "Respect your elders," "Don't talk with your mouth full"—Sharon had often heard her parents say these things.

(3) To indicate an interruption

A dash is sometimes used in dialogue to mark a sudden interruption—for example, a correction, a hesitation, a sudden shift in tone, or an unfinished thought.

> Groucho told the steward, "I'll have three hard-boiled eggs—make that four hard-boiled eggs."

> "I think—no, I know—this is the worst day of my life," Julie sighed.

(4) Editing misused or overused dashes

Dashes can make a passage seem disorganized and out of control. For this reason, dashes should not be overused in academic writing and should not be used in place of periods or commas.

> **FAULTY (overuse of dashes):** Registration was a nightmare—most of the courses I wanted to take—geology and conversational Spanish, for instance—met at inconvenient times—or were closed by the time I tried to sign up for them—it was really depressing—even for registration.

> **REVISED (moderate use of dashes):** Registration was a nightmare. Most of the courses I wanted to take—geology and conversational Spanish, for instance—met at inconvenient times or were closed by the time I tried to sign up for them. It was really depressing—even for registration.

EXERCISE 2

Add dashes where needed in the following sentences. If a sentence is correct, mark with a *C*.

> **EXAMPLE:** World War I called "the war to end all wars" was, unfortunately, no such thing.
>
> World War I—called "the war to end all wars"—was, unfortunately, no such thing.

ANSWERS TO EXERCISE 2

1. Tulips, daffodils, hyacinths, lilies—all of these flowers grow from bulbs.
2. St. Kitts and Nevis—two tiny island nations—are now independent after 360 years of British rule.
3. "But it's not—" She paused and reconsidered her next words.
4. He considered several different majors—history, English, political science, and business—before deciding on journalism.
5. The two words added to the Pledge of Allegiance in the 1950s—"under God"—remain part of the Pledge today.

1. Tulips, daffodils, hyacinths, lilies all of these flowers grow from bulbs.
2. St. Kitts and Nevis two tiny island nations are now independent after 360 years of British rule.
3. "But it's not" She paused and reconsidered her next words.
4. He considered several different majors history, English, political science, and business before deciding on journalism.
5. The two words added to the Pledge of Allegiance in the 1950s "under God" remain part of the Pledge today.

32c Using Parentheses

Like commas and dashes, **parentheses** may be used to set off interruptions within a sentence. Parentheses, however, indicate that the enclosed material is of secondary importance.

(1) To set off nonessential material

Parentheses may be used to set off nonessential material that expands, clarifies, defines, illustrates, or supplements an idea.

> A compound may be used in any grammatical function: as noun *(wishbone)*, adjective *(foolproof)*, adverb *(overhead)*, verb *(gainsay)*, or preposition *(without)*. (Thomas Pyles, *The Origin and Development of the English Language*)

> It took Gilbert Fairchild two years at Harvard College **(**two academic years, from September, 1955, to June, 1957**)** to learn everything he needed to know. (Judith Martin, *Gilbert: A Comedy of Manners*)

When a complete sentence set off by parentheses falls within another sentence, it should not begin with a capital letter or end with a period.

> Born in 1893, four years before Queen Victoria's Diamond Jubilee, at Cathedral Choir School, Oxford, where her father was headmaster, Sayers became a first-rate medievalist **(**she translated Dante**)** and a theologian **(**her miracle play, *The Man Born to Be King*, outsold her mystery novels in her lifetime**)**; she died in 1957. (Barbara Grizzuti Harrison, *Off Center*)

If the parenthetical sentence does not interrupt another sentence, it must begin with a capital letter and end with a period, question mark, or exclamation point that falls within the closing parenthesis.

> A few days later he called and asked me to come in and bring anything else I had written. **(**The only thing I had was a notebook full of

isolated sentences like "She walked across the room wearing her wedding ring like a shield.") (Jeremy Bernstein, *New York Times Book Review*)

CLOSE-UP

USING PARENTHESES

When a parenthetical element falls within a sentence, punctuation never precedes the opening parenthesis. Punctuation may follow the closing parenthesis, however.

(2) In other conventional situations

Parentheses are used to set off letters and numbers that identify points on a list and around dates, cross-references, and documentation.

All reports must include the following components: **(1)** an opening summary; **(2)** a background statement; and **(3)** a list of conclusions and recommendations.

Russia defeated Sweden in the Great Northern War **(**1700–1721**)**.

Other historians also make this point **(**see p. 54**)**.

One critic has called the novel "peurile" **(**Arvin 72**)**.

EXERCISE 3

Add parentheses where necessary in the following sentences. If a sentence is correct, mark it with a *C*.

> **EXAMPLE:** The greatest battle of the War of 1812 the Battle of New Orleans was fought after the war was declared over.
>
> The greatest battle of the War of 1812 **(**the Battle of New Orleans**)** was fought after the war was declared over.

1. George Orwell's *1984* 1949 focuses on the dangers of a totalitarian society.
2. The final score 45–0 was a devastating blow for the Eagles.
3. Belize formerly British Honduras is a country in Central America.
4. The first phonics book *Phonics Is Fun* has a light blue cover.
5. Some high school students have so many extracurricular activities band, sports, drama club, and school newspaper, for instance that they have little time to study.

1. George Orwell's *1984* (1949) focuses on the dangers of a totalitarian society.
2. The final score (45–0) was a devastating blow for the Eagles.
3. Belize (formerly British Honduras) is a country in Central America.
4. The first phonics book *(Phonics Is Fun)* has a light blue cover.
5. Some high school students have so many extracurricular activities (band, sports, drama club, and school newspaper, for instance) that they have little time to study.

32d Using Brackets

Brackets are used in two situations.

(1) To set off comments within quotations

Brackets are used within quotations to tell readers that the words enclosed are yours and not those of your source. You can bracket an explanation, a clarification, a correction, or an opinion.

"Dues are being raised $1.00 per week [to $5.00]," the treasurer announced.

"The use of caricature by Dickens is reminiscent of the satiric sketches done by [Joseph] Addison and [Richard] Steele [in *The Spectator*]."

"Even as a student at Princeton he [F. Scott Fitzgerald] felt like an outsider."

"The miles of excellent trails are perfect for [cross-country] skiing."

If a quotation contains an error, indicate that the error is not yours by following the error with the italicized Latin word *sic* ("thus") in brackets.

"The octopuss [*sic*] is a cephalopod mollusk with eight arms."

CLOSE-UP

USING BRACKETS

See ◀
39c1

Brackets are also used to indicate changes you make to tailor a quotation so that it fits the context of your sentence.

(2) In place of parentheses within parentheses

When one set of parentheses falls within another, substitute brackets for the inner set.

In her study of American education between 1945 and 1960 (*The Trouble Crusade* [New York: Basic Books, 1963]), Diane Ravitch addresses issues like progressive education, race, educational reforms, and campus unrest.

32e Using Slashes

The **slash** is used in three situations.

(1) To separate one option from another

The either/or fallacy assumes that a given question has only two possible answers.

Will pass/fail courses be accepted for transfer credit?

The producer/director attracted more attention at the film festival than the actors.

When you use a slash to separate one option from another, do not leave a space before or after the slash.

CLOSE-UP

USING SLASHES

Unless you are actually presenting three alternatives (Bring a pencil or pen or both to the exam), avoid the construction *and/or* (Bring a pencil and/or a pen to the exam). Instead use *and* or *or.*

Bring a pencil and pen to the exam.

Bring a pencil or pen to the exam.

(2) To separate lines of poetry run into the text

The poet James Schevill writes, "I study my defects / And learn how to perfect them."

▶ See 31d3

When you use a slash to separate lines of poetry, leave a space both before and after the slash.

(3) To separate the numerator from the denominator in fractions

7/8

1 4/5

If your typewriter or computer has a special key for a particular fraction ($\frac{1}{2}$, $\frac{1}{4}$) use that instead of the slash.

TEACHING STRATEGY

Ellipses
You may want to use samples of student writing (from another class or year) to illustrate the correct use of ellipses.

Discuss the problem of "moth-eaten" quotations. Student writers sometimes try to pare long quotations down to size by chopping them to bits—and then gluing them together with ellipses. Such efforts often leave the reader more confused than enlightened. Rather than string phrases and half-sentences of a long quotation into a whole that is less than the sum of its parts, students might try one of the following strategies.

- Cut the quotation down to one vital, central sentence (or even a few key words)—and then lead into it with their own words.

- Look for another quotation. If the quotation seems too long or ponderous to be used in its entirety, it might be better to scrap the whole thing and look for a more appropriate quotation.

- Be brave and use the whole passage (ellipsis-free). This response is especially appropriate where the quotation is used because of its beauty of style, wit, or humor—all of which are likely to be lost if the passage is dotted with ellipses.

32f Using Ellipses

(1) To indicate an omission in a quotation

Use an **ellipsis**—three *spaced* periods—to indicate words or entire sentences omitted from a quotation. When deleting material, be careful not to change the meaning of the original passage.

ORIGINAL: "When I was a young man, being anxious to distinguish myself, I was perpetually starting new propositions. But I soon gave this over; for I found that generally what was new was false." (Samuel Johnson)

WITH OMISSIONS: "When I was a young man, . . . I was perpetually starting new propositions. But I soon . . . found that generally what was new was false." (Three spaced periods indicate omissions.)

If you delete words immediately after an internal punctuation mark (such as a comma), retain the punctuation mark before the ellipsis.

CLOSE-UP

USING ELLIPSES

Do not begin a quoted passage with an ellipsis.

FAULTY: Barzun notes that Dickens ". . . might just as well have been the common schoolboy. . . ."

REVISED: Barzun notes that Dickens "might just as well have been the common schoolboy. . . ."

Do not use an ellipsis following a quotation unless you have deleted words at the end of that sentence.

Deleting Words at the Beginning of a Sentence When you delete *words at the beginning of a sentence* within a quoted passage, retain the previous sentence's end punctuation, followed by the ellipsis.

In the final paragraph, Jaynes poses—and answers—her central question: "What is power? . . . the option not only of saying *no* but also of saying *yes*."

Deleting Words at the End of a Sentence   When you delete *words at the end of a sentence* within a quoted passage, retain the sentence's end punctuation, followed by the ellipsis.

> According to humorist Dave Barry, "From outer space Europe appears to be shaped like a large ketchup stain. . . ."

> The play ends with George sadly crooning, "Who's afraid of Virginia Woolf, Virginia Woolf, Virginia Woolf? . . ."

CLOSE-UP

USING ELLIPSES

If a quotation ending with an ellipsis is followed by parenthetical documentation, the end punctuation follows the documentation.

> As Jarman argues, "There was no willingness to compromise . . ." (161).

Deleting One or More Complete Sentences   If you omit *one or more complete sentences* from a quoted passage, retain the previous sentence's end punctuation, followed by an ellipsis.

> According to Donald Hall, "Everywhere one meets the idea that reading is an activity desirable in itself. . . . People surround the idea of reading with piety, and do not take into account the purpose of reading. . . ."

Note that complete sentences must precede and follow the period plus ellipsis.

(2)   To indicate an omission within verse

When you omit one or more lines of poetry (or a paragraph or more of prose), use a complete line of spaced periods.

ORIGINAL: *Stitch! Stitch! Stitch!*
In poverty, hunger, and dirt,
And still with a voice of dolorous pitch,
Would that its tone could reach the Rich,
She sang this "Song of the Shirt!"
                                –Thomas Hood

529

ANSWERS TO EXERCISE 4

Your students' answers will vary. Some possibilities appear below. This is a good exercise to discuss in connection with a research paper assignment. Students must determine what information is essential here and must be prepared to explain their choices. An additional instruction might incorporate part of the quotation into a hypothetical research paper paragraph.

1. "When I was eighteen . . . my mother told me that when out with a young man I should always leave a half-hour before I wanted to."
2. "When I was eighteen or thereabouts, my mother told me that when out with a young man I should always leave a half-hour before I wanted to. . . . I recognized this advice as sound, and exactly the same rule applies to research."
3. "One must stop *before* one has finished. . . ."
4. "The most important thing about research is to know when to stop. . . . When I was eighteen or thereabouts, my mother told me that when out with a young man I should always leave a half-hour before I wanted to."

ANSWERS TO EXERCISE 5

1. Mark Twain (Samuel L. Clemens) made the following statement: "I can live for two months on a good compliment."
2. Liza Minelli, the actress/singer who starred in several films, is the daughter of Judy Garland. For emphasis, dashes may replace the commas.

continued on following page

WITH OMISSION: *Stitch! Stitch! Stitch!*
In poverty, hunger, and dirt,
. . . .
She sang this "Song of the Shirt!"

(3) To indicate unfinished statements

An ellipsis can also indicate an interrupted statement.

"If only. . . ." He sighed and turned away.

This use is generally not appropriate in college writing unless you are reproducing dialogue.

EXERCISE 4

Read this paragraph and follow the instructions below it, taking care in each case not to delete essential information.

The most important thing about research is to know when to stop. How does one recognize the moment? When I was eighteen or thereabouts, my mother told me that when out with a young man I should always leave a half-hour before I wanted to. Although I was not sure how this might be accomplished, I recognized the advice as sound, and exactly the same rule applies to research. One must stop *before* one has finished; otherwise, one will never stop and never finish. (Barbara Tuchman, *Practicing History*)

1. Delete a phrase from the middle of one sentence and mark the omission with ellipses.
2. Delete words at the beginning of any sentence and mark the omission with ellipses.
3. Delete words at the end of any sentence and mark the omission with ellipses.
4. Delete one complete sentence from the middle of the passage and mark the omission with ellipses.

EXERCISE 5

Add appropriate punctuation—colons, dashes, parentheses, brackets, or slashes—to the following sentences. Be prepared to explain why you chose the punctuation marks you did. If a sentence is correct, mark it with a *C*.

EXAMPLE: There was one thing she was sure of if she did well at the interview, the job would be hers.

There was one thing she was sure of: If she did well at the interview, the job would be hers.

1. Mark Twain Samuel L. Clemens made the following statement "I can live for two months on a good compliment."
2. Liza Minnelli, the actress singer who starred in several films, is the daughter of Judy Garland.
3. Saudi Arabia, Oman, Yemen, Qatar, and the United Arab Emirates all these are located on the Arabian peninsula.
4. John Adams 1735–1826 was the second president of the United States; John Quincy Adams 1767–1848 was the sixth.
5. The sign said "No tresspassing *sic*."
6. *Checkmate* a term derived from the Persian phrase meaning "the king is dead" announces victory in chess.
7. The following people were present at the meeting the president of the board of trustees, three trustees, and twenty reporters.
8. Before the introduction of the potato in Europe, the parsnip was a major source of carbohydrates in fact, it was a dietary staple.
9. In this well-researched book (*Crime Movies* New York Norton, 1980), Carlos Clarens studies the gangster genre in film.
10. I remember reading though I can't remember where that Upton Sinclair sold plots to Jack London.

STUDENT WRITER AT WORK

PUNCTUATION

Review Chapters 27–32; then read this student essay. Commas, semicolons, quotation marks, apostrophes, parentheses, and dashes have been intentionally deleted; only the end punctuation has been retained. When you have read the essay carefully, add all appropriate punctuation marks.

The dry pine needles crunched like eggshells under our thick boots.

Wont the noise scare them away Dad?

He smiled knowingly and said No deer rely mostly on smell and sight.

I thought That must be why were wearing fluorescent orange jumpsuits but I didn't feel like arguing the point.

It was a perfect day for my first hunting experience. The biting winds were caught by the thick bushy arms of the tall pines and I could feel a numbing redness in my face. Now I realized why Dad always grew that ugly gray beard that made him look ten years older. A few sunbeams managed to carve their way through the layers of branches and leaves creating pools of white light on the dark earth.

How far have we come? I asked.

continued on the following page

continued from previous page

3. Saudi Arabia, Oman, Yemen, Qatar, and the United Arab Emirates—all these are located on the Arabian Peninsula.
4. John Adams (1735–1826) was the second president of the United States; John Quincy Adams (1767–1848) was the sixth.
5. The sign said, "No tresspassing [*sic*]."
6. *Checkmate*—a term derived from the Persian phrase meaning "the king is dead"—announces victory in chess. (Parentheses are another option.)
7. The following people were present at the meeting: the president of the board of trustees, three trustees, and twenty reporters.
8. Before the introduction of the potato in Europe, the parsnip was a major source of carbohydrates—in fact, it was a dietary staple.
9. In this well-researched book (*Crime Movies* [N.Y.: Norton 1980]), Carlos Clarens studies the gangster genre in film.
10. I remember reading—though I can't remember where—that Upton Sinclair sold plots to Jack London.

STUDENT WRITER AT WORK

Punctuation

Revision

The dry pine needles crunched like eggshells under our thick boots.

"Won't the noise scare them away, Dad?"

He smiled knowingly and said, "No, deer rely mostly on smell and sight."

I thought, "That must be why we're wearing fluorescent orange jumpsuits," but I didn't feel like arguing the point.

continued on following page

continued from previous page

It was a perfect day for my first hunting experience. The biting winds were caught by the thick, bushy arms of the tall pines, and I could feel a numbing redness in my face. Now I realized why Dad always grew that ugly, gray beard that made him look ten years older. A few sunbeams managed to carve their way through the layers of branches and leaves, creating pools of white light on the dark earth.

"How far have we come?" I asked.

"Oh, only a couple of miles. We should be meeting Joe up ahead."

Joe was one of Dad's hunting buddies. He always managed to go off on his own for a few hours and come back with at least a four-pointer. Dad was envious of Joe and liked to tell people what he called "the real story."

"You know, Joe paid a fortune for that buck at the checking station," he'd tell his friends. Dad was sure that this would be his lucky year.

We trudged up a densely wooded hill for what seemed like hours. The sharp, needled branches whipped my bare face as I followed close behind my father; occasionally I wiped my cheeks to discover a new cut in my frozen flesh.

"All this for a deer," I thought.

The still pine air was suddenly shattered by four rapid gunshots echoing across the vast green valley below us.

"Joe's got another one. Come on!" Dad yelled. It seemed as if I were following a young kid as I watched my father take leaping strides down the path we had just ascended. I had never seen him so enthusiastic before. I plodded breathlessly along, trying to keep up with my father.

Suddenly—out of the corner of my eye—I caught sight of an object that didn't fit in with the monotony of trunks and branches and leaves and needles. I froze and observed the largest, most majestic buck I had ever seen. It too stood motionless, apparently grazing on some leaves or berries. Its coloring was beautiful, with alternating patches of tan, brown, and snow-white fur. The massive antlers towered proudly above its head as

continued from the previous page

Oh only a couple of miles. We should be meeting Joe up ahead.

Joe was one of Dads hunting buddies. He always managed to go off on his own for a few hours and come back with at least a four-pointer. Dad was envious of Joe and liked to tell people what he called the real story.

You know Joe paid a fortune for that buck at the checking station hed tell his friends. Dad was sure that this would be his lucky year.

We trudged up a densely wooded hill for what seemed like hours. The sharp needled branches whipped my bare face as I followed close behind my father occasionally I wiped my cheeks to discover a new cut in my frozen flesh.

All this for a deer I thought.

The still pine air was suddenly shattered by four rapid gunshots echoing across the vast green valley below us.

Joes got another one. Come on! Dad yelled. It seemed as if I were following a young kid as I watched my father take leaping strides down the path we had just ascended. I had never seen him so enthusiastic before. I plodded breathlessly along trying to keep up with my father.

Suddenly out of the corner of my eye I caught sight of an object that didnt fit in with the monotony of trunks and branches and leaves and needles. I froze and observed the largest most majestic buck I had ever seen. It too stood motionless apparently grazing on some leaves or berries. Its coloring was beautiful with alternating patches of tan brown and snow-white fur. The massive antlers towered proudly above its head as it looked up and took notice of me. What struck me most were the tearful brown eyes almost feminine in their gaze.

Once again the silence was smashed this time by my fathers thundering call and I watched as the huge deer scampered gracefully off through the trees. I turned and scurried down the path after my father. I decided not to mention a word of my encounter to him. I hoped the deer was far away by now.

Finally I reached the clearing from where the shots had rung out. There stood Dad and Joe smiling over a fallen six-point buck. The purple-red blood dripped from the wounds to form a puddle in the dirt. The bucks sad brown eyes gleamed in the sun but no longer smiled and blinked.

Where have you been? asked Dad. Before I could answer he continued Do you believe this guy? Every year he bags the biggest deer in the whole state!

While they laughed and talked I sat on a tree stump to rest my aching legs. Maybe now we can go home I thought.

But before long I heard Dad say Come on Bob I know theres one out there for us.

We headed right back up that same path and sure enough that same big beautiful buck was grazing in that same spot on the same berry bush. The only difference was that this time Dad saw him.

This is our lucky day he whispered.

I froze as Dad lifted the barrel of his rifle and took careful aim at the silently grazing deer. I closed my eyes as he squeezed the trigger but instead of the deadly gun blast I heard only a harmless click. His rifle had jammed.

Use your rifle quick he whispered.

As I took aim through my scope the deer looked up at me. Its soulful brown eyes were magnified in my sight like two glassy bulls-eyes. My finger froze on the trigger.

Shoot him! Shoot him!

But instead I aimed for the clouds and fired. The deer vanished along with my fathers dreams. Dad never understood why this was the proudest moment of my life.

---

it looked up and took notice of me. What struck me most were the tearful brown eyes—almost feminine in their gaze.

Once again the silence was smashed—this time by my father's thundering call—and I watched as the huge deer scampered gracefully off through the trees. I turned and scurried down the path after my father. I decided not to mention a word of my encounter to him. I hoped the deer was far away by now.

Finally, I reached the clearing from where the shots had rung out. There stood Dad and Joe, smiling over a fallen six-point buck. The purple-red blood dripped from the wounds to form a puddle in the dirt. The buck's sad brown eyes gleamed in the sun but no longer smiled and blinked.

"Where have you been?" asked Dad. Before I could answer, he continued, "Do you believe this guy? Every year he bags the biggest deer in the whole state!"

While they laughed and talked, I sat on a tree stump to rest my aching legs. "Maybe now we can go home," I thought.

But before long I heard Dad say, "Come on, Bob, I know there's one out there for us."

We headed right back up that same path, and—sure enough—that same big, beautiful buck was grazing in that same spot on that same berry bush. The only difference was that this time Dad saw him.

"This is our lucky day," he whispered.

I froze as Dad lifted the barrel of his rifle and took aim at the silent, grazing deer. I closed my eyes as he squeezed the trigger, but instead of the deadly gun blast, I heard only a harmless click. His rifle had jammed.

"Use your rifle—quick!" he whispered.

As I took aim through my scope, the deer looked up at me. Its soulful brown eyes were magnified in my sight like two glassy bull's-eyes. My finger froze on the trigger.

"Shoot him! Shoot him!"

But instead I aimed for the clouds and fired. The deer vanished along with my father's dreams. Dad never understood why this was the proudest moment of my life.

CHAPTER 33

Capitalization

Familiarizing yourself with the conventions of capitalization is important. Conventions change, however, and if you are not certain whether a word should be capitalized, consult the latest edition of a dictionary.

33a Capitalize the First Word of a Sentence or of a Line of Poetry

The first word of a sentence, including a sentence of directly quoted speech or writing, should start with a capital letter.

> The square of the hypotenuse is equal to the sum of the squares of the other two sides.

> As Shakespeare wrote, "Who steals my purse steals trash."

Do not capitalize a sentence set off within another sentence by dashes or parentheses.

> **FAULTY:** Finding the store closed—It was a holiday—they went home.

> **REVISED:** Finding the store closed—it was a holiday—they went home.

> **FAULTY:** The candidates are Frank Lester and Jane Lester (They are not related).

> **REVISED:** The candidates are Frank Lester and Jane Lester (they are not related).

See ◄ 32a2 When a complete sentence is introduced by a colon, capitalization is optional.

CLOSE-UP

CAPITALIZATION

When you quote poetry, remember that the first word of a line of poetry is generally capitalized. If the poet uses a lowercase letter to begin a line, however, that style should be followed when you quote the line.

33b **Capitalize Proper Nouns, Titles Accompanying Them, and Adjectives Formed from Them**

Proper nouns—the names of specific persons, places, or things (Diana Ross, Madras, the Enoch Pratt Free Library)—are capitalized, and so are adjectives formed from proper nouns.

(1) Specific people's names

Eleanor Roosevelt	Elvis Presley
Jackie Robinson	William the Conqueror

When a title precedes a person's name or is used instead of the name, it, too, is capitalized.

Dad	Pope John XXIII
Count Dracula	Justice Ruth Bader Ginsburg

Titles that *follow* names or those that refer to the general position, not the particular person who holds it, are usually not capitalized. A title denoting a family relationship is never capitalized when it follows an article or a possessive pronoun.

CAPITALIZE:	**DO NOT CAPITALIZE:**
Grandma	my grandmother
Private Hargrove	Mr. Hargrove, a private in the army
Queen Mother Elizabeth	a popular queen mother
Senator Carol Moseley-Braun	Carol Moseley-Braun, the senator from Illinois
Uncle Harry	my uncle
General Patton	a four-star general

Titles or abbreviations of academic degrees are always capitalized, even when they follow a name.

More on Capitalization

Charles Macintosh invented a raincoat—but over the centuries, the "macintosh" lost its capital letter. And so it went with Amelia Bloomer's underwear, Monsieur Silhouette's outline drawings, the Earl of Sandwich's favorite snack, and the elastic tights worn by French trapeze star Jules Léotard.

Today's corporations, mindful of lost royalties past, spend a great deal of money fighting to keep the capital letters on *their* trademark names. Ask students to name some trademarks that seem to be in transition—on the way to becoming "common" nouns. (Some of the possibilities for discussion: Levi's, Frisbee, Jello, Bic, Brillo, Velcro.)

Perry Mason, Attorney at Law

Benjamin Spock, M.D.

Titles that indicate high-ranking positions may be capitalized even when they are used alone or when they follow a name.

the Pope

William Jefferson Clinton, President of the United States

(2) Names of particular structures, special events, monuments, vehicles, and so on

the *Titanic*	the Taj Mahal
the Brooklyn Bridge	Mount Rushmore
the World Series	the Eiffel Tower

NOTE: When a common noun such as *bridge*, *river*, *county*, or *lake* is part of a proper noun, it, too, is capitalized. Do not, however, capitalize such words when they complete the names of more than one thing (as in Kings and Queens counties).

(3) Places, geographical regions, and directions

Saturn	the Straits of Magellan
Budapest	the Western Hemisphere
Walden Pond	the Fiji Islands

North, *east*, *south*, and *west* are capitalized when they denote particular geographical regions, but not when they designate directions.

The Middle West seemed like a wasteland to F. Scott Fitzgerald's Nick Carraway, so he decided to come East. (Capital letters necessary because *Middle West* and *East* refer to specific regions.)

Turn west at the corner of Broad Street and continue north until you reach Market. (No capitals used because *west* and *north* refer to directions, not specific regions.)

(4) Days of the week, months of the year, and holidays

Saturday	Ash Wednesday
January	Rosh Hashanah
Veterans Day	Kwanzaa

(5) Historical periods, events, and documents; names of legal cases and awards

the Battle of Gettysburg	the Treaty of Versailles

the Industrial Revolution   the Voting Rights Act
the Reformation   *Brown v. Board of Education*

NOTE: Names of court cases are italicized except in bibliographic entries.

(6) Philosophic, literary, and artistic movements

Naturalism	Dadaism
Romanticism	Fauvism
Neoclassicism	Expressionism

NOTE: Dictionaries and manuals of style vary in their advice on capitalizing these nouns.

(7) Races, ethnic groups, nationalities, and languages

African American	Korean
Latino	Dutch
Caucasian	Turkish

NOTE: When the words *black* and *white* refer to races, they have traditionally not been capitalized. Current usage is divided on whether or not to capitalize *black*.

(8) Religions and their followers; sacred books and figures

Muslims	the Talmud	Buddha
Jews	the Koran	the Virgin Mary
Islam	God	the Messiah
Judaism	the Lord	the Scriptures

NOTE: It is not necessary to capitalize pronouns referring to God unless the pronoun might also refer to another antecedent in the sentence.

    CONFUSING: Alex's grandfather taught him to trust in God and to love all <u>his</u> creatures. (The word *his* could refer either to God or to Alex's grandfather.)

    REVISED: Alex's grandfather taught him to trust in God and to love all <u>His</u> creatures.

(9) Political, social, athletic, civic, and other groups and their members

the New York Yankees

the Democratic Party

the International Brotherhood of Electrical Workers

the American Civil Liberties Union

the National Council of Teachers of English

Pearl Jam

(10) Businesses; government agencies; and medical, educational, and other institutions

Congress	Lincoln High School
Environmental Protection Agency	the University of Maryland

NOTE: When the name of a group or institution is abbreviated, the abbreviation uses capital letters in place of the capitalized words.

IBEW
ACLU
NCTE

(11) Trade names and words formed from them

Pontiac	Coke
Sanka	Pampers
Kleenex	Xeroxing

NOTE: Trade names that have been used so often and for so long that they have become synonymous with the product—for example, aspirin and nylon—are no longer capitalized. (Consult a dictionary to determine whether or not to capitalize a familiar trade name.)

CLOSE-UP

CAPITALIZATION

In general, use generic references, not brand names, in college writing—*photocopy*, not *Xerox*, for example.

(12) Specific academic courses

Sociology 201   English 101

NOTE: Do not capitalize a general subject area unless it is the name of a language.

Although his major was <u>engineering</u>, he registered for courses in <u>sociology</u>, English, and <u>zoology</u>.

(13) Adjectives formed from proper nouns

Keynesian economics	Elizabethan era
Freudian slip	Shakespearean sonnet
Platonic ideal	Marxist ideology
Aristotelian logic	Shavian wit

When words derived from proper nouns have lost their specialized meanings, do not capitalize them.

The <u>china</u> pattern was very elaborate.

We need a 75-<u>watt</u> bulb.

33c Capitalize Important Words in Titles

In general, all words in titles of books, articles, essays, films, and the like—including your own papers—are capitalized, with the exception of articles (*a*, *an*, and *the*), prepositions, conjunctions, and the *to* in infinitives. If an article, preposition, or conjunction is the *first* or *last* word in the title, however, it is capitalized.

"Dover Beach"	*On the Waterfront*
The Declaration of Independence	*The Skin of Our Teeth*
Across the River and into the Trees	*Of Human Bondage*
Two Years before the Mast	"Politics and the English Language"

33d Capitalize the Pronoun I and the Interjection O

Even if the pronoun *I* is part of a contraction (*I'm, I'll, I've*), it is always capitalized.

Sam and <u>I</u> finally went to the Grand Canyon, and <u>I</u>'m glad we did.

The interjection *O* is also always capitalized.

Give us peace in our time, <u>O</u> Lord.

The interjection *oh,* however, is capitalized only when it begins a sentence.

1. The Brontë sisters wrote *Jane Eyre* and *Wuthering Heights*, two nineteenth-century novels that are required reading in many English classes that study Victorian literature.

2. It was a beautiful day in the spring—it was April 15, to be exact—but all Ted could think about was the check he had to write to the Internal Revenue Service and the bills he had to pay by Friday.

3. Traveling north, they hiked through British Columbia, planning a leisurely return on the cruise ship *Canadian Princess*.

4. Alice liked her mom's apple pie better than Aunt Nellie's rhubarb pie; but she liked Grandpa's punch best of all.

5. A new elective, Political Science 30, covers the Vietnam War from the Gulf of Tonkin to the fall of Saigon, including the roles of Ho Chi Minh, the Viet Cong, and the Buddhist monks; the positions of Presidents Johnson and Nixon; and the influence of groups like the Student Mobilization Committee and Vietnam Veterans Against the War.

6. When the Central High School Drama Club put on a production of Shaw's *Pygmalion*, the director Xeroxed extra copies of the parts for Eliza Doolittle and Professor Henry Higgins so he could give them to the understudies.

7. Shaking all over, Bill admitted, "Driving on the Los Angeles Freeway is a frightening experience for a kid from the Bronx, even in a BMW."

continued on following page

CLOSE-UP

CAPITALIZATION

Many other single letters are capitalized in certain usages. Check your dictionary to determine whether to use a capital letter.

U-boat	Vitamin B
D-Day	an A in history
Model T	C major

33e Capitalize Salutations and Closings of Letters

In salutations of business or personal letters, always capitalize the first word.

See ◄ 48a2

Dear Mr. Reynolds: Dear Fred,

The first word of the complimentary close is also always capitalized.

Sincerely,   Very truly yours,

33f Editing Misused or Overused Capitals

(1) Seasons

Do not capitalize the names of the seasons—summer, fall, winter, spring—unless they are strongly personified, as in Old Man Winter.

(2) Centuries and loosely defined historical periods

Do not capitalize the names of centuries or general historical periods.

seventeenth-century poetry   the automobile age

But do capitalize names of specific historical, anthropological, and geological periods.

Iron Age   Paleozoic Era

(3) Diseases and other medical terms

Do not capitalize names of diseases or medical tests or conditions unless a proper noun is part of the name or unless the disease is an acronym.

polio	Apgar test	AIDS
Reye's syndrome	mumps	SIDS

EXERCISE

Capitalize words where necessary in these sentences.

> **EXAMPLE:** John F. Kennedy won the pulitzer prize for his book *profiles in courage.*
>
> John F. Kennedy won the Pulitzer Prize for his book *Profiles in Courage.*

1. The brontë sisters wrote *jane eyre* and *wuthering heights,* two nineteenth-century novels that are required reading in many english classes that study victorian literature.
2. It was a beautiful day in the spring—it was april 15, to be exact—but all Ted could think about was the check he had to write to the internal revenue service and the bills he had to pay by friday.
3. Traveling north, they hiked through british columbia, planning a leisurely return on the cruise ship *canadian princess.*
4. Alice liked her mom's apple pie better than aunt nellie's rhubarb pie; but she liked grandpa's punch best of all.
5. A new elective, political science 30, covers the vietnam war from the gulf of tonkin to the fall of saigon, including the roles of ho chi minh, the viet cong, and the buddhist monks; the positions of presidents johnson and nixon; and the influence of groups like the student mobilization committee and vietnam veterans against the war.
6. When the central high school drama club put on a production of shaw's *pygmalion,* the director xeroxed extra copies of the parts for eliza doolittle and professor henry higgins so he could give them to the understudies.
7. Shaking all over, Bill admitted, "driving on the los angeles freeway is a frightening experience for a kid from the bronx, even in a bmw."
8. The new united federation of teachers contract guarantees teachers many paid holidays, including columbus day, veterans day, and washington's birthday; a week each at christmas and easter; and two full months (july and august) in the summer.
9. The sociology syllabus included the books *beyond the best interests of the child, regulating the poor,* and *a welfare mother;* in anthropology we were to begin by studying the stone age; and in geology we were to focus on the mesozoic era.
10. Winners of the nobel peace prize include lech walesa, former leader of the polish trade union solidarity; the reverend dr. martin luther king, jr., founder of the southern christian leadership conference; and bishop desmond tutu of south africa.

continued from previous page

8. The new United Federation of Teachers contract guarantees teachers many paid holidays, including Columbus Day, Veterans Day, and Washington's Birthday; a week each at Christmas and Easter; and two full months (July and August) in the summer.
9. The sociology syllabus included the books *Beyond the Best Interests of the Child, Regulating the Poor,* and *A Welfare Mother;* in anthropology we were to begin by studying the Stone Age; and in geology we were to focus on the Mesozoic Era.
10. Winners of the Nobel Peace Prize include Lech Walesa, former leader of the Polish trade union Solidarity; the Reverend Dr. Martin Luther King, Jr., founder of the Southern Christian Leadership Conference; and Bishop Desmond Tutu of South Africa.

CHAPTER 34

Italics

34a Setting Off Titles and Names

Italicize the titles of books, newspapers, magazines, and journals; pamphlets; films; television and radio programs; long poems; plays; long musical works; and paintings and sculpture. Also italicize names of ships, trains, aircraft, and spacecraft. All other titles are set off with quotation marks. If your typewriter or computer does not have an italic typeface, indicate italics by underlining.

See ◄
31b

CLOSE-UP

USING ITALICS

Titles of your own essays, typed at the top of the first page or on a title page, are neither italicized nor placed within quotation marks. Names of sacred books, such as the Bible, and well-known documents, such as the Constitution and the Declaration of Independence, are also neither italicized nor placed within quotation marks.

TITLES AND NAMES SET IN ITALICS

Books

    David Copperfield   *The Bluest Eye*

Newspapers

    the *Washington Post*   *The Philadelphia Inquirer*

continued on the following page

continued from the previous page

Articles and names of cities are italicized only when they are part of a title.

Magazines

Rolling Stone   *Scientific American*

Journals

New England Journal of Medicine

American Sociological Review

Pamphlets

Common Sense

Films

Casablanca   *Who's Afraid of Virginia Woolf?*

Punctuation is italicized when it is part of a title.

Television programs

Sesame Street   *Roseanne*

Radio programs

All Things Considered   *Prairie Home Companion*

Long poems

John Brown's Body   *The Faerie Queen*

Plays

Macbeth   *A Raisin in the Sun*

Long musical works

Rigoletto   *Eroica*

Paintings and sculpture

Pietà   *Guërnica*

Ships

Lusitania   USS *Saratoga*

SS and USS are not italicized when they precede the name of a ship.

Trains

City of New Orleans   *The Orient Express*

Aircraft

the *Hindenburg*   *Enola Gay*

Only particular aircraft, not makes or types such as Piper Cub or Boeing 707, are italicized.

Spacecraft

Sputnik   *Enterprise*

Foreign Words

William Strunk, Jr., upheld the notion that unless a foreign word was so much a part of contemporary English that it didn't *need* italics, good writers needn't use it. Strunk's law: "Some writers, however, from sheer exuberance or a desire to show off, sprinkle their work liberally with foreign expressions, with no regard for the reader's comfort. It is a bad habit. Write in English."

Do students agree entirely with Strunk's injunction? Are some students willing to accept the use of foreign phrases as a stylistic device? What, for example, is Tom Wolfe "getting at" when he uses the German word for *soldier* in the quotation below?

"Everybody at the craps table was staring in consternation to think that anybody would try to needle a tough, hip, elite *soldat* like a Las Vegas craps dealer."

–from "Las Vegas (What?)"

34b Setting Off Foreign Words and Phrases

Thousands of foreign words and phrases are now considered part of the English language. These words—*naive, lasso, chaperon,* and *catharsis,* for example—receive no special treatment. Foreign words and phrases not yet fully assimilated into the English language, however, should be set in italics. (Foreign proper nouns are not italicized.)

> "*C'est la vie,*" Madeline said when she noticed the new, lower sales price.

> *Spirochaeta plicatilis, Treponema pallidum,* and *Spirilbum minus* are all bacteria with corkscrew-like shapes.

If you are not sure whether a foreign word has been assimilated into English, consult a dictionary.

34c Setting Off Elements Spoken of as Themselves and Terms Being Defined

Italicize letters, numerals, words, and phrases when they refer to the letters, numerals, words, and phrases themselves.

> Is that a *p* or a *g*?

> I forget the exact address, but I know it has a *3* in it.

> Does *through* rhyme with *cough*?

> His pronunciation of the phrase *Mary was contrary* told us he was from the Midwest.

Italicize to set off words and phrases that you go on to define.

> A *closet drama* is a play meant to be read, not performed.

34d Using Italics for Emphasis

Italics lend unusually strong weight to a word or phrase and should, therefore, be used in moderation. Overuse of italics interferes with the tone and even the meaning of what you write. Whenever possible, then, indicate emphasis with word choice and sentence structure. The following sentences illustrate acceptable use of italics for emphasis.

As for protecting the children from exploitation, the chief and indeed only exploiters of children these days *are* schools. (John Holt, "School Is Bad for Children")

Initially, poetry might be defined as a kind of language that says *more* and says it *more intensely* than does ordinary language. (Lawrence Perrine, *Sound and Sense*)

 34e Using Italics for Clarity

Occasionally, it is necessary to italicize a word to avoid confusion or ambiguity when a sentence may have more than one meaning.

This time Jill forgot the *key*. (Last time Jill forgot something else.)

This time *Jill* forgot the key. (Last time someone else forgot the key.)

EXERCISE

Underline to indicate italics where necessary, and delete any italics that are incorrectly used. If a sentence is correct, mark it with a *C*.

> **EXAMPLE:** However is a conjunctive adverb, not a coordinating conjunction.
>
> <u>However</u> is a conjunctive adverb, not a coordinating conjunction.

1. I said Carol, not Darryl.
2. A *deus ex machina*, an improbable device used to resolve the plot of a fictional work, is used in Charles Dickens's novel Oliver Twist.
3. He dotted every i and crossed every t.
4. The Metropolitan Opera's production of Carmen was a tour de force for the principal performers.
5. *Laissez-faire* is a doctrine holding that government should not interfere with trade.
6. Antidote and anecdote are often confused because their pronunciations are similar.
7. Hawthorne's novels include Fanshawe, The House of the Seven Gables, The Blithedale Romance, and The Scarlet Letter.
8. Words like mailman, policeman, and fireman are rapidly being replaced by nonsexist terms like letter carrier, police officer, and firefighter.
9. A classic black tuxedo was considered de rigueur at the charity ball, but Jason preferred to wear his *dashiki*.
10. Thomas Mann's novel Buddenbrooks is a Bildungsroman.

ANSWERS TO EXERCISE

1. I said <u>Carol</u>, not <u>Darryl</u>.
2. A <u>deus ex machina</u>, an improbable device used to resolve the plot of a fictional work, is used in Charles Dickens's novel <u>Oliver Twist</u>.
3. He dotted every <u>i</u> and crossed every <u>t</u>.
4. The Metropolitan Opera's production of <u>Carmen</u> was a <u>tour de force</u> for the principal performers.
5. C
6. <u>Antidote</u> and <u>anecdote</u> are often confused because their pronunciations are similar.
7. Hawthorne's novels include <u>Fanshawe</u>, <u>The House of the Seven Gables</u>, <u>The Blithedale Romance</u>, and <u>The Scarlet Letter</u>.
8. Words like <u>mailman</u>, <u>policeman</u>, and <u>fireman</u> are rapidly being replaced by nonsexist terms like <u>letter carrier</u>, <u>police officer</u>, and <u>firefighter</u>.
9. A classic black tuxedo was considered <u>de rigueur</u> at the charity ball, but Jason preferred to wear his <u>dashiki</u>.
10. Thomas Mann's novel <u>Buddenbrooks</u> is a <u>Bildungsroman</u>.

CHAPTER 35

Hyphens

Hyphens have two conventional uses: to break words at the end of a typed or handwritten line and to link words in certain compounds.

35a Breaking Words at the End of a Line

Whenever possible, avoid breaking a word at the end of a line; if you must do so, divide words only between syllables. Consult a dictionary if necessary to determine correct syllabication. When you can, divide a word between prefix and root (sus • pending) or between root and suffix (develop • ment). Divide words that contain doubled consonants between the doubled letters (let • ter) unless the doubled letters are part of the root (cross • ing) or unless the doubled letters do not break into two syllables (ex • pelled). Do not end two consecutive lines with hyphens, and never divide a word at the end of a page.

Additional guidelines for determining how words should be divided are listed below. Note, however, that if you use a computer, you do not have to make decisions about how to hyphenate at the end of a line. A computer does not hyphenate; instead, the entire word is brought down to the next line.

(1) One-syllable words

Never hyphenate one-syllable words. Keep a one-syllable word intact even if it is relatively long *(thought, blocked, French, laughed)*. If you cannot fit the whole word at the end of the line, move it to the next line.

FAULTY:	Mark Twain's novel *The Prin- ce and the Pauper* considers the effects of environment on personality.
REVISED:	Mark Twain's novel *The Prince and the Pauper* considers the effects of environment on personality.

(2) Short syllables

Never leave a single letter at the end of a line or carry one or two letters to the beginning of a line. One-letter prefixes (like the *a-* in *away*) or short suffixes (like *-y, -ly, -er,* and *-ed*) should not be separated from the rest of the word. The suffixes *-able* and *-ible* cannot be broken into two syllables.

FAULTY:	Nadia walked very slowly a- long the balance beam.
REVISED:	Nadia walked very slowly along the balance beam.
FAULTY:	Amy's parents wondered whether the terrib- le twos would ever end.
REVISED:	Amy's parents wondered whether the ter- rible twos would ever end.

(3) Compound words

If you must hyphenate a compound word, put the hyphen be-tween the elements of the compound.

▶ **See
35b**

FAULTY:	Environmentalists believe snowmo- biles produce air and noise pollution.
REVISED:	Environmentalists believe snow- mobiles produce air and noise pollution.

If the compound already contains a hyphen, divide it at the exist-ing hyphen.

FAULTY:	She met her hus- band-to-be on a blind date.
REVISED:	She met her husband- to-be on a blind date.

(4) Illogical or confusing hyphenation

Some words contain letter combinations that look like other words. To avoid confusing readers, do not isolate a part of a word that can be read as a separate word.

"Hyphens should be used when not using them causes confusion. As Fowler wrote in his *Modern English Usage*: 'A little used car is not necessarily the same as a little-used car.' . . . Similarly, a small-business man could be a huge fellow who runs a mom-and-pop enterprise, but a small businessman is always a runt."

–William Safire
On Language

"Hooray for the hyphen. Or, rather, Hoo-ray!"

–George Plimpton

1. tran/scen/den/tal/ism
   tran-scendentalism or
   transcen-dentalism
2. cal/li/o/pe
   calli-ope
3. mar/tyr
   mar-tyr
4. lon/gi/tude
   longi-tude
5. book/keep/er
   book-keeper
6. side-/split/ting
   side-splitting
7. mark/ed/ly
   markedly [no break]
8. a/maz/ing
   amaz-ing
9. un/like/ly
   un-likely
10. thor/ough
    thor-ough

CONFUSING: The supervisor did not appreciate the facetious remark.

REVISED: The supervisor did not appreciate the
facetious remark.

(5) Numerals, contractions, and abbreviations

Numerals, contractions, and abbreviations (including acronyms)
should not be divided. A hyphen is not used between a numeral and
an abbreviation.

FAULTY: The special program on child abuse was presented to over 23,-
000 schoolchildren.

REVISED: The special program on child abuse was presented to over
23,000 schoolchildren.

FAULTY: Whether or not the meeting began on time was-
n't important.

REVISED: Whether or not the meeting began on time
wasn't important.

FAULTY: During the 1960s, participation in RO-
TC declined on many college campuses.

REVISED: During the 1960s, participation in ROTC
declined on many college campuses.

FAULTY: The balloon was launched at precisely 8-
P.M.

REVISED: The balloon was launched at precisely 8 P.M.

EXERCISE 1

Divide each of these words into syllables, consulting a dictionary if necessary; then, indicate with a hyphen where you would divide each word at
the end of a line.

EXAMPLE: underground
un • der • ground
under-ground

1. transcendentalism
2. calliope
3. martyr
4. longitude
5. bookkeeper
6. side-splitting
7. markedly
8. amazing
9. unlikely
10. thorough

35b Dividing Compound Words

A **compound word** is composed of two or more words. Some familiar compound words are always hyphenated.

no-hitter   helter-skelter

Other compounds are always written as one word.

fireplace   peacetime   sunset

Finally, some compounds are always written as two separate words.

medical doctor   labor relations   bunk bed

Your dictionary can tell you whether a particular compound requires a hyphen: *snow job*, for instance, is two unhyphenated words; *snowsuit* is one word; and *snow-white* is hyphenated. Usage changes, however, so it is important to have an up-to-date dictionary.

Although hyphenization of compound words is not uniform, a few reliable rules do apply.

(1) In compound adjectives

A **compound adjective** is two or more words combined into a single grammatical unit that modifies a noun. When a compound adjective *precedes* the noun it modifies, its elements are joined by hyphens.

He stayed tuned to his favorite listener-supported radio station, waiting for a hard-hitting editorial.

The research team tried to use nineteenth-century technology to design a space-age project.

However, when a compound adjective *follows* the noun it modifies, it does not include a hyphen.

The three government-operated programs were run smoothly, but the one that was not government operated was short of funds.

Use **suspended hyphens**—hyphens followed by space or by the appropriate punctuation and space—in a series of compounds that have the same principal elements.

The three-, four-, and five-year-old children take daily naps.

CLOSE-UP

USING HYPHENS

Compound adjectives that contain words ending in *-ly* are not hyphenated, even when they precede the noun.

Many <u>upwardly mobile</u> families consider items like home computers, microwave ovens, and videocassette recorders to be necessities.

(2) With certain prefixes or suffixes

Use a hyphen between a prefix and a proper noun or an adjective formed from a proper noun.

mid-July   pre-Columbian

Use a hyphen to connect the prefixes *all-, ex-, half-, quarter-, quasi-,* and *self-* and the suffixes *-elect* and *-odd* to a noun.

all-pro	quasi-serious
ex-senator	self-centered
half-pint	president-elect
quarter-moon	thirty-odd

NOTE: The words *selfhood, selfish,* and *selfless* do not include hyphens. In these cases *self* is the root, not a prefix.

(3) For clarity

Hyphenate to prevent misreading one word for another.

In order to <u>reform</u> criminals, we must <u>re-form</u> our ideas about prisons.

Hyphenate to avoid combinations that are hard to read such as two *i*'s (*semi-illiterate*) or more than two of the same consonant (*shell-less*).

Hyphenate in most cases between a capital initial and a word when the two combine to form a compound.

A-frame   T-shirt

But check your dictionary; some letter- or numeral-plus-word compounds do not require hyphens.

B flat   F major

(4) In compound numerals and fractions

Hyphenate compounds that represent numbers below one hundred, even if they are part of a large number.

the *twenty-first* century   three hundred *sixty-five* days

Compounds that represent numbers over ninety-nine *(two thousand, thirty million, two hundred fifty)* are not hyphenated. Therefore, in the expression *three hundred sixty-five days,* the compound *sixty-five* is hyphenated because it represents a number below one hundred, but no hyphens connect *three* to *hundred.*

Hyphenate the written form of a fraction when it modifies a noun.

a two-thirds share of the business

a three-fourths majority

Hyphens are not required in other cases, but most writers do use them.

PREFERRED: seven-eighths of the circle

ACCEPTABLE: seven eighths of the circle

(5) In newly created compounds

A coined compound, one that uses a new combination of words as a unit, requires hyphens.

He looked up with a who-do-you-think-you-are expression on his face.

EXERCISE 2

Form compound adjectives from the following word groups, inserting hyphens where necessary.

EXAMPLE: A contract for three years

a three-year contract

1. a relative who has long been lost
2. someone who is addicted to video games
3. a salesperson who goes from door to door
4. a display calculated to catch the eye
5. friends who are dearly beloved
6. a household that is centered on a child
7. a line of reasoning that is hard to follow
8. the border between New York and New Jersey
9. a candidate who is thirty-two years old
10. a computer that is friendly to its users

1. long-lost relative
2. video-game addict
3. door-to-door salesman
4. eye-catching display
5. dearly beloved friends
6. child-centered household
7. hard-to-follow line of reasoning
8. New York-New Jersey border
9. thirty-two-year-old candidate
10. user-friendly computer

ANSWERS TO EXERCISE 3

1. One of the restaurant's blue-plate specials is chicken-fried steak.
2. Virginia and Texas are both right-to-work states.
3. He stood on tiptoe to see the near-perfect statue, which was well hidden by the security fence.
4. The five- and ten-cent store had a self-service make-up counter and many up-to-the-minute gadgets.
5. The so-called Saturday night special is opposed by pro-gun-control groups.
6. He ordered two all-beef patties with special sauce, lettuce, cheese, pickles, and onions on a sesame seed bun.
7. The material was extremely thought provoking, but it hardly presented any earth-shattering conclusions.
8. The Dodgers-Phillies game was rained out, so the long-suffering fans left for home.
9. Bone-marrow transplants carry the risk of what is known as a graft-versus-host reaction.
10. The state-funded child care program was considered a highly desirable alternative to family day care.

EXERCISE 3

Add hyphens to the compounds in these sentences wherever they are required. Consult a dictionary if necessary.

EXAMPLE: Alaska was the forty ninth state to join the United States.

Alaska was the forty-ninth state to join the United States.

1. One of the restaurant's blue plate specials is chicken fried steak.
2. Virginia and Texas are both right to work states.
3. He stood on tiptoe to see the near perfect statue, which was well hidden by the security fence.
4. The five and ten cent store had a self service makeup counter and stocked many up to the minute gadgets.
5. The so called Saturday night special is opposed by pro gun control groups.
6. He ordered two all beef patties with special sauce, lettuce, cheese, pickles, and onions on a sesame seed bun.
7. The material was extremely thought provoking, but it hardly presented any earth shattering conclusions.
8. The Dodgers Phillies game was rained out, so the long suffering fans left for home.
9. Bone marrow transplants carry the risk of what is known as a graft versus host reaction.
10. The state funded child care program was considered a highly desirable alternative to family day care.

CHAPTER 36

Abbreviations

Abbreviations save time and space. They also communicate meaning quickly and efficiently—but only when they are familiar to your readers.

Many abbreviations are acceptable only in informal writing and are not appropriate in college writing. Others are acceptable in scientific, technical, or business writing, or only in a particular discipline. If you are unsure whether to use a particular abbreviation, check a style manual in your field.

▶ See 40f

36a Abbreviating Titles

Titles before and after proper names are usually abbreviated.

Mr. John Singleton   Dr. Mathilde Krim
Henry Kissinger, Ph.D.   St. Jude

But military, religious, academic, and government titles are not abbreviated.

General George Patton   Professor Kenneth G. Schaefer
the Reverend William Gray   Senator Ben Nighthorse
                                                   Campbell

36b Abbreviating Organization Names and Technical Terms

Certain abbreviations are used in speech and in college writing to designate groups, institutions, people, substances, and so on.

553

See ◀
27a2

QUOTATION

After discovering that the plaque left on the moon by Neil Armstrong contains the error "July 1969, A.D.," William Safire commented:

"[I hope] some descendant of mine will take a sharp stylus on some weekend rocket to the moon and, while awaiting a transfer rocket to Mars, will draw a little arrow placing it [A.D.] in front of the word *July*. This will show that human beings in the early days of space were grammatically fallible; that mankind . . . is forever editing, and that a little precision is a dangerous thing."

For example, well-known businesses and government, social, and civic organizations are commonly referred to by initials. These abbreviations fall into two categories: abbreviations formed from capitalized initials (CIA) and those that are acronyms (CORE).

Accepted abbreviations for terms that are not well known may also be used, but only if you spell out the full term the first time you mention it, followed by the abbreviation in parentheses.

Citrus farmers have been injecting ethylene dibromide (EDB), a chemical pesticide, into the soil for more than twenty years. Now, however, EDB has seeped into wells and contaminated water supplies, and it is a suspected carcinogen.

CLOSE-UP

USING ABBREVIATIONS

The extent to which abbreviations are used varies from discipline to discipline. Regardless of the discipline, however, excessive use of abbreviations can be confusing to your readers, so use them sparingly.

36c Abbreviating Designations of Dates, Times of Day, Temperatures, and Numbers

50 B.C. (B.C. follows the date)	A.D. 432 (A.D. precedes the date)
525 B.C.E.	730 C.E.
6 A.M.	3:03 P.M.
20° C (Centigrade or Celsius)	180° F (Fahrenheit)

Always capitalize B.C. and A.D. The alternatives B.C.E., for "before common era," and C.E., for "common era," are also capitalized. You may, however, use either uppercase or lowercase letters for a.m. and p.m. (A.M., a.m., P.M., p.m.). Printers conventionally set A.M., P.M., B.C., and A.D. in small capital letters (A.M., P.M., B.C., A.D.). These abbreviations are used only when they are accompanied by numbers.

FAULTY: We will see you in the A.M.

REVISED: We will see you in the morning.

REVISED: We will see you at 8 A.M.

Avoid the abbreviation *no.*, except in technical writing, and then only before a specific number. This abbreviation may be written either *no.* or *No.*

FAULTY: The no. on the label of the unidentified substance was 52.

REVISED: The unidentified substance was labeled no. 52.

In nontechnical writing *no.* is acceptable only in certain documentation formats.

CORRECT: *American Anthropologist* 4, no. 2 (1973): 65–84.

 36d **Editing Misused or Overused Abbreviations**

In college writing abbreviations are not used in the following cases.

(1) Certain familiar Latin expressions

Abbreviations of the common Latin phrases *i.e.* ("that is"), *e.g.* ("for example"), and *etc.* ("and so forth") are sometimes appropriate for informal writing, and they may occasionally be acceptable in a parenthetical note. In most college writing, however, an equivalent phrase should be written out in full.

INFORMAL: Poe wrote "The Gold Bug," "The Tell-Tale Heart," etc.

PREFERABLE: Poe wrote "The Gold Bug, " "The Tell-Tale Heart," and other stories.

INFORMAL: Other musicians (e.g., Bruce Springsteen) have been influenced by Dylan.

PREFERABLE: Other musicians (for example, Bruce Springsteen) have been influenced by Dylan.

The Latin abbreviations *et al.* ("and others") and *cf.* ("compare") are used only in footnotes and bibliographic entries.

Davidson, Harley, et al. *You and Your Motorcycle*. New York: Ten Speed, 1968.

(2) The names of days, months, or holidays

FAULTY: Sat., Aug. 9, was the hottest day of the year.

REVISED: Saturday, August 9, was the hottest day of the year.

FAULTY: Only twenty-three shopping days remain until Xmas.

REVISED: Only twenty-three shopping days remain until Christmas.

(3) Units of measurement

In informal and technical writing, some units of measurement are abbreviated when preceded by a numeral.

The hurricane had winds of 35 m.p.h.

The new Honda gets over 50 m.p.g.

In general, however, write out such expressions, and spell out words such as *inches, feet, years, miles, pints, quarts,* and *gallons*. NOTE: Abbreviations for units of measurement are not used in the absence of a numeral.

> **FAULTY:** The laboratory equipment included pt. and qt. measures and a beaker that could hold a gal. of liquid.

> **REVISED:** The laboratory equipment included pint and quart measures and a beaker that could hold a gallon of liquid.

(4) Names of places, streets, and the like

Abbreviations of names of streets, cities, states, countries, and geographical regions are common in informal writing and in correspondence. For college assignments these words should be spelled out.

FAULTY:	REVISED:
B'way	Broadway
Riverside Dr.	Riverside Drive
Phila.	Philadelphia
Calif. or CA	California
Catskill Mts.	Catskill Mountains

EXCEPTIONS: The abbreviation *U.S.* is often acceptable (U.S. Coast Guard), as is *D.C.* in *Washington, D.C.* It is also permissible to use the abbreviation *Mt.* before the name of a mountain *(Mt. Etna)* and *St.* in a place name *(St. Albans)*.

(5) Names of academic subjects

Names of academic subjects are not abbreviated.

> **FAULTY:** Psych., soc., and English lit. are all required courses.

> **REVISED:** Psychology, sociology, and English literature are all required courses.

(6) Parts of books

Abbreviations that designate parts of written works *(Pt. II, Ch. 3, Vol. IV)* should not be used within the body of a paper except in parenthetical documentation.

(7) People's names

FAULTY: Mr. Harris's five children were named Robt., Eliz., Jas., Chas., and Wm.

REVISED: Mr. Harris's five children were named Robert, Elizabeth, James, Charles, and William.

(8) Company names

The abbreviations *Inc., Bros., Co.,* or *Corp.* and the **ampersand** *(&)* are not used unless they are part of a firm's official name.

Company names are written exactly as the firms themselves write them.

Western Union Telegraph Company	AT&T
Santini Bros.	Charles Schwab & Co., Inc.

CLOSE-UP

USING ABBREVIATIONS

Shortened forms of publishers' names—Harcourt, not Harcourt Brace College Publishers—are preferred in MLA bibliographic citations.

▶ See 40b2

Abbreviations for *company, corporation,* and the like are not used in the absence of a company name.

FAULTY: The corp. merged with a small co. in Pittsburgh.

REVISED: The corporation merged with a small company in Pittsburgh.

The ampersand (&) is used in college writing only in the name of a company that requires it or in citations that follow APA documentation style.

▶ See 44c

1. The committee meeting, attended by representatives from Action for Children's Television (ACT) and the National Organization for Women (NOW), Senator Putnam, and the president of ABC, convened at 8 AM on Monday, February 24, at the YWCA on Germantown Avenue.

2. An economics professor was suspended after he encouraged his students to speculate on securities issued by corporations under investigation by the Securities and Exchange Commission (SEC).

3. Benjamin Spock, the M.D. who wrote *Baby and Child Care,* is a respected doctor known throughout the USA.

4. [If this sentence can be defined as "technical writing" (see 35b).]

5. The Reverend Dr. Martin Luther King, Jr., leader of the Southern Christian Leadership Conference (SCLC), led the famous Selma, Alabama, march.

6. William Golding, a novelist from the United Kingdom, won the Noble Prize in literature.

7. The adult education center, financed by a major computer corporation, offers courses in basic subjects like introductory biology and technical writing as well as teaching programming languages such as PASCAL.

8. All the brothers in the fraternity agreed to write to President Dexter appealing their disciplinary probation under Chapter 4,

continued on following page

continued from previous page

Section 3, of the Inter-Fraternity Council constitution.

9. A four-quart (that is, one-gallon) container is needed to hold the salt solution.

10. According to Professor Morrison, all those taking the MCATs should bring two sharpened no. 2 pencils to the St. Joseph's University auditorium on Saturday.

(9) Symbols

The symbols %, =, +, #, and ¢ are acceptable in technical and scientific writing but not in nontechnical college writing.

FAULTY: The cost of admission has increased 50%.

REVISED: The cost of admission has increased fifty percent.

The symbol $ is acceptable before specific numbers, but not as a substitute for the words *money* or *dollars.*

CORRECT: The first of her books of poetry cost $4.25 per copy.

FAULTY: The value of the $ has declined steadily in the last two decades.

REVISED: The value of the dollar has declined steadily in the last two decades.

EXERCISE

Correct any incorrectly used abbreviations in the following sentences, assuming that all are intended for an academic audience. If a sentence is correct, mark it with a *C.*

EXAMPLE: *Romeo & Juliet* is a play by Wm. Shakespeare.

Romeo and Juliet is a play by William Shakespeare.

1. The committee meeting, attended by representatives from Action for Children's Television (ACT) and NOW, Sen. Putnam, & the pres. of ABC, convened at 8 A.M. on Mon. Feb. 24 at the YWCA on Germantown Ave.
2. An econ. prof. was suspended after he encouraged his students to speculate on securities issued by corps. under investigation by the SEC.
3. Benjamin Spock, the M.D. who wrote *Baby and Child Care,* is a respected dr. known throughout the USA.
4. The FDA banned the use of Red Dye no. 2 in food in 1976, but other food additives are still in use.
5. The Rev. Dr. Martin Luther King, Jr., leader of the S.C.L.C., led the famous Selma, Ala., march.
6. Wm. Golding, a novelist from the U.K., won the Nobel Prize in lit.
7. The adult education center, financed by a major computer corp., offers courses in basic subjects like introductory bio. and tech. writing as well as teaching programming languages, such as PASCAL.
8. All the bros. in the fraternity agreed to write to Pres. Dexter appealing their disciplinary probation under Ch. 4, Sec. 3, of the IFC constitution.
9. A 4 qt. (i.e., 1 gal.) container is needed to hold the salt solution.
10. According to Prof. Morrison, all those taking the MCATs should bring two sharpened no. 2 pencils to the St. Joseph's University auditorium on Sat.

CHAPTER 37

Numbers

Convention determines when to use a **numeral** (22) and when to spell out a number (twenty-two). Numerals are generally more common in scientific and technical writing, in journalism, and in informal writing. Numbers are more often spelled out in academic or literary writing although there are some exceptions.

CLOSE-UP

USING NUMBERS

In some disciplines, such as the social sciences and engineering, the rule is to spell out all numbers less than ten and to use numerals for ten and above. The conventions that follow apply to the humanities. For information on conventions that apply to other disciplines, consult a specific style manual.

37a   Spelling Out Numbers That Begin Sentences

Never begin a sentence with a numeral. If a number begins a sentence, spell the number out.

FAULTY: 250 students are currently enrolled in English composition courses.

REVISED: Two hundred fifty students are currently enrolled in English composition courses.

TEACHING STRATEGY

Numbers
Ask students to look through a newspaper, textbook, or journal and examine its use of numerals. From the examples they find, ask the students to infer principles of usage. How, if at all, do these principles differ from those in the text?

Bring a collection of stylebooks to class (the *Associated Press Stylebook*, *The New York Times Manual of Style and Usage*, and others) and discuss any differences in rules concerning numerals. Remind students that academic and research institutions, corporations, news organizations, and other groups frequently have very specific rules that may not match those discussed in the text.

Read aloud the following passage from Stephen Jay Gould's book *Wonderful Life,* pausing at each use of numerals or spelled-out forms. At each pause, ask students to tell you which of the text's rules governs the writer's choice.

In 1986, a year after his monograph on *Wiwaxia,* Simon Conway Morris published a "blockbuster" of another type—a comprehensive ecological analysis of the entire Burgess community. He began with some interesting facts and figures. About 73,300 specimens on 33,520 slabs have been collected from Burgess Shale. Ninety percent of this material resides in Washington, in Walcott's collection; 87.9 percent of these specimens are animals, and nearly all the rest are algae. Fourteen percent of the animals have shelly skeletons; the remainder are soft-bodied.

The fauna contains 119 genera in 140 species; 37 percent of these genera are arthropods. . . .

. . . He recognized four major groups: (1) Deposit-feeding collectors (mostly arthropods)—60 percent of the total number of individuals; 25–30 percent of the genera. (This category includes *Marrella* and *Canadaspis,* the two most common Burgess animals, hence the high representation for individuals.) (2) Deposit-feeding swallowers (mostly ordinary mollusks with hard parts)—1 percent of individuals; 5 percent of genera. (3) Suspension feeders (mostly sponges, taking food directly from the water column)—30 percent of individuals; 45 percent of genera. (4) Carnivores and scavengers (mostly arthropods)—10 percent of individuals; 20 percent of genera.

You may also reword the sentence.

Current enrollment in English composition courses is 250 students.

37b Spelling Out Numbers Expressed in One or Two Words

Unless a number falls into one of the categories listed in 37d, spell it out if you can do so in one or two words.

The Hawaiian alphabet has only twelve letters.

Class size stabilized at twenty-eight students.

Approximate numbers can often be expressed in one or two words.

Guards turned away more than ten thousand disappointed fans.

The subsidies are expected to total about two million dollars.

CLOSE-UP

USING NUMBERS

Use a hyphen when you spell out compound numbers from twenty-one to ninety-nine.

37c Using Numerals for Numbers Expressed in More Than Two Words

Numbers more than two words long are expressed in figures.

The pollster interviewed 3,250 voters before the election.

The dietitian prepared 125 sample menus.

When Levittown, Pennsylvania, was built in the early 1950s, the builder's purchases included 300,000 doorknobs, 153,000 faucets, 53,600 ice cube trays, and 4,000 manhole covers.

NOTE: Numerals and spelled-out forms denoting the same kind of item should not be mixed in the same passage. For consistency, then, the number 4,000 in the last sentence above is expressed in figures even though it could be written in only two words.

<table>
<tr><td>**37d**</td><td>## Using Numerals Where Convention Requires Their Use</td></tr>
</table>

(1) Addresses

1600 Pennsylvania Avenue

10 Downing Street

111 Fifth Avenue, New York, New York 10003

(2) Dates

January 15, 1929	62 B.C.
November 22, 1963	1914–1919

(3) Exact times

9:16   10 A.M. (or 10:00 A.M.)   6:50

EXCEPTIONS: Spell out times of day when they are used with *o'clock*: *eleven o'clock,* not *11 o'clock.* Also spell out times expressed as round numbers: *They were in bed by ten.*

(4) Exact sums of money

$25.11

$6752.00 (or $6,752.00)

$25.5 million (or $25,500,000)

EXCEPTION: You may write out a round sum of money if the number can be expressed in one or two words.

five dollars	two thousand dollars
fifty-three cents	six hundred dollars

(5) Pages and divisions of written works

Numerals are used for chapter and volume numbers; acts, scenes, and lines of plays; chapters and verses of the Bible; and line numbers of long poems.

The "Out, out brief candle" speech appears in Act 5, scene 5, of *Macbeth* (lines 17–28); in Kittredge's *Complete Works of Shakespeare* it appears on page 1142.

(6) Measurements

When a measurement is expressed by a number accompanied by a symbol or an abbreviation, use figures.

55 mph   12″
32°   15 cc

(7) Numbers containing percentages, decimals, or fractions

80% (or 80 percent)   6 3/4
98.6   3.14

(8) Ratios, scores, and statistics

Children preferred Crispy Crunchies over Total Bran by a ratio of 20 to 1.

The Knicks defeated the Bullets 92 to 84.

The median age of the voters was 42; the mean age was 40.

(9) Identification numbers

Route 66   Track 8
Channel 12   Social Security number 146-07-3846

37e Using Numerals with Spelled-Out Numbers

Even when all the numbers in a particular passage are short enough to be spelled out, figures are sometimes used along with spelled-out numbers to distinguish one number from another.

CONFUSING: The team was divided into twenty two-person squads.

CLEAR: The team was divided into twenty 2-person squads.

EXERCISE

Revise the use of numbers in these sentences, being sure usage is correct and consistent. If a sentence uses numbers correctly, mark it with a *C*.

EXAMPLE: The Empire State Building is one hundred and two stories high.

The Empire State Building is 102 stories high.

Numerals with Spelled-Out Numbers

1. *1984,* a novel by George Orwell, is set in a totalitarian society.
2. The English placement examination included a 30-minute personal-experience essay, a 45-minute expository essay, and a 150-item objective test of grammar and usage.
3. In a control group of two hundred forty-seven patients, almost three out of four suffered serious adverse reactions to the new drug.
4. Before the Thirteenth Amendment to the Constitution, slaves were counted as 3/5 of a person.
5. The intensive membership drive netted 2,608 new members and additional dues of over 5 thousand dollars.
6. They had only 2 choices: Either they could take the yacht at Pier Fourteen, or they could return home to the penthouse at Twenty-seven Harbor View Drive.
7. The atomic number of lithium is three.
8. Approximately 3 hundred thousand schoolchildren in District 6 were given hearing and vision examinations between May third and June 26.
9. The United States was drawn into the war by the Japanese attack on Pearl Harbor on December seventh, 1941.
10. An upper-middle-class family can spend over 250,000 dollars to raise each child up to age 18.

ANSWERS TO EXERCISE

1. C (*1984* is a book title.)
2. C
3. In a control group of 247 patients, almost 3 out of 4 suffered serious adverse reactions to the new drug.
4. Before the Thirteenth Amendment to the Constitution, slaves were counted as three-fifths of a person.
5. The intensive membership drive netted 2,608 new members and additional dues of over five thousand dollars.
6. They had only two choices: Either they could take the yacht at Pier 14, or they could return home to the penthouse at 27 Harbor View Drive.
7. C
8. Approximately 300,000 schoolchildren in District 6 were given hearing and vision examinations between May 3 and June 26.
9. The United States was drawn into World War II by the Japanese attack on Pearl Harbor on December 7, 1941.
10. An upper-middle-class family can spend over 250,000 dollars to raise each child up to age eighteen.

Writing with Sources

Research for Writing

The four chapters included in Part 8, Writing with Sources, all deal with the research process. Each chapter is a mini resource manual that can be used alone or in combination with the other chapters. Chapter 38 focuses on gathering information for writing a research paper and discusses library and nonlibrary research. Chapter 39 deals with evaluating sources as well as with taking notes—summarizing, paraphrasing, and recording quotations—and avoiding plagiarism. Chapter 40, which concentrates on documentation, includes MLA, Chicago, APA, and CBE formats. Finally, Chapter 41 presents an overview of the process of writing a research paper—including a detailed case study of a student's research process.

Before you begin to cover the material in the text, ask students to choose a topic for research—something that genuinely interests them. Ask students to do a limited amount of general reading on the topic and to draw up an informal proposal that includes a tentative research question and also a "plan of action"—a search strategy following the guidelines on page 567. Because so many students feel overwhelmed by their first research projects, you may want to check these proposals in brief student conference sessions and to offer any advice you feel might help (additional resources, suggestions on direction, pacing, etc.)

CHAPTER 38

Research for Writing

Research is the systematic study and investigation of a topic outside your own experience and knowledge. When you do research you move from what you know about a topic to what you do not know. You may do **primary research**—working with sources such as letters or novels, historical documents, and interviews—and you may also do **secondary research**—examining other researchers' interpretations or analyses of sources. Many of your research papers will rely on both primary and secondary source material.

See ◀ 39a1

38a   Mapping Out a Preliminary Search Strategy

A **search strategy** is a systematic process of collecting and evaluating source material, moving in a logical, orderly progression from general sources to more specific ones. A search strategy reflects the way research works: As your research becomes more focused, so do your ideas. Like the writing process, though, the research process does not progress in a straight line. For example, even after you have read many articles about your topic, you might find you need to consult a general encyclopedia or an almanac to answer a specific question.

The following diagram presents an overview of a preliminary search strategy. Although you will not go through every stage each time you do research, you will follow the same general order, moving from exploratory to focused research. However you map out your plans, though, you should remain flexible and willing to modify your search strategy as your research progresses.

A PRELIMINARY SEARCH STRATEGY

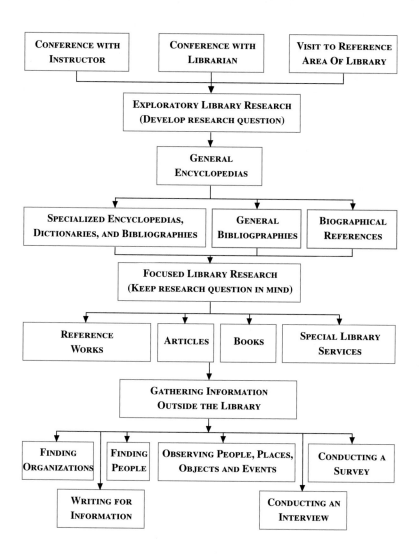

A preliminary search strategy

QUOTATIONS

"Perhaps there is no valid distinction between being clever and being well-informed. The two are so often confused as to be indistinguishable. In many situations information is so great a part of effectiveness that without information a really clever person cannot get started. With information a much less clever person can get very far."

–Edward De Bono

"It is important that students bring a certain ragamuffin barefoot irreverence to their studies; they are not here to worship what is known, but to question it."

–Jacob Bronowski
The Ascent of Man

TEACHING STRATEGY

Library Tour
Arrange a tour of the library and its resources, preferably with the assistance of a librarian familiar with the library's reference materials and any special collections or services. To gauge student understanding of how to use library resources, have the class complete the activity suggested in the marginal notes on the next page.

As a follow-up to the library tour suggested in "Teaching Strategies" on page 567, divide the class into pairs and assign topics for a quick "practicum" to measure students' understanding of how to use the library. (A list of possible topics appears below.) Ask student partners to draw up an informal bibliography that includes one book, one article, etc., found by using the following research tools: 1) a catalog; 2) a general periodical; 3) a specialized periodical index; 4) a general encyclopedia; 5) a suject-specific encyclopedia; 6) *New York Times Index;* and 7) two other resources of the students' choice.

Possible Topics:

Abortion

Alcoholism

College Athletics

Jazz Musicians

The Single Parent

Medical Ethics

Presidential Elections

Capital Punishment

Computer Fraud

Postmodern Architecture

Religion in India

Affirmative Action

Labor Unions

Japanese Art

Psychology of Adolescents

Or, give each small group a section of the library to investigate and report on for the student newspaper or for a collaborative report. One group might have the reference room, another the catalog, etc. Groups could

continued on following page

38b Doing Library Research

> **STRATEGIES FOR DOING LIBRARY RESEARCH**
> **Before you start:**
> - Know the library's physical layout.
> - Take a tour of the library if one is offered.
> - Familiarize yourself with the library's holdings.
> - Find out if your college library has a printed guide to its resources.
> - Meet with a reference librarian or your instructor if necessary.
> - Be aware of the library's hours.
>
> **As you do research:**
> - Be sure to copy down or print out the complete publication information—author, title, volume number, date of publication, and page numbers—that you will need to locate a particular source.
> - Remember how to distinguish between a book and a periodical citation: A periodical citation includes an article title set in quotation marks.

When you use the library as a research tool, you begin by doing **exploratory research**, looking at general reference works that give you a broad overview of your topic and its possibilities. Once you have finished you exploratory research, you do **focused research**—returning to the library with more narrowly focused objectives and making use of specialized reference sources, articles, books, and special library services.

(1) Exploratory research

See ◀ 41b

During exploratory research your goal is to find a **research question** for your paper. At this early stage you want to consult works that give useful overviews and that are not too specialized or technical. The exploration you do now can help you familiarize yourself with key terms, people, and events relevant to your topic. The following reference works are most useful for exploratory research.

General Encyclopedias   General multivolume encyclopedias, such as *Encyclopedia Americana* (30 vols.), *Collier's Encyclopedia* (24 vols.), and *The New Encyclopaedia Britannica* (30 vols.) contain information about many different subjects. Therefore, they are a good place to begin your exploratory research.

To get a quick overview of your topic, use a one-volume general encyclopedia. Two of the best are *The New Columbia Encyclopedia* and *The Random House Encyclopedia*, both of which contain many short entries listed in alphabetical order by subject. However, because the cross-references and in-depth bibliographical information provided by the multivolume encyclopedias do not appear in these short volumes, you may find that you have to consult a multivolume encyclopedia.

continued from previous page

compose a section—even do research in books and periodicals if necessary—and copy it for the whole class to edit.

CLOSE-UP

EXPLORATORY RESEARCH

Although a general encyclopedia often contains detailed discussions of selected subjects, it is *not* a substitute for a more specialized research source. Even when written by experts in a particular field, encyclopedia articles are aimed at general readers. Moreover, articles are often written a year or two before the publication date of the encyclopedia and, therefore, do not contain the most up-to-date information.

Encyclopedias, Dictionaries, and Bibliographies in Special Subjects   These specialized reference works contain in-depth articles focusing on a single field or subject. Articles are frequently more detailed than those in general encyclopedias and sometimes include annotated bibliographies and cross-references as well. Specialized reference works are listed in Eugene P. Sheehy's *Guide to Reference Books,* available at the reference desk in most libraries. For information on specialized reference works used in specific disciplines, **see Part 9**.

General Bibliographies   General bibliographies list books available in a wide variety of fields.

Books in Print. A helpful index of authors and titles of every book in print in the United States. *The Subject Guide to Books in Print* indexes books according to subject area. *Paperbound Books in Print* is an index to all currently available paperbacks.

The Bibliographic Index. A tool for locating bibliographies. This index is particularly useful for researching a subject that is not well covered in other indexes.

Biographical References Biographical reference books provide valuable information about people's lives and times as well as bibliographic listings.

Living Persons

Who's Who in America. Gives very brief biographical data and addresses of prominent living Americans.
Who's Who. Collects concise biographical facts about notable living British men and women.
Current Biography. Includes informal articles on living people of many nationalities.

Deceased Persons

Dictionary of American Biography. Considered the best of American biographical dictionaries. Includes articles on over thirteen thousand deceased Americans who have made contributions in all fields.
Dictionary of National Biography. The most important reference work for British biography.
Webster's Biographical Dictionary. Perhaps the most widely used biographical reference work. Includes people from all periods and places.
Who Was When? A Dictionary of Contemporaries. A reference source for historical biography; covers 500 B.C. through the early 1970s.
Who Was Who in America. Consists of nine volumes covering 1607 to 1985, a historical volume covering 1607 to 1896, and an index volume.

(2) Focused research: Locating reference works

See ◄
41e When you do **focused research**, you keep a specific research question in mind as you look for information. Focusing on your research question, you fill in specific details—facts, examples, statistics, definitions, quotations—to support your ideas.

The following reference works are most useful for focused research.

Unabridged Dictionaries   **Unabridged dictionaries**, such as ▶ See
the *Oxford English Dictionary*, are comprehensive works that give   19d
detailed information about words.

Special Dictionaries   These dictionaries focus on topics such
as usage, synonyms, slang and idioms, etymologies, and foreign
terms. For lists of selected specialized dictionaries arranged according to discipline, **see Part 9**.

Yearbooks and Almanacs   A **yearbook** is an annual publication that updates factual and statistical information already published in a reference source. An **almanac** provides lists, charts,
and statistics about a wide variety of subjects.

World Almanac. Includes statistics about government, population,
   sports, and many other subjects. Includes a chronology of
   events of the previous year. Published annually since 1868.
Information Please Almanac. Could be used to supplement the
   World Almanac (each work includes information unavailable in
   the other). Published annually since 1947.
Facts on File. A world news digest with index. Covering 1940 to
   the present, this work offers digests of important news stories
   from metropolitan newspapers.
Editorials on File. Reprints important editorials from American
   and Canadian newspapers. Editorials represent both sides of
   controversial issues and are preceded by a summary of the
   principles involved.
Statistical Abstract of the United States. Summarizes the innumerable statistics gathered by the U.S. government.

Atlases   An **atlas** contains maps and charts and often a wealth
of supplementary historical, cultural, political, and economic information.

National Geographic Society. *National Geographic Atlas of the
   World.* The most up-to-date atlas available.
Rand McNally Cosmopolitan World Atlas. A modern and extremely
   legible medium-sized atlas.
Times, London. *The Times Atlas of the World*, five volumes, John
   Bartholomew, ed. Considered one of the best large world at-
   lases.
Allen, James Paul and Eugene James Turner. *We the People: An
   Atlas of America's Ethnic Diversity.* Presents information about

Americans of every ethnic background, immigration and relocation history, economic status, and employment patterns. Maps show immigration routes and settlement patterns.

Shepherd, William Robert. *Historical Atlas,* 9th ed. Covers period from 2000 B.C. to 1955. Excellent maps showing war campaigns and development of commerce.

Quotation Books A **quotation book** contains numerous quotations on a wide variety of subjects, often by well-known persons. Such quotations can be especially useful for your paper's introductory and concluding paragraphs.

Bartlett's Familiar Quotations. Quotations are arranged chronologically by author.

The Home Book of Quotations. Quotations are arranged by subject. An author index and a key work index are also included.

(3)   Focused research: Locating articles

Periodicals are magazines, newspapers, or scholarly journals that are published regularly throughout the year. A **magazine** is a serial publication—such as *Time, MacUser,* or *National Geographic*—aimed at general readers. A **journal** is a serial publication—such as *Nature, American Anthropologist,* or *College English*—written for specialists in a particular field. **Periodical indexes**, which may be in electronic form or in bound volumes, list articles from a selected group of magazines, newspapers, or scholarly journals. In some bound index volumes, entries are arranged according to subject; in others, entries are arranged according to author. The key to the abbreviations at the front of the volume enables you to use the index and gives all the information you need to find the articles you select. The figure below illustrates the information usually contained in a citation from a bound periodical index.

Citation from periodical index

LOCATING ARTICLES

Bibliographic information about periodical articles (and in some cases, the articles themselves,) may be available in print, on film, or in a **database**—a collection of related information that you access with a computer. Often, information about a single periodical article will be available in more than one of these formats. Information can, therefore, be retrieved in a number of different ways, some of which are considerably faster and more flexible than others. Consult a librarian to determine the available format for the indexes you need.

- **Print Indexes:** Bibliographic information about periodical articles may be collected in print, with listings collected annually in bound volumes. The *Readers' Guide to Periodical Literature*, for example, is available in bound volumes as well as on-line.

- **Microfilm:** Extremely small images of the pages of a periodical may be stored on microfilm. You need a microfilm scanner to read or photocopy the pages.

- **Microfiche:** Microfiche is similar to microfilm, but images are on a 5 × 7 sheet of film and scanned with a microfiche reader.

- **Computerized Databases:** Bibliographic information may be obtained from commercial information services— such as DIALOG and WILSONLINE—that contain databases listing thousands of sources. You access these databases with a computer and a modem. Charges can be minimal or quite high, depending on the database you use. Be sure to ask a librarian to estimate the cost of the search before you begin.

- **CD-ROM (Compact Disk—Read Only Memory):** Another electronic format uses small CDs similar to the ones on which music is recorded. One CD can contain an entire database such as the *Readers' Guide* or *Magazine Index*. Some college libraries subscribe to CD-ROM services— such as INFOTRAC—and make them available to students at no charge. You access CD-ROMs with a computer equipped with a CD-ROM reader.

Databases

If you have access to database-search facilities, ask a librarian (or someone else with hands-on experience) to show your students what a database has to offer. You may want to mention the many specialized databases: AgriStar Network (agriculture), Lexis (law), InnerLine (banking and finance), Videolog (engineering), Artsnet (performing arts), and Educom's BITNET (an electronic link with the computer facilities of more than 400 colleges and universities). Emphasize that database searches can be expensive and tricky; the techniques are worth learning, but students should not (for a while, at least) try to do all their research on-line.

There are three levels of periodical indexes: *general indexes*, *specialized indexes*, and *abstracting services*.

General Indexes   General indexes lead you to articles in newspapers and popular magazines.

Readers' Guide to Periodical Literature. This index lists articles that appear in more than 150 magazines for general readers. Articles in the *Readers' Guide* are listed and cross-referenced under subject headings. You have probably used this index before, and the fact that you are familiar with it may encourage you to consult it first. However, you should be aware of its primary limitation: The *Readers' Guide* indexes only popular periodicals, whose articles may oversimplify complex issues. Because many of the periodicals indexed here are not suitable for college research, you must supplement information derived from such sources with material from more scholarly sources.

Magazine Index. Now widely used in college libraries, this database, available on-line and on microfilm, indexes popular periodicals. With *Magazine Index* you can search more than four years of periodicals at a time.

INFOTRAC. Widely used in college libraries, INFOTRAC is available on CD-ROM or from an information service. It contains many databases, listing entries for over a thousand general-interest periodicals as well as entries for business, technical, and government periodicals.

```
Heading: COLLEGE SPORTS
        -Analysis
    2. Covering the bases, (female athletes in college)
       (panel discussion) by Katherine Ingram j1  v13 Women's Sports
       and Fitness Sept '91 p62 (2)
         6103445
         ABSTRACT/HEADINGS/HOLDINGS
    1. Women's college programs are enjoying boom times.
       (28 new soccer programs for women in Division I
       colleges) by Alex Yannis 14 col  in. v142 The New York Times
       Sept 15 '93 pB21 (L) col. 4
         HEADINGS/HOLDINGS
```

INFOTRAC *printout*

New York Times Index. The *New York Times Index* lists major articles and features of the *Times* since 1851 by year. To find articles on a subject, locate the volume for the appropriate year. Articles are listed alphabetically by subject with short summaries. Supplements are published every two weeks. If your library subscribes to the *New York Times* microfilm service, you have access to every issue of the *Times* back to 1851.

Specialized Indexes Specialized indexes, such as the *Humanities Index*, lead you to articles in professional journals. The most commonly used index scholarly periodicals are largely American and fairly easy to obtain. Many of the articles listed in such indexes assume expert knowledge, but some are accessible to general readers. For information on specialized indexes used in specific disciplines, **see Part 9**.

Abstracting Services Abstracting services have a wider scope than general or specialized indexes, with comprehensive, often international, listings of published articles in a discipline. In addition to providing citations for journal articles, abstracting services also include **abstracts**, brief summaries of the articles' major points. For information on abstracting services used in specific disciplines, **see Part 9**. (Even if your library does not receive printed copies of these abstracts, you can gain access to many of them if your library has electronic database searching.)

GUIDELINES FOR CARRYING OUT A DATABASE SEARCH

The fastest way to locate periodicals is to use a computer to do a database search. The following general guidelines for carrying out a database search will make retrieving information easier.

- *Choose Appropriate Databases* Not all databases will be equally useful. Some, for example, will be too advanced for beginning researchers. Consult with your librarian to decide which ones are appropriate for your research.

- *Narrow Your Topic to Key Words* In order to find the information in a database, you must narrow your topic to one or more **descriptors** or key words; the computer will call up articles that contain these words in their titles or abstracts. The more precise your descriptors are, the more specific (and useful) the information you get will be. For example, the descriptors *affirmative* or *action* would separately yield thousands of references. *Affirmative action*, however, yields about thirty—a manageable number. When you enter the descriptors, the computer will indicate the number of citations they have elicited. You can then decide whether you want to view or print the citations or whether you want to narrow or widen your search.

---

continued from the previous page

- *Select the Format for Citations* With many databases, you can view just the bibliographic citation, which gives author, article title, and journal title, or you can view additional material—bibliographic information, an abstract, publication information, reprint costs, and related key words, for example.

- *Make a Printout* You can print out all the citations the descriptors elicit or just the ones that seem most useful. Once you have printed out the citations, you can then locate the articles in your college library. (See Figure 2 for an example of a database printout.)

Volume   Pages   Date

Title

Author

Journal

Key Words

```
    1/5/1
    09316150    DIALOG File 647: MAGAZINE ASAP   *Use Format 9 for FULL TEXT*
    Why did we take so long? (college athletics regulations reformed) (column)
    Rawlings, Hunter R., III
    Sports Illustrated v74 p72 (1) Jan 21, 1991
    SOURCE FILE: MI File 47
    ISSN: 0038-822X
    ARTICLE TYPE: column
    AVAILABILITY: FULL TEXT Online LINE COUNT: 00070
    DESCRIPTORS: College sports—Rules and regulations;
        NCAA—Rules and regulations
    ?*t  s1/5/2-5

    1/5/2
    09302592    DIALOG file 647: MAGAZINE ASAP *Use Format 9 for FULL TEXT*
    The bottom line in college athletics. (recruiting reforms)
    U.S. News & World Report v110 p12 (1) Jan 21, 1991
    SOURCE FILE: MI File 47
    ISSN: 0041-5537
    illustration; photograph
    AVAILABILITY: FULL TEXT  Online   LINE COUNT: 00016
    DESCRIPTORS: College athletes—Recruiting;   NCAA—Rules and regulations;
        Scholarships—Rules and regulations

    1/5/3
    08960802    DIALOG File 647: MAGAZINE ASAP    *Use Format 9 for FULL TEXT*
    Win at Any Cost: the Sell Out of College Athletics. (book reviews)
    Stuttaford, Genevieve
    Publishers Weekly   v237 p51(2) Oct 12, 1990
    SOURCE FILE: MI File 47
    ISSN: 0000-0019
    ARTICLE TYPE:  review
    AVAILABILITY:  FULL TEXT Online   LINE COUNT:  00017
    GRADE: B
    REVIEWEE: Dealy, Francis X., Jr.
```

Database printout

(4) Focused research: Locating books

Some libraries list their holdings by author, title, and subject in print form on cards arranged in drawers of **card catalogs**. Other libraries have **on-line catalogs**, computerized systems that allow students to call up catalog entries by author, title, or subject on video terminals.

CLOSE-UP

ON-LINE CATALOGS

The on-line catalogs of various libraries differ, and so the techniques you use to retrieve information from one on-line catalog will not necessarily work with another. As with any technical task, however, the more you use the on-line catalog, the more adept you will get. In addition, most college libraries have printed instructions designed to help you use their on-line catalog.

To locate a book, use its **call number**, which refers you to the area of the library that houses books on your subject. (The same call number that appears in the catalog is written on the spine of the book.) When you find a book you need, look through the books shelved nearby. In itself, browsing is not an effective research technique, but as part of a focused search strategy, it can sometimes yield good results. Also keep in mind that the catalog itself contains cross-references that refer you to related subject headings. By using them, you can find additional books related to your subject.

If you cannot find a book, go to the circulation desk for help. The book may be out, on reserve, or shelved in a different section of the library.

(5) Focused research: Using special library services

Your most valuable resource is your librarian, a trained professional whose business it is to know how to locate information. Before you begin any complicated research project, you should ask your librarian about any of the following special services you plan to use.

Government Documents

Many students find the arcane rules of government documents research somewhat overwhelming. Yet government documents are a rich source of facts, figures, and quotations for essays on current political or social issues and many other topics. Consider arranging for small-group student tours of the nearest government documents room or library. In addition, you might do a government documents search for information on a topic of current interest and spend some class time showing the class how you did the research and what you found.

The answers to this exercise can be found in this chapter or in 43a, 44a, or 45a.

1. *Book Review Index*
2. *Dictionary of American Biography, Encyclopedia Americana, Webster's Biographical Dictionary*
3. catalog
4. *Encyclopaedia Britannica, Encyclopedic Dictionary of Physics, Encyclopedia of Physics*
5. catalog
6. If the book is in your library, the information will be in the Library of Congress catalog; if not, use *Bibliographic Index*.
7. catalog, *Subject Guide to Books in Print, Bibliographical Index*

continued on following page

IF YOUR SOURCE IS NOT AVAILABLE

Problem	Possible Solution
Book checked out of library	Consult person at circulation desk.
Book not in library's collection	Call other nearby libraries. Ask instructor if he or she owns a copy. Arrange for interlibrary loan (if time permits).
Journal not in library's collection/article ripped out of journal	Arrange for interlibrary copy (if time permits). Check to see whether article is available in a database. Ask librarian whether article has been reprinted as part of a collection.

Interlibrary Loans   If you need a book or periodical the library does not own, you can ask your librarian to arrange an **interlibrary loan**. Libraries can arrange to borrow books or acquire copies of articles from other college libraries or from the holdings of municipal and state libraries. However, because interlibrary loans may take several weeks, you may not be able to take advantage of this service unless you initiate the loan early in your research.

Special Collections   Your library may include **special collections** of books, manuscripts, or documents. In addition, churches, ethnic societies, historical trusts, and museums sometimes have materials you cannot find anywhere else, and your librarian may be able to get you permission to use these.

Government Documents   Federal, state, and local governments publish a variety of print materials, ranging from consumer information to detailed statistical reports. Although some **government documents** may be too technical—or too simplistic—for your research needs, others can be valuable. Government documents are particularly useful sources of statistics.

Some government documents, such as *Statistical Abstract of the United States,* may be housed in your library's reference room. A large university library may have a separate government documents area with its own catalog or index. The *Monthly Catalog of U.S. Government Publications,* an index of many, but not all, publications of the federal government, may be located either there or among the indexes in the reference room.

Vertical File   The **vertical file** is the place where libraries keep miscellaneous source materials, which may or may not be suitable for research. This file contains pamphlets from a variety of organizations and interest groups, some requested by your library and others unsolicited. The file may also include newspaper clippings and other material collected by your librarians because of its relevance to the research interests of your college's population.

*EXERCISE 1

What library research sources would you consult to find the following information?

1. A book review of Maxine Hong Kinston's *China Men* (1980)
2. Biographical information about the American anthropologist Margaret Mead
3. Books about Margaret Mead and her work
4. Information about the theories of Albert Einstein
5. Whether your college library has *The Human Use of Human Beings* by Norbert Wiener
6. How many pages there are in *On Death and Dying* by Elisabeth Kübler-Ross and how long the bibliography is
7. Where you could find other books on death and dying
8. Where you could find other books by Elisabeth Kübler-Ross
9. Articles about Walt Whitman's *Leaves of Grass*
10. Whether *Leaves of Grass* is presently available in a Norton Critical Edition

*EXERCISE 2

Use the resources of your library to help you answer the following questions. Cite the source or sources of your answers.

1. What government publication could give you information about how to solar heat your home?
2. What government agency could you contact to find out what is being done to help the aging receive proper nutrition?
3. At what address could you contact Bruce Evans, an American artist?

continued from previous page

8. catalog, perhaps *Books in Print*
9. *Literary History of the United States, Essay and General Literature Index, Humanities Index, MLA International Bibliography*
10. *Books in Print*

ANSWERS TO EXERCISE 2

Sending all your students to the library to do the same exercise will strain your library's resources without helping your students. Assigning them to do this exercise in groups will be more effective. You may wish to encourage students to try to find the information in more than one source.

1. *Consumer Information Catalog* or *Monthly Catalog of US Government Documents* will list available publications.
2. The *US Government Manual* lists all federal agencies and programs.
3. The best place to start is with *Who's Who in America*. You could also consult the microfiche *Biobase,* which will direct you to the appropriate sources.
4. Both the *National Faculty Directory* and *Who's Who* list Sagan as a member of the Department of Astronomy and Geology at Cornell University.
5. The *Encyclopedia of Associations* lists organizations by subject; there are several with the word *wolves* in the title.
6. There is a *Directory of Lobbyists,* but the *Congressional Quarterly Directory of Lobbyists* is probably

continued on following page

continued from previous page

more recent and more complete.

7. The *US Government Manual* will direct you to the Department of Immigration and Naturalization.

8. *Who's Who in America* lists a home address.

9. See the *Consumer Information Catalog* for a list.

10. The *US Government Manual* lists the address for the Peace Corps as 806 Connecticut Avenue NW, Washington, D.C. 20506.

4. At what university does the astronomer Carl Sagan teach?

5. What organizations could you contact to find out what is being done to prevent the killing of wolves in North America?

6. How could you get current information about the tobacco lobby?

7. What government agency could tell you what government services are available to resident aliens?

8. At what address could you contact Harold Bloom, a scholar who does work on nineteenth-century English literature?

9. Is there a government pamphlet that gives information about buying a new car?

10. How could you get current information about the Peace Corps?

38c Gathering Information Outside the Library

By relying exclusively on the library for research materials, you ignore important sources of current information. Public service organizations, lobbying organizations, and government agencies can supply you with a good deal of current data. People who work in a particular field or who have done research on your subject are also excellent sources.

(1) Finding organizations

Numerous organizations offer literature, often free of charge, to interested parties. Your instructor or reference librarian may suggest the most appropriate organizations for you to contact, and local businesses, chambers of commerce, corporate public information departments, and government offices or publications may also refer you to helpful groups. The most useful source of information, however, is the *Encyclopedia of Associations*, which lists thousands of organizations by subject area. It also has a key word index, which enables you to find out whether or not an organization in a particular subject area exists. In addition to writing or calling for information, you can sometimes arrange to visit an organization, perhaps to observe a meeting or to interview a member.

(2) Writing for information

Once you have identified people or organizations that may be able to supply useful information, your next step is to contact them directly and ask for their help. If the organization is local or has a toll-free number, call to request information. If not, you may want to write a letter—but first check with your instructor to

See ◀
48d

make sure that you have enough time to wait for an answer. Whether you call or write, be sure you are prepared with the following information.

- The name of the person or department to contact
- The specific information you would like
- Why you want the information
- When you need it

(3) Finding people

An important step in your research is locating people who can suggest reliable and up-to-date sources of information—or even provide you with some of that information. A meeting with your instructor may be all that you need to get started, or your instructor may refer you to someone else more familiar with your topic. Just one or two good contacts can help you to establish a **research network**: Your first contact suggests another, who in turn suggests two more.

Many excellent guides to experts in various fields are available in the reference section of your library. Some of these guides are listed below.

Who's Who Among Black Americans
Who's Who and Where in Women's Studies
Who's Who in American Art
Who's Who in American Education
Who's Who in American Politics
American Men and Women of Science
Biographical Directory of the American Psychological Association
Contemporary Authors
Directory of The Modern Language Association

Remember too that your own college or university is likely to have a number of experts in particular subjects on its faculty. Your instructor or librarian can identify such experts for you.

EXERCISE 3

You are beginning to gather information for the research projects outlined below. Where in your college or your community might you turn if you wanted to establish a research network? Write five questions that you would ask each person.

1. A paper for a biology course examining new developments in DNA recombinant research

STUDENTS WORKING TOGETHER

Divide the class into groups of three or four students. Assign each of the topics in Exercise 3 to one or two groups. (If your class is a large one, you may need to come up with an expanded list of topics.) Each group should submit a brief written report on their research, including the five questions suggested in the exercise. Discuss the project. Students are sometimes not aware of how much information is available on request—or how many experts can be found with a simple telephone call. Encourage your students to look for expert opinion to support their conclusions.

IN THE CLASSROOM

Bring to class a collection of question-and-answer-format interviews from current magazines and newspapers. Have students find and read aloud examples of leading questions, open- and closed-ended questions, questions designed to provoke an emotional response from the subject, and so on. Discuss the interviewers' techniques for drawing stories, detailed memories, and vivid quotations from the subject.

2. A short paper for a history course in which you examine the usefulness of slave narratives for historical research
3. A research paper for a composition course in which you explore the possible future uses of virtual reality
4. A paper for a political science course about issues in a local election

(4) Observing people, places, objects, and events

Your own observations can be a useful source of information. For example, an art or music paper may be enriched by information gathered during a visit to a museum or concert; an education paper may include a report of a classroom observation; and a psychology or sociology paper may include observations of an individual's behavior or of group dynamics.

GUIDELINES FOR MAKING OBSERVATIONS

- Determine in advance what information you hope to gain from your observations. This material should not be available from the library or other research sources.
- Decide exactly what you want to observe and where you are going to observe it.
- Make an appointment for your visit.
- Bring a small notepad or tape recorder so you can keep a record of your observations.
- Bring any additional materials you may need—a camera, videocassette recorder, or stopwatch, for example.
- Copy your observations onto note cards. Include the date, place, and time of your observations.

---

(5) Conducting an interview

See ◀ 41e

Interviews often give you material that you cannot get by any other means—for instance, biographical information, a firsthand account of an event, or the views of an expert on a particular subject.

Your interview questions should elicit detailed, useful responses. **Leading questions**—those phrased to elicit a certain answer—tend to make some people defensive, and **vague questions**—those too general to elicit useful responses—can be confusing. **Dead-end questions**—questions that call for yes or no answers—yield limited information.

The city intends to spend 2 billion dollars over the next ten years to re-habilitate the port area. Do you support this plan? Why or why not? (good question)

Are you against the city's shortsighted plans to rehabilitate the port area? (leading question)

How do you feel about the port area? (vague question)

Do you think the port area should be developed? (dead-end question)

The kinds of questions you ask in an interview depend on the information you are seeking. **Open-ended questions**—those intended to elicit general information—allow a respondent great flexibility in answering.

Do you think students today are motivated? Why or why not?

If you could change something in your life, what would it be?

Closed-ended questions—those designed to elicit specific information—enable you to zero in on a particular aspect of a subject.

Has your family become more or less religious over the past ten years? How do you account for this?

How much money did the government's cost-cutting programs actually save?

Prepare students for interviewing by having them conduct interviews in pairs. Divide the class into pairs. Ask each person to write down one or two specific subjects on which he or she might be an interesting "interview." Partners should trade their suggestion lists and then bring to the next class a list of questions (both open- and closed-ended) on one of the topics suggested. After the interviews (some students may want to tape-record, others to write notes), conduct a class discussion on the experience. You may also ask students to turn in a brief self-evaluation of their strengths and weaknesses as interviewers.

As an additional exercise, consider assigning teams of students to do interviews for an informal "oral history" of the college or university, or of the city or town in which the class is held.

GUIDELINES FOR CONDUCTING AN INTERVIEW

- Always make an appointment.
- Prepare a list of specific questions tailored to the subject matter and time limit of your interview.
- Do background reading about your topic. Be sure you do not ask for information you can easily find elsewhere.
- Have a pen and paper with you. If you want to tape the interview, get the respondent's permission in advance.
- Allow the person you are interviewing to complete an answer before you ask another question.
- Take notes, but continue to pay attention as you do so. Occasionally nod or make comments to show you are interested and to encourage the respondent to continue.
- Pay attention to the reactions of the respondent.
- Do not hesitate to deviate from your prepared list of questions to ask follow-up questions.
- At the end of the interview, thank the respondent for his or her time and cooperation.
- Send a brief note of thanks.

(6) Conducting a survey

If you are examining a contemporary social, psychological, or economic issue—the rise of racism on college campuses, for instance—a **survey** of attitudes or opinions could be indispensable. Before you start, remember that conducting a survey requires a good deal of advance planning. If time is short, plan to use another approach.

Begin by identifying the group of people you will poll. This group can be a **convenient sample**—for example, people in your chemistry lecture—or a **random sample**—names chosen from a telephone directory, for instance. When you choose a sample, your goal is to designate a population that is *representative*. You do not want, for example, a sample composed of all sophomores, all females, or all business majors—unless, of course, the goal of your survey is to poll only these groups.

The number of people to poll is another consideration: You must have enough respondents to convince readers that your sample is *significant*. If you poll ten people in your French class about an issue of college policy, and your university has ten thousand students, you cannot expect your readers to be convinced by your results. Scientific formulas exist to determine exactly what constitutes a significant sample, so if you plan to conduct a survey, you should consult a social science instructor for further information.

You should also be sure your questions are worded clearly and specifically designed to elicit the information you wish to get. Short-answer or multiple-choice questions, whose responses can be quantified, are much easier to handle than questions that call for paragraph-length answers. Test the questions on a friend to be sure they are understandable. Be sure you do not ask so many questions that respondents lose interest and stop answering. Also, be sure you do not ask biased or leading questions.

You have an advantage if you are surveying a population to which you have easy access. If, for example, your population is your fellow students, you can slip questionnaires under their doors in the residence hall, or you can distribute them in the cafeteria during lunch, or (with the instructor's permission) you can give them out during a lecture. If your questionnaire is fairly brief, the best way to ensure a high response rate is to allow respondents a specific amount of time and collect the forms yourself. If you believe filling out forms on the spot will be time consuming or annoying, or if you think respondents may not want to be identified, you can request responses be returned to you—placed in a box set up in a central location, for instance. If you do not collect the forms yourself, be sure to set a deadline for their completion well in advance of when you need to begin analyzing responses.

If you have a manageable number of questions and respondents, you should be able to classify and categorize your responses fairly easily. If your questionnaire is complex, you may want to get a few friends to help you with this step. In any case, determining exactly what your results suggest is the most challenging (and the most unpredictable) part of the process.

GUIDELINES FOR CONDUCTING A SURVEY

- Determine what you want to know.
- Select your sample.
- Design your questions.
- Type and duplicate the questionnaire.
- Distribute the questionnaires.
- Collect the questionnaires.
- Analyze your responses.
- Decide how to use the results in your paper.

CHAPTER 39

Working with Sources

See ◄
5a

Once you have located sources of information for your paper, your next step is to evaluate their usefulness to you so you can begin to take notes. At this point, your emphasis shifts from looking *for* sources to looking *at* sources. In the process you use **active reading** strategies to identify and highlight key ideas and to record significant relationships. As you do so, you make judgments about the writer's ideas so you can formulate a critical response. When you write your paper, you will use the information you gather to develop your own ideas about your topic and to support the points you make in your discussion.

39a Evaluating Sources

Whenever you read, you should evaluate the potential usefulness of a source. You should do this as soon as possible so you will not waste time reading irrelevant material. As you look for sources, you should explore as many different viewpoints as possible. Focusing on sources that present a single view of your topic will not give you the perspective you need to develop your own ideas. You should also locate more sources than you actually intend to use in your paper because one or more of the sources you find may be one-sided, outdated, unreliable, biased, superficial, or irrelevant—and therefore unusable. To be safe, then, you should collect about twice the number of sources you think you will need.

(1) Distinguishing between primary and secondary sources

To evaluate a source, you should first be certain that you know whether you are reading a **primary** or a **secondary source**—that is, whether you are considering original documents and observations or interpretations of those documents and observations.

> **PRIMARY SOURCE:** United States Constitution, Amendment XIV (Ratified July 9, 1868). Section I.
> All persons born or naturalized in the United States, and subject to the jurisdiction thereof, are citizens of the United States and the state wherein they reside. No state shall make or enforce any law which shall abridge the privileges or immunities of citizens of the United States; nor shall any state deprive any person of life, liberty, or property, without the process of law; nor deny to any person within its jurisdiction the equal protection of the laws.

> **SECONDARY SOURCE:** Paula S. Rothenberg, *Racism and Sexism: An Integrated Study*
> Congress passed The Fourteenth Amendment . . . in July 1868. This amendment, which continues to play a major role in contemporary legal battles over discrimination, includes a number of important provisions. It explicitly extends citizenship to all those born or naturalized in the United States and guarantees all citizens due process and "equal protection" of the law.

For many research projects, primary sources such as letters, speeches, and data from questionnaires are essential. However, secondary sources, which provide the critical comments of scholars who know a good deal about the area you are studying, can also be

---

PRIMARY AND SECONDARY SOURCES

Primary Source	Secondary Source
Novel, poem	Literary criticism
Diary, autobiography	Biography
Letters, historical documents, oral testimony	Historical commentary
Newspaper report	Editorial
Raw data from questionnaires	Social science article
Observations/experiment	Scientific article
Television show/film	Critical analysis
Interview	Case study

Previewing
Your students may be unfamiliar with the time-saving strategy of previewing resource material. As a demonstration, you might choose a research topic and bring to class several books that might be good resources. Preview the books as you would in your own research work (use an overhead projector where appropriate), looking at tables of contents, index listings, pictures and graphs, and so on.

Topic Overview
It can be difficult for students to judge whether an authority is unbiased, or at least presenting a fair argument. As one strategy for dealing with research on controversial topics, suggest that students always begin their reading with a broad-ranging general article on the subject. An encyclopedia entry, textbook treatment, or popular magazine cover story (in *Time* or *Newsweek*, for instance) is likely to touch upon any controversies or different viewpoints relating to the topic. This background knowledge will help students detect an author's too-selective or biased use of argument or fact.

valuable. Keep in mind, though, that the further from the primary source you get, the more chances exist for inaccuracies caused by researchers' inadvertent distortion and misinterpretation of material.

(2) Evaluating the usefulness of print sources

One efficient way to evaluate a source and its author is to ask a librarian or your instructor for an opinion. But even if a source is highly recommended, it may not suit your needs.

To measure the usefulness of a print source, determine how *relevant* it is to your topic. How detailed is its treatment of your subject? Skim a book's table of contents and index for references to your topic. To be of any real help, a book should devote a section or chapter to your topic, not simply a footnote or a brief mention. For articles, read the abstract, or skim the entire article for key facts, looking closely at section headings, information set in boldface type, and topic sentences. An article should have your topic as its central subject, or at least as a major concern.

The date of publication tells you whether the information in a book or article is *current*. A discussion of computer languages written in 1966, for instance, will now be obsolete. Scientific and technological subjects usually demand up-to-date treatment. Even in the humanities, new discoveries and new ways of thinking lead scholars to reevaluate and modify their ideas over time.

Some classic works, however, never lose their usefulness. For example, although Edward Gibbon wrote *The History of the Decline and Fall of the Roman Empire* in the eighteenth century, the book still offers a valuable overview of the events it describes. Contemporary historians may interpret events differently, but Gibbon's information is sound, and the book is required reading for anyone studying Roman history. If a number of your sources cite certain earlier works, you should consult those works, regardless of their publication dates. Do, however, be alert for out-of-date information in such sources.

Another factor to consider is the *reliability* of your source. Is a piece of writing intended to inform or to persuade? Does the author have an ulterior motive? One way to judge the objectivity of a source is to find out something about its author. The source itself may contain biographical information, sometimes in a separate section, or you can consult a biographical dictionary. Skim the preface to see what the author says about his or her purpose. What do other sources say about the author? Do they consider the author fair? Biased? Compare a few statements with a fairly neutral source—a textbook or an encyclopedia, for instance—to see whether an author seems to be slanting facts.

In assessing your source's reliability, you should also try to determine how respected it is. A contemporary review of a source can help you make this assessment. *Book Review Digest*, available in the reference section of your library, lists popular books that have been reviewed in at least three newspapers or magazines and includes excerpts from representative reviews. Scholarly books are indexed in *Book Review Index*. Although this index contains no excerpts, it does include citations that refer you to the periodicals in which books were reviewed.

You can also find out about the standing of a source in the scholarly community by consulting a special class of indexes called **citation indexes**. These books list all scholarly articles published in a given year that mention a particular source. Information is listed under the original article, the author of the article in which the original article is mentioned, or the subject. Seeing how often an article is mentioned can help you determine how influential it is. Citation indexes are available for the humanities, the sciences, and the social sciences.

CLOSE-UP

EVALUATING PRINT SOURCES

Do not use articles from popular periodicals as sources without your instructor's permission. You may consult popular sources for background or for ideas, but remember that they are aimed at a general audience and do not follow the same rigorous standards as scholarly publications. Although some popular periodicals—such as *Atlantic* and *Harper's*—often contain articles that are reliable and carefully researched, others do not. Keep in mind that popular magazines are commercial; their main aim is to sell as many copies as they can. For this reason, their treatment of a topic may be superficial, one-sided, or sensational.

(3) Evaluating the usefulness of nonprint sources

Nonprint sources—interviews, telephone calls, films, and so on—must also be evaluated. Here, too, you should consider the *relevance* of the source—the extent to which it addresses your needs. An expert on family planning who knows little about adolescent health problems may be an excellent source if your paper will focus on changing trends in birth control methods, but not if your paper is about teenage pregnancy.

The *currency* of a nonprint source is also a factor. A 1970 television documentary on the topography of a Pacific island may still be accurate, but a documentary on the lives of its people may not reflect today's conditions at all.

Reliability is important, too. Is a radio feature on energy conservation part of a balanced news program or a thinly veiled commercial sponsored by a public utility? Is the material presented by experts in the field or by actors? Check the credits and acknowledgments and read reviews to see which sources were consulted. Do the participants in a panel discussion on acid rain agree on the magnitude of the problem, or do they represent different points of view? Is the person you plan to interview fair and impartial or biased on some issues? Try to find out by consulting an instructor in a related field or by reading the person's writings before the interview.

GUIDELINES FOR EVALUATING SOURCES

- Are you considering a primary or a secondary source?
- How relevant is your source to your needs?
    How detailed is its treatment of your subject?
    Is your topic a major focus of your source?
- How current is your source?
    Have recent developments made any parts of your source dated?
    Do any of your sources cite a classic work?
- How reliable is your source?
    Is your source meant to inform or to persuade?
    Does the author of your source show any bias?
    How respected is your source?
    Do other scholars mention your source favorably?

---

*EXERCISE 1

Read the following paragraphs carefully, paying close attention to the information provided about their sources and authors as well as to their content. Decide which sources would be most useful and reliable in supporting the thesis "Winning the right to vote has (or has not) significantly changed the role of women in national politics." Which sources, if any, should be disregarded? Which would you examine first? Be prepared to discuss your decisions.

1. Almost forty years after the adoption of the Nineteenth Amendment, a number of promised or threatened events have failed to materialize. The millennium has not arrived, but neither has the country's social fabric been destroyed. Nor have women organized a political party to elect only women candidates to public office. . . . Instead, women have shown the same tendency to divide along orthodox party lines as male voters. (Eleanor Flexner, *Century of Struggle*, Atheneum 1968. *A scholarly treatment of women's roles in America since the* Mayflower, *this book was well reviewed by historians*.)

2. Woman has been the great unpaid laborer of the world, and although within the last two decades a vast number of new employments have been opened to her, statistics prove that in the great majority of these, she is not paid according to the value of the work done, but according to sex. The opening of all industries to women, and the wage question as connected with her, are the most subtle and profound questions of political economy, closely interwoven with the rights of self-government. (Susan B. Anthony; first appeared in Vol. I of *The History of Woman Suffrage*; reprinted in *Voices from Women's Liberation*, ed. Leslie B. Tanner, NAL 1970. *An important figure in the battle for women's suffrage, Susan B. Anthony* [1820–1906] *also lectured and wrote on abolition and temperance*.)

3. Women . . . have never been prepared to assume responsibility; we have never been prepared to make demands upon ourselves; we have never been taught to expect the development of what is best in ourselves because no one has ever expected *anything* of us—or for us. Because no one has ever had any intention of turning over any serious work to us. (Vivian Gornick, "The Next Great Moment in History is Ours," *Village Voice* 1969. *The* Voice *is a liberal New York City weekly*.)

4. With women as half the country's elected representatives, and a woman President once in a while, the country's *machismo* problems would be greatly reduced. The old-fashioned idea that manhood depends on violence and victory is, after all, an important part of our troubles. . . . I'm not saying that women leaders would eliminate violence. We are not more moral than men; we are only uncorrupted by power so far. When we do acquire power, we might turn out to have an equal impulse toward aggression. (Gloria Steinem, "What It Would Be Like If Women Win," *Time* 1970. *Steinem, a well-known feminist and journalist, is one of the founders of* Ms. *magazine*.)

5. Nineteen eighty-two was the year that time ran out for the proposed equal rights amendment. Eleanor Smeal, president of the National Organization for Women, the group that headed the intense 10-year struggle for the ERA, conceded defeat on June 24. Only 24 words in all, the ERA read simply: "Equality of rights under the law shall not be denied or abridged by the United States or by any state on account of sex." Two major opinion polls had reported just weeks before the ERA's defeat that a majority of Americans continued to favor the amendment. (June Foley, "Women 1982: The Year that Time Ran Out," *The World Almanac & Book of Facts*, 1983)

ANSWERS TO EXERCISE 1

1. Informative source for thesis. Although published in 1968, its issues remain relevant today.
2. Old but classic source; information may establish role of women before they were given right to vote. Note importance of author in history of women's suffrage.
3. Not comprehensive or current; popular source may not be suitable for research paper.
4. Note author's feminist bias. Also note popular source may not be appropriate. Essay may also be dated.
5. Acceptable source for basic facts.
6. Although such popular sources may not be suitable for research paper, examples may be useful. Note possible feminist bias of author.
7. Useful source for factual information; presents specific current example of women's political power.

6. The president of the National Women's Political Caucus, Sharon Rodine, grinned as she pronounced, "From Connecticut to Missouri to Oregon, women won great victories . . . 1990 is a clear rehearsal for the decade ahead." Ellen Malcolm, president of Emily's List, a fund-raising network for pro-choice Democratic women candidates, was almost effusive. "We have twenty Democratic women in the new Congress, an increase of two-thirds since 1987," she said. Most significantly, Cardiss Collins, congresswoman from Chicago, will be joined by three more African Americans: Barbara-Rose Collins from Detroit; the eminent feminist civil rights activist Eleanor Holmes Norton from Washington D.C.; and the redoubtable Maxine Waters, who graduates to Congress after fourteen years as a state assemblywoman from Los Angeles. Returned to the House in this election were two other minority women: Ileana Ros-Lehtinen, a Florida Republican who is Cuban American, and Democrat Patsy T. Mink, an Asian American from Hawaii. (Jane O'Reilly, "Running for Our Lives," *Glamour* Jan. 1991. *O'Reilly is a founding editor of* Ms. *as well as a newspaper columnist and a political correspondent for* Time. Glamour *is a fashion, beauty, and lifestyle magazine for young women.*)

7. A chastened Senate yesterday bowed to Senator Carol Moseley-Braun, its only African-American member, and reversed a vote that would have given what an infuriated Moseley-Braun describes as an "imprimatur" to an insignia that features the confederate flag.

   After an extraordinary emotional debate in which senators talked with candor about racism in America, the Senate voted 75–25 to kill a proposal to renew a design patent for the insignia of the United Daughters of the Confederacy (UDC). It featured the original flag of the Confederacy encased in a wreath.

   Only a few hours before, the Senate signaled its intention to approve the proposal, sponsored by Senator Jesse Helms (R–NC), on a procedural vote of 50–48.

   Helms contended the 24,000 UDC members were "delightful gentleladies" engaged in charitable endeavors and deserving of the largely honorific patent protection that Congress almost routinely confers on national groups.

   What happened in between the two votes was Moseley-Braun (D–IL), the first black woman in the history of the Senate, whose voice was eloquent and angry as she spoke of the legacy of slavery and the Confederacy's fight to preserve it. (*Helen Dewar*, "Senate Bows to Pressure," Philadelphia Inquirer 7 *Jan. 1993. Dewar is a columnist for the* Washington Post, *a daily newspaper published in Washington, D.C.*)

39b Taking Notes

Although it may seem like a sensible strategy, simply copying down the words of a source is the least efficient way to take notes.

Experienced researchers know it makes more sense to take notes that combine direct quotation with *paraphrase* and *summary*. By doing so, they make sure they understand both the material and its relevance to their research. In fact, the very act of putting the ideas of others into your own words helps you gain a better understanding of what they are saying.

▶ **See 41e2**

(1) Writing a summary

You summarize when you want to capture the general idea of a source. A **summary**, sometimes called a *précis*, is a brief restatement in your own words of the main idea of a passage, article, or entire book. When you write a summary, you condense the author's ideas into a few concise sentences, taking care not to misrepresent his or her views. You do not, however, include your own ideas or observation.

A summary is always much shorter than the original because it omits the examples, asides, analogies, and rhetorical strategies that writers use to add emphasis and interest. Just how short a summary should be depends on the length and complexity of the source and on the use you are going to make of it in your paper. You should try, if possible, to condense each paragraph of the original to a sentence or two—although complicated passages might demand fuller treatment, and simple, direct passages will require briefer treatment.

Before you write your summary, you should understand the source material. Because the purpose of a summary is to communicate the main idea of the source, you should read the original carefully, paying particular attention to its thesis, topic sentences, and supporting details. Sometimes—especially with complicated material—you will find it helpful to highlight the source or list its key points.

Once you have isolated a source's key points, you should try to summarize its main idea in a single sentence. This is not always easy. Your sentence must be general enough to capture the sense of the entire source, not just a part of it. You may have to write several versions of this sentence before you find one that accurately reflects the original source.

Next, draft the summary, using your one-sentence restatement as your first sentence. Your summary should present a general overview of the original and should not include the author's specific examples or analogies. As you write, use your own words, not the exact language or phrasing of the source. If you think it is necessary to quote a distinctive word or phrase, place it within quotation marks. Finally, remember that your summary should include only the ideas of your source, not your own interpretations or opinions.

After you have written a draft of your summary, check to make sure that it conveys both the meaning and spirit of the original. Make certain you have not inadvertently used any of the author's exact words without quotation marks. (Of course, you can use *some* words from a source. As a rule you can use the proper nouns, simple words, or technical terms found in the original source without enclosing them within quotation marks.) As you reread your summary, check to see that you have included smooth transitions. When you are finished, add documentation to identify the source you have summarized.

STEPS FOR WRITING A SUMMARY

1. Reread your source until you understand it.
2. Write a one-sentence restatement of the main idea.
3. Draft your summary, using the one-sentence restatement as your first sentence. Use your own words, not the words or phrasing of the source. Include quotation marks where necessary.
4. Revise your summary, making sure that it accurately reflects the source and that you have not included any of your own ideas or opinions.
5. Add transitions where necessary.
6. Add appropriate documentation.

Compare the following three passages. The first is an original source, the second is an unacceptable summary, and the third is an acceptable summary.

ORIGINAL SOURCE: Sudo, Phil, "Freedom of Hate Speech?" *Scholastic Update* 124.14 (1992) 17-20.

Today, the First Amendment faces challenges from groups who seek to limit expressions of racism and bigotry. A growing number of legislatures have passed rules against "hate speech"—[speech] that is offensive on the basis of race, ethnicity, gender, or sexual orientation. The rules are intended to promote respect for all people and protect the targets of hurtful words, gestures, or actions.

Legal experts fear these rules may wind up diminishing the rights of all citizens. "The bedrock principle [of our society] is that government may never suppress free speech simply because it goes against what the community would like to hear," says Nadine Strossen, president of the American Civil Liberties Union and professor of constitutional law at New York University Law School. In recent years, for example, the courts have upheld the right of neo-Nazis to march in Jewish neighborhoods; protected cross-burning as a form of free expression; and allowed protesters to burn the American flag. The offensive, ugly, distasteful, or repugnant nature of expression is not reason enough to ban it, courts have said.

But advocates of limits on hate speech note that certain kinds of expression fall outside of First Amendment protection. Courts have ruled that "fighting words"—words intended to provoke immediate violence—or speech that creates a clear and present danger are not protected forms of expression. As the classic argument goes, freedom of speech does not give you the right to yell "Fire!" in a crowded theater.

> **UNACCEPTABLE SUMMARY:** Today, the First Amendment faces challenges from lots of people. Some of these people are legal experts who want to let Nazis march in Jewish neighborhoods. Other people have the sense to realize that some kinds of speech create a clear and present danger and therefore should not be protected (Sudo 17).

The unacceptable summary above uses words and phrases from the original without placing them in quotation marks and also uses some of the source's phrasing. This use constitutes plagiarism. In addition, the summary expresses its writer's opinion ("Other people have the sense to realize . . ."). Compare this unacceptable summary with the following acceptable summary. Notice that the acceptable summary presents an accurate, objective overview of the original without using its exact language or phrasing. (The one distinctive phrase from the source is placed within quotation marks.)

▶ See
39d

> **ACCEPTABLE SUMMARY:** The right to freedom of speech, guaranteed by the First Amendment, is becoming more difficult to defend. Some people think that stronger laws against the use of "hate speech" weaken the First Amendment. But others argue that some kinds of speech are not protected (Sudo 17).

(2) Writing a paraphrase

A summary conveys just the essence of a source; a **paraphrase** gives a *detailed* restatement of all a source's important ideas. It not only indicates the source's key points, but also reflects its order, tone, and emphasis. Often a paraphrase will quote key words or phrases to suggest the flavor of the original. For this reason, a paraphrase can be of any length—even the same length as the source. Like a summary, a paraphrase should present the *source's* ideas, not your own. The difference between a summary and a paraphrase is illustrated in the following side-by-side comparison of the passage discussed in the previous section.

SUMMARY	PARAPHRASE
The right to freedom of speech, guaranteed by the First Amendment, is becoming more difficult to defend. Some people feel that stronger laws against "hate speech" weaken the First Amendment. But others argue that some kinds of speech are not protected (Sudo 17).	Many groups want to limit the right of free speech guaranteed by the First Amendment to the Constitution. They do this to protect certain groups of people from "hate speech." Women, people of color, and gay men and lesbians, for example, may find that hate speech is used to intimidate them. Legal scholars are afraid that even though the rules against hate speech are well intentioned, they undermine our freedom of speech. As Nadine Strossen, president of the American Civil Liberties Union says, "The bedrock principle [of our society] is that government may never suppress free speech simply because it goes against what the community would like to hear" (Sudo 17). People who support speech codes point out, however, that certain types of speech are not protected by the First Amendment—for example, words that create a "clear and present danger" or that would lead directly to violence (Sudo 17).

The purpose of a paraphrase is to restate the source in order to clarify its meaning. You use a paraphrase only when you want to explain a source in detail to readers. For this reason, you would summarize—not paraphrase—extremely long passages or entire books or articles.

Before you begin to draft your paraphrase, carefully read the original. Because a paraphrase reflects the *exact* points of the source, you may wish to outline the source before you start to write.

Next, draft your paraphrase, following the order, tone, and emphasis of the original. Make certain you use your own words, except when you want to quote to give readers a sense of the original.

If you do include quotations, circle the quotation marks so you will not forget to include them when you revise your paraphrase. Try not to look at the source when you write—use the language and syntax that come naturally to you—and avoid duplicating the wording or sentence structure of the original. Whenever possible, use synonyms that accurately convey the meaning of the original word or phrase. If you cannot think of a synonym for an important term, quote. Notice in the paraphrase above that the student chose to quote "hate speech" and "clear and present danger" rather than use less distinctive terms.

Once you finish writing a draft of your paraphrase, revise it carefully, making sure that it is accurate and complete and that you have not inadvertently used the language, phrasing, or sentence patterns of the source. Make certain you have covered all important points and included quotation marks where they are required. As you revise, add transitions where necessary to make your paraphrase a clear and coherent whole. Remember to document all direct quotations from your source as well as the entire paraphrase.

STEPS FOR WRITING A PARAPHRASE

1. Reread your source until you understand it.
2. Outline your source if necessary.
3. Draft your paraphrase, following the order, tone, and emphasis of the original.
4. Revise your paraphrase, making sure it reflects the order and emphasis of the original. Be sure you do not use the words or phasing of the original without enclosing the borrowed material within quotation marks.
5. Add transitions where necessary.
6. Add appropriate documentation.

Following are an original passage, an unacceptable paraphrase, and an acceptable paraphrase.

ORIGINAL SOURCE: Turkle, Sherry. *The Second Self: Computers and the Human Spirit.* New York: Simon & Schuster, 1984: 83–84.

When you play a video game, you enter into the world of the programmers who made it. You have to do more than identify with a character on a screen. You must act for it. Identification through action has a special kind of hold. Like playing a sport, it puts people into a highly focused and highly charged state of mind. For many people, what is being pursued in the video game is not just a score, but an altered state.

The pilot of a race car does not dare to take . . . attention off the road. The imperative of total concentration is part of the high. Video games demand the same level of attention. They can give people the feeling of being close to the edge because, as in a dangerous situation, there is no time for rest and the consequences of wandering attention [are] dire. With pinball, a false move can be recuperated. The machine can be shaken, the ball repositioned. In a video game, the program has no tolerance for error, no margin for safety. Players experience their every movement as instantly translated into game action. The game is relentless in its demand that all other time stop and in its demand that the player take full responsibility for every act, a point that players often sum up [with] the phrase "One false move and you're dead."

UNACCEPTABLE PARAPHRASE: Playing a video game, you enter into a new world—one the programmer of the game made. You can't just play a video game; you have to identify with it. Your mind goes to a new level and you are put into a highly focused state of mind.

Just as you would if you were driving a race car or piloting a plane, you must not let your mind wander. Video games demand complete attention. But the sense that at any time you could make one false move and lose is their attraction—at least for me. That is why I like video games more than pinball. Pinball is just too easy. You can always recover. By shaking the machine or quickly operating the flippers, you can save the ball. Video games, however, are not so easy to control. Usually, one slip means that you lose (Turkle 83–84).

The unacceptable paraphrase above does little more than echo the phrasing and syntax of the original, borrowing words and expressions without enclosing them in quotation marks. This con-

See ◄
39d

stitutes plagiarism. In addition, the paraphrase digresses into a discussion of the writer's own views about the relative merits of pinball and video games. Although the acceptable paraphrase below follows the order and emphasis of the original—and even quotes a key phrase—its wording and sentence structure are very different from those of the source. It not only conveys the major ideas of the source, but it also maintains an objective stance.

ACCEPTABLE PARAPHRASE: The programmer defines the reality of the video game. The game forces a player to merge with the character who is part of the game. The character becomes an extension of the player, who determines how he or she will think and act. Like sports, video games put a player into a very intense "altered state" of mind that is the most important part of the activity (Turkle 83).

The total involvement they demand is what attracts many people to video games. These games can simulate the thrill of participating in a dangerous activity without any of the risks. There is no time for rest and no opportunity to correct errors of judgment. Unlike video games, pinball games are forgiving. A player can—within certain limits—manipulate a pinball game to correct minor mistakes. With video games, however, every move has immediate consequences. The game forces a player to adapt to its rules and to act carefully. One mistake can cause the death of the character on the screen and the end of the game (Turkle 83–84).

(3) Recording quotations

When you **quote**, you copy an author's remarks exactly as they appear in a source, word for word and punctuation mark for punctuation mark.

CLOSE-UP

WORKING WITH SOURCES

When recording quotations in your notes, be sure you do not inadvertently leave out quotation marks. You may want to circle them to be sure you notice them when you transfer them from your notes to your paper.

As a rule, avoid including numerous direct quotations in your papers. The use of one quotation after another interrupts the flow of your discussion and gives readers the impression that your paper is just an unassimilated collection of other people's ideas. Quote only when something vital would be lost otherwise. When you do quote, use only those words and phrases that support your points and provide a perspective that contributes something vital to your paper. Before you include any quotation, ask yourself whether your purpose would be better served if you used your own words.

WHEN TO QUOTE

- Quote when a source's wording or phrasing is so distinctive that a summary or paraphrase would diminish its impact. In such cases it is best to let the source speak for itself.

- Quote when a source's words lend authority to your presentation. If an author is a recognized expert on your subject, his or her words are as convincing as expert testimony at a trial.

continued on the following page

continued from the previos page

- Quote when an author's words are so concise that paraphrase would create a long, clumsy, or incoherent phrase or would change the meaning of the original.
- Quote when you are going to disagree with a source. Using a source's exact words assures readers you are being fair to those on the other side of the issue and not setting up a **straw man**.

See ◄
7a5

39c Integrating Your Notes into Your Writing

You cannot simply drop the summaries, paraphrases, and quotations from your notes into your paper; you should weave them smoothly into your discussion, adding analysis or explanation to increase coherence and to show why you are using each source's words or ideas. As you integrate borrowed material into your paper, be sure to differentiate your ideas from those of your sources.

(1) Integrating quotations

Quotations should be smoothly worked into your sentences and introduced by identifying phrases. They should never be awkwardly dropped into the paper, leaving the exact relationship between the quotation and the writer's point unclear, as in the following example.

> **UNACCEPTABLE:** For the Amish, the public school system represents a problem. "A serious problem confronting Amish society from the viewpoint of the Amish themselves is the threat of absorption into mass society through the values promoted in the public school system" (Hostetler 193).

The next example shows how a brief introductory remark provides a context for a quotation, giving readers the information they need to make sense of the quoted material. In addition, the writer uses only the part of the quotation she needs to make her point.

> **ACCEPTABLE:** For the Amish, the public school system is a problem because it represents "the threat of absorption into mass society" (Hostetler 193).

CLOSE-UP

INTEGRATING YOUR NOTES INTO YOUR WRITING

To avoid monotonous sentence structure, experiment with different methods of integrating source material into your paper.

- Vary the verbs you use for attribution.

acknowledges	*discloses*	*implies*
suggests	*observes*	*notes*
concludes	*believes*	*comments*
insists	*explains*	*claims*
predicts	*summarizes*	*illustrates*
reports	*finds*	*proposes*
warns	*concurs*	*speculates*
admits	*affirms*	*indicates*

- Vary the placement of the identifying phrase, putting it at the beginning or at the end of the quoted material, or even in the middle.

QUOTATION WITH ATTRIBUTION IN MIDDLE: "A serious problem confronting Amish society from the viewpoint of the Amish themselves," observes Hostetler, "is the threat of absorption into mass society through the values promoted in the public school system" (193).

PARAPHRASE WITH ATTRIBUTION AT END: The Amish are also concerned about their children's exposure to the public school system's values, notes Hostetler (193).

Sometimes, for the sake of clarity, you might want to use a **running acknowledgment**, which enables you to introduce the source of the quotation into the text. Running acknowledgments are particularly helpful when you are using several sources in the same section of your paper.

> **RUNNING ACKNOWLEDGMENT:** As John Hostetler points out, the Amish see the public school system as a problem because it represents "the threat of absorption into mass society" (193).

When you use a quotation by an expert to support your opinion, you can lend additional authority to your statement by establishing his or her expertise in the running acknowledgment.

RUNNING ACKNOWLEDGMENT ESTABLISHING EXPERTISE: <u>According to John A. Hostetler, a noted authority on Amish life</u>, one of the most serious problems the Amish face is "the threat of absorption" into the dominant culture represented by the public schools (193).

Another option for integrating quotations smoothly into your writing is to combine quotation and paraphrase, quoting only a significant word or two and paraphrasing the rest.

QUOTATION AND PARAPHRASE: According to John A. Hostetler, one of the most serious problems the Amish face is "the threat of absorption" into the dominant culture represented by the public schools (193).

Integrating quotations seamlessly into your writing can create problems. Solutions for some of these problems are listed below.

Substitutions or Additions within Quotations   When you have to change or add a word to make a quotation fit your paper—for example, to change the verb tense of the original or to supply a missing antecedent for a pronoun—acknowledge your changes by enclosing them in brackets (not parentheses).

ORIGINAL QUOTATION: "Immediately after her wedding, she and her husband followed tradition and went to visit almost everyone who attended the wedding" (Hostetler 122).

QUOTATION REVISED TO MAKE VERB TENSES CONSISTENT: Nowhere is the Amish dedication to tradition more obvious than in the events surrounding marriage. Right after the wedding celebration the Amish bride and groom "visit almost everyone who [has attended] the wedding" (Hostetler 122).

QUOTATION REVISED TO SUPPLY AN ANTECEDENT FOR A PRONOUN: "Immediately after her wedding, [Sarah] and her husband followed tradition and went to visit almost everyone who attended the wedding" (Hostetler 122).

QUOTATION REVISED TO CHANGE A CAPITAL TO A LOWERCASE LETTER: The strength of the Amish community is illustrated by the fact that "[i]mmediately after her wedding, she and her husband followed tradition and went to visit almost everyone who attended the wedding" (Hostetler 122).

Omissions within Quotations   You can reduce the length of quotations by deleting unnecessary or irrelevant words, substituting an ellipsis (three spaced periods) for the deleted words.

See ◄
32f1

ORIGINAL: "Not only have the Amish built and staffed their own elementary and vocational schools, but they have gradually organized on local, state, and national levels to cope with the task of educating their children" (Hostetler 206).

QUOTATION REVISED TO ELIMINATE UNNECESSARY WORDS: "Not only have the Amish built and staffed their own elementary and vocational schools, but they have gradually organized . . . to cope with the task of educating their children" (Hostetler 206).

NOTE: When you omit a word or phrase at the *beginning* of a quoted passage, you do not use ellipses to indicate the omission.

FAULTY (ELLIPSES AT THE BEGINNING OF A QUOTATION): For the Amish, the calendar year reflects agricultural activities and ". . . the kind of leisure and the customs that tend to become associated with the seasons of the year" (Hostetler 94).

REVISED: For the Amish, the calendar year reflects agricultural activities and "the kind of leisure and the customs that tend to become associated with the seasons of the year" (Hostetler 94).

CLOSE-UP

OMISSIONS WITHIN QUOTATIONS

Be sure you do not misrepresent quoted material when you shorten it. The material you delete should not affect the meaning of the passage you are using. For example, do not say, "the Amish have managed to maintain . . . their culture" when the original quotation is "the Amish have managed to maintain *parts of* their culture."

Long Quotations   Occasionally, you may want to quote more than four lines from a source. Set off such a quotation by indenting it ten spaces from the margin. Double-space, do not use quotation marks, and introduce the long quotation with a colon. If you are quoting a single paragraph, do not indent the first line. If you are quoting more than one paragraph, indent the first line of each paragraph three spaces.

▶ See 31d2

According to Hostetler, the Amish were not always hostile to public education:

> The one-room rural elementary school served the Amish community well in a number of ways. As long as it was a public school, it stood midway between the Amish community and the world. Its influence was tolerable, depending upon the degree of influence the Amish were able to bring to the situation. As long as it was small, rural, and near the community, a reasonable influence could be maintained over its worldly character. (196)

Assign a mini research paper. Ask students to write a short essay on a topic of general interest, using a short quotation, a long quotation, a summary, and a paraphrase to support their thesis statements. You might suggest that they write the paper first, using their own experiences and ideas as support; then, as they revise, they can add the source material to develop their arguments further.

CLOSE-UP

USING LONG QUOTATIONS

Use long quotations when you want to convey a sense of an author's style or thought process. Keep in mind, however, that long quotations can be distracting. They interrupt your discussion and, when used excessively, give the impression that you have relied too heavily on the words of others. For this reason, keep the use of long quotations to a minimum.

(2) Integrating paraphrases and summaries

See ◀
39d
Introduce your paraphrases and summaries with running acknowledgments and end them with appropriate documentation. By surrounding your paraphrases and summaries with documentation, you make certain that your readers are able to differentiate your own ideas from the ideas of your sources. If you do not do this, you risk misleading your readers and possibly being accused of plagiarism.

MISLEADING (IDEAS OF SOURCE BLEND WITH IDEAS OF WRITER): Art can be used to uncover many problems that children have at home, in school, or with their friends. For this reason, many therapists use art therapy extensively. Children's view of themselves in society is often reflected by their art style. A cramped, crowded art style using only a portion of the paper shows their limited role (Alschuler 260).

REVISED WITH RUNNING ACKNOWLEDGMENT (IDEAS OF SOURCE DIFFERENTIATED FROM IDEAS OF WRITER): Art can be used to uncover many problems that children have at home, in school, or with their friends. For this reason, many therapists use art therapy extensively. According to William Alschuler in *Art and Self-Image*, children's view of themselves in society is often reflected by their art style. A cramped, crowded art style using only a portion of the paper shows their limited role (260).

In some cases running acknowledgments can be distracting—particularly if you are weaving a number of examples from various sources into your discussion. In such situations, you must be careful to connect examples smoothly to avoid awkwardness.

AWKWARD: The extent of colonial sympathy for the British cause has been widely debated, but even so certain conclusions can be drawn. William Appleton Williams says that at least thirty-five and possibly sixty percent of the colonists initially sided with the British (235). Daniel Boorstin says after the war started, that number probably dropped to twenty percent (78). John Hope Franklin says that at no time did the number of colonists loyal to England fall below ten percent (45).

REVISED: The extent of colonial sympathy for the British cause has been widely debated, but even so, certain conclusions can be drawn. According to William Appleton Williams, for example, at least thirty-five and possibly sixty percent of the colonists initially sided with the British (235). After the war started, as Daniel Boorstin notes, that number probably dropped to twenty percent (78). According to John Hope Franklin, however, at no time did the number of colonists loyal to England fall below ten percent (45).

As you integrate paraphrases and summaries into your paper, make the relationship between your ideas and the ideas of your sources clear by using appropriate transitional words and phrases. Not only will such words and phrases ensure that your paraphrases and summaries are smoothly integrated into your discussion, they will also clarify your purpose in using them.

CONFUSING: The chain of events that led to Richard Nixon's resignation is well known. According to Woodward and Bernstein, both Alexander Haig and Henry Kissinger urged Nixon to cut his ties with his aides (366).

REVISED: The chain of events that led to Richard Nixon's resignation is well known. First, according to Woodward and Bernstein, both Alexander Haig and Henry Kissinger urged Nixon to cut his ties with his aides (366). Next, . . .

EXERCISE 2

Assume that in preparation for a paper on the topic "The effects of the rise of suburbia," you read the following passage from the book *Great Expectations: America and the Baby Boom Generation* by Landon Y. Jones. Reread the passage and then write a brief summary. Next, paraphrase one paragraph. Finally, take notes that combine paraphrase with quotation, making certain to quote only when appropriate.

As an internal migration, the settling of the suburbs was phenomenal. In the twenty years from 1950 to 1970, the population of the suburbs doubled from 36 million to 72 million. No less than 83 percent of the total population growth in the United States during the 1950s was in the suburbs, which were growing fifteen times faster than any other segment of the country. As people packed and moved, the national mobility rate leaped by 50 percent. The only other comparable influx was the wave of European immigrants to the United States around the turn of the century. But as *Fortune* pointed out, more people moved to the suburbs every year than had ever arrived on Ellis Island.

Your students' answers will vary. One sample of each activity follows.

Summary
The movement of people into the suburbs was the largest resettlement in United States history. From 1950 to 1970 the number of people living in the suburbs doubled, and the national mobility rate increased.

To accomodate these people, home-builders turned farmlands into developments of three-bedroom houses owned by young couples holding government loans. By 1960, the percentage of homeowners had risen above that of renters; most homeowners now lived in the new suburbs.

Abraham Levitt built the three largest suburbs. The first Levittown had 17,447 identical houses, occupied by couples who either had or wanted to have children.

Paraphrase (Paragraph 3)
The suburbs became "gigantic nurseries." Abraham Levitt, the son of Russian-Jewish immigrants and the builder of military barracks in World War II, constructed the largest developments. A Long Island potato field became the first Levittown, which housed 82,000 people in 17,447 homes. Two more communities were built on the East Coast. Each offered identical four-room houses with built-in television, outdoor grill, washer, and attic for $7,900, or $60 per month. The common denominator among Levittown residents was the desire to have children.

continued on following page

continued from previous page

Notes (Paragraph 3)

- largest suburban developments, which Jones calls "gigantic nurseries," built by Abraham Levitt
- first Levittown on Long Island; two others on East Coast
- $7,900 or $60 a month for "Monopoly-board bungalow with four rooms, attic, washing machine, outdoor barbecue, and a television set built into the wall"
- 17,447 units/82,000 people, lots of children

TEACHING STRATEGY

Plagiarism

Prevention is the best way to deal with plagiarism. Discuss what plagiarism is, why it is wrong, and what the penalties may be. Make your standards—and your department's or college's policies—clear. Students should see that improving their own skills is a survival matter both at school and at work. Usually the student who cheats is worried about improving, is lazy about study activities, or has given up trying to do well in a particular course or for a particular instructor. Your attitude can be a preventive, not only by encouraging a student to grow, but also by demonstrating that you will identify any work stolen from another source and disguised as a student's own and that you will penalize the student.

Types of Plagiarism

Plagiarism generally falls into one of four categories:

continued on following page

By now, bulldozers were churning up dust storms as they cleared the land for housing developments. More than a million acres of farmland were plowed under every year during the 1950s. Millions of apartment-dwelling parents with two children were suddenly realizing that two children could be doubled up in a spare bedroom, but a third child cried loudly for something more. The proportion of new houses with three or more bedrooms, in fact, rose from one-third in 1947 to three-quarters in 1954. The necessary *Lebensraum* could only be found in the suburbs. There was a housing shortage, but young couples armed with VA and FHA loans built their dream homes with easy credit and free spending habits that were unthinkable to the baby-boom grandparents, who shook their heads with the Depression still fresh in their memories. Of the 13 million homes built in the decade before 1958, 11 million of them—or 85 percent—were built in the suburbs. Home ownership rose 50 percent between 1940 and 1950, and another 50 percent by 1960. By then, one-fourth of *all* housing in the United States had been built in the fifties. For the first time, more Americans owned homes than rented them.

We were becoming a land of gigantic nurseries. The biggest were built by Abraham Levitt, the son of poor Russian-Jewish immigrants, who had originally built houses for the Navy during the war. The first of three East Coast Levittowns went up on the potato fields of Long Island. Exactly $7900—or $60 a month and no money down—bought you a Monopoly-board bungalow with four rooms, attic, washing machine, outdoor barbecue, and a television set built into the wall. The 17,447 units eventually became home to 82,000 people, many of whom were pregnant or wanted to be. In a typical story on the suburban explosion, one magazine breathlessly described a volleyball game of nine couples in which no less than five of the women were expecting.

39d Avoiding Plagiarism

(1) Defining plagiarism

Plagiarism is presenting another person's words or ideas as if they are your own. By not acknowledging a source, you mislead readers into thinking that the material you are presenting is yours when, in fact, it is the result of someone else's time and effort.

Some writers plagiarize deliberately, copying passages word for word or even presenting another person's entire work as their own. Students who do this are doing themselves and their classmates a great disservice. They are undercutting the learning process, thereby sacrificing the education that they are in college to obtain. If found out, they are usually punished severely. Many students have failed courses and some have even had degrees withheld because of plagiarism.

Most plagiarism, however, is accidental. It occurs when students are not aware of what constitutes plagiarism, or when they forget that a note they jotted down is really a direct quotation or that an idea they are using is actually someone else's. Still, accidental plagiarism is often dealt with just as harshly as intentional plagiarism. Plagiarism is not taken lightly in education, business, or anyplace else. *Plagiarism is theft.*

In general, you must document all direct quotations, opinions, judgments, and insights of others that you summarize or paraphrase. You must also document information that is not well known, is open to dispute, or is not commonly accepted. Finally, document tables, graphs, charts, and statistics taken from a source.

▶ See 40a

Common knowledge, information that you would expect most educated readers to know, need not be documented. Thus, you can use facts that are widely available in encyclopedias, textbooks, newspapers, and magazines without citing a source. Even if the information is new to you, if it seems to be generally known—for instance, if it appears in several of your sources—you need not document it. Information that is in dispute, however, or that a particular person has discovered or theorized about, must be acknowledged. For example, you need not document the fact that John F. Kennedy graduated from Harvard in 1940 or that he was elected president in 1960. You must, however, document a historian's analysis of Kennedy's performance as president or a researcher's recent discoveries about his private life.

(2) Revising to eliminate plagiarism

You can avoid plagiarism by using documentation wherever it is required and by watching for the situations that cause the most common types of unintentional plagiarism.

COMMON TYPES OF UNINTENTIONAL PLAGIARISM

- Borrowed words not enclosed in quotation marks
- Paraphrase too close to its source
- Statistics not attributed to a source
- Writer's words and ideas not differentiated from those of the source

---

continued from previous page

The Unauthorized Help Paper Many students are used to getting informal (or formal) help from peers. Make clear what your guidelines for seeking such help are, especially if your college has a writing lab in which students can consult peer tutors.

The Copied Paper Most often the copied paper is one that earned someone else a good grade and is being turned in by a panicked student. You should be suspicious of work radically different from your student's usual work and of sudden topic changes. Ask that students turn in all materials generated in the composition of the paper. (Students who copy usually cannot manufacture the notes and drafts.) Requiring some writing to be done in class not only gives you a better idea of what constitutes your student's work but makes it more difficult for students to pass off copied papers as their own.

The Bought Paper Again, the clue to such a paper is work radically different from the student's norm. In addition, bought papers rarely fulfill the exact specifications of the assignment. Companies selling papers probably advertise in your campus newspaper. Requiring working materials—and, in the case of research papers, photocopies of sources used—should prevent this kind of plagiarism.

The Undocumented or Poorly Documented Paper Doing the research project in stages and requiring all

continued on following page

continued from previous page

materials to be turned in should ensure appropriate work and eliminate the last-minute production in which a student commits plagiarism. Check note cards for distinctions between quotations, paraphrases, and summaries and check final drafts for suddenly elevated vocabulary or Ciceronian sentence structure.

TEACHING STRATEGY

Plagiarism

Discuss real-life incidents of plagiarism with students. A good place to begin is the speech made by Senator Joseph Biden during the 1987 presidential primaries in which Biden plagiarized Britain's Labour Party leader, Neil Kinnock. Another case to consider is writer John Hersey's article on James Agee, which appeared in *The New Yorker* and failed to give credit to author Laurence Bergreen for material taken from Bergreen's book *James Agee: A Life*. (See "A Critic at Large," *The New Yorker*, July 18, 1988; "Hersey Apologizes to a Writer Over an Article on Agee," *New York Times*, July 22, 1988; and "Department of Amplification," *The New Yorker*, August 1, 1988).

Borrowed Words Not Enclosed in Quotation Marks

ORIGINAL: Historically, only a handful of families have dominated the fireworks industry in the West. Details such as chemical recipes and mixing procedures were cloaked in secrecy and passed down from one generation to the next. . . . One effect of familial secretiveness is that, until recent decades, basic pyrotechnic research was rarely performed, and even when it was, the results were not generally reported in scientific journals. (Conkling, John A. "Pyrotechnics." *Scientific American* July 1990: 96.)

PLAGIARISM: John A. Conkling points out that until recently, little scientific research was done on the chemical properties of fireworks, and when it was, the results were not generally reported in scientific journals (96).

Even though the student who wrote this passage documented the source of his information, he did not acknowledge that he borrowed the source's exact wording. To correct this problem, the student should paraphrase the source's words or use quotation marks to acknowledge his borrowing.

CORRECT (BORROWED WORDS IN QUOTATION MARKS): John A. Conkling points out that until recently, little scientific research was done on the chemical properties of fireworks, and when it was, "the results were generally not reported in scientific journals" (96).

CORRECT (PARAPHRASE): John A. Conkling points out that research conducted on the chemical composition of fireworks was seldom reported in the scientific literature (96).

Paraphrase Too Close to Its Source

ORIGINAL: Let's be clear: this wish for politically correct casting goes only one way, the way designed to redress the injuries of centuries. When Pat Carroll, who is a woman, plays Falstaff, who is not, casting is considered a stroke of brilliance. When Josette Simon, who is black, plays Maggie in *After the Fall*, a part Arthur Miller patterned after Marilyn Monroe and which has traditionally been played not by white women, but by blonde white women, it is hailed as a breakthrough.

But when the pendulum moves the other way, the actors' union balks. (Quindlen, Anna. "Error, Stage Left." *New York Times* 12 Aug. 1990, sec. 1:21)

PLAGIARISM: Let us be honest. The desire for politically appropriate casting only goes in one direction, the direction intended to make up for the damage done over hundreds of years. When Pat Carroll, a female, is cast as Falstaff, a male, the decision is a brilliant one. When Josette Simon, a black woman, is cast as Maggie in *After the Fall*, a role Arthur Miller based on Marilyn Monroe and which has usually been played by a woman who is not only white but also blonde, it is considered a major advance.

But when the shoe is on the other foot, the actors' union resists (Quindlen 21).

Although this student documents the passage and does not use the exact words of her source, she closely imitates the original's syntax and phrasing. In fact, all she has really done is substitute synonyms for the author's words; the distinctive style of the passage is still the author's. The student could have avoided plagiarism by changing the syntax as well as the words of the original.

> **CORRECT (PARAPHRASE IN STUDENT'S OWN WORDS; ONE DISTINCTIVE PHRASE PLACED IN QUOTATION MARKS):** According to Anna Quindlen, the actors' union supports "politically correct casting" (21) only when it means casting a woman or or minority group member in a role created for a male or a Caucasian. Thus, it is acceptable for actress Pat Carroll to play Falstaff or for black actress Josette Simon to play Marilyn Monroe; in fact, casting decisions such as these are praised. But when it comes to casting a Caucasian in a role intended for an African American, Asian, or Hispanic, the union objects (21).

Statistics Not Attributed to a Source

> **ORIGINAL:** From the time they [male drivers between 16 and 24] started to drive, 187 of these drivers (almost two-thirds) reported one or more accidents, with an average of 1.6 per involved driver. Features of 303 accidents are tabulated in Table 2. Almost half of all first accidents occurred before the legal driving age of 18, and the median age of all accidents was 19. (Schuman, Stanley, et al. "Young Male Drivers: Accidents and Violations." *JAMA* 50 (1983): 1027)

> **PLAGIARISM:** By and large male drivers between the ages of 16 and 24 accounted for the majority of accidents. Of 303 accidents recorded in Michigan, almost one half took place before the drivers were legally allowed to drive at 18.

The student who used this information assumed statistics are common knowledge—because they may be found in many sources and are accepted as accurate by many experts in a field. Statistics, however, are virtually always the result of original research that deserves acknowledgment. Moreover, readers will be interested in the source of any statistics in order to determine their reliability. For these reasons, you should always document any use of statistics.

> **CORRECT:** According to one study, male drivers between the ages of 16 and 24 accounted for the majority of accidents. Of 303 accidents recorded, almost one half took place before the drivers were legally allowed to drive at 18 (Schuman et al., 1027).

Writer's Words and Ideas Not Differentiated from Those of the Source

ORIGINAL: At some colleges and universities traditional survey courses of world and English literature . . . have been scrapped or diluted. At others they are in peril. At still others they will be. What replaces them is sometimes a mere option of electives, sometimes "multicultural" courses introducing material from Third World cultures and thinning out an already thin sampling of Western writings, and sometimes courses geared especially to issues of class, race, and gender. Given the notorious lethargy of academic decision-making, there has probably been more clamor than change; but if there's enough clamor, there will be change. (Howe, Irving. "The Value of the Canon." *The New Republic* 2 Feb. 1991: 40–47)

PLAGIARISM: Debates about expanding the literary canon take place at many colleges and universities across the United States. At many universities the Western literature survey courses have been edged out by courses that emphasize minority concerns. These courses are "thinning out an already thin sampling of Western writings" in favor of courses geared especially to issues of "class, race, and gender" (Howe 40).

GUIDELINES FOR AVOIDING PLAGIARISM

- **Take careful notes.** Make certain you have recorded information from your sources carefully and accurately.
- **In your notes, put all words taken from sources inside circled quotation marks** and enclose your own comments within brackets.
- **In your paper**, **differentiate your ideas from those of your sources** by clearly introducing borrowed material with the author's name and by following it with documentation.
- **Enclose all direct quotations** used in your paper within quotation marks.
- **Review paraphrases and summaries in your paper** to make certain they are in your own words and that any distinctive words and phrases from a source are quoted.

See ◄
Ch.40

- **Document all direct quotations and all paraphrases and summaries** of your sources.
- **Document all facts** that are open to dispute or are not common knowledge.
- **Document all opinions, conclusions, figures, tables, graphs, and charts** taken from a source.

Because the student who wrote this passage does not differentiate his ideas from those of his source, it appears he borrowed only the quotations in the last sentence. Actually, the student is indebted to his source for the second sentence of the passage as well. By blending his ideas with Howe's, the student passes off some of his source's ideas as his own and unwittingly commits plagiarism. He should have clearly defined the boundaries of the borrowed material by placing a running acknowledgment *before*—and documentation *after*—the borrowed material. In the following correct example, notice that both the summary and the quotation are documented. (A quotation always requires separate documentation.)

> **CORRECT:** Debates about expanding the literary canon take place at many colleges and universities across the United States. According to the noted critic Irving Howe, at many universities the Western literature survey courses have been edged out by courses that emphasize minority concerns (40). These courses, says Howe, are "thinning out an already thin sampling of Western writings in favor of courses geared especially to issues of class, race, and gender" (40).

STUDENT WRITER AT WORK

WORKING WITH SOURCE MATERIAL

This student paragraph uses material from three sources, but its author has neglected to cite them. After reading the paragraph and the three sources that follow it, identify material that has been quoted directly from a source. Compare the wording against the original for accuracy and insert quotation marks where necessary, being sure the quoted passages fit smoothly into the paragraph. Next, paraphrase any passages the student did not need to quote, and, after consulting Chapter 40, document each piece of information that requires it.

Student Paragraph: Oral History

Oral history became a legitimate field of study in 1948, when the Oral History Research Office was established by Allan Nevins. Like recordings of presidents' fireside chats and declarations of war, oral history is both oral and historical. But it is more: Oral history is the creation of new historical documentation, not the recording or preserving of documentation that already exists. Oral history also tends to be more spontaneous and personal and less formal than ordinary tape recordings. Nevins's purpose was to collect and prepare materials to help future historians to better

continued on the following page

STUDENT WRITER AT WORK

Working with Source Material
Material taken from a source has been underlined, with comments about it bracketed.

Oral history became a legitimate field of study in 1948, when the Oral History Research Office was established by Allan Nevins. [factual, but not readily accessible; source 1] Like recordings of presidents' fireside chats and declarations of war, oral history is both oral and historical. [slightly garbled, but essentially direct from source 3] But it is more: oral history is the creation of new historical documentation, not the recording or preserving of documentation that already exists. [patches together, in exact words, two sentences from source 3] Oral history also tends to be more spontaneous and personal and less formal than ordinary tape recordings. [paraphrases source 2] Nevins's purpose was to collect and prepare materials to help future historians to better understand the past. [paraphrase—not a very good one—of source 1] Oral history has enormous potential to do just this, for it draws on people's memories of their own lives and deeds and of their associations with particular people, periods, and events. [mixed paraphrase and quotation from source 2] The result, when it is recorded and transcribed, is a valuable new source. [general knowledge; idea in all sources]

Commentary
Note that much of this student's work is sloppy—if it is meant to be quotation, quotation marks should be used; if it is meant to be paraphrase,

continued on following page

continued from previous page

it is not revised enough from the exact words and structure of the original to be anything but plagiarism, even if there had been documentation. And there is no attribution.

Possible Revision

Oral history became a legitimate field of study in 1948, when Allan Nevins established the Oral History Research Office (Benison 71). [source cited for fact not easily accessible] Like historical recordings, oral history is both oral and historical, but it is more: It requires, according to Norman Hoyle, developing "a new historical document" instead of saving already existing materials (61). [new paraphrase, incorporating quotation, requires documentation and should have attribution] Oral history also tends to be more spontaneous and personal and less formal than ordinary tape recordings (Rumics 602). [paraphrase documented] Nevins's purpose in establishing this office was to provide future historians with material which would enable them to understand "contemporary historical moments" (Benison 71). [mixed paraphrase and quotation requires documentation] Oral history has enormous potential to foster this understanding, for such history draws on people's "reminiscences . . . and their associations with a particular person, period, or event" (Rumics 602). [quotation, which uses ellipses to indicate omitted material, requires documentation] Consequently, when the interview is recorded and transcribed, the result is a valuable new source of historical information. [no documentation as this information is common to all sources]

continued on following page

continued from the previous page

understand the past. Oral history has enormous potential to do just this, for it draws on people's memories of their own lives and deeds and of their associations with particular people, periods, or events. The result, when it is recorded and transcribed, is a valuable new source.

Source 1

When Allan Nevins set up the Oral History Research Office in 1948, he looked upon it as an organization that in a systematic way could obtain from the lips and papers of living Americans who had led significant lives a full record of their participation in the political, economic, and cultural affairs of the nation. His purpose was to prepare such material for the use of future historians. It was his conviction that the individual played an important role in history and that an individual's autobiography might in the future serve as a key to an understanding of contemporary historical movements. (Excerpted from Benison, Saul. "Reflections on Oral History." The American Archivist 28.1 [January 1965]: 71)

Source 2

Typically, an oral history project comprises an organized series of interviews with selected individuals or groups in order to create new source materials from the reminiscences of their own life and acts or from their association with a particular person, period, or event. These recollections are recorded on tape and transcribed on a typewriter into sheets of transcript. . . . Such oral history may be distinguished from more conventional tape recordings of speeches, lectures, symposia, etc., by the fact that the former creates new sources through the more spontaneous, personal, multitopical, extended narrative, while the latter utilizes sources in a more formal mode for a specific occasion. (Excerpted from Rumics, Elizabeth. "Oral History: Defining the Term." Wilson Library Bulletin 40 [1966]: 602)

Source 3

Oral history, as the term came to be used, is the creation of new historical documentation, not the recording or preserving of documenation—even oral documentation—that already exists. Its purpose is not, like that of the National Voice Library at Michigan

Source 39d

State University, to preserve the recordings of fireside chats or presidential declarations of war or James Whitcomb Riley reciting "Little Orphan Annie." These are surely oral and just as surely the stuff of history; but they are not oral history. For this there must be the creation of a new historical document by means of a personal interview. (Excerpted from Hoyle, Norman. "Oral History." Library Trends [July 1972]: 61)

continued from previous page

Commentary on Possible Revision
Note changes which have been made to strengthen coherence and unity.

These sources, including inclusive page numbers for the complete articles, would appear in a list of works cited at the end of the paper.

Documentation

Life, as a wit once said, is an open-book test. Encourage your students to learn as much as they can in your overview of this chapter, but remind them that this chapter will be most valuable as an "open-book" reference and guide for future research papers.

Help students understand the importance of careful documentation—not simply to protect against charges of plagiarism, but to make things easier for readers. Documentation enables readers to check the writer's sources and statistics or to locate cited source material for use in another related research project.

Some students will overdocument and some will underdocument. You might spend some class time going over examples of the types of information typically found in student papers and asking students to tell you what needs documentation.

CHAPTER 40

Documentation

Documentation, the formal acknowledgment of the sources you use in your paper, enables your readers to judge the quality and originality of your work and to determine how authoritative and relevant each work you cite is. Different academic disciplines use different documentation styles. This chapter explains and illustrates the documentation styles recommended by the Modern Language Association (MLA), *The Chicago Manual of Style* (CMS), the American Psychological Association (APA), and the Council of Biology Editors (CBE).

40a Knowing What to Document

You should document any information you borrow from a source, except information that is common knowledge. In addition to printed material, sources may include interviews, conversations, films, records, or radio or television programs.

WHAT TO DOCUMENT
Do Document

- Direct quotations
- Opinions, judgments, and insights of others that you summarize or paraphrase
- Information that is not widely known
- Information that is open to dispute
- Information that is not commonly accepted
- Tables, charts, graphs, and statistics taken from a source

continued on the following page

---

continued from the previous page

> **Do Not Document**
>
> - Your own ideas, observations, and conclusions
> - Common knowledge: information that is widely available in reference books, newspapers, and magazines
> - Familiar quotations

Your documentation should clearly identify the source of each piece of information. For this reason, you should avoid using a single reference to cover several pieces of information from a variety of different sources. Instead, place documentation after each quotation and at the end of each passage of paraphrase or summary. Be sure to place documentation so that it will not interrupt your ideas—ideally, at the end of a sentence. Finally, be sure to differentiate your ideas from those of your sources by placing running acknowledgments before, and documentation after, all borrowed material.

▶ See 39c

40b Using MLA Format*

MLA format is recommended by the Modern Language Association and is required by many teachers of English and other languages as well as teachers of other humanities disciplines. (Student papers illustrating the use of MLA format appear in **41j** and **43d1**.)

(1) Parenthetical references in the text

MLA documentation uses parenthetical references within the text keyed to a list of works cited at the end of the paper. A typical reference consists of the author's last name and a page number.

▶ See 40b2

The colony's religious and political freedom appealed to many idealists in Europe (Ripley 132).

To distinguish two or more sources by the same author, shorten the title of each work to one or two key words, and include the appropriate shortened title in the parenthetical reference after the author's name.

*MLA documentation format follows the guidelines set in the *MLA Handbook for Writers of Research Papers*, 3rd ed. New York: MLA, 1988.

Penn emphasized his religious motivation (Kelley, William Penn 116)

If you state the author's name or the title of the work in your sentence, do not include it in the parenthetical reference that follows.

Penn's political motivation is discussed by Joseph P. Kelley in Pennsylvania, The Colonial Years, 1681–1776 (44).

GUIDELINES FOR PUNCTUATING WITH PARENTHETICAL REFERENCES: MLA

Paraphrases and summaries Parenthetical references are placed *before* the end punctuation.

> Penn's writings epitomize seventeenth-century religious thought (Dengler and Curtis 72).

Quotations run in with the text Parenthetical references are placed *after* the quotation but *before* the end punctuation.

> As Ross says, "Penn followed this conscience in all matters" (127).

> We must now ask, as Ross does, "Did Penn follow Quaker dictates in his dealings with Native Americans" (128)?

> According to Williams, Penn's utopian vision was informed by his Quaker beliefs . . ." (72).

Quotations set off from the text Parenthetical references are placed two spaces *after* the end punctuation.

> According to Arthur Smith, William Penn envisioned a state based on his religious principles:
>
> > Pennsylvania would be a commonwealth in which all individuals would follow God's truth and develop according to God's law. For Penn this concept of government was self-evident. It would be a mistake to see Pennsylvania as anything but an expression of Penn's religious beliefs. (314)

DIRECTORY OF MLA PARENTHETICAL REFERENCES

1. A work by one author
2. A work by two or three authors
3. A work by more than three authors
4. A book with a volume and page number
5. A work without a listed author
6. A work that is one page long
7. An indirect source
8. More than one work
9. A literary work
10. An entire work
11. Two authors with the same last name
12. A government document or a corporate author
13. Tables and illustrations

Sample MLA Parenthetical References

1. A Work by One Author

Far from cheapening it, the death penalty actually affirms the value of human life (Koch 395).

2. A Work by Two or Three Authors

For works with two or three authors, list the authors in the order in which they appear on the title page, with *and* before the last name.

One group of physicists questioned many of the assumptions of relativity (Harbeck and Johnson 31).

With the advent of behaviorism, psychology began a new phase of inquiry (Cowen, Barbo, and Crum 31–34).

3. A Work by More Than Three Authors

For works with more than three authors, list only the first author, followed by *et al.* ("and others") in place of the rest.

A number of important discoveries were made off the coast of Crete in 1960 (Dugan et al. 63).

4. A Book with a Volume and Page Number

If you list more than one volume of a multivolume work in your Works Cited list, include the appropriate volume and page number, separated by a colon, in the parenthetical reference. Do not use the words or abbreviations for *volume* and *page*.

> In 1912 Virginia Stephen married Leonard Woolf, with whom she founded the Hogarth Press (Woolf 1:17).

If you use only one volume of a multivolume work and have included the volume number in the Works Cited list, include just the page number in the parenthetical reference (Woolf 17).

5. *A Work without a Listed Author*

Use a shortened version of the title in the parenthetical reference, beginning with the word by which it is alphabetized in the Works Cited list.

> Television ratings wars have escalated during the past ten years ("Leaving the Cellar" 102).

6. *A Work That Is One Page Long*

Do not include a page reference for a one-page article.

> Sixty percent of Arab-Americans work in white-collar jobs (El-Badru).

7. *An Indirect Source*

If you must use a statement by one author quoted in the work of another author, indicate that the material is from an indirect source with the abbreviation *qtd. in* ("quoted in").

> Wagner stated that myth and history stood before him "with opposing claims" (qtd. in Winkler 10).

8. *More Than One Work*

Cite each work as you normally would, separating one from another with semicolons.

> The Brooklyn Bridge has been used as a subject by many American artists (McCullough 144; Tashjian 58).

See ◀
40b3 Long parenthetical references distract readers. Whenever possible, present them as content notes.

9. *A Literary Work*

When citing literary works it is often helpful to include more than just the author and page number. For example, the chapter number of a novel enables readers to locate your reference in any edition of the work. For the same reason, biblical citations include chapter and verse as well as an abbreviated title of the book (Gen. 5.12).

In a parenthetical reference to a prose work, begin with the page number, follow it with a semicolon, and then add any additional information that might be necessary.

> In Moby-Dick Melville refers to a whaling expedition funded by Louis XIV of France (151; ch. 24).

In parenthetical references to long poems, cite both division and line numbers, separating them with periods.

> In The Aeneid Virgil describes the ships as cleaving the "green woods reflected in the calm water" (8.124). (In this citation the reference is to book 8, line 124 of *The Aeneid*.)

In citing classic verse plays, include the act, scene, and line numbers separated by periods (*Macbeth* 2.2.14–16 or II.ii. 14–16).

10. An Entire Work

When citing an entire work, include just the author's name in the text of your paper or in a parenthetical reference.

> Northrup Frye's Fearful Symmetry presents a complex critical interpretation of Blake's poetry.

> Fearful Symmetry presents a complex critical interpretation of Blake's poetry (Frye).

11. Two Authors with the Same Last Name

To distinguish authors with the same last name, include their initials in the parenthetical references.

> Recent increases in crime have probably caused thousands of urban homeowners to install alarms (Weishoff, R. 115). Some of these alarms use sophisticated sensors that were developed by the U.S. Army (Weishoff, C. 76).

12. A Government Document or a Corporate Author

If the author of a work is listed as a government agency or corporation, cite the work using the organization's name followed by the page number (American Automobile Association 34). You can avoid long parenthetical references by working the organization's name into the text.

According to the President's Commission for the Study of Ethical Problems in Medicine and Biomedical and Behavioral Research, the legal issues relating to euthanasia are complicated (76).

13. Tables and Illustrations

Place the documentation for tables and illustrations below the illustrative material. (**See Appendix A** for the format for tables and illustrations.)

(2)   Works Cited list

The **Works Cited list**, which appears at the end of your paper, lists all the research materials you cite. If your instructor tells you to list all the sources you read, whether you actually cited them or not, use the title *Works Consulted*.

GUIDELINES FOR PREPARING THE WORKS CITED SECTION: MLA

- Begin the list of works cited on a new page after the last page of text or content notes.
- Number the page on which the list of works cited begins as the next page of the paper—for example, if your paper ends on page 8, the Works Cited section begins on page 9.
- List entries alphabetically according to the author's last name. List the author's full name as it appears on the title page. Alphabetize unsigned sources by the first main word of the title (excluding *a, an,* or *the*).
- Type the first line of each entry flush with the left-hand margin; indent subsequent lines five spaces.
- Separate the major divisions (author, title, publication data) within each entry with a period and two spaces.
- Double-space within and between entries.

Works Cited Format: MLA

Author's last name   First name   Underlined title (all major words capitalized)
         ↓ [1]   ↓ [2]   ↓   [2]   [1]
     Dyson, Freeman. Disturbing the Universe. New York:

Double →   [1]   ↑   ↑
space   Harper, 1979.
                        ↑   ↑
                   Publisher   Date   Period   City

DIRECTORY OF MLA WORKS CITED ENTRIES

Citations for Books

1. A book by one author
2. A book by two or three authors
3. A book by more than three authors
4. Two or more books by the same author
5. An edited book
6. An essay in an anthology
7. More than one essay from the same anthology
8. A multivolume work
9. The foreword, preface, or afterword of a book
10. A short story, play, or poem in an anthology
11. A short story, play, or poem in a collection of an author's work
12. A book-length poem
13. A book whose title contains a title that is normally enclosed within quotation marks
14. A book whose title contains a title that is normally underlined
15. A translation
16. A reprint of an older edition
17. A dissertation (published/unpublished)
18. An article in an encyclopedia (signed/unsigned)
19. A pamphlet
20. A government publication

Citations for Articles

21. An article in a scholarly journal with continuous pagination through an annual volume
22. An article in a scholarly journal with separate pagination in each issue
23. An article in a weekly magazine (signed/unsigned)
24. An article in a monthly magazine
25. An article that does not appear on consecutive pages
26. An article in a newspaper (signed/unsigned)
27. An editorial
28. A letter to the editor
29. A book review (titled/untitled)
30. An article whose title includes a quotation or a title within quotation marks
31. An article whose title includes a title that is normally underlined

continued on the following page

continued from the previous page

Citations for Nonprint Sources

32. Computer software
33. Material from an on-line information service
34. A lecture
35. A personal interview
36. A published interview
37. A personal letter
38. A published letter
39. A letter in a library's archives
40. A film
41. A videotape
42. A radio or television program
43. A recording

Sample MLA Works Cited Entries: Books   Book citations include the author's name; the book title (underlined); and the publication information (place, publisher, and date). Capitalize all major words of the title except articles, prepositions, conjunctions, and the *to* of an infinitive (unless such a word is the first or last word of the title or subtitle). Use a short form of the publisher's name; do not include *Incorporated, Publishers,* or *Company* after the name of the publisher. *Alfred A. Knopf, Inc.,* for example, is shortened to *Knopf,* and *Oxford University Press* becomes *Oxford UP.*

1. A Book by One Author

Bettelheim, Bruno.   The Uses of Enchantment: The Meaning
    and Importance of Fairy Tales.   New York: Knopf, 1976.

When citing an edition other than the first, indicate the edition number as it appears on the work's title page.

Gans, Herbert J.   The Urban Villagers.   2nd ed.   New York:
    Free, 1982.

2. A Book by Two or Three Authors

List the first author last name first. Subsequent authors are listed first name first in the order in which they appear on the book's title page.

Davidson, James West, and Mark Hamilton Lytle.   After the
    Fact: The Art of Historical Detection.   New York: Knopf,
    1982.

3. A Book by More Than Three Authors

For more than three authors, list only the first author, followed by *et al.* ("and others").

> Spiller, Robert E., et al., eds. Literary History of the United
> States. New York: Macmillan, 1974.

4. Two or More Books by the Same Author

List books by the same author in alphabetical order by title. Three hyphens followed by a period take the place of the author's name after the first entry.

> Thomas, Lewis. The Lives of a Cell: Notes of a Biology
> Watcher. New York: Viking, 1974.
>
> ---. The Medusa and the Snail: More Notes of a Biology
> Watcher. New York: Viking, 1979.

If the author is the editor or translator of the second entry, place a comma and the appropriate abbreviation after the hyphens (---, ed.).

5. An Edited Book

An edited book is a work that has been prepared for publication by a person other than the author. If your emphasis is on the author's work, begin your citation with the author's name.

> Bartram, William. The Travels of William Bartram. Ed. Mark
> Van Doren. New York: Dover, 1955.

If your emphasis is on the editor's work, begin your citation with the editor's name.

> Van Doren, Mark, ed. The Travels of William Bartram. By
> William Bartram. New York: Dover, 1955.

6. An Essay in an Anthology

When citing an essay appearing in an anthology, include the *full* span of pages on which the whole essay appears, even though you may cite only one page in your paper.

> Lloyd, G. E. R. "Science and Mathematics." The Legacy of
> Greece. Ed. Moses I. Finley. New York: Oxford UP,
> 1981. 256–300.

If the essay you cite has been published previously, include publishing data for the first publication followed by the current information along with the abbreviation *Rpt. in* ("Reprinted in").

Warren, Austin. "Emily Dickinson." The Sewanee Review 17
(1957): 132–77. Rpt. in Emily Dickinson: A Collection
of Critical Essays. Ed. Richard B. Sewall. Englewood
Cliffs: Prentice, 1963. 101–16.

NOTE: If you cite an essay that appears in a collection of the author's work, use the format illustrated in number 11.

7. *More Than One Essay from the Same Anthology*

If you cite more than one essay from the same anthology, list each essay separately, followed by a cross-reference to the entire anthology. List complete publication information for the anthology itself.

Bolgar, Robert R. "The Greek Legacy." Finley 429–72.

Finley, Moses I., ed. The Legacy of Greece. New York: Oxford
UP, 1981.

Williams, Bernard. "Philosophy." Finley 202–55.

8. *A Multivolume Work*

When all volumes of a multivolume work have the same title, include the number of the volume you are using.

Raine, Kathleen. Blake and Tradition. Vol. 1. Princeton:
Princeton UP, 1968. 2 vols.

When using two or more volumes, cite the entire work.

Raine, Kathleen. Blake and Tradition. 2 vols. Princeton:
Princeton UP, 1968.

If the volume you are using has an individual title, supply the title after the author's name, followed by the number of the volume, the title of the entire work, the total number of volumes, and the inclusive publication dates.

Durant, Will, and Ariel Durant. The Age of Napoleon: A History of European Civilization from 1789 to 1815. New
York: Simon, 1975. Vol. 11 of The Story of Civilization.
11 vols. 1935–75.

9. *The Foreword, Preface, or Afterword of a Book*

Taylor, Telford. Preface. Less Than Slaves. By Benjamin B.
Ferencz. Cambridge: Harvard UP, 1979. xiii–xxii.

If the author of the foreword, preface, or afterword is also the
author of the book, use *By* and the author's last name after the
title of the book.

Killingray, David. Introduction. A Plague of Europeans: West-
erners in Africa Since the Fifteenth Century. By
Killingray. Baltimore: Penguin, 1973. 6–12.

10. *A Short Story, Play, or Poem in an Anthology*

Singer, Isaac Bashevis. "The Spinoza of Market Street." The
Norton Anthology of Short Fiction. Ed. R. V. Cassill. 3rd
ed. New York: Norton, 1986. 1210–24.

Shakespeare, William. Othello, The Moor of Venice. Shakespeare: Six Plays and the Sonnets. Ed. Thomas Marc
Parrott and Edward Hubler. New York: Scribner's, 1956.
145–91.

**11. *A Short Story, Play, or Poem in a Collection of an Author's
Work***

Singer, Isaac Bashevis. "The Spinoza of Market Street." The
Collected Stories of Isaac Bashevis Singer. New York:
Farrar, 1983. 79–93.

Walcott, Derek. "Nearing La Guaira." Selected Poems. New
York: Farrar, 1944. 47–48.

12. *A Book-Length Poem*

Eliot, T. S. The Waste Land. T. S. Eliot: Collected Poems
1909–1962. New York: Harcourt, 1963. 51–70. (The title
of a book-length poem is underlined.)

**13. *A Book Whose Title Contains a Title That Is Normally
Enclosed within Quotation Marks***

If the book you are citing contains a title that is normally enclosed within quotation marks, keep the quotation marks.

Herzog, Alan, ed. Twentieth Century Interpretations of "To a
Skylark." Englewood Cliffs: Prentice, 1975.

14. A Book Whose Title Contains a Title That Is Normally Underlined

If the work you are citing contains another title that you would normally underline (a novel, play, or long poem, for example), do not underline the interior title . (Do not use italics if your are writing with a computer.)

> Knoll, Robert E. ed. Storm Over The Waste Land. Chicago: Scott, 1964.

15. A Translation

> García Márquez, Gabriel. One Hundred Years of Solitude. Trans. Gregory Rabassa. New York: Avon, 1991.

16. A Reprint of an Older Edition

Include the original publication date after the title.

> Wharton, Edith. The House of Mirth. 1905. New York: Scribner's, 1975. (1905 is the original publication date.)

17. A Dissertation (Published/Unpublished)

Enter a published dissertation the way you would a book. For dissertations published by University Microfilms International (UMI), include the order number.

> Peterson, Shawn. Loving Mothers and Lost Daughters: Images of Female Kinship Relations in Selected Novels of Toni Morrison. Diss. U of Oregon, 1993. Ann Arbor: UMI, 1994. DA 9322935.

Use quotation marks for the title of an unpublished dissertation.

> Romero, Yolanda Garcia. "The American Frontier Experience in Twentieth-Century Northwest Texas." Diss. Texas Tech U, 1993.

18. An Article in an Encyclopedia (Signed/Unsigned)

Enter the title of an unsigned article just as it is listed in the encyclopedia. No volume number or page numbers are needed.

> "Cubism." Encyclopaedia Britannica: Micropaedia. 1974.

For a signed article, begin with the author's name.

Monro, D. H. "Humor." The Encyclopedia of Philosophy. 1974 ed.

When citing relatively unfamiliar encyclopedias, especially those that have appeared in only one edition, give full publication information.

Hendrickson, Robert. "Melting Pot." The Facts on File Encyclopedia of Word and Phrase Origins. New York: Facts on File, 1987.

19. A Pamphlet

Enter pamphlets as if they were books. If no author is listed, enter the underlined title first.

Existing Light Photography. Rochester: Kodak, 1989.

20. A Government Publication

If the publication has no author, begin with the name of the government, followed by the name of the agency.

United States. President's Commission for the Study of Ethical Problems in Medicine and Biomedical and Behavioral Research. Deciding to Forgo Life-Sustaining Treatment: Ethical, Medical, and Legal Issues in Treatment Decisions. Washington: GPO, 1989.

If the publication has an author, begin with either the author's name or the agency.

Wilson, J. R. Hearing on Minority Unemployment. U.S. 91st Cong., 2nd sess. Washington: GPO, 1970.

United States. Cong. House. Hearing on Minority Unemployment. By J. R. Willson. 91st Cong., 2nd sess. Washington: GPO, 1970.

Sample MLA Works Cited Entries: Articles Article citations include the author's name; the title of the article, in quotation marks; the underlined name of the periodical; and the pages on which the full article appears, without the abbreviations *p.* or *pp.*

21. An Article in a Scholarly Journal with Continuous Pagination through an Annual Volume

For an article in a journal with continuous pagination—for example, one in which an issue ends on page 172 and the next issue begins with page 173—include the volume number followed by the date of publication in parentheses. Follow the publication date with a colon, a space, and the page numbers.

Huntington, John. "Science Fiction and the Future." College
     English 37 (1975): 340–58.

22. An Article in a Scholarly Journal with Separate Pagination in Each Issue

For a journal in which each issue begins with page 1, add a period and the issue number after the volume number.

Sipes, R. G. "War, Sports, and Aggression: An Empirical Test of
     Two Rival Theories." American Anthropologist 4.2
     (1973): 65–84.

23. An Article in a Weekly Magazine (Signed/Unsigned)

In dates, the day precedes the month. Abbreviate all months except May, June, and July.

Traub, James. "The Hearts and Minds of City College." The
     New Yorker 7 June 1993: 42–53.

"Solzhenitsyn: A Candle in the Wind." Time 23 Mar. 1970: 70.

24. An Article in a Monthly Magazine

Roll, Lori. "Careers in Engineering." Working Woman Nov.
     1982: 62.

25. An Article That Does Not Appear on Consecutive Pages

When an article does not appear on consecutive pages—that is, when it begins on page 58, continues on page 59, and then skips to page 112—include only the first page number and a plus (+) sign.

Rodman, Selden. "Where Art Is Joy." Caribbean Travel and
     Life Oct. 1987: 58+.

26. An Article in a Newspaper (Signed/Unsigned)

Stipp, David. "Japanese Firms Find Little Success in the U.S.
Small Computer Market." <u>Wall Street Journal</u> 11 Sept.
1991, late ed.: 6.

"Soviet Television." <u>Los Angeles Times</u> 13 Dec. 1990, sec. 2: 3+.

27. An Editorial

"Tough Cops, Not Brutal Cops." Editorial. <u>New York Times</u>
5 May 1994, late ed.: A26.

28. A Letter to the Editor

Bishop, Jennifer. Letter. <u>Philadelphia Inquirer</u> 10 Dec. 1992:
A17.

29. A Book Review (Titled/Untitled)

Begin with the reviewer's name, followed by the title of the review (if any), the title and author of the book reviewed, and the
date on which the review appeared.

Fox-Genovese, Elizabeth. "Big Mess on Campus." Rev. of
<u>Illiberal Education: The Politics of Race and Sex on
Campus</u>, by Dinesh D'Souza. <u>Washington Post</u> 15 Apr.
1991, national weekly ed.: 32.

Harris, Joseph. Rev. of <u>Perspectives on Research and Scholarship in Composition</u>, eds. Ben W. McLelland and Timothy
R. Donovan. <u>College Composition and Communication</u> 38
(1987): 101–02.

NOTE: A citation for an unsigned review begins with the title of
the review, if any, or the title of the book reviewed.

30. An Article Whose Title Includes a Quotation or a Title within Quotation Marks

If a title of an article includes material that is normally enclosed within quotation marks, enclose the material within single
quotation marks.

Nash, Robert. "About 'The Emperor of Ice-Cream.'" <u>Perspectives</u>
7 (1954): 122–24.

31. An Article Whose Title Includes a Title That is Normally Underlined

If an article includes a title that would normally be underlined (to indicate italics), underline it in your Works Cited entry.

Leicester, H. Marshall, Jr. "The Art of Impersonation: A General Prologue to The Canterbury Tales." PMLA 95 (1980): 213–24.

Sample MLA Works Cited Entries: Other Sources Citations for other kinds of sources typically include author, title, publication information, and other pertinent material.

32. Computer Software

Citations for computer software include the writer of the program, the title of the program, the version (preceded by *vers.*), the descriptive label *computer software,* the distributor, the year of publication, and any relevant information about the operating system.

Norton, Peter. The Norton Utilities for Macintosh. Vers. 2.0. Computer software. Symantec, 1992. System 6.0 or higher, 1.7 MB, disk.

33. Material from an On-Line Information Service

Enter material from an on-line information service just as you would printed material. Include the name of the information service and the numbers identifying the material.

Baer, Walter S. "Telecommunications Technology in the 1990s." Computer Science June 1994: 152+. Dialog file 102, item 0346142.

34. A Lecture

Sandman, Peter. "Communicating Scientific Information." Communications Seminar, Dept. of Humanities and Communications. Drexel U, 26 Oct. 1994.

35. A Personal Interview

West, Cornel. Personal interview. 28 Dec. 1993.

Tannen, Deborah. Telephone interview. 8 June 1994.

36. A Published Interview

Stravos, George. "An Interview with Gwendolyn Brooks." Contemporary Literature 11.1 (Winter 1970): 1–20.

37. A Personal Letter

Tan, Amy. Letter to the author. 7 April 1993.

38. A Published Letter

Joyce, James. "Letter to Louis Gillet." 20 Aug. 1931. James Joyce. By Richard Ellmann. New York: Oxford, 1965. 631.

39. A Letter in a Library's Archives

Stieglitz, Alfred. Letter to Paul Rosenberg. 5 Sept. 1923. Stieglitz Archive. Yale, New Haven.

40. A Film

Include the title of the film (underlined), the distributor, and the date, along with other information of use to a reader, such as the performers, the director, and the writer.

Citizen Kane. Dir. Orson Welles. With Orson Welles, Joseph Cotton, Dorothy Comingore, and Agnes Moorehead. RKO, 1941.

If you are focusing on the contribution of a particular person, begin with that person's name (Welles, Orson, dir. Citizen Kane . . .).

41. A Videotape

Interview with Arthur Miller. Videocassette. Dir. William Schiff. The Mosaic Group, 1987. 20 min.

42. A Radio or Television Program

"Prime Suspect 3." Writ. Lynda LaPlante. With Helen Mirren. Mystery! PBS. WNET, New York. 28 April 1994.

43. A Recording

Begin with the composer, conductor, or performer (whichever you are emphasizing), followed by the title, manufacturer, catalog number, and year of issue.

Boubill, Alain, and Claude-Michel Schönberg. <u>Miss Saigon</u>.
    With Lea Salonga, Claire Moore, and Jonathan Pryce.
    Cond. Martin Koch. Geffen, DIDX006369, 1989.

When citing jacket notes, include the author's name, the title, and a description of the material—lyrics, for example.

Marley, Bob. "Crisis." Lyrics. <u>Bob Marley and the Wailers</u>.
    Kava Island Records, 422–846, 209–2, 1978.

(3) Content notes

Content notes—multiple bibliographic citations or other material that does not fit smoothly into your paper—may be used to supplement parenthetical documentation. Content notes are indicated by a raised number (superscript) in the paper, which is keyed to the note. These notes can appear either as footnotes at the bottom of the page or endnotes on a separate page, entitled *Notes,* placed after the last page of the paper and before the list of works cited. Content notes are double-spaced within and between entries.

For Multiple Citations   Use content notes instead of combining several bibliographic citations in a single parenthetical reference. (Including many references within a single pair of parentheses can distract readers.)

IN THE PAPER

Many researchers emphasize the necessity of having dying patients share their experiences.[1]

IN THE NOTE

[1]Kübler-Ross 27; Stinnette 43; Poston 70; Cohen and Cohen 31–34; Burke 1:91–95.

For Explanations   Use content notes to provide comments, explanations, or other information that is needed to clarify a point in the text.

IN THE PAPER

The massacre of the Armenians during World War I is an event the survivors could not easily forget.[2]

IN THE NOTE

²Many accounts of the Armenian massacre have been published. For a firsthand account of this event, see Bedoukian 17–81. For a fictional account, see Werfel.

40c Using the Chicago Format

*The Chicago Manual of Style** (CMS) is used in history and some social science and humanities disciplines. (Excerpts from a student paper illustrating the use of CMS format appear in **43d3**.) **CMS format** has two parts: notes at the end of the paper (endnotes) and a list of bibliographic citiations. Although Chicago style encourages the use of endnotes, it also allows the use of footnotes at the bottom of the page. If you are required to use footnotes rather than endnotes, be sure the notes at the bottom of a particular page of your paper correspond to the note numbers on that page.

(1) Endnotes and footnotes

The notes format calls for a raised numeral (superscript) in the text after source material you have either quoted or referred to. This numeral is placed after all punctuation marks except the dash. (If you are documenting a quotation, the note comes directly after the quoted material, not after the author's name or an introductory phrase.) This numeral corresponds to a number placed before the note.

GUIDELINES FOR PREPARING THE LIST OF ENDNOTES: CMS

- Begin endnotes on a new page after the last page of the paper.
- Number the page on which the endnotes appear as the next page of the paper.
- Type and number notes in the order in which they appear in the paper, beginning with number one.

continued on the following page

---

*The Chicago format follows the guidelines set in *The Chicago Manual of Style*. 14th ed. Chicago: University of Chicago Press, 1993.

continued from the previous page

- Type the note number on the line followed by a period and one space.
- Indent the first line of each note three spaces; type subsequent lines flush with the left-hand margin
- Double-space within and between entries.

Endnote and Footnote Formats: CMS

IN THE TEXT

By November of 1942, the Allies had proof that the Nazis were engaged in the systemic killing of Jews.[1]

IN THE NOTE

1. David S. Wyman, The Abandonment of the Jews: America and the Holocaust 1941–1945 (New York: Pantheon Books, 1984), 65.

DIRECTORY OF CMS FOOTNOTES AND ENDNOTES

1. A book by one author
2. A book by two or three authors
3. A book by more than three authors
4. A multivolume work
5. An edited book
6. An essay in an anthology
7. An article in an encyclopedia (signed/unsigned)
8. An article in a scholarly journal with continuous pagination through an annual volume
9. An article in a scholarly journal with separate pagination in each issue
10. An article in a weekly magazine (signed/unsigned)
11. An article in a monthly magazine
12. An article in a newspaper

Sample CMS Footnotes and Endnotes: Books Enter the full title of the book, and use the publisher's full name as it appears on the title page. Although underscoring to indicate italics is acceptable, CMS style encourages the use of italics for titles.

1. A Book by One Author

1. Herbert J. Gans, The Urban Villagers, 2d ed. (New York: Free Press, 1982), 100.

2. A Book by Two or Three Authors

2. James West Davidson and Mark Hamilton Lytle, After the Fact: The Art of Historical Detection (New York: Alfred A. Knopf, 1982), 54.

3. A Book by More Than Three Authors

3. Robert E. Spiller et al., eds., Literary History of the United States (New York: Macmillan, 1974), 24.

4. A Multivolume Work

When all the volumes of a multivolume work have the same title, include the number of the volume you are using.

4. Kathleen Raine, Blake and Tradition (Princeton: Princeton University Press, 1968), 1: 100.

When the volume you are using has an individual title, give the title as well as the volume number.

5. Will Durant and Ariel Durant, The Age of Napoleon: A History of European Civilization from 1789 to 1815, vol. 11, The Story of Civilization (New York: Simon & Schuster, 1975), 90.

5. An Edited Book

6. William Bartram, The Travels of William Bartram, ed. Mark Van Doren (New York: Dover Press, 1955), 85.

6. An Essay in an Anthology

7. Peter Kidson, "Architecture and City Planning," in The Legacy of Greece, ed. M. I. Finley (New York: Oxford University Press, 1981), 376–400.

7. An Article in an Encyclopedia (Signed/Unsigned)

8. The Encyclopedia of Philosophy, 1967 ed., s.v. "Hobbes, Thomas," by R. S. Peters.

9. <u>The Focal Encyclopedia of Photography</u>, 1965 ed., s.v. "daguerreotype."

NOTE: The abbreviation *s.v.* stands for *sub verbo*—under the word.

Sample CMS Footnotes and Endnotes: Articles Article citations include the author's name; the title of the article, in quotation marks; the name of the periodical, underlined or italicized; and the pages on which the full article appears, without the abbreviations *p.* or *pp.*

8. *An Article in a Scholarly Journal with Continuous Pagination through an Annual Volume*

1. John Huntington, "Science Fiction and the Future," <u>College English</u> 37 (fall 1975): 340.

9. *An Article in a Scholarly Journal with Separate Pagination in Each Issue*

2. R. G. Sipes, "War, Sports, and Aggression: An Empirical Test of Two Rival Theories," <u>American Anthropologist</u> 4, no. 2 (1973): 84.

10. *An Article in a Weekly Magazine (Signed/Unsigned)*

3. James Traub, "The Hearts and Minds of City College," <u>The New Yorker</u>, 7 June 1993, 43.

4. "Solzhenitsyn: A Candle in the Wind," <u>Time</u>, 23 March 1970, 70.

11. *An Article in a Monthly Magazine*

5. Lori Roll, "Careers in Engineering," <u>Working Woman</u>, November 1982, 62.

12. *An Article in a Newspaper*

6. Raymond Bonner, "A Guatemalan General's Rise to Power," <u>New York Times</u>, 21 July 1982, 3(A).

Sample CMS Footnotes and Endnotes: Subsequent References The first time you make reference to a work, use the full citation; subsequent references to the same work, however, should list only the author's last name, followed by a comma and a page number.

FIRST NOTE ON ESPINOZA

> 1. J. M. Espinoza, The First Expedition of Vargas in New Mexico, 1962 (Albuquerque: University of New Mexico Press, 1940), 10–15.

SUBSEQUENT NOTE

> 4. Espinoza, 69.

The Chicago Manual of Style allows the use of the abbreviation *ibid.* ("in the same place") for subsequent references to the same work as long as there are no intervening references. *Ibid.* takes the place of the author's name and the work's title—but not the page number.

FIRST NOTE ON ESPINOZA

> 1. J. M. Espinoza, The First Expedition of Vargas in New Mexico, 1692 (Albuquerque: University of New Mexico Press, 1940), 10–15.

SUBSEQUENT NOTE

> 2. Ibid., 69.

NOTE: Keep in mind that use of *Ibid.* is rapidly giving way to the style that uses the author's name in subsequent notes.

(2)   Bibliography

The bibliography lists the sources you have used in your paper. List entries in alphabetical order on a separate page after the last page of your paper or after the endnotes. In addition to the heading *Bibliography,* Chicago style allows *Selected Bibliography, Works Cited, Literature Cited, References,* and *Sources Consulted,* depending on which most accurately describes the works listed.

GUIDELINES FOR PREPARING THE BIBLIOGRAPHY: CMS

- Begin the bibliography on a new page after the list of endnotes or, if footnotes are used instead of endnotes, after the last page of the paper.

continued on the following page

---

continued from the previous page

- Number the page on which the bibliography appears as the next page of the paper.
- List entries alphabetically according to the author's last name. Give the author's full name as it appears on the title page. Alphabetize unsigned sources by the first main word of the title (excluding *a, an,* or *the*).
- Type the first line of each entry flush with the left-hand margin; indent subsequent lines three spaces.
- Separate the major divisions within each entry with a period and one space.
- Double-space within and between entries.

Bibliography Format: CMS

```
Last name   First                    Underlined title
    ↓    [1]  ↓  [1]                       ↓                    [1]
Hilberg, Raul. The Destruction of the European Jews. New
         [1]                    [1]
    York: Harper & Row, 1961.
      ↑            ↑           ↑
     City      Publisher     Date      Double-space
```

DIRECTORY OF CMS BIBLIOGRAPHIC ENTRIES

1. A book by one author
2. A book by two or three authors
3. A book by more than three authors
4. A multivolume work
5. An edited book
6. An essay in an anthology
7. An article in an encyclopedia (signed/unsigned)
8. An article in a scholarly journal with continuous pagination through an annual volume
9. An article in a scholarly journal with separate pagination in each issue
10. An article in a weekly magazine (signed/unsigned)
11. An article in a monthly magazine
12. An article in a newspaper

Sample CMS Bibliographic Entries: Books

1. A Book by One Author

Gans, Herbert J. The Urban Villagers. 2d ed. New York: Free
    Press, 1982.

2. A Book by Two or Three Authors

Davidson, James West, and Mark Hamilton Lytle. After the
Fact: The Art of Historical Detection. New York: Alfred A.
Knopf, 1982.

3. A Book by More Than Three Authors

Spiller, Robert E., et al., eds. Literary History of the United
States. New York: Macmillan, 1974.

4. A Multivolume Work

When all volumes of a multivolume work have the same title,
include the number of the volume you are using.

Raine, Kathleen. Blake and Tradition. Vol. 1. Princeton: Princeton University Press, 1968.

When the volume you are using has an individual title, give the
title after the author's name.

Durant, Will, and Ariel Durant. The Age of Napoleon: A History
of European Civilization from 1789 to 1815. Vol. 11, The
Story of Civilization. New York: Simon & Schuster, 1975.

5. An Edited Book

Bartram, William. The Travels of William Bartran, edited by
Mark Van Doren. New York: Dover Press, 1955.

6. An Essay in an Anthology

Kidson, Peter. "Architecture and City Planning." In The Legacy
of Greece, edited by M. I. Finley, 376–400. New York: Oxford University Press, 1981.

7. An Article in an Encyclopedia (Signed/Unsigned)

The Encyclopedia of Philosophy. 1967 ed., s.v. "Hobbes,
Thomas," by R. S. Peters.

The Focal Encyclopedia of Photography. 1965 ed., s.v.
"Daguerreotype."

Sample CMS Bibliographic Entries: Articles

8. An Article in a Scholarly Journal with Continuous Pagination through an Annual Volume

Huntington, John. "Science Fiction and the Future." <u>College</u>
<u>English</u> 37 (fall 1975): 340–58.

9. *An Article in a Scholarly Journal with Separate Pagination in Each Issue*

Sipes, R. G. "War, Sports, and Aggression: An Empirical Test of
Two Rival Theories." <u>American Anthropologist</u> 4, no. 2
(1973): 65–84.

10. *An Article in a Weekly Magazine (Signed/Unsigned)*

Traub, James. "The Hearts and Minds of City College." <u>The New</u>
<u>Yorker</u>, 7 June 1993, 43.

"Solzhenitsyn: A Candle in the Wind." <u>Time</u>, 23 March 1970, 70.

11. *An Article in a Monthly Magazine*

Roll, Lori. "Careers in Engineering." <u>Working Woman</u>, November 1982, 62.

12. *An Article in a Newspaper*

Bonner, Raymond. "A Guatemalan General's Rise to Power."
<u>New York Times</u>, 21 July 1982, 3(A).

40d   Using APA Format*

APA format, which is used extensively in the social sciences,
relies on short parenthetical citations consisting of the last name
of the author, the year of publication, and—for direct quotations—the page number. These references are keyed to an alphabetical list of references that follows the paper. APA format also
permits content notes placed after the last page of the text. (A student paper illustrating the use of APA format appears in **44d**.)

(1)   Parenthetical references in the text

APA style requires a comma between the name and the date.

---

*APA documentation format follows the guidelines set in the *Publication Manual of
the American Psychological Association*. 4th ed. Washington, DC: APA, 1994.

One study of stress in the workplace (Weisberg, 1983) shows a correlation between. . . .

You should not include in the parenthetical reference any information that appears in the text of your paper.

In his study Weisberg (1983) shows a correlation. . . .
(author's name in text)

In Weisberg's 1983 study of stress in the workplace. . . .
(author's name and date in text)

DIRECTORY OF APA PARENTHETICAL REFERENCES

1. A work by one author
2. Two works by the same author(s), same year
3. A work by two or more authors
4. A work by a corporate author
5. A work with no listed author
6. A personal communication
7. An indirect source
8. A specific part of a source
9. Two or more works within the same parenthetical reference
10. A long quotation
11. Tables

Sample APA Parenthetical References

1. A Work by One Author

Supporters of bilingual education programs often speak about the psychological well-being of the child (Bakka, 1992).

2. Two Works by the Same Author(s), Same Year

If you cite two or more works by the same author that appeared the same year, the one whose title comes first alphabetically is designated *a,* the second, *b* (e.g., Weisberg 1983a and Weisberg 1983b), and so on. These letter designations also appear in the reference list that follows the text of your paper.

▶ See
40d2

The effects of stress in the workplace are measurable (Weisberg, 1983a, 1983b). . . .

3. A Work by Two or More Authors

When a work has two authors, cite both names every time you refer to it.

> There is growing concern over the use of psychological testing in elementary schools (Albright & Glennon, 1982).

If a work has three, four, or five authors, cite all names in the first reference, and in subsequent references cite the first author followed by *et al.* and the year (Sparks et al., 1984).

When a work has six or more authors, cite the last name of the first author followed by *et al.* and the year in the first and subsequent references.

When referring to multiple authors in the text of your paper, join the last two names with *and*: According to Rosen, Wolfe, and Ziff (1988). . . . In parenthetical documentation, however, use an ampersand (Rosen, Wolfe, & Ziff, 1988).

4. A Work by a Corporate Author

The name of a corporate author is usually spelled out each time it appears in a citation. If a name is cumbersome, you may abbreviate it in subsequent citations.

First reference:

(National Institute of Mental Health [NIMH], 1987)

Subsequent reference:

(NIMH, 1987)

5. A Work with No Listed Author

If a work has no listed author, cite the first two or three words of the title (capitalized) and the year. Use quotation marks around the title of an article and underline the title of a periodical or book.

> This explains the fears of some Asian immigrants ("New Immigration," 1994).

6. A Personal Communication

Cite letters, memos, telephone conversations, E-mail, messages from electronic bulletin boards, and the like only in the text (not in the reference list). Give the initial of the first name and the complete last name of the person with whom you communicated and the date.

Another researcher confirms these findings (R. Takaki, personal communication, October 17, 1993)

7. *An Indirect Source*

Identify statements by one author mentioned in the work of another with the phrase *as cited in.*

Cogan and Howe offered very different interpretations of the problem (as cited in Swenson, 1990).

8. *A Specific Part of a Source*

When citing a specific part of a source, use abbreviations for the words *page* (p.), *chapter* (chap.), and *section* (sec.). Always supply the page number for a quotation.

These theories have an interesting history (Lee, 1966, chap. 2).

9. *Two or More Works within the Same Parenthetical Reference*

List works by different authors in alphabetical order by last name. Use semicolons between citations.

. . . among several studies (Barson & Roth, 1985; Rose, 1987; Tedesco, 1982).

List works by the same author or authors in order of date of publication.

. . . among several studies (Weiss & Elliot, 1982, 1984, 1985).

Distinguish works by the same author that appeared in the same year by designating the first *a,* the second *b,* and so on; repeat the year with each citation.

. . . among several studies (Hossack, 1985a, 1985b, 1985c, in press).

(*In press* designates a work about to be published.)

10. *A Long Quotation*

Parenthetical documentation for a quotation of forty words or more should be placed two spaces after the final punctuation.

Todd Gitlin (1987) sees a change in left-wing politics beginning in 1969:

> Women had been the cement of the male-run movement; their "desertion" into their own circles completed the

dissolution of the old boys' clan. While men outside the hard-line factions were miserable with the crumbling of their one-time movement, women were riding high. (1987, p. 374)

NOTE: Long quotations are double-spaced and indented five to seven spaces from the left-hand margin. If the quotation is more than one paragraph long, indent the first line of the second and subsequent paragraphs five spaces. APA style allows triple- or quadruple-spacing before and after long quotations.

11. Tables

Number tables consecutively in the order in which they appear in your paper (Table 1, Table 2, and so on). Double-space each table, and begin each table on a separate page after the reference list. Indicate the location of each table with a specific reference in the text of your paper: *Table 3 summarizes this study*.

(2)   Reference list

The list of all the sources cited in your paper falls at the end on a new numbered page headed *References*. If you are listing all the works you consulted, regardless of whether they are cited in your paper, label your list *Bibliography*.

GUIDELINES FOR PREPARING THE REFERENCE LIST: APA

- Begin the reference list on a new page after the last page of the paper or after the content notes.
- Number the page on which the reference list appears as the next page of the paper.
- List the items in the reference list alphabetically. Spell out the author's last name, and use initials for the first and middle names.
- Indent the first line of each entry five to seven spaces; type subsequent lines flush with the left-hand margin.
- Separate the major divisions within each entry with a period and one space.
- Double-space within and between entries.

GUIDELINES FOR ARRANGING WORKS IN THE REFERENCE LIST

- Single-author entries precede multiple-author entries that begin with the same name.

  Field, S. (1987). . . .

  Field, S., & Levitt, M. P. (1984). . . .

- Entries by the same author or authors are arranged according to date of publication, starting with the earliest date.

  Ruthenberg, H., & Rubin, R. (1985). . . .

  Ruthenberg, H., & Rubin, R. (1987). . . .

- Entries with the same author(s) and date of publication are arranged alphabetically according to title, with the first designated *a*, the second *b*, and so on.

  Wolk, E. M. (1986a). Analysis. . . .

  Wolk, E. M. (1986b). Hormonal. . . .

- Entries by the same first author and different second author are arranged alphabetically according to the last name of the second author.

  Chisolm, T., & Abel, J. (1993). . . .

  Chisolm, T., & Gainor, C. (1993). . . .

Reference List Format: APA

Last name Initials Date Underlined title (only first word capitalized)
[1] [1] [1]

Morgan, C. T. (1986). Introduction to psychology. New

York: Knopf.
[1]
City Publisher Double-space

DIRECTORY OF APA REFERENCE LIST ENTRIES

1. A book with one author
2. A book with more than one author
3. An edited book

continued on the following page

continued from the previous page

4. A book with no listed author or editor
5. A work in several volumes
6. A book with a corporate author
7. A government report
8. One selection from an anthology
9. Two selections from the same anthology
10. An article in a reference book
11. The foreword, preface, or afterword of a book
12. An article in a scholarly journal with continuous pagination through an annual volume
13. An article in a scholarly journal with separate pagination in each issue
14. A magazine article
15. A newspaper article (signed/unsigned)
16. A letter to the editor
17. A personal communication
18. An interview
19. Electronic media

Sample APA Reference List Entries: Books Capitalize only the first word of the title and the first word of the subtitle. Underline the entire title (including end punctuation) and enclose the date, volume number, and edition number in parentheses.

1. A Book with One Author

Maslow, A. H. (1974). <u>Toward a psychology of being.</u> Princeton: Van Nostrand.

2. A Book with More Than One Author

List all the authors—by last name and initials—regardless of how many there are. All authors are cited last name first. Use an ampersand (&), not *and,* before the last author's name.

Wolfinger, D., Knable, P., Richards, H. L., & Silberger, R. (1990). <u>The chronically unemployed.</u> New York: Berman Press.

3. An Edited Book

Lewin, K., Lippitt, R., & White, R. K. (Eds.). (1985). <u>Social learning and imitation.</u> New York: Basic Books.

4. A Book with No Listed Author or Editor

Writing with a computer. (1993). Philadelphia: Drexel Press.

5. A Work in Several Volumes

Jones, P. R., & Williams, T. C. (Eds.). (1990–1993). Handbook of therapy (Vols. 1–2). Princeton: Princeton University Press.

NOTE: The parenthetical citation in the text would be (Jones & Williams, 1990–1993).

6. A Book with a Corporate Author

League of Women Voters of the United States. (1991). Local league handbook. Washington, DC: Author.

NOTE: When the author and publisher are the same, use the word *Author* at the end of the citation instead of repeating the publisher's name.

7. A Government Report

National Institute of Mental Health. (1987). Motion pictures and violence: A summary report of research (DHHS Publication No. ADM 91-22187). Washington, DC: U.S. Government Printing Office.

8. One Selection from an Anthology

Lorde, A. (1984). Age, race, and class. In P. S. Rothenberg (Ed.), Racism and sexism: An integrated study (pp. 352–360). New York: St. Martin's Press.

9. Two Selections from the Same Anthology

Give the full citation for the anthology in each entry.

Lorde, A. (1984). Age, race, and class. In P. S. Rothenberg (Ed.), Racism and sexism: An integrated study (pp. 352–360). New York: St. Martin's Press.

Rimer, S. (1986). At a welfare hotel, mothers find support in weekly talks. In P. S. Rothenberg (Ed.), Racism and sexism: An integrated study (pp. 435–360). New York: St. Martin's Press.

10. An Article in a Reference Book

Edwards, P. (Ed.). (1987). The encyclopedia of philosophy (4th ed., Vols. 1–8). New York: Macmillian.

11. The Foreword, Preface, or Afterword of a Book

Taylor, T. (1979). Preface. In Less than slaves by Benjamin B. Ferencz. Cambridge: Harvard University Press.

Sample APA Reference List Entries: Articles   Capitalize only the first word of the title and the first word of the subtitle. Do not underline the title of the article or enclose it in quotation marks. Give the periodical title in full; underline it and capitalize all major words. Underline the volume number and include the issue number in parentheses. Give inclusive page numbers. Use *pp.* when referring to page numbers in newpapers, but omit this abbreviation when referring to page numbers in periodicals with volume numbers.

12. An Article in a Scholarly Journal with Continuous Pagination through an Annual Volume

Miller, W. (1969). Violent crimes in city gangs. Journal of Social Issues, 27, 581–593.

13. An Article in a Scholarly Journal with Separate Pagination in Each Issue

Williams, S., & Cohen, L. R. (1984). Child stress in early learning situations. American Psychologist, 21 (10), 1–28.

14. A Magazine Article

McCurdy, H. G. (1983, June). Brain mechanisms and intelligence. Psychology Today, 13, 61–63.

15. A Newspaper Article (Signed/Unsigned)

James, W. R. (1993, November 16). The uninsured and health care. The Wall Street Journal, pp. A1, A12. (article appears on two separate pages)

Study finds many street people mentally ill. (1994, June 7). New York Times, p. A7.

16. *A Letter to the Editor*

> Williams, P. (1993, July 19). Self-fulfilling stereotypes [Letter to the editor]. Los Angeles Times, p. A22.

17. *A Personal Communication*

Personal communications are not included in the reference list but are cited only in the text of your paper. Published letters, however, are included in the reference list.

> Joyce, J. (1931). Letter to Louis Gillet. In Richard Ellmann, James Joyce (p. 631). New York: Oxford University Press.

18. *An Interview*

Like personal communications, personal interviews are cited only in the text of your paper.

19. *Electronic Media*

APA recommends the following generic formats for referencing on-line information.

> Author, I. (date). Title of article. Name of Periodical [On-line], XX. Available: Specify path.

> Author, I. (date). Title of article. [CD-ROM]. Title of Journal, XX, XXX–XXX. Abstract from: Source and retrieval number.

(3) Content notes

APA format allows—but does not encourage—the use of content notes, which are indicated by raised numbers (superscripts) in the text. The notes are listed on a separate numbered page, entitled *Footnotes,* following the last page of text. Double-space all notes, indenting the first line of each note five to seven spaces and beginning subsequent lines at the left-hand margin.

IN THE TEXT

Skinner's behaviorist theories fell into disfavor and were displaced by the ideas of cognitive psychologists.[1]

IN THE NOTE

[1]Skinner himself remained largely unconvinced by the cognitive theorists. In a New York Times interview just two weeks before his death in 1990, he affirmed his belief in his model of human behavior.

40e Using CBE Format*

Documentation formats recommended by the Council of Biology Editors (CBE) and distributed by the American Institute of Biological Sciences are used by authors, editors, and publishers in biology, botany, zoology, physiology, anatomy, and genetics. The *CBE Style Manual* recommends several documentation styles, including the **number-reference** format described here. Numbers inserted parenthetically in the text correspond to a list entitled *References* or *Literature Cited* at the end of the paper.

GUIDELINES FOR PREPARING THE REFERENCE LIST: CBE

- Begin the reference list on a new page after the last page of the paper.
- Number the page on which the reference list appears as the next page of the paper.
- List the items *either* in the order in which they are mentioned in the paper *or* alphabetically, with last name first. Spell out the author's last name, and use initials for the first and middle names.
- Number entries consecutively; type note numbers on the line followed by a period and two spaces.
- Type the first line of each entry flush with the left-hand margin; align subsequent lines directly beneath the first letter of the author's last name.
- Separate the major divisions within each entry with a period and two spaces.
- Double-space within and between entries.

Reference List Format: CBE

IN THE PAPER

One study (1) has demonstrated the effect of low dissolved oxygen. Cell walls of. . . .

---

*CBE documentation format follows the guidelines set in the *CBE Style Manual*. 5th ed. Bethesda: Council of Biology Editors, 1983.

650

I N THE R EFERENCE L IST

Name   Initials   Title not underlined, only first word capitalized
     [2]   [2]

1. White, R. P. An introduction to biochemistry.

    Philadelphia: W. B. Saunders; 1974.

    City   Publisher   Date   Double-space

DIRECTORY OF CBE REFERENCE LIST ENTRIES

1. A book with one author
2. A book with more than one author
3. An edited book
4. A specific section of a book
5. An article in a scholarly journal with continuous pagination through an annual volume
6. An article in a scholarly journal with separate pagination in each issue
7. An article with a subtitle
8. An article with no listed author
9. An article with discontinuous pagination

Sample CBE Reference List Entries: Books List the author(s), the title (with only the first word capitalized), the city of publication (followed by a colon), the name of the publisher (followed by a semicolon), and the year (followed by a period). Do not underline book titles.

1. A Book with One Author

1. Rathmil, P. D. The synthesis of milk and related products.
   Madison, WI: Hugo Summer; 1985.

2. A Book with More Than One Author

2. Krause, K. F.; Paterson, M. K., Jr. Tissue culture: methods
   and application. New York: Academic Press, Inc.; 1973.

3. An Edited Book

3. Marzacco, M. P., editor. A survey of biochemistry. New
   York: R. R. Bowker Co.; 1985.

4. A Specific Section of a Book

> 4. Baldwin, L. D.; Rigby, C. V. A study of animal virology. 2nd ed. New York: John Wiley & Sons; 1984: 121–133.

Sample CBE Reference List Entries: Articles List the author(s), the title of the article (with only the first word capitalized), an abbreviated title of the journal (with all major words capitalized), the volume number (followed by a colon), the inclusive page numbers of the article (followed by a semicolon), and the year (followed by a period). Do not place titles of articles in quotation marks or underline titles of journals.

5. An Article in a Scholarly Journal with Continuous Pagination through an Annual Volume

> 1. Bensley, K. Profiling women physicians. Medica 1:140–145; 1985.

6. An Article in a Scholarly Journal with Separate Pagination in Each Issue

> 2. Wilen, W. W. The biological clock of insects. Sci. Amer. 234(2):114–121; 1991.

7. An Article with a Subtitle

> 3. Schindler, A; Donner, K. B. On DNA: the evolution of an amino acid sequence. J. Mol. Evol. 8:94–101; 1980.

8. An Article with No Listed Author

> 4. Anonymous. Developments in microbiology. Int. J. Microbiol. 6:234–248; 1987.

9. An Article with Discontinuous Pagination

> 5. Williams, S.; Heller, G. A. Special dietary foods and their importance for diabetics. Food Prod. Dev. 44:54–62, 68–73; 1984.

40f Using Other Documentation Styles

The following style manuals describe documentation formats different from the ones already discussed. If you are writing in a field that does not have a style manual, ask your instructor what documentation style you should use before you begin your research assignment.

Chemistry

American Chemical Society. *Handbook for Authors of Papers in American Chemical Society Publications.* Washington: American Chemical Soc., 1986.

Geology

Unites States Geological Survey. *Suggestions to Authors of the Reports of the United States Geological Survey.* 6th ed. Washington DC: Dept. of the Interior, 1978.

Mathematics

American Mathematical Society. *A Manual for Authors of Mathematical Papers.* 8th ed. Providence: American Mathematical Soc., 1984.

Medical Sciences

International Steering Committee for Medical Editors. "Uniform Requirements for Manuscripts Submitted to Biomedical Journals." *Annals of Internal Medicine* 90 (Jan. 1979): 95–99.

Physics

American Institute of Physics. Publications Board. *Style Manual for Guidance in the Preparation of Papers.* 4th ed. New York: American Inst. of Physics, 1990.

40g   Using Abbreviations

Many of the abbreviations that made documenting your sources so tedious have been eliminated from the latest editions of style manuals. Some, however, are still in use. Because you may still encounter such abbreviations, however, you should be familiar with them.

STUDENTS WORKING TOGETHER

Have students work in pairs to create bibliography problems for other pairs. In other words, let the students create the exercise (in essence, the data to be put in a format), and then let them do the exercise they create, using the pair who created the exercise as the evaluators. This activity gets students involved in creation, writing, and evaluation.

STUDENTS WORKING TOGETHER

Ask students to complete the Exercise and then exchange papers. Students should mark any questionable use of MLA format and work together to correct problems. If a pair of students can't agree on a correction, ask them to hold it for class discussion. Use the board to show the correct format, and discuss any problem points.

In the following exercise, the parenthetical references are listed by number, followed by the list of works cited in the proper order. Note that when the author's name is already mentioned in the text, it does not have to appear in the reference.

1. (2)
2. (Lindemann, "Approaches")
3. (34–36)
4. (D'Angelo 80)
5. (Tate vii)
6. (Jones 73)
7. (61)
8. (Best D1)
9. (41)
10. (Lindemann, *Rhetoric* 236–37)
11. (Bain)
12. (Jones 78)
13. (179–185)
14. (Larson 635)

Works Cited

Bain, Robert. Personal interview. 5 Nov. 1990.

Best, Kim. "Using a Newspaper in the Classroom." Durham Morning Herald 14 Nov. 1982: D1.

D'Angelo, Frank. "The Search for Intelligible Structure." The Writing Teacher's Sourcebook. Eds. Gary Tate and Edward P. J. Corbett. New York: Oxford UP, 1981. 80–88.

Irmscher, William F. Teaching Expository Writing. New York: Holt, Rinehart and Winston, 1979.

Jones, M. Hardy. "Teaching Johnny to Write." Time 25 Oct. 1980: 72–79.

continued on following page

COMMONLY USED SCHOLARLY ABBREVIATIONS

anon.	anonymous
bk.	book
c., ca.	circa ("about"). Used with dates that are approximate, as in c. 1920 (approximately 1920).
cf.	*confer* ("compare")
ch.	chapter
col.	column
colloq.	colloquial
comp., comps.	compiled by, compiler(s)
diss.	dissertation
ed.	edition, editor, edited by
e.g.	*exempli gratia* ("for example")
et al.	*et alia* ("and others")
ff.	and the following pages, as in pp. 88 ff.
i.e.	*id est* ("that is")
illus.	illustrated by, illustration
l., ll.	line(s)
ms., mss.	manuscript(s)
n., nn.	note(s), as in p. 12, n. 1
NB	*nota bene* ("take notice")
n.d.	no date (of publication)
n.p.	no place (of publication), no publisher
n. pag.	no pagination
p., pp.	page(s)
rev.	revision, revised by; review, reviewed by
rpt.	reprint, reprinted by
sec.	section
supp.	supplement
trans.	translated by, translator, translation
vol.	volume

*EXERCISE

The following are notes for a paper on approaches to teaching composition. Put them in the proper format for MLA parenthetical documentation, and then arrange them in the proper format for the list of works cited. (If your instructor requires a different method of documentation, use that style instead.)

1. Page 2 in a book called Teaching Expository Writing by William F. Irmscher. The book has 188 pages and was published by Holt, Rinehart and Winston, which at that time was located in New York, in 1979. (author's name mentioned in the text of your paper)
2. Something Erika Lindemann said in a lecture on November 9, 1982. She called the talk Approaches to Teaching. (Lindemann's name is not mentioned in the text)
3. Irmscher's book again, this time pages 34, 35, and 36. (author's name mentioned in the text)
4. The Search for Intelligible Structure, an essay written by Frank J. D'Angelo in a book by Gary Tate and Edward P. J. Corbett that is a collection of essays. Oxford University Press in New York published the book, and its copyright date is 1981. Your quotation is from the first page of the essay, which runs from page 80 to page 88. The book is called The Writing Teacher's Sourcebook. (no author's name mentioned in the text)
5. Page vii of the introduction to a book called Teaching Composition: 10 Bibliographical Essays, which Texas Christian University Press published in 1976 in Fort Worth. It was edited by Gary Tate. (no author's name mentioned in the text)
6. An article in *Time* on October 25, 1980, called Teaching Johnny to Write. You paraphrased a paragraph on page 73. The article ran from page 72 to page 79, and the last page was signed M. Hardy Jones. (no author's name mentioned in the text)
7. Page 61 in Teaching Expository Writing. (author's name mentioned in the text)
8. Using a Newspaper in the Classroom, which appeared in the Durham Morning Herald on page 1 of section D in the Sunday paper on November 14, 1982. It was written by Kim Best. (no author's name mentioned in the text)
9. Lee Odell's article in the February 1979 issue of College Composition and Communication. The article, called Teachers of Composition and Needed Research in Discourse Theory, ran from page 39 to page 45, and you got your information from page 41. That was volume 30 of the journal. (author's name mentioned in the text)
10. You found a book written by Erika Lindemann that you want to quote. You use material on pages 236 and 237 of the book, titled A Rhetoric for Writing Teachers, which was published in 1982 by the New York office of Oxford University Press. (no name mentioned in the text)
11. You decide you need to talk to someone with some experience, so you interview English professor Robert Bain at the University of North Carolina. You talked to him on November 5 and now you are quoting something he said. (Bain's name is not mentioned in the text)
12. This material is from page 78 of that *Time* article. (no author's name mentioned in the text)
13. You summarize pages 179–185 of Irmscher. (author's name mentioned in the text)

continued from previous page

Larson, Richard. "Problem-Solving, Composing, and Liberal Education." College English 33 (March 1972): 628–35.

Lindemann, Erika. A Rhetoric for Writing Teachers. New York: Oxford UP, 1982.

---. "Approaches to Teaching." Lecture. 9 Nov. 1982.

Odell, Lee. "Teachers of Composition and Needed Research in Discourse Theory." College Composition and Communication 30 (Feb. 1979): 39–45.

Tate, Gary. Introduction. Teaching Composition: 10 Bibliographical Essays. Ed. Tate. Fort Worth: Texas Christian UP, 1976, vii.

14. You find the perfect conclusion on page 635 of volume 33 of College English in Richard Larson's article Problem-Solving, Composing and Liberal Education. This is the March 1972 issue and the article begins on page 635. (no author's name mentioned in the text)

CHAPTER 41

Writing a Research Paper

Doing research involves more than just absorbing the ideas of others; it requires you to think critically, evaluating and interpreting the ideas presented in your sources and developing ideas of your own. In addition, it requires strategic planning, careful time management, and the willingness to rethink, reformulate, and reshape ideas.

The following schedule will give you an overview of the research process and help you manage your time.

THE RESEARCH PROCESS

	Activity	Date Due	Date Completed
Planning Your Paper	*Moving from Assignment to Topic*		
	• understanding your assignment **41a1**	_____	_____
	• choosing a topic **41a2**	_____	_____
	• starting a research notebook **41a3**	_____	_____
	Focusing on a Research Question		
	• mapping out a preliminary search strategy **41b1**	_____	_____
	• doing exploratory research **41b2**	_____	_____

Activity	Date Due	Date Completed
Assembling a Working Bibliography and Making Bibliography Cards		
• assembling a working bibliography **41c1**	_____	_____
• making bibliography cards **41c2**	_____	_____
Developing a Tentative Thesis **41d**		
Doing Focused Research and Taking Notes		
• reading sources **41e1**	_____	_____
• taking notes **41e2**	_____	_____
Shaping Your Material — *Deciding on a Thesis* **41f**	_____	_____
Preparing a Formal Outline **41g**	_____	_____
Writing and Revising — *Writing a Rough Draft* **41h**	_____	_____
Revising Your Drafts **41i**	_____	_____
Preparing a Final Draft **41j**	_____	_____

41a   Moving from Assignment to Topic

(1)   Understanding your assignment

Every research paper begins with an assignment. Before you can find a direction for your research, you must be sure you understand the exact requirements of this assignment.

WRITING CHECKLIST: UNDERSTANDING YOUR ASSIGNMENT

- Has your instructor provided a list of possible topics?
- Has your instructor asked you to develop a topic from a general subject area?
- Does your instructor expect you to select a topic on your own?
- Is your purpose to explain or to persuade?
- Is your audience your instructor? Your fellow students? Both?
- Can you assume your audience knows a lot (or just a little) about your topic?
- When is the completed research paper due?
- About how long should it be?
- Will you be given a specific research schedule to follow, or are you expected to set your own schedule?
- Is collaborative work encouraged? If so, at what stages of the research process?
- Does your instructor expect you to take notes on note cards? To prepare a formal outline?
- Are instructor–student conferences required?
- Will your instructor review note cards, outlines, or drafts with you at regular intervals?
- Does your instructor require you to do research only in the library, or can you also gather information outside the library?
- Does your instructor require you to keep a research notebook?
- What paper format and documentation style are you to use?
- What help is available to you—from your instructor, other students, experts in the field your paper will explore, community resources, your library?

(2)   Choosing a topic

Once you understand the requirements and scope of your assignment, you can look for a direction for your research. You begin this task by narrowing your focus to a topic you can explore within the boundaries of your assignment.

In many cases, your instructor will help you to choose a topic, either by providing a list of suitable topics or by suggesting a general subject area—a famous trial, an event that happened on the day you were born, a problem on your college campus. Even in

659

these instances, you will still need to choose one of the topics or narrow the subject area: decide on one trial, one event, one problem.

If your instructor prefers that you select a topic entirely on your own, your task is somewhat more difficult: You must consider various topics and weigh both their suitability for research and your interest in researching them. You find a focus for your paper in much the same way you decide on a topic for a short essay: You brainstorm, ask questions, talk to people, and read. With a research paper, however, you know from the start that you will examine not only your own ideas on a topic, but also the ideas of others.

An effective research topic has four characteristics.

A Research Topic Should Be Neither Too Broad Nor Too Narrow   The subject areas listed below are too general for research. The narrowed topics are more suitable starting points.

Subject Area	Topic
Computer technology	The possible negative effects of computer games on adolescents
Feminism	The relationship between the feminist movement and the use of sexist language
Mood-altering drugs	The use of mood-altering drugs in state mental hospitals

Exactly how broad or narrow should a topic be? There is no easy answer. For one thing, a topic must fit within the boundaries of your assignment. "Julius and Ethel Rosenberg: Atomic Spies or FBI Scapegoats?" is far too broad for a ten-page—or even a hundred-page—treatment. However, "One piece of evidence that played a decisive role in establishing the Rosenbergs' guilt" would probably be too narrow for a ten-page research paper—even if you could gain access to significant information. But how one newspaper reported the Rosenbergs' espionage trial or how a particular group of people—government employees, peace activists, or college students, for example—reacted at the time to the couple's 1953 execution could work.

A Research Topic Should Be Suitable for Research   Topics based exclusively on personal experience or on value judgments are not suitable for research. For example, "How attending an integrated high school has made me a more tolerant person" is a topic that can be developed only by self-analysis. Similarly, "The superiority of Freud's work to that of Jung" is a topic that might interest you, but no amount of research can establish that one person's work is better than another's.

A Research Topic Should Be One That Can Be Researched in a Library to Which You Have Access　If your instructor gives you a specific topic or subject area to write about, he or she will probably have made sure that your library has the resources you need. If you choose your own topic, however, you will first have to learn something about your library's strengths and weaknesses. For instance, the library of an engineering or business college may not have a large collection of books of literary criticism; the library of a small liberal arts college may not have extensive resources for researching technical or medical topics.

Finally, a Research Topic Should Be One in Which You Are Genuinely Interested　You will be deeply involved with the topic you select for several weeks—in some cases, even for an entire semester—and your research will be most productive if you care about your subject. If you are interested in your topic, you are likely to see research less as a tedious chore and more as an opportunity for discovering new ideas and new connections among ideas.

(3)　Starting a research notebook

As soon as you have your assignment, you should start a **research notebook**, a combination journal of your reactions and log of your progress. A research notebook maps out your direction and keeps you on track; throughout the research process it helps you define and redefine the boundaries of your assignment.

In this notebook you can record lists of things to do, sources to check, leads to follow up on, appointments, possible community contacts, questions to which you would like to find answers, stray ideas, possible thesis statements or titles, and so on. Be sure to date your entries and to check off and date work completed; this will save you from repeating steps.

Some students use a spiral notebook that includes pockets to hold note and bibliography cards. Others find a small assignment book more convenient. Whatever kind of book you use, the research notebook can serve as a useful record of what has been done and what is left to do.

TEACHING STRATEGY

Research Notebook
Structuring a paper is, by nature, a process of trial and error. Suggest that students keep a research notebook for each paper. The notebook is a kind of writer's journal or companion to the formal writing assignment. It will record structural techniques and strategies, problems encountered, and solutions crafted. Of course, the students may use the notebook to record all details (not only structural) of the writing process. This is an effective way to help students analyze what they do as they write, think about their successes and failures, and consider ways to control future writing assignments.

MARION DUCHAC'S WRITING PROCESS

Marion Duchac, a student in a composition class, was given this assignment: "Write an eight- to ten-page research paper that takes a persuasive stance on one issue that affects students on our campus." This was to be a full-semester project, so Marion had fourteen weeks in which to research and write the paper. Throughout this chapter shaded boxes will trace her progress and reproduce the comments she made in her research notebook at each stage of her work.

Marion's instructor, Dr. Mary Ann Potter, told the class she would require regular conferences at which she would review each student's progress and check his or her research notebook; a segment of the assignment would be due at each meeting. At various points in the process, she would require collaborative work, and she expected them to use nonlibrary as well as library sources. With these general guidelines in mind, Marion began to think about her assignment.

MOVING FROM ASSIGNMENT TO TOPIC

For my composition class, I was assigned to do research and to write an argumentative paper about an issue that affects students on my campus. At first I planned to write about need vs. merit in financial aid to college students, but that topic didn't have the right argumentative edge. Besides, according to the two articles I located in the library, the federal government is overhauling the system, and the facts keep changing. I talked to the financial aid officer at my school and to various students, whose most common complaint was that financial aid forms are too hard to fill out. This topic just wasn't producing a thesis I could support, so I had to abandon it. Eventually, I talked to a classmate whose athletic scholarship may be cancelled next year because of new NCAA regulations, and as a result I became interested in exploring the topic of athletic scholarships.

Assignment	Topic
Defend a position on an issue that affects students on campus.	Athletic scholarships

continued on the following page

continued from the previous page

> *I started this research notebook by copying down my assignment on the first page and taping Dr. Potter's research schedule to the inside front cover, so I could check off each stage of my research as I completed it. Next I jotted down the assignment's other requirements and the time of my first conference with Dr. Potter. Then I recorded my topic and listed a few people I thought might be able to help me get started. At this point I felt I was ready to move on to the next stage of my assignment: focusing on a research question.*

EXERCISE 1

Using your own instructor's guidelines for selecting a research topic, choose a topic for your paper. Begin your research notebook by entering information about your assignment, schedule, and topic.

41b Focusing on a Research Question

(1) Mapping out a preliminary search strategy

The key to successful research lies in finding out what questions to ask. Your research should be guided by a **search strategy**, a plan for systematically gathering and evaluating potential source material, moving from general to specific sources. At your first meeting your instructor can help you map out a tentative search strategy for the project you have in mind and direct you to appropriate sources. Keep in mind, though, that as you continue your research, you will probably modify your search strategy to fit your changing priorities.

▶ See 38a

(2) Doing exploratory research

Exploratory research helps you get an overview of your topic and an understanding of its possibilities. At this early stage, you want to explore the boundaries of your topic. One way to do this is to discuss your ideas with others. Teachers, librarians, family, and friends may all suggest possible sources—sometimes unexpected or unconventional ones—for your paper. Another way to explore your topic is to skim general reference works in your college library—encyclopedias, for example—and jot down the names of sources these works cite in their bibliographies. At this

▶ See 38b1

stage, your notes should identify potential sources, not the specific information each source contains.

As you do exploratory research, your goal is to formulate a **research question**, the question you want your research paper to answer. This question will help you focus your exploratory research and guide you as you assemble a working bibliography. By suggesting a direction for your research, a research question helps you to decide which sources to seek out, which to examine first, and which to skip. The answer to your research question will be your paper's **thesis**, the statement the body of your paper will support.

See ◄
2b

Whatever type of research you do, the paper you eventually write will either *explain* something to readers or *persuade* them. Your assignment determines whether your paper will be informative or persuasive, and your research question should reflect this purpose. For example, the question "What characteristics do horror movies of the 1990s have in common?" calls for an informative paper. However, the question "Does the explicit violence in horror movies of the 1990s affect the behavior of adolescent viewers?" calls for a persuasive paper.

FOCUSING ON A RESEARCH QUESTION

Mapping Out a Preliminary Search Strategy

Before I started writing about athletic scholarships, I spent some time in the library. Since I wanted to write a paper that would take a stand in favor of athletic scholarships, I looked for articles and books that would help me support my position. Unfortunately, I didn't find many printed sources indicating that athletic scholarships should be retained, at least in their present form. This led me to start thinking about qualifying my support of athletic scholarships rather than wholeheartedly endorsing them.

My library research was somewhat frustrating because I couldn't find much objective information about athletic scholarships. I found plenty of research and reporting about the controversial aspects of athletic scholarships and their place in the world of academics, so I ended up focusing my search in that direction.

Consulting Experts

When I felt comfortable with some of the basic facts and issues of my subject, I talked to Dr. Potter about how to continue my re-

continued on the following page

continued from the previous page

search. I had already planned to interview student athletes on campus, but she suggested that I also try to arrange a few personal and telephone interviews with coaches from local schools. Unfortunately, only one of the coaches I contacted was willing to let me use his name. Another coach said he would talk to me but preferred to remain anonymous. Dr. Potter didn't think it would be ethical to cite an anonymous source, so I decided I wouldn't be able to rely on information from coaches to the extent I had hoped to. But I did begin to feel more confident that I was clarifying the direction of my search and developing my own opinions on my subject.

Asking Questions

My preliminary reading, my conference with Dr. Potter, and my attempts to interview local coaches suggested a number of questions I thought I should explore as I continued my research.

- What is the purpose of athletic scholarships?
- What percentage of schools offer them?
- What percentage of college students receive them?
- What kinds of students receive athletic scholarships?
- In what sports are scholarships awarded?
- What is the relationship between athletics and academics?
- What should this relationship be?
- Who should determine what this relationship should be?
- What are the advantages and disadvantages of athletic scholarships?
- What are some of the problems and controversies associated with athletic scholarships?

Focusing on a Research Question

As I was thinking about these questions, I realized my primary question, the one I wanted my paper to answer, was really quite simple: Given the problems they create, should athletic scholarships be continued?

Establishing Research Priorities

Now that I knew what I wanted to find out, I had to plan my research. I decided to do a database search, looking for articles on my topic. I also thought I might try to talk to someone in our financial aid or admissions office who might help me get access to statistics, and I wanted to try to get hold of a recruitment brochure, so I could see how schools sell themselves to athletes. I really felt, though, that

continued on the following page

When to Focus

Students tend to believe they should have the "right question" in mind from the beginning. Reinforce the text's assertion that most writers develop the central question of a paper only *after* exploratory research. Read and discuss the quotation on page 663. Whether they are choosing a topic or focusing on a research question, writers must be willing to adapt their ideas as research suggests new directions or angles on the subject.

continued from the previous page

> *my first priority should be to arrange an interview with Coach*
> *Walker, who had agreed to talk to me about my paper. Meanwhile, I*
> *planned to talk informally with students, including student athletes,*
> *about my research question.*

41c Assembling a Working Bibliography and Making Bibliography Cards

(1) Assembling a working bibliography

Whenever you encounter a promising source, jot down complete bibliographic information and a brief evaluation of the source on an index card. You will use these cards to compile a **working bibliography**, which includes all the sources (print and nonprint) you will examine later when you do your focused research and take notes. This preliminary bibliography is neither permanent nor complete: You will probably discard some of the sources and add others as your research progresses.

(2) Making bibliography cards

When you make a bibliography card for a source that looks promising, include the following information.

MAKING BIBLIOGRAPHY CARDS

Book

Author(s)
Title (underlined)
Call number (for future
  reference)
City of publication
Publisher
Date of publication
Brief evaluation

Article

Author(s)
Title of article (in quotation
  marks)
Title of journal (underlined)
Volume
Date
Inclusive page numbers
Brief evaluation

See ◀
40b2

Using the required bibliographic format on your cards now will decrease your chance of error later on and thus make your job easier in the long run. This bibliographic information should be *full* and *accurate*; if it is not, you may be unable to find sources

later. Your brief evaluation of each source should note the kind of
information the source contains, the amount of information offered, its relevance to your topic, and its limitations—whether it
is biased or outdated, for instance.

▶ **See 39a**

CLOSE-UP

MAKING BIBLIOGRAPHY CARDS

Be sure to make cards not only for books and articles but
also for interviews (including telephone interviews), meetings, lectures, films, and other nonprint sources of information.

As your research progresses, review your bibliography cards
regularly. Keeping your research question in mind, look over your
cards to reevaluate your sources. Select the cards that seem most
useful, and make plans to reexamine the sources they describe.
(Retain *all* the cards you have made, however, even those for
sources that do not seem very promising. You may decide to use a
rejected source later on when you have a more definite direction
for your research.) If you identify areas where additional sources
are needed, plan to do further research.

ASSEMBLING A WORKING BIBLIOGRAPHY AND MAKING BIBLIOGRAPHY CARDS

*As I did my exploratory research, I recorded bibliographic data
and evaluations of all source material—including books, articles,
and interviews—on index cards.*

call number ⌐

author ⟶
title ⟶
publication ⟶
information

evaluation ⟶

> Dealy, Francis K.   GV 351. D43 1990
> <u>Win at Any Cost</u>
> (New York : Carol 1990)
> A recent book that treats the history
> of collegiate athletics and the ways the
> pressure to win affects coaches, students,
> and athletic programs. Dealy emphasizes
> exploitation of student athletes in the
> quest for victory.

Bibliography card: Print source   *continued on the following page*

continued from the previous page

ANSWERS TO EXERCISE 2

Responses will vary. You will probably want to check students' work.

subject ⟶

evaluation ⟶

date ⟶

Walker, Skippy "Tiptoe"
Telephone interview   1 Mar. 1994

Assistant coach at large Texas high school. (Doesn't want school named in my paper.) The only local coach who was willing to talk to me on the record. Confirms some of Dealy's statistics. Denies unethical recruitment of athletes he has known.

Bibliography card: Interview

EXERCISE 2

Do exploratory research to find a research question for your paper, carefully evaluating the relevance and usefulness of each source. Make a bibliography card for each source, thus compiling a working bibliography for your research paper in progress. When you have finished, re-evaluate your sources and plan additional research if necessary.

41d   Developing a Tentative Thesis

See ◀
2b

Your **tentative thesis** is a preliminary statement of what you think your research will support. This statement, which you will eventually refine into a thesis, should be the answer to your research question.

As you move through the research process, your tentative thesis and even your research question may change considerably. A line of inquiry may lead to a dead end, a key source may not be available, or a lead you uncover in your research may encourage you to branch out in a new direction. But whether or not you make major adjustments to your tentative thesis, it should grow increasingly more precise, eventually leading you to a thesis your research can support.

DEVELOPING A TENTATIVE THESIS

Subject Area
Computer technology

Topic	Research Question	Tentative Thesis
The possible negative effects of computer games on adolescents	Do computer games have any negative effects on adolescents?	Computer games interfere with adolescents' ability to learn

Subject Area
Feminism

Topic	Research Question	Tentative Thesis
The relationship between the feminist movement and the use of sexist language.	What is the relationship between the feminist movement and the use of sexist language?	The feminist movement is largely responsible for the decline of sexist language.

Subject Area
Mood-altering drugs

Topic	Research Question	Tentative Thesis
The use of mood-altering drugs in state mental hospitals	How has the use of mood-altering drugs affected patients in state mental hospitals?	The use of mood-altering drugs has changed the population of state mental hospitals.

TEACHING STRATEGY

Incubation
If your students have the luxury of considering a topic or assignment for several days or more, they may find *incubation*, a strategy closely related to meditation, useful. After carefully thinking about the problem or writing task, the writer "deposits" it in the subconscious and leaves it alone to grow and develop. The subject has not been forgotten, but merely deposited on the "back burner" of the mind. This process, valued by many creative and critical thinkers, is one that Isaac Asimov says "invariably worked" for him. He describes the technique he used when he found himself stumped by a writing problem: "I went to the movies. Not just any movie. I had to pick a movie which was loaded with action but which made no demands on the intellect. As I watched, I did my best to avoid any conscious thinking concerning my problem, and when I came out of the movie I knew exactly what I would have to do. . . . It never failed."

from *The Left Hand of the Electron*

You will probably want your students to submit their tentative thesis statements to you for approval. It takes practice to develop a concise statement of purpose. Sometimes after students reread their notes and rearrange their ideas, they produce better thesis statements, ones that are more accurate and more focused.

DEVELOPING A TENTATIVE THESIS

My answer to my research question became the tentative thesis for my paper.

Athletic scholarships should be retained because athletic teams add something vital to college life.

EXERCISE 3

Following your own instructor's guidelines, develop a tentative thesis for your research paper.

41e Doing Focused Research and Taking Notes

Once you have decided on a tentative thesis, you are ready to begin your focused research and note taking.

(1) Reading sources

See ◄
5a-b

As you read, use **active reading** strategies: Preview each source, skimming it quickly; then read it carefully, highlighting potentially useful material. Be sure to approach your text with a critical eye, distinguishing fact from opinion and evaluating writers' support carefully. Remain alert to bias, faulty reasoning, logi-

See ◄
5c

cal fallacies, and unfair appeals. Eventually, you will **annotate** the source and then go on to take notes in the form of summary, paraphrase, or direct quotation.

Your limited time makes it impossible for you to read all your print sources thoroughly, so read only those sections of a work that pertain to your topic. Before you begin, survey the work carefully, checking a book's index and the headings and subheadings in the table of contents to determine which pages to read thoroughly and which to skim. Look carefully at abstracts and head-

See ◄
39a

ings of articles. These strategies will enable you to evaluate each source's usefulness to you as efficiently as possible.

(2) Taking notes

Take careful notes as you do library research and be sure to take notes on nonprint sources—interviews, lectures, films, and so on—too. And remember, if you encounter a promising new source while you are taking notes, make a bibliography card for the new source immediately.

Note Cards As you do library research, take notes on index cards. The advantages of index cards become obvious when you start arranging and rearranging your material. You seldom know where a particular piece of information belongs at first—or even whether you will use it. You will rearrange your ideas many times, and index cards make it easy for you to add and delete information and to experiment with different sequences—something you cannot do with notes you make on photocopies or on loose sheets of paper.

Each card should include a short descriptive *heading* at the top that indicates the relevance of the note to some aspect of your topic. Because you will use these headings to make your outline and organize your notes, they should be as specific and descriptive as possible. Labeling every note for a paper on athletic scholarships *scholarships* or *financial aid*, for example, will not prove very helpful to you later on. More focused headings—*recruitment of athletes* or *benefits of athletic scholarships*, for instance—will be much more useful.

Each card should accurately identify the source of the information you are recording. You need not include the complete citation, but you must include enough information to identify your source. *Durrant 62* would be enough to send you back to the bibliography card carrying the complete documentation for Sue M. Durrant's "Title IX—Its Powers and Its Limitations." If you use more than one source by the same author, however, you need a more complete reference. For example, *Blum, Graduation Rate 42* would be necessary if you were using more than one source by Debra Blum. Be sure to provide information to identify each nonprint source as well—for example, *Interview with Coach Walker*.

The note itself—in the form of **summary**, **paraphrase**, or **quotation**—will naturally be the focus of your note card. In this note you should record the information you will include in your paper.

▶ See 39b

Finally, each card should include a brief comment that makes clear your rationale for making the note and identifies what you think it will add to your paper. This comment (placed in brackets so you will know it represents your own ideas, not those of your source) should establish the purpose of your note—what you think it can explain, support, clarify, describe, or contradict—and perhaps suggest its relationship to other notes or other sources.

IN THE CLASSROOM

A class discussion on note-taking techniques might be valuable at this point. Who uses 4 × 6 cards—and wouldn't dream of using anything else? Who likes legal pads? Post-it®-notes? Photocopies? Some researchers think that writing information down in longhand fixes it in the mind in a way that photocopying does not. Most researchers, in fact, use a combination of all these methods. Let students discuss favorite note-taking methods as well as any problems (forgetting to note a source, for example) and let other students offer suggestions.

Any questions you have about the note can also be included in your comment.

GUIDELINES FOR WORKING WITH NOTE CARDS

- **Put only one note on each card.** If you do not keep your notes distinct from one another, you lose the flexibility that is the whole point of this method of note taking.

- **Include everything now that you will need later** to understand your note—names, dates, places, connections with other notes. After a few weeks you will not remember the significance of any but the most explicit notes.

- **Distinguish quotations from paraphrase and summary, and your own ideas from those of your sources.** If you copy a source's words, use quotation marks. If you use a source's ideas but not its words, do not use quotation marks. If you write down your own ideas, enclose them in brackets. This system will help you to avoid mixups—and plagiarism **(see 39d)**.

- **Put an author's comments into your own words whenever possible.** Word-for-word copying is probably the most inefficient way to take notes. You may use quotations in your final paper, but for the most part you will summarize and paraphrase your source material **(see 39b)**, adding your own observations and analysis. Thus, it makes no sense to postpone the inevitable; summarize and paraphrase *now*, as you read each source. Not only will this strategy save you time later on, but it will also help you understand your sources and evaluate their usefulness to you *now*, when you still have time to find additional sources to substitute for or complement them if necessary. If you do expect to quote an author's words in your paper, be sure to copy them accurately, transferring the *exact* words, spelling, punctuation marks, and capitalization to your note card.

Following are two useful note card formats, the first for a print source, the second for a nonprint source.

Author and page
number
↓

Descriptive ———→
heading
Note ————→
(Summary,
paraphrase, or
quotation)

Your comment →
(opinions, reac-
tions, purpose
of note, con-
nections with
other sources,
etc.)

Recruitment   Dealy 180

Surprisingly, "of the 265,000 seniors playing high school football, only 2,253, one out
of every 118 high school senior football
players, receive scholarships."

[This is an important statistic. It helps to explain
the pressure and competition for these
scholarships among high school athletes— and
suggests how important recruitment is.
It might also explain why there are so many
recruiting violations.]

Recruitment   Interview with Coach Walker

"When our players are recruited we don't see
any methods that people might consider unethical. . . . Most of our players don't get
scholarships. Actually, we've only had
two receive scholarships in the last 10
years — and remember, those were 10
years in which we scored high in our
division."

[His statistic about scholarship students is
consistent with Dealy's numbers.]

Note card formats

Photocopies   With the availability of photocopy machines in
libraries, many researchers routinely copy useful portions of
sources. As long as you are careful to accurately record the bibliographic information for the source, this is a useful and time-
saving strategy. As you prepare to photocopy material, however,
keep the following guidelines in mind.

673

Students are so accustomed to using secondary research sources that generating original research (through observation, interviews, analysis of data, etc.) may not occur to them as a realistic possibility. Do what you can to encourage students to include among their various writing assignments at least one research project including fieldwork.

As a preliminary exercise, divide the class into groups of three or four. Using a list you have developed before class, assign each group a research topic or question that could be developed in part through fieldwork. Ask students to brainstorm for fieldwork ideas and jot down notes for a research strategy. Be available to help if groups are stumped for ideas. It might be a good idea to bring to class some local telephone books and any available lists (from newspaper, chamber of commerce, and so forth) of local organizations, agencies, research institutes, etc. Discuss the results. Are students excited by the idea of generating fresh material?

Have students go all the way if possible. Why list fieldwork questions without trying them out? How else will they really know if questions are any good? Even if students just get to role play in class—one group answering another group's questions and then reporting on their effectiveness in writing—this exercise should be valuable.

GUIDELINES FOR WORKING WITH PHOTOCOPIES

- Bring exact change for the machine.
- Record full and accurate source information, including the page numbers, on the first page of each copy.
- Clip or staple together consecutive pages of a single source.
- Do not copy a source without reminding yourself—*in writing*—why you are doing so. In pencil or on removeable self-stick notes, record your initial responses to the source's ideas, jot down cross-references to other works or notes, and highlight important sections.
- Photocopying can be time-consuming and expensive, so try to avoid copying material that is only marginally relevant to your paper.

Photocopied information is not a substitute for notes. In fact, photocopying is only the first step in the process of taking thorough, careful notes on a source. In some cases—for example, for a short paper that uses only one or two sources—the annotating and highlighting you do on copies may be enough to guide your draft. But in most research assignments, it is unwise to use photocopies as a substitute for detailed notes. You should be especially careful not to allow the ease and efficiency of photocopying to encourage you to postpone decisions about the usefulness of your information. Remember, you can accumulate so many photocopied pages that it will be almost impossible to keep track of all your information.

You should also keep in mind that photocopies do not have the flexibility of note cards: A single page may include notes that should be earmarked for several different sections of your paper. This lack of flexibility makes it virtually impossible for you to arrange your source material into any meaningful order.

Finally, remember that the annotations you make on photocopies are usually not focused or polished enough to be incorporated into your paper. You will still have to paraphrase and summarize your source's ideas and make connections among them. Therefore, you should approach a photocopy just as you approach any other print source—as material that you read, highlight, annotate, and take notes about.

> ### LIBRARY WORK
>
> *I now felt ready to choose the most useful books and articles from the database search I had done earlier. As I began to take detailed notes from these sources, I discovered that many of them stressed the reform of existing athletic programs. The problems discussed most often are unfair recruiting practices; inequities in support of men's and women's sports, "major" and "minor" sports, and winning and losing teams; and neglect of student athletes' academic obligations. I began to think I might need to revise my tentative thesis to emphasize reform of college athletic programs.*
>
> ### FIELDWORK
>
> *One of the student athletes I spoke to, a tennis player, mentioned that one of the teams on our campus won a national championship a few years ago. I hadn't been aware of the wheelchair basketball team at all, since I wasn't a student here the year they won the championship. But I was intrigued by the whole idea and decided to attend one of their games. The team exhibited tremendous skill and was supported by a gym full of very enthusiastic fans. I later found out that crowds of local people follow this team wherever they play, but the players are not on scholarship. I think they should be. I don't know if I can use this experience in my paper, but it's certainly affected my thinking about athletic scholarships.*

Some teachers check note cards and require photocopies of text to check paraphrases against sources. To be sure your students understand paraphrasing before they write their research papers, you may wish to do this as well.

EXERCISE 4

Begin focused research for your paper, taking careful notes from your sources. Remember, your notes should include paraphrase, summary, and your own observations and analysis as well as direct quotations.

41f Deciding on a Thesis

Even after you have finished your focused research and note taking, your thesis remains tentative. Now you must refine this tentative thesis into a **thesis**, a carefully worded statement that expresses a conclusion your research can support. This thesis should be more detailed than your tentative thesis, accurately conveying the direction, emphasis, and scope of your paper.

▶ See 2b2

Arrange a roundtable discussion of tentative thesis statements students have chosen for their papers. Explore any limitations, potential problems with supporting the central idea, opposing arguments to be handled, possible resources, and so on. If any students are writing about the same topic—even from very different viewpoints—they might use the exercise to share information or brainstorm for ideas.

If students are to work on the format of the thesis, the most effective exercise might be to copy all thesis statements beforehand, put students in groups, and have them evaluate them before discussion. If the interest is also in content, groups might take each other's statements and generate either questions or revisions for each thesis. Then both groups might get together and have a more in-depth discussion than a whole class could do (and in the same amount of time). A summary report to the instructor discussing the conclusions of this discussion might be composed by each group.

DECIDING ON A THESIS

Tentative Thesis	**Thesis**
Computer games interfere with adolescents' ability to learn.	Because they interfere with concentration and teach players to expect immediate gratification, computer games interfere with adolescents' ability to learn.
The feminist movement is responsible for the decline of sexist language.	By raising public awareness of careless language habits and changing the image of women, the feminist movement has helped to bring about a decline of sexist language.
The development of mood-altering drugs has changed the population of state mental hospitals.	It is the development of psychotropic (mood-altering) drugs, not advances in psychotherapy, that has made possible the release of large numbers of patients from state hospitals into the community.

If your thesis does not express a conclusion your research can support, you may need to revise it. Reviewing your notes carefully, perhaps grouping your note cards in different ways, may help you to decide on a suitable thesis. Or you may try other techniques—for instance, brainstorming or freewriting with your research question as a starting point, or asking questions about your topic. The thesis you finally decide on should be consistent with the kind and amount of source material you have collected and the ideas you have developed in response to that material.

See ◀ 1c2

DECIDING ON A THESIS

After I attended the wheelchair basketball game, I knew I had to revise my thesis. When I began my research, I had wanted to write a

continued on the following page

paper in support of retaining athletic scholarships, but now I wanted to call for reforming the system to provide equal support for athletic and academic efforts, men's and women's sports, major and minor sports, and winning and losing teams. This change, in my opinion, would mean shifting the focus of athletic programs away from winning at all costs toward enriching campus life and the lives of individual student athletes. As a result, I decided to use the following thesis statement.

> Athletic scholarships should be continued, but college athletic programs should be reformed to deemphasize winning at all costs and to focus instead on fairness to all students athletes.

*EXERCISE 5

Read the following passages from various sources. Assume you are writing a research paper on the influences that shaped young writers in the 1920s. What possible thesis statements could be supported by the information in these passages?

1. Yet in spite of their opportunities and their achievements the **generation** deserved for a long time the adjective [lost] that Gertrude Stein had applied to it. The reasons aren't hard to find. It was lost, first of all, because it was uprooted, schooled away and almost wrenched away from its attachment to any region or tradition. It was lost because its training had prepared it for another world that existed after the war (and because the war prepared it only for travel and excitement). It was lost because it tried to live in exile. It was lost because it accepted no older guides to conduct and because it formed a false picture of society and the writer's place in it. The generation belonged to a period of transition from values already fixed to values that had to be created. (Malcolm Cowley, *Exile's Return*)

2. The 1920s were a time least likely to produce substantial support among intellectuals for any sound, rational, and logical program. Prewar stability and convention were condemned because all evidences of stability seemed illusory and artificial. The very lively and active interest in science was perhaps the decade's most substantial contribution to modern civilization. Yet in this case as well, achievement became a symbol of disorder and a source for disenchantment. (Frederick J. Hoffman, *The 20's*)

3. Societies do not give up old ideals and attitudes easily; the conflicts between the representatives of the older elements of traditional American culture and the prophets of the new day were at times as bitter as they were extensive. Such matters as religion, marriage, and moral standards, as well as the issues over race, prohibition, and immigration were at the heart of the conflict. (Introduction to *The Twenties*, ed. George E. Mowry)

Some possibilities are:

The erosion of traditional values and intellectual support in the 1920s produced that decade's disenchantment with anything not concrete and factual.

As a result of postwar changes in values, Americans born in the 1920s were called the "lost generation."

Writers in the 1920s often found themselves at odds with traditional values.

677

You will probably want your students to submit their thesis statements to you for approval.

Outlining

If your students seem unfamiliar or uncomfortable with formal outlining conventions, try one or both of the following reinforcement activities.

1. "Shuffle" the elements of a formal outline, and give students an unorganized list of topics, subtopics, and supporting points. Ask them to arrange the items into a meaningful formal outline and to compare the results to the original.

2. Ask students to work with a piece of their own writing (an old essay or paper) for which no outline was developed. Ask the students to develop a formal outline and discuss the results. Did the exercise reveal any problems with organization or the flow of ideas? In light of the newly developed outline, would they significantly change the piece if they had it to do again?

EXERCISE 6

Carefully read over all the notes you have collected during your focused research and develop a thesis for your paper.

41g Preparing a Formal Outline

Although most students will not prepare one for a short essay, a formal outline is almost essential for a longer or more complex writing project. A **formal outline** indicates the order in which you will present your ideas and the relationship of main ideas to supporting details. A formal outline is more polished than an informal one. It is more strictly parallel and more precise, pays more attention to form, and presents points in the exact order in which you plan to present them in your draft.

A formal outline may be a **topic outline**, in which each entry is a single word or a short phrase, or a **sentence outline**, in which each entry is a complete sentence. Each of these outline forms has advantages and disadvantages. A sentence outline is a more fully developed guide for your paper: You have a head start on your paper when you are able to use the sentences of your outline in your draft. Because it is so polished and complete, however, the sentence outline is more difficult and time consuming to construct, especially at an early stage of the writing process. A topic outline provides less precise guidance, but it is easier to prepare.

Formal outlines conform to specific conventions of structure, content, and style. If you follow the conventions of outlining carefully, your formal outline can help you to plan a paper in which you cover all relevant ideas in an effective order, with appropriate emphasis, within a logical system of subordination.

THE CONVENTIONS OF OUTLINING
Structure

• Outline format should be followed strictly.

    I. First major point of your paper
       A. First subpoint
       B. Next subpoint
          1. First supporting example

continued on the following page

---

continued from the previous page

> 2. Next supporting example
>   a. First specific detail
>   b. Next specific detail
> II. Second major point

- Headings should not overlap.
- Each heading should have at least two subheadings.
- Each entry should be preceded by an appropriate letter or number, followed by a period.
- The first word of each entry should be capitalized.

Content

- Outline should include the paper's thesis statement.
- Outline should cover only the body of the essay.
- Headings should be specific and concrete.
- Headings should be descriptive, clearly related to the topic to which they refer.

Style

- Headings of the same rank should be grammatically parallel.
- Sentence outlines should use complete sentences, with all sentences in the same tense.
- In a sentence outline, each entry should end with a period.
- Topic outlines should use words or short phrases, with all headings of the same rank using the same parts of speech.
- In a topic outline, each entry should not end with a period.

By the time you have completed your focused research and note taking, you will have accumulated a good many note cards. These cards will probably be arranged haphazardly, perhaps in the order in which you took notes—and perhaps in no order at all. Before you can write a rough draft, you will need to make some sense out of all these cards, and you do this by sorting and organizing them. By identifying categories and subcategories of information, you begin to see the emerging shape of your paper and are able to construct a formal outline that reflects this shape.

GUIDELINES FOR PREPARING A FORMAL OUTLINE

- Make sure each note card contains only one general idea or one brief related group of facts. If it does not, *carefully* recopy any unrelated information onto another card. If the information on two cards overlaps, combine it on one card.
- Check to be sure the heading on each card specifically and accurately characterizes the information on the card. If it does not, change the heading.
- Lay out your note cards on a big table—or on the floor—and sort them into piles, guided by their headings. (Keep a miscellaneous pile for notes that do not seem to fit any-place. You may discard these notes later, or you may dis-cover—or create—a place for them in your paper when you construct your formal outline.)
- Check your categories for balance. If most of your notes fall into one or two categories, your categories are proba-bly too broad. Try rewriting some of your headings to cre-ate narrower, more focused categories. If you have only one or two cards in a category you now see as very im-portant to your paper's thesis, you should do additional research so that you can expand the undeveloped cate-gory. If time is short, however, you really have no choice but to treat the topic only briefly or to drop it entirely, re-vising your thesis accordingly.
- When you are satisfied that the categories you have iden-tified make sense in terms of the scope of your thesis and the major topics you plan to cover in your paper, you can begin to sort and organize the cards *within* each group. Within each group, put related cards together, adding more specific subheads to the cards' headings. Now arrange ideas in an order that highlights the most impor-tant points and subordinates lesser ones. Once again, do some discarding, setting aside note cards that do not fit into your emerging scheme. (And remember to retain these cards. They may fit a new line of inquiry as you ex-periment with different arrangements.)
- Decide on a logical order in which to discuss your paper's major points.
- When you are satisfied with the arrangement of major categories of information and supporting details within

continued on the following page

continued from the previous page

> each category, make a topic or sentence outline with divisions and subdivisions corresponding to the headings on your note cards.
> - Review your completed outline for potential problems—for instance, whether you need more supporting information in a particular area or have placed too much emphasis on a relatively unimportant idea, whether ideas are illogically or ineffectively placed, or whether overlapping discussions turn up in different parts of your outline.

Remember that the outline you construct at this stage is only a guide for you to follow as you draft your paper; it is likely to change as you write and revise. The final outline, written after your paper is complete, will serve as a guide for your readers.

PREPARING A FORMAL OUTLINE

After deciding exactly what I would write about and what position I would take, I reread my note cards and sorted them into related groups in order to break my topic down into parts I could handle. Then, to see how all the parts were related and to decide on a logical order for the parts, I made a topic outline. (Later, I'd have to do a more detailed sentence outline to hand in with my paper.)

▶ See 41j

Thesis: Athletic scholarships should be continued, but college athletic programs should be reformed to deemphasize winning at all costs and to focus instead on fairness to <u>all</u> student athletes.

I. Advantages of Athletic Programs
   A. School Spirit
   B. Money
   C. Balanced Education

II. Disadvantages of Athletic Programs
   A. Unequal Support of Sports
      1. Football and basketball vs. other sports
      2. Men's sports vs. women's sports

continued on the following page

Answers will vary. You will probably
want to check students' work.

continued from the previous page

B. Treatment of Student Athletes
  1. Acceptance of unqualified students
  2. Little academic support
  3. Heavy schedules

III. Causes of Problems
  A. Dishonest and Exploitive Athletic Directors
  B. Concentration on Winning

IV. Historical Background
  A. Harvard–Yale Contest
  B. Competition for Students
  C. Efforts to Control Violence and Standardize Rules
    1. Early organizations
    2. NCAA

V. Abuses in Athletic Programs and Scholarships
  A. Recruitment
    1. Before 1980s
    2. NCAA intervention
    3. Continuing violations

  B. Sexism
    1. Concentration on men
    2. Title IX
    3. NCAA study

  C. Academics
    1. Proposition 48
    2. Controversy
    3. Recent Developments

VI. Suggested Changes
  A. Fair Recruitment and Academic Support Programs
  B. Clear Priorities
  C. Balanced Treatment of Students
  D. Responsible Recruitment

EXERCISE 7

Review your note cards carefully. Then, sort and group your cards into
categories and construct a formal topic outline for your paper.

41h Writing a Rough Draft

When you are ready to write your rough draft, arrange your note cards in the order in which you intend to use them. Follow your outline as you write, moving from one entry to the next and using your notes as you need them.

To make it easier for you to revise your rough draft later on, follow the guidelines recommended in Chapter 3: Double-space, write or type on only one side of your paper, and so on. Be careful to recopy your source information accurately on this and every subsequent draft, placing the documentation as close as possible to the material it identifies.

Once you begin drafting, you will find that the time you spent taking careful, accurate notes and preparing an outline will pay off. Even so, do not expect to write the whole draft in a single sitting. Developing one major heading from your outline is a realistic goal for a morning or afternoon of writing.

Your paragraphs will probably correspond to the points of your outline, at least in this draft. (Later on, you may make changes.) As you write, make an effort to supply transitions between sentences and paragraphs. These transitions need not be polished; you will refine them in subsequent drafts. But if you leave them out entirely at this stage, you may lose track of the logical and sequential links between ideas, and this will make revision difficult.

If words do not come easily, freewriting for a short period may help. Sometimes leaving your paper for five or ten minutes gives you a fresh view of your material. Another good strategy for avoiding writer's block is beginning your drafting with the section for which you have the most material. ▶ See 1c2

Remember, the purpose of the first draft is to get ideas down on paper so that you can react to them. You should *expect* to revise, so postpone making precise word choices and refining style. As you draft, jot down questions to yourself and note points that need further clarification; leave space for material you plan to add; and bracket phrases or whole sections that you may later decide to move or delete. In other words, lay the groundwork for a major revision. Remember that even though you are guided by an outline and notes, you are not bound to follow their content or sequence exactly. As you write, new ideas or new connections among ideas may occur to you. Make a note of such ideas as they come to mind (perhaps bracketing them or printing them in ▶ See 3b boldface on a typed draft or jotting them down on self-stick notes), and incorporate them into your next draft. If you find yourself deviating from your thesis or outline, reexamine them to see whether the departure is justified. (For information about drafting your essay on a computer, **see Appendix B**.)

(1) Shaping the parts of the paper

Like other essays, the research paper has an *introduction*, a *body*, and a *conclusion*, but in the rough draft you should focus on developing the body of your paper. You should not spend time planning an introduction or conclusion at this stage. Your ideas will change as you write, and you will want to revise your opening and closing paragraphs later to reflect your revisions.

See ◀
4g2

Introduction   In your **introduction** you identify your topic and establish how you will approach it. Here, you also include your thesis—the position you will support in the rest of the paper. Sometimes the introductory paragraphs briefly summarize your major supporting points (the major divisions of your outline) in the order in which you will present them. Such an overview of your thesis and support provides a smooth transition into the body of your paper. In your rough draft, however, your thesis alone can serve as a placeholder for a more polished introduction.

See ◀
4c1

Body   As you draft the **body** of your paper, indicate its direction with strong topic sentences that correspond to the divisions of your outline.

> Despite their obvious advantages, college athletic programs have problems.

You can also use headings if they are a convention of the discipline in which you are writing.

Problems of Athletic Programs
> Despite their obvious advantages, college athletic programs have problems.

Even in your first draft, descriptive headings and topic sentences will help you keep your discussion under control.

See ◀
4e

See ◀
4g1

Use different **patterns of development** to shape the individual sections of your paper, and be sure to connect ideas with clear transitions. If necessary, connect two sections of your paper with a **transitional paragraph** that shows their relationship.

Conclusion The **conclusion** of a research paper often re-states the thesis. This is especially important in a long paper be-cause by the time your readers get to the end, they may have lost sight of your paper's main point. You can close your paper with a summary of your major points, a call for action, or perhaps an apt quotation. In your rough draft, however, your concluding para-graph can be very tentative.

▶ See 4g3

(2) Working source material into your paper

In the body of your paper, you evaluate and interpret your sources, comparing different ideas and synthesizing conflicting points of view. As a writer, your job is to draw your own conclu-sions, consolidating information from various sources into a paper that presents a coherent, original view of your topic to your readers.

Your source material must be smoothly integrated into your paper, with the relationships among various sources (and be-tween those sources' ideas and your own) clearly and accurately identified. If two sources present conflicting interpretations, you must be especially careful to use precise language and accurate transitions to make the contrast apparent (For instance, "Al-though Durrant believes the situation has changed, a later study suggests . . ."). Even if two sources agree, you should make this clear (For example, "Like Blum, Durrant believes" or "Coach Walker's statistics confirm those reported in Dealy"). Such phras-ing will provide a context for your own comments and conclu-sions. If different sources present complementary information about a subject, blend details from the sources *carefully,* keeping track of which details come from which source, to reveal the com-plete picture.

CLOSE-UP

WRITING A RESEARCH PAPER

As you write your rough draft, be sure to record all source information fully and accurately. Include every source and page number, as Marion Duchac does in the paragraphs from her draft on pages 687–88.

If you decide to look at rough drafts, remember that you cannot cover everything. Offer suggestions, but avoid doing your students' editing for them.

"Don't be afraid to seize whatever you have written and cut it to ribbons. It can always be restored to its original condition in the morning."

–E. B. White

"Good writing is good manners. You can both please and help your public only when you learn how to be the first victim of your writing, how to anticipate a reader's difficulties and to hear yourself as others hear you."

–Ritchie R. Ward

"I have rewritten—often several times—every word I have ever published. My pencils outlast their erasers."

–Vladimir Nabokov

EXERCISE 8

Write a draft of your paper, being careful to incorporate source material smoothly and to record source information accurately. Begin with the section for which you have the most material.

41i   Revising Your Drafts

A good way to start revising is to check to see that your thesis is still appropriate for your paper. Make an outline of your completed draft, and compare it with the outline you made before you began the draft. If you find significant differences, you will have to revise your thesis or rewrite sections of your paper.

See ◄
3b-d

As you review your drafts, follow the revision procedures that apply to any paper. (For information about revising on a computer, **see Appendix B**.) In addition, you should consider the following questions that apply specifically to research papers.

REVISING A RESEARCH PAPER

- Should you do more research to find support for certain points?
- Do you need to reorder the major divisions of your paper?
- Should you rearrange the order in which you present your points within those divisions?
- Do you need to add section headings? Transitional paragraphs?
- Are sources smoothly integrated into your paper?
- Are direct quotations blended with paraphrase, summary, and your own observations and reactions?
- Do you introduce source material with running acknowledgments?
- Are all borrowed ideas carefully documented?
- Have you analyzed and interpreted the ideas of others rather than simply stringing those ideas together?

See ◄
3c2

If your instructor allows **collaborative revision**, take advantage of it. As you move from rough to final draft, you should think more and more about your readers' reactions. Testing out others' reactions to a draft can be extremely helpful.

CLOSE-UP

REVISING YOUR DRAFTS

You will probably take your paper through several drafts, changing different parts of it each time or working on one part over and over again. After revising each draft thoroughly, type or print out a corrected version and make additional corrections by hand on that draft before typing your next version.

▶ See 3d

WRITING AND REVISING

In the first draft of my paper I tried to explain the effects of Title IX by using my own words to paraphrase and summarize Sue M. Durrant. But a classmate who read my draft in a collaborative revision session asked me what the regulation actually says, so I decided to add the exact wording of a portion of Title IX (quoted in Durrant) in my final draft.

First Draft

In general, the money-making teams are the men's teams. Because the emphasis is on winning and making money, it is not surprising that colleges and recruiters concentrate on men when building and maintaining their sports programs. Since the introduction of Title IX in 1972, however, such a focus could be interpreted as illegal. Sue M. Durrant reports that although Title IX encompasses nearly all facets of education, it is mainly associated with increased opportunities for women in the area of athletics (60).

Revised

In general... since the introduction of Title IX in 1971, however, such a focus could be interpreted as illegal. Title IX states, "No person in the United States shall, on the basis of sex, be excluded from participation in, be

continued on the following page

COMPUTER TIP

Embedding Writer-to-Self Comments Within Drafts
Tell students to write notes to themselves within the text as they compose—reminders of material they need to include or questions that occur to them. Students can put these notes within brackets, or they can format them in bold, uppercase, or *hidden text* characters (if their software offers that feature).

QUOTATION

"No matter how carefully you write each sentence as you assemble a piece of writing, the final product is bound to have some ragged edges. . . . I've been writing this book by the bricklayer method, slowly and carefully. That's all very well as far as it goes—at the end of every chapter the individual bricks may look fine. But what about the wall? The only way to check your piece for unity is to go over it one more time from start to finish, preferably reading it aloud. See if you have executed all the decisions that you made before you started writing.)

–William Zinsser
Writing With a Word Processor

continued from the previous page

denied the benefits of, or be subjected to discrimination under any program or activity receiving Federal financial assistance" (qtd. in Durrant 60).

Sue M. Durrant reports that although....

Another classmate pointed out that one of my paragraphs seemed unfocused. He said the first two sentences of the paragraph didn't seem to go with the rest of the sentences. I decided to delete the first two sentences of the paragraph and insert a new topic sentence that accurately stated the main idea.

First Draft

Francis X. Dealy points out that as long as there have been human civilizations, there have been athletes and their competitions. Spanning from the ancient Greek games to last year's Super Bowl, human beings have organized contests to celebrate athletes and sports. Dealy reports that the first American intercollegiate spectacle was held in 1852 in New Hampshire when a rowing contest between Harvard and Yale was staged. Harvard won, and so began a fierce rivalry between the two schools (56). As Dealy observes, "Judging from the intensity of the spectators and the participants, the stakes included which school had the more beautiful campus, the smarter faculty, the brighter student body, and the more successful alumni" (59).

Revised

Athletic contests have had a long historical tradition in American colleges. Francis X. Dealy reports that the first American intercollegiate spectacle was held in 1852 in New Hampshire when a rowing contest between Harvard and Yale was staged. Harvard won, and so began a fierce rivalry between the two schools (56). As Dealy observes, "Judging from the intensity of the spectators and the participants, the stakes included which school had the more beautiful campus, the smarter faculty, the brighter student body, and the more successful alumni" (59).

EXERCISE 9

Following the guidelines in 41i and 3c, revise your research paper until you are ready to prepare your final draft.

41j   Preparing a Final Draft

Before you type or print out the final version of your paper, you will have to prepare a detailed formal outline (usually a sentence outline) to hand in with your paper, and you will also have to prepare your list of works cited. After you have finished these tasks, you will *edit* all your material—paper, outline, documentation, Works Cited list, and so on.

After this point, stop for a moment to consider your title. It should be descriptive enough to tell your readers what your paper is about, and, ideally, it should create interest in your subject. Your title should also be consistent with the purpose and tone of your paper. (You would hardly want a humorous title for a paper about the death penalty or world hunger.) Finally, your title should be engaging and to the point—and perhaps even provocative. Often a quotation from one of your sources suggests a likely title: Marion Duchac's title was suggested by the title of one of her sources, Dealy's *Win at Any Cost*.

▶ See 3d3

Now you can proceed to type your final draft. Before you hand in your manuscript, read it through one last time, proofreading for grammar, spelling, or typing errors you may have missed. Pay particular attention to parenthetical documentation and Works Cited entries. Remember that every error undermines your credibility. If your instructor gives you permission, you may make *minor* corrections on your final draft with correction fluid or by neatly crossing out a word or two and writing the correct word or phrase above the line. If you find a mistake that you cannot correct neatly, however, retype or reprint the page. Once you are satisfied that your manuscript is as accurate as you can make it, you are ready to hand it in.

▶ See App. A

EXERCISE 10

Prepare a sentence outline and Works Cited list for your research paper. Then, edit your paper, outline, and Works Cited list; decide on a title; and type your paper according to the format your instructor requires. Proofread your typed copy carefully before you hand it in.

COMPUTER TIP

Dealing with the Mask of Neatness
Students sometimes fail to perceive errors or problems in word-processed text. Because word processing makes it easy to correct spelling mistakes and typographical errors and because laser printers produce clear, crisp copy, student writing (in the words of one experienced teacher) always looks neat and publishable—even in early or conceptually messy drafts.

QUOTATIONS

"A good title is the title of a successful book."
–Raymond Chandler

"I have never been a title man. I don't give a damn what it is called."
–John Steinbeck

Collect a variety of titles (from another class of yours)—some of which are effective and appropriate and some of which are ineffective and inappropriate. Discuss in class how titles are made and what distinguishes a good title from a poor one.

Critical Thinking

To remind students that even a finished piece of writing remains a vital, lively document subject to musing, speculation, and questioning, ask them to read the final draft of "Athletic Scholarships: Who Wins?" and write a brief comment to the author. They might keep the following guidelines in mind.

1. If the author were here, what questions would you ask?
2. What ideas in the paper interested you most?
3. Are there any ideas in the essay that you would challenge? Explain.
4. Choose two or three sentences in the essay that you think exemplify its strong points. Explain why.
5. If you think the essay still has some weak points, choose two or three sentences to illustrate your reasons.
6. Use this essay as a taking-off point for finding a new topic. Which of this writer's points offers you a possible topic to explore in an essay of your own?

Allow 10 to 15 minutes for writing; then ask students to let their writing form the basis for class discussion on the essay's strengths and weaknesses.

THE COMPLETED PAPER

Marion Duchac's completed research paper, "Athletic Scholarships: Who Wins?" appears on the pages that follow. The paper, which uses MLA documentation style, is accompanied by a title page, a sentence outline, a notes page, and a list of works cited. Annotations opposite each page of the manuscript comment on stylistic and structural aspects of the paper; explain the format for proper documentation; illustrate various methods of incorporating source material into the paper; and highlight some of the choices Marion made as she moved from note cards to rough draft to completed paper.

If your instructor does not require a title page, include all identifying information—your name, the name of the course, your instructor's name, and the date—in the upper-left-hand corner of your paper's first page, one inch from the top and flush with the left-hand margin. Your title should be centered two spaces below the last line of this heading. Type your name and the number *1* in the upper right-hand corner, one-half inch from the top.

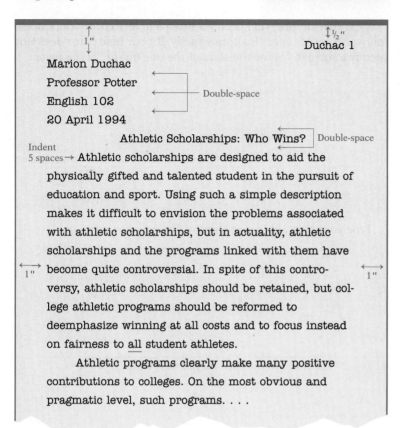

↕ ½ "
Duchac 1

↑
1"
↓

Marion Duchac

Professor Potter

English 102

20 April 1994

— Double-space

Athletic Scholarships: Who Wins? ← Double-space

Indent
5 spaces → Athletic scholarships are designed to aid the

physically gifted and talented student in the pursuit of

education and sport. Using such a simple description

makes it difficult to envision the problems associated

with athletic scholarships, but in actuality, athletic

scholarships and the programs linked with them have

←→ become quite controversial. In spite of this contro- ←→
1"   1"

versy, athletic scholarships should be retained, but col-

lege athletic programs should be reformed to

deemphasize winning at all costs and to focus instead

on fairness to all student athletes.

Athletic programs clearly make many positive

contributions to colleges. On the most obvious and

pragmatic level, such programs. . . .

Most instructors will specify a title-page format. If yours does, follow his or her specifications exactly. If your instructor does not specify a format, follow the style of the one on the facing page.

Type your title about one-third of the way down the page.

Type your name about two inches below the title.

Type your course number, instructor's name, and date about two inches below your paper's title. Double-space between these lines.

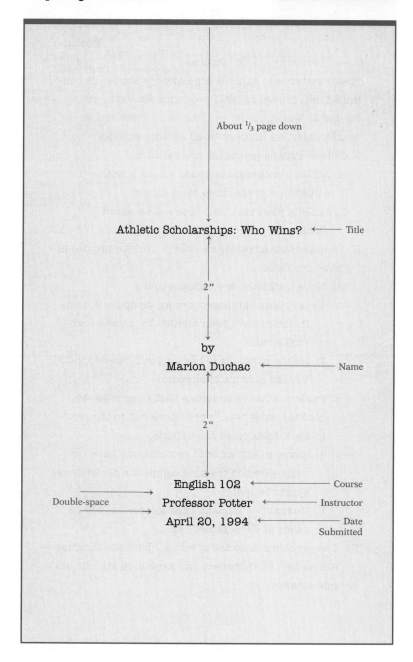

About ⅓ page down

Athletic Scholarships: Who Wins? ← Title

2"

by

Marion Duchac ← Name

2"

English 102 ← Course

Double-space → Professor Potter ← Instructor

April 20, 1994 ← Date Submitted

Duchac i

Center ⟶ Outline ⟵ Double-space

Thesis statement: Athletic scholarships should be continued, but college athletic programs should be reformed to deemphasize winning at all costs and to focus instead on fairness to all student athletes.

I. College athletic programs are valuable.

   A. Athletic programs increase school spirit.

   B. Athletic programs help raise money.

   C. Athletic programs help create a balanced education.

II. Despite their advantages, college athletic programs have problems.

   A. Not all athletes are valued equally.

      1. On many campuses money, equipment, and facilities have been allotted to football and basketball.

      2. Men's sports have been given a disproportionate amount of support.

   B. Student athletes complain that their roles as athletes have been overemphasized to the detriment of their roles as students.

      1. Some college athletic departments have become glorified training camps for professional sports teams.

      2. Unqualified students are accepted solely because of their athletic skills.

III. The problems associated with athletic scholarships are caused by dishonest and exploitive athletic administrators.

Duchac ii

    A. It is understandable that the main focus of most collegiate sports programs is winning.

    B. The concentration on winning leads to the worst abuses in athletic programs and scholarships.

IV. Athletic contests have had a long historical tradition in American colleges.

    A. American intercollegiate sports began in 1852 with a rowing contest between Harvard and Yale, which began a fierce rivalry.

    B. The emphasis on winning encouraged the recruitment of the best athletes, regardless of cost.

    C. The history of college athletics has been marked by violence and controversy.

        1. Organizations formed to solve these problems were short-lived.

        2. In 1905 the NCAA was formed to deal with the violence and to standardize rules.

        3. One of the NCAA's main concerns has always been the equitable distribution of financial aid and scholarships.

V. Today, the NCAA continues to address abuses associated with athletic programs and scholarships.

    A. Recruitment of student athletes is a large and controversial part of the athletic scholarship process.

        1. Until the late 1980s, recruiters enticed high school football players with generous financial aid.

        2. After several instances of unethical recruitment practices became public, the NCAA intervened, but recruitment violations continue.

Duchac iii

3. Recruitment is still the principal means of matching students with available funds.

B. Sexism is another serious problem in college athletic programs.

1. Because the emphasis is on winning and making money, colleges and recruiters concentrate on men.

2. Since the introduction of Title IX, such a focus could be interpreted as illegal.

   a. Title IX granted acceptability and status to female athletes in elementary schools, high schools, and colleges.

   b. Initially, there was a rapid growth in female athletic programs.

   c. More recently, there has been an obvious slowing of the movement toward equality between men and women in collegiate sports programs.

3. A March 1992 NCAA study of gender equality in colleges found that more resources were allotted to men's teams than to women's.

C. The relationship between sports and academics has always been tempestuous.

1. Proposition 48 was instituted when it was made public that some of America's star athletes were unable to read.

   a. Proposition 48 contains a "partial qualifier" clause.

   b. The NCAA voted to close this loophole in 1989 with the passage of a series of reforms.

Duchac iv

2. The proposed reforms are extremely controversial.

    a. John Chaney, the men's basketball coach at Temple University, criticizes them.

    b. Arthur Ashe, the tennis star, believed that athletes would get the message.

3. Changes are occuring.

    a. Athletes' graduation rates have increased.

    b. Fewer athletes are academically unprepared.

4. Changes may or may not indicate improvement.

VI. Colleges should retain athletic scholarships--with certain changes.

    A. Academic support programs should be fair to all student athletes.

    B. Academics--not sports--must be given first priority.

    C. Students who receive athletic scholarships should not be exploited.

    D. Recruitment should be responsible.

Beginning with the first page of your paper, type your last name and the page number one-half inch from the top in the upper right-hand corner. Do not use punctuation before or after the page number. Leave one-inch margins.

Title typed one inch from top of page

Two spaces between title and first line of paper

¶1: Introduction, presenting background information
Thesis statement

¶2: (Outline point I): Paragraph presents background on value of athletic programs and need for athletic scholarships. First four sentences of paragraph support student's opinion with information available in several sources; therefore, no documentation is required. In the last sentence, "as Alvin Sanoff observes" introduces material from a source. Introducing the summary with the author's name and following it with parenthetical documentation clearly identifies the boundaries of the borrowed material.

↕½"

Duchac 1

Athletic Scholarships: Who Wins?
←—— Double-space

¶1 Athletic scholarships are designed to aid the physically gifted and talented student in the pursuit of education and sport. Using such a simple description
makes it difficult to envision the problems associated
with athletic scholarships, but actually, athletic scholarships and the programs linked with them have become quite controversial. In spite of this controversy,
athletic scholarships should be retained, but college
athletic programs should be reformed to deemphasize
winning at all costs and to focus instead on fairness to
all student athletes.

¶2 Athletic programs clearly make many positive
contributions to colleges. On the most obvious and
pragmatic level, such programs increase school spirit
and help to create a sense of community. They also
help to raise money: Winning teams spark alumni
contributions, and athletic events raise funds
through ticket sales. More important, athletic pro-
grams--like programs in performing arts and
music--help to create a rewarding, balanced education for all students. In addition, student athletes
make important academic, social, and cultural contributions to their schools and thus enrich the college
experience for others. Without athletic scholarships,
many of these students would be denied the opportunity to attend college because, as Alvin Sanoff observes, the aid for which many economically
deprived student athletes are eligible does not cover

¶3: (Outline point IIA): Paragraph's discussion of disparities in athletic programs is supported by source material, which combines paraphrase with direct quotation.

¶4 (Outline point IIB): Student's point is supported by her own observations as well as by a source whose ideas she summarizes (to convey general ideas rather than specific details). Because the author's name is mentioned in the paper, it is not included in the parenthetical documentation.

Duchac 2

the expense of a college education the way athletic
scholarships do (68).

¶3 Despite their obvious advantages, college athletic
programs have problems. First, not all athletes--or all
programs--are valued equally. According to Sue M.
Durrant, on many campuses money, equipment, and
facilities have traditionally been allotted to football
and basketball at the expense of less visible sports
such as swimming, tennis, and field hockey. Men's
sports have been given a disproportionate amount of
support, and "winning" teams and coaches have been
compensated accordingly. In fact, until recently it was
not unusual for women's teams to use "hand-me-
down" gear while men's teams played with new "state
of the art" equipment or for women's teams to travel
by bus while men's teams traveled by air (60).

¶4 Another problem is that college athletes at all levels
complain their roles as athletes are overemphasized, to
the detriment of their roles as students. According to
Francis X. Dealy, some college athletic departments
have become little more than glorified training camps
for professional sports teams, and student athletes have
been pressured to put academics last. This problem is
compounded by overzealous recruiting practices, with
academically unqualified students accepted to the college
solely because of their athletic skills. Once at college,
these students are exploited and overworked, treated as
commodities rather than as students, and given little
academic support; many fail to graduate (106). With
the demands of heavy travel and practice schedules,

¶5 (Outline points IIIA and IIIB): Two quotations identify emphasis on winning, introducing student's point that this emphasis is a significant cause of the abuses her paper will discuss.

Note that the Lombardi quotation is so well known it does not require documentation.

Qtd. in indicates that the Santayana quotation appears in Dealy.

¶6 (Outline points IVA and IVB): Paragraph provides background to explain roots of today's problems. Two parenthetical references cite the same source (Dealy).

Duchac 3

many student athletes, even those with strong academic backgrounds, run the risk of falling behind in their studies. Moreover, their grueling schedules tend to isolate them from other students, excluding them from the college community. Given these difficulties, college athletic programs are under considerable pressure to reform their recruitment and financial aid packages.

¶5   The problems associated with athletic scholarships are particularly numerous and complex, but they have less to do with the scholarships themselves than with the way dishonest and exploitive athletic administrators run their programs. It is understandable that the main focus of most collegiate sports programs is winning. According to Vince Lombardi, the famous football coach, "Winning isn't everything; it's the only thing." Even the philosopher George Santayana has observed, "In athletics, as in all performances, only winning is interesting. The rest has value only as leading to it or reflecting it" (qtd. in Dealy 61). However, this concentration on winning leads to the worst abuses in athletic programs and the scholarships they award.

¶6   Athletic contests have had a long historical tradition in American colleges. Francis X. Dealy reports that the first American intercollegiate spectacle was held in 1852 in New Hampshire when a rowing contest between Harvard and Yale was staged. Harvard won, and so began a fierce rivalry between the two schools (56). As Dealy observes, "Judging from the intensity of the spectators and the participants, the stakes included which school had the more beautiful

Note that a direct quotation from a source requires its own reference.

¶7 (Outline point IVC): Paragraph continues tracing background, summarizing problems historically associated with collegiate athletics and steps taken to address those problems.

Quotation is introduced by a running acknowledgment.

Fleisher, Goff, and Tollison 38–41 indicates summary of four pages of material from a source.

¶8 (Outline point V): Transitional paragraph identifies the three principal abuses on which the paper will focus (unethical recruitment practices, sexism, and admissions irregularities).

Duchac 4

campus, the smarter faculty, the brighter student body, and the more successful alumni" (59). The emphasis on winning encouraged the recruitment of the best athletes, no matter what the cost. In fact, Dealy observes that the first athletic scholarships were in the form of salaries paid to professional athletes to perform in the name of a particular school. Without regulation, athletic scholarships were much like shady financial deals arranged in smoky back rooms: Athletes became commodities to be bought and sold.

¶7   Fleisher, Goff, and Tollison report that up until the late 1870s, collegiate games "were marked by violence, rough play, haphazard rules, and controversy over eligibility requirements. Athletes moved from school to school with no loss of eligibility, and club members hired professional athletes to participate in intercollegiate events" (37). Several organizations were formed to help control violence and to standardize rules, but all had spotty participation and were short-lived. In December 1905, in order to deal with violence and to standardize rules of play, the National Collegiate Athletic Association (NCAA) was formed in response to the concerns of Theodore Roosevelt, then president of the United States. The scope of the NCAA has widened tremendously in the last ninety years, but one of its main concerns has always been the equitable distribution of financial aid and scholarships (Fleisher, Goff, and Tollison 38–41).

¶8   Today the NCAA continues to address abuses associated with athletic programs and scholarships,

¶9 (Outline points VA, VA1, and VA2): Paragraph introduces first abuse (unethical recruitment practices).

Dealy 173–80 indicates that paragraph summarizes several pages of a source. Concluding citation includes Dealy's name because parenthetical reference is far from running acknowledgment.

¶10 (Outline point VA3): Paragraph introduces discussion of recruitment practices, blending information from three sources, one of which is a telephone interview. (Interview subject is listed with other sources in the Works Cited section of the paper.)

Superscript (raised numeral) at end of first sentence of paragraph refers readers to a content note. (See notes following text of paper.)

Student's note cards for information from Dealy and from interview with Coach Walker appear on page 673.

Duchac 5

including aggressive and often unethical recruitment techniques, a disproportionate amount of money being awarded to men over women, and academically under-prepared athletes being admitted to and retained by colleges and universities. The organization's task is a difficult one, however, because the problems have deep roots.

¶9   Recruitment of student athletes is a large and con-troversial part of the athletic scholarship process. Un-derstandably, colleges and universities want to recruit the finest athletes for their teams, but sometimes this quest for the best has led to unethical and aggressive re-cruitment practices. Dealy reports, for example, that up until the late 1980s, recruiters openly enticed talented high school football players with promises of generous financial aid and merchandise including cars or expen-sive athletic clothing and shoes. After several instances of unethical recruitment practices became public, most notably the fact that one university had been paying its football players salaries just to play ball, the NCAA in-tervened and became more vigorous in its attempt to regulate the recruitment process. Recruiting violations continue to account for 60 to 70 percent of all NCAA in-fractions, however, and the NCAA has had great diffi-culty enforcing guidelines in this area (Dealy 173–80).

¶10 Skippy "Tiptoe" Walker, assistant football coach at a large Texas high school, reports that none of his athletes has been wooed in any manner that could be considered unethical.[1] Walker is quick to point out, however, that most of his athletes do not receive scholarships. In fact,

Title alone (abbreviated) is used in parenthetical reference because source does not specify an author's name.

¶11 (Outline points VB1 and VB2): Paragraph introduces second major abuse to be discussed in paper (sexism in college sports programs).

Note that quotation is used rather than paraphrase or summary. Using the exact wording from a key piece of legislation is important because it conveys the law's ideas clearly and eliminates any possibility of readers' misinterpreting its intent.

Duchac 6

only two football players from his high school have received athletic scholarships in the past ten years despite
his team's consistently high rankings in their division.
This statistic is in line with the rest of the country. As
reported by Dealy, only one out of every 118 high
school football seniors receives an athletic scholarship
(180). It is possible for students to try to locate their
own athletic scholarships by paying a nominal fee to an
independent search service, which enters the student's
name into a national database and also provides the student with a list of available scholarships and schools
seeking recipients ("You C.A.N." 56). Recruitment, however, remains the principal form of matching a student
with available funds and positions. Recruitment builds
the strongest teams, because recruiters can keep their
eyes on promising students who are still in the early
part of their high school careers.

¶11 Sexism is another serious problem in college athletic organizations. The strongest teams, in the view of
colleges, are the ones that generate the greatest
amount of interest (and revenue) both on and off campus. In general, the money-making teams are the
men's teams. Because the emphasis is on winning and
making money, it is not surprising that colleges and
recruiters concentrate on men when building and
maintaining their sports programs. Since the introduction of Title IX in 1972, however, such a focus could
be interpreted as illegal. Title IX states, "No person in
the United States shall, on the basis of sex, be excluded from participation in, be denied the benefits of,

¶12 (Outline points VB2a-c): Paragraph continues discussion of sexism in college sports programs.

Paragraph includes three parenthetical references to a single source (Durrant). Note that each quotation requires its own parenthetical reference.

¶13 (Outline point VB3): Paragraph continues discussion of sexism in college sports programs, blending paraphrase and summary from three sources.

Note that two sources by a single author (Lederman) are differentiated by title to avoid confusion.

Duchac 7

or be subjected to discrimination under any program or activity receiving Federal financial assistance" (qtd. in Durrant 60).

¶12   Sue M. Durrant reports that although Title IX encompasses nearly all facets of education, it is mainly associated with increased opportunities for women in the area of athletics (60). In fact, Durrant notes, "Title IX tilted the balance of power. Title IX granted acceptability and status to elementary school, high school, and college female athletes" (61). During the first decade that Title IX was in place, there was a huge response, and reform was immediate and readily apparent. The number of women athletes in colleges doubled, and there was rapid growth in female athletic programs at all levels of education, but particularly in colleges and universities (Durrant 61). Since this ten-year span of compliance to the law, there has been an obvious slowing of the movement toward equality between men and women in collegiate sports programs. One of the areas in which this problem is most visible and measurable is in athletic financial aid and scholarships.

¶13 The NCAA's study of gender equality in colleges that play big-time sports, completed in March 1992, found that men's teams received almost 70 percent of the athletic scholarship money, 77 percent of the operating money, and 83 percent of the recruiting money (Lederman, "Men Get Money" A45). Proponents for women argue that the distribution of money should be based on enrollment, which, as reported in a Chronicle of Higher Education study of gender equality, would

¶14 (Outline points VC, VC1, and VC1a): Paragraph introduces discussion of third major abuse (admissions irregularities in college sports programs).

Paragraph begins with a transitional sentence that moves readers from one problem (sexism) to another (admissions irregularities).

Running acknowledgment ("Alvin Sanoff reports that . . .") introduces paraphrase of a source.

Note that paraphrase includes quotation of a key term ("partial qualifier") which cannot be effectively paraphrased.

¶15 (Outline point VC1b): Paragraph continues discussion of admissions irregularities, introducing a controversy associated with athletes' admissions and scholarships and a discussion of the NCAA's role in the controversy.

Duchac 8

give women a slight edge over men (Lederman, "Men Outnumber" A1). In order for future progress to be made in the area of gender equality in college sports, it is important that the NCAA and other independent organizations continue surveys like the NCAA gender equality study. And, as Durrant points out, it is also important that complaints continue to be filed when discrimination is suspected or encountered (63).

¶14 Although the relationship between women and organized sports might be stormy at times, the relationship between sports and academics has always been tempestuous under the best of circumstances. When it was made public that some of America's star college athletes were unable to read (Dealy 111), the NCAA was forced into action. Proposition 48, the result of much compromise and maneuvering during the NCAA's 1983 convention, required that athletes meet some basic academic requirements before they could receive athletic scholarships. Alvin Sanoff reports that the potential recipients had to score at least 700 out of a possible 1600 points on the Scholastic Aptitude Test (or 15 out of 36 on the American College Test) or attain a C average in eleven core academic courses. If the student achieved only one of these goals and not the other, he or she was considered a "partial qualifier" and, although eligible for an athletic scholarship, would not be allowed to participate in sports during his or her freshman year (68).

¶15 Since Proposition 48 went into effect in 1986, approximately 600 students per year have received athletic scholarships under the "partial qualifier"

Author Lederman is named earlier in the paragraph; an abbreviated title is included in the parenthetical reference to distinguish source from other sources by Lederman.

¶16 (Outline points VC2, VC3, and VC4): Paragraph continues discussion of admissions irregularities, blending direct quotation and paraphrase from two sources. *Qtd in* indicates that Chaney and Ashe quotations were found in Sanoff.

Parenthetical reference includes title because paper cites two articles by Blum.

Duchac 9

umbrella. Ninety percent of these students were African-American football or basketball players (Sanoff 68). In 1989, however, the NCAA voted to remove this loophole of opportunity with the passage of a series of reforms, the most stringent of which is scheduled to take effect in August 1995. The most significant change, as reported by Lederman, will be that freshmen athletes must achieve a 2.5 grade point average in thirteen academic core courses rather than 2.0 in eleven courses as previously required. Students will also have to score a minimum of 700 on the SAT in addition to the GPA requirement ("NCAA Votes" A1).

¶16 Because underprivileged athletes are most affected by these rule changes, the proposed reforms are extremely controversial. John Chaney, the men's basketball coach at Temple University, calls the new rule "an insane, inhuman piece of legislation that will fill the streets with more of the disadvantaged" (qtd. in Sanoff 68). The late tennis player Arthur Ashe believed, however, that "any time educational standards have been raised, the athletes have gotten the message" (qtd. in Sanoff 68). Preliminary results of ongoing studies have indicated that the athletes are indeed getting the message: The graduation rates of Division I scholarship athletes entering college in 1986 were six percentage points higher than the average graduation rates of athletes who enrolled at those same colleges three years before Proposition 48 took effect (Blum "Graduation" A42). Other study results show that the number of academically underprepared athletes

Over four lines long, this quotation is typed as a block, indented ten spaces from the left margin, double-spaced, with two spaces above and below. No quotation marks are used. Because the quotation is a single paragraph, no paragraph indentation is needed. (In a quotation of more than one paragraph, all paragraphs are indented three spaces.)

Note that in a long quotation the parenthetical reference is placed two spaces after the final punctuation.

Paragraph 16 closes with the student's original idea, leading to her paper's concluding paragraphs.

¶17 (Outline Point VI): Paragraph introduces concluding remarks, which make recommendations about reforms. Because all ideas in this paragraph represent the student's original conclusions, no documentation is necessary.

Duchac 10

enrolling in Division I colleges dropped in 1991. As reported by Debra Blum, however, these statistics can be read in two diametrically opposed ways:

> The decline in the number of academically underqualified athletes going to Division I and II colleges may mean that more athletes are meeting the standard, as supporters of the standard contend. On the other hand, the decline may suggest that the underprepared students are simply moving in greater numbers into junior colleges or preparatory schools or, as some critics fear, that they are not continuing their education at all. ("More Freshmen" A39)

The legislation will go into effect as scheduled regardless of the raging argument. It is important, though, to remember that athletic scholarships are given by academic institutions of higher learning and not by clearinghouses for potentially profitable athlete-products.

¶17 If athletic departments insist on treating their student athletes as commodities, colleges should end the charade and treat them as the professionals they are--eliminating all academic requirements and paying them a percentage of what they earn for the college. If, however, they want their athletic programs to be student programs, colleges should retain athletic scholarships--but not in their present form. Instead, they must move to reform recruitment and academic support programs so that they are fair to all student athletes---men and women, tennis players and football

¶18: (Conclusion): Student presents further recommendations for reforming abuses in college athletic programs.

Long quotation from a source, identifying purpose of college athletic programs, serves as paragraph's focal point.

Paper closes with an emphatic final sentence that echoes the title, reminding readers of what the student has identified as the cause of the problems in college athletics programs: the focus on winning.

Duchac 11

players, winners and losers. Academics--not sports--
must be given first priority. Students who receive athletic scholarships should not be exploited; they should
be treated like all other scholarship recipients. Recruitment should be responsible, academic standards
should be maintained, and promises made to athletes
should be realistic.

¶18 What should the true role of athletic programs on
college campuses be? Durrant suggests that it goes far
beyond competition for trophies:

> Participation in sport and physical activity
> may be for health and fitness, the aesthetic as-
> pects, social interaction, the pride of accom-
> plishment, or merely for the joy of moving.
> Through this participation, the development of
> identity and self-esteem occurs. To be able to
> move ourselves in desired ways increases our
> confidence, competence, and efficiency. (61)

If we accept this interpretation, it follows that the
scholarship athlete should be treated like any other
exceptional student on campus who loves his or her
subject and takes joy in the process of learning and
mastery. Athletic programs clearly benefit their institutions, and athletic scholarships should certainly be a
part of any college system: However, the focus of sports
programs should expand to encompass the personal
enrichment of the whole student. Shifting the focus of
athletics away from winning will ultimately benefit not
only college athletes and the scholarship programs that
support them, but also the colleges themselves.

This page is numbered.

A content note provides supplementary information to the reader. Because the information is a digression, including it in the body of the paper would be distracting.

↕ ½"
Duchac 12

↑
1"
↓

Center ────────────→ Notes
←──────── Double-space

Indent ──→ [1]I also interviewed another high school coach
5 spaces

who asked that he not be identified. He admitted that
several of his star athletes over the past few years
had been lured with expensive dinners and generous
financial packages into big-time colleges. When I asked
the coach if he had reported these infractions to the
proper authorities, he said he had not.

Every page of the works cited section is numbered. List entries in alphabetical order, indenting the second and subsequent lines of every entry 5 spaces. Use double spacing between and within entries.

The first entry illustrates the correct form for a signed newspaper article by a single author. Note that it indicates both section and page numbers and provides inclusive pagination.

The list of works cited includes two works by Debra E. Blum. Note that the author's name is not repeated in the second entry; instead, three unspaced hyphens, followed by a period, are used.

Entry illustrates form for a book by a single author.

Entry identifies a journal article by a single author.

Entry illustrates form for a book with more than one author.

Entry identifies a newspaper article by a single author. (The two entries that follow list additional articles by the same author.)

Entry identifies a signed periodical article.

Entry identifies the subject of a telephone interview.

Entry identifies an unsigned periodical article (alphabetized by title).

Duchac 13

Center ────────→ Works Cited

Blum, Debra E. "Graduation Rate of Scholarship Athletes
Rose after Proposition 48 Was Adopted, NCAA Reports." The Chronicle of Higher Education 7 July
1993: A42–44.

———. "More Freshmen Meet Academic Standards Set by
NCAA." The Chronicle of Higher Education 21 Apr.
1993: A38–40.

Dealy, Francis X. Win at Any Cost. New York: Carol, 1990.

Durrant, Sue M. "Title IX--Its Power and Its Limitations."
Journal of Physical Education, Recreation and Dance
45 (1992): 60-64.

Fleisher, Arthur A., Brian L. Goff, and Robert D. Tollison.
The National Collegiate Athletic Association. Chicago:
U of Chicago P, 1992.

Lederman, Douglas. "Men Get 70% of Money Available for
Athletic Scholarships and Colleges That Play Big-Time
Sports, New Study Finds." The Chronicle of Higher Ed-
ucation 18 Mar. 1992: A1+.

———. "Men Outnumber Women and Get Most of Money in
Big-Time Sports Programs." The Chronicle of Higher
Education 8 Apr. 1992: A1+.

———. "NCAA Votes Higher Academics Standards for College
Athletes." The Chronicle of Higher Education 15 Jan.
1992: A1+.

Sanoff, Alvin P. "When Is the Playing Field Too Level?" U.S.
News and World Report 30 Jan. 1989: 88-89.

Walker, Skippy "Tiptoe." Telephone interview. 1 March
1994.

"You C.A.N. Get Help with a Scholarship." Scholastic Coach
Aug. 1992: 56.

PART 9

Writing in the Disciplines

FURTHER READING

Toby Fulwiler. "How Well Does Writing across the Curriculum Work?" *College English* **46** (February 1984): 113–25.

This essay is a report on and assessment of a six-year writing across the curriculum program at Michigan Tech. The author, a leading figure in the development of WAC, divides the article into two sections: *problems* and *unexpected benefits*.

One problem arose with terminology. The program was designed around James Britton's discourse taxonomy of expressive, transactional, and poetic. Expressive writing—what it meant—caused endless problems; it was often interpreted by workshop participants as encouraging the students to write whatever they wanted. This problem lasted three years, until expressive writing was reconceived as "exploratory" or "speculative" writing.

Another problem was the resistance by unmotivated or suspicious faculty members. Some members were ordered by department chairs to attend the writing workshops. Participation in the workshop must be on a voluntary basis. Yet another problem was over turf, with philosophers and English teachers being the most difficult. While philosophers constantly questioned everything, English teachers generally believed that only they could teach writing.

There was an inevitable problem with translation, e.g., finding ways for writing to be included in advanced mathematics classes. Other problems included class enrollments that were too large to include writing assignments and the faculty's lack of trust in students' abilities in peer tutoring.

continued on following page

CHAPTER 42

Understanding the Disciplines

All instructors, regardless of academic discipline, have certain basic concerns when they read a paper. They expect to see standard English, correct grammar and spelling, logical thinking, and clear documentation of sources. In addition, they expect sensible organization, convincing support, and careful editing. Despite these similarities, however, instructors in various disciplines have different ideas about what they expect in a paper.

One way of putting these differences into perspective is to think of the various disciplines as communities of individuals who exchange ideas about issues that concern them. Just as in any community, scholars who write within a discipline have agreed to follow certain practices—conventions of style and vocabulary, for example. Without these conventions it would be difficult or even impossible for them to communicate effectively with one another. If, for example, everyone writing about literature used a different documentation format or a different specialized vocabulary, the result would be chaos. To a large extent, then, learning to write in a particular discipline involves learning the conventions that govern a discourse community.

 42a **Research Sources**

Gathering information is basic to all disciplines, but not all disciplines rely on the same kinds of resources. In the humanities, for example, library research is an important part of most studies. Although some historians will conduct interviews and some literary scholars will collect quantifiable data, most people who do research in the humanities spend a great amount of time reading primary and secondary sources.

▶ See 39a1

Those who work in the social sciences also spend a lot of time examining the literature on a particular topic. But they also rely heavily on nonprint sources of information—observation of behavior, interviews, and surveys, for example. Because of the kind of data they generate, social scientists use statistical methodology and record their results in charts, graphs, and tables. Those who work in the natural sciences (such as biology, chemistry, and physics) and the applied sciences (engineering and computer science, for example) rely almost exclusively on **empirical data**—information obtained through controlled laboratory experiments or from mathematical models. They use the data they collect to formulate theories that try to explain their observations.

The information you gather in the library, in the field, or in the laboratory will help you support the conclusions you formulate when you write. Whether you want to make a point about the color gray in a work by Herman Melville or about the effect of a particular amino acid on the respiratory system, your intent is the same: to find the support you need to explain something to readers or convince them of something.

The kind of support that is acceptable and persuasive, however, varies from discipline to discipline. Students of literature often use quotations from fiction or poetry to support their statements, whereas historians are likely to refer to documents, letters, and court or church records. Social scientists frequently rely on statistics to support their conclusions; those in the natural sciences use the empirical data they derive from controlled experiments.

Although all scholars may use the opinions of recognized experts to support their ideas, the weight given to these opinions varies from discipline to discipline. A historian developing a Marxist analysis of Western industrialism, for example, will use the ideas of Karl Marx to support his or her conclusions; a feminist literary critic will cite the work of other well-respected feminist critics. However, while scientists frequently begin papers with a literature survey, they base their conclusions not on interpretations of expert opinion, but on empirical data. This does not mean that scientific research somehow yields more objective conclusions than does research in other disciplines. The history of science is filled with examples of researchers seeing what they want to see, whether it be canals on Mars, a cure for cancer, or room-temperature fusion.

continued from previous page

In contrast to these problems were some unexpected benefits. For example, there eventually developed a community of scholars, whereas earlier, most of the faculty had been isolated in their own departments. The environment across campus changed subtly; the faculty and the students were working on a common project that they could discuss, say, in the cafeteria. Moreover, the faculty's attitude toward writing began to improve, in both the teaching of writing and their own writing abilities. Also, many teachers learned that they could still learn something about teaching.

COMPUTER TIP

Learning in Collaboration with Others
Because much of the writing students will do in the working world will be collaborative—especially if they are engineering or business majors—consider using computers to facilitate collaborative work and to teach students how to work together on writing projects.

- Ask students to write separate sections of a report, merge the sections into a single file, and make the style consistent. One individual can be in charge of polishing the final version, or the group can work together in front of a single monitor.

- After students have written the separate sections of a report, ask them to exchange files and edit someone else's work before they merge the sections and make the style consistent.

- Take advantage of intercollegiate computer networks by having students design a project that requires

continued on following page

continued from previous page

them to work with individuals from another school. In compiling the final document, have students follow the general guidelines above.

FURTHER READING

Gary A. Olson (interview conducted by). "Social Construction and Composition Theory: A Conversation with Richard Rorty." *Journal of Advanced Composition* 9.1–2 (1989): 1–9.
This interview, though brief, is important, for many composition theorists (e.g., Kenneth Bruffee and Patricia Bizzell) have been influenced by Richard Rorty. (His important works include *Philosophy and the Mirror of Nature* and *Contingency, Irony, and Solidarity.*) Specifically, his work has had an influence on developers of Writing Across the Curriculum (or WAC) and Collaborative Learning. These individuals have stated that they "founded" their programs on Rorty's thinking. It is especially interesting in this interview, therefore, to observe how Rorty does not always give back to these theorists/developers what they have given to him: while they praise him as their father, he comes close to disowning them. Olson, the interviewer, consequently concludes that "we" need to reassess our thinking about Rorty.
One of the phrases often used in the field of rhetoric and composition is "social construction of reality" (a sociological concept made famous by Peter Berger). It became an especially

continued on following page

42b   Writing Assignments

Because each discipline has a different set of concerns, writing assignments vary from course to course. A sociology course, for example, may require a statistical analysis; a literature course will ask for a literary analysis. Therefore, it is not enough simply to know the material about which you are asked to write. You must also be aware of what the instructor in a particular discipline expects of you. For example, when your art history instructor asks you to write a paper on the Brooklyn Bridge, she does not expect you to write an analysis of Hart Crane's poem "The Bridge"—a topic suitable for a literature class—nor does she want a detailed discussion of the steel cables used in bridge building—a topic suitable for a materials engineering class. What she might expect is for you to discuss the use of the bridge as an artistic subject—as it is in the paintings of Joseph Stella and the photographs of Alfred Steiglitz, for example.

When you get any assignment, make certain you understand exactly what you are being asked to do; if you have any doubt, ask your instructor for clarification. Then, try to acquaint yourself with the ideas that are of interest to those who publish on your topic. Take the time to look through your text and class notes or do some exploratory research in the library to get a sense of the theories, concerns, and controversies related to your topic. (See Chapters 43 through 45 for lists of specialized research sources used in each discipline.) Keep in mind, however, that you should not limit yourself to areas that researchers have already explored. On the contrary, papers that explore new ideas, move in new directions, or make new connections between disciplines—for example, quoting a few lines from the poem "The Bridge" in your art history paper to shed some light on the power of the bridge as a literary and artistic symbol—are often the most valuable and enlightening.

42c   Conventions of Style and Documentation

(1)   Style

Specialized Vocabulary   Learning the vocabulary of a discipline is like learning a new language. At first, you observe native speakers from a distance. Eventually, as you learn a few words, you begin to communicate—if only slightly—with the natives. Finally, after you can speak the language well, you become actively involved with those around you and, if you are lucky, participate in the life of the community. Only by learning the specialized vocabulary of a field can you communicate with those who work in it. Once you know this vocabulary, you can begin to participate in the discussions and debates that define the discipline.

When you write a paper for a course, you use the specialized vocabulary of people who publish in the discipline. When you write a paper for a literature course, for example, use the literary terms you hear in class and read in your textbook—*point of view*, *persona*, and *imagery*, for example. Do the same for assignments in your other classes. It makes no sense to use inexact words or colloquial phrases when the discipline offers a vocabulary that will enable you to express concepts accurately and concisely. What other words, for example, would you use to denote *ecosystem*, *conditioned reflex*, or *metaphor* without using these specific terms?

CLOSE-UP 🔍

WRITING IN THE DISCIPLINES

Although technical terms facilitate communication within a discipline, they can do exactly the opposite when used outside the discipline.

▶ **See 18c1**

Level of Diction　　Within any field, particular assignments have different degrees of formality and, therefore, different stylistic requirements. Regardless of discipline, research papers tend to be formal: They contain learned words, are grammatically correct, avoid contractions and colloquialisms, and use third-person pronouns. Proposals—whether they are in the humanities, social sciences, or natural science—also are relatively formal.

▶ **See 18a**

Other assignments, by their very nature, are less formal than proposals and research papers. Because its purpose is to present an individual's personal reactions, a response statement uses subjective language, first person, and active voice ("I see a lot of weaknesses I didn't notice last time"). A lab report is also informal, but because its purpose is to report the observations themselves—not the observer's reactions—it frequently relies on objective language and passive voice ("The acid was poured" rather than "I poured the acid"). Perhaps the only thing that can be said with certainty about the levels of diction used in various

continued from previous page

important concept to composition theorists when they realized that there are no firmly rooted foundations on which to build a thinking, writing, reading program. Although many composition researchers and theorists, such as Flower and Hayes, are foundationalists, many composition theorists are antifoundationalists, as Rorty himself is (and Clifford Geertz, Thomas Kuhn, and Stanley Fish are). If, as the antifoundationalists say, reality cannot be known as a thing in itself, or known directly and thereby agreed upon by all, then, at best, what is left is a socially constructed reality. Hence, *rhetoric*, which has been in favor of such a view (or views) of reality and which has been an advocate for versions of reality, becomes greatly important again. This antifoundationalist view has become important for such social-epistemic rhetoricians in composition as Bizzell and James Berlin. Hence, the *apparent* bond that exists between Rorty and antifoundationalists in rhetoric and composition.]

In the interview, Rorty drops a bomb when he says that he is "suspicious" of socially justified programs of the teaching of writing. He calls WAC "a terrible idea." The teaching of writing, he says, is all about teaching students how to "write complete sentences" and how to "get the commas in the right place." He questions, though he has sympathy with, Bruffee's concept of collaborative writing as based on social construction of reality, for he believes that a "program" cannot be built on it. He also differs with the leftists (Meyers) who argue that peer groups in collaborative writing can lead to "group think." At

continued on following page

continued from previous page

worst, peer groups can lead to "committee prose." He also says that WAC in the U. S. is only perpetuating the individual jargons of the disciplines.

He agrees with E. D. Hirsch that students need to gain knowledge and fluency in the dominant culture and discourse. In Rorty's own term, students need to learn "normal discourse" (i.e., the particular disciplinary discourse) *before* they can hope for "abnormal discourse" (i.e., discourse that will change the status quo of a particular discipline or challenge a discipline altogether). He believes that abnormal discourse cannot be taught but can be studied. Specifically, he says that normal discourse should be taught at the elementary and secondary and undergraduate levels and that "abnormal discourse" should then be studied at the graduate level. He disagrees with those critics (M. Cooper) who say that it is extremely difficult, if not impossible, for women to get into the so-called (male) normal discourse and that therefore, they must speak from their own feminine experience, which would count as abnormal discourse. Rorty says that all change by marginals has been brought about by their entering the dominant, normal discourse, from which he believes women are quite capable of speaking and writing.

He says that he, unlike the leftists, believes that "America [is] still a spectacular success story of the growth of democratic freedom."

disciplines is that all of them use different degrees of formality at different times, depending on audience and purpose.

Format   Each discipline has certain distinctive formats that govern the way in which written information is presented. A **format** is a way of arranging material that has become accepted practice. A format may govern the arrangement of an entire piece of writing, such as a lab report, which has certain prescribed sections. For this reason, in some disciplines—particularly in the sciences and social sciences—writers use internal headings in their papers, and in others—particularly in the humanities—they seldom do. A format may also determine how certain kinds of information are presented within a paper. For example, social scientists expect statistical data to be presented in tables or graphs. See ◀ Ch. 37 Specific mechanical concerns, such as whether to spell out a number or use numerals, also differ from discipline to discipline.

Professional organizations define the guidelines that govern paper formats within their disciplines. Typically, a professional organization will issue or recommend a handbook or style sheet that defines the standards for spelling, mechanics, punctuation, and capitalization. This style sheet also gives explicit guidelines for the use and placement of information in charts, graphs, tables, and illustrations within a paper as well as for typing conventions, such as the placement of page numbers on a paper and the arrangement of information on a title page. In addition, these style sheets discuss and illustrate the documentation format they recommend for research papers. Because significant differences exist between disciplines, you should consult the appropriate style sheet for the discipline before you begin your paper.

(2)   Documentation

Different disciplines use different styles of documentation. Four of the most widely used formats are those recommended by the Modern Language Association (MLA), the American Psychological Association (APA), *The Chicago Manual of Style* (CMS), See ◀ Ch. 40 and the Council of Biology Editors (CBE).

Instructors in the humanities usually prefer MLA style, which uses parenthetical references within the text to refer to a Works Cited list at the end of the paper, or CMS style, which uses footnotes or endnotes keyed to bibliographic citations at the end of the paper. Instructors in some of the social sciences, such as psychology and education, prefer APA style, which uses parenthetical references that differ slightly from MLA style. Other disciplines—the physical and biological sciences and medicine, for example—prefer a number-reference format, such as CBE style, which uses numbers in parentheses in the text that refer to a numbered list of works at the end of the paper.

Because of the lack of uniformity among the disciplines, it is especially important that you consult your instructor to see which documentation style he or she requires. Your instructor may expect you to use a certain style sheet that he or she will place on reserve in the library or ask you to purchase. If this is not the case, however, it is your responsibility to determine which style to use—possibly by looking at an important journal in the field and following its documentation format.

GUIDELINES FOR DOCUMENTATION

- Make sure you understand what information must be documented **(see 40a)**.
- Do not assume the documentation style you use in one class is appropriate for another.
- Use one documentation style consistently throughout the paper.
- Make sure you have a copy of the appropriate style sheet or journal so that you can consult it as you write.
- Follow *exactly* the conventions of the format you decide to use.
- When you proofread the final draft of your paper, make certain that you have documented all information that needs documentation and that you have punctuated all entries correctly.

WRITING IN THE DISCIPLINES

	Discipline	Research Sources	Assignment
Humanities	Languages Literature Philosophy History Linguistics Religion Art history Music	Library sources Interviews Observations (museums, concerts) Oral history	Response statement Book review Art, music, dance, or film review Bibliographic essay Annotated biblio- graphy Literary analysis Research paper
Social Sciences	Anthropology Psychology Economics Business Education Sociology Political science Social work Criminal justice	Library sources Surveys Observation (behavior of groups and individuals)	Experience paper Case study Literature review Proposal Research paper
Natural and Applied Sciences	Natural sciences Biology Chemistry Physics Astronomy Geology Mathematics Applied sciences Engineering Computer science Nursing Pharmacy	Library sources Observations Experiments Surveys	Abstract Literature survey Laboratory report Research paper

Style and Format	Documentation
Style Specialized vocabulary Direct quotations *Format* Little use of internal headings, tables, etc.	English, languages, philosophy: MLA History, art: CMS
Style Specialized vocabulary, including statistical terminology *Format* Internal headings Use of charts and figures (graphs, maps, flow charts, photographs) Numerical data (in tabular form)	Psychology, education, etc.: APA Social work, etc.: CMS
Style Frequent use of passive voice Few direct quotations *Format* Internal headings Use of tables, graphs, and illustrations (exact formats vary)	Biology: CBE Other scientific disciplines use a variety of different documentation styles

▶ See 40f

FURTHER READING

James L. Kinneavy. "Restoring the Humanities: The Return of Rhetoric from Exile." In *The Rhetorical Tradition and Modern Writing*. Ed. James J. Murphy. New York: The Modern Language Association, 1982. 19–28.

The author begins by making reference to and quoting from *The Humanities in American Life*, the Report of the Commission on the Humanities. He concludes that the report is "disheartening." It is more of a lament than a productive look toward the future. The question unanswered by the report is "Why the decline of the humanities in schools?" The author attempts to answer this question and to make recommendations.

Specifically, he is interested in the "alienation of rhetoric from the humanities" and its restoration. The author limits his definition of rhetoric to the definition and place that it had in the *trivium*, along with grammar and logic. It is unquestionable, the author says, that "the humanities were born in a rhetorical manger." Humanities and humanism mean "freedom" to speak with one's neighbors to determine the life of the *polis*. The author gives an overview of the life and work of both Socrates and Cicero.

In America in 1914, however, teachers of speech and elocution, who were then members of English departments, walked out of the National Conference of Teachers of English in order to form speech departments. With them, they took rhetoric. Similarly, logic was removed to departments of philosophy or mathematics. Hence, the trivium was torn asunder.

continued on following page

CHAPTER 43

Writing in the Humanities

The humanities include a variety of disciplines, such as art, music, literature, languages, religion, and philosophy. In these fields research is often conducted in order to analyze or interpret a **primary source**—a literary work, a historical document, a musical composition, or a painting or piece of sculpture—or to make connections between one work and another. Scholars in humanities disciplines may also examine **secondary sources**—commentaries on primary sources—in order to make critical judgments and, sometimes, to develop new theories.

See ◀
39a1

43a   Research Sources

Library research is an important part of study in many humanities disciplines. When you begin your research in any subject area, the *Humanities Index* is one general source you can use. This periodical index lists articles from more than two hundred scholarly journals in areas such as history, language, literary criticism, and religion. It is available in print—with entries arranged alphabetically in yearly volumes according to author and subject—and on-line.

Many specialized sources are also available to help you with your research.

(1) Specialized library sources

The following reference sources are used in various humanities disciplines.

Art

Art Index
Encyclopedia of World Art
Index to Art Reproductions in Books
New Dictionary of Modern Sculpture
Oxford Companion to Art
Praeger Encyclopedia of Art

Drama

The Crown Guide to the World's Great Plays from Ancient Greece
    to Modern Times
A Guide to Critical Reviews
McGraw-Hill Encyclopedia of World Drama
Modern World Drama: An Encyclopedia
Oxford Companion to the Theatre

Film

The Film Encyclopedia
Guide to Critical Reviews
International Index to Film Periodicals
International Index to Multimedia Information
Lander's Film Reviews
Magill's Survey of Cinema
New York Times Film Reviews

History

America: History and Life (United States)
Cambridge Ancient History
Cambridge Medieval History
CRIS (Combined Retrospective Index to Journals in History,
    1838–1974)
Dictionary of American History
Great Events in History
Guide to Historical Literature
Harvard Guide to American History
Historical Abstracts (Europe)
New Cambridge Modern History

continued from previous page

Today, the author claims, we must put things back together. We must reintroduce a concern for persuasion along with aesthetics and science in all departments, not only in English departments. While the author does not directly address "writing (or rhetoric) across the curriculum/disciplines," what he has to say is of great importance to this movement.

FURTHER READING

Richard E. Young. "Concepts of Art and the Teaching of Writing." In *The Rhetorical Tradition and Modern Writing*. Ed. James J. Murphy. New York: The Modern Language Association of America, 1982. 130–41.

The key concepts the author is dealing with in relation to the teaching of writing are "craft" and "art." He opens with two representatives of the concepts, John Genung (*The Practical Elements of Rhetoric* [1892]) and Gordon Rohman ("Pre-Writing: The Stage of Discovery in the Writing Process" [1965]). Genung's view is that "art" cannot be taught but must be learned. What can be taught is "craft"—that is, organizing and editing. Rohman, in contrast, argues that "craft" and "art" can be taught. Rohman is specifically concerned with the "dynamic of conceptualizing" and "creative discovery." He, therefore, is rightly associated with the new rhetoric.

The author sees these two contrasting views as the coordinates of the present theorization about and teaching of composition. He sees the former as a threat to the teaching of

continued on following page

continued from previous page

composition, for it claims that writing cannot be taught and, as a result, impoverishes the whole enterprise. What the author argues for is a "new classicism," one that reinaugurates a view of art as teachable in the form of rhetorical invention. He refers to Aristotle and to such contemporary neoclassical rhetoricians as Richard Weaver, Edward Corbett, Ross Winterowd, and Francis Christensen in speaking of this classicism.

If we are to teach the new classicism, then what are we to teach? The author answers "heuristics," that is, inventional or discovery procedures. A heuristic strategy is a procedure that allows a rhetor to guess effectively at the reasonable answers to a rhetorical problem. The author here, as elsewhere, argues for specific use of heuristic procedures such as Ciceronian *topoi* (as put forth in Corbett's *Classical Rhetoric for the Modern Student*) or tagmemics (as espoused in Young, Becker, and Pike's *Rhetoric: Discovery and Change*).

Language and Literature

Annual Bibliography of English Language and Literature
Biography Index
Book Review Digest
Book Review Index
Cassell's Encyclopedia of World Literature
Children's Literature Abstracts
Contemporary Authors
Current Biography
Essay and General Literature Index
Language and Language Behavior Abstracts (LLBA)
Literary History of the United States (LHUS)
MLA International Bibliography
Oxford Companion to American Literature
Oxford Companion to Classical Literature
Oxford Companion to English Literature
Oxford English Dictionary
PMLA General Index
Princeton Encyclopedia of Poetry and Poetics
Salem Press Critical Surveys of Poetry, Fiction, Long Fiction, and Drama
Short Story Index
Twentieth Century Authors
Webster's Biographical Dictionary

Music

Harvard Dictionary of Music
Music Article Guide
Music Index
The New Grove Dictionary of Music and Musicians
The New Oxford Companion to Music

Philosophy

The Consise Encyclopedia of Western Philosophy and Philosophers
Dictionary of the History of Ideas
Encyclopedia of Philosophy
Philosopher's Index

Religion

> *Encyclopedia Judaica*
> *Encyclopedia of Islam*
> *The Hutchinson Encyclopedia of Living Faiths*
> *New Catholic Encyclopedia*
> *Oxford Dictionary of the Christian Church*

(2)   Specialized databases for computer searches

Many of the indexes that appear on the preceding list of specialized library sources are available on-line as well as in bound volumes. Some of the most helpful databases for humanities disciplines include *Art Index, MLA Bibliography, Religion Index, Philosopher's Index, Music Literature International (RRM), Essay and General Literature Index, Artbibliographies Modern, Historical Abstracts, the LLBA Index, Dissertation Abstracts, Arts and Humanities Search,* and *OnLine.*

(3)   Other sources of information

Research in the humanities is not limited to the library. Historians may do interviews or archival work or consult records collected in town halls, churches, or courthouses. Art historians may visit museums and galleries, and music scholars may attend concerts.

Nonprint sources, such as the oral history interview excerpted below, can be important additions to a paper in any humanities discipline. ▶ **See 38c5**

Excerpt from Interview with Arturo Tapia, a registered Tigua Indian

My daddy never used to say he was Tigua Indian. . . . We never talked about it. . . . Other Indians never liked us, and the white people never allowed us in their bars or stores. I have gone up to people and told them I am Tigua and they say, "What a low class Indian," or "Them down there, the Mexicans, they sold out."

The student who recorded this interview chose to use it in her paper's conclusion.

As many of the sources examined in this paper have demonstrated, the history of the Tiguas is full of misconceptions. The New Mexico version of the Tiguas' migration is that they fled with a Spanish party to El Paso during the Indian uprising of August

10, 1680, but the Tigua version of their migration is quite different. The New Mexico Indians have portrayed the Tiguas of Isleta as a "Judas Tribe" who turned against their own people to ally themselves with the Spanish. As A. Tapia, a registered Tigua Indian, points out, even today the Tiguas face discrimination from other Indians as well as from whites and feel they are considered "low class."

 Assignments in the Humanities

(1) The response statement

In some disciplines, particularly literature, you may be asked to write a **response statement**, in which you express (sometimes informally) your reactions to a literary work, a painting, a film, a dance performance, or a concert. You might focus on an assigned question, take a particular critical perspective, or react to a character's behavior. Such an assignment requires you to write a first-person account of your feelings and to explore the factors that influenced your reactions. The following response statement was written for a journal kept for an Introduction to Literature course.

Sample Response Statement

Rereading The Catcher in the Rye after two years, I see a lot of weaknesses I didn't notice last time. The style is too cute and too repetitive, and it calls attention to itself. Salinger has Holden Caulfield say things like "I mean" and "if you know what I mean" too many times, and he seems to use bad language for no particular reason. Also, I don't like Holden as much as I did the first time I read the book. Before I saw him as isolated and misunderstood, a pathetic character who could have been happy if only he had stayed a child forever. Now he gets on my nerves. I keep thinking he could do something to help himself if he didn't have to blame everything on the "phonies." All this probably says more about how I have changed in two years than about how good or bad the book is.

(2) The book review

A **book review**, which may be assigned in any humanities discipline, asks you to respond critically to a book, judging it according to a particular standard. Before you begin your evaluation of the work, you should provide some context to enable readers to follow your discussion—for example, a brief summary or overview of the work or an outline of the author's key points or arguments. After you summarize the book, you should evaluate its content (the writer's central idea and supporting information) and the way in which the material is presented (the book's style and organization). End your book review with a statement that summarizes your critical evaluation of the text. (Reviews of performances or nonprint works are similar to book reviews in that they present judgments that assess the worth of a work or an artist.) The book review that follows was written for a cross-disciplinary composition class.

Excerpt from a Book Review

In his thought-provoking book Chaos: Making a New Science, James Gleick presents the events of recent years that have shaped this new science and introduces the individuals responsible for those events. He begins with what is considered the starting point of the new science, Edward Lorenz's Butterfly Effect, and ends with a comprehensive discussion of the newest discoveries and the future of chaos. Most importantly, he explains the equations, theories, and concepts that are the heart of chaos. One example of how this new science is applied is Mitchell Feigenbaum's theory of universality. Using only a hand-held calculator, this physicist proved that simple equations from a simple system can be applied to a totally unrelated system to produce a complicated solution. Although Feigenbaum's theory was at first greeted with skepticism, it soon became the basis for finding order in otherwise unrelated irregularities. Gleick's clear analysis of this and other theories makes his explanation of chaos appropriate for readers with little scientific background as well as for readers familiar with the field.

(3)   The bibliographic essay

A **bibliographic essay** surveys research in a field and compares and contrasts the usefulness of various sources on a particular subject. Several publications in various humanities specialties publish annual bibliographic essays to inform scholars of recent developments in a field. Students in advanced literature or history classes may also prepare this type of essay. The following excerpt, written by a student in an American literature class, comes from a bibliographic essay on Mark Twain's novel *Pudd'nhead Wilson*.

▶ See
43a

Excerpt from a Bibliographic Essay

Most early critics analyze the novel in terms of racial issues and the doctrine of environmental determinism. Langston Hughes, for instance, writes that "the basic theme [of Pudd'nhead Wilson] is slavery . . . and its main thread concerns the absurdity of man-made differentials, whether of caste or 'race'" (viii). James M. Cox also focuses on the racial issue, specifically on miscegenation, but in a more symbolic vein than does Hughes. Cox's analysis of the novel is very similar to Fiedler's. Like Fiedler, Cox believes the novel deals with American guilt; it is a "final unmasking of the heart of darkness beneath the American dream" (361). Tom serves as the white man's nemesis, spawned by the guilt of miscegenation, a violation of the black race. According to Cox, he is "the instrument of an avenging destiny which has overtaken Dawson's Landing" (353).

(4)   The annotated bibliography

Each entry in an **annotated bibliography** includes full source information and a *brief* summary of the source's main points or arguments. The following example is from a student's annotated bibliography of *Pudd'nhead Wilson.*

Excerpt from an Annotated Bibliography

Chellis, Barbara A. "Those Extraordinary Twins: Negroes and Whites." American Quarterly 21 (1969): 100–12.

> Chellis sees Pudd'nhead Wilson as an exposure of "the fiction of law and custom" that has justified distinctions between blacks and whites. Twain develops his theme through his characterizations of Roxy, Tom, and Chambers. According to Chellis, Roxy's "crime"—condemning the real Tom to slavery—stems not from the influence of her race but from that of her white values. Tom is spoiled and selfish—the kind of person produced by white society's values. The invalidity of race distinctions is further pointed out through the servility of Chambers, who is white. Black servility is ultimately seen to be nothing more than the result of training.

(5)   The literary analysis

Students in literature classes are frequently called on to analyze poems, plays, short stories, or novels. For a detailed discussion of how to write a **literary analysis** as well as two sample papers—illustrating an analysis of a short story and of a poem—**see Chapter 47**.

43c   Conventions of Style and Documentation

(1)   Style

Although papers in the humanities may include abstracts and internal headings, they usually do not, nor do they generally present material in tables, charts, or graphs. Although each humanities discipline has its own specialized vocabulary, you should avoid overusing technical terminology. Using the first person (*I*) is acceptable when you are expressing your own reactions and convictions—for example, in a response statement. In other cases, however, you should use a third-person point of view.

When you write papers about literature, you should follow certain conventions of literary analysis. These conventions—for example, the use of present tense—are listed and explained in **47c**.

(2)   Documentation

Literature and modern and classical language scholars use MLA format; history scholars use *The Chicago Manual of Style*. ▶ **See 40b,c**

43d   Sample Humanities Research Papers

(1)   Modern Language Association (MLA) format

The following literary research paper, "Assertive Men and Passive Women: A Comparison of Adrienne Rich's 'Aunt Jennifer's Tigers' and 'Mathilde in Normandy,'" is based on two of Adrienne Rich's poems. It uses MLA documentation style. Because the instructor did not require a separate title page, the student included identifying information on the first page of her paper. For a detailed discussion of the process of writing a research paper in the humanities (accompanied by an annotated student paper), **see Chapter 41**.

Cathy Thomason

Composition 102

Dr. Alvarez-Goldstein

9 May 1994

Assertive Men and Passive Women: A Comparison
of Adrienne Rich's "Aunt Jennifer's Tigers" and
"Mathilde in Normandy"

Adrienne Rich began her poetic career in the
1950s as an undergraduate at Radcliffe College. The
product of a conservative southern family, Rich was
greatly influenced by her father, who encouraged the
young poet to seek higher standards for her work.
However, her early work shows evidence of Rich's future struggle with this upbringing. Although she used
styles and subjects that gained approval from both her
father and the almost all-male literary world, some of
her early works suggest an inner struggle with the
status quo of male dominance (Bennett 178). "Aunt
Jennifer's Tigers" and "Mathilde in Normandy" clearly
reflect this struggle.

Both "Aunt Jennifer's Tigers" and "Mathilde in
Normandy" tell the story of women living in the
shadow of men's accomplishments. The central figures
of the two poems exhibit their creativity and intelligence through a traditional female occupation, embroidery. Although the poems are set in different time
periods ("Mathilde in Normandy" in the Middle Ages
and "Aunt Jennifer's Tigers" in the 1950s), their focus
is the same: Both are about women who recognize
their subordinate positions to men but are powerless

Thomason 2

to change their situations. The difference between the two, however, lies in the fact that Mathilde does not suffer the inner conflict that Aunt Jennifer does.

"Aunt Jennifer's Tigers" is the story of a woman trapped in a conventional and apparently boring marriage. Her outlet is her needlework; through her embroidery she creates a fantasy world of tigers as an escape from her real life. In 1972, in her essay "When We Dead Awaken," Rich said, "In writing this poem, composed and apparently cool as it is, I thought I was creating a portrait of an imaginary woman" (469). What she later discovered was that Aunt Jennifer was probably typical of many married women in the 1950s. These women, Rich observed, "didn't talk to each other much in the fifties—not about their secret emptiness, their frustrations . . . " (470).

Standing in opposition to the timid and frustrated Aunt Jennifer are the tigers in the first stanza:

> Aunt Jennifer's tigers prance across a screen,
> Bright topaz denizens of a world of green.
> They do not fear the men beneath the tree;
> They pace in sleek chivalric certainty. (1–4)

Unlike Aunt Jennifer, the tigers move with confidence and "chivalric certainty." They are not dominated by men, as Aunt Jennifer is. According to Claire Keyes, this stanza contains the dominant voice. Although Aunt Jennifer is not assertive or forceful, her creations--the tigers--are. In other words, Aunt Jennifer's work is more interesting than she is. Even so, the tigers are a part of Aunt Jennifer's imagination, and

Thomason 3

they show there is more to Aunt Jennifer than her existence as a housewife would suggest.

"Mathilde in Normandy" is based on the folktale that William the Conqueror's wife, Queen Mathilde, created the Bordeaux tapestry to tell the story of the Norman invasion of England (Keyes 24). The men in this poem are as powerful as the men who dominate Aunt Jennifer, showing their masculinity in their attack on another country. Although Rich acknowledges Mathilde's position as a woman who may never see her husband again--"Say what you will, anxiety there too / Played havoc with the skein" (21-22)--Mathilde, unlike Aunt Jennifer, has no problem with the notion that women should stay home. According to the speaker, "Yours was a time when women sat at home" (17). Therefore, Mathilde does not suffer inner turmoil about her role in life as a subordinate wife.

Mathilde is separated from her husband's world because she is female and considered unable to fight, to compete in a male world. Aunt Jennifer, however, is held back from the male world by a different barrier:

> Aunt Jennifer's fingers fluttering through
> her wool
> Find even the ivory needle hard to pull.
> The massive weight of Uncle's wedding band
> Sits heavily upon Aunt Jennifer's hand. (5-8)

Her marriage is not a happy one; her wedding band holds her down (Keyes 23). In accepting her role as housewife, Aunt Jennifer is denied a satisfying existence. She is reduced to maintaining a household for a

Thomason 4

husband who dominates all aspects of her life. Even Aunt Jennifer's pursuit of art is made difficult by her marriage. The "weight" of her wedding band makes it difficult to embroider.

For Mathilde, life is objectionable only because she fears that her husband and the men of the kingdom will not return. She identifies with her husband's cause and accepts the fact that the life men lead is dangerous, "Harsher hunting on the opposite coast" (16). Unlike Mathilde, Aunt Jennifer sees her marriage as imprisonment; she will only be free when she dies:

> When Aunt is dead, her terrified hands will
> lie
> Still ringed with the ordeals she was
> mastered by.
> The tigers in the panel that she made
> Will go on prancing, proud and un-
> afraid. (9-12)

Aunt Jennifer's hands are "terrified," overwhelmed by the power that her husband, and society, have over her. Aunt Jennifer, who is "mastered" by her situation, stands in opposition to the tigers, who flaunt their independence, forever "prancing, proud, and unafraid." Mathilde's experience with male power is also terrifying. She is in the vulnerable position of being a queen whose husband and court are away at war. If her husband does not return, she can be subjugated by another man, one wishing to take over the kingdom.

Both "Aunt Jennifer's Tigers" and "Mathilde in Normandy" present stories of women bound to men.

Although both poems suggest that conventional relationships between men and women are unsatisfactory, Mathilde does not struggle against her role as Aunt Jennifer does. Claire Keyes points out that although these poems are "well mannered and feminine on the surface," they "speak differently in their muted stories" (28). Although only "Aunt Jennifer's Tigers" openly criticizes the status quo, both poems reflect the inner conflicts of women and express their buried desires subtly, but nonetheless powerfully.

Thomason 6

Works Cited

Bennett, Paula. "Dutiful Daughter." My Life a Loaded
    Gun: Female Creativity and Feminist Poetics.
    Boston: Beacon, 1986. 171–76.

Keyes, Claire. The Aesthetics of Power: The Story of
    Adrienne Rich. Athens: U of Georgia P, 1986.

Rich, Adrienne. "Aunt Jennifer's Tigers." Poems: Se-
    lected and New. New York: Norton, 1974. 81.

–––. "Mathilde in Normandy." Poems: Selected and
    New. New York: Norton, 1974. 94–95.

–––. "When We Dead Awaken." Ways of Reading: An
    Anthology for Writers. Ed. David Bartholomae and
    Anthony Petrowsky. Boston: Bedford, 1993.
    461–76.

(2) *Chicago Manual of Style* (CMS) format

See ◄
40c

The following pages are excerpted from a history research paper, "Native Americans and the Reservation System," which uses Chicago style. Because the instructor did not require a separate title page, the student included identifying information on the first page of her paper.

1

Angela M. Womack
American History 301
December 3, 1994

Native Americans and the Reservation System

It is July 7th, and 10,000 Navajo Indians make ready to leave land in Arizona that they have called home for generations. This land has been assigned to the Hopi tribe by the U.S. government to settle a boundary dispute between the two tribes.[1] Ella Bedonie, a member of the Navajo tribe, says, "The Navajo and the Hopi people have no dispute. It's the government that's doing this to us. I think the Hopis may have the land for a while, but then the government . . . will step in."[2] The Hopis are receiving 250,000 acres to compensate them for the 900,000 acres they will lose in this land deal; however, the groundwater on this land is questionable because of possible contamination by a uranium mine upstream. To offset this unappealing aspect the government has sweetened the deal with incentives of livestock.[3]

This was the fate many Native Americans faced as western expansion swept across the North American continent. Now consider that the incident mentioned occurred not on July 7, 1886, but on July 7, 1986. Indian relations with the U.S. government are as problematic today as ever before, for the federal

2

government's administration of the reservation system both promotes and restricts the development of the Native American culture.

A reservation is an area of land reserved for Indian use. There are approximately 260 reservations in the United States at present.[4] The term reservation can be traced back to the time when land was "reserved" for Indian use in treaties between whites and Native Americans. Figures from 1978 by the Bureau of Indian Affairs show that 51,789,249 acres of land are in trust for Native Americans; 41,678,875 acres of this are for tribes, and 10,110,374 acres are for individuals.[5] The Native Americans are by no means restricted to these areas, although this assumption is commonly made. They are as free as any other citizen to leave these areas.

10

Notes

1. Trebbe Johnson, "Indian Land, White Greed," Nation, 4 July 1987, 15.

2. Johnson, 17.

3. Johnson, 16.

4. Ted Williams, "On the Reservation: America's Apartheid," National Review, 8 May 1987, 28.

5. U.S. Bureau of Indian Affairs, "Information About . . . The Indian People" (Washington, D.C.: Government Printing Office, 1981), 6, mimeographed.

11

Bibliography

"Adrift in Their Own Land." Time, 6 July 1987: 89.

Arrandale, Tom. "American Indian Economic Development." Editorial Research Reports, 17 Feb. 1984: 127–142.

Battise, Carol. Personal interview, 25 Sept. 1987.

Cook J. "Help Wanted--Work, Not Handouts." Forbes, 4 May 1987: 68–71.

Horswell, Cindy. "Alabama-Coushattas See Hope In U.S. Guardianship." Houston Chronicle, 26 May 1987: 11.

Johnson, Trebbe. "Indian Land, White Greed." Nation, 4 July 1987: 15–18.

Martin, Howard N. "Alabama-Coushatta Indians of Texas: Alabama-Coushatta Historical Highlights." Brochure, Alabama-Coushatta Indian Reservation: Livingston, Texas, n.d.

"A New Brand of Tribal Tycoons." Time, 16 March 1987: 56.

Philp, K. R. "Dillon S. Myer and the Advent of Termination: 1950–1953." Western Historical Quarterly 3 (Jan. 1988): 37–59.

U.S. Bureau of Indian Affairs. "Information About . . . The Indian People." Washington, D.C.: Government Printing Office, 1981. Mimeographed.

Williams, Ted. "On the Reservation: America's Apartheid." National Review, 8 May 1987: 28–30.

Young, J., and Williams. T. American Realities: Historical Realities from the First Settlements to the Civil War. Boston: Little, Brown, 1981.

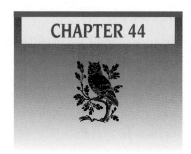

CHAPTER 44

Writing in the Social Sciences

The social sciences include anthropology, business, criminal justice, economics, education, political science, psychology, social work, and sociology. When you approach an assignment in the social sciences, your purpose is often to study individuals or groups in order to make generalizations about their behavior. You may be seeking to understand causes; predict results; define a policy, habit, or trend; draw an analogy between one group and another; or analyze a problem. Before you can approach a problem in the social sciences, you must develop a **hypothesis**, an educated guess about what you believe your research will suggest. Then you can go on to gather data you hope will support that hypothesis. Data may be quantitative or qualitative. **Quantitative data** are essentially numerical—the "countable" results of surveys and polls. **Qualitative data** are less exact and more descriptive—the results of interviews or observations, for example.

44a Research Sources

Although library research is an important component of research in the social sciences, researchers also frequently engage in field work. In the library, social scientists consult compilations of statistics, government documents, and newspaper articles in addition to scholarly books and articles. Outside the library, social scientists conduct interviews and surveys and observe individuals and groups. Because so much of their data are quantitative, social scientists must know how to analyze statistics and how to read and interpret tables.

(1) Specialized library sources

The following reference sources are useful in a variety of social science disciplines.

ASI Index (American Statistics Institute)
Bibliografia Chicana: A Guide to Information Sources
Dictionary of Mexican American History
Editorial Research Reports
Encyclopedia of Black America
Handbook Of North American Indians
Harvard Encyclopedia of American Ethnic Groups
Human Resources Abstracts
International Bibliography of the Social Sciences
International Encyclopedia of the Social Sciences
The Negro Almanac: A Reference Work on the Afro-American
PAIS (Public Affairs Information Service)
Population Index
Social Sciences Citation Index
Social Sciences Index
Women's Studies: A Recommended Core Bibliography

The following are the reference sources most often used for research in specific disciplines.

Anthropology

Abstracts in Anthropology
Anthropological Literature
Dictionary of Anthropology

Business and Economics

Accountants' Index
Business Periodicals Index
The Encyclopedia of Banking and Finance
The Encyclopedia of Management
Journal of Economic Literature
The McGraw-Hill Dictionary of Modern Economics
Personnel Management Abstracts
Wall Street Journal Index

Criminal Justice

Criminology and Penology Abstracts
Criminal Justice Abstracts
Criminal Justice Periodicals Index
Encyclopedia of Crime and Justice
Police Science Abstracts

Education

Current Index to Journals in Education
Dictionary of Education
Education Index
Encyclopedia of Educational Research

Political Science

ABC Political Science
American Political Dictionary
CIS Index (Congressional Information Service)
Combined Retrospective Index to Journals in Political Science
Dictionary of Political Thought
Encyclopedia of Modern World Politics
Encyclopedia of the Third World
Encyclopedia of the United Nations and International Agreements
Europa Year Book
Foreign Affairs Bibliography
Information Services on Latin America
International Political Science Abstracts

Psychology

Biographical Dictionary of Psychology
Contemporary Psychology
Encyclopedia of Psychology
International Encyclopedia of Psychiatry, Psychology, Psychoanalysis and Neurology
Psychological Abstracts

Sociology and Social Work

> *Encyclopedia of Social Work*
> *Encyclopedia of Sociology*
> *Poverty and Human Resources Abstracts*
> *Rural Sociology Abstracts*
> *Sage Family Studies Abstracts*
> *Social Work Research and Abstracts*
> *Sociological Abstracts*

Government Documents Government documents are important resources for social scientists because they contain complete and up-to-date facts and figures on a wide variety of subjects.

Government documents can be located through the *Monthly Catalog,* which contains the list of documents published each month together with a subject index. Other useful indexes include *The Congressional Information Service Index, The American Statistics Index,* and *The Index to U.S. Government Periodicals.*

Newspaper Articles Newspaper articles are particularly useful sources for researching subjects in political science, economics, and business. Students usually consult the *New York Times.* For information from newspapers from across the country, a useful source is *Newsbank,* which provides subject headings under the appropriate government agencies. For instance, articles on child abuse are likely to be listed under Health and Human Services. Another useful source of information from newspapers is *Lexis/Nexus.* This powerful database enables you not only to survey thousands of newspapers from across the country but also to view and reprint the articles. *Lexis/Nexus* also has a special library that contains articles focusing on legal matters and tax issues.

(2)   Specialized databases for computer searches

Many of the print sources cited above are also available in computer databases. Some of the more widely used databases for social science disciplines include *Cendata, Business Periodicals Index, Social Sciences Index, PsycINFO, ERIC, Social Scisearch, Sociological Abstracts, Information Science Abstracts, PAIS International, Population Bibliography, Economic Literature Index, ABI/INFORM, Legal Resource Index, Management Contents, Trade & Industry Index, PTSF + S Indexes* and *Facts on File.*

(3)   Other sources of information

Interviews, surveys, and observation of the behavior of various groups and individuals are important nonlibrary sources for social science research. Assignments may ask you to use your classmates as subjects for surveys or interviews. For example, in a political science class, your teacher may ask you to interview a sample of college students and classify them as conservative, liberal, or moderate. You may be asked to poll each group to find out college students' attitudes on issues such as acquaintance rape, affirmative action, or the problems of the homeless. If you were writing a paper on educational programs for the mentally gifted, in addition to library research you might want to observe two classes—one of gifted students and one of average students. You might also want to interview students, teachers, or parents. Similarly, research in psychology and social work may rely on your observations of clients, patients, or their families.

▶ See 38c

44b   Assignments in the Social Sciences

(1)   The experience paper

Instructors in the social sciences often ask students to do hands-on research outside the library; one typical assignment is the **experience paper**, in which students are asked to record their observations and reactions to a field trip or site visit. For example, students in an education class might write up their observations of a class of hearing-impaired students, criminal justice students might record their reactions to a trial, business majors might write about their impressions of how a particular small business operates, and students in a psychology class could write an experience paper about a visit to a state-run psychiatric facility.

The following excerpt, written by a student in a sociology of religion class, describes visits to two different churches.

Excerpt from an Experience Paper

The Pentecostal church service I observed was full of self-expression, motion, and emotion. People sang and praised the Lord in loud voices. In an atmosphere similar to that of a revival, people spontaneously expressed their joy and their reactions to the sermon. Each word the preacher spoke elicited responses like

"Praise the Lord," "Hallelujah," "Thank the Lord," and "Amen." Rather than focusing on religious doctrine, the sermon concentrated on the problems of everyday life. The Presbyterian service I observed was very different. Compared to the Pentecostal service it seemed orderly, structured, and traditional. The sanctuary was quiet; organ music was the only sound. Worshippers did not shout or clap; the most they did during the service was stand up. They conducted silent, almost whispered, prayers; even their hymns were quiet and solemn. Finally, the sermon was less practical than the Pentecostal sermon; it focused on the interpretation of theological doctrine and only tangentially discussed the doctrine's relevance to life.

(2) The case study

The **case study** is important for the presentation of information in psychology, sociology, anthropology, and political science, where it can examine an individual case, the dynamics of a group, or the operations of a political organization. Case studies are usually informative, describing a problem and suggesting solutions or treatments. They generally follow a set format: the statement of the problem, the background of the problem, the observations of the behavior of the individual or group being studied, the conclusions arrived at, and suggestions for improvement or future recommendations.

Different disciplines use case studies in different ways. In political science, case studies can focus on deliberations in policy making and decision making. Foreign policy negotiations, for instance, may be written up as case studies, and so can issue analyses such as "Should government control the media?" In psychology, social work, and educational psychology or counseling, the case study typically focuses on an individual and his or her interaction with peers or with agency professionals. Such a case study usually describes behavior and outlines the steps to be taken in solving the problem that the caseworker or researcher observes.

Excerpt from a Case Study (Social Work)

Mona Freeman, a 14-year-old girl, was brought to the Denver Children's Residential Treatment Center by her 70-year-old, devoutly religious adoptive mother. Both were personable, verbal, and neatly groomed. The presenting problem was seen differently by various members of the client system. Mrs. Freeman described Mona's "several years of behavior problems," including "lying, stealing, and being boy crazy." Mona viewed herself as a "disappointment" and wanted "time to think." She had been expelled from the local Seventh Day Adventist School for being truant and defiant several months earlier and had been attending public school. The examining psychiatrist diagnosed a conduct disorder but saw no intellectual, physical, or emotional disabilities. He predicted that Mona probably would not be able to continue to live in "such an extreme disciplinary environment" as the home of Mrs. Freeman because she had lived for the years from seven until twelve with her natural father in Boston, Massachusetts—a situation which was described as a "kidnapping" by Mrs. Freeman. The psychiatrist mentioned some "depression" and attributed it to Mona's inability to fit in her current environment and the loss of her life with her father in Boston.

(3)   The literature review

The **literature review**, similar in purpose and format to the bibliographic essay assigned in the humanities, is often part of the background section of a social science research paper. By reading, summarizing, and commenting on recent scholarship on a particular topic, students demonstrate a knowledge of the topic as well as an understanding of different critical approaches to that topic. The following excerpt, written for an introductory psychology class, is from a literature review on depression among college students.

▶ See
43b3

Excerpt from a Literature Review

Negative events and outlook are not the only causes of depression among college students (Brown & Silberschatz, 1988; Cochran & Hammen, 1985). In fact, some students do not become depressed in the presence of one or more negative events, while others are depressed even if no negative event takes place (Billings et al., 1983). Depression can appear to arise from a range of social and environmental factors, which can appear as a combination of stressful life events, poor coping style, and a lack of social resources (Cochran & Hammen, 1985; Vrendenburg et al., 1985).

(4) The proposal

A **proposal**, often the first stage of a research project, can help clarify and focus the project's direction and goals. In a proposal, you make a convincing case for your idea. To do so, you define your research project and defend it. In the process you must adhere strictly to any specifications outlined by your instructor or in the request for proposals issued by the grant-giving agency.

See ◀
Ch. 48 Along with a proposal you usually send a cover letter, called a *letter of transmittal,* and a **résumé**, which lists your specific qualifications for the project. This résumé summarizes your relevant work experience and accomplishments and reinforces your qualifications.

Many proposals include some or all of the following components.

Cover Sheet   The cover sheet contains your name, the title of your project, and the person or agency to which your proposal is submitted. It also provides a short title that expresses your subject concisely. Usually another line on this sheet states the reason for the submission of the proposal—for example, to satisfy a course requirement or to request funding or facilities.

Advantages of the Maquiladora Project
in El Paso
Submitted to: Professor Lawrence Howley
For: Fulfillment of Research Requirement
for Sociology 412
by Laura Talamantes

Abstract   Usually on a separate page, the abstract provides a short summary of your proposal. (See **45b1** for information on writing abstracts; for a sample, see the abstract accompanying the social science research paper in **44d**.)

Statement of Purpose   Your statement of purpose tells why you are conducting your research and what you hope to accomplish—for example, "The Maquiladora Project is an industrial development program that relies on international cooperation with Mexican industries to utilize Mexican labor while boosting the employment of U.S. white-collar workers."

Background of the Problem   This section summarizes previous research and indicates the need for your specific study.

Rationale In this section, you explain as persuasively as possible why your research project is necessary and what makes it important at this time.

Statement of Qualification This section demonstrates why you are qualified to carry out the research and enumerates the special qualifications you bring to your work.

Literature Survey This section contains a brief survey of each source you have consulted. Because it helps to establish your credibility as a researcher, this section should be thorough.

Research Methods This section describes the exact methods you will use in carrying out your research and the materials you will need; its purpose is to demonstrate the soundness of your method.

Timetable This section states the time you will need to carry out the project.

Budget Where applicable, this section estimates the costs for carrying out the research.

Conclusion In this section you restate the importance of your project.

44c Conventions of Style and Documentation

(1) Style

Social science writing uses a technical vocabulary. For instance, the social work case study excerpted in 44b2 identifies "the presenting problem"—that is, the reason the "subject," Mona, was brought to the Denver Children's Facility. Because you are addressing specialists when you write papers in individual social science disciplines, you should use the specialized vocabulary of each field and, when you describe charts and tables, you should use statistical terms, such as *mean, percentage,* and *chi square.* Keep in mind, however, that you should use plain English to explain what percentages, means, and standard deviations signify in terms of your analysis.

A social science research paper includes a **running head**, an abbreviated title printed at the top of each page of a paper. It also uses **internal headings** (for example, *Method, Results, Background of Problem, Description of Problem, Solutions,* and *Conclusion*). Unlike a humanities paper, each section of a social science paper is a complete unit with a beginning and an end so that it can be read separately, out of context, and still make sense. The body of the paper may present and discuss graphs, maps, photographs or flow charts. Finally, a social science paper frequently presents numerical data in tabular form.

(2) Documentation

See ◄
40d
Many of the journals in the various social science disciplines use the documentation style of the American Psychological Association's *Publication Manual.*

Sample Social Science Research Paper: American Psychological Association (APA) format

The following psychology research paper, "Impression Formation in Customer-Clerk Interactions," uses APA documentation style.

Running head: IMPRESSION FORMATION

Impression Formation in Customer-Clerk Interactions

Jennifer Humble

Dr. Barbara Bremer

December 11, 1994

Impression Formation   2

Abstract

The present study examined the extent to which physical appearance and dress affect the quality of customer-clerk interactions. An observational study was conducted in which a well-dressed actor and a poorly dressed actor posed as customers and engaged in customer-clerk interactions at nine different stores in a suburban shopping mall. The sociability of the clerk, time of initiation of interaction, duration of interaction, and prices of the first two watches shown were recorded. In support of the proposed hypothesis that dress and physical appearance will affect the quality of social interactions, the results indicated that the sociability of each clerk was significantly higher when interacting with the well-dressed actor than when interacting with the poorly dressed actor.

Impression Formation in Customer-Clerk Interactions

     In an attempt to understand the factors that influence the quality of social interactions between customers and clerks, some theorists have proposed that the sociability of the customer is the critical factor in determining the sociability of the interaction (Hester, Koger, & McCauley, 1985). Hester et al. (1985) reported that the customer's sociability will determine the sociability of the salesperson, for the salesperson appears to adapt to and mimic the sociability of the customer. Furthermore, Segal and McCauley (1986) reported that sociability of customer-clerk interactions is only minimally affected by factors such as urbanism of location of interaction and business of the location. Thus, they too indicated that customer sociability plays a crucial role in determining the quality of customer-clerk interactions.

     Although customer sociability is believed to be a key factor in determining the quality of social interactions, one must also consider the effect of first impressions as a determinant of the quality of these interactions. Past research indicates that individuals tend to form impressions and make judgments of others on the basis of cues, including facial expression, gestures, and dress (Hamid, 1972). Hamid examined the effects of glasses and makeup on impression formation and judgments and discovered that female actors who wore glasses and no makeup were perceived as being conservative while actors who wore makeup and no glasses were perceived as intelligent, neat, and self-confident. These stereotypical responses occurred despite the fact that the subjects had neither seen nor communicated with the actors before. This indicates that individuals tend to make intrinsic judgments about a person based on external cues. Francis and Evans (1987) further demonstrated the significant affects of personal coloring and garment style on the assessment of personality trait factors such as emotional, sociable, adaptable, and scientific. Additional research has also confirmed that there is an overall general tendency to form impressions of strangers primarily on the basis of physical/biological traits (e.g., being well dressed and physically attractive) (Lennon & Davis, 1989).

The impressions that individuals form of others, while at times accurate judgments of intrinsic characteristics, nevertheless often prove to be inaccurate. In addition, once individuals integrate external information into their impression of a person it is very difficult to discount such information (Tetlock 1983). One can see how impression formation can play a critical role in determining the quality of social interactions, for individuals tend to make intrinsic trait assumptions of others based on external cues. This tendency in turn alters their behavior toward the individual according to their preconceived perception of the individual.

Furthermore, social interactions can be greatly hindered when the impressions one forms are inaccurate.

The purpose of the present study is to examine the effects of physical appearance and dress on impression formation and to determine how the formed impression of a customer will consequently affect the quality of customer-clerk interactions. Based on the previous research indicating the effect of dress cues on impression formation, it is proposed that the manipulation of dress and physical appearance will have a significant effect on the sociability of customer-clerk interactions.

Method

Subjects

The study included nine salespersons employed at various stores located in a large suburban shopping center. The subjects were unaware of the fact that an observational study was in progress.

Materials

A previously developed observational measure of sociability or friendliness of public interaction was used to determine the sociability of each clerk (Segal & McCauley, 1986). The clerk's sociability was scored based on the following six behaviors: (a) greeting; (b) conversation; (c) farewell (0 = none; 1 = routine, conversational; 2 = friendly, personal recognition); (d) smiles; (e) facial regard (0 = none; 1 = one or two briefly; 2 = three or more; 3 = more or less continuous); and (f) overall tone (1 = unfriendly; 2 = functional, routine; 3 = friendly; 4 = personal recognition, willingness to go beyond business at hand). In addition, the time taken to initiate the interaction (minutes), the duration of the interaction (minutes), and the prices of the first two watches shown were recorded.

Procedure

The study was conducted by three experimenters (one male and two females) at a large suburban shopping center on a Sunday afternoon between 11 a.m. and 4 p.m. The two female experimenters (Actor 1 and Actor 2) posed as customers in a customer-clerk interaction while the male experimenter served as the Stable Observer of the interaction. Additionally, Actor 1 and Actor 2 served as observers when not directly participating in the interaction.

To test the effects of dress and appearance cues on impression formation and the quality of customer-clerk interactions, Actor 1 was at first dressed in a beige tailored suit, wore high-heeled shoes, and carried a leather handbag. In addition, she wore gold jewelry and makeup and had her hair neatly arranged. Actor 2 wore an old blue hooded sweatshirt, blue sweatpants, and old running sneakers. She additionally wore glasses, no makeup, and no jewelry, and had her hair combed straight back. In each interaction Actor 1 entered a store, approached a jewelry counter, and began to look at watches. A jewelry counter was chosen as a site for the interaction because of the ease of observation of the interaction and the high probability of obtaining the same clerk for both Actor 1 and Actor 2. Actor 2 and the Stable Observer also approached the jewelry counter or surrounding areas (within 10-20 feet of Actor 1) in order to record the sociability of the clerk.

The basic scenario involved the salesperson's approaching Actor 1 and inquiring if she needed assistance, to which Actor 1 was instructed to respond that she was interested in purchasing a watch. If Actor 1 were asked if she had a price range in mind, she was to respond that she had no price range. This measure was taken to allow for the clerk to make a decision as to the price of the watch that Actor 1 could afford based on his or her impression of Actor 1. After being shown a minimum of two watches, Actor 1 thanked the salesperson for his or her help and departed.

After the interaction, Actor 2 and the Stable Observer scored the sociability of the clerk on small notepads that had been concealed in their pockets during the interaction. Actor 2 and the Stable Observer were on opposite sides of Actor 1; thus, each was unaware of the degree of sociability recorded by the other. After the interaction involving Actor 1 was complete and the data recorded, approximately five minutes elapsed before Actor 2 approached the same

Impression Formation   8

jewelry counter. Actor 2 then engaged in an interaction with the clerk using the same dialogue as Actor 1. The Stable Observer and Actor 1 scored the sociability of the clerk. The Stable Observer also timed the initiation of each interaction and the duration of each interaction. This procedure was repeated in five large department stores and four smaller jewelry stores. This measure was taken to determine whether the clerk's sociability would vary according to store type.

Results

Analysis of the results of the sociability scale indicated significant differences in each of the six behavioral ratings of the clerk's sociability with respect to Actor 1 and Actor 2. In Actor 1/Actor 2 evaluation of the clerk's interaction with both Actor 1 and Actor 2, significant differences were seen with respect to the clerk's greeting ($t(16) = 4.81$, $p<.000$), conversation ($t(16) = 2.98$, $p<.009$), farewell ($t(16) = 3.58$, $p<.003$), smile ($t(16) = 5.41$, $p<.000$), facial regard ($t(16) = 8.50$, $p<.000$), and overall tone ($t(16) = 3.50$, $p<.008$), using two-tailed t-tests for independent samples. Each clerk's sociability rating tended to be higher when interacting with Actor 1 than when interacting with Actor 2. Means and standard deviations of the six behavioral scores are presented in Table 1.

The results of the Stable Observer's evaluation of the clerk's behavior when interacting with Actor 1 and Actor 2 also indicated significant differences in the clerk's greeting ($\bar{t}$(16) = 4.81, $\bar{p}$<.001), conversation ($\bar{t}$(16) = 2.98, $\bar{p}$<.009), farewell ($\bar{t}$(16) = 4.37, $\bar{p}$<.000), smiles ($\bar{t}$(16) = 6.43, $\bar{p}$<.000), facial regard ($\bar{t}$(16) = 8.50, $\bar{p}$<.000), and overall tone ($\bar{t}$(16) = 4.38, $\bar{p}$<.000) using two-tailed $\bar{t}$-tests for independent samples. The Stable Observer also rated the clerk as being more sociable when interacting with Actor 1 than when interacting with Actor 2. Means and standard deviations of the six behavioral scores are presented in Table 1.

Interrater reliability was shown to be significant in each of the six behaviors rated: greeting ($\bar{r}$ = .94) conversation ($\bar{r}$ = 1.00), farewell ($\bar{r}$ = .94), smile ($\bar{r}$ = .98), facial regard ($\bar{r}$ = 1.00), and overall tone ($\bar{r}$ = .93), $\bar{p}$<.001, 1-tailed, for all correlations.

The overall evaluations by Actor 1/Actor 2 and the Stable Observer of clerk sociability in six behaviors over 18 interactions were shown to be strikingly similar. Means and standard deviations are presented in Table 2.

Significant differences were also noted in the prices of the first two watches shown to Actor 1 and Actor 2 ($\bar{t}$(14) = 5.73, and 2.95, respectively, $\underline{p}$<.01, two-tailed $\underline{t}$-test for independent samples). Actor 1 tended to be shown higher priced watches (watch #1, $\bar{x}$ = \$433.33; S.D. = 98.68; watch #2, $\bar{x}$ = \$730.00; S.D. = 457.24) than Actor 2 (watch #1, $\bar{x}$ = \$196.42; S.D. = 51.94; watch #2, $\bar{x}$ = \$215.00; S.D. = 38.30).

Impression Formation   10

The amount of time until the initiation of the interaction (minutes) was also shown to be significantly shorter ($\underline{t}(16) = -3.91$, $\underline{p} < .001$) and the duration of the interaction significantly longer ($\underline{t}(16) = 2.98$, $\underline{p} < .009$; two-tailed $\underline{t}$-tests for independent samples) for interactions involving Actor 1 as opposed to Actor 2. Means and standard deviations are presented in Table 3.

Discussion

The results of the study clearly support the proposed hypothesis that physical appearance and dress influence impression formation, which will consequently affect the quality of social interactions, namely customer-clerk interactions. It was demonstrated that each salesperson tended to be more sociable to the well-dressed Actor 1 than he or she was to the poorly dressed Actor 2 despite the lack of variation in behavior or dialogue between the two actors. Both the Stable Observer and Actor 1/Actor 2 rated each salesperson as generally exhibiting a more friendly greeting ($\overline{x} = 1.78$; S.D. = .44) and conversation ($\overline{x} = 1.67$; S.D. = .50) when interacting with Actor 1 as opposed to Actor 2. In addition, a higher degree of smiles ($\overline{x} = 2.67$; S.D. = .50) and facial regard ($\overline{x} = 2.67$; S.D. = .50) was recorded in each clerk's interaction with Actor 1. This suggests that each salesperson tended to form different impressions about Actor 1 and Actor 2 based on dress and physical appearance cues, for these two factors were the only intended differences between the interactions. The role of dress could be confirmed in future studies through the use of a single actor posing as both a poorly dressed customer and a well-dressed customer, thus eliminating the influence of differing personality factors of each actor on the clerk's sociability. Nevertheless, the differences in impressions formed led each clerk to behave in a more sociable manner to the well-dressed, physically attractive Actor 1.

The degree to which dress cues can influence impression formation and judgment is further demonstrated by the prices of the watches shown to Actor 1 (watch #1, $\overline{x}$ = $433.33; watch #2, $\overline{x}$ = $730.00) and Actor 2 (watch #1, $\overline{x}$ = $196.43; watch #2, $\overline{x}$ = $215.00). The salesperson in each case clearly made the assumption that a well-dressed customer would be interested in a more expensive watch, whereas a poorly dressed customer would be interested in a less expensive watch. The impression formation was so strong in one case that the clerk recommended Actor 2 visit a store that sold less expensive watches. Thus, each clerk made judgments about each Actor despite a lack of information about the person's socioeconomic status. The appearance of each Actor also influenced the amount of time taken for service. In all cases the well-dressed Actor was waited on sooner. In fact, in two stores the poorly dressed Actor could not get waited on for 15 minutes and therefore left the store. The duration of the interaction was also shorter for Actor 2, and in most cases the conversation was very routine, involving no friendliness or personal recognition.

The results further indicated a high interrater reliability in the subjective measurement of each clerk's behavior. It must be noted that a potential weakness in the results exists due to the fact that each observer was previously aware of the hypothesis being tested. Thus, the potential for biased observations does exist. However, this factor appears not to have significantly influenced the results, for the objective measurements (i.e., price of watches shown, time of initiation of interaction, and duration of interaction) also indicated the tendency of each clerk to form different impressions of each Actor based on appearance.

The results of the present study support previous research indicating the effects of physical appearance and dress on impression formation (Francis & Evans, 1987; Hamid, 1972). This noted importance of dress cues on impression formation and resulting social interactions can have important implications in situations other than customer-clerk interactions, for dress is an integral part of one's appearance; thus, one must pay particular attention to mode of dress when trying to convey a given impression, particularly in situations such as job interviews.

Impression Formation   13

References

Francis, S. K., & Evans, P. K. (1987). Effects of hue, value, and style of garment and personal coloring of model on person perception. Perceptual and Motor Skills, 64, 383–390.

Hamid, P. N. (1972). Some effects of dress cues on observational accuracy, a perceptual estimate, and impression formation. Journal of Social Psychology, 86, 279–289.

Hester, L., Koger, P., & McCauley, C. (1985). Individual differences in customer sociability. European Journal of Social Psychology, 15, 453–456.

Lennon, S. J., & Davis, L. L. (1989). Categorization in first impressions. Journal of Psychology, 123(5), 439–446.

Segal, M. E., & McCauley, C. R. (1986). The sociability of commercial exchange in rural, suburban, and urban locations: A test of the urban overload hypothesis. Basic and Applied Social Psychology, 7(2), 115–135.

Tetlock, P. E. (1983). Accountability and the perseverance of first impressions. Social Psychology Quarterly, 46(4), 285–292.

Table 1

Mean (S.D.) Scores and Correlations of Interrater Reliability of Actor 1/Actor 2 and Stable Observer's Rating of Salesperson Sociability

	Actor 1/Actor 2 Rating				Stable Observer's Rating				Interrater Reliability
	Clerk 1		Clerk 2		Clerk 1		Clerk 2		Correlation Coefficient
	$\bar{x}$	S.D.	$\bar{x}$	S.D.	$\bar{x}$	S.D.	$\bar{x}$	S.D.	
Greeting	1.78	0.44	0.78	0.44	1.78	0.44	0.78	0.44	0.87
Conversation	1.67	0.50	0.89	0.60	1.67	0.50	0.89	0.60	1.00
Farewell	1.78	0.44	0.89	0.60	1.89	0.33	0.89	0.60	0.94
Smile	2.56	0.53	0.78	0.83	2.56	0.53	0.67	0.71	0.98
Facial Regard	2.67	0.50	0.78	0.44	2.67	0.50	0.78	0.44	1.00
Overall tone	2.78	0.67	1.44	1.13	2.89	0.60	1.22	0.97	0.93

N = 9 for both Clerk 1 and Clerk 2.

Impression Formation   15

Table 2

Mean (S.D.) Scores of Actor 1/Actor 2 and Stable Observer's Overall Evaluation of Clerk Sociability

	Actor 1/Actor 2 Evaluation		Stable Observer's Evaluation	
	$\overline{x}$	S.D.	$\overline{x}$	S.D.
Greeting	1.28	0.67	1.28	0.67
Conversation	1.28	0.67	1.28	0.67
Farewell	1.33	0.67	1.39	0.70
Smile	1.67	1.14	1.61	1.14
Facial Regard	1.72	1.07	1.72	1.07
Overall Tone	2.11	1.13	2.06	1.16

$N = 18$

All means are significantly different, $p<.001$, by two-tailed t-test for paired samples.

Table 3

Mean (S.D.) Scores of Prices of Watches, Time to Approach, and Time of Interaction

	Price of Watch #1		Price of Watch #2		Time to Approach (minutes)		Time of Interaction (minutes)	
	$\bar{x}$	S.D.	$\bar{x}$	S.D.	$\bar{x}$	S.D.	$\bar{x}$	S.D.
Actor 1	$433.33	98.68	$730.00	457.24	5.11	2.03	3.11	0.78
Actor 2	$196.43	51.94	$215.00	38.30	10.11	3.26	1.78	1.09

N = 9 for all cases except for Actor 2 price of watch #1 and watch #2 (n = 7).

All means differ significantly, $\underline{p}$<.01, by two-tailed $\underline{t}$-test for independent samples.

CHAPTER 45

Writing in the Natural and Applied Sciences

Writing in the natural and applied sciences relies on **empirical data**—information derived from observations or experiments. Although science writing is usually expository, concerned with accurately reporting observations and experimental data, it may also be persuasive.

Basic to research in the natural and applied sciences is the **scientific method**—a process by which scientists gather and interpret information.

THE SCIENTIFIC METHOD

1. Define a problem you want to solve or an event you want to explain. Conduct a search of the literature to find out what previous work has been done on the problem.
2. Formulate a hypothesis that attempts to explain the problem.
3. Plan a method of investigation that will allow you to test your hypothesis.
4. Carry out your experiment. Make careful observations and record your data.
5. Analyze the results of your experiment, and determine whether or not they support your initial hypothesis. Revise your hypothesis, if you can, to account for any discrepancies. If you cannot, plan further research that may help you explain the phenomena you have observed.

FURTHER READING

David Bartholomae. "Inventing the University." *When a Writer Can't Write: Studies in Writer's Block and Other Composing Process Problems*. Ed. Mike Rose. New York: Guilford P, 1985. 134–65.

The title of this article is one key to understanding its importance. When students write, they "invent" the university—that is, the various disciplines and their discourses. But while students "invent" the university, the university "invents" the students. In other words, while students are learning by appropriating the various discourses of the disciplines, they are simultaneously appropriated by those discourse communities. As students become fluent in the discourses, however, they can change the university and its discourse communities.

Bartholomae points out that the primary problem basic writing students have is constructing a role (voice, persona) to address the audience of each discourse. Student writers must understand how the conventions of each discourse shape, or invent, both the writer and the audience. It is a matter of understanding how these conventions "control" the self-concepts of basic writers vis-à-vis their teacher or wider audience.

Bartholomae takes issue with Linda Flower and John Hayes's "cognitive process theory of writing," in which a writer's persona and audience is "solely" invented by the writer. Bartholomae sees this view as insensitive to basic writers. He, instead, favors a view of writers constructed across a social, cultural context.

continued on following page

45a   Research Sources

The methods of data collection used in the sciences frequently require observation and experimental research. In addition to being discussed, most results are tabulated and presented graphically. In addition, scientists often carry out literature searches to determine what other work has been done in their areas of interest.

(1)   Specialized library sources

Because scientists are interested in the number of times and the variety of sources in which a study is cited, they frequently use the *Science Citation Index*. The following specific sources are also useful.

General Science

Applied Science and Technology Index
CRC Handbook of Chemistry and Physics (and other titles in the CRC series of handbooks)
Current Contents
General Science Index
McGraw-Hill Encyclopedia of Science and Technology
Scientific and Technical Information Sources

Chemistry

Analytical Abstracts
Chemical Abstracts
Chemical Technology
Dictionary of Organic Compounds
Kirk-Othmer Encyclopedia of Chemical Technology
Van Nostrand Reinhold Encyclopedia of Chemistry

Engineering

Engineering Index
Environment Abstracts
Government Reports Announcements (NTIS)
HRIS Abstracts (highway engineering)
Pollution Abstracts
Selected Water Resources Abstracts

Research Sources

Earth Sciences

Abstracts of North American Geology
Annotated Bibliography of Economic Geology
Bibliography and Index of Geology
Bibliography of North American Geology
Climatology and Data (U.S. Environmental Data Service)
Encyclopedia of Earth Sciences
Geophysical Abstracts
Publications of the USGS

Life Sciences

G. F. Zimek's Animal Life Encylcopedia
Cumulative Index to Nursing and Allied Health Literature
Biological Abstracts
Biological and Agricultural Index
Biology Digest
A Dictionary of Genetics
Encyclopedia of Bioethics
Encyclopedia of the Biological Sciences
Environmental Abstracts Annual
Hospital Literature Index
Index Medicus
International Dictionary of Medicine and Biology
International Nursing Index

Mathematics

Current Index to Statistics
Encyclopedia of Statistical Sciences
Encyclopedic Dictionary of Mathematics
Mathematical Reviews
Universal Encyclopedia of Mathematics

Physics

Astronomy and Astrophysics Abstracts
Encyclopedia of Physics
Encyclopedic Dictionary of Physics
Physics Abstracts
Solid State Abstracts Journal

continued from previous page

Beginning writers must "both imagine and write from a position of privilege to be successful." That is, they must understand that writers within particular discourse communities are "privileged." With an awareness of "privilege," Bartholomae argues, the students may be ready "to transform the political and social relationships between students and teachers."

The remaining section of the essay is a quasi-case study of the problems that basic writers have in relation to cognitive and social contexts. The study is limited to first-year university students.

(2)   Specialized databases for computer searches

As in other disciplines, many print indexes are available on-line. Helpful databases for research in the sciences include *BIO-SIS Previews, CASearch, SCISEARCH, Agvicola, CAB Abstracts, Compendex, NTIS, Inspec, MEDLINE, MATHSCI, Life Sciences Collection, GEOREF, Zoological Record Online,* and *World Patents Index.*

(3)   Other sources of information

Opportunities for research outside the library vary widely because of the many ways in which scientists can gather information. In agronomy, for example, researchers collect soil samples; in toxicology, they test air or water quality. In marine biology, they might conduct research in a particular aquatic environment, while in chemistry they conduct experiments to identify an unknown substance. Scientists also conduct surveys: Epidemiologists study the spread of communicable diseases, and cancer researchers question populations to determine how environmental or dietary factors influence the likelihood of contracting cancer.

45b   Assignments in the Sciences

Many writing assignments in a science class are similar to those assigned in other disciplines. Three additional assignments that are common in (but not limited to) scientific disciplines are the *abstract,* the *literature survey,* and the *laboratory report.*

(1)   The abstract

Most scientific articles begin with **abstracts**, highly condensed summaries that serve as road maps or guides for readers. Many scientific indexes also provide abstracts of articles so researchers can determine whether an article is of use to them. An **indicative abstract** gives a sense of the content of an article. It helps readers decide whether the article will be useful and whether they want to read it in full. An annotated bibliography includes short indicative abstracts following each complete citation. An **informative abstract** is more complete; it includes enough detail so readers can obtain essential information without reading the article itself.

When you write an abstract, follow the organization of your paper, devoting a sentence or two to each of its major sections. In 200 to 500 words, state the purpose, method of research, results, and conclusion in the order in which they appear in the paper, but include only essential information. Avoid quoting from your paper or repeating the title. An abstract should provide clear information for a wide audience; therefore, use only the technical terms you are sure your readers will understand.

Abstract: Biology

"Purification to Near Homogeneity of Bovine Transforming Epithelial Growth Factor," by Stephen McManus, Cooperative Education Student, Smith, Kline & French.

The control of cellular proliferation is known to be mediated at an extra-cellular level by polypeptide growth factors; examples include epidermal growth factor (EGF), platelet-derived growth factor (PDGF), and transforming growth factors alpha and beta (TGF-a, TGF-β). The transforming growth factors are so called because of their ability to induce anchorage-independent growth of selected target cell lines. Our studies have identified an apparently novel growth factor activity associated with epithelial cells and tissues. This activity, termed epithelial transforming growth factor (TGFe), is identified by the anchorage-independent growth of the SW13 epithelial cell line, derived from human adrenocortical carcinoma. The purification of this factor was accomplished by a multi-step chromatography and electrophoretic process. The total purification was estimated as 6×10^5-fold with 1% recovery, corresponding to a yield of 0.1µg TGf-e/kg bovine kidney.

(2)   The literature survey

Literature surveys are common in the sciences, most often as a section of a proposal or as part of a research paper. Unlike an abstract, which summarizes a single source, a literature survey summarizes the work of a number of studies and sometimes compares and contrasts them. By doing so, the literature survey provides a theoretical context for the discussion.

Literature Survey: Parasitology

Ultrastructural studies of micro- and macrogametes have included relatively few of the numerous Eimerian species. Major early studies include the following (hosts are listed in parentheses): micro- and macrogametes of E. performans (rabbits), E. stiedae (rabbits), E. bovis (cattle), and E. auburnensis (cattle) (Hammond et al., 1967; Scholtyseck et al., 1966), macrogametogenesis in E. Magna (rabbits) and E. intestinalis (rabbits) (Kheysin, 1965), macrogametogony of E. tenella (chickens) (McLaren, 1969), and the microgametocytes and macrogametes of E. neischulzi (rats) (Colley, 1967). More recent investigations have included macrogametogony of E. acervulina (chickens) (Pitillo and Ball, 1984).

(3) The laboratory report

A laboratory report is the most common assignment for students taking courses in the sciences. It is divided into sections that reflect the stages of the scientific method and generally conform to the specifications for the laboratory report outlined below. Not every section will be necessary for every experiment, and some experiments may call for additional components, such as an abstract or a reference list. In addition, a lab experiment may include tables, charts, graphs, and illustrations. The exact format of a student lab report is usually defined by the specific course's lab manual.

A lab report is an explanation of a process. To enable readers to follow and understand what may be a complex series of tasks, it is important for you to explain stages clearly and completely, to present steps in exact chronological order, and to illustrate the purpose of each step. In addition, you must provide clear descriptions of the equipment used in an experiment.

Laboratory Report: Chemistry

Purpose In this section you describe the goal of the experiment, presenting the hypothesis you are going to test or examine.

The purpose of this lab is to determine the iron content of an unknown mixture containing an iron salt by titration with potassium permanganate solution.

Equation to find % of Fe: $5Fe^{2+} + MnO_{4-} + 8H^+ = 5Fe^{3+} + Mn^{2+} + 4H_2O$

Equipment In this section you list the equipment you will use in the experiment. Often this section also identifies and explains your methodology.

Equipment includes two 60 ml beakers, a graduated cylinder, a scale, 600 ml of distilled water, 2 grams of H_2SO_4, 5 grams of $KMnO_4$, 100 ml of $H_2C_2O_4 \cdot 2H_2O$, and a Bunsen burner.

Procedure   In this section you describe the steps of the experiment in the order in which they occur, usually numbering the steps.

1) A $KMnO_4$ solution was prepared by dissolving 1.5 grams of $KMnO_4$ in 500 ml of distilled water.
2) Two samples $H_2C_2O_4 \cdot 2H_2O$ of about 0.2 grams each were weighed.
3) Each sample was dissolved in 60 ml of H_2O and 30 ml of H_2SO_4 in a 250 ml beaker.
4) The mixture was heated to 80°C and titrated slowly with $KMnO_4$ until the mixture turned pink.
5) The procedure was repeated twice.

Results   In this section you present the results that you obtained from your experiments. These results can be observations, measurements, or equations.

Percentage of iron: 1st run = 12.51 ml

2nd run = 11.2 ml

Conclusion or Discussion of Results   In this section you explain your results or justify them in terms of the initial questions asked in the *Purpose* section.

Calculation for % of iron

$$\frac{12.5 \text{ ml} \times .0894 \text{ M}}{100 \text{ ml}/1} \times \frac{5 \text{ moles Fe}}{1 \text{ mole MnO}_4} \times \frac{55.85 \text{ g/mol}}{.5\text{g}} \times 100$$

$$= \frac{66.11}{500} = 13.22\% \text{ Fe}$$

 Conventions of Style and Documentation

(1)   Style

Because writing in the sciences focuses on the experiment, not on those conducting the experiment, writers often use the passive voice. For example, in a lab report, you would say "The mixture was heated for forty-five minutes" rather than "I heated the mixture for forty-five minutes." You would use the first person, however, if you were asked to write a reaction statement or to give your opinion about something. Another stylistic convention to remember concerns verb tense: A conclusion or a statement of generally accepted fact should be in the present tense ("Objects in motion *tend* to stay in motion"); a summary of a study, however, should be in the past tense ("Watson and Crick *discovered* the structure of DNA"). Finally, note that direct quotations are seldom used in scientific papers.

See ◀ 23cl

Because you are writing to inform or persuade other scientists, you should write clearly and concisely. Remember that you should use technical terms only when they are necessary to convey your meaning. Too much technical jargon can make your paper difficult to understand—even for scientists familiar with your discipline. Often a scientific paper will include a glossary that lists and defines terms that may be unfamiliar to readers.

See ◀ 45d

Tables and illustrations are an important part of most scientific papers. Be careful to place tables as close to your discussion of them as possible and to number and label any type of illustration or diagram so you can refer to it in your text. Keep in mind that each professional society prescribes formats for tables and illustrations and the way they are to be presented. Therefore, you cannot use a single format for all your scientific writing.

Remember that different scientific journals follow different conventions of style and use different paper formats. For example, although the *CBE Style Manual* governs the overall presentation of papers in biology, the *Journal of Immunology* might have a different format from the *Journal of Parasitology*. (The *CBE Style Manual* lists the different journals that use their own paper formats.) Your instructor may ask you to prepare your paper according to the style sheet of the journal to which you might wish to submit your work. Although publication may seem a remote possibility to you, following a style sheet helps to remind you that writing in the sciences may involve writing for a variety of different audiences.

You should also learn the various abbreviations with which journals are referred to in the reference sections of science papers. For example, *The American Journal of Physiology* is abbreviated Amer. J. Physiol., and *The Journal of Physiological Chemistry* is abbreviated J. of Physiol. Chemistry. Note that in CBE style, the abbreviated forms of journal titles are *not* underlined in the reference list.

(2)   Documentation

Documentation style varies from one scientific discipline to another; even within each discipline, documentation style may vary from one journal to another. For this reason, you should ask your instructor which documentation format is required. Most disciplines in the sciences use a number-reference format prescribed by their professional societies. For instance, electrical engineers use the format of the Institute for Electronics and Electrical Engineers, chemists use the format of the American Chemical Society, physicists use the format of the American Institute of Physics, and mathematicians use the format of the American Mathematical Society.

▶ See 40f

Sample Science Research Paper: Council of Biology Editors (CBE) format

The following excerpts from a biology research paper, "Maternal Smoking: Deleterious Effects on the Fetus," uses a number-reference format recommended by the *CBE Style Manual*.

June M. Fahrman
Biology 306
April 17, 1994

Maternal Smoking: Deleterious
Effects on the Fetus

Introduction

The placenta, lifeline between fetus and mother, has been the subject of various studies aimed at determining the mechanisms by which substances in the mother's bloodstream affect the fetus. For example, cigarette smoking is clearly associated with an increased risk in the incidence of low birthweight infants, due both to prematurity and to intrauterine growth retardation. (1)

Development of the Placenta

At the morula stage of development, less than one week after fertilization, two types of cells can be distinguished. . . .

Maternal Smoking

10

Conclusion

In summary, abundant evidence exists as to the harmful effects maternal smoking may have on the fetus. These include low birthweight, low IQ scores, minimal brain dysfunction, shorter stature, perinatal mortality, and premature birth. . . .

Maternal Smoking

11

GLOSSARY

1. ABRUPTO PLACENTA--Partial or complete premature separation of a normally implanted placenta.
2. CATECHOLAMINES--Pyrocatechols with an alkylamine side chain; examples of biological interest are epinephrine, norepinephrine, and dopa.
3. HYPOXIA--Decrease below normal levels of oxygen in air, blood, or tissue, short of anoxia.
4. MINIMAL BRAIN DYSFUNCTION--Also referred to as hyperactivity and/or attention deficit disorder.
5. NEONATAL PERIOD--Newborn; related to period immediately succeeding birth through the first 28 days.
6. PERINATAL PERIOD--Period before delivery from the twenty-eighth week of gestation to the first seven days after delivery.

Maternal Smoking

12

References

1. Rakel, Robert E. Conn's current therapy 1988.
   Philadelphia: W. B. Saunders; 1988.
2. Meberg, A.; Sande, H.; Foss. O. P.; Stenwig, J. T.
   Smoking during pregnancy--effects on the fetus and
   on thiocyanate levels in mother and baby. Acta.
   Paediatr. Scand 68:547–552; 1979.
3. Lehtovirta, P.; Forss, M. The acute effect of smoking on intervillous blood flow of the placenta. Brit.
   Obs. Gyn. 85:729–731; 1978.
4. Phelan, Jeffrey P. Diminished fetal reactivity with
   smoking. Amer. Obs. Gyn. 136:230–233; 1980.
5. VanDerVelde, W. J. Structural changes in the placenta of smoking mothers: a quantitative study.
   Placenta 4:231–240; 1983.
6. Asmussen, I. Ultrastructure of the villi and fetal
   capillaries in placentas from smoking and non-
   smoking mothers. Brit. Obs. Gyn. 87:239–245;
   1980.
7. Meyer, M. Perinatal events associated with maternal smoking during pregnancy. Amer. Epid.
   103(5):464–476; 1976.

Writing Essay Examinations

One of the most difficult parts of essay examinations for many students is writing against the clock. Writing ungraded first drafts of essays in class helps students learn to manage their ideas and their time. Some instructors find it helpful to pace the students' progress further, allowing ten minutes for an introduction, thirty for the body, and five for the conclusion, with the last minutes devoted to proofreading and minor revisions.

Occasionally, you may find it useful to set up an essay examination that mimics the research and writing process. Several days before the exam, distribute a list of topics (enough of them to ensure that students doing similar research won't swamp the library or other resource centers). Ask students to select a topic, do exploratory research, develop a thesis statement, and search for material to support it. Students will turn in this preliminary material no later than the day prior to the exam and be asked to come to the exam with nothing but a pen and a supply of paper. Return each student's research and notes, and begin. This kind of essay examination is more a test of students' writing and reasoning skills than a test of sheer memory.

Collect from instructors examples of essay examinations their students have written and the instructor's comments. Share these answers with your students, discussing the writing problems that are evident and how the writing of the examination might have been improved.

CHAPTER 46

Writing Essay Examinations

Taking examinations is a skill, one you have been developing throughout your life as a student. Although both short-answer and essay examinations require you to study, to recall what you know, and to budget your time carefully as you write your answers, only essay questions ask you to synthesize information and to arrange ideas in a series of clear, logically connected sentences. To write an essay examination, or even a paragraph-length answer, you must do more than memorize facts; you must see the relationships among them. In other words, you must think critically about your subject.

 46a **Planning an Essay Examination Answer**

Because you are under pressure during an examination and tend to write quickly, you may be tempted to skip the planning and revision stages. But if you write in a frenzy and hand in your examination without a second glance, you are likely to produce a

---

PLANNING AN ESSAY EXAMINATION ANSWER

1. Review your material.
2. Consider your audience and purpose.
3. Read through the entire examination.
4. Read each question very carefully.
5. Brainstorm to find ideas.

disorganized or even incoherent answer. With advance planning and sensible editing, you can write an answer that demonstrates your understanding of the material.

(1)   Review your material

Be sure you know beforehand the scope and format of the examination. How much of your text and class notes will the examination cover—the entire semester's work or only the material covered since the last test? Will you have to answer every question, or will you be able to choose among alternatives? Will the examination be composed entirely of short-answer questions, or will it include one-sentence, one-paragraph, or essay-length answers? Will the examination emphasize your recall of specific facts or your ability to demonstrate your understanding of the course material by drawing conclusions?

Examinations challenge you to recall and express in writing what you already know—what you have read, what you have heard in class, what you have reviewed in your notes. Before you even begin any examination, then, you must study for it: Reread your text and class notes, highlight key points, and perhaps outline particularly important sections of your notes. When you prepare for a short-answer examination, you may memorize facts without analyzing their relationship to one another or to a body of knowledge as a whole—the definition of pointillism, the date of Queen Victoria's death, the formula for a quadratic equation, three reasons for the fall of Rome, two examples of conditioned reflexes, four features of a feudal economy, six steps in the process of synthesizing Vitamin C. When you prepare for an essay examination, however, you must do more than remember information; you must also make connections among ideas.

When you are certain you know what to expect, see if you can anticipate the essay questions your instructor might ask. Try out likely questions on classmates, and see whether you can do some collaborative brainstorming to outline answers to possible questions. You might even practice answering one or two.

(2)   Consider your audience and purpose

The audience for any examination is usually the instructor who prepared it. He or she already knows the answers to the questions and a great deal more about the subject. As you read the questions, think about what your instructor has emphasized in class.

QUOTATION

"As the weeks dwindled toward the Honors exams, I became numbly philosophical. Either I would do well, or I wouldn't. Chances were I would. I had looked at past tests on file in the library, and I could answer most of those questions. After all, we would have choices. I would be able to study on the two in-between days that separated the three exams. I would certainly be able to write something on Restoration drama or seventeenth-century prose, and I could stumble through the rest with a bit of luck. I could define terms, and reorient the questions to what I knew. Four hours weren't really so long, if I took a break in the middle to get a drink of water. I could do it, of course I could do it."

–Susan Allen Toth
Ivy Days: Making My Way Out East

Although you may certainly arrange material in a new way or use it to make your own point, keep in mind that your purpose is to demonstrate that you understand the material, not to make clever remarks or introduce irrelevant information.

CLOSE-UP

WRITING ESSAY EXAMINATIONS

See ◄
42c

When you prepare to write a response to an examination question, you should consider the wider academic audience your instructor represents and make every effort to use the specific vocabulary of the field and to follow any discipline-specific stylistic conventions your instructor has discussed.

(3) Read through the entire examination

Your time is usually limited when you take an examination, so you must plan carefully. How long should a "short-answer" or "one-paragraph" or "essay-length" answer be? How much time should you devote to answering each question? The question itself may specify the time allotted for each answer, so look for that information. More often the point value of each question or the number of questions on the examination will determine how much time to spend on each answer. If an essay question is worth 50 out of 100 points, for example, you will probably have to spend at least half and perhaps more of your time planning, writing, and proofreading your answer.

Before you begin to write, read the entire examination carefully to determine your priorities and your strategy. First, be sure your copy of the test is complete and that you understand the format each question requires. If you need clarification, ask your instructor or proctor for help. Then, decide where to start. Responding first to short answers (or to questions whose answers you are sure of) is usually a good strategy. This tactic ensures you will not become bogged down in a question that baffles you, left with too little time to write a strong answer to a question you understand well. Moreover, starting with the questions you are sure of can help build your confidence.

(4)   Read each question very carefully

To write an effective answer, you need to understand the question. As you read any essay question, you may find it helpful to underline key words and important terms.

> **SOCIOLOGY:** Distinguish among Social Darwinism, instinct theory, and sociobiology, giving examples of each.
>
> **MUSIC:** Explain how Milton Babbitt used the computer to expand Schoenberg's twelve-tone method.
>
> **PHILOSOPHY:** Define existentialism and identify three influential existentialist works, explaining why they are important.

Look carefully at the wording of each examination question. If the question calls for a *comparison and contrast* of *two* styles of management, a *description* or *analysis* of *one* style, no matter how comprehensive, will not be acceptable. If the question asks for causes *and* effects, a discussion of causes alone will not do.

The wording of the question suggests what you should emphasize. For instance, an American history instructor would expect very different answers to the following two examination questions.

1. Give a detailed explanation of the major causes of the Great Depression, noting briefly some of the effects of the economic collapse on the United States. (1 hour)
2. Give a detailed summary of the effects of the Great Depression on the United States, briefly discussing the major causes of the economic collapse. (1 hour)

Although the preceding questions look somewhat alike, the first calls for an essay that focuses on *causes*, whereas the second calls for one that stresses *effects*.

The following question requires a very specific treatment of the topic.

> **LITERATURE:** Identify three differences between the hard-boiled detective story and the classical detective story.

The following, a response to this question that simply *identifies* three characteristics of *one* kind of detective story, is not acceptable.

> **UNACCEPTABLE ANSWER:** The hard-boiled detective story, popularized in Black Mask magazine in the 1930s and 1940s, is very different from the classical detective story of Edgar Allan Poe or Agatha Christie. The hard-boiled story features a down-on-his-luck detective

IN THE CLASSROOM

Collect essay exam questions from your own past tests and those of colleagues. Copy a variety of essay questions on transparencies. Ask students to tell you what kind of question each one is: Does it ask for comparison and contrast? For description? Analysis? Causes? Effects? A personal reaction?

IN THE CLASSROOM

Giving students practice in writing ungraded paragraph-length examination answers can help build students' confidence in their ability to write longer ones. From time to time, distribute copies of a short (one- to two-page) article. Give students time to read the passage carefully. Collect the articles, write an exam question that can be answered in one paragraph, and give students fifteen minutes to write. Discuss the results.

who is constantly tempted and betrayed. His world is dark and chaotic, and the crimes he tries to solve are not out-of-the-ordinary occurrences but the norm. These stories have no happy endings; even when the crime is solved, their world is still corrupt.

The answer below, which *contrasts* the *two* kinds of detective stories, is acceptable.

> **REVISED:** The hard-boiled detective story differs from the classical detective story in its characters, its setting, and its plot. The classical detective is usually well educated and well off; he is aloof from the other characters and therefore can remain in total control of the situation. The hard-boiled detective, on the other hand, is typically a decent but down-on-his-luck man who is drawn into the chaos around him, constantly tempted and betrayed. In the orderly world of the classical detective, the crime is a temporary disruption; in the hard-boiled detective's dark and chaotic world, the crimes he tries to solve are not out-of-the-ordinary occurrences but the norm. In the classical detective story, order is restored at the end, but hard-boiled stories have no happy endings; even when the crime is solved, their world is still corrupt.

KEY WORDS IN EXAMINATION QUESTIONS

- Explain
- Compare
- Contrast
- Trace
- Evaluate
- Discuss

- Clarify
- Relate
- Justify
- Analyze
- Interpret
- Describe

- Classify
- Identify
- Illustrate
- Define
- Support
- Summarize

---

(5) Brainstorm to find ideas

See ◄ 1c2

Once you understand the question, begin **brainstorming**, quickly listing all the relevant ideas you can remember. Then, identify significant points on your list, and delete less promising ones. A quick review of the exam question and your supporting ideas should lead you toward a workable thesis for your essay answer.

46b Shaping an Essay Examination Answer

(1) Finding a thesis

Often you can rephrase the examination question as a thesis statement. For example, the American history examination question "Give a detailed summary of the effects of the Great Depression on the United States, briefly discussing the major causes of the economic collapse" suggests the following thesis.

> **EFFECTIVE THESIS:** The Great Depression, caused by the American government's economic policies, had major political, economic, and social effects on the United States.

An effective thesis addresses all aspects of the question but highlights only relevant concerns. The following thesis statements are not effective.

> **VAGUE:** The Great Depression, caused largely by profligate spending patterns, had a number of very important results.

> **INCOMPLETE:** The Great Depression caused major upheaval in the United States.

> **IRRELEVANT:** The Great Depression, caused largely by America's poor response to the 1929 stock market crash, had more important consequences than World War II.

SHAPING AN ESSAY EXAMINATION ANSWER

1. Find a thesis.
2. Make a scratch outline.

---

(2) Making a scratch outline

Because time is limited, you should plan your answer before you write it. Therefore, once you have decided on a suitable thesis, you should make a scratch outline of your major points.

Write on the inside cover of your exam book or on its last sheet. Use the pattern of development suggested by the question— ▸ See definition, comparison and contrast, or cause and effect, for in- 4e stance—to shape your outline, and list your supporting points in the order in which you plan to discuss them. Once you have completed your outline, check it against the exam question to make certain it covers everything the question requires—and *only* what the question requires.

A scratch outline for an answer to the American history question ("Give a detailed summary of the effects of the Great Depression on the United States, briefly discussing the major causes of the economic collapse.") might look like this.

THESIS: The Great Depression, caused by the American government's economic policies, had major political, economic, and social effects on the United States.

SUPPORTING POINTS:

Causes

American economic policies: income poorly distributed, factories expanded too much, more goods produced than could be purchased.

Effects

1. Economic situation worsened—farmers, businesses, workers, and stock market all affected.
2. Roosevelt elected—closed banks, worked with Congress to enact emergency measures.
3. Reform—TVA, AAA, NIRA, etc.
4. Social Security Act, WPA, PWA

See ◀
4e5

An answer based on this outline will follow a **cause-and-effect** pattern, with an emphasis on effects, not causes.

46c Writing and Revising an Essay Examination Answer

Referring to your thesis and your outline, you can now draft your answer. A simple statement of your thesis that summarizes your answer is your best *introduction,* for it shows the reader you are addressing the question directly. Do not bother crafting an elaborate or unusual introduction; your time is precious, and so is your reader's.

To develop the *body* of the essay, follow your outline point by point, using clear topic sentences and transitions to indicate your progression and to help the reader see you are answering the question in full. Such signals, along with parallel sentence structure and repeated key words, make your answer easy to follow.

The most effective *conclusion* for an essay examination is a clear, simple restatement of the thesis or a summary of the essay's main points.

Essay answers should be complete and detailed, but they should not contain irrelevant material. Every unnecessary fact or opinion only increases your chance of error. Do not repeat yourself or volunteer unrequested information. Do not express your own feelings or opinions unless such information is specifically called for, and be sure to support all your general statements with specific examples.

Finally, leave enough time to reread and revise what you have written. Try to view your answer from a fresh perspective. Have you left out words or written illegibly? Is your thesis clearly worded? Does your essay support your thesis and answer the question? Are your facts correct, and are your ideas presented in a logical order? Review your topic sentences and transitions. Check sentence structure and word choice, spelling and punctuation. If a sentence—or even a whole paragraph—seems irrelevant, cross it out. If you suddenly remember something you want to add, you can insert a few additional words with a caret (ˆ). Neatly insert a longer addition at the end of your answer, box it, and label it so your instructor will know where it belongs.

The following one-hour essay answer was written in response to the question outlined in **46b2**. Notice how the student restates the question in her thesis and keeps the question in focus by repeating key words like *cause, effect, result, response,* and *impact.*

QUESTION: Give a detailed summary of the effects of the Great Depression on the United States, briefly discussing the major causes of the economic collapse.

EFFECTIVE ESSAY EXAM ANSWER

Introduction—Thesis rephrases exam question

1   The Great Depression, caused by the American government's economic policies, had major political, economic, and social effects on the United States.

Summarizes policies leading to Depression (causes)

2   The Depression was precipitated by the stock market crash of October 1929. But its actual causes were more subtle; they lay in the U.S. government's economic policies. First, personal income was not well distributed. Although production rose during the 1920s, the farmers and other workers got too little of the profits; instead, a disproportionate amount of income went to the richest 5 percent of the population. The tax policies at this time made inequalities in income even worse. A good deal of income also went into development of new manufacturing plants. This expansion stimulated the economy but encouraged the production of more goods than consumers could purchase. Finally, during the economic boom of the 1920s the government did not attempt to limit speculation or impose regulations on the securities market; it also did little to help

build up farmers' buying power. Even after the crash began, the government made mistakes: Instead of trying to counter the country's deflationary economy, the government focused on keeping the budget balanced and making sure the United States adhered to the gold standard.

3   The Depression, devastating to millions of individuals, had a tremendous impact on the nation as a whole. Its political, economic, and social consequences were great.

4   Between October 1929 and Roosevelt's inauguration on March 4, 1932, the economic situation grew worse. Businesses were going bankrupt, banks were failing, and stock prices were falling. Farm prices fell drastically, and hungry farmers were forced to burn their corn to heat their homes. There was massive unemployment, with millions of workers jobless and humiliated, losing skills and self-respect. President Hoover's Reconstruction Finance Corporation made loans available to banks, railroads, and businesses, but he felt state and local funds (not the federal government) should finance public works programs and relief. Confidence in the president declined as the country's economic situation worsened.

5   One result of the Depression, then, was the election of Franklin Delano Roosevelt. By the time of his inauguration, most American banks had closed, 13 million workers were unemployed, and millions of farmers were threatened by foreclosure. Roosevelt's response was immediate: Two days after he took office, he closed all banks and took steps to support the stronger ones with loans and prevent weak ones from reopening. During the first hundred days of his administration, he kept Congress in special session. Under his leadership, Congress enacted emergency measures designed to provide "Relief, Recovery, and Reform."

6   In response to the problems caused by the Depression, Roosevelt set up agencies to reform some of the conditions that had helped to cause the Depression in the first place. The Tennessee Valley Authority, created in May of 1933, was one of these. Its purposes were to control floods by building new

dams and improving old ones and to provide cheap, plentiful electricity. The TVA improved the standard of living of area farmers and drove down the price of power all over the country. The Agricultural Adjustment Administration, created the same month as the TVA, provided for taxes on basic commodities, with the tax revenues used to subsidize farmers to produce less. This reform measure caused prices to rise.

More effects: NIRA, other laws, etc.

7   Another response to the problems of the Depression was the National Industrial Recovery Act. This act established the National Recovery Administration, an agency that set minimum wages and maximum hours for workers and set limits on production and prices. Other laws passed by Congress between 1935 and 1940 strengthened federal regulation of power, interstate commerce, and air traffic. Roosevelt also changed the federal tax structure to redistribute American income.

More effects: Social Security, etc.

8   One of the most important results of the Depression was the Social Security Act of 1935, which established unemployment insurance and provided financial aid for the blind and disabled and for dependent children and their mothers. The Works Progress Administration (WPA) gave jobs to over 2 million workers, who built public buildings, roads, streets, bridges, and sewers. The WPA also employed artists, musicians, actors, and writers. The Public Works Administration (PWA) cleared slums and created public housing. In the National Labor Relations Act (1935), workers received a guarantee of government protection for their unions against unfair labor practices by management.

Conclusion

9   As a result of the economic collapse known as the Great Depression, Americans saw their government take responsibility for providing immediate relief, for helping the economy recover, and for taking steps to ensure that the situation would not be repeated. The economic, political, and social impact of the laws passed during the 1930s are still with us today, helping to keep our government and our economy stable.

Notice that the student does not digress: She does not describe the conditions of people's lives in detail, blame anyone in particular, discuss the president's friends and enemies, or consider parallel events in other countries. She covers only what the question asks for. Notice, too, how topic sentences ("One result of the Depression. . . ."; "In response to the problems caused by the Depression . . . ."; "One of the most important results of the Depression. . . .") keep the primary purpose of the discussion in focus and guide the reader through the essay.

A well-planned essay like the preceding one is not easy to write. Consider the following ineffective answer to the same question.

INEFFECTIVE ESSAY EXAM ANSWER

1   The Great Depression is generally considered to have begun with the Stock Market Crash of October 1929 and to have lasted until the defense buildup for World War II. It was a terrible time for millions of Americans, who were not used to being hungry or out of work. Perhaps the worst economic disaster in our history, the Depression left its scars on millions of once-proud workers and farmers who found themselves reduced to poverty. We have all heard stories of businessmen committing suicide when their investments failed, of people selling apples on the street, and of farmers and their families leaving the dust bowl in desperate search of work. My own grandfather, laid off from his job, had to support my grandmother and their four children on what he could make from odd carpentry jobs. This was the Depression at its worst.

No clear thesis Vague, subjective impressions of the Depression

2   What else did the Depression produce? One result of the Depression was the election of Franklin Delano Roosevelt. Roosevelt immediately closed all banks. Then Congress set up the Federal Emergency Relief Administration, the Civilian Conservation Corps, the Farm Credit Administration, and the Home Owners' Loan Corporation. The Reconstruction Finance Corporation and the Civil Works Administration were two other agencies designed to provide Relief, Recovery, and Reform. All these agencies

800

Gratuitous
summary

helped Roosevelt in his efforts to lead the nation to recovery while providing relief and reform.

3   Along with these emergency measures, Roosevelt set out to reform some of the conditions he felt were responsible for the economic collapse. Accordingly, he created the Tennessee Valley Authority (TVA) to control floods and provide electricity in the Tennessee Valley. The Agricultural Adjustment Agency levied taxes and got the farmers to grow less, causing prices to rise. Thus, these agencies, the TVA and the AAA, helped to ease things for the farmers.

Unsupported
generalization

4   The National Industrial Recovery Act established the National Recovery Administration, which was designed to help workers. It established minimum wages and maximum hours, both of which made conditions better for workers. Other important agencies included the Federal Power Commission, the Interstate Commerce Commission, the Maritime Commission, and the Civil Aeronautics Authority. Changes in the tax structure at about this time made the tax system fairer and eliminated some inequities. Roosevelt, working smoothly with his cabinet and with Congress, took many important steps to ease the nation's economic burden.

Why were
these agencies
important?
What did
they do?

5   Despite the fact that he was handicapped by polio, Roosevelt was a dynamic president. His fireside chats, which millions of Americans heard on the radio every week, helped to reassure Americans that things would be fine. This increased his popularity. But he had problems, too. Not everyone agreed with him. Private electric companies opposed the TVA, big business disagreed with his support of labor unions, the rich did not like the way he restructured the tax system, and many people saw him as dangerously radical. Still, he was one of our most popular presidents ever.

Discussion of
Roosevelt
irrelevant to
topic

6   Social Security Act—unemployment insurance, aid to blind and disabled and children

WPA—built public projects

PWA—public housing

National Labor Relations Act—strengthened labor unions

Undeveloped
information

This essay only indirectly answers the examination question. It devotes too much space to unnecessary elements—an emotional introduction, needlessly repeated words and phrases, gratuitous summaries, and unsupported generalizations. Without a thesis to guide her, the writer slips into a discussion of only the immediate impact of the Depression and never discusses its causes or long-term effects. Although the body paragraphs do provide the names of many agencies created by the Roosevelt administration, they do not explain the purpose of most of them. Consequently, the student seems to consider the formation of the agencies, not their contributions, to be the Depression's most significant result.

Because the student took a time-consuming detour, she had to list points at the end of the essay without discussing them fully; moreover, she was left with no time to sum up her main points, even in a one-sentence conclusion. Although it is better to include undeveloped information than to skip it altogether, an undeveloped list has shortcomings. Many instructors will not give credit if you do not write out your answer in full. More important, you cannot effectively show logical or causal relationships in a list.

46d   Writing Paragraph-Length Essay Examination Answers

Some essay questions ask for a paragraph-length answer, not a full essay. A paragraph should be just that: not one or two sentences, not a list of points, not more than one paragraph.

See ◀ 4c A paragraph-length answer should be **unified** by a clear topic sentence. Just as an essay answer begins with a thesis statement, a paragraph answer opens with a topic sentence that summarizes what the paragraph will cover. You should generally word this sentence to echo the examination question. The paragraph should See ◀ 4d also be **coherent**—that is, its statements should be linked by transitions that move the reader along. And the paragraph should be See ◀ 4f **well developed**, with enough relevant detail to convince your reader you know what you are talking about.

A typical question on a business examination, reproduced here, asks for a paragraph-length response.

> **QUESTION:** In one paragraph, define the term *management by objectives,* give an example of how it works, and briefly discuss an advantage of this approach.

EFFECTIVE PARAGRAPH-LENGTH EXAMINATION ANSWER

Definition

Example

Advantage

As defined by Horngren, management by objectives is an approach by which a manager and his or her superior together formulate goals, and plans by which they can achieve these goals, for a forthcoming period. For example, a manager and a superior can formulate a responsibility accounting budget, and the manager's performance can then be measured according to how well he or she meets the objectives defined by the budget. The advantage of this approach is that the goals set are attainable because they are not formulated in a vacuum. Rather, the objectives are based on what the entire team reasonably expects to accomplish. As a result, the burden of responsibility is shifted from the superior to the team: The goal itself defines all the steps needed for its completion.

In this answer, key phrases ("As *defined* by. . . ."; "For *example*. . . ."; "The *advantage* of this approach. . . .") clearly identify the various parts of the question being addressed. The writer volunteers no more than the question asks for, and his use of the wording of the question helps make the paragraph orderly, coherent, and emphatic.

The student who wrote the following response may know what *management by objectives* is, but his paragraph sounds more like a casual explanation to a friend than an answer to an examination question.

INEFFECTIVE PARAGRAPH-LENGTH EXAMINATION ANSWER

Sketchy, casual definition

No example given

Vague

Management by objectives is when managers and their bosses get together to formulate their goals. This is a good system of management because it cuts down on hard feelings between managers and their superiors. Since they set the goals together, they can make sure they're attainable by considering all possible influences, constraints, etc., that might occur. This way neither the manager nor the superior gets all the blame when things go wrong.

Just as with an essay-length answer, a paragraph answer will not be effective unless you take the time to read the question carefully, plan your response, and outline your answer before you begin to write. It is always a good idea to use the wording of the question in your answer and to reread your answer to make sure it explicitly addresses the question.

Writing about Literature
When student essays about literature go astray, they tend to go in opposite directions. Some students present little more than personal reactions to the work. Others try too hard for a completely objective analysis. Ask students to read and discuss the quotations in the margin on page 799. These writers speak of "talking back" to books, making art a "personal matter." They focus on the need for readers and critics to interact with the text, to take a stand. Discuss the following proposition: Good critics ultimately go beyond their personal response to a work—but that "gut level" response is where they *begin* the critical process.

CHAPTER 47

Writing about Literature

47a Approaching Literature

Writing about literature is different from other kinds of writing. When writers create works of imaginative literature, they work within certain **genres**: short stories, novels, plays, poems, and the like. Each of these types of literature has its own special characteristics. If you recognize these special forms and understand their features, literary works will be more accessible to you.

A writer of imaginative literature may create a mood by experimenting with language and form. In experimental modern fiction, for example, a short story can consist entirely of an alumni magazine's class notes or a short series of diary-like entries moving backward in time to the narrator's birth; it can even consist of just a single paragraph. A poem may sound like prose, and may be just a few lines—or just a few words—long. It may be written entirely in lowercase letters, or its lines may be arranged in the shape of an animal. A play may have only a single character, or it may have a narrator who speaks directly to the audience. A novel may switch narrators with each section or chapter, and it can skip from one time period to another. In other words, literature has the power to surprise readers by doing the unexpected: breaking the rules.

47b Reading Literature

When you read a literary work about which you plan to write, use the same critical thinking skills and active reading strategies you apply to other works you read: **Preview** the work, and

See ◀
5a

highlight it to identify key ideas and cues to meaning; then, **annotate** it carefully.

▶ See 5c

As you read and take notes, focus on the special concerns of literary analysis, considering elements like a short story's plot, a poem's rhyme or meter, or a play's characters. Look for *patterns*, related groups of words, images, or ideas that run through a work. Look for *anomalies*, unusual forms, unique uses of language, unexpected actions by characters, or unusual treatments of topics. And look for *connections*, links with other literary works, with historical events, or with biographical information.

When you have finished your reading and annotating, **brainstorm** to discover material to write about; then, organize your material. As you arrange related material into categories, a structure for your paper should emerge.

▶ See 1c2

These strategies, tailored to the special demands of writing about literature, help you detect relationships among ideas, uncover links to other works, and find a central focus and shape for your essay.

When you read a work of literature, keep in mind that you do not read to magically discover the one correct meaning the writer had in mind. The "meaning" of a literary work is created by the interaction between a text and its readers. Do not assume, however, that a work can mean whatever you want it to mean; ultimately, your interpretation must be consistent with the stylistic signals, thematic suggestions, or patterns of imagery in the text. These elements may be subject to interpretation, but they cannot be ignored. Ideally, your reading should balance what you know or can learn about the special qualities of literature with your unique reactions to it.

47c Writing about Literature

Just as literature is different from other kinds of writing, the writing you do *about* literature is different. When you write about literature, you respond to the possibilities created by a work's form, content, and style. As you write, you observe the conventions of literary criticism, which has its own vocabulary and formats. You also respond to certain discipline-specific assignments. For instance, you may be asked to **analyze** a work, to take it apart and consider one or more of its elements—perhaps the plot or characters in a story or the use of language in a poem. Or, you

---

QUOTATION

Writing in the *New York Times Book Review*, editor/critic Anatole Broyard quotes a friend who says the following.

"I used to have too much respect for books, or the wrong kind of respect. . . . I made sure my hands were clean before taking up a book, as if literature had to be kept sterile . . . [Now] I can talk back to the book. I remind myself that I bought and paid for the book—it's mine to do as I please with. . . . As soon as I open it, I *occupy* the book, I stomp around in it. I underline passages, scribble in the margins, leave my mark. . . .

"I read the book the way I want to read it—I gawk, stare, dawdle, digress, retrace my steps. If the author says something I don't like or can't believe, I argue with him. . . . A good book has a voice and a good reading has a voice too. I like to be able to *hear* myself responding to a book, answering it, agreeing and disagreeing in a manner I recognize as peculiarly my own. . . .

"You can't remain an innocent bystander."

QUOTATION

"It is the capacity for making good or bad art a *personal matter* that makes . . . a critic."

–George Bernard Shaw

Although students understand (generally) what it means to research a general topic, they often seem puzzled about how to research a novel, story, poem, or play. A special library tour in which the librarian shows students helpful reference books, indexes, and journals is obviously in order. However, it seems that some meaningful activity might make the process clearer. In a fiction class, for example, students might each choose one story from a text and find one article about the story. Once the student has found the article he or she might either make an oral report to the class about the article or write a summary of the article to turn in (attaching a photocopy of the article, of course).

Distribute copies of several book reviews from the *New York Times Book Review* or other publications. Ask students to read and discuss the reviews. The reviewer *evaluates* the work—but does he/she also *analyze* and *interpret* it? Does the reviewer try to connect the work to other literary works and traditions? Is the work evaluated in the context of the author's life and previous works? Does the reviewer give readers a sense of his or her personal response to the work?

may be asked to **interpret** a work, to try to discover its possible meanings. Less often, you may be called on to **evaluate** a work, to judge its strengths and weaknesses.

More specifically, you may be asked to trace the critical or popular reception to a work; to compare two works by a single writer (or by two different writers); to consider the relationship between a work of literature and a literary movement or historical period. You may be asked to analyze a character's motives or the relationship between two characters, or to comment on a story's setting or tone. In any case, understanding exactly what you are expected to do will make your writing task easier.

CLOSE-UP

WRITING ABOUT LITERATURE

Some papers that analyze, interpret, or evaluate works of literature use secondary sources—books and articles *about* the works—to support their ideas. For an example of a literary research paper that uses secondary sources, **see 43d**.

When you write about literature, you use all the skills you bring to any writing assignment. Your goal is to make a point and support it with appropriate references to the work under discussion or to related works or secondary sources.

CONVENTIONS OF WRITING ABOUT LITERATURE

- Use present tense verbs when discussing works of literature: "The character of Mrs. Mallard's husband *is* not developed. . . ."
- Use past tense verbs only when discussing historical events ("Owen's poem conveys the destructiveness of World War I, which at the time the poem *was* written *was* considered to be. . . ."), presenting historical or biographical data ("Her first novel, *published* in 1811 when Austen *was* 36, . . ."), or identifying events that occurred prior to the time of the story's main action ("Miss Emily is a recluse; since her father *died* she has lived alone except for a servant").
- Support all points with specific, concrete examples from the work you are discussing: Briefly summarize key

continued on the following page

continued from the previous page

events, quote dialogue or description, describe characters or setting, or paraphrase ideas.

- Combine paraphrase, summary, and quotation with your own interpretations, weaving quotations smoothly into your paper **(see 39a–b)**.

- Be careful to acknowledge all sources, including the work or works under discussion. Introduce the words or ideas of others with a reference to the source and follow borrowed material with appropriate parenthetical documentation. Enclose the words of others in quotation marks.

- In accordance with MLA documentation style, use parenthetical documentation and include a Works Cited list **(see 40b)**.

- When citing a part of a short story or novel, supply the page number (168); for a poem, give the line numbers (2–4); for a classic verse play, include act, scene, and line numbers (1.4.29–31 or I.iv.29–31). For other plays, supply act and/ or scene numbers. When quoting more than four lines of prose or three lines of poetry, follow the guidelines outlined in **31d2–3**.

- Avoid subjective expressions like *I feel, I believe, it seems to me,* and *in my opinion.* These weaken your paper by suggesting that its ideas are "only" your opinion and have no validity in themselves.

- Do not rely on excessive plot summary. Your goal is to draw a conclusion about one or more works and to support that conclusion with pertinent details. If a plot detail supports a point you wish to make, a *brief* summary alluding to a particular event or arrangement of events is acceptable. But plot summary is no substitute for analysis.

- Use literary terms accurately **(see 47f)**. For example, be careful to avoid confusing *narrator* or *speaker* with *author;* feelings or opinions expressed by a narrator or character do not necessarily represent those of the author. You should not say, "In the poem's last stanza, Frost expresses his indecision" when you mean that the poem's *speaker* is indecisive.

- Underline titles of novels and plays; set titles of short stories and poems within quotation marks.

Reviewer's Checklist

At First Glance
Do I like it? Does it appeal to my emotions, to my intellect, or to both? Can I name the emotions I feel about this work? Does it strike me as a professional job, or amateurish? Does it come alive for me in some way? Am I entertained, amused, challenged, startled, or _____ (fill in the blank) by this work?

Thinking It Over
What is the artist/creator trying to do? How well does he or she succeed? Does the style of the work seem right for the subject? How effectively do the parts of the work (plot, characters, mood—or color, shape, technique) fit together? Does anything stick out, seem out of place or jarring? Is the work "too much" or "too little" in some way?

Outside Sources
Where does the work fit into its field (film, painting, drama, etc.)? How does it connect with earlier works by the same artist/creator? With other current works? To earlier works with similar themes? What can comparing/contrasting this work with other works tell me about it?

Evaluation
What's the best I can say about this work? Within its field, is this a major or minor effort? Is it successful in what it tries to do? Will it last; does it have long-standing value, or a "today only" appeal? Is it likely to appeal to most readers/viewers or only to a limited group?

–"Reviewer's Checklist"
from *Writing!* Magazine

47d   **Writing about Fiction**

When you write a **literary analysis** of a work of fiction, you follow the same process you use when you write any paper about literature. However, you concentrate on elements—such as plot, character, setting, and point of view—characteristic of works of fiction.

WRITING ABOUT FICTION

Carla Watts, a student in an introductory literature course, was asked to select a short story from a list supplied by her instructor and to write an essay about it. The following story, written in 1983 by Gary Gildner, is the one she decided to write about.

Sleepy Time Gal

In the small town in northern Michigan where my father lived as a young man, he had an Italian friend who worked in a restaurant. I will call his friend Phil. Phil's job in the restaurant was as ordinary as you can imagine—from making coffee in the morning to sweeping up at night. But what was not ordinary about Phil was his piano playing. On Saturday nights my father and Phil and their girlfriends would drive ten or fifteen miles to a roadhouse by a lake where they would drink beer from schooners and dance and Phil would play an old beat-up piano. He could play any song you named, my father said, but the song everyone waited for was the one he wrote, which he would always play at the end before they left to go back to the town. And everyone knew of course that he had written the song for his girl, who was as pretty as she was rich. Her father was the banker in their town, and he was a tough old German, and he didn't like Phil going around with his daughter.

My father, when he told the story, which was not often, would tell it in an offhand way and emphasize the Depression and not having much, instead of the important parts. I will try to tell it the way he did, if I can.

So they would go to the roadhouse by the lake, and finally Phil would play his song, and everyone would say, Phil, that's a great song, you could make a lot of money from it. But Phil would only shake his head and smile and look at his girl. I have to break in here and say that my father, a gentle but practical man, was not inclined to emphasize the part about Phil looking at his girl.

It was my mother who said the girl would rest her head on Phil's shoulder while he played, and that he got the idea for the song from the pretty way she looked when she got sleepy. My mother was not part of the story, but she had heard it when she and my father were younger and therefore had that information. I would like to intrude further and add something about Phil writing the song, maybe show him whistling the tune and going over the words slowly and carefully to get the best ones, while peeling onions or potatoes in the restaurant; but my father is already driving them home from the roadhouse, and saying how patched up his tires were, and how his car's engine was a gingerbread of parts from different makes, and some parts were his own invention as well. And my mother is saying that the old German had made his daughter promise not to get involved with any man until after college, and they couldn't be late. Also my mother likes the sad parts and is eager to get to their last night before the girl goes away to college.

So they all went out to the roadhouse, and it was sad. The women got tears in their eyes when Phil played her song, my mother said. My father said that Phil spent his week's pay on a new shirt and tie, the first tie he ever owned, and people kidded him. Somebody piped up and said, Phil, you ought to take that song down to Bay City—which was like saying New York City to them, only more realistic—and sell it and take the money and go to college too. Which was not meant to be cruel, but that was the result because Phil had never even got to high school. But you can see people were trying to cheer him up, my mother said.

Well, she'd come home for Thanksgiving and Christmas and Easter and they'd all sneak out to the roadhouse and drink beer from schooners and dance and everything would be like always. And of course there were the summers. And everyone knew Phil and the girl would get married after she made good her promise to her father because you could see it in their eyes when he sat at the old beat-up piano and played her song.

That last part about their eyes was not, of course, in my father's telling, but I couldn't help putting it in there even though I know it is making some of you impatient. Remember that this happened many years ago in the woods by a lake in northern Michigan, before television. I wish I could put more in, especially about the song and how it felt to Phil to sing it and how the girl felt when hearing it and knowing it was hers, but I've already intruded too much in a simple story that isn't even mine.

Well, here's the kicker part. Probably by now many of you have guessed that one vacation near the end she doesn't come home to see Phil, because she meets some guy at college who is good-looking and as rich as she is and, because her father knew about Phil all along and was pressuring her into forgetting about him, she gives in to this new guy and goes to his hometown during the vacation and falls in love with him. That's how the people in town figured it, because after she graduates they turn up, already married, and right away he takes over the old German's bank—and buys a new Pontiac at the place where my father is the mechanic and pays cash for it. The paying cash always made my father pause and shake his head and mention again that times were tough, but here comes this guy in a spiffy white shirt (with French cuffs, my mother said) and pays the full price in cash.

And this made my father shake his head too: Phil took the song down to Bay City and sold it for twenty-five dollars, the only money he ever got for it. It was the same song we'd just heard on the radio and which reminded my father of the story I just told you. What happened to Phil? Well, he stayed in Bay City and got a job managing a movie theater. My father saw him there after the Depression when he was on his way to Detroit to work for Ford. He stopped and Phil gave him a box of popcorn. The song he wrote for the girl has sold many millions of records, and if I told you the name of it you could probably sing it, or at least whistle the tune. I wonder what the girl thinks when she hears it. Oh yes, my father met Phil's wife too. She worked in the movie theater with him, selling tickets and cleaning the carpet after the show with one of those sweepers you push. She was also big and loud and nothing like the other one, my mother said.

Carla began by reading the story through quickly. Then she reread it more carefully, highlighting and annotating as she read. A portion of the highlighted and annotated story appears below.

When do events take place?

In the small town in northern Michigan where my father lived as a young man, he had an Italian friend who worked in a restaurant. I will call his friend Phil. Phil's job in the restaurant was as ordinary as you can imagine--from making coffee in the morning to sweeping up at night. But what was not ordinary about Phil *[handwritten left margin:]* Sat. nights = special — dancing, beer, etc.

was his piano playing. On Saurday nights my father and Phil and the girlfriends would drive ten or fifteen miles to a roadhouse by a lake where they would drink beer from schooners and dance and Phil would play a old beat-up piano. He could play any song you named, my father said, but the song everyone waited for was the one he wrote, which he would always play at the end before they left to go back to the town. And everyone knew of course that he had written the song for his girl,

[handwritten left margin:] fairy tale style

who was as pretty as she was rich. Her father was the banker in their town, and he was a tough old German, and he didn't like Phil going around with his daughter.

My father, when he told the story, which was not often, would tell it in an offhand way and emphasize the Depression and not having much, instead of the important parts. I will try to tell it the way he did, if I can.

Carla's next task was to brainstorm to find ideas. As she searched for a topic for her essay, she found it helpful to brainstorm separately on plot, character, setting, point of view, tone and style, and theme, to see which suggested the most promising possibilities.

Brainstorming List

[handwritten:]

Plot

Flashback — narrator remembers story father told.

Story: Phil loved rich banker's daughter, wrote song for her, girl married someone else, Phil sold song for $25.00, married another woman.

Ordinary, predictable story of "star-crossed lovers" from different backgrounds ("Probably by now many of you have guessed. ..."), but what actually happened isn't important.

Character

Phil — Italian, never went to high school, ordinary job in restaurant, extraordinary piano player.
Girl — no name, pretty, rich, educated
Narrator —?
Mother — romantic
Father — mechanic; gentle, practical

Setting

"Small town in northern Michigan"
Past — when narrator's father was a young man
In woods — near lake
Roadhouse — dancing, drinking, beat-up piano

Point of View

Narrator tells story to reader but there's a story inside the story.
Father tells his story, mother qualifies his version (she's "not part of the story" but has heard it), narrator tells how they told it.
Point of view keeps shifting — characters compete to tell the story ("I would like to intrude further...").
Father's version: stresses Depression, hard times
Mother's version: stresses relationship, "sad parts"
Readers encouraged to find own point of view; narrator of story addresses readers.
Three characters invent and reinvent and embellish story each time they tell it.

Tone and Style

Conversational style — narrator talks to reader ("Well, here's the kicker part.")

Like a fairy tale (girl = "as pretty as she
   was rich"; father = "a gentle but
   practical man")
Casual speech: contractions; "well,"
   "some guy," etc.

Theme
Which is "real" story?
   Subject of Phil's story = missed chances,
   failure.
   Subject of narrator's story = the past?
   Values of different characters? Conflict
   between real events and memory?

When Carla looked over her brainstorming list, she saw at once
that character and point of view suggested the most interesting
possibilities for her paper. Still, she found herself unwilling to
start drafting her essay until she could find out more about the
story's title, which she thought must be significant. She asked
around until she found someone who told her that the title was
the name of an actual song—and supplied the lyrics. She recorded
her reactions to this information in a journal entry.

Journal Entry

"Sleepy Time Gal" = name of song
Mother says Phil got inspiration for song
   from the way his girl looked when she got
   sleepy. ** Does title of story refer to
   girl or to song? **
Song = fantasy about the perfect married life
   that should follow the evenings of dancing:
   in a "cottage for two" wife will be happy
   cooking and sewing for her husband and
   will end her evenings early. She'll be
   happy to forget about dancing and be a
   stay-at-home wife.
Maybe lyrics describe what Phil wants
   and never gets?

At this point Carla decided to arrange some of the most useful material from her brainstorming list, journal entry, and annotations into categories. She gave these categories headings that corresponded to the three versions of Phil's story presented in "Sleepy Time Gal," and she added related supporting details as they occurred to her.

Three Versions of Phil's Story

Mother's Version

("Likes the sad parts") and the details of the romance: the way the father made the daughter promise not to get involved with a man until she finished college, the way the women got tears in their eyes when Phil played his song.

Remembers girl's husband had French cuffs.

Remembers Phil's wife = "big and loud"

Notes people were trying to cheer Phil up

Remembers girl resting head on Phil's shoulder, and how he got idea for song.

Father's Version

Depression/money: mentions Phil's patched tires and engine, how he spent a week's pay on new clothes, how girl's husband pays cash for a new Pontiac

("Times were tough")

Narrator's Version

Facts of story — but wants to add more about Phil's process of writing song (because he, like Phil, = artist?), more about romance ("you could see it in their eyes"). Wants to embellish story.

("I wish I could put more in . . . .")

Carla's notes and lists eventually suggested the following thesis for her paper: "'Sleepy Time Gal' is a story that is not about the 'gal' of the title or about the man the narrator calls Phil but about the different viewpoints of its three narrators." Guided by this tentative thesis, she went on to write and revise her paper, following the process detailed in Chapter 3. The final draft of Carla's paper begins on the following page. Annotations have been added to identify the conventions that apply to writing essays about works of fiction.

Carla Watts   Watts 1

Professor Sierra

English 1001

12 March 1994

Paper title
centered
                                     Whose Story?

Title in quota-
tion marks
        Midway through Gary Gildner's short story "Sleepy

Time Gal" the narrator acknowledges, "I've already in-

Quotation in-
troduced
truded too much in a simple story that isn't even mine"

Parenthetical
documenta-
tion identifies
page on which
quotation ap-
peared
(215). But whose story is "Sleepy Time Gal"? It is pre-

sented as the tale of Phil, an ordinary young man of

modest means who falls in love with a rich young

woman, writes a song for her, and loses both the woman

and the song, as well as the fame and fortune the song

could have brought him, apparently because he is un-

willing to fight for either. But actually, "Sleepy Time

Thesis
Gal" is not Phil's story, and it is not the story of the girl

he loves; the story belongs to the three characters who

compete to tell it.

Brief plot
summary
combined
with interpre-
tation
        The story these characters tell is a simple one; it is

also familiar. Phil is a young man with an ordinary job.

He has little education and no real prospects of doing any-

thing beyond working in a restaurant doing menial jobs.

He is in love with a girl whose father is a rich banker, a

girl who goes to college. Phil has no more chance of mar-

rying the girl than he has of becoming educated or becom-

ing a millionaire. He has written a song for her, but he is

doomed to sell the rights to it for twenty-five dollars. Phil

may be a man with dreams and expectations that go be-

yond the small Michigan town and the roadhouse, but he

does not seem to be willing to struggle to make his dreams

come true. Ironically, he never achieves the happy

Watts 2

married life his song describes; his dreams remain just dreams, and he settles for life in the dream world of a movie theater.

Father's perspective

Past tense used to identify events that occurred before story's main action

The character who seems to be the author of Phil's story is the narrator's father: He is the only one who knew Phil and witnessed the story's events, and he has told it again and again to his family. But the story he tells reveals more than just what happened to Phil; it says a lot about his own life, too. The father is a mechanic who eventually leaves his small Michigan town for Detroit. As the narrator observes, he is "a gentle but practical man" (214). We can assume he has seen some hard times; he sees Phil's story only in the context of the times, and "times were tough" (216). The narrator says, "My father, when he told the story, . . . would tell it in an offhand way and emphasize the Depression and not having much, instead of the important parts" (214). In the father's version, seemingly minor details are important: Phil's often-mended car engine, "a gingerbread of parts from different makes" (215), and incidents like how Phil spent a week's pay on a new shirt and tie, "the first tie he ever owned" (215), and how the girl's husband paid cash for a new Pontiac. These details are important to the father because they have to do with money. He sees Phil's story as more about a particular time (the Depression era) than about particular people. Whenever he hears Phil's song on the radio, he remembers that time.

Ellipses indicate words omitted from quotation

Mother's perspective

The narrator's mother, however, sees Phil's story as a romantic, timeless story of hopelessly doomed lovers. She did not witness the story's events, but she has

Point supported by specific references to story

heard the story often. According to the narrator, she "likes the sad parts and is eager to get to their last night before the girl goes away to college" (215). She remembers how the women in the roadhouse got tears in their eyes when Phil played the song he wrote. The mother's selective memory helps to characterize her as somewhat romantic and sentimental, interested in people and their relationships (the way the girl's father made her promise to avoid romantic entanglements until after college; the way Phil's friends tried to cheer him up) and in visual details (the way the girl rested her head on Phil's shoulder; the French cuffs on her husband's shirt). In the interaction between the characters she sees drama and even tragedy. The sentimental story of lost love appeals to her just as the story of lost opportunity appeals to the father.

Narrator's perspective

The narrator knows the story only through his father's telling and retelling of it, and he says, "I will try to tell it the way he did, if I can" (214). But this is impossible: He embellishes the story, and he makes it his own. He is the one who communicates the story to readers, and he ultimately decides what to include and what to leave out. His story reflects both his parents' points of view: the focus on both characters and events, on romance and history. In telling Phil's story, he tells the story of a time, recreating a Depression-era struggle of a man who could have made it big but wound up a failure; however, he also recounts a story about people, a romantic, sentimentalized story of lost love. And, he tells a story about his own parents.

Watts 4

The narrator, like Phil, is creative; he needs to convey the facts of the story, but he must struggle to resist the temptation to add to them: to add more about how Phil went about writing the song, "maybe show him whistling the tune and going over the words slowly and carefully to get the best ones" (214–15), more about the romance itself. The narrator is clearly embellishing the story---for instance, when he says everyone knew Phil and the girl would get married because "you could see it in their eyes" (215), he admits that this detail is not in his father's version of the story---but he is careful to identify his own contributions, explaining, "I couldn't help putting it in there" (215). The narrator cannot help wondering about the parts his father does not tell, and he struggles to avoid rewriting the story to include them. Sometimes he cannot help himself, and he apologizes for his lapses with a phrase like "I have to break in here . . ." (214). But, for the most part, the narrator knows his place, knows it is not really his story to tell: "I wish I could put more in, especially about the song and how it felt to Phil to sing it and how the girl felt when hearing it and knowing it was hers, but I've already intruded too much in a simple story that isn't even mine" (215).

Phil's story is, as the narrator acknowledges, a simple one, almost a cliché. But Gary Gildner's story, "Sleepy Time Gal," is much more complex. In it, three characters create and recreate a story of love and loss, ambition and failure, each contributing the details they feel should be stressed and, in the process, revealing something about themselves and about their own hopes and dreams.

Watts 5

Work Cited

Gildner, Gary. "Sleepy Time Gal." Sudden Fiction:
American Short-Short Stories. Ed. Robert
Shapard and James Thomas. Salt Lake City:
G. M. Smith, 1986. 214–16.

Carla's paper focuses on the story's shifting point of view and the contributions of the three central characters to Phil's story. She supports her thesis with specific references to "Sleepy Time Gal"—in the form of quotation, summary, and paraphrase—and interprets the story's events in light of the points she is making. Her paper does not include every item in her notes, nor should it: She selects only those details that support her thesis.

47e   Writing about Poetry

When you write a paper about poetry, you follow the process you use when you write any paper about literature. However, you concentrate on the elements poets use to create and enrich their work—for example, voice, form, sound, meter, language, and tone.

WRITING ABOUT POETRY

Daniel Johanssen, a student in an introductory literature course, followed this process as he planned an essay about Delmore Schwartz's 1959 poem "The True-Blue American."

Daniel's assignment was a general one; his instructor had asked only that each student choose a poem and react to it. After he chose his poem, Daniel read it several times. Then he read it aloud, paying special attention to the sound of the poem. As he read, he highlighted and annotated the poem, as illustrated here.

The True-Blue American

Jeremiah Dickson was a (true-blue) American,   *= loyal; faithful*
For he <u>was a little boy who understood America</u>, for   *(also = "red, white*
   he felt that he must   *& blue"?)*
Think about *everything*; because that's *all* there is to
   think about,
Knowing immediately the intimacy of truth and
   comedy,

QUOTATION

"Analysis is never in any sense a substitute for the poem. The best any analysis can do is to prepare the reader to enter the poem more perceptively. By isolating for special consideration some of the many simultaneous elements of the poem, analysis makes them more visible in one sense, and less interesting in another. It is up to the reader, once the analysis is completed, to re-read the poem in a way that will restore the simultaneity and therefore the liveliness of the poetic structure. The only reason for taking a poem apart is that it may then be put back together more richly. . . .

"The point is that *the language of experience is not the language of classification.* A boy burning with ambition to become a jockey does not study a text on zoology. . . . He may never learn how many incisors a horse has, nor how many yards of intestines. What does it matter? He is concerned with *feel*, a response-to, a sense of the character and reaction of the living animal. And zoology cannot give him that. Not all the anatomizing of all the world's horses could teach a man horse-sense.

"So for poetry. The concern is not to arrive at a definition and to close the book, but to arrive at an experience. There will never be a complete system for 'understanding' or for 'judging' poetry. Understanding and critical judgment are admirable goals, but neither can take place until the poem has been experienced, and even then there is always some part of every good work of art that can never be fully explained or categorized. It
continued on following page

continued from previous page

still remains true that the reader who has experienced most fully will finally be the best judge."

–John Ciardi,
How Does a Poem Mean?

5 Knowing intuitively how a sense of humor was a
why? necessity
For one and for all who live in America. Thus, natively, and
Naturally when on an April Sunday in an ice cream parlor Jeremiah
Was requested to choose between a chocolate sundae and a banana split
He answered unhesitatingly, having no need to think of it
10 Being a true-blue American, determined to continue as he began:
Rejecting the either-or of Kierkegaard, and any another European; *?*
Refusing to accept alternatives, refusing to believe the choice of between;
Rejecting selection; denying dilemma; electing absolute affirmation: *?*
knowing
15   in his breast
The infinite and the gold
Of the endless frontier, the deathless West.   *rhyme*
"Both: I will have them both!" declared this true-blue
American   *^ambition; refusal to settle for half. Or just greedy?*
In Cambridge, Massachusetts, on an April Sunday, instructed
20   By the great department stores, by the Five-and-Ten,   *What do all these things teach?*
Taught by Christmas, by the circus, by the vulgarity and grandeur of
Niagara Falls and the Grand Canyon,
Tutored by the grandeur, vulgarity, and infinite appetite gratified and
Shining in the darkness of the light
25 On Saturdays at the double bills of the moon pictures, *= movies*
The consummation of the advertisements of the imagination of the light
Which is as it was—the infinite belief in infinite hope—   *Why part of this series? Irony?*
exploration— of Columbus, Barnum, Edison, and Jeremiah
Dickson.   *circus   light*

On first reading, Daniel saw the poem as a patriotic catalog of all the things that made America great; in fact, its seemingly patriotic theme was what had made him select this poem to write about. The closer he looked, however, the more clearly he saw his initial impression as inaccurate. As he studied the poem, he saw how all its parts contributed to one impression: a critical look at American greed and materialism. Even though he did not like what he found, he could not ignore the clues to the poem's central theme; in writing his paper he would have to interpret the poem in light of these clues.

After Daniel read and reread "The True-Blue American," he decided to do some brainstorming in a systematic fashion, considering voice, form, sound, meter, language, tone, and theme one by one. His brainstorming on the first four of these categories was not very productive: He concluded only that the poem's speaker was an anonymous voice, not identified as a particular person; that line length varied quite widely and did not seem to follow a particular pattern; and that the poem did not seem to have a regular rhyme scheme or meter. When he brainstormed about language, tone, and theme, however, he was able to discover some interesting ideas.

Brainstorming List

Language
Repetition: true-blue American (3x + title),
    Jeremiah Dickson (first and last words
    of poem), vulgarity, grandeur, infinite,
    light
Parallelism: "Knowing immediately...
    knowing intuitively"; "Rejecting....
    Refusing to accept...refusing to believe...
    Rejecting...denying...electing...Knowing;"
    "instructed by...taught by...tutored by..."
Imagery: "moon pictures"—glowing in
    the dark

Tone
Speaker's attitude seems angry, bitter,
    disillusioned — why?
What does it mean to be "American"? To
    understand America? Is being "true-blue"
    a positive or negative goal?

Is title ironic?

Theme

Poem's stated subject = true-blue American,
    but what is true-blue American?

Is poem's real subject what it means
    to be an American?

What is meaning of America? Is America
    great, full of possibilities? or corrupted
    by materialism and greed?

What is significance of names of people
    and places? How do they fit in with the
    poem's theme?

After he had finished his brainstorming, Daniel made a more focused list to organize some of the poem's key ideas and allusions.

Key Ideas and Allusions

Places

America: ice cream parlor, West (frontier),
    Cambridge Mass, department stores, Five-
    and-Ten, circus, Niagara Falls, Grand
    Canyon, movie theater

People

Kierkegaard (?), Columbus, Barnum, Edison,
    (+ Jeremiah Dickson, typical average
    American)

Miscellaneous

Chocolate sundae, banana split, Christmas,
    advertisements

Contrasting ideas

Positive —
    Promise of western frontier (infinite/
    endless/deathless)

grandeur
gold
imagination
light
hope
Negative —
    vulgarity
    greed
    appetite
darkness

Daniel's notes suggested many interesting paper topics. For example, he could contrast the poem's superficial patriotism with its actual pessimism or explore the relationship between language and theme. He could pursue the idea of making choices or the contrast between positive and negative images. He could do some research in order to consider the poem in the context of the United States in 1959, when it was written, or to consider how Delmore Schwartz's life or other works, or the works of his contemporaries, might be pertinent. He could examine the specific people and places mentioned in the poem and consider their possible significance. Finally, he could compare the poem to another poem with a similar—or contrasting—theme. Any of these possibilities would be perfectly appropriate. However, Daniel knew that it made sense to focus on a thesis his notes could support, and this eliminated some possible topics—for example, those focusing on the poem's sound or form.

As Daniel proceeded to develop a thesis and an effective arrangement for his ideas, and, eventually, to write and revise his essay about "The True-Blue American," he followed the writing process outlined in Chapters 2 and 3. The final draft of his paper follows.

Daniel Johanssen   Johanssen 1

Professor Stang

English 1001

8 April 1994

Irony in Schwartz's "The True-Blue American"

The poem "The True-Blue American" by Delmore
Schwartz is not as simple and direct as its title suggests. In fact, the title is extremely ironic. At first,
the poem seems patriotic, but actually the flag-waving
strengthens the speaker's criticism. Even though the
poem seems to support and celebrate America, it is

actually a bitter critique of the negative aspects of
American culture.

According to the speaker, the primary problem
with America is that its citizens falsely believe themselves to be authorities on everything. The following
lines introduce the theme of the "know-it-all" Ameri-

can: "For he was a little boy who understood America, for he felt that he must / Think about everything;
because that's all there is to think about" (2–3). This
theme is developed later in a series of parallel
phrases that seem to celebrate the value of immediate intuitive knowledge and a refusal to accept or to
believe anything other than what is American (4–6).

Americans are ambitious and determined, but
these qualities are not seen in the poem as virtues.
According to the speaker, Americans reject sophisticated "European" concepts like doubt and choices and
alternatives and instead insist on "absolute affirmation" (13)--simple solutions to complex problems. This
unwillingness to compromise translates into stubborn-

Johanssen 2

ness and materialistic greed. This tendency is illustrated by the boy's asking for <u>both</u> a chocolate sundae
<u>and</u> a banana split at the ice cream parlor---not "either-
or" (11). Americans are characterized as pioneers who
want it all, who will stop at nothing to achieve "The infinite and the gold / Of the endless frontier, the deathless West (16–17). For the speaker, the pioneers and
their "endless frontier" are not noble or self-sacrificing;
they are like a greedy little boy at an ice cream parlor.

According to the speaker, the greed and materialism of America began as grandeur but ultimately
became mere vulgarity. Similarly, the "true-blue
American" is not born a vulgar parody of grandeur;
he learns it from his true-blue fellows:

> instructed
> By the great department stores,
> by the Five-and-Ten,
> Taught by Christmas, by the circus, by
> the vulgarity and grandeur of
> Niagara Falls and the Grand Canyon
> Tutored by the grandeur, vulgarity, and
> infinite appetite gratified. . . . (19–23)

Long verse quotation (over 3 lines) set off from text. Quotation is indented 10 spaces from left margin; no quotation marks are used.

Among the "tutors" the speaker lists are American institutions such as department stores and
national monuments. Within these institutions,
grandeur and vulgarity coexist; in a sense, they are
one and the same.

The speaker's negativity climaxes in the phrase
"Shining in the darkness, of the light" (24). This
paradoxical statement suggests the negative truths

hidden beneath America's glamorous surface. All the grand and illustrious things of which Americans are so proud are personified by Jeremiah Dickson, the spoiled brat in the ice cream parlor.

Conclusion reinforces thesis

Like America, Jeremiah has unlimited potential. He has native intuition, curiosity, courage, and a pioneer spirit. Unfortunately, however, both America and Jeremiah Dickson are limited by their willingness to be led by others, by their greed and impatience, and by their preference for quick, easy, unambiguous answers rather than careful philosophical analysis. Regardless of his---and America's---potential, Jeremiah Dickson is doomed to be hypnotized and seduced by glittering superficialities, light without substance, and to settle for the "double bill of the moon pictures" (25) rather than the enduring truths of a philosopher like Kierkegaard.

Work Cited

Schwartz, Delmore. "The True-Blue American."
   Summer Knowledge: New and Selected Poems
   1938–1958. Garden City: Doubleday, 1959. 163.

TEACHING STRATEGY

Literary Terms

Help students fix literary terms in their minds by assigning them to find "something interesting" written about some of the terms contained in the glossary beginning on page G1. The "something" could be an example, a literary anecdote, a quotation, or a passage from a book about writing (John Gardner's *The Art of Fiction* contains some memorable and useful passages, for example). Ask students to read some of their finds to the class. You might begin with the following reflection on *metaphor*.

> Metaphors are the icing on the pound cake of ordinary prose. Metaphors also are a touch of brandy in the sauce of style. A literary cook can get too heavy with the icing, or too heavy with the booze. Metaphors can be overdone, as I am overdoing this one. . . .
>
> What are the rules for using the tool of metaphor? The first rule is to keep metaphors short. Metaphors are rope-slung bridges over Andean gorges: They are fragile affairs, incapable of bearing great weight. . . . The second [rule] is to keep our metaphors internally consistent. If we fail to sustain a metaphor, we get into what has been termed a mixaphore. . . . Former [California] Governor Jerry Brown once was quoted in the *San Francisco Chronicle* with a lively criticism of President Carter: "The president has been riding the waves of his current saber-rattling, but his balloon is about to be deflated. . . ."

continued on following page

REVISION CHECKLIST: WRITING ABOUT LITERATURE

- Reconsider your topic. Is it specific enough? Do you focus on the concerns of your assignment?
- Does your introduction provide readers with the background or context they need to understand the discussion to follow?
- Do you clearly state your thesis? Does your thesis reflect your purpose? Does it clearly identify the aspects of the work you will discuss?
- How effective is the organization of your essay? Would your discussion be more effective if you arranged your points in a different order?
- Do you support your interpretations and judgments about the work? Would more support strengthen your case? Do you include too many points? Do you include a wide enough range of examples from the work and from your research? Do your examples actually support your points?
- Have you supplied the transitional words and phrases you need to reinforce the logical and sequential connections among your ideas?
- Have you included the plot details and definitions of literary terms that your readers need to understand your discussion?
- Does your conclusion reinforce your main points? Would a different strategy improve your ending?
- Have you placed all words that are not your own in quotation marks? Have you quoted accurately?
- Are parenthetical references to page and line numbers accurate?

47f Using Literary Terms

When you write about literature, you use a vocabulary appropriate to the discipline. The following glossary defines some of the terms you will use.

alliteration   repetition of initial sounds in a series of words, as in "dark, damp dungeon."

allusion   an unacknowledged reference to a historical event, work of literature, biblical passage, or the like that the author expects readers to recognize.

antagonist   the character who is in conflict with or in opposition to the *protagonist*. Sometimes the antagonist is a force or situation, such as war or poverty.

assonance   repetition of vowel sounds in a series of words, as in "f<u>i</u>ne sl<u>i</u>de on the <u>i</u>ce."

blank verse   lines of unrhymed iambic pentameter in no particular stanzaic form; approximates the rhythms of ordinary English speech.

character   the fictional representation of a person. Characters may be *round* (well developed) or *flat* (undeveloped stereotypes), *dynamic* (changing and growing during the course of the story), or *static* (remaining essentially unchanged by the story's events).

climax   the point of greatest tension or importance in a play or story; the point at which the story's decisive action takes place.

closed form   a kind of poetic structure characterized by a consistent pattern of rhyme, meter, or stanzaic form.

conflict   the opposition between two or more characters, between a character and a natural force, or between contrasting tendencies or motives or ideas within one character.

consonance   repetition of consonant sounds in a series of words, as in "the g<u>n</u>arled fi<u>n</u>gers of his <u>n</u>ervous ha<u>n</u>ds."

denouement   the point in the plot of a work of fiction or drama at which the action comes to an end and loose ends are tied up.

end-stopped line   a line of poetry that ends with a full stop, usually at the end of a sentence.

enjambment   a line of poetry ending with no punctuation or natural pause so that it runs over into the next line.

exposition   the initial stage of the plot of a work of fiction or drama, in which the author presents basic information readers need to understand the story's charcters and events.

figurative language   language whose meaning is not to be taken literally. The most commonly used figures of speech are *hyperbole, metaphor, personification, simile,* and *understatement.*

free verse   poetry that does not follow a fixed meter or rhyme scheme.

hyperbole   a figure of speech characterized by intentional overstatement or exaggeration.

continued from previous page

The third rule for metaphors is to make them like breakfast biscuits, fresh every morning. Forget that steady hand at the helm of the ship of state. If we can find nothing fresher than that, let us write, 'The president seemed in firm control of his budget bill,' and let it go at that."

–James J. Kilpatrick,
The Writer's Art

imagery   use of sensory description (description that relies on sight, sound, smell, taste, or touch) to make what is being described more vivid. A *pattern of imagery* combines a group of related images in order to create a single effect.

irony   language that suggests a discrepancy or incongruity between what is said and what is meant *(verbal irony)*, between what actually happens and what we expected to happen *(situational irony)*, or between what a character knows or believes and what the reader knows *(dramatic irony; also called tragic irony)*.

lyric poetry   poetry that expresses a speaker's mood or feelings. Lyric poems are usually short.

metaphor   a comparison that equates two things that are essentially unlike. Unlike a *simile*, a metaphor does not use *like* or *as*.

meter   the pattern of stressed and unstressed syllables in a line of poetry; each repeated unit of meter is called a *foot*. An *anapest* has three syllables, the first two unstressed and the third stressed; a *dactyl* has three syllables, the first stressed and subsequent ones unstressed; an *iamb* has two syllables, of which the second is stressed; a *spondee* has two syllables, both stressed; and a *trochee* has two syllables, the first stressed and the second unstressed. A poem's meter is described by the kind of foot (iamb, dactyl, and so on) and the number of feet in each line (one foot per line = monometer, two feet per line = dimeter, three feet = trimeter, four = tetrameter, five = pentameter, and so on). Thus, a poetic line containing five feet, each of which contains an unstressed syllable followed by a stressed syllable, would be described as *iambic pentameter.*

monologue   an extended speech by one character.

narration   the recounting of events in a work of fiction. When an event that has already occurred is recounted in a later sequence of events, it is called a *flashback*; when something that will occur later in a narration is suggested earlier, the suggestion is called a *foreshadowing.*

open form   a kind of poetic structure not characterized by any consistent pattern of rhyme, meter, or stanzaic form.

paradox   a seemingly contradictory statement.

persona   the narrator or speaker of a story or poem; the persona's attitudes and opinions are not necessarily those of the author.

personification   the assigning of human qualities to nonhuman things.

832

plot   the arrangement of events in a work of literature.

point of view   the perspective from which a story is told. A story may have a *first-person narrator*, who may be a major or minor character in the story. Alternatively, a story may have a *third-person narrator*, who is not a character in the story. Such a narrator may be an *omniscient narrator*, who knows the thoughts and motives of all the story's characters, or a *limited omniscient narrator*, who sees into the minds of only some of the characters. A narrator who cannot be trusted—because he or she is naive, evil, stupid, or self-serving—is called an *unreliable narrator*. The objective perspective that presents only information an audience would get from watching the action unfold on stage is called the *dramatic* point of view.

protagonist   the principal character of a work of drama or fiction.

rhyme   the repetition of the last stressed vowel sound and all subsequent sounds. *End rhyme* occurs at the ends of poetic lines; *internal rhyme* is the rhyming of words within a line of poetry.

rhythm   the regular repetition of stresses and pauses.

setting   the background against which the action of a work of literature takes place: the historical period, locale, season, time of day, etc.

simile   a comparison that equates two essentially unlike things using the word *like* or *as*.

soliloquy   a convention of drama in which a character speaks directly to the audience, revealing thoughts and feelings that the play's other characters, even if they are present on the stage, are assumed not to hear.

stanza   a group of lines in a poem, separated from others by a blank space on the page, which forms a unit of thought, mood, or meter. Common stanzaic forms include the *couplet* (two lines), *tercet* (three lines), *quatrain* (four lines), *sestet* (six lines), and *octave* (eight lines).

stock character   a character who behaves consistently and predictably and who is instantly recognizable and familiar to the audience.

symbol   an image whose meaning transcends its literal or denotative sense in a complex way. Its multiple associations give it significance beyond what it could carry on its own.

theme   an idea expressed by a work of literature.

tone   the attitude of the speaker toward a work's subject, characters, or audience, conveyed by the work's word choice and arrangement of words.

understatement   intentional downplaying of a situation's significance, often for ironic effect.

CHAPTER 48

Practical Writing

48a Composing Business Letters

As a student you will probably have occasion to write letters—for example, to request information for a research paper, to appeal a decision or policy, to complain about a product or service, or to apply for employment.

(1) Planning, writing, and revising your letter

Before you write a business letter, you should think carefully about your purpose, audience, and tone. Many large organizations receive hundreds of letters each day, so your letter should be brief and to the point, with important information placed early in the letter. Be concise, avoid digressions, and try to sound as natural as possible. Stilted or flowery language gets in the way of clear communication, and so does legalistic terminology *(The person named* supra [above]; *in re: your letter)* and business jargon *(in regards to, herewith enclosed).*

▶ **See 1b**

The first paragraph of your letter should introduce your subject and mention any pertinent previous correspondence. The rest of your letter should present the facts readers will need to understand your points. If the ideas you are communicating are complicated, you may want to present points in a numbered list. Your conclusion should reinforce your message, and the entire letter should communicate your good will.

After you revise and edit your letter, proofread it carefully before you type or print the final copy. Type your business letter on good quality 8 1/2" × 11" paper. Leave wide margins, at least an inch all around, and center between the top and bottom of the page. Type your letter single-spaced and use a conventional format. One of the most common, the block format, is illustrated on page 838. (The semiblock format appears on page 840, and the indented format on page 842.)

Remember, the appearance of your letter affects your reader's response to it. A neatly typed letter, free of smudges and errors, makes a favorable first impression on your readers. A sloppy letter or one with misspellings or corrections made by hand presents you and your case badly.

(2) Understanding the conventions of a business letter

The format of a business letter may seem arbitrary and prescriptive, but remember that this format has evolved in response to the special needs of the business audience. An inside address, for example, seems pointless until you consider that business letters often circulate to people other than the recipient, and for these readers knowing the identity of the original recipient may be important. Dates are also necessary because letters frequently become part of a permanent record, filed for some future use.

If you conform to the following conventions, your readers will know where to find each piece of information in your letter.

The Heading The **heading** of a business letter consists of the sender's return address (but not his or her name) and the date on which the letter is written. If you use letterhead stationery, supply only the date, typing it at least two spaces below the letterhead. Each line of the heading falls under the one above it.

Spell out words like *Street, Avenue, Road, Place, East,* and *West.* You may, however, use postal abbreviations to abbreviate the names of states.

CLOSE-UP **COMPOSING BUSINESS LETTERS**

Be sure to punctuate the heading correctly. Commas separate the city from the state and the day from the year. However, no punctuation is used before the zip code, at the ends of lines, or in postal abbreviations for states.

The Inside Address The **inside address** identifies the *recipient's* name and address. It begins at the left margin, four to six lines below the heading (depending on the need to use space to create a balanced page). Include an appropriate title (Mr., Ms.,

Dr.) with the recipient's name and the recipient's full address. Previous correspondence should be your guide to your recipient's correct name and title.

The Salutation　The **salutation** ("Dear _____") appears two spaces below the inside address at the left margin. It consists of the person's title followed by the last name as it appears in the inside address. In business letters the salutation is almost always followed by a colon, not a comma.

Even if you are on a first-name basis with someone, use his or her full name and title in the inside address. (In this case, however, you may use the first name, followed by a comma, in the salutation). If you are writing to someone whose name you do not know—the Director of Personnel, for example—you can avoid the awkward phrase *To Whom It May Concern* by routing your letter to a specific department or by referring to a particular subject.

```
Marketing Department
Harcourt Brace College Publishers
301 Commerce Street
Fort Worth, TX 76102

Attention: Director of Personnel

        or

Subject: Sales Position
```

CLOSE-UP

COMPOSING BUSINESS LETTERS

If you are writing to a woman, consult previous correspondence and use the title she uses. If you do not know her preference, use *Ms.* If you know someone's initials but do not know whether the person is a man or a woman, you can call the company switchboard and ask. You can also use a neutral form of address—*Dear Editor* or *Dear Supervisor,* for example.

Keep in mind that salutations such as *Gentlemen* and *Dear Sirs* should be used only when you are certain your audience is male. As general forms of address these salutations should be avoided.

Sample Letter: Block Format

Heading   6732 Wyncote Avenue
Houston, TX 77004
May 3, 1994

Inside
address   Mr. William S. Price, Jr., Director
Division of Archives and History
Department of Cultural Resources
109 East Jones Street
Raleigh, NC 27611

Salutation   Dear Mr. Price:

Thank you for sending me the material I requested
about pirates in colonial North Carolina.

Body   Both the pamphlets and the bibliography were
extremely useful for my research. My instructor
told me I had presented information in my paper
that he had never seen before. Without your help,
I am sure my paper would not have been so well
received.

I have enclosed a copy of my paper, and I would
appreciate any comments you may have. Again,
thank you for your time and trouble.

Compli-
mentary
close   Sincerely yours,

Written
signature   *Kevin Wolk*

Typed
signature   Kevin Wolk

Copy sent   cc: Dr. N. Provisor, Professor of History

Additional
data   Enc.: Research paper

The Body   The **body** of your letter contains your message. Begin this section two spaces below the salutation and single-space the text. (In a short letter of two or three sentences, you may double-space.) In block and semi-block formats you do not use paragraph indentations; in some other formats you indent the first line of each paragraph five spaces from the left-hand margin.

If your letter is more than one page long, place the recipient's name, the date, and the page number in the upper left-hand corner of the second page.

The Complimentary Close   The **complimentary close** appears two spaces below the body of the letter. The most common complimentary closes are *Sincerely yours, Yours truly,* and *Yours very truly.* If you are on friendly terms with the recipient, *Best wishes* or *Cordially* is appropriate. Note that only the first word of the complimentary close is capitalized.

The Signature   Leave four spaces below the complimentary close and type your name and title in full. Sign your name, without a title, above the typewritten line.

Additional Data   Indicate any additional information below the signature.

Enc.: (Material enclosed along with letter)

cc: Eric Brody (copy sent to the person mentioned)

SJL/lew (The initials of the writer/the initials of the typist)

48b   **Writing Letters of Application**

The two kinds of letters of application you are most likely to write are letters requesting employment and letters requesting admission to graduate school.

(1)   Letters requesting employment

When you apply for employment, your primary objective is to interest a prospective employer enough to interview you. After carefully considering why you want the job and what about you might interest a prospective employer, make an informal outline, consulting your **résumé** and the advertisement to which you are responding. Then write a rough draft of your letter.

▶ See
48c

Sample Letter Requesting Employment: Semiblock Format

246 Hillside Drive
Urbana, IL 61801
October 20, 1994

Mr. Maurice Snyder, Personnel Director
Guilford, Fox, and Morris
Eckerd Building
22 Hamilton Street
Urbana, IL 61822

Dear Mr. Snyder:

My adviser, Dr. Raymond Walsh, told me you are interested in hiring a part-time accounting assistant. I feel that my academic background and my work experience qualify me for this position.

I am presently a junior accounting major at the University of Illinois. During the past year, I have taken courses in taxation, trusts, and business law. I have also worked with a personal computer to develop my own tax program. Last spring, I gained practical accounting experience by working in the Accounting Department's tax clinic. After I graduate, I hope to get a master's degree in taxation and then return to the Urbana area.

I believe my experience in taxation as well as my familiarity with the local business community would enable me to contribute to your firm's current needs. I have enclosed a résumé for your examination. I will be available for an interview any time after midterm examinations, which end October 25.

I look forward to hearing from you.

Sincerely yours,

Sandra Kraft

Sandra Kraft

Enc.: Résumé

Pract 48b

CLOSE-UP

COMPOSING BUSINESS LETTERS

When you apply for a job, address your letter to a specific person. If the job advertisement does not include a name, call the company to find out the name of the personnel director or the individual who is responsible for hiring. Be certain to spell this person's name correctly in the salutation and on the envelope. Do not forget to type your name below your signature and to type *Enc.:* followed by a list of any documents you are enclosing—your résumé, for example.

Begin your letter by identifying the job you are applying for and stating where you heard about it—in a newspaper, in a journal, from a professor, or from your school's job placement service, for instance. Be sure to include the date of the advertisement and the exact title of the position. End your introduction with your thesis: a statement of your ability to do the job.

The body of your letter provides the information that will convince your reader of your qualifications—for example, relevant courses you have taken and pertinent job experience. Be sure to address any specific concerns mentioned in the advertisement. Above all, emphasize your strengths and explain how they relate to the specific job for which you are applying.

Conclude by saying that you have enclosed your résumé. State that you are available for an interview, noting any dates on which you cannot be available.

EXERCISE 1

Look through the employment advertisements in your local paper or in the files of your college placement service. Choose one job and write a letter of application in which you outline your achievements and discuss your qualifications for the position.

(2) Letters to graduate or professional schools

Graduate and professional schools routinely ask applicants for autobiographical statements explaining why they have applied. These personal statements reveal a great deal about you, including your goals, your level of maturity, and your ability to communicate. They are read carefully, and they help determine whether or not you will be accepted.

Sample Letter to Professional School: Indented Format

33513 Capstan Drive
Laguna Niguel, CA 92677
April 7, 1994

Marylouise Esten, Dean of Admissions
Temple University School of Law
1719 North Broad Street
Philadelphia, PA 19122

Dear Ms. Esten:

During my sophomore and junior years in college, I worked part time for a law firm in Cincinnati. I began as a file clerk and eventually was promoted to the firm's research department. This experience was a turning point in my life. It helped me decide to become a lawyer and to take courses that would prepare me for this goal.

The firm for which I worked does general practice and also a good deal of community legal work. After eight months, I was given the responsibility of screening legal-aid clients and assigning them to one of three lawyers. I consulted with attorneys and, at times, discussed specific cases with them. Eventually they allowed me to be present when depositions were taken.

My experiences with this law firm have given me a realistic picture of the legal profession and have enabled me to make a mature, informed decision to become a lawyer. I am certain that as a result, I will be an understanding and compassionate attorney, one who puts her clients' interests before her own.

To prepare for my legal career, I have majored in political science and minored in business, earning a grade point average of 3.35 on a 4.0 scale. I have taken courses in political theory, government, accounting, and constitutional law as well as in sociology, psychology, and business writing. In addition, I am a member of my school's pre-law society, and I won honorable mention in a regional moot-court competition held last spring.

M. Esten
April 7, 1994
Page 2

    My background has made me realize my
responsibility to the legal profession and to society. My
involvement with the legal profession has provided me
with the motivation to pursue a career in law. My
academic record and my experience with legal work
make me certain that a career in law is a realistic goal
for me.

                Sincerely,

                Jacqueline Reyes
                Jacqueline Reyes

Again, emphasize specific examples to bolster any generalizations students may make about themselves. Encourage students to use plain English, not jargon. Admissions committees are seldom—if ever—impressed by an applicant's ability to use what he or she thinks is the lingo of the field.

IN THE CLASSROOM

If the class is particularly interested in résumé skills, you may want to assign a few students to read résumé handbooks and report briefly on various formats. For example, Yana Parker's *Damn Good Résumé Guide* (Ten Speed Press) focuses on a *skill-based* résumé format that is especially useful for students who need to play up their skills and play down their lack of job experience.

As you plan, write, and revise such a letter, concentrate on what distinguishes you from others who are applying to the school. Be specific, offering examples from your academic, employment, and personal experience to illustrate the points you make. Everything in your statement should underscore your thesis: that you are committed to the field and should be admitted to the program.

EXERCISE 2

Assume that you are applying to a graduate school of law, medicine, business, journalism, or social work, or to a graduate program in another field. Write a personal statement in which you tell the admissions officer what led you to choose your field. Be specific in describing your motivation, your experience, and your aspirations.

48c Writing Résumés

The letter of application summarizes your qualifications for a specific position; the **résumé** provides an overview of your accomplishments, focusing on your education and your work experience.

Before you compose your résumé, list all general information about your education, your job experience, your goals, and your personal interests. Then select the information that is most appropriate for the job you want, emphasizing the accomplishments that differentiate you from other candidates. For example, if you have received academic honors or awards, or if you have financed your own education, include this information as well.

There is no single correct format for a résumé. You may decide to arrange your résumé in **chronological order**, listing your education and work experience in sequence (beginning with the most recent) or in **emphatic order**, presenting your material in the order that stresses the experience that will be of most interest to an employer. Whatever a résumé's arrangement, it should be brief—one page is sufficient for an undergraduate—easy to read, and well organized. An employer should be able to see at a glance what your qualifications are.

SECTIONS OF A RÉSUMÉ

- The **heading** includes your name, school address, home address, and telephone number.

- A statement of your **career objective** (optional), placed at top of the page, identifies your professional goals.

- The **education section** includes the schools you have attended, starting with the most recent one and moving back in time. (After graduation from college, do not list your high school unless you have a compelling reason to do so.)

- The **summary of work experience** generally starts with your most recent job and moves backward.

- The **background** or **interests section** lists your most important (or most relevant) special interests and community activities.

- The **honors section** lists academic achievements and awards.

- The **references section** lists the full names and addresses of at least three references. If your résumé is already one full page long, a line saying that your references will be sent upon request is sufficient.

Caution students that preparing a résumé can be very time consuming. This assignment requires absolute accuracy. Badly typed résumés should not be accepted. Typeset résumés generally have the best appearance, so you may want to look into typesetting facilities available to students in your area.

CLOSE-UP

WRITING RÉSUMÉS

Remember that federal law prohibits employers from discriminating on the basis of age, sex, or race. You should not include such information in your résumé.

EXERCISE 3

Prepare a résumé to include with the letter requesting employment you wrote for Exercise 1.

Sample Résumé: Chronological Order

Michael D. Fuller

Address	Home	Campus
	1203 Hampton Road	27 College Avenue
	Joppa, MD 21085	College Park, MD 20742
	Telephone:	Telephone:
	(301) 877-1437	(301) 357-0732

Education
1992–94   University of Maryland, College Park, MD
(sophomore). Biology major. Expected date of graduation: June 1996. Presently maintain a 3.3 average on a 4.0 scale.

1988–92   Forest Park High School, Baltimore, MD. Basketball team, track team, debating society, class president, mathematics tutor. Graduated in top fifth of class.

Experience
1993–94   University of Maryland Library, College Park, MD.
Assistant to reference librarian. Filed, sorted, typed, shelved, and catalogued. Earnings offset college expenses.

1992–93   University of Maryland Cafeteria, College Park, MD.
Busboy. Cleaned tables, set up cafeteria, and prepared hot trays.

1992
Summer   McDonald's Restaurant, Pikesville, MD. Cook. Prepared hamburgers. Acted as assistant manager for two weeks while manager was on vacation.

Interests   Member of University Debating Society.
Tutor in University's Academic Enrichment Program.

References   Ms. Stephanie Young, Librarian
Library
University of Maryland
College Park, MD 20742

Mr. William Czernick, Manager
Cafeteria
University of Maryland
College Park, MD 20742

Mr. Arthur Sanducci, Manager
McDonald's Restaurant
5712 Avery Road
Pikesville, MD 22513

Sample Résumé: Emphatic Order

Michael D. Fuller

27 College Avenue
University of Maryland
College Park, MD 20742
(301) 357-0732

1203 Hampton Road
Joppa, MD 21085
(301) 877-1437

Restaurant Experience

McDonald's Restaurant, Pikesville, MD. Cook. Prepared
hamburgers. Acted as assistant manager for two weeks
while manager was on vacation. Supervised employees,
helped prepare payroll and work schedules. Was named
employee of the month. Summer 1992.

University of Maryland Cafeteria, College Park, MD. Busboy.
Cleaned tables, set up cafeteria, and prepared hot trays.
September 1992–May 1993.

Other Work Experience

University of Maryland Library, College Park, MD.
Assistant to reference librarian: filed, sorted, typed, shelved,
and catalogued. Earnings offset college expenses. September
1993–May 1994.

Education

University of Maryland, College Park, MD (sophomore).
Biology major. Expected date of graduation: June 1996.
Forest Park High School, Baltimore, MD.

Interests

Member of University Debating Society.
Tutor in University's Academic Enrichment Program.

References

Mr. Arthur Sanducci, Manager
McDonald's Restaurant
5712 Avery Road
Pikesville, MD 22513

Mr. William Czernick, Manager
Cafeteria
University of Maryland
College Park, MD 20742

Ms. Stephanie Young, Librarian
Library
University of Maryland
College Park, MD 20742

Sample Letter Requesting Information: Semiblock Format

> 17 Maple Drive
> Clinton, MS 39058
> December 2, 1994
>
> Dr. Norman Murphy
> English Department
> Louisiana State University at Shreveport
> Shreveport, LA 75115
>
> Dear Dr. Murphy:
>
> I am a third-year English major at Clinton College, and I am interested in pursuing a career in scientific and technical writing. Dr. Stewart Lage, my adviser, thought you could give me advice about schools that have strong graduate programs in this field.
>
> Although I prefer to stay in the South, I am willing to go to school in any part of the country. I am particularly interested in schools with internship programs that would allow me to gain practical experience in industry. I have already written to Stanford University and am waiting for its catalog.
>
> I am currently at home and will be back at school on January 6th. I would appreciate any information you could communicate to me or to Dr. Lage.
>
> I hope to hear from you soon.
>
> > Yours truly,
> >
> > *Daniel Howell, Jr.*
> >
> > Daniel Howell, Jr.

48d   Writing Letters Requesting Information

Students sometimes have to send letters requesting information from a person or a business. For instance, you might need to ask an instructor for a recommendation or write to an expert in a field to gather information for a research project.

Before you write such a letter, decide exactly what information you need. Make a list, if necessary, and eliminate any questions that you can answer yourself. Think about what you need the information for and how much time you have to get it. Usually you can limit your letter to a few questions that can be answered quickly and easily.

Write a courteous and concise letter. Begin by introducing yourself, and then specify clearly what information you want and why you want it. Be specific; your reader will be doing you a favor by responding, and you should not waste his or her time. If you have several requests, number them—but keep them brief.

After you have received the information you have requested, you should write a letter thanking the person you have contacted.

▶ See 48a

▶ See
48a

EXERCISE 4

For a research paper on television situation comedies, write a letter requesting information from Dr. Alan Friedman (38 University Place, New York University, New York, NY 10011), a noted authority on the subject. Ask him four questions you would like him to answer. Remember that Dr. Friedman is very busy, but he will probably answer a short, businesslike letter.

48e   Writing Letters of Appeal

Students sometimes have occasion to write letters asking for a clarification of or change in college policy. For example, you may need to request permission to take a two-credit overload, to waive or change a particular requirement for graduation, to request a leave of absence, to appeal a faculty or administrative decision, or to request permission to live off campus. In many cases your audience will be inclined to hold a view different from yours, and you will have to convince them to change their minds.

STUDENTS WORKING TOGETHER

Have students submit the letter of application and résumé (prepared for Exercises 1 and 3) for collaborative revision and evaluation. Peer critics should comment on writing skills and format but also try to evaluate how convincing this application "package" might be to the employer.

ANSWERS TO EXERCISE 4

Remind students that their letters should be short. Their questions should be open-ended but should not require long answers. You may wish to refer them to "Conducting an Interview" **(38c5)**. Discussing possible questions in class would also be useful.

Sample Letter of Appeal: Block Format

Room 405A
Building A
October 15, 1994

Ms. Andrea Perry, Director
Residential Life

Dear Ms. Perry:

I am writing this letter because I would like to keep my pet
Burmese Python (Python molurus bivittatus) in my dorm room.
This snake is nonpoisonous and harmless to humans; like all
snakes, pythons make no noise, do not smell, and do not need to
be walked. Because of my background, I am uniquely qualified to
attest to the desirability of having pythons as pets.

The first quality that makes Burmese pythons good pets is that
they are safe. For two years, I worked at a large pet store in
Philadelphia where I cared for more than two hundred snakes and
other reptiles. During my employment, I was responsible for their
care and feeding. I also kept many snakes as pets. Not one of the
pets I kept ever caused any problems for either my parents or
me. In addition, I was never harmed by any of the animals in the
store or by any of the snakes I kept at home.

Another quality that makes Burmese pythons good pets is that
they are inactive much of the day. The only time they become
animated is when they are fed. After they eat their meal—one
small mouse a week—they settle down to digest their food. As a
result, they lie dormant for long periods of time, moving only to
drink water or to change their position. This characteristic makes
pythons ideally suited for the dorms.

I realize that there is a rule against having pets in the dorm, but I
think that because this pet is so easy to keep, an exception should
be made. Finally, I believe that due to my years of experience
working with snakes, I have the ability to care properly for my
pet.

I would like to set up an appointment with you to discuss this
matter further. Thank you for your time and trouble.

Sincerely yours,

Christopher Mauro

Christopher Mauro

As you write a letter of appeal, keep in mind the principles of writing an argumentative essay. Remember that the purpose of your letter is to convince readers to accept your arguments. To achieve this end, be sure you maintain a rational tone. Support your points with facts and reasonable arguments, and avoid arguing against a policy simply because "it's not fair." Present yourself as a reasonable person who sees both sides of an issue, and remember that anger or sarcasm will only undercut your case. Write your letter with your reader in mind: If your reader is unfamiliar with your case, begin with an overview, not an involved discussion of your problem. Arrange events in logical order, using transitions to make their sequence apparent, and be sure that your tone is firm but reasonable. Finally, try to end on a positive note—at the very least asserting your belief in the fairness or goodwill of the reader.

▶ See
Ch. 7

▶ See Ch. 7

ANSWERS TO EXERCISE 5

Although it might make the students feel better to write a tirade, remind them that they are more likely to get a positive response if they present themselves as reasonable people.

EXERCISE 5

Your college or university owes you a $250 refund from your tuition. Apparently they charged you twice for a student activities fee. After discovering the error you go to the registrar, who tells you that as a matter of policy all refunds are credited to the next semester's tuition. After a day and a half of hearing the same story at one office after another, you decide to write a letter to the president of your school, explaining why you think the school should reimburse you now. Make a strong case and present the facts clearly and logically.

 48f **Composing Memos**

The process of composing memos is much the same as the process of composing business letters. Unlike letters, however, memos communicate information *within* a business organization, transmitting brief messages of a paragraph or two or short reports or proposals. Their function is either to convey information or to persuade. Regardless of their function, most memos have the following general structure.

The Opening Component The **opening component**—*To, From, Subject,* and *Date*—replaces the heading and inside address of a letter. This section establishes the audience and the subject of your communication. Because a memo often circulates beyond its original audience, all names and titles should be stated in full. The *subject line,* which exists to give your reader a clear idea of

For this exercise, students may wish to look at the discussion of sexist language in **18f2**.

what your memo is about, should include more than one word. For example, "Housing" will mean very little to readers; even though they are familiar with the subject of housing, they will not know what your memo will be about. "Changes in Student Housing Policy" states the subject more precisely.

The Body   The **body** of your memo should begin with a purpose statement containing key words that immediately convey your message. Some people like to present the purpose statement as a separate component with its own heading. In any case, the purpose statement should include a word that clearly defines your intention—for example, *evaluates, proposes, questions, reports, describes,* or *presents.*

The first paragraph of the body summarizes your conclusions; the rest of your memo tells readers how you arrived at your conclusions, backing them up with facts and figures. Often each paragraph has a heading that identifies its subject and helps to guide readers through the body of the memo.

The Conclusion   The **conclusion** of your memo should contain a detailed restatement of your points. If the memo's purpose is to persuade your readers of something, you should include a list of recommendations. Because readers remember best what comes last, you should end your memo with a summary of your conclusions or recommendations.

In the memo on pages 853–54, the writer, a student who works in a tutoring program, uses headings to identify the major divisions of her discussion. Because her purpose is to persuade her audience, she addresses her reader's major concerns—cost, ease of construction, and projected benefits—and ends with a list of recommendations.

EXERCISE 6

Your duties at your summer job with a public utility include reading correspondence sent from your division to the public. While reading a pamphlet that discusses energy conservation, you notice repeated use of the word *repairmen*, and you come across the sentences "Each consumer must do *his* part" and "*Mothers* should teach their children about conservation." With the approval of your supervisor, you decide to write a memo to John Durand, public relations manager, explaining that some customers might perceive this language as sexist—and therefore offensive. In your memo, explain to Mr. Durand why the language should be changed and suggest some words and phrases he could use in their place. Mr. Durand is your superior, so maintain a reasonable and respectful tone.

TO: Ina Ellen, Senior Counselor
FROM: Kim Williams, Student Tutor Supervisor
SUBJECT: Construction of a Tutoring Center
DATE: November 10, 1994

The purpose of this memo is to propose the establishment of a tutoring center in the Office of Student Affairs.

BACKGROUND
Under the present system, tutors must work with students in a number of facilities scattered across the university campus. This situation has a number of drawbacks, including a lack of contact among tutors and the inability of tutors to get immediate help with problems. As a result, tutors waste a lot of time running from one facility to another—and are often late for appointments. Most tutors agree that the present system is unwieldy and ineffective.

NEW FACILITY
I propose that we establish a tutoring facility adjacent to the Office of Student Affairs. The two empty classrooms adjacent to the office, presently used for storage of office furniture, would be ideal for this use. Incurring a minimum of expense and using its own maintenance workers, the university could use dividers to convert these rooms into ten small offices. We could furnish these offices with the desks and file cabinets already stored in these rooms.

BENEFITS
The benefits of this facility would be the centralizing of the tutoring service and the proximity of the facility to the Office of Student Affairs. The tutoring facility could also use the secretarial services of the Office of Student Affairs, ensuring that student tutors get messages from the students with whom they work.

Memo to I. Ellen
November 10, 1994
Page 2

RECOMMENDATIONS
To implement this project we would need to do the
following:

1. Clean up and paint rooms 331 and 333 and
   connect them to the Office of Student Affairs
2. Use folding partitions to divide each room into
   five single-desk offices
3. Use stored office equipment to furnish the
   center

I am certain that these changes would do much to improve the tutoring service that the Office of Student
Affairs now offers. I look forward to discussing this
matter with you in more detail.

APPENDIX A

Preparing Your Papers

A clean, neatly typed paper is a courtesy that you owe your readers. Sloppily typed or smudged papers not only make reading difficult but also detract from your credibility. Some of your instructors will give you specific guidelines for preparing a paper—and, of course, you should follow them. But others will expect you to be familiar with the conventions for preparing a paper. The following conventions are generally consistent with those prescribed by the MLA style sheet (3rd ed).

A1   Paper Format

Submit typed or computer printed papers whenever possible. Because typewritten papers are easier to read, they are worth the extra effort—even if you are a slow typist. Before you type, be sure that your keys are clean and that you have a fresh black ribbon in your typewriter or a new cartridge in your printer. Do not use "fancy" type, such as script, that could distract your readers. Be sure to keep a copy of your paper in case your instructor misplaces it.

Use white, twenty-pound weight 8½″ × 11″ bond paper. Avoid both erasable paper, which smudges, and "onionskin," which is difficult to read. Never use paper that is not white or that is smaller than 8½″ × 11″. If you are using a computer, make sure to separate pages from one another and detach the pinhole feed strips on the sides of the pages.

Double-space your paper throughout. Single-spacing is hard to read and does not leave enough room for instructors' comments or corrections. Leave a one-inch margin at the top and bottom and on both sides of your paper. Indent five spaces for each new paragraph and ten spaces for a long quotation set off from the text.

(1) Paper without a separate title page

Many instructors do not require a separate title page. If yours does not, type your name, the course number, your instructor's name, and the date (all double-spaced) one inch from the top of the first page of the paper, flush with the left-hand margin. Double-space again and center the title. If the title is longer than a single line, double-space and center the second line below the first. Capitalize all important words in the title, but not prepositions, conjunctions, articles, or the *to* in infinitives, unless they begin or end the title. Do not underline the title or enclose it in quotation marks, but do underline words in the title if they are underlined in your paper (for example, book titles). Never put a period after a title, even if it is a sentence. Double-space between the last line of the title and the first line of your paper.

CLOSE-UP

PREPARING YOUR PAPERS

If you have permission to submit a handwritten paper, use 8½″ × 11″ wide-lined paper. Do not use paper that leaves a ragged edge when torn from a spiral-bound notebook. Leave wide margins, write on every other line, and be sure to use only one side of each paper. Use black or dark blue ink, never colored ink or pencil. If your handwriting is sloppy, try printing.

Number all pages of your paper consecutively in the upper right-hand corner, one-half inch from the top, flush right. Do not put *p.* before the page numbers, and do not put periods or any other punctuation after them. To ensure that your instructor will be able to reconstruct a paper whose pages are separated, put your name and the page number on *every* page, including the first.

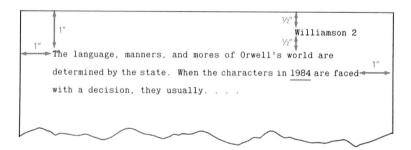

Sample of First Page of Paper without a Title Page

Sample Page Format

(2) Paper with a separate title page

Some instructors prefer a separate title page and an outline like the ones appearing with the paper at the end of Chapter 40 (page 691). The title page includes the title; your name, course, and section number; your instructor's name; and the date you submitted your paper.

When you use a title page, repeat your title on the first page of your manuscript. Subsequent pages follow the format for a paper without a title page. If your paper includes an outline of more than one page, number each page with a lowercase Roman numeral (i, ii, etc.). A single-page outline need not be numbered. All pages of the text include your last name and the page number.

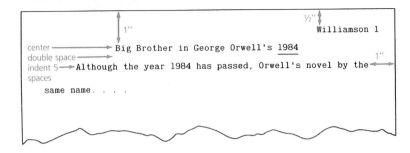

Sample of First Page of Paper with a Separate Title Page

A2   Typing Punctuation

The conventions for the spacing of typed punctuation are illustrated below. Be sure to follow them consistently throughout the paper.

TYPED PUNCTUATION		
Type of Punctuation	*Typing Conventions*	*Examples*
1. apostrophes	Leave no space before or after the apostrophe unless the apostrophe ends a word.	Turner's landscapes The writer's local color novels
2. commas and semicolons	Leave no spaces before commas or semicolons; leave one space after commas and semicolons.	In fact, the story was untrue; the slave revolt failed.
3. hyphens and dashes	A hyphen is one stroke; a dash is two unspaced hyphens.	quick-witted Ford was born poor--not rich--on a farm.

continued on the following page

continued from the previous page

4. colons, question marks, and exclamation points	Leave no spaces before these punctuation marks; leave one space after a colon and two spaces after a question mark or exclamation point.	Discuss the following: plot, theme, and character. Could he be correct? Perhaps. Never! Tyranny can never triumph.
5. quotation marks	Leave no spaces between quotation marks and the words or punctuation marks they enclose.	The short story "The Gold Bug" is by Poe. "I come to bury Caesar," said Mark Antony, "not to praise him."
6. periods	Leave no space before a period at the end of a sentence. Leave two spaces after the period.	It has been an issue since 1938. The fight for intellectual freedom continues.
7. ellipses	Separate all periods of an ellipsis by one space. Do not leave a space before the period at the end of a sentence that is followed by an ellipsis.	It was . . . a bitter satire of the motion picture industry. The report examines many classes of people. . . .
8. italics	Underline words with an unbroken line to indicate italics.	The Sun Also Rises

continued on the following page

continued from the previous page

9. parentheses, brackets	Leave one space before opening a parenthesis and one after closing a parenthesis. Do not space between the enclosed words and the parentheses. If the parentheses contain an entire sentence, leave two spaces before and after the parentheses. Brackets follow the same conventions as parentheses.	Charles Lindbergh attended the University of Wisconsin (1920–22) but left to learn how to fly. In 1907, Luther Burbank wrote *Training of the Human Plant.* (The pamphlet dealt with environment and human development.) In 1909. . .
10. slashes	Leave no space before or after a slash except if the slash separates lines of poetry. In this case leave one space before and one after the slash.	The object travels at 800 m/sec in a parallel direction. "Nature's first green is gold. / Her hardest hue to hold." (Robert Frost)

A3 Tables and Illustrations

Tables and illustrations aid readers by summarizing material. In order to carry out this function, tables and illustrations should be placed as close as possible to the part of the paper in which they are discussed, and they should be integrated into the text, not just dropped into it. To show readers the function and meaning of a table or illustration, you might include a sentence such as "Table 3 shows the year-by-year increase in the national debt from 1980 to 1988."

Tables should be headed *Table* and given an Arabic numeral and a descriptive caption. Type the heading and the descriptive caption flush left on separate lines above the title. Capitalize the heading and the caption as if they were titles. Type the full citation for the table (if it is borrowed from a source) and for any notes below the table beginning at the left-hand margin. Double-space the table's text and the notes.

Table 2

Reading Miscues Based on the Reading Comprehension Test

Reading Miscue Inventory	Percentile
Total miscues	53
Semantically inappropriate miscues	60
Miscues which alter meaning	51
Overall loss of comprehension	40
Retelling score	20

Source: Adapted from Alice S. Horning, "The Trouble with Writing Is the Trouble with Reading," Journal of Basic Writing 6(1987):46.

Illustrations—graphs, charts, line drawings, and photographs—should be labeled *Figure* (abbreviated *Fig.*) and given an Arabic number. The label, a descriptive caption, and the full citation are typed below the illustration beginning at the left-hand margin.

Fig. 1. Etching of the Globe Playhouse from Thomas Mark Parrott and Edward Hubler, Six Plays and the Sonnets (New York: Scribner's, 1956) 2.

A4 Editing Your Final Draft

After finishing the final draft of your paper, proofread it carefully and correct any errors. If a page is messy or if you have to make extensive corrections, retype or reprint the entire page. Make minor corrections with correction fluid, but never make corrections in the margin or below the line. If your instructor gives you permission, you may make some corrections in black ink. Use the proofreader's marks on the next page to indicate your changes.

A5 Submitting Your Paper

When you are ready to submit your paper, fasten the pages together with a paper clip in the upper left-hand corner. Do not staple the pages together like a book or use a binder that cannot easily be removed. Remember that your instructor cannot write comments on your paper if he or she cannot separate the pages.

Proofreader's Marks

Mark	Mark in text
∧	add; insert a word here
ℊ	delete; take a a word out
‿	close up; one w ord
ℊ̂	delete; and close uup
# ∧	add a # space
(stet)	disregard change (stet)
¶	¶ Begin new paragraph
/	lower-case Letter
≡	capital letter
∽	transpose lettres
(ital)	italics (ital)
⌇ (bf)	boldface (bf)
⟋	You can also indicate that you do not want to begin a paragraph. Draw a line to connect the two sentences.

Choosing Software

Although research on teaching writing with computers shows that the quality and kind of instruction instructors give is much more important than the software students use, you can give students the following advice about choosing writing tools.

1. Choosing a Word-Processing Program

Make sure the word-processing software you choose is easy to use; a poorly designed program can interfere with the composing process. An early version of one well-known word processor, for example, inhibited students from carrying out revisions because they had to move awkwardly back and forth between separate writing and editing modes.

Make sure the word processor is powerful enough to meet your needs and the needs of your students. Although you can ignore features, you cannot make a program do more than it was intended to do.

If you are are satisfied with the word processing program you use yourself, consider choosing it. Because you know it backwards and forwards (or ought to), you can easily demonstrate to students your own struggles with writing and your own strategies for coping.

2. Choosing Computer Software

If your school or department has decided to purchase software, and if you have a voice in selecting it, there are guidelines you can follow. When considering software that integrates developing, composing, revising, and editing aids, avoid any program that

continued on following page

APPENDIX B

Guide to Writing with Computers

Why should you take valuable time from your other classes or your social life to learn how to use a computer? A long-range answer is that when you enter the working world, your employer will expect you to be computer literate. A more immediate answer is that many of your classes will demand writing—sometimes a lot of it—and a computer will make writing tasks easier for you.

GENERAL ADVICE FOR WRITING WITH COMPUTERS

- Learn to type quickly and accurately.
- Know your software's capabilities.
- Proofread carefully.
- Print your work frequently so you can see it in its entirety.
- Save your work frequently, perhaps every page or so—certainly after you have done anything that would make rewriting difficult if lost.
- Make a backup copy of every file on another disk.

Of course, computers will not do the writing for you, and word processing alone will not make you a better writer; classroom instruction in writing skills and hard work on your part are absolutely essential. Any word-processing program, however, will take much of the drudgery out of writing; in fact, many reluctant writers report that **word processing**—the use of a computer to write and edit text—makes writing more enjoyable.

USING A COMPUTER

As important as it is to your work in college and outside, word processing is not the only benefit of learning to use a computer for writing. Electronic mail; collaborative composing; worldwide networks such as Bitnet, the Internet, the national "information highway"; and other advances in technology can be useful, educational, and entertaining. Bitnet listserve lists and Internet news groups can be valuable sources of information for research topics. Internet file transfers (ftp) and other research tools, such as *gopher* and *archie* can help you find information about thousands of topics. And Internet users can search university libraries around the world—all from the convenience of a computer on their own campus or in their dorm room.

B1   How Writing with Computers Can Help You

Experienced writers know that rewriting is central to the writing process. A computer will allow you to revise papers with a minimum of time and effort—and, because working on a computer makes revising and editing so easy, you may be more inclined to make revisions than you might have been before.

(1)   Planning

Whether you are writing an essay or a research paper, you can type your preliminary notes or research ideas and store them in a file you have created especially for that purpose. Or, you can take notes with pen and paper and enter them later into the computer. When you begin to write your paper, you can review these notes as you develop ideas and begin to organize them; you can even move these notes directly into your paper as you write.

When you plan your essay, you may find it helpful to carry out pre-writing strategies at the computer. Freewriting, for example, is much easier and more efficient with a computer. You can type

continued from previous page

locks students into a writing routine or interrupts the flow of words, thoughts, or the writing process itself. For example, if students are revising a draft, they ought to be able to move quickly and easily, back and forth, between the composing mode and the program's prewriting and brainstorming aids.

Although some style checkers are better than others, be wary of them. Many tend to serve experienced writers better than inexperienced ones. Differences between conventions and rules—and between advice and commands—are not always apparent to novice writers, who may be intimidated by the comments many programs make and who may assume that tentative suggestions are absolute directives.

If you encourage or require students to use a style checker, train them in its use and emphasize that *they* are the final authority. If students do not understand what the program is supposed to do and what its limitations are, they may fail to look critically at their writing because they assume the program is doing that for them.

Finally, keep in mind that writing *software* needs to be integrated into a writing *program*. Whether the writing program is process oriented may not be as important as whether it is a coherent whole into which the software naturally fits. Both research and experience show that writing software will work only to the extent that students take it seriously and teachers train them in the basics of writing and in the use of the software.

Teaching Word Processing

1. Mastering the Program Itself
The most productive time you can spend prior to entering class on the first day is getting to know your software and *using* it. Investigate *all* of its features in order to discover its potential as a writing tool. Only if you are an enthusiastic and competent user of the program will your students take seriously your testimony about the advantages of writing with it. Only if you are familiar with this tool can you ask, "Which of its word-processing features can I adapt as teaching and writing aids?"

2. Deciding Who Should Teach Word Processing
Although your school may have a computer center to teach word processing, there are advantages in teaching it yourself. Although some of your colleagues may believe that using valuable class time to teach a nonacademic skill like word processing is at best unproductive, computer-center personnel may not stress the intimate connection between word processing and the writing process.

3. Deciding How You Should Teach Word Processing
If, as some researchers claim, inexperienced writers are likely to find themselves suffering from cognitive overload when they try to learn word processing techniques and writing strategies at the same time, it is your job as a writing instructor to help students learn the former in ways that will enhance the latter. With imagination, you can link instruction in word processing to instruction in writing.

continued on following page

See ◀
1c2

quickly and neatly; you can move the writing you produce directly into an essay without having to retype it; and you can easily group and organize your typed notes.

Word processing makes it easy to be spontaneous and create the free-flowing writing that brainstorming and journals contain. This does not mean that you cannot do your brainstorming, free writing, journal entries, and note taking by hand if you want to. But you should try experimenting with the computer at all stages of the writing process so that you can find out what works for you.

(2) Shaping

Your computer's ability to *move* parts of text—whether a single word or a number of paragraphs—from one location to another can help you organize your notes. For example, if you have created brainstorming or note-taking files, you can use the computer to group similar notes together into "idea chunks." Experiment with putting the groups in different orders until you discover the best way to present your information.

You can use word processing in this way to arrange your notes into something approaching an informal outline or even a very rough draft. You can also make an outline by typing words or phrases representing main and supporting ideas, arranging and rearranging them until the order begins to make sense, and indenting them with the <Tab> key to indicate the hierarchy of your ideas.

Following is an example of how Marion Duchac organized her first thoughts on college athletics (note that the arrows indicate tabs).

College athletics
→ Problems
→ → Too much emphasis on winning
→ → → Effects on programs
→ → → Effects on individuals
→ Solutions
→ → Less emphasis on winning
→ → → How it would affect programs
→ → → How it would affect individual athletes

At a later stage in the writing process, Marion Duchac created a more formal topic outline by using standard outline format. (Note that she still needs to express her thoughts in parallel terms.)

A12

$\rightarrow$ I. Advantages of Athletic Programs
$\rightarrow$ $\rightarrow$ A. School Spirit
$\rightarrow$ $\rightarrow$ B. Money
$\rightarrow$ $\rightarrow$ C. Balanced Education
$\rightarrow$ II. Disadvantages of Athletic Programs
$\rightarrow$ $\rightarrow$ A. Unequal Support of Sports
$\rightarrow$ $\rightarrow$ $\rightarrow$ 1. Football and basketball vs. other sports
$\rightarrow$ $\rightarrow$ $\rightarrow$ 2. Men's sports vs. women's sports
$\rightarrow$ $\rightarrow$ B. Treatment of Student Athletes
$\rightarrow$ $\rightarrow$ $\rightarrow$ 1. Acceptance of unqualified students
$\rightarrow$ $\rightarrow$ $\rightarrow$ 2. Little academic support
$\rightarrow$ $\rightarrow$ $\rightarrow$ 3. Heavy schedules

CLOSE-UP   USING A COMPUTER

Be sure you keep your planning and shaping material separate from your first draft. You can do this by creating two or more separate documents within a file. For example, if you are writing a paper on environmental issues in your community, you can call your planning and shaping file ENVIR-PS and your draft file ENVIR-D. In your ENVIR-PS file, you can store journal entries, brainstorming lists, notes on observations and readings, and/or informal outlines.

(3) Writing

As you write, keep in mind that writing usually does not proceed neatly and directly from a beginning right through to a conclusion. As writers write—sometimes while in the middle of developing a paragraph or even writing a sentence—they go back to reread and rethink what they have written and then edit or revise before going on.

Because writing is not a neat, orderly process, the computer has definite advantages. Text can easily be reshaped when you make changes and reformatted when you add new text or delete old text; thus, it is possible for you to begin in the middle or "end" at the beginning. As ideas come to you, you can type them immediately and insert them wherever you think they ought to go—and, all the while, you can be certain that you can change their location (or even delete them) later on with little effort.

continued from previous page

At the same time you are teaching the basics of word processing, you can pass along your own favorite composing and revising strategies. In other words, make sure students learn word-processing techniques at the point where those techniques really matter to them—for example, learning to execute block moves when they actually need (and want) to move sentences and paragraphs.

4. Deciding How Much Word Processing to Teach

Walking the thin line between too much and not enough training in computers and word processing is difficult. On the one hand, you should keep in mind that the course is a *writing* course, not a computer course. On the other hand, students—especially novices—need to feel comfortable with the computer and the word processor and need to know how to perform the basic operations, like a block move, which will enable them to carry out global revision strategies.

Among teachers experienced in word processor use in the writing class, there is little agreement on how much word processing to teach. What is clear, however, is that you ought to avoid overloading students with huge chunks of abstract word-processing instruction. Since students generally resist theory and information that is presented abstractly, give them only as much information as they need in order to do what they have to do *now*. For example, they are likely to become adept at using a split screen only when they see the need to copy text from one document to another, and they will execute search-and-replace operations

continued on following page

continued from previous page

only when they want to avoid the tedium of replacing portions of text that occur many times in a document.

Writing instructors who teach word processing find that many students do not easily master operations such as deleting, inserting, and moving blocks of text—not to mention more advanced operations—because they have a difficult time seeing computers as anything more than glorified typewriters. You will probably have greater success teaching the basics of word processing if you have students work with "real" writing—that is, with writing they have produced and in which they presumably have some investment.

If, however, it is early in the term and your students have not yet produced anything substantial of their own, turn an article (like the one in Exercise 4 on p. 125–26 of *The Holt Handbook*) into a word-processing document and use the exercise as a tool to teach basic word-processing skills. Instead of simply asking students to discuss what needs to be omitted, added, or rearranged, have them make the changes in the text.

Brainstorming is a useful technique for finding ideas, and it is not something you do only at the beginning of a writing assignment. Experienced writers often brainstorm whenever they need to generate ideas, no matter what their current "stage" (*planning, shaping,* or *writing and revision*) in the writing process. Indeed, with a computer, the divisions between "stages" in the writing process tend to disappear.

CLOSE-UP

USING A COMPUTER

As you write, enclose any comments or questions that occur to you in brackets or format them in bold or uppercase letters. Later, as you revise, you will easily be able to identify and react to these notes.

(4)   Revising

Revising on a computer will not necessarily save time, but your time will be spent effectively. Instead of taking hours to type a neater version of an essay, you will be able to spend your time doing real revision—developing new paragraphs, rewording sentences, or rearranging ideas, for example.

Although some people revise exclusively on the computer screen, this is not a good strategy for everyone. For one thing, you can only see a small portion of your text on the screen at any given time. For another, it is difficult for you to compare drafts. For these reasons, you will probably want to print out a hard copy of each draft. This strategy enables you to see the entire paper and thus to make global revisions. You should make your handwritten changes on this hard copy and then enter them into your computer.

REVISING ON A COMPUTER

- Move paragraphs around in your essay so that the order of the paragraphs is scrambled. Ask another student to try to determine the original order of the paragraphs. If he or she has trouble doing so, you may not have organized the material in a logical way or you may not have used transitions between paragraphs effectively.

continued on the following page

continued from the previous page

- "Explode" a paragraph you have written by putting a <Return> after each sentence; this will format each sentence as a separate paragraph. Then, as you did with the paragraphs in your essay, scramble the text. (You can also use the search-and-replace function to replace spaces between sentences with paragraph returns.) Ask a classmate to try to put the sentences back in their original order. If he or she has difficulty doing so, your paragraph may not be coherent—that is, the sentences may not be in a logical order or you may not have used transitions effectively.

- Take a typical paragraph you have written in which the topic sentence occurs at the beginning. Copy the paragraph twice so you have a total of three identical copies. Leave the topic sentence in the first copy in its original position at the beginning; move the topic sentence in the second copy to some suitable point near the middle; finally, move the topic sentence in the third copy to the end. In each case, ask yourself what advantages or disadvantages there are in having the topic sentence in that particular location. When you place the topic sentence in a new location, do you have to make other changes in the rest of the paragraph? What changes? Why?

- As you revise, put material you might want to delete in **bold** characters, material you might want to relocate in UPPERCASE characters, and material you might want to add in <u>underlined</u> characters.

- Be careful about permanently discarding material that does not seem useful at the time. Get in the habit of moving unwanted material to the end of the document on which you are working until you are certain you really do not need it.

(5) Editing

You may find it easier to spot surface-level problems—typing mistakes or grammar, punctuation, and spelling errors—if you can see what you have written in a different way or from a different perspective. Several strategies can help you accomplish this while you are editing your paper.

Since in a lab setting you can give all students similar tasks, encourage (or even require) them to share texts either by working in teams at the same computer or by exchanging disks in order to act as peer editors for one another.

Meeting in a Computer Center

Before planning your computer and composition course, find out how often you will be able to meet your entire class in either a computerized classroom or a computer lab—all the time, frequently, infrequently, or never. If you have a choice, meet your class as often as possible in the setting that places students in front of computers when you teach, even if they have to share computers.

There are a number of advantages in doing so. Instead of passively observing while you demonstrate word-processing techniques and writing strategies, and only later experimenting with them, students will be able to try them out immediately. In addition, since writing done in a computer lab setting is, in a sense, public writing, students are more likely to form a community of writers through sharing ideas and strategies than they would if they were writing in isolation.

Ultimately, however, since you will be communicating attitudes toward writing as well as writing theory and instruction, the success of your class will greatly depend on how comfortable you are and how you function in a lab setting. Teaching in a lab or a computerized classroom is not like teaching in a "normal" classroom.

continued on following page

continued from previous page

Students sitting in front of computers tend to act independently—for example, they are likely to form informal groups and share ideas and to play with the machines in front of them. They will be even less tolerant of lectures and will want to *do something.* These attitudes and behaviors can be somewhat unnerving if you are not prepared.

If you can not arrange to meet your class on a regular basis in a computer lab or classroom, or if your students have difficulty obtaining access to a computer, you will have to decide at what point logistics impinge on, or perhaps overwhelm, pedagogy. Even if you have devised "perfect" computerized exercises, you must ask yourself whether it makes sense to give computer instruction to students who have limited access to computers.

TEACHING STRATEGY

Creating Exercise Files
The simplest computer-assisted-instruction aid is a word-processing file. You can easily adapt exercises from textbooks or create exercises of your own that students can complete on the screen.

Using the Power of Networking
A *network* exists when two or more computers are connected so that users can "talk" with one another and exchange data. The connection can be a local-area network (LAN), connecting computers within a classroom or a building or across a campus, or it can be a distance network, linking one campus or (geographical location) to another.

continued on following page

EDITING ON A COMPUTER

- Try looking at only a small portion of text at a time. If your software allows you to split the screen and create another *window,* create one so small that you can see only one or two lines of text. Try this technique also while reading your essay backwards—last sentence to first—and you may find you can dramatically reduce the number of surface-level errors in your essays.

- Print a hard copy and work with a pen or pencil to mark final changes you want to make before entering the changes with the computer. Seeing your work on paper rather than on a computer screen can help you find new ways to revise.

- Use the *search* or *find* command to look for words or phrases about which your instructor has warned you, or for errors you commonly make—for instance, confusing *it's* with *its, lay* with *lie, effect* with *affect, their* with *there, its* with *it's,* or *too* with *to.* You can also locate inadvertant use of sexist language by searching for words like *he, his, him,* or *man.*

Finally, learn how to use the spell checker. Remember that it does not find "mistakes"; it simply identifies character strings it does not recognize. Thus, a spell checker will not recognize *there* in "They forgot there books on the table" as incorrect, nor will it spot the typographical error in "Whatever he wanted to do, he dad." You still must proofread your papers carefully.

(6) Formatting

When you write with a computer, you can easily tailor the format of your document for special audiences and occasions. With a little practice, you can make paragraphs look any way you wish for special kinds of documents.

> Here is a paragraph—the kind you will use most of the time for your essays—formatted in a conventional way, with the first line indented five spaces.

And here is a paragraph—the kind you might use in business letters—formatted with no indent.

With a computer you can quickly change the size of margins and the amount of space between lines. You might want to make such changes when you give a draft of an essay to a peer editor or to your instructor, printing it triple-spaced, for example, or with extra-wide margins to leave room for written comments.

continued from previous page

A further distinction exists between *synchronous* and *asynchronous* communication. When synchronous communication takes place, users transmit and receive messages simultaneously; when asynchronous communication takes place, there is a lapse of anywhere from a few minutes to a few hours between the sending and the receiving of a message.

CLOSE-UP

USING A COMPUTER

Word processing can give your writing a neat, "finished" appearance. Keep in mind, however, that neatness does not equal correctness. Ironically, the ability of a computer to produce neat-looking text can disguise flaws that might otherwise be readily apparent. Because your writing has a professional look, you will have to take special care to ensure that spelling errors and typographical errors do not slip by.

(7)   Documenting

Word processing can eliminate much of the work, frustration, and confusion involved in documenting sources for research papers. Many programs will, with a few simple keystrokes, create notes, number the notes correctly, and automatically change the reference numbers of all subsequent notes if you add a new one or if you delete text that contains a reference number. Most software programs allow you either to place the notes at the bottom of the appropriate pages (footnotes) or to collect them at the end of the document (endnotes).

▶ See Ch. 40

(8)   Other kinds of software

Style checkers are special programs designed to perform operations such as identifying possible mistakes in grammar or punctuation, critiquing your writing style (for example, the length and complexity of your sentences), or commenting on proper word usage (slang terms or sexist language). However, style checkers are most helpful to experienced writers who know how to interpret a program's advice. The danger for inexperienced writers is that they will depend too much on the style checker and not look critically at their own writing. Remember that style checkers are not "smart." Although they may point out weaknesses or problems in your writing, they can recognize only a limited number of writing patterns and are only as good as the programming behind them. You are—and should be—the final authority.

1. Establishing the Virtual Classroom

Some instructors have their students "hand in" papers over a campus network, either to a computer in the instructor's office or, by means of a modem (a device which connects computers via telephone lines), to a computer in the instructor's home. Instructors then have the option of evaluating, commenting on, and correcting the essays on-line and sending them back to students' computers. Carried to an extreme, such an operation can take the place of some, or even all, regular class meetings—creating, in effect, a *virtual classroom* where none physically exists.

2. Engaging in Distance Learning

By means of various intercollegiate computer networks (for example, Bitnet), you can have students exchange data or drafts and critique letters with students in writing classes at other institutions.

1. Turn Exercise 1 on page 277 of *The Holt Handbook* ("Identify the parallel elements in these sentences by underlining parallel words and bracketing parallel phrases and clauses") into a word-processing file and have students complete it on-line.

2. Turn Exercise 3 on page 249, of *The Holt Handbook* ("Identify the sentence fragments in the following paragraph, and correct each [by either] attach[ing] the fragment to a related independent clause, or add[ing] a subject and a finite verb to create a new independent clause") into a word-processing document and have students complete it on-line.

3. Create a word-processing document from a paragraph or a short essay and ask students to alter it for a different audience, tone, or point of view.

Tutorial programs are designed to provide practice in punctuation, mechanics, grammar, and usage. You can use them at your own pace for extra practice in specific areas.

Other software allows you to collaborate with other students and to send and receive electronic mail.

B2   Important Terms

archive   A collection of several files compressed and combined into one file. Used to save storage space and file transfer time.

ASCII   Acronym for American Standard Code for Information Interchange. The dominant character set encoding of modern computers.

baud   In popular use (although not technically correct) as data transfer speed measured in bits per second. A 2400 baud connection will transfer 2,400 bits per second, a 9600 baud connection will transfer 9600 bits per second, etc.

BBS   Acronym for Bulletin Board System, an electronic bulletin board system; that is, a message database where people can log in and leave messages for others grouped into topic groups. The thousands of local BBS systems in operation throughout the U. S. are usually operated as a hobby.

bells and whistles   Features added to a program or system to make it more fun to use.

bit   Binary digit, the smallest unit of information in a binary computer, usually represented by on or off, or numerically by 0 or 1.

BITNET   Acronym for Because It's Time NETwork, a collection of IBM mainframe computers located mostly at universities and connected via a wide area network.

boot   To start a computer and initialize the operating system.

bug   A problem in a software application or system program that causes malfunctions.

copy   A command that makes it possible for the user to reproduce either a whole file or part of a file in a different place.

crash   A sudden, catastrophic system or program failure.

cursor   The blinking marker, usually a small bar or rectangle, on a computer screen. The cursor is the entry point for all text you create on a computer screen. As you type, text will appear immediately to the left of the cursor.

delete   A command that permanently removes a portion of text from a document or a complete *file* from a *disk*.

document   A series of computer-generated characters—usually letters and numbers—stored on a *disk*. Each document is given a name so the computer can find it when commanded to do so. When you write an essay (or part of an essay) and *save* it, you create a *document* (sometimes called a *file*) that remains on the floppy or hard disk and can be accessed again later to change, display, or print.

download   To transfer a file or other data from a remote system to the local system; a means of receiving data or files.

e-mail   Abbreviation for electronic mail.

emoticons   Symbols used in E-Mail communication to indicate an emotion.

Fidonet   A worldwide network of personal computers for the exchange of E-Mail.

forum   A BBS discussion group.

hard copy   The printed version of material produced on a computer. A hard copy enables you to reconstruct your work if you lose or damage your floppy disk or accidentally erase or destroy your file.

insert   A command that adds a portion of text to a document.

Internet   A worldwide network of computers at educational, corporate, and government and military sites.

modem   Modulator/demodulator; a device used to transmit data over telephone lines.

mouse   A hand-held pointing device for moving the cursor around the screen and executing certain commands.

move   A command that relocates text from one place to another in the same document or from one document to another document.

Dealing with the Computer as Plaything
One of the pitfalls awaiting the inexperienced when they write with a word processor is succumbing to what has been called the "lure of the gadgetry." To that we can attribute, at least to some extent, the tendency to concentrate prematurely on relatively minor matters. Because it takes their minds off the difficult task of writing, it is easy for writers working on early drafts to slip into correcting typos or fidgeting with formatting. It is also easy for writers to overwrite by trying for the "perfect" text.

If experienced writers admit (and they frequently do) that word processing tends to focus their attention on surface-level features, imagine its potential effect on inexperienced writers. Warn your students about the dangers of "playing around" too much, too early in the writing process, with punctuation, spelling, and word choice, and do not insist upon or reward neat early drafts. Challenge students to take chances and experiment with messiness.

Be aware, also, that style checkers and spell checkers both tend to focus attention on minutiae.

MS-DOS   Acronym for MicroSoft Disk Operating System. The operating system that underlies virtually all IBM compatible and DOS systems.

network   Two or more computers connected to one another so users can exchange information.

RAM   Acronym for Random Access Memory.

PC   Abbreviation for Personal Computer.

post   To send a public message to a mailing list or newsgroup on a public bulletin board system.

replace   The command—usually used along with the *search* or *find* command—that substitutes one character string for another. The *replace* command allows you to make quick, systematic changes in your writing. For example, if you decided to change the name of a character in a short story, you could instruct the computer to replace every instance of *Jim* with *Frederick*.

save   The command that makes a permanent copy of a document on either a floppy or hard disk. Until you save a document, it exists only as electronic impulses in the memory of your computer. If there should be an accidental power failure, all text produced since the last time you saved will be permanently lost. Learn the *save* command immediately and get in the habit of saving regularly.

search   A command that looks for a designated character string—usually a word or a group of words chosen by the user—in a file. The command can help you find places in your writing where you might want to make changes or check for errors. For example, you can tell your word processor to search for the pronouns, *he, him,* and *his* so that you can be sure you have avoided sexist language.

server   A computer that performs a service for other computers, such as file storage and distribution of E-Mail.

software   A program, written in a computer language, that tells a computer what to do and how to do it.

upload   To transfer a file or other data from the local system to a remote system; a means of sending data or files.

WYSIWYG   Acronym for What You See Is What You Get. A graphic interface that shows documents as they will appear printed, with all special characters and fonts, on the screen.

Some Emoticons in Current Use on E-Mail

:-o	Wow! *or* Uh oh!
:-c	Real unhappy
:-|	Grim
:-C	Just totally unbelieving *or* User is really bummed
:=|	Baboon
:-B	Drooling
:-v	Speaking
:-,	Smirk
:-V	Shout
:-||	Anger
:-r	Sticking tongue out
:-*	Oops!
'-)	Wink
:-T	Keeping a straight face
:-D	Said with a smile *or* User is laughing (at you!)
:-[	Pouting
:-#	My lips are sealed
:-Y	A quiet aside
8-|	Eyes wide with surprise
&-|	Tearful
8-]	"Wow, maaan"
:^D	"Great! I like it!"
:-)	Basic smiley; used to inflect a sarcastic or joking statement since we can't hear voice inflection
;-)	Winky smiley; user just made a flirtatious and/or sarcastic remark, more a "don't hit me for what I just said" smiley *or* User is so happy, s/he is crying
:-(	Frowning smiley; user did not like that last statement or is upset or depressed about something
[:]	User is a robot
8-)	User is wearing sunglasses
B:-)	Sunglasses on head
::-)	User wears normal glasses
B-)	User wears horn-rimmed glasses
:-7	User just made a wry statement
;-(	User is crying
:-@	User is screaming
:<)	User is from an Ivy League School
:-&	User is tongue tied
:-P	Nyahhhh!
:-S	User just made in incoherent statement

continued on the following page

continued from the previous page

:-/	User is skeptical
@=	User is pro-nuclear war
*<:-)	User is wearing a Santa Claus hat
(8-o	It's Mr Bill!
*:0)	And Bozo the Clown!
3:]	Pet smiley
3:[	Mean Pet smiley
:-0	No yelling! (quiet lab)
:-:	Mutant smiley
X-(	User just died

Some Abbreviations Currently Used on E-Mail

CIS	Consumer Information Service (of CompuServe)
RTFM	Read the Forgotten Manual
PITA	Pain In The "Acronym"
BTW	By The Way
BFN	Bye, For Now
TSR	Terminate and Stay Resident program
OTOH	On The Other Hand
FWIW	For What It's Worth
RSN	Real Soon Now
OIC	Oh, I see!
IMHO	In My Humble Opinion [the speaker is never humble]
IMCO	In My Considered Opinion
g,d&r	grinning, ducking, and running
OOTB	Out Of The Box

Glossary of Usage

This glossary of usage lists words and phrases that often give trouble to writers. As you use it, remember that this glossary is intended only as a guide. Language is constantly changing, so throughout this glossary an effort has been made to reflect current usage in college, business, and technical writing. When a usage is in dispute or in flux, the alternatives are discussed along with the advantages and disadvantages of each. In addition to the advice you receive from this glossary, your own sense of the language, as well as your assessment of your audience and purpose, should help you decide whether a particular usage is appropriate.

a, an   Use *a* before words that begin with consonants or words that have initial vowels that sound like consonants.

> *a* primitive artifact   *a* one-horse carriage

Use *an* before words that begin with vowels and words that begin with a silent *h*.

> *an* aqueous solution   *an* honest person

accept, except   *Accept* is a verb that means to "to receive." *Except* is a preposition or conjunction that means "other than." As a verb *except* means "to leave out."

> The auditors will *accept* all your claims *except* the last two.

> Aliens who have lived in the United States for more than five years are *excepted* from the regulation.

advice, advise   *Advice* is a noun meaning "opinion or information offered." *Advise* is a verb that means "to offer advice to."

> The king sent a messenger to the oracle to ask for *advice*.

> The broker *advised* her client to stay away from speculative stocks.

affect, effect　*Affect* is a verb meaning "to influence." *Effect* can be a verb or a noun. As a verb it means "to bring about," and as a noun it means "result."

A severe cutback in federal funds for student loans could *affect* his plans for graduate school.

The arbitrator tried to *effect* a settlement that would satisfy both the teachers and the school board.

The most notable *effect* of the German bombing of London was to strengthen the resolve of the British.

afraid, frightened　See **frightened, afraid**.

aggravate, irritate　*Aggravate* means "to worsen." *Irritate* means "to annoy." Avoid using *aggravate* as a colloquial term for *irritate*.

The malfunction of the computer system *irritated* the project leaders, who worried that the breakdown would *aggravate* an already tense situation.

all ready, already　*All ready* means "wholly prepared." *Already* means "by or before this or that time."

During the thirties President Roosevelt made the country feel that it was *all ready* for any challenge that might confront it.

By the time Horatius decided to call for help, it was *already* too late.

all right, alright　Although there is a tendency in the direction of *alright*, current usage calls for *all right*.

allusion, illusion　An *allusion* is a reference or hint. In literature it is a brief reference to a person, place, historical event, or other literary work with which a reader is expected to be familiar. An *illusion* is something that is not what it seems.

In *The Catcher in the Rye* the main character makes an *allusion* to *The Return of the Native,* a novel by Thomas Hardy.

The Viking landing proved that the canals of Mars are an *illusion* caused by atmospheric and topographical conditions.

a lot　*A lot* is always two words. It is used colloquially as a substitute for "many" or "a great deal."

among, between　*Among* refers to groups of more than two things. *Between* refers to just two things. The distinction between these terms seems to be fading, and it is becoming increasingly acceptable when speaking to use *between* for three or more things. In formal writing situations, however, you should maintain the distinction.

The three parties agreed *among* themselves to settle the question out of court.

By the time of his death in 323 B.C., Alexander's empire encompassed all the territory *between* Macedon and India.

amount, number *Amount* refers to a quantity that cannot be counted. *Number* refers to things that can be counted. Always use *number* when referring to groups of people.

Because he had missed several payments, the bank called in the full *amount* of the loan.

Seeing their commander fall, a large *number* of troops ran to his aid.

an, a See **a**, **an**.

and/or In business or technical writing, use *and/or* when either or both of the items it connects can apply. In college writing, however, the use of *and/or* should generally be avoided.

The data recorder can print *and/or* display the temperatures of the cooling vats.

apt to See **likely to, liable to, apt to**.

as . . . as . . . In such constructions, *as* signals a comparison; therefore, you must use the second *as*.

AWKWARD: John Steinbeck's *East of Eden* is as long if not longer than *The Grapes of Wrath*.

CLEAR: John Steinbeck's *East of Eden* is *as* long *as* if not longer than *The Grapes of Wrath*.

as, like Current usage accepts *as* as a conjunction or a preposition. *Like*, however, should be used as a preposition only. If a full clause is introduced, *as* is preferred.

In his novel *The Scarlet Letter* Hawthorne uses imagery *as* he does in his other works.

In its use of imagery *The Scarlet Letter* is *like The House of the Seven Gables*.

When used as a preposition, *as* indicates equivalency or identity.

After classes he works *as* a manager of a fast-food restaurant.

Like, however, indicates resemblance but never identity.

Writers *like* Carl Sandburg appear once in a generation.

as, than When making comparisons, either objective or subjective case pronouns can follow *as* or *than*. To determine case, you must know whether the things being compared are subjects or objects of verbs. A simple way to test this is to add the missing verb.

Nassim was as tall *as* he (is tall).

I have walked farther *than* he (has walked).

I like Jim more *than* (I like) *him*. (*Him* is the object of the missing verb *like*.)

assure, ensure, insure *Assure* means "to tell confidently or to promise." *Ensure* and *insure* can be used interchangeably to mean "to make certain." Always use *insure* (not *ensure*) to mean "to protect people or property against loss."

Caesar wished to *assure* the people that if they surrendered, he would not plunder their city.

To *ensure* (or *insure*) the smooth operation of the mechanism, you should oil it every six months.

It is extremely expensive for physicians to *insure* themselves against malpractice suits.

at, to   Many people use the prepositions *at* and *to* after *where* in conversation. This usage is redundant and should not be used in college writing.

> **COLLOQUIAL:** *Where* are you working *at*?
> *Where* are you going *to*?

> **REVISED:** Where are you working?
> Where are you going?

awhile, a while   *Awhile* is an adverb. *A while*, which consists of an article and a noun, is used as the object of a preposition.

Before we continue we will rest *awhile*. (modifies the verb *rest*)

Before we continue we will rest for *a while*. (object of the preposition *for*)

bad, badly   *Bad* is an adjective and *badly* is an adverb.

The school board decided that *The Tin Drum* by Günter Grass was a *bad* book and deleted it from the high school reading list.

For the past five years American automobile makers have been doing *badly*.

After verbs that refer to any of the senses or any other linking verb, use the adjective form.

He looked *bad*. He felt *bad*. It tasted *bad*.

Bad meaning "very much" is colloquial and should be avoided in college writing.

> **COLLOQUIAL:** Jake Barnes felt that he needed a vacation in Spain real *bad*.

> **REVISED:** Jake Barnes felt that he *badly* needed a vacation in Spain.

being as, being that   Colloquial for *because*. These awkward phrases add unnecessary words and weaken your sentences.

Because (not *being that*) the climate was getting colder, a great number of animals migrated south.

beside, besides   *Beside* is a preposition meaning "next to" and occasionally "apart from." *Besides* can be either a preposition or an adverb. As a preposition, *besides* means "except" or "other than." As an adverb it means "in addition to."

Beside the tower was a wall that ran the length of the old section of the city.

The judge pointed out to the lawyer that his argument was *beside* the point.

Besides its industrial uses, laser technology has many other applications.

Edison not only invented the light bulb and the ticker tape, but the phonograph *besides*.

between, among   See **among, between**.

bring, take   *Bring* means to transport from a farther place to a nearer place. *Take* means to carry or convey from a nearer place to a farther one.

In the late nineteenth century many Russian Jewish immigrants were able to *bring* to this country only the clothes they wore.

Take this message to the general and wait for a reply.

but, however, yet   *But, however,* and *yet* should be used alone, not in combination.

She thought her essay was adequate, *but* (not *but yet* or *but however*) she continued to revise.

can, may   *Can* denotes ability and *may* indicates permission.

Can (are they *able* to?) first-year students participate in the work-study program?

May (do they have permission?) registered aliens collect unemployment benefits?

censor, censure   To *censor* is to label as undesirable passages of books, plays, films, news broadcasts, essays, etc. To *censure* is to condemn or criticize harshly.

Many recording artists are concerned that their albums will be *censored*.

In 1633 Galileo was *censured* by the Inquisition for holding that the sun was the center of the universe.

center around   This common colloquialism is acceptable in speech and informal writing but not in college writing.

The report *centers on* (not *around*) the effects of cigarette smoking on the circulatory system.

complement, compliment   *Complement* means "to complete or add to." *Compliment* means "to give praise."

A double-blind study would *complement* their preliminary work on this anticancer drug.

Before accepting the 1949 Nobel Price for literature, William Faulkner *complimented* the people of Sweden for their courtesy and kindness.

conscious, conscience   *Conscious* means "having one's mental faculties awake." *Conscience* is the moral sense of right and wrong.

With a local anesthetic a patient remains *conscious* during the procedure.

G5

During the American Civil War, the Copperheads followed the dictates of *conscience* and refused to fight.

consensus "Consensus of opinion" is redundant because *consensus* means an "agreement of the majority." Write "they reached a consensus," or use "they agreed" or "the majority view was."

continual, continuous *Continual* means "recurring at intervals." *Continuous* refers to an action that occurs without interruption.

A pulsar is a star that emits a *continual* stream of electromagnetic radiation. (It emits radiation at regular intervals.)

A small battery allows the watch to run *continuously* for five years. (It runs without stopping.)

could of, would of In speech, the contractions *could've* and *would've* sound like the nonstandard constructions *could of* and *would of*. Spell out *could have* and *would have* in college writing.

Macbeth *would have* (not *would of*) defied his wife if he *could have* (not *could of*).

couple of *Couple* means "a pair," but *couple of* may mean "several" or "a few." When you designate quantities, avoid ambiguity. Write "four points," "three reasons," or "two examples" rather than "a couple of."

criterion, criteria Although many people use these singular and plural words interchangeably, *criteria,* from the Greek, is the plural of *criterion,* meaning "standard for judgment."

Of all the *criteria* for hiring graduating seniors, class rank is the most important *criterion*.

data *Data* is the plural of the Latin *datum,* meaning "fact." In everyday speech and writing *data* is used for both singular and plural. In college writing you should generally preserve the distinction.

The *data* discussed in this section *are* summarized in the graph in Appendix A.

different from, different than *Different than* is used extensively in American speech. Stylists, who point out that *different than* indicates a comparison where none is intended, prefer *different from*. In college writing, use *different from*.

His test scores were not much *different from* (not *than*) mine.

discreet, discrete *Discreet* means "careful or prudent." *Discrete* means "separate or individually distinct."

Because Madame Bovary was not *discreet* with her lover, her reputation suffered.

Current research has demonstrated that atoms can be broken into hundreds of *discrete* particles.

disinterested, uninterested *Disinterested* means "objective" or "capable of making an impartial judgment." *Uninterested* means "indifferent or unconcerned."

The narrator of Ernest Hemingway's "A Clean, Well-Lighted Place" is a *disinterested* observer of the action.

Finding no treasure after leading an expedition from Florida to Oklahoma, Hernando de Soto was *uninterested* in going farther.

due to *Due to* is always correct when used as an adjective following a form of the verb *be*. Many object to the use of *due to* as a preposition meaning "because of."

The cancellation of classes was *due to* a sudden snow storm.

Classes were cancelled *because of* (not *due to*) a sudden snow storm.

effect, affect See **affect**, **effect**.

emigrate from, immigrate to *To emigrate* is "to leave one's country and settle in another." *To immigrate* is "to come to another country and reside there." The noun forms of these words are *emigrant* and *immigrant*.

In 1887 by great-grandfather *emigrated from* the Russian city of Minsk and traveled by ship to Boston. During that year many other *emigrants* made the same trip.

The potato famine of 1846–47 caused many Irish to *immigrate* to the United States. These *immigrants* became builders, politicians, and storekeepers.

ensure, assure, insure See **assure**, **ensure**, **insure**.

enthused *Enthused*, a colloquial form of *enthusiastic*, should never be used in college writing.

President John F. Kennedy was *enthusiastic* (not *enthused*) about the U. S. space program.

especially, specially *Especially* means "particularly" or "very." *Specially* means "for a particular reason or purpose."

He was *especially* proud of his daughter's athletic abilities when he learned a scholarship had been created in her honor. The *specially* created college scholarship was earmarked for athletically gifted women.

etc. *Etc.*, the abbreviation of *et cetera*, means "and the rest." Although *etc.* is common in popular writing, do not use it in your college writing. Say "and so on" or, better, specify exactly what *etc.* stands for.

UNCLEAR: Before beginning to draft your research paper, you should have paper and pencil, *etc.*

REVISED: Before beginning to draft your research paper, you should have a pencil or pen, paper, and your note cards.

everyday, every day *Everyday* is an adjective that means "ordinary" or "commonplace." *Every day* means "occurring daily."

In the Gettysburg Address, Lincoln used *everyday* words to create a model of clarity and conciseness.

In *The Canterbury Tales* Chaucer describes a group of pilgrims who tell stories *every day* as they ride from London to Canterbury.

except, accept   See **accept**, **except**.

explicit, implicit   *Explicit* means "expressed or stated directly." *Implicit* means "implied" or "expressed or stated indirectly."

The director *explicitly* warned the actors to be on time for rehearsals. Her *implicit* message was that lateness would be grounds for dismissal from the play.

farther, further   In speech and in informal writing, the distinction between these words as adjectives has all but disappeared. In formal writing, however, *farther* is preferred to designate distance and *further* to designate degree.

I have traveled *farther* from my home town than any of my relatives.

Critics of the welfare system charge that government subsidies to the poor encourage *further* dependence.

Further has two additional uses. As a conjunctive adverb, *further* means "besides." As a transitive verb, *to further* means "to promote" or "to advance."

Napoleon I was one of the greatest generals in history; *further,* he promoted liberalism through widespread legal reforms.

Tom Jones, the hero of Fielding's novel, is a poor boy who is able to *further* himself with luck and good looks.

fewer, less   Use *fewer* with nouns that can be counted: *fewer* books, *fewer* people, *fewer* dollars. User *less* with quantities that cannot be counted: *less* pain, *less* power, *less* enthusiasm.

figuratively, literally   See **literally**, **figuratively**.

firstly (secondly, thirdly, . . .)   Archaic forms meaning "in the first . . . second . . . third place." Use *first, second, third.*

former   *Former* as an adjective means "preceding" or "previous." As a noun it means "the first of two things mentioned previously." It is often used in conjunction with *latter.*

The *former* residents of this area, the Delaware Indians, were forced to cede their land in 1795.

Two books mark the extremes of Herman Melville's career: *Typee* and *Moby-Dick.* The *former* was a best seller; the *latter* was generally ignored by the public.

freshman, freshmen   *Freshman* is singular and *freshmen* is plural. Even so, only *freshman* is used as the adjective form: *freshman* composition, *freshman* registration, *freshman* dormitories. Some consider *freshman* sexist and prefer the term *first-year student.*

frightened, afraid   *Frightened* should be accompanied by the prepositions *at* or *by; afraid,* by *of.*

Dolley Madison, wife of President James Madison, was *frightened at* the thought of the British burning Washington.

In Charles Dickens's *A Christmas Carol,* Scrooge is *frightened by* three ghosts.

Young children are often *afraid of* the dark.

further, farther   See **farther**, **further**.

good, well   *Good* is an adjective, never an adverb.

The townspeople thought the proposal for a new municipal water plant was a *good* one.

Well can function as an adverb or an adjective. As an adverb it means "in a good manner:" "He did *well* (not *good*) on the test" and "She swam *well* (not *good*) in the meet."

Well is used as an adjective with verbs that denote a state of being or feeling. Here *well* can mean "in good health": "I feel *well.*"

good and . . .   This colloquial phrase meaning "very" is not appropriate in college writing.

After escaping from the Iroquois, Natty Bumppo was *very* (not *good and*) tired.

got to   *Got to* is not suitable in college writing. To indicate obligation use *have to, has to,* or *must.*

COLLOQUIAL: Anyone who takes a literature course has *got to* get a copy of *A Glossary of Literary Terms* by M. H. Abrams.

REVISED: Anyone who takes a literature course *has* to get a copy of *A Glossary of Literary Terms* by M. H. Abrams.

hanged, hung   Both *hanged* and *hung* are past participles of *hang. Hanged* is used to refer to executions. *Hung* is used in all other senses meaning "suspended" or "held up."

Billy Budd was *hanged* from the mainyard of the ship for killing the master-at-arms.

The pictures in the National Gallery were *hung* to take advantage of the natural lighting in the various rooms.

he, she   Traditionally *he* has been used in the generic sense to refer to both males and females. To acknowledge the equality of the sexes, however, avoid the generic *he.* Constructions such as *he or she* or *he/she* can be cumbersome, especially when used a number of times in a paragraph. To avoid problems, use the first and third person plural pronouns when possible.

TRADITIONAL: Before registering, *each student* should be sure *he* has received *his* student number.

REVISION: Before registering, *we* should receive *our* student numbers.

REVISION: Before registering, *students* should receive *their* student numbers.

hopefully   The adverb *hopefully* should modify a verb, an adjective, or another adverb.

During the 1930s many of the nation's jobless looked *hopefully* to the federal government for relief. (*Hopefully* modifies *looked*.)

Increasingly, however, *hopefully* is being used as a sentence modifier meaning "it is hoped." In college writing, use *hopefully* in its traditional sense to avoid ambiguity.

AMBIGUOUS: *Hopefully*, scientists will discover a cure for the common cold within the next five years. (Who is hopeful? Scientists or the writer?)

REVISED: Scientists *hope* they will discover a cure for the common cold within the next five years.

however   See **but, however, yet**.

if, whether   When asking indirect questions or expressing doubt, use *whether*.

He asked *whether* (not *if*) the flight would be delayed because of the fog.

The attendant was not sure *whether* (not *if*) the fog would delay the flight.

Use *whether or not* when expressing alternatives.

He did not know *whether or not* to change his travel plans.

illusion, allusion   See **allusion, illusion**.

immigrate to, emigrate from   See **emigrate from, immigrate to**.

implicit, explicit   See **explicit, implicit**.

imply, infer   *Imply* means "to hint" or "to suggest." *Infer* means "to conclude from." When you *imply*, you *send out* a suggestion; when you *infer*, you *receive* or draw a conclusion.

Mark Antony *implied* that Brutus and the other conspirators had wrongfully killed Julius Caesar. The crowd *inferred* his meaning and called for the punishment of the conspirators.

in, into   Use *in* when you want to indicate position. Use *into* when you want to indicate motion to a point within a thing.

As he stood *in* the main burial vault of the tomb of Tutankhamen, Howard Carter saw a wealth of artifacts.

Before he walked *into* the cave, Tom Sawyer grasped Becky Thatcher's hand.

In 1828 Russia and Persia entered *into* the Treaty of Turkmanchai.

infer, imply   See **imply, infer**.

ingenious, ingenuous   *Ingenious* means "clever at inventing or organizing." *Ingenuous* means "open" or "artless."

> Ludwig van Beethoven is recognized as one of the most *ingenious* composers who ever lived.

> For a politician the mayor was surprisingly *ingenuous*.

inside of, outside of   *Of* is unnecessary when *inside* and *outside* are used as prepositions.

> He waited *inside* (not *inside of*) the coffee shop.

Inside of is colloquial in references to time.

> He could run a mile in *under* (not *inside of*) eight minutes.

insure, ensure, assure   See **assure**, **ensure**, **insure**.

irregardless, regardless   See **regardless**, **irregardless**.

irritate, aggravate   See **aggravate**, **irritate**.

its, it's   *Its* is a possessive pronoun. *It's* is a contraction of *it is*.

> The most obvious characteristic of a modern corporation is the separation of *its* management from *its* ownership.

> *It's* not often that you see a collection of rare books such as the one housed in the Library of Congress.

-ize, -wise   The suffix *-ize* is used to change nouns and adjectives into verbs: *civilize, industrialize, immunize*. The suffix *-wise* is used to change a noun or adjective into an adverb: *likewise, otherwise*. Unfortunately, some writers, particularly in advertising and government, use these suffixes carelessly, making up words as they please: *prioritize, taste-wise, weather-wise*, and *policy-wise*, for example. Be sure to look up suspect *-ize* and *-wise* words in the dictionary to be sure they are standard forms.

kind of, sort of   *Kind of* and *sort of* to mean "rather" or "somewhat" are colloquial and should not appear in college writing.

> **COLLOQUIAL:** The countess was surprised to see that Napoleon was *kind of* short.

> **REVISED:** The countess was surprised to see that Napoleon was *rather* short.

Reserve *kind of* and *sort of* for occasions when you categorize.

> Willie Stark, a character in Robert Penn Warren's *All the King's Men*, is the *kind of* man who begins by meaning well and ends by being corrupted by his success.

latter   See **former**.

lay, lie   See **lie**, **lay**.

leave, let   *Leave* means "to go way from" or "to let remain." *Let* means "to allow" or "to permit."

Many missionaries were forced to *leave* China after the Communist revolution in 1948.

As the liquid boils away, it will *leave* a dark brown precipitate at the bottom of the flask.

In London it is illegal to *let* dogs foul the footpath.

less, fewer See **fewer**, **less**.

let, leave See **leave**, **let**.

liable to See **likely to**, **liable to**, **apt to**.

lie, lay *Lie* is an intransitive verb (one that does not take an object) that means "to recline." *Lay* is a transitive verb (one that takes an object) meaning "to put" or "to place."

Base Form	Past	Past Participle	Present Participle
lie	lay	lain	lying

Each afternoon she would *lie* in the sun and listen to the surf.

As I Lay Dying is a novel by William Faulkner.

In 1871 Heinrich Schliemann unearthed the city of Troy, which had *lain* undisturbed for two thousand years.

The painting *Odalisque* by Eugene Delacroix shows a nude *lying* on a couch.

Base Form	Past	Past Participle	Present Participle
lay	laid	laid	laying

The Federalist Papers *lay* the foundation for the American conservative movement.

In October of 1781 the British *laid* down their arms and surrendered to George Washington at Yorktown.

After he had *laid* his money on the counter, he walked out of the restaurant.

We watched the Amish stone masons *laying* a wall without using mortar.

like, as See **as**, **like**.

likely to, liable to, apt to *Likely to* implies a strong chance that something might happen. *Liable to* implies that something undesirable is about to occur. *Apt to* implies a natural tendency.

Medical researchers feel that in fifty years human beings are *likely* to have a life span of more than a hundred years.

If we do not do something to correct the poor drainage in this area, we are *liable to* have a repeat of last year's flooding.

Old books are *apt to* increase in value if you protect them from heat and moisture.

literally, figuratively *Literally* means "following the letter" or "in a strict sense." *Figuratively* means "metaphorically" or "not literally."

Literally, the Declaration of Independence is a list of grievances that the English colonists had against their king. *Figuratively,* the

Declaration of Independence is a document that elevates the rights of common people above the divine right of kings.

loose, lose   *Loose* is an adjective meaning "not rigidly fastened or securely attached." *Lose* is a verb meaning "to misplace."

The marble facing of the building became *loose* and fell to the sidewalk.

After only two drinks, most people *lose* their ability to judge distance.

lots, lots of, a lot of   These words are colloquial substitutes for "many," "much," or "a great deal of." Avoid their use in college writing.

The students had several (not *lots of* or *a lot of*) options for essay topics.

When using these words informally, be careful to use correct subject-verb agreement.

There are (not *is*) *lots of* possible topics.

majority, plurality   These words are often confused. *Majority* denotes more than half. *Plurality* means a larger number but not necessarily a majority. A candidate with a *majority* has over 50 percent of the votes cast. A candidate with a *plurality* has more votes than any of the other candidates, but not over 50 percent of the total. Use *most* rather than *majority* when you do not know the exact numbers.

INCORRECT: The soprano got the *majority* of the applause.

REVISED: The soprano got *most* of the applause.

man   Like the generic pronoun *he, man* has been used in English to denote members of both sexes. This usage is being replaced by *human beings, people,* or similar terms that do not specify gender.

The dinosaur was extinct long before *human beings* (not *man*) walked the earth.

may, can   See **can**, **may**.

may be, maybe   *May be* is a verb phrase. *Maybe* is an adverb meaning "perhaps."

She *may be* older than the other students, but she is more enthusiastic than they are.

Maybe her experience in the corporate world will give her an advantage in the management courses.

media, medium   *Medium,* meaning a "means of conveying or broadcasting something," is singular. *Media* is the plural form.

Television has replaced print and film as the *medium* of communication that has the most profound effect on our lives.

A good business presentation uses a number of different *media* to make its point.

might have, might of   *Might of* is a nonstandard construction, not the written form for the contraction of *might have.*

G13

John F. Kennedy *might have* (not *might of*) been a great president had he not been assassinated.

number, amount   See **amount**, **number**.

OK, O.K., okay   While all three spellings are acceptable, this term should be avoided in college writing. Replace this term with a more specific word or words.

The instructor's lecture was *adequate* (not *okay*), if uninspiring.

on account of   Use *because of*.

The computer malfunctioned *because of* (not *on account of*) a faulty circuit board.

outside of, inside of   See **inside of**, **outside of**.

percent, percentage   *Percent* indicates a part of a hundred when a specific number is referred to: "10 *percent* of his weekly salary"; "5 *percent* of the monthly rent." *Percentage* is used when no specific number is referred to: "a *percentage* of the people"; "a *percentage* of next year's receipts." In technical and business writing it is permissible to use the % sign after percentages you are comparing. Write out *percent* in college writing.

phenomenon, phenomena   A *phenomenon* is a single observable fact or event. It can also refer to a rare or significant occurrence. *Phenomena* is the plural form.

Metamorphosis is a *phenomenon* that occurs in many insects, mollusks, amphibians, and fish.

John Stuart Mill was a *phenomenon*. He could read classical Greek at the age of five.

Comets are celestial *phenomena* that have been regarded with awe and terror and were once seen as omens of unfavorable events.

plenty   *Plenty*, when used as a noun followed by "of," means "abundance" or "a large amount." Avoid using *plenty* as a colloquial substitute for "very" or "quite."

There are *plenty* of benefits to recycling plastic.

Recycling can be *quite* (not *plenty*) time consuming, but it is well worth the trouble.

plus   As a preposition, *plus* means "in addition to." Avoid using *plus* as a substitute for *and*.

Include the sum of the principal, *plus* the interest, in your calculations.

The amount you quoted was too high. Moreover (not *plus*), it was inaccurate.

precede, proceed   *Precede* means "to go or come before." *Proceed* means "to go forward in an orderly way."

Robert Frost's *North of Boston* was *preceded* by another volume of poetry, *A Boy's Will*.

In 1532 Francisco Pizarro landed at Tumbes and *proceeded* south until he encountered the Incas.

principal, principle As a noun, *principal* means "a sum of money (minus interest) invested or lent" or "a person in the leading position." As an adjective it means "most important."

If you cash the bond before maturity, a penalty can be subtracted from the *principal* as well as the interest.

The *principal* of the high school is a talented administrator who has instituted a number of changes.

Women are the *principal* wage earners in many American households.

A *principle* is a rule of conduct or a basic truth.

The Constitution embodies the fundamental *principles* upon which the American republic if founded.

raise, rise *Raise* is a transitive verb, and *rise* is an intransitive verb—that is, *raise* takes an object and *rise* does not.

A famous photograph taken during World War II shows American Marines *raising* the flag on Iwo Jima.

The planet Venus is called the morning star because when it *rises*, it is brighter than any light in the sky except the sun or moon.

real, really *Real* means "genuine" or "authentic." *Really* means "actually." In your college writing, do not use *real* as an adjective meaning "very."

COLLOQUIAL: The planarian is a *real* flat worm that we studied in biology class.

REVISED: The planarian is a *very* flat worm that we studied in biology class.

reason is that, reason is because *Reason* should be used with *that* and not with *because*, which is redundant.

The *reason* he moved out of the city *is that* (not *is because*) property taxes rose sharply.

regardless, irregardless *Irregardless* is a nonstandard version of *regardless*. The suffix *-less* means "without" or "free from," so the negative prefix *ir-* is unnecessary.

Regardless (not *irregardless*) of what some people might think, drunk drivers kill more than twenty-five thousand people a year.

respectively, respectfully, respectably *Respectively* means "in the order given." *Respectfully* means "giving honor or deference." *Respectably* means "worthy of respect."

In this paper I will discuss "The Sisters" and "The Dead," which are the first and the last stories, *respectively*, in James Joyce's collection *Dubliners*.

When being presented to Queen Elizabeth of England, foreigners are asked to bow *respectfully*.

G15

Even though Abraham Lincoln ran his campaign for the United States Senate quite *respectably,* he was defeated by Stephen Douglas in 1858.

rise, raise See **raise, rise**.

set, sit To *set* means "to put down" or "to lay." To *sit* means "to assume a sitting position."

Base Form	Past	Past Participle	Present Participle
set	set	set	setting
sit	sat	sat	sitting

After rocking the baby, he *set* her down carefully in her crib.

Research has shown that many children *sit* in front of the television five to six hours a day.

shall, will *Will* has all but replaced *shall* to express all future action.

should of See **could of, would of**.

sit, set See **set, sit**.

so Avoid using *so* alone as a vague intensifier meaning "very" or "extremely." Follow *so* with *that* and a clause that describes the result.

She was *so* pleased with their work *that* she took them out to lunch (not *She was so pleased with their work.*)

sometime, sometimes, some time *Sometime* means "at some time in the future." *Sometimes* means "now and then." *Some time* means "a period of time."

In his essay "The Case Against Man," Isaac Asimov says that *sometime,* far in the future, human beings will not be able to produce enough food to sustain themselves.

All automobiles, no matter how well constructed, *sometimes* need repairs.

At the battle of Gettysburg, General Meade's failure to counterattack gave Lee *some time* to regroup his troops.

sort of, kind of See **kind of, sort of**.

specially, especially See **especially, specially**.

stationary, stationery *Stationary* means "staying in one place." *Stationery* means "materials for writing" or "letter paper."

When viewed from the earth, a communications satellite traveling at the same speed as the earth appears to be *stationary* in the sky.

The secretaries are responsible for keeping departmental offices supplied with *stationery*.

supposed to, used to Both *supposed to* and *used to* require the final *d* to indicate past tense.

She was *supposed to* (not *suppose to*) turn in her paper yesterday.

She always *used to* (not *use to*) turn in her papers on time.

take, bring See **bring, take**.

than, as   See **as**, **than**.

than, then   *Than* is a conjunction used to indicate a comparison, and *then* is an adverb indicating time.

The new shopping center is bigger *than* the old one.

He did his research; *then* he wrote a report.

that, which, who   Use *that* or *which* when referring to a thing. Use *who* when referring to a person.

In *How the Other Half Lives,* Jacob Riis described the conditions *that* existed in working-class slums in nineteenth-century America.

The Wonderful Wizard of Oz, which was published in 1900, was originally entitled *From Kansas to Fairyland.*

Anyone *who* (not *that*) visits Maine cannot help being impressed by the beauty of the scenery and the ruggedness of the landscape.

themselves, theirselves, theirself   *Theirselves* and *theirself* are non-standard variants of *themselves* and are not acceptable in college writing.

Pioneer families had to build their homes and clear their land by *themselves* (not *theirself* or *theirselves*).

then, than   See **than**, **then**.

there, their, they're   Use *there* to indicate place and in the expressions *there is* and *there are.*

I have always wanted to visit the Marine Biological Laboratory in Woods Hole, Massachusetts, but I have never gotten *there.*

There is nothing we can do to resurrect a species once it becomes extinct.

Their is a possessive pronoun.

James Watson and Francis Crick did *their* work on the molecular structure of DNA at the Cavendish Laboratory at Cambridge University.

They're is a contraction of *they are.*

White sharks and Mako sharks are dangerous to human beings because *they're* good swimmers and especially sensitive to the scent of blood.

thus, therefore   *Thus* means "in this way," not "therefore" or "so."

In Joseph Conrad's *Heart of Darkness*, Kurtz becomes a man-god to the natives. *Thus*, he is able to collect a fortune in ivory.

INCORRECT: Throughout the past year, interest rates have dropped dramatically. *Thus,* businesses are able to buy the equipment they need to modernize their operations.

REVISED: Throughout the past year, interest rates have dropped dramatically, *so* businesses are able to buy the equipment they need to modernize their operations.

till, until, 'til   *Till* and *until* have the same meaning, and both are acceptable. *Until* is preferred in college writing. *'Til*, a contraction of *until*, should be avoided.

to, at   See **at, to**.

to, too, two   *To* is a preposition that indicates direction.

> Last year we flew from New York *to* California.

Too is an adverb that means "also" or "more than is needed."

> "Tippecanoe and Tyler *too*" was William Henry Harrison's campaign slogan during the 1840 presidential election.

> The plot was *too* complicated for the average reader.

Two expresses the number 2.

> Just north of *Two* Rivers, Wisconsin, is a petrified forest.

try to, try and   *Try and* is the colloquial equivalent of the more formal *try to*.

> **COLLOQUIAL:**   Throughout most of his career E. R. Rutherford was determined to *try and* discover the structure of the atom.

> **REVISED:**   Throughout most of his career E. R. Rutherford *tried to* discover the structure of the atom.

-type   Deleting this empty suffix eliminates clutter and clarifies meaning.

> **COLLOQUIAL:**   Found in the wreckage of the house was an *incendiary-type* device.

> **REVISED:**   Found in the wreckage of the house was an *incendiary* device.

uninterested, disinterested   See **disinterested, uninterested**.

unique   *Unique* means "the only one," not "remarkable" or "unusual."

> **COLLOQUIAL:**   Its undershot lower jaw makes the English bulldog *unique* among dogs.

> **REVISED:**   Its undershot lower jaw makes the English bulldog unusual among dogs.

> **CORRECT USAGE:**   In their scope and unity, Michelangelo's paintings are *unique*.

Because *unique* means "the only one," it can take no intensifiers. Never use constructions like "the most unique" or "very unique."

until   See **till, until, 'til**.

used to   See **supposed to, used to**.

wait for, wait on   To *wait for* means "to defer action until something occurs." To *wait on* means "to act as a waiter."

> **COLLOQUIAL:**   I am *waiting on* dinner.

G18

REVISED: I am *waiting for* dinner.

CORRECT: The captain *waited on* the head table himself.

well, good   See **good**, **well**.

were, we're   Some people pronounce these words alike, and so they confuse them when they write. *Were* is a verb; *we're* is the contraction of *we are*.

The Trojans *were* asleep when the Greeks climbed out of the wooden horse and took the city.

We Americans are affected by the advertising we see. *We're* motivated by the ads we see to buy billions of dollars worth of products each year.

whether, if   See **if**, **whether**.

which, who, that   See **that**, **which**, **who**.

who, whom   When a pronoun serves as the subject of its clause, use *who* or *whoever;* when it functions in a clause as an object, use *whom* or *whomever*.

Sarah, *who* is studying ancient civilizations, would like to visit Greece.

Sarah, *whom* I haven't seen in a year, wants me to travel to Greece with her.

To determine which to use at the beginning of a question, use a personal pronoun to answer the question.

Who tried to call me? *He* called. (subject)

For *whom* is the package? It is for *her.* (object of a preposition)

Whom do you want for the job? I want *her.* (object)

who's, whose   Use *who's* when you mean *who is*.

Who's going to take calculus?

Use *whose* when you want to indicate possession.

The writer *whose* book was in the window was autographing copies in the store.

will, shall   See **shall**, **will**.

-wise, -ize   See **-ize**, **-wise**.

would of, could of   See **could of**, **would of**.

yet   See **but**, **however**, **yet**.

your, you're   Because these words are pronounced alike they are often confused. *Your* indicates possession, and *you're* is the contraction of *you are*.

You can improve *your* stamina by jogging two miles a day.

You're certain to be impressed the first time you see the Golden Gate Bridge spanning San Francisco Bay.

GLOSSARY

Glossary of Grammatical and Rhetorical Terms

absolute phrase   See **phrase**.

abstract noun   See **noun**.

acronym   A word formed from the first letters or initial sounds of a group of words: NATO = North Atlantic Treaty Organization.

active voice   See **voice**.

adjective   A word that describes, limits, qualifies, or in any other way modifies nouns or pronouns. A **descriptive adjective** names a quality of the noun or pronoun it modifies: *junior year*. A **proper adjective** is formed from a proper noun: *Hegelian philosophy*. Other kinds of words may be used to limit or qualify nouns, and they are then considered adjectives: **articles** (*a, an, the*): *the book, a peanut*; **possessive pronouns** (*my, your, his*, and so on): *their apartment, my house*; **demonstrative pronouns** (*this, these, that, those*): *that table, these chairs*; **interrogative pronouns** (*what, which, whose*, and so on): *Which car is yours?*; **indefinite pronouns** (*another, each, both, many*, and so on): *any minute, some day*; **relative pronouns** (*what, whatever, which, whichever, whose, whosever*): *Bed rest was what the doctor ordered.*; **numbers** (*one, two, first, second*, and so on): *Claire saw two robins.* **21d**; **25a**

adjective clause   See **clause**.

adverb   A word that describes the action of verbs or modifies adjectives, other adverbs, or complete phrases, clauses, or sentences. Adverbs answer the questions "How?" "Why?" "Where?" "When?" and "To what extent?" Adverbs are formed from adjectives, many by adding *-ly* to the adjective form (*dark/darkly, solemn/solemnly*), and may also be derived from prepositions (*Joe carried on.*). Other adverbs that indicate time, place, condition, cause, or degree are not derived from other parts of speech: *then*, *never*, *very*, and *often*, for example. The words *how*, *why*, *where*, and *when* are classified as **interrogative adverbs** when they ask questions (*How did we get into this mess?*). See also **conjunctive adverb**. **21e**; **25c**

adverb clause   See **clause**.

adverbial conjunction   See **conjunctive adverb**.

agreement   The correspondence between words in number, person, and gender. Subjects and verbs must agree in number (singular or plural) and person (first, second, or third): *Soccer is a popular European sport.*; *I play soccer too.* **24a** Pronouns and their antecedents must agree in number, person, and gender (masculine, feminine, neuter); *Lucy loaned Charlie her car.* **24b**

allusion   A reference to a famous historical, literary, or biblical person or event which the reader is expected to recognize. **18d5**

analogy   A form of **figurative language** in which the writer explains an unfamiliar idea or object by comparing it to a more familiar one: *Sensory pathways of the central nervous system are bundles of nerves rather like telephone cables that feed information about the outside world into the brain for processing.* **18d3**

antecedent   The word or group of words to which a pronoun refers: *Brian finally bought the stereo he had always wanted.* (*Brian* is the antecedent of the pronoun *he*.)

appositive   A noun or noun phrase that identifies, in different words, the noun or pronoun it follows: *Columbus, the capital of Ohio, is in the central part of the state.* Appositives may be used without special introductory phrases, as in the preceding example, or they may be introduced by *such as, or, that is, for example*, or *in other words*: *Japanese cars, such as Hondas, now have a large share of the U.S. automobile market.* In a restrictive appositive, the appositive precedes the noun or pronoun it modifies: *Singing cowboy Gene Autry became the owner of the California Angels.* **8e4**; **28d1**

article   The word *a*, *an*, or *the*. Articles signal that a noun follows and are usually classified as adjectives. See also **adjective**. **21d**

auxiliary verb   See **verb**.

balanced sentence   A sentence neatly divided between two parallel structures. Balanced sentences are typically **compound sentences** made up of two parallel clauses (*The telephone rang, and I answered.*), but the parallel clauses of a **complex sentence** can also be balanced. **10c**

cardinal number   A number that expresses quantity—*seven, thirty, one hundred*. (Contrast **ordinal**.)

case   The form a noun or pronoun takes to indicate how it functions in a sentence. English has three cases. A pronoun takes the **subjective** (or **nominative**) **case** when it acts as the subject of a sentence or a clause: *I am an American*. **22a1** A pronoun takes the **objective case** when it acts as the object of a verb or of a preposition: *Fran gave me her dog*. **22a2** Both nouns and pronouns take the **possessive** (or **genitive**) **case** when they indicate ownership: *My house is brick, Brandon's T-shirt is red*. This is the only case in which nouns change form. **22a3**

clause   A group of related words that includes a subject and a predicate. An **independent** (main) **clause** may stand alone as a sentence (*Yellowstone is a national park in the West.*), but a **dependent** (**subordinate**) clause must always be accompanied by an independent clause (*Yellowstone is a national park in the West that is known for its geysers.*). Dependent clauses are classified according to their function in a sentence. An **adjective clause** (sometimes called a **relative clause**) modifies nouns or pronouns: *The philodendron, which grew to be twelve feet tall, finally died* (the clause modifies *philodendron*). An **adverb clause** modifies single words (verbs, adjectives, or adverbs) or an entire phrase or clause: *The film was exposed when Bill opened the camera* (the clause modifies *exposed*). A **noun clause** acts as a noun (as subject, direct object, indirect object, or complement) in a sentence: *Whoever arrives first wins the prize* (the clause is the subject of the sentence). An **elliptical clause** is grammatically incomplete—that is, part or all of the subject or predicate is missing. If the missing part can be easily inferred from the context of the sentence, such a construction is acceptable: *When (they are) pressed, the committee will act*. **8c2**

climactic word order   The writing strategy of moving from the least important to the most important point in a sentence and ending with the key idea. **10a2**

collective noun   See **noun**.

comma splice   An error created when two independent clauses are incorrectly joined by a comma. **14a–d**

COMMA SPLICE: The Mississippi River flows south, the Nile River flows north.
REVISED: The Mississippi River flows south. The Nile River flows north.
REVISED: The Mississippi River flows south; the Nile River flow north.
REVISED: The Mississippi River flows south, and the Nile River flows north.
REVISED: Although the Mississippi River flows south, the Nile River flows north.

common noun   See **noun**.

comparative/superlative The forms taken by an adjective or an adverb to indicate degree. The **positive degree** describes a quality without indicating comparison (*Frank is tall*.). The **comparative degree** indicates comparison between two persons or things (*Frank is taller than John*.). The **superlative degree** indicates comparison between one person or thing and two or more others (*Frank is the tallest boy in his scout troop*.). **25d**

complement A word or word group that describes or renames a subject, an object, or a verb. A **subject complement** is a word or phrase that follows a linking verb and renames the subject. It can be an adjective (called a **predicate adjective**) or a noun (called a **predicate nominative**): *Clark Gable was a movie star*. An **object complement** is a word or phrase that describes or renames a direct object. Object complements can be either adjectives or nouns: *We call the treehouse the hideout*.

complex sentence See **sentence**.

compound Two or more words that function as a unit, such as **compound nouns**: *attorney at law; boardwalk*; **compound adjectives**: *hardhitting editorial*; **compound prepositions**: *by way of, in addition to*; **compound subjects**: *April and May are spring months*.; **compound predicates**: *Many try and fail to climb Mt. Everest*.

compound adjective See **compound**.

compound noun See **compound**.

compound predicate See **compound**.

compound preposition See **compound**.

compound sentence See **sentence**.

compound subject See **compound**.

compound-complex sentence See **sentence**.

conjunction A word or words used to connect single words, phrases, clauses, and sentences. **Coordinating conjunctions** (*and, or, but, nor, for, so, yet*) connect words, phrases, or clauses of equal weight: *crime and punishment* (coordinating conjunction *and* connects two words). **Correlative conjunctions** (*both . . . and, either . . . or, neither . . . nor*, and so on), always used in pairs, also link items of equal weight: *Neither Texas nor Florida crosses the Tropic of Cancer* (correlative conjunction *neither . . . nor* connects two words). **Subordinating conjunctions** (*since, because, although, if, after*, and so on) introduce adverb clauses: *You will have to pay for the tickets now because I will not be here later* (subordinating conjunction *because* introduces adverb clause). **21g** *conjunctive adverb*   An adverb that joins and relates independent clauses in a sentence (*also, anyway, besides, hence, however, nevertheless, still,* and so on): *Howard tried out for the Yankees; however, he didn't make the team.* **21e**

connotation   The emotional associations that surround a word. (Contrast **denotation**.) **18b1**

contraction   The combination of two words with an apostrophe replacing the missing letters: *We + will = we'll; was + not = wasn't.*

coordinate adjective   One of a series of adjectives that modify the same word or word group: *The glen was quiet, shady,* and *cool.*

coordinating conjunction   See **conjunction**.

coordination   The pairing of similar elements (words, phrases, or clauses) to give equal weight to each. Coordination is used in simple sentences to link similar elements into compound subjects, predicates, complements, or modifiers. If can also link two independent clauses to form a compound sentence: *The sky was cloudy, and it looked like rain.* (Contrast **subordination**.) **8a**

correlative conjunction   See **conjunction**.

cumulative sentence   A sentence that begins with a main clause followed by additional words, phrases, or clauses that expand or develop it: *On the hill stood a schoolhouse, paint peeling, windows boarded, playground overgrown with weeds.*

dangling modifier   A modifier for which no logical headword appears in the sentence. To correct dangling modifiers, either change the subject of the sentence's main clause, creating a subject that can logically service as the headword of the dangling modifier, or add words that transform the dangling modifier into a dependent clause. **15b**

DANGLING: Pumping up the tire, the trip continued.
REVISED: After pumping up the tire, they continued the trip.

dead metaphor   A metaphor so overused that it has become a meaningless cliché. **18e1**

declarative sentence   See **sentence**.

deductive argument   An argument that begins with a general statement or proposition and establishes a chain of reasoning that leads to a conclusion. **6b**

demonstrative adjective   See **adjective**.

demonstrative pronoun   See **pronoun**.

denotation   The dictionary meaning of a word. (Contrast **connotation**.) **18b1**

dependent clause   See **clause**.

descriptive adjective   See **adjective**.

G24

direct object　See **object**.

direct quotation　See **quotation**.

documentation　The formal acknowledgement of the sources used in a piece of writing. **40a**

documentation style　A format for providing information about the sources used in a piece of writing. Documentation styles vary from discipline to discipline. **40b–e**

double negative　The use of two negative words within a single sentence: *She didn't have no time*. Such constructions are nonstandard English. Revised: *She had no time* or *She didn't have any time*.

ellipsis　Three spaced periods used to indicate the omission of a word or words from a quotation: *"The time has come . . . and we must part."* **32f**

elliptical clause　See **clause**.

embedding　A strategy for varying sentence structure that involves changing some sentences into modifying phrases and working them into other sentences. **12b3**

enthymeme　A syllogism in which one of the premises—usually the major premise—is implied rather than stated. **6b2**

expletive　A construction in which *there* or *it* is used with a form of the verb *be*: *There is no one here by that name*.

faulty parallelism　See **parallelism**.

figurative language　Imaginative comparisons between different ideas or objects using common figures of speech—**simile**, **metaphor**, **analogy**, **personification**, **allusion**, **hyperbole**, and **understatement**. **18d**

figure of speech　See **figurative language**.

finite verb　A verb that can stand as the main verb of a sentence. Unlike **participles**, **gerunds**, and **infinitives** (see **verbal**), finite verbs do not require an auxiliary in order to function as the main verb: *The rooster crowed*.

fragment　See **sentence fragment**.

function word　An article, preposition, conjunction, or auxiliary verb that indicates the function of and the grammatical relationship among the nouns, verbs, and modifiers in a sentence.

fused sentence　A **run-on sentence** that occurs when two independent clauses are joined either without suitable punctuation or without a coordinating conjunction. Correct fused sentences by separating the independent clauses with a period, a semicolon, or a comma and a coordinating conjunction, or by using subordination. **14a–d**

FUSED SENTENCE: Protein is needed for good nutrition lipids and carbohydrates are too.

REVISED: Protein is needed for good nutrition. Lipids and carbohydrates are too.

REVISED: Protein is needed for good nutrition; lipids and carbohydrates are too.

REVISED: Protein is needed for good nutrition, but lipids and carbohydrates are too.

REVISED: Although protein is needed for good nutrition, lipids and carbohydrates are too.

gender   The classification of nouns and pronouns as masculine (*father, boy, he*), feminine (*mother, girl, she*), or neuter (*radio, kitten, them*). See also **case**.

gerund   A special form of verb ending in *-ing* that is always used as a noun: *Fishing is relaxing* (gerund *fishing* serves as subject; gerund *relaxing* serves as subject complement). Note: When the *-ing* form of a verb is used as a modifier, it is considered a **present participle**. (See also **verbal**.)

gerund phrase   See **phrase**.

headword   The word or phrase in a sentence that is described, defined, or limited by a modifier.

helping verb   See **auxiliary verb**.

idiom   An expression that is characteristic of a particular language and whose meaning cannot be predicted from the meaning of its individual words: *lend a hand*.

imperative mood   See **mood**.

indefinite adjective   See **adjective**.

indefinite pronoun   See **pronoun**.

independent clause   See **clause**.

indicative mood   See **mood**.

indirect object   See **object**.

indirect question   A question that tells what has been asked but, because it does not use the speaker's exact words, does not take a question mark: *He asked whether he could use the family car*.

indirect quotation   See **quotation**.

inductive argument   An argument that begins with observations or experiences and moves toward a conclusion. **6a**

infinitive   The base form of the verb preceded by *to*, an infinitive can serve as an adjective (*He is the man to watch.*), an adverb (*Chris hoped to break the record.*), or a noun (*To err is human.*). See also **verbal**.

infinitive phrase   See **phrase**.

intensifier   A word that adds emphasis but not additional meaning to words it modifies. *Much, really, too, very,* and *so* are typical intensifiers.

intensive pronoun   See **pronoun**.

interjection   A grammatically independent word, expressing emotion, that is used as an exclamation. An interjection can be set off by a comma, or, for greater emphasis, it can be punctuated as an independent unit, set off by an exclamation point: *Ouch! That hurt.* **21h**

interrogative adjective   See **adjective**.

interrogative adverb   See **adverb**.

interrogative pronoun   See **pronoun**.

intransitive verb   See **verb**.

inverted appositive   See **appositive**.

irregular verb   A verb that does not form both its past tense and past participle by the addition of *-d* or *-ed* to the base form of the verb. **23a2**

isolate   Any word, including **interjections**, that can be used in isolation: *Yes. No. Hello. Good-bye. Please. Thank you.*

linking verb   A verb that connects a subject to its complement: *The crowd became quiet.* Words that can be used as linking verbs include *seem, appear, believe, become, grown, turn, remain, prove, look, sound, smell, taste, feel,* and forms of the verb *be.*

main clause   See **clause**.

main verb   See **verb**.

mass noun   See **noun**.

metaphor   A form of **figurative language** in which the writer makes an implied comparison between two unlike items, equating them in an unexpected way: *The subway coursed through the arteries of the city.* **18d2**

misplaced modifier   A modifier that has no clear relationship with its headword, usually because it is placed too far from it. **15a**

MISPLACED: By changing his diapers, Dan learned much about his new baby son.
REVISED: Dan learned much about his new baby son by changing his diapers.

mixed construction   A sentence made up of two or more parts that do not fit together grammatically, causing readers to have trouble determining meaning. **17f**

MIXED: The Great Chicago Fire caused terrible destruction was what prompted changes in the fire code. (independent clause used as a subject)

REVISED: The terrible destruction of the Great Chicago Fire prompted changes in the fire code.

REVISED: Because of the terrible destruction of the Great Chicago Fire, the fire code was changed.

mixed metaphor   The combination of two or more incompatible images in a single figure of speech: *During the race John kept a stiff upper lip as he ran like the wind.* **18e2**

modal auxiliary   See **verb**.

modifier   A word that adds information to a sentence and shows connections between ideas.

mood   The verb form that indicates the writer's basic attitude. There are three moods in English. The **indicative mood** is used for statements and questions: *Nebraska became a state in 1867.* **23h** The **imperative mood** specifies commands or requests and is often used without a subject: *(You) Pay the rent.* **23i** The **subjunctive mood** expresses wishes or hypothetical conditions: *I wish the sun were shining.* **23j**

nominal   A word, phrase, or clause that functions as a noun.

nominative case   See **case**.

nonfinite verb   See **verbal**.

nonrestrictive modifier   A modifying phrase or clause that does not limit or particularize the words it modifies, but rather supplies additional information about them. Nonrestrictive modifiers are set off by commas: *Oregano, also known as marjoram and suganda, is a member of the mint family.* **28d1** (Contrast **restrictive modifier**.)

noun   A word that names people, places, things, ideas, actions, or qualities. A **common noun** names any of a class of people, places, or things: *Lawyer, town, bicycle.* A **proper noun**, always capitalized, refers to a particular person, place, or thing: *Anita Hill, Chicago, Schwinn.* A **mass noun** names a quantity that is not countable: *sand, time, work.* An **abstract noun** refers to an intangible idea or quality: *bravery, equality, hunger.* A **collective noun** designates a group of people, places, or things thought of as a unit: *Congress, police, family.* **21a**

noun clause   See **clause**.

noun phrase   See **phrase**.

number   The form taken by a noun, pronoun, demonstrative adjective, or verb to indicate one (**singular**): *car, he, this book, boast*; or many (**plural**): *cars, they, those books, boasts.* **17d**

numerical adjective   See **adjective**.

object   A noun, pronoun, or other noun substitute that receives the action of a **transitive verb**, **verbal**, or **preposition**. A **direct object** indicates where the verb's action is directed and who or what is affected by it: *John caught a butterfly*. An **indirect object** tells to or for whom the verb's action was done: *John gave Nancy his butterfly*. An **object of a preposition** is a word or word group introduced by a preposition: *John gave Nancy his butterfly for an hour*.

object complement   See **complement**.

object of a preposition   See **object**.

objective case   See **case**.

ordinal number   A number that indicates position in a series: *seventh, thirtieth, one-hundredth*.

parallelism   The use of similar grammatical elements in sentences or parts of sentences: *We serve beer, wine, and soft drinks*. Words, phrases, clauses, or complete sentences may be parallel, and parallel items may be paired or presented in a series. When elements that have the same function in a sentence are not presented in the same terms, the sentence is flawed by **faulty parallelism**. **10c**; **16a–b**

participial phrase   See **phrase**.

participle   A verb form that functions in a sentence as an adjective. Virtually every verb has a **present participle**, which ends in *-ing* (*breaking, leaking, taking*), and a **past participle**, which usually ends in *-d* or *-ed* (*agreed, walked, taken*). (See also **verbal**.) **Present participle**: *The heaving seas swamped the dinghy*. (present participle *heaving* modifies noun *seas*); **past participle**: *Aged people deserve respect*. (past participle *aged* modifies noun *people*.)

parts of speech   The eight basic building blocks for all English sentences: *nouns, pronouns, verbs, adjectives, adverbs, prepositions, conjunctions*, and *interjections*.

passive voice   See **voice**.

past participle   See **participle**.

periodic sentence   A sentence that moves from a number of specific examples to a conclusion, gradually building in intensity until a climax is reached in the main clause: *Wan and pale and looking ready to crumble, the marathoner headed into the last mile of the race*.

person   The form a pronoun or verb takes to indicate the speaker (**first person**): *I am/we are*; those spoken to (**second person**): *you are*; and those spoken about (**third person**): *He/she/it is*; *they are*. **17d**

personal pronoun   See **pronoun**.

personification   A form of **figurative language** in which the writer describes an idea or inanimate object in terms that imply human attributes, feelings, or powers: *The big feather bed beckoned to my tired body*. **18d4** *phrase*   A grammatically ordered group of related words that lacks a subject or a predicate or both and functions as a single part of speech. A **verb phrase** consists of an auxiliary (helping) verb and a main verb: *The wind was blowing hard*. A **noun phrase** includes a noun or pronoun plus all related modifiers: *She broke the track record*. A **prepositional phrase** consists of a preposition, its object, and any modifiers of that object: *The errant ball sailed over the fence*. A **verbal phrase** consists of a verbal and its related objects, modifiers, or complements. A verbal phrase may be a **participial phrase** (*Undaunted by the sheer cliff, the climber scaled the rock*.), a **gerund phrase** (*Swinging from trees is a monkey's favorite way to travel*.), or an **infinitive phrase** (*Wednesday is Bill's night to cook spaghetti*.). An **absolute phrase** usually consists of a noun or pronoun and a participle, accompanied by modifiers: *His heart racing, he dialed her number*. **8c1**

positive degree   See **comparison**.

possessive adjective   See **adjective**.

possessive case   See **case**.

predicate   A verb or verb phrase that tells or asks something about the subject of a sentence is called a **simple predicate**: *Well-tended lawns grow green and thick*. (*grow* is the simple predicate.) A **complete predicate** includes all the words associated with the predicate: *Well-tended lawns grow green and thick*. (*grow green and thick* is the complete predicate.) **8a**

predicate adjective   See **complement**.

prefix   A letter or group of letters put before a root or word that adds to, changes, or modifies it. **19f3**

preposition   A part of speech that introduces a word or word group consisting of one or more nouns or pronouns or of a phrase or clause functioning in the sentence as a noun: *Jeremy crawled under the bed*. **21f**

prepositional phrase   See **phrase**.

present participle   See **participle**.

principal parts   The forms of a verb from which all other forms can be derived. The principal parts are the **base form** (*give*), the **present participle** (*giving*), the **past tense** (*gave*), and the **past participle** (*given*).

pronoun   A word that may be used in place of a noun in a sentence. The noun for which a pronoun stands is called its **antecedent**. There are eight different types of pronouns. Some have the same form but are distinguished by their function in the sentence. A **personal pronoun** stands for a person or thing: *I*, *me*, *we*, *us*, *my*, and so on. (*They broke his window*.) A **reflexive pronoun** ends in

-self, or *-selves* and refers to the subject of the sentence or clause: *myself, yourself, himself*, and so on. (*They painted the house themselves.*) An **intensive pronoun** ends in *-self* or *-selves* and emphasizes a preceding noun or pronoun. (*Custer himself died in the battle.*) A **relative pronoun** introduces an adjective or noun clause in a sentence: *which, who, whom*, and so on. (*Sitting Bull was the Sioux chief who defeated Custer.*) An **interrogative pronoun** introduces a question: *who, which, what, whom*, and so on. (*Who won the lottery?*) A **demonstrative pronoun** points to a particular thing or group of things: *this, that, these, those.* (*Who was that masked man?*) A **reciprocal pronoun** denotes a mutual relationship: *each other, one another.* (*We still have each other.*) An **indefinite pronoun** refers to persons or things in general, not to specific individuals. Most indefinite pronouns are singular—*anyone, everyone, one, each*—but some are always plural—*both, many, several.* (*Many are called, but few are chosen.*) **21b**

proper adjective   See **adjective**.

proper noun   See **noun**.

quotation   The use of the written or spoken words of others. A **direct quotation** occurs when a passage is borrowed word for word from another source. Quotation marks (" ") establish the boundaries of a direct quotation: *"These tortillas taste like cardboard,"* *complained Beth*. **31a** An **indirect quotation** reports someone else's written or spoken words without quoting that person directly. Quotation marks are not used: *Beth complained that the tortillas tasted like cardboard*.

reciprocal pronoun   See **pronoun**.

reflexive pronoun   See **pronoun**.

regular verb   A verb that forms both its past tense and past participle by the addition of *-d* or *-ed* to the base form of the verb. **23a**

relative adverb   See **adverb**.

relative clause   See **clause**.

relative pronoun   See **pronoun**.

restrictive modifier   A modifying phrase or clause that limits the meaning of the word or word group it modifies. Restrictive modifiers are not set off by commas: *The Ferrari that ran over the fireplug was red*. **28d1** (Contrast **nonrestrictive modifier**.)

root   A word from which other words are formed. An understanding of a root word increases a reader's ability to understand unfamiliar words that incorporate the root.

run-on sentence   An incorrect construction that results when the proper connective or punctuation does not appear between independent clauses. A run-on occurs either as a **comma splice** or as a **fused sentence**.

sentence An independent grammatical unit that contains a subject and a predicate and expresses a complete thought: *Carolyn sold her car*. **8a** A **simple sentence** consists of one subject and one predicate: *The season ended*. **8a**; a **compound sentence** is formed when two or more simple sentences are connected with coordinating conjunctions, conjunctive adverbs, semicolons, or colons: *The rain stopped, and the sun began to shine*. **9a**; a **complex sentence** consists of one simple sentence, which functions as an independent clause in the complex sentence, and at least one dependent clause, which is introduced by a subordinating conjunction or a relative pronoun: *When he had sold three boxes* [dependent clause], *he was halfway to his goal.* [independent clause] **9b**; and a **compound-complex sentence** consists of two or more independent clauses and at least one dependent clause: *After he prepared a shopping list* [dependent clause], *he went to the store* [independent clause], *but it was closed.* [independent clause]. **9c**

sentence fragment An incomplete sentence, phrase, or clause that is punctuated as if it were a complete sentence. **13a–g**

shift A change of *tense, voice, mood, person, number,* or *type of discourse* within or between sentences. Some shifts are necessary, but problems occur with unnecessary or illogical shifts. **17a–e**

simile A form of **figurative language** in which the writer makes a comparison, introduced by *like* or *as*, between two unlike items on the basis of a shared quality: *Like sands through the hourglass, so are the days of our lives*.; *The wind was as biting as his neighbor's doberman*. **18d1**

simple predicate See **predicate**.

simple sentence See **sentence**.

simple subject See **subject**.

split infinitive An infinitive whose parts are separated by a modifier. **15a4**

   SPLIT: She expected *to* ultimately *swim* the channel.
   REVISED: She expected ultimately *to swim* the channel.

squinting modifier A modifier that seems to modify either a word before it or one after it and that conveys a different meaning in each case. **15a**

   SQUINTING: The task completed simply delighted him.
   REVISED: He was delighted to have the task completed simply.
   REVISED: He was simply delighted to have the task completed.

subject A noun or noun substitute that tells who or what a sentence is about is called a **simple subject**: *Healthy thoroughbred horses run like the wind*. (*Horses* is the simple subject.) The **complete subject** of a sentence includes all the words associated with the subject: *Healthy thoroughbred horses run like the wind*. (*Healthy thoroughbred horses* is the complete subject.) **8a**

G32

subject complement   See **complement**.

subjective case   See **case**.

subjunctive mood   See **mood**.

subordinate clause   See **clause**.

subordinating conjunction   See **conjunction**.

subordination   Making one or more clauses of a sentence grammatically dependent upon another element in a sentence: *Preston was only eighteen when he joined the firm.* (Contrast **coordination**.) **9b**

suffix   A syllable added at the end of a word or root that changes its part of speech. **19f3**

superlative degree   See **comparison**.

suspended hyphen   A hyphen followed by a space or by the appropriate punctuation and a space: *The wagon was pulled by a two-, four-, or six-horse team.*

syllogism   A three-part set of statements or propositions, devised by Aristotle, that contains a major premise, a minor premise, and a conclusion. **6b1**

tag question   A question, consisting of an auxiliary verb plus a pronoun, that is added to a statement and set off by a comma: *You know it's going to rain, don't you?*

tense   The form of a verb that indicates when an action occurred or when a condition existed. **23b–f**

transitive verb   See **verb**.

verb   A word or phrase that expresses action (*He painted the fence.*) or a state of being (*Henry believes in equality.*). A **main verb** carries most of the meaning in the sentence or clause in which it appears: *Winston Churchill smoked long, thick cigars.* A main verb is a **linking verb** when it is followed by a **subject complement**: *Dogs are good pets.* An **auxiliary verb** (sometimes called a **helping verb**) combines with the main verb to form a **verb phrase**: *Graduation day has arrived.* The auxiliaries *be* and *have* are used to indicate the tense and voice of the main verb. The auxiliary *do* is used for asking questions and forming negative statements. Other auxiliary verbs, known as **modal auxiliaries** (*must, will, can, could, may, might, ought (to), should,* and *would*), indicate necessity, possibility, willingness, obligation, and ability: *It might rain next Tuesday.* A **transitive verb** requires an **object** to complete its meaning in the sentence: *Pete drank all the wine* (*wine* is the direct object). An **intransitive verb** has no direct object: *The candle glowed.* **21c1**; **23a–n**

verb phrase   See **phrase**.

verbal   (nonfinite verb) Verb forms—**participles**, **infinitives**, and **gerunds**—that are used as nouns, adjectives, or adverbs. Verbals

G33

do not behave like verbs. Only when used with an auxiliary can such verb forms serve as the main verb of a sentence. *The wall painted* is not a sentence; *The wall was painted* is. **21c2**

verbal phrase   See **phrase**.

voice   The form that determines whether the subject of a verb is acting or is acted upon. When the subject of a verb performs the action, the verb is in the **active voice**: *Palmer sank a thirty-foot putt*. When the subject of a verb receives the action—that is, is acted upon—the verb is in the **passive voice**: *A thirty-foot putt was sunk by Palmer*. **10e**; **17b**; **23k–n**

Acknowledgments

Rachel Phillips Belash, "Why Girls Schools Remain Necessary," by Rachel Phillips Belash. From *The New York Times,* February 22, 1988, Op-Ed. Copyright © 1988 by The New York Times. Reprinted by Permission.

John Berger, Excerpt from *About Looking* by John Berger. Copyright © 1980 by John Berger. Reprinted by permission of Pantheon Books, a division of Random House, Inc.

Duane Gall, "Multicultural Education," by Duane Gall. From *Utne Reader,* November/December 1990. Copyright © Duane Gall. Reprinted by permission of the author.

Ellen Goodman, "Sexual Harassment in Schools: Nearly Everybody 'Is Doing It" by Ellen Goodman. From *The Philadelphia Inquirer,* June 8, 1993. Copyright © 1993, The Boston Globe Company. Reprinted by permission.

Michael Finkel, "Undecided—and Proud of It," by Michael Finkel. From *The New York Times,* November 24, 1989, Op-Ed. Copyright © 1989 by The New York Times. Reprinted by permission.

Victoria Fromkin and Robert Rodman, Excerpt from *An Introduction to Language,* Fourth Edition by Victoria Fromkin and Robert Rodman, copyright © 1988 by Holt, Rinehart and Winston, Inc., adapted and reprinted by permission of the publisher.

Ron Kovic, Excerpt from *Born on the Fourth of July* by Ron Kovic. Copyright © 1976 by Ron Kovic. Reprinted by permission of McGraw-Hill, Inc.

Jake Lamar, "Whose Legacy Is It, Anyway?" by Jake Lamar. From *The New York Times,* September 9, 1991, Op-Ed. Copyright © 1991 by the New York Times. Reprinted by permission.

Richard D. Lamm, "English Comes First," by Richard D. Lamm. From *The New York Times,* July 1, 1986, Op-Ed. Copyright © 1986 by the New York Times. Reprinted by permission.

Joseph Nocera, "How the Middle Class Has Helped Ruin the Public Schools," by Joseph Nocera. From *The Washington Monthly,* September/October 1989. Reprinted with permission from The Washington Monthly. Copyright by The Washington Monthly Company, 1611 Connecticut Avenue, NW, Washington, D. C. 20009. (202) 462-0128.

William R. Quimby, Excerpt from article, "The World of Hunting," by William R. Quimby. From *Safari Magazine,* July/August 1992. Copyright © 1992 by William R. Quimby. Reprinted by permission of The Safari Club International.

Anna Quindlen, "Error, Stage Left," by Anna Quindlen. From *The New York Times,* August 12, 1990, Op-Ed. Copyright © 1990 by The New York Times. Reprinted by permission.

Phyllis Schafly, Excerpt from article "Comparable Worth Unfair to Men and Women," by Phyllis Schafly, *The Humanist,* May/June 1986. Reprinted with permission of The American Humanist Association, copyright © 1986.

Index

Index to Instructor's Annotations